The Pearl Oyster

The Pearl Oyster

Edited By

Paul C. Southgate
John S. Lucas

ELSEVIER

Amsterdam • Boston • Heidelberg • London • New York • Oxford
Paris • San Diego • San Francisco • Singapore • Sydney • Tokyo

Elsevier
The Boulevard, Langford Lane, Kidlington, Oxford OX5 1GB, UK
Radarweg 29, PO Box 211, 1000 AE Amsterdam, The Netherlands

First edition 2008

Notice
No responsibility is assumed by the publisher for any injury and/or damage to persons or property as a
matter of products liability, negligence or otherwise, or from any use or operation of any methods, products,
instructions or ideas contained in the material herein. Because of rapid advances in the medical sciences, in
particular, independent verification of diagnoses and drug dosages should be made.

British Library Cataloguing in Publication Data
A catalogue record for this book is available from the British Library

Library of Congress Cataloging-in-Publication Data
A catalog record for this book is available from the Library of Congress

ISBN–13: 978-0-44-452976-3

For information on all Elsevier publications
visit our website at elsevierdirect.com

Printed and bound by CPI Group (UK) Ltd, Croydon, CR0 4YY

Transferred to Digital Print 2012

Working together to grow
libraries in developing countries

www.elsevier.com | www.bookaid.org | www.sabre.org

ELSEVIER   BOOK AID
International   Sabre Foundation

CONTENTS

PREFACE

The plain outer surfaces of pearl oyster shells conceal the lustrous beauty of their mother-of-pearl lining. Pearl oysters of the family Pteriidae have been fished for thousands of years to satisfy human fascination with mother-of-pearl, an activity which occasionally uncovered rare and highly prized natural pearls. In the early 1900s, the ability of pearl oyster tissue to secrete mother-of-pearl was harnessed for production of cultured pearls. This process, which has changed little since, provides the basis for a cultured marine pearl industry with a current value of around US$ 500 million per annum.

Exploitation of pearl oysters by fishermen seeking natural pearls and mother-of-pearl, or by pearl farmers, has a rich and fascinating history and the modern pearling industry does not necessarily reflect the past glories of some countries. While many of us associate pearls with fashionable jewelry-shop windows and glossy magazines, it is an industry which also offers economic opportunities to coastal communities in less developed countries; an industry which involves individuals, co-operatives and families, as well as large multinational companies. This book presents an overview of the utilization of pearl oysters through history, from ancient pearl fisheries to modern pearl farming.

A large volume of research on the biology and culture of pearl oysters has been published, predominantly since the 1950s. Major databases show a rapid growth from 55 publications in the 1950s to around 400 in the 1980s and to 860 in the 1990s. Over the past 20–25 years pearl oyster biology has well and truly become a science. Now there are new focuses for research in this field including genetics and inheritance, structure and properties of mother-of-pearl (nacre), mechanisms of nacre formation and the potential of this process in medical science.

In seeking to be comprehensive, this book brings together contributors from a broad range of fields, including commercial pearl farming, economics, fisheries biology, gemology and jewelry retail, humanities and science, to provide an overview of pearl oyster biology, culture techniques, global pearl production, and the current pearl market. We trust that it will be of much interest and assistance to all those who share our fascination with pearl oysters and their products.

Paul Southgate
John Lucas

LIST OF CONTRIBUTORS

Héctor Acosta-Salmón, Center for Aquaculture and Stock Enhancement, Harbor Branch Oceanographic Institute at Florida Atlantic University, 5600 US 1 North, Fort Pierce, FL 34946, USA

Micheline Cariño, Department of Humanities, Universidad Autónoma de Baja California Sur, La Paz, B.C.S., Mexico

Angelique Fougerouse, Pearl Culture Agency, P.O. Box 9047, Motu Uta 98715, Tahiti, French Polynesia

Scott P. Gifford, School of Environmental and Life Sciences, University of Newcastle, Callaghan, NSW, 2308, Australia

Anthony Hart, West Australian Fisheries Research Laboratories, P.O. Box 20, North Beach, WA 6920, Australia

John D. Humphrey, Aquatic Animal Health, Department of Primary Industry, Fisheries and Mines, GPO Box 3000, Darwin, NT 0801, Australia

Odette Ison, School of Marine and Tropical Biology, James Cook University, Townsville, Qld. 4811, Australia

Dean R. Jerry, School of Marine and Tropical Biology, James Cook University, Townsville, Qld. 4811, Australia

Sandra Langy, Ministry of Sea Fisheries Aquaculture and Research, P.O. Box 2551, Papeete 98713, Tahiti, French Polynesia

Cedrik Lo, Pearl Culture Agency, P.O. Box 9047, Motnuka 98715, Tahiti, French Polynesia

John S. Lucas, Centre for Marine Studies, University of Queensland, St. Lucia, Qld 4072, Australia

Mario Monteforte, Centro de Investigaciones Biologicas del Noroeste, S.C.,, Mar Bermejo 195, Col. Playa Palo de Santa Rita, La Paz, B.C.S., Mexico

Rocky de Nys, School of Marine and Tropical Biology, James Cook University, Townsville, Qld. 4811, Australia

Wayne A. O'Connor, NSW Department of Primary Industries, Port Stephens Fisheries Centre, Locked Bag 1, Nelson Bay, NSW 2315

Bernard Poirine, Université de la Polynésie Française, BP 5670, Tahiti, French Polynesia

Marthe Rousseau, MNHN Département Milieux et Peuplements, Aquatiques, UMR 5178, Biologie des Organismes Marins et Ecosystèmes, CP26 43, rue Cuvier, 75005 Paris, France

Pedro E. Saucedo, Centro de Investigaciones Biologicas del Noroeste, S.C., Mar Bermejo 195, Col. Playa Palo de Santa Rita, La Paz, B.C.S., Mexico

Brigitte Sheung, Golay Buchel & Cie SA, Avenue De Rhodanie 60, CH-1000 Lausanne 3 Cour, Switzerland

Paul C. Southgate, School of Marine and Tropical Biology, James Cook University, Townsville, Qld. 4811, Australia

Elisabeth Strack, Gemmologisches Institut Hamburg, Gerhofstr 19, 20354 Hamburg, Germany

Joseph Taylor, Pertokoan Sanur Raya N0. 10–19, Jalan Bypass Ngurah Rai, Sanur – Denpasar Selatan 80227, Bali, Republic of Indonesia

Ilya Tëmkin, Division of Invertebrate Zoology, American Museum of Natural History, Central Park West at 79th Street, New York, NY 10024, USA

Clem Tisdell, School of Economics, University of Queensland, Brisbane, Qld. 4072, Australia

Richard D. Torrey, *Pearl World: The International Pearling Journal*, 302 West Kaler Drive, Phoenix, Arizona 85021-724302, USA

Katsuhiko T. Wada, Research Planning and Coordination Division, National Research Institute of Aquaculture (NRIA), Fisheries Research Agency (FRA), Nansei, Mie 5160193, Japan

Aimin Wang, Ocean College, Hainan University, Haikou 570228, China

DEDICATION

We dedicate this book to our "pearls" Dawn and Helen

ACKNOWLEDGMENTS

We acknowledge that many people have assisted the editors and authors of chapters with photographs, translations, information, and in other ways. We express our sincere gratitude for their contributions.

CHAPTER 1

Introduction

Elisabeth Strack

1.1. HISTORICAL OVERVIEW

The delicate beauty of the pearl has fascinated mankind since the dawn of time and it has been cherished by the most uncivilized peoples and highly refined civilizations in equal measures. Pearls were treasures, symbols of wealth, power and prestige, and they were met with devotion and respect.

The first pearls may have been found accidentally while searching for food, and it is not difficult to imagine that both the iridescent shells and the round, white and shining objects found within them attracted the admiration of our early ancestors.

Shells were themselves worked into jewelry and decorative objects. The oldest findings, discovered in early 2004, are 75,000 years old; they come from the Blombos cave in South Africa. Forty-one shells of the marine snail *Nassarius kraussianus*, only 7–8 mm in size, were pierced in an identical manner and were painted with ocher on the inside.

1.1.1. Antiquity

Mother-of-pearl (MOP) inlays made around 4500 BC were found within the ruins of Bismaya in Mesopotamia, and sculptures dating from the Babylonian era from excavations in Nineveh and Nimrud confirm that pearls had been known as decorative objects (Strack, 2006). Although no direct findings are available, historians take it for granted that the early civilizations knew of the pearl grounds in the Persian Gulf and the Red Sea. The oldest discoveries of pearl shells in Egypt date back to the 4th millennium BC, while it is only after the Persian conquest in the 5th century that interest in pearls became larger, leading to extravagant use in Alexandria during the Ptolemaic period.

Quotations in the Talmud and the Bible are an indication of the significance the ancients attributed to pearls. The Talmud mentions in Genesis that the dresses God provided for Adam and Eve were "as beautiful as pearls" and speaks of "Manna as white as pearls". The Old Testament refers to pearls within the Proverbs of Solomon (8: 11):

> Wisdom is more precious than coral or pearls,
> and it is not even equalled by rubies.

Later translations do not mention pearls anymore and speak instead of rubies as treasures. The reason for this is that the old Hebrew word for pearls has not yet been defined precisely and different interpretations are encountered today. The New Testament mentions pearls in a parable of Jesus (Matthew 13: 45–46):

> Again, the kingdom of heaven is like a merchant looking for fine pearls. When he
> found one of great value, he went away and sold everything he had and bought it.

And Matthew 7: 6 mentions in the famous Sermon on the Mount:

> Do not give dogs what is sacred;
> do not throw your pearls to pigs.
> If you do, they may trample them under their feet,
> and then turn and tear you to pieces.

The Revelation of John (21: 21) speaks of the 12 pearly gates in the wall encircling New Jerusalem, made entirely of precious stones:

> And the twelve gates were twelve pearls,
> each gate was made of a single pearl.

The Koran mentions pearls as a symbol for precious objects. The trees in paradise bear fruit made of pearls and emeralds; and the believer who enters paradise will be crowned with pearls of incomparable beauty.

Indian history makes first mention of pearls in the Veda, the holy books of the Brahmans, who speak of "krisana". A Hindu legend recounts that Lord Krishna offered the first pearl that he found in the deepest place of the sea to his daughter Pandaia as a wedding gift, and until today Indian brides traditionally wear pearls on their wedding days. But pearls are also a symbol of tears, going back to the story of the daughter of the Great Mogul of Delhi whose daughter wept bitter tears on her lost love and when she died, the God of Love and Passion transformed her tears into pearls.

In classical Sanskrit pearls were named "mukta", which means as much as purity or escape, alluding to the spirit of the mollusc that wants to escape and solidify as a pearl. In the East, as far away as the Sulu Islands, the names for pearls will later be derived from the Sanskrit name. Sanskrit also uses the word "maracata" from which the Persian "morvarid" and the Greek "margarita" were later derived.

The ancient Greeks dedicated the pearl to Aphrodite, the goddess of love who was herself compared to a precious pearl as she had risen out of the water from a shell.

Homer rarely mentions pearls, although the following line from the Ilias (XIV, 183) may refer to pearls:

> In three shining drops the glittering gems were hanging from her ears.

The origin of pearls was explained by the help of legends and the idea was favoured that pearls were being formed by lightning striking the sea. In the 4th century BC, Theophrastus mentions for the first time the word "margarita" and makes the first scientific attempt to explain why pearls form in molluscs. The word "margarita", still used in modern Greek language for pearls, was used by 18th and 19th century zoologists when they created the first scientific names for pearl-producing molluscs, for example, *Mytilus margaritiferus, Meleagrina margaritifera*. "Margaritology" denotes the science of pearls. Most European languages use "Margarita" as a name for girls and in English language even the word "Pearl" itself is used as a name for girls.

It was not until the conquests of Alexander the Great (356–323 BC) that lavish quantities of pearls from the East poured onto the Western markets. Alexandria, capital of the Ptolemaic empire since 304 BC, developed into a rich metropolis where even freshwater pearls, derived from mussels from Britain, were offered for sale. They were brought by the Phoenicians who came to England as early as the 7th century BC.

The early trade with pearls reached its climax during the Roman Empire. The riches originated from Pompeius' conquest of the East, and consequently pearls flowed into Rome from all countries and all major cities of the old world. The pearl trade was grouped in the hands of the so-called "margaritarii", a name given to divers but also to pearl traders and jewelers as well. They kept their "officinae margaritariorum" in the Forum Romanum. Pearls, called "margarita", rose to become the top luxury item. In his "Historia naturalis", Gaius Plinius Secundus (24–79 AD) writes of a number of other names that were used for pearls, among them the word "perla"; while "margarita" was also used for other cherished objects or even for beloved persons, especially children.

Plinius gives a vivid description of the luxurious use of pearls of his time. Caesar paid six million sestertia for one single pearl in the year 59 BC, which he gave to Servilia, mother of Brutus. The amount would equal about US$650,000 today. Plinius also writes of seeing Lollia Paulina, the second wife of Emperor Caligula, wearing lavish pearl and emerald jewelry; and he complains about the vanity of the Roman ladies of his time. Plinius also recounts the legend of Cleopatra's bet with Marc Anthony, that she would be able to eat in his presence a meal worth 60 million sestertia (ca. US$6.5 million today). She dissolved the pearl from an earring in a glass of vinegar and proceeded to drink. The pearl from the second earring, saved from suffering the same fate as the first by Lucius Plancus, the adjudicator of the bet, was allegedly later brought to Rome, divided into two halves, and used to adorn the ears of the state of Venus in the Pantheon.

1.1.2. Medieval times

The luxurious use of pearls continued long into the later period of antiquity and into the early Middle Ages. When Rome was taken by Alarich in 410 AD, the Roman pearl treasures fell into the hands of the Goths and passed in the 6th and 7th century to the Franconian rulers. Byzantium became the capital of the Eastern Roman empire and replaced Rome as the center of the ancient world. Pearls continued to be used luxuriously, and their lavish use is best reflected in the famous mosaic picture of Ravenna's San Vitale cathedral that shows the early Byzantine Emperor Justinian I (483–585) and his wife Theodora in shining regalia.

The Franconian period saw a continuing use of pearls in sacred goldsmith art, and the tradition was continued until into the 15th century. Reliquaries, crosses, chalices and evangelaries, scepters, coats for priests and antependia were decorated and embroidered with pearls.

In the early Middle Ages pearls became a symbol of purity for worshiping the Madonna, and Christ came to be seen as the exquisite pearl carried by Mary. Images of love and femininity had by now been successfully integrated into Christian symbolism. The Orthodox Church later adopted the traditions and used pearls to decorate icons.

The sacred use of pearls reached its climax during the time of Charlemagne (768–814), and Albrecht Dürer's painting of the emperor in his coronation robes demonstrates the Franconian pearl luxury. While the last traces of pearls from antiquity were gradually lost, local river pearls replaced them. The Romans had already known pearls from freshwater mussels in the River Moselle and its tributaries in the Eifel and Hunsrück mountains in southwest Germany. Ausonius, a tutor to Emperor Valentian I, speaks of them in his hymn to the Mosella.

Pearls remained a symbol of power and wealth even after the empire of Charlemagne was divided up upon his death. The so-called Imperial Crown of Konrad II (1024–1039), kept today at the Kunsthistorisches Museum in Vienna, was made during the 10th and 11th centuries (Fig. 1.1).The crown is embellished with enameled pictures of Byzantine influence and a variety of colored gemstones, interspersed with pearls and signed with a pearl-studded inscription of the emperor's name.

The 11th and 12th centuries saw the art of pearl embroideries develop in European monasteries and courts. The origin goes back to the East Roman Empire, and Crusaders brought the art to Europe. The church soon discovered the new art as a means of representation and had altar cloths, garments for priests, bishops' caps, corporalia bags and tiny coats for statuettes of the Virgin Mary embroidered with pearls. Pearl embroideries were most popular from the 12th century onwards in the Heath Monasteries on Lüneburg Heath in North Germany. The "Kasel Cross" of Ebstorf Monastery on Lüneburg Heath, originating from around 1500, is an especially beautiful example (Fig. 1.2). The art did not survive the reformation period.

Crusaders brought out the last treasures from Alexandria and Byzantium during the 12th and 13th century and the new merchant towns in Germany and Northern Italy became the new centers for the pearl trade. The city of Venice excelled in the luxurious use of pearls. The return of the crusaders and the consequent formation of the Knight's Orders contributed to the popularization of pearls, which remained, however, a privilege of the ruling classes who showed them off at all possible occasions. The Dukes of Burgundy became among others famous for their pearl luxury. A few decades later, Henry VIII (1491–1547) rose to fame for similar excesses in England.

Variations of the word "pearl" emerged in European languages from the 14th century onwards from the Latin "perla" or "perula". German and French languages transformed it into "perle" and Danish and Dutch into "paarl".

FIGURE 1.1 The so-called Imperial Crown of Konrad II. The front plate is 14.9 cm high. Weltliche und Geistliche Schatzkammer, Kunsthistorisches Museum, Vienna. (Photo: Kunsthistorisches Museum, Vienna.) (See Color Plate 1)

1.1.3. The early modern ages

When Christopher Columbus made his third journey to the New World in 1498, he discovered the first pearls along the coasts of what is today Venezuela. The early 16th century saw the exploration of pearl grounds also along the coast of present-day Colombia; and Mexican pearls were discovered after Hernando Cortez had reached Baja California in 1535.

Pearls from the palaces and temples of the Aztecs found their way into the treasuries of the Spanish court and paved the way for a new "Pearl Age", a second pearl rush after the excesses of the Roman Empire. The pearl riches spread from Spain to Europe and were by now called "occidental pearls" in order to distinguish them from the "oriental pearls" from the classical finding places in the East that had been known until then.

Under the reign of Queen Elizabeth I (1558–1603) England rose to become a world power equal to Spain. Several portrait painters to the queen, daughter of Henry VIII,

FIGURE 1.2 The "Kasel Cross" was embroidered around 1500 and is today part of on antependium at Ebstorf Monastery in North Germany. It measures 95 cm in height and 73.5 cm in width. The frames around the pictorial representations, the gowns of the persons depicted and the various symbols are embroidered with river pearls. The photograph shows a section only. (Photo: Elisabeth Strack. Reproduction with the permission of The Abbess, Kloster Ebstorf.) (See Color Plate 2)

depicted her in fanciful dresses adorned with thousands of pearls (Fig. 1.3). After the death of Maria Stuart, she came into possession of the so-called Hanoverian pearls that had originally belonged to Catherine de' Medici. Queen Alexandra was seen wearing these pearls again around 1900, and Queen Elizabeth II wore them on her wedding day in 1947.

The Habsburg family owned the largest collection of pearls and displayed the most beautiful specimens in the crown that was made in 1602 for Emperor Rudolf II in Prague, using the pearls as a symbol of the secular power of Austria.

Goldsmiths of the Renaissance period extensively used the pearl riches of the New World, and fanciful figurine pendants, representing mermaids, sea beasts, dragons, swans and pelicans, were set with large pearls and decorated with colored enamel, gemstones and the first table-cut diamonds (Fig. 1.4). The Portuguese word "barocco" meaning something like strange or peculiar, lent its name to the baroque pearls. In Florence, Benvenuto Cellini created figurines of high perfection for the Medici family. The goldsmith's art of the late 16th century also favored cups made of exotic shells from bivalve molluscs, snails and cephalopods like nautilus.

Jean Baptiste Tavernier, born in Paris in 1605, wrote of the pearls he witnessed during his extensive travels to the East in the Arabian sultanates and at the courts of the Shah of Persia and the Great Mogul Aureng-Zeb in Delhi, where he had seen the large pearl of about 50 carats suspended from the Peacock Throne.

FIGURE 1.3   Queen Elizabeth I. (*Source*: Kunz and Stevenson, 1908.) (See Color Plate 3)

FIGURE 1.4   Pendant: Sea dragon, late 16th century. Spanish or German, height 10 cm. (Photo: British Museum Publications Ltd. The Waddeston Bequest 3, Renaissance Jewellery, ML 82.) (See Color Plate 4)

1.1.4. The modern ages

By the 18th century, the pearl age was over. Pearls from America and India had become rare and there was less interest in local river pearls. The Brazilian diamond deposits had been discovered in 1727, and diamonds rose in popularity above pearls. Court etiquette now reserved pearls for occasions of semi-official mourning, a tradition that has been kept at European courts until today and in which one can see another source of the superstitious belief that associates pearls with tears. The topic is not new; it is encountered in various historical contexts from Oriental and Greek mythology up to the symbolic languages of the Middle Ages. Dew drops or even rain drops were also interpreted as tears causing the formation of pearls. In his "Westöstlicher Divan" the German poet Johann Wolfgang von Goethe draws on a parable of Saadi from Shiraz, which he presents at the beginning of his "Book of Parables":

> From heaven there fell upon the foaming wave
> A timid drop; the flood with anger roared,
> But God, its modet boldness to reward,
> Strength to the drop and firm endurance gave.
> Its form the mussel captive took,
> And to its lasting glory and renown,
> The pearl now glistens in our monarch's crown,
> With gentle gleam and loving look.

However, there is another not quite so legendary reason why pearls were taken to represent tears that does deserve mentioning. It is thought that during the "Pearl Age", the excessive use of pearls literally ruined a number of smaller courts in Europe, stripping them of their fortunes and causing them to shed tears.

Interest in scientific enlightenment did not start until the 18th century. The Swedish scientist Carl von Linné researched the formation of pearls and in his Systema Naturae of 1758 he gives the first scientific name to pearl-producing bivalve molluscs, *Mytilus margaritiferus*. He even conducted experiments on pearl culture.

During the Baroque and Rococo periods the pendants of the Renaissance period evolved into exaggerated, grotesque and comical figurines with pearls of bizarre shapes at their center. Extraordinary pieces were created at the various European courts; the most famous ones belong to the collection at the Grüne Gewölbe Museum in Dresden, dating back to King August the Strong, who purchased them before 1725. The 18th century also saw an increase in the number of buyers from the wealthy bourgeoisie classes.

At the end of the century, *Pinctada* was used for the genus of pearl-producing species within the Family Pteriidae by the Danish scientist Johann Hieronymus Chemnitz, who published his eight-volume "Conchylien-Cabinet" between 1779 and 1795. The word may have been derived from the Portuguese "pintada" (=painted, stained), while in French language "pintade" means guinea fowl or a dark gray wild chicken whose feathers are stained with roundish "pearls" of light coloring.

The German Peter Friedrich Röding used *Pinctada* in 1798 when he wrote the so-called Bolten catalog, the sales catalog of the collection of Joachim Friedrich Bolten, a well-known physician in Hamburg. The Bolten catalog fell, however, into neglect for over a century and with it a number of names which were Bolten's or Röding's own classification. The generic name *Pinctada* itself did not resurface until 1915. During the 19th century the *Pinctada* genus was referred to as *Meleagrina* (since 1819), *Margaritophora* (since 1811) or *Margaritifera* (since 1857), and the names were used in a rather haphazard way, together with *Pinctada*, until into the 1960s.

The 19th century was the last century of the natural pearl. The start of the century saw for some years a veritable flood of pearls from the Ceylonese (Sri Lankan) grounds. Pearls were still a privilege of the aristocratic classes and the Napoleonic court used pearls to boast of its new stature. Josephine, the first empress, possessed several large pearls and the emperor's sisters even tried to outdo her. Jean-Louis David depicted the heavy pearl jewelry worn by the Napoleonic ladies on his coronation picture. At the end of the 18th century, the French crown jewels contained some of the most beautiful and largest pearls. They included "La Reine des Perles", a pearl of 110 grains (27.5 carats) that disappeared when the crown jewels were stolen in 1792. The remaining pearls were sold during the Third Republic at the end of the 19th century.

The bourgeoisie classes gained increasing political power and influence during the first decades of the 19th century, but the jewelry remained at first modest. Due to increasing industrialization, the second half of the century led to the economic boom that brought growing wealth to the middle classes who were then able to afford to wear pieces of jewelry that had previously been the privilege of aristocratic families.

The European pearl trade reached its final high period after 1850, lasting until the end of the 1920s. Jewelry designed for the middle classes used small pearls and half pearls. To an extent, pearls now started to replace diamonds in everyday jewelry and they were used to accentuate colored stones like amethyst and peridot, which were typical features of Victorian brooches and pendants. Tiny pearls, so-called seed pearls, were frequently used to produce parures (set of matching jewelry) that Victorians used as wedding jewelry. They were drilled and sewn onto terrace-shaped mother-of-pearl (MOP) rosettes that were worn as complements with broad ribbons of seed pearls.

The first dark pearls from the Tuamotu Archipelago in French Polynesia appeared around the 1850s. They became only fashionable after Empress Eugenie of France developed a liking for them.

After the Civil War, pearl jewelry became even more popular in the United States than in Europe. Around 1900, pearls were considered to be the highest status symbol. The pearl boom was substantially influenced by rich findings of freshwater pearls from the Mississippi River and its tributaries.

The Paris World Fair of 1900 was to lead to the last golden age for natural pearls, and it established the city as the second international pearl center after London. This was the Paris of the Belle Epoque where the wealthy and mighty from all parts of the world were seen together with the demimonde in the new luxury hotels and restaurants. Pearls were the favorite object of adornment. The London fashion scene was led by Queen Alexandra, who loved pearls and is herself the most exquisite example for the refined, dainty luxury of the time.

Pearl prices climbed to astronomic heights between 1893 and 1907. The pearl boom differed from previous boom periods insofar that it used pearls from the whole world, while the Roman pearl riches and the pearls from the New World had come from individual finding places only. In 1916, Pierre Cartier used a natural pearl necklace to pay for his house on 5th Avenue in New York. The successful conclusion of the sale prompted him to change the American saying "Never underestimate the power of a woman" into "Never underestimate the power of pearls".

The life and work of Leonard Rosenthal fell in the last golden age of pearls. He came in the year 1900 from the Caucasian city of Vladikavkaz to Paris where he rose to become the most important pearl dealer of all time. At the height of his fame he controlled the whole market from Venezuela, Panama and Mexico to the Persian Gulf, India and China. He also witnessed the emergence of cultured pearls and was a member of the group of Paris dealers who appealed to a French court against the cultured pearls produced in Japan by Mikimoto (see Section 8.9.1).

The Art Nouveau jewelry of the early 20th century preferred pastel colors that went well with the subdued white color of pearls, resembling "milk and snow". Focusing on artistic design, attention turned once more to baroque pearls. Lalique was the first to use the pink conch pearl produced by the marine gastropod *Stombus gigas*. Art Deco used pearl sautoirs and long, multiple-string necklaces, but pearls do not seem to go well with the geometric shapes and strong color accentuations of the period. Coco Chanel initiated the use of white pearls on a black dress, using imitation pearls, either as a fashion item or as a personal accessory. MOP was also used in elegant items of the time such a cigarette boxes and table clocks.

Pearls did not become available to women on all levels of society until some decades later, towards the end of the 20th century. But today's market is nearly exclusively a market for cultured pearls. In 1916, Kokichi Mikimoto started production of round cultured pearls in Japan, after having harvested his first cultured blister pearl in 1893 (see Section 8.9.1). The new cultured pearls gained a foothold on the world market during the 1920s.

The world economic crisis of 1929 slashed the prices for natural pearls overnight but they continued to have a regular market until the 1950s. After this, until today, they have had a small but fine market for themselves and have even achieved astonishingly high prices on international auctions again since 2004. The natural pearl was, however, unable to halt the victorious rising of the cultured pearl, which represented the ideal pearl for the changing post-war democratic societies in America and Europe. Starting in the 1950s, the new pearls from Japan offered women of all classes the fulfillment of their dreams of owning round, white and shiny pearls. In addition, the last 40 years have seen the rise of freshwater and South Sea cultured pearls.

At the beginning of the 21st century, cultured pearls were much sought after jewelry objects, able to satisfy almost all needs and ideals. Never before has the market offered so many shapes, colors and qualities from so many different countries (see Chapter 10).

1.2. COUNTRIES OF ORIGIN

Several *Pinctada* species, and to a much lesser extent *Pteria* species, produce the pearls of jewelry quality that seem to have sustained from ancient times the human definition of beauty and rarity. Pearls have become known from the Indo-Pacific, the Eastern Pacific and Western Atlantic tropical zones. The tropics (Tropic of Cancer and Tropic of Capricorn) mark the northern and southern boundaries. Pearl grounds stretch, however, at certain points to outside these boundaries. It is interesting to note that the oldest known, so-to-say classical places of origin for pearls are situated within the oldest areas of recorded human history and civilizations. The following sections of this chapter outline the history of fisheries and exploitation of pearl oysters stocks in major producing counties. Further details of fisheries production on a species-by-species basis are presented in Chapter 9 along with details of cultured pearl production from the major species.

1.3. INDIAN OCEAN

1.3.1. Persian Gulf

The first reference is found in an Assyrian inscription that dates back to around 2000 BC, speaking of "a parcel of fisheyes from Dilmun", probably pearls from Bahrain, where the Dilmun civilization was centered from 3,200 to 1,600 BC. In 1989, a 4 mm pearl, estimated to be about 4,000 years old, was found in a Dilmun settlement.

Probably the oldest known pearl necklace comes from an Achemenid grave at the Acropolis of the Winter Palace of the Persian kings in Susa, dating back to the middle of the 4th century BC. The necklace came in 1908 to the Persian Gallery at the Louvre in Paris. It was discovered during an expedition by Jacques de Morgan in 1901. The regularly arranged pearls are sized between 3.5–7 mm, with baroque to barrel shapes. Some pearls of a white to grayish color have been preserved well over the period of nearly 2,500 years and reveal a near perfect luster.

Pearl grounds originally stretched on the Arabian side from Kuwait along the coast of Saudi Arabia to Bahrain, Qatar, the United Arab Emirates and the Sultanate of Oman (Fig. 1.5). They also run along nearly the whole coast of the Persian side of the Gulf, from near Bandar-e-Busehr (island of Khark) to Bandar-e-Lengeh (island of Kish) in the south, and even further south into the Strait of Hormuz (Fig. 1.5).

The Phoenicians, who probably held the first monopoly of the pearl trade, were succeeded by Arab seafarers, undertaking long and difficult journeys with their sailing ships of simple construction, the so-called dhows. The city of Masqat (Muscat) in the Gulf of Oman was the main trade center. After 1515, when Alfonso d'Albuquerque came to the Strait of Hormuz, the Portuguese brought the pearl fisheries under their control. In 1622, the Persian Shah Abbas took victory over the Portuguese, and the new harbor town Bandar-e-Abbas on the Persian side became the new center. The pearl fisheries returned to the control of the local sheikhs who generously allowed their subjects to fish for pearls for a small fee. Jean Baptiste Tavernier, who visited the city of Ormuz in 1670, described pearl trading in the Gulf in his memoirs.

The last boom time was between about 1850 and 1930 and when it was over, partly due to the world economic crisis of 1929 and partly due to the advent of the cultured pearl, pearl fishing went into a decline. The countries at the Gulf were saved just in time from falling into poverty by the discovery of oil in 1932.

There are Akoya pearl oysters[1], known locally as "mohar", in the Persian Gulf and most of the so-called oriental pearls were obtained from these. In the past, the species was often described as *Pinctada vulgaris* or the "common *Pinctada*" in gemmological literature. Akoya pearl oysters grow to shell length 60–80 mm in the Gulf, with a life-span of 7–8 years. They live on sand banks and coral rocks. The thin shells show a white, cream or pale yellow MOP, and pale pink to somewhat violet is occasionally observed. The MOP industry did not start using the shells until the 19th century. Until World War II, large quantities were imported to London and Hamburg under the name "Lingah-Persian Gulf". They were mainly processed into cheap buttons used for

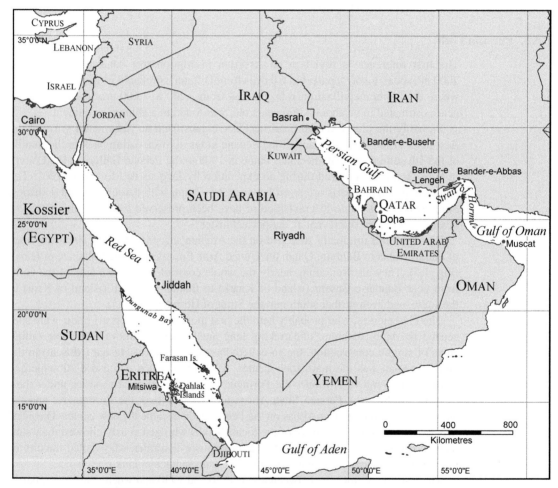

FIGURE 1.5   The Persian Gulf and Red Sea coasts.

[1]Throughout this text the term 'Akoya' is used for pearl oysters in the currently unresolved complex of *Pinctada fucata-martensii-radiata-imbricata* (see Section 2.3.1.13)

uniforms and undergarments. The Japanese called the shells "shinju" (pearls). Exports were mainly handled from Basra, which even before the 19th century had been a center for the export of shells to other Arabian countries and India, where they were used for intarsia and also in the production of pills and betels.

The "Lingah Shells" were named after Lingah, a small place on the Arabian side of the Gulf. The term was soon used generally for small and thin shells, and traders started referring to "Ceylonese Lingah", "Australian Lingah" or "Venezuelan Lingah", depending on the place of origin. Other names used for the shells from the Gulf were "Bastard Shell" and "Lapi Shell".

Pinctada margaritifera is also found in the Persian Gulf, hence the geographical name *P. margaritifera persica* from Jameson (1901). The Arabian names are "sudaifee" or "zinni". They may grow to 200 mm shell length in the Gulf. The color of the MOP is cream to light gray. Indian dealers probably purchased the shells since early times, as the 19th century MOP trade called them "Bombay Shell". Over the centuries, the shells have probably been of greater commercial significance than the pearls. In 1908, during the high time of the MOP industry, ca. 3,000 of shells were exported. Around the middle of the 19th century, approximately 60,000 people, nearly the whole population of the Arabian side of the Gulf, worked in the pearl fisheries. The fishing season was usually concentrated between April and September. At first the shallow coastal waters were fished, leaving the deeper grounds for the warmer season. The grounds were fished in rotation in order to ensure they were left fallow for several years. The pearling fleets were centered in Bahrain and in Doha in Qatar (Fig. 1.5). They consisted of about 150 ships in the tradition of the dhows that were usually owned by rich merchants, often from India, the so-called "bunnias". An important person aboard the ships was the so-called "nahhams" who sang the ritual prayers that determined the schedule of the day. The captain, called "nokhata", had absolute command of the divers, the so-called "ghai ghawwas" who lived and worked on the ships. The divers used characteristic utensils like a wooden nose clamp (fitaam), cotton tampoons trenched in oil for the ears, a basket (dadjin) to hold the molluscs and a knife for cutting them off from the bottom. On their fingers and toes they wore small leather caps, the "khabbal". A diver descended on two ropes that his assistant, the "saib", held and controlled from the ship (Fig. 1.6). He remained underwater for 60–90 seconds, reaching a depth of 6–20 m. Each diver would usually descend 30 to 40 times in one day and there were usually 8 to 40 divers on one ship. A crew of 20 to 30 divers could harvest 7,000 to 8,000 pearl oysters per day. The whole pearling fleet harvested a total of approximately 150 million pearl oysters during certain seasons. Modern diving suits, coming into fashion around 1900, were not used initially and were later prohibited by the sheikhs in order to further avoid overfishing.

The pearl oysters were opened on deck with a knife; large pearls were removed immediately by hand and the rest were harvested using 25 brass sieves, the so-called "tasah". As a rule there were several small pearls in each 500 pearl oysters. Pistol shots heralded the lucky find of a large pearl, the sound of which could be heard in far away places on the coast. All pearls were eventually collected by the captain in the traditional red cloth and after he had sold the first pearls he paid the divers in cash. They received the lowest part as they were on the lowest end of the hierarchy and were usually caught in a vicious circle of poverty from which they could not free themselves.

FIGURE 1.6 Pearl at work in the Persian Gulf. (Photo: Falcon Cinefoto Bahrain.)

The pearls were usually purchased by middlemen, the so-called "tawwash", who came directly to the ships and sold the pearls on to larger dealers. The pearls dominated life in the Gulf for centuries and they once dominated the world trade as "oriental pearls". Currently, pearls from the Gulf are sold to Indian pearl dealers who clean, bleach and drill them in India, from where they are sold internationally and are also sold back the Gulf states.

New pearls from present fisheries account for only 15% of the trade and are still called "Basra pearls". All pearls entering Bahrain are examined by the Gemstone and Pearl Testing Laboratory, which was established by the Ministry of Commerce in 1990.

About 4,000 to 5,000 pearls of good quality were fished each season during the boom years between 1850 and 1930. Colors range from white to dark cream and very light pink. There are even yellowish-pink pearls that resemble the apricot hue of today's Chinese freshwater cultured pearls. Arabian language calls them "nabatee", the word for sugar of the same color. Nevertheless white pearls achieve the highest prices. The majority of pearls are small and irregular in shape, mainly 4–6 mm in size. Pearls of 8–9 mm in size are exceptions, and supposedly originate from *P. margaritifera*. Round or nearly round shapes are rare, usually the pearls are slightly to distinctly off-round or slightly baroque.

1.3.2. Red Sea

Pearls from the Red Sea, a center of mankind's oldest settlements, also have a long history like those from the Persian Gulf, although pearl fishing did not start to flourish until under the Ptolemaic rule (302–30 BC). While European seafaring nations and colonial powers began to impinge from the early 16th century onwards, the pearl fisheries were always controlled exclusively by the local people. The actual diving was often done by black slaves. Canoes were used for shallow waters and the diver initially searched the seabed using a form of bucket with a glass bottom. It seems that the few fisheries left today still use this method.

FIGURE 1.7 Carved, so-called "Jerusalem Shell", depicting the resurrection of Jesus Christ, ca. 1900. (*Source*: Kunz and Stevenson, 1908.)

Originally, pearls were found all over the Red Sea. Two areas were important: the grounds around Farasan Island in the south of the Saudi Arabian coast and the southwestern coast around the Dahlak Islands, just off the city of Mitsiwa in what is now Eritrea (Fig. 1.5). The lagoon of Djibouti in the Gulf of Aden was also fished for pearls, as early as in the third century BC.

During the pearl boom of the 19th century, fishing centered around Kosseir (Al Kusair) on the Egyptian coast and in the southern half on the western coast from Jiddah to the Gulf of Aden. The pearl trade was dominated by Indian dealers who often financed the pearling fleets. Pearls from the Red Sea are still found on the natural pearl market today, although extremely rarely. Both the Akoya pearl oyster and *Pinctada margaritifera* occur in the Red Sea.

The latter was often termed *P. margaritifera erythrensis* and its shells were an important trade item until the 1920s. Approximately 1,000t were exported each year around 1900. A small part went directly to Bombay, where they were traded as "Bombay Shell" but the London market also spoke of "Egyptian Shells". They were nearly exclusively used to make buttons due to their dark color. The MOP industry distinguished three main color qualities: "Jiddah" was dark, "Massowah" had a medium color and "Aden" had a light color. Some of the shells went directly to Jerusalem and Bethlehem, hence the name "Jerusalem Shell". The shells were processed in the Holy Land (Palestine, at the time under British mandatory rule) to make rosaries, crosses and many other souvenirs for pilgrims. Nearly the entire population of Bethlehem found work in the small domestic industry. The true "Jerusalem Shells" were full-sized shells that were engraved with Bible scenes of the life of Jesus Christ. They are collectors' items today (Fig. 1.7).

A project supported by the British colonial government in what is now Sudan dealt with the artificial breeding of *P. margaritifera*, it aimed at increasing the MOP shell production. The scientist in charge was Dr. Cyril Crossland and a hatchery station was established in Dungunab Bay on Rawaya Peninsula in 1912 (Fig. 1.5). Three million larvae were ultimately grown to adult size. A successful pearl culture project was started in the late 1920s.

Pearls from the Red Sea come nearly exclusively from Akoya pearl oysters and have a striking similarity to pearls from the Persian Gulf in both shape and size. Arabian dealers called pearls from Dahlak "durr" or "durra", meaning "shining drop", and traded them to African countries, while pearls from the Arabian side went directly via Alexandria to the European market where they were known as "Egyptian pearls". The local Arabian name for pearls is "lulu". Figure 1.8 shows pearls from the Red Sea that came to Germany via India in the 1990s. Intense pink and light to dark violet, gray-violet and orangey-violet colors seem to come only from the Red Sea. The colors are natural, and the shells of Akoya pearl oysters from the Red Sea may already show a violet or even dark red rim. They have, however, never played a role in the MOP industry.

1.3.3. East Africa and Madagascar

Individual grounds of Akoya pearl oysters and *Pinctada margaritifera* (also often described as *Pinctada margaritifera erythrensis*) stretch from the Gulf of Aden along the coast of East Africa to the Mozambique Channel (Fig. 1.9). The many islands along the East African coast, especially on the stretch between Pemba Island in northern Tanzania and Ibo Island in northern Mozambique, provide sheltered lagoons. Another center of pearl fisheries, recorded as early as around the year 1600, was the Bazaruto Islands in southern Mozambique (Fig. 1.9). The fisheries of the 19th century were intended to export shells to the world-wide MOP industry. A number of

FIGURE 1.8 Pearls with slightly varying hues from the Red Sea. Maximum size is 7 mm. (Photo: Elisabeth Strack. Pearls: Martin Wolf KG.)

German companies held fishing licenses in what was then German East Africa. Pearls, named "African pearls" were considered more of a by-product. They had probably been traded to the Persian Gulf and India by Arabian seafarers from the 7th century onwards. Around 1900, an American company and the "East African Pearl Society" established themselves on the Bazaruto Islands and Ibo Island. This was probably with the aim of building hatchery stations in order to create new sources for the American MOP industry.

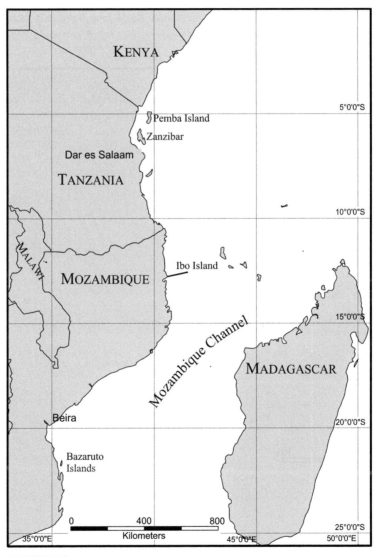

FIGURE 1.9   The coasts of east Africa and Madagascar.

At the beginning of the 20th century, Madagascar, which has rich grounds of several *Pinctada* species, was discovered by the French colonial government as a new source for MOP. Fisheries with modern diving equipment were conducted on the western coast and any pearls discovered went directly to the French market.

1.3.4. Strait of Manaar

The pearl grounds in the Strait of Manaar between India and Sri Lanka (Ceylon) (Fig. 1.10) have been known for about 2,500 years. Pearls are deeply rooted in both Sri Lankan and Indian culture, both Singhalese and Tamil languages call them "mootoo", going back to the Sanskrit name "mukta".

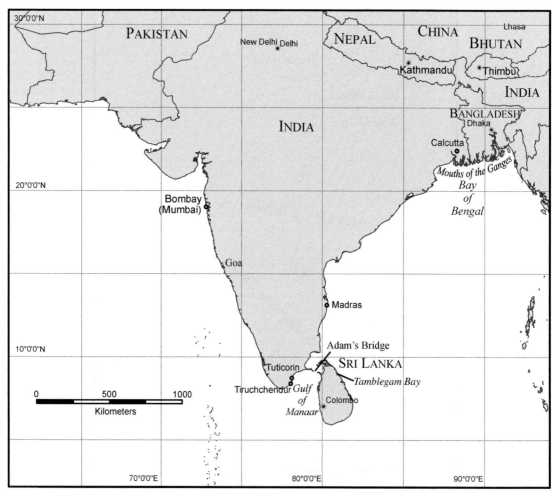

FIGURE 1.10   The coasts of India, Sri Lanka (Ceylon) and Bangladesh.

Phoenician traders were the first to bring pearls from the island of Ceylon, the "Taprobane" of European antiquity, to the Mediterranean area. They were followed by the Arabian seafarers who brought Indian pearls with their dhows to the Persian Gulf. In the last quarter of the 13th century, Marco Polo, son of a merchant family from Venice, traveled for 25 years to China and Southeast Asia. In his memoirs ("Il Milione") he provides a lively description of the pearling fleets that were then hardly different to those of the 19th century.

The legendary pearl grounds are situated on both sides of the passage between Sri Lanka and the southern tip of India, made up of the Gulf of Manaar, Palk Bay and Palk Strait and divided by the small islands of the so-called Adam's Bridge (Fig. 1.10).

Rich pearl grounds are concentrated in the northwest of Sri Lanka, where they stretch along the Bay of Kondatschy for about 50 km at a distance from the coast. They may periodically become densely populated with Akoya pearl oysters. There are also grounds within Palk Bay and on the eastern coast of Sri Lanka in Tamblegam Bay, which have been fished occasionally. On the Indian side, the pearl grounds lie also within the Gulf of Manaar, on the Tinnevelly coast between Tuticorin and Tiruchchendur.

Pearl fishing in the Gulf of Manaar followed the same pattern as in the Persian Gulf, but fisheries were not regular and only lasted 4–6 weeks. They were under the control of different colonial powers and during the British time supervision was conducted by expert colonial officers. Pearl fishing was mainly in the hands of a certain caste, the "parawa" who came from the Tinnevelly coast in South India; they were Tamils.

The Portuguese came to Ceylon in 1506 and proceeded immediately into a treaty with the Ceylonese king that guaranteed them an annual tribute of pearls. Pearl fishing took place regularly, even to the extent of probably destroying a large part of the pearl grounds. Dealers from Goa, a Portuguese settlement on the western coast of India, bought the pearls directly from the fishermen.

The Dutch came in 1640 and organized their first pearl fishery in 1666. Regular fisheries followed and Amsterdam began to supersede Lisbon as a new trading center for pearls. There was no fishing between 1732 and 1746, and from 1768 onwards it stopped altogether. When Ceylon became an English colony in 1796, the grounds had remained unfished for 28 years. The first fishery, conducted in 1797, was extremely successful. The same was true for the years to follow. This meant, the pearl grounds had turned into a remarkable source of income for the colonial power. The government was, however, wise enough to introduce long dormant periods. As 1903 and 1905 had been very good years, the Ceylon Company of Pearl Fishers was founded with the aim of introducing scientific hatchery methods, inspired by European oyster culture, in order to improve the situation of the pearl grounds. The project ended as a financial disaster. The same was true for a second project that aimed at increasing pearl production by infecting 50,000 pearl oysters with parasitic cestode larvae from the feces of certain fish. This project arose from the belief that pearls resulted from a response to invertebrate parasites imbedding in and irritating the flesh of pearl oysters.

Only three new fishing periods of any significance took place after 1923, the last being in 1956. They were followed by two more in 1979 and 1984 (probably due to a renewed interest in natural pearls by Arab countries).

During the British times, two thirds of all pearl oysters went to the government and the rest was divided between the ship owner, the divers and the rest of the team. The government auctioned its part of the molluscs in lots; they were left to rot and were later washed out. The thin shells were usually thrown back into the sea, and only towards the end of the 19th century, they were sold to the MOP industry, becoming known in Europe as "Eastern Lingah Shells".

The pearl merchants were organized in a rather complicated hierarchy of middle-men up to the representatives of the richest Indian merchants, the so-called "chetties" who came from Madras and Bombay and who bought at the government auctions. Nearly all the pearls went to Bombay and the city, now called Mumbai, is still a center for the trade in natural pearls.

The majority of pearls were sold to major and minor rajahs until the 1950s. In addition to this, extraordinary objects such as pearl carpets were produced. The Baroda Pearl Carpet, measuring 2.72×1.62 m and being made up of approximately 1.4 million seed pearls, is a fine example. It had been commissioned by the Gaekwar of Baroda.

The situation has changed today when many Indian dealers buy back pearls from the Maharajas who are forced to sell parts of their wealth. The urban Indian middle class has meanwhile discovered pearls for itself, although it will mainly buy cultured pearls. On an international level, India is still the number one trading place for natural pearls.

Pearls from the Gulf of Manaar have nevertheless nearly become a part of history. Called Indian pearls in the trade, their quality is generally comparable with those from the Persian Gulf, but there are more seed and dust pearls, and the yield from harvests is lower.

The Indian dealers are still grand masters of all methods for processing pearls. The pearls were, for instance, boiled in water for several days, after which they were kept in tiny glass bottles and placed in the sun. More than a century ago, Indian dealers adopted the European carat weight unit. "Chov" and "tank" are also still used. One tank equals 24 rati (1 rati = 7/8 carat). The chov unit, still used in Arab countries, specifies the volume of pearls and is an evaluation unit. The chov is calculated using tables on the basis of the weight. Another Indian weight unit is the tola (=58.32 carats).

1.3.5. India, Pakistan and Bangladesh

Banks of Akoya pearl oysters are found along the whole coast of the Indian subcontinent. In addition, there are more than 50 species of freshwater mussels that have yielded pearls over the centuries. In India, the Malabar Coast to the south of Mumbai (Bombay) and the Gulf of Kutch further to the north are notable. On the eastern coast, pearl grounds spread from Tuticorin up to the delta of the Ganges River (Fig. 1.10). Fishing has never reached the scale of the Gulf of Manaar.

In Pakistan, marine pearl fishing took place to the south of Karachi. The rivers yield colorful freshwater pearls that are all exported to India.

The pearl grounds of Bangladesh in the Bay of Bengal stretch into the border area with Myanmar, and they are still worked today. The pearls are all sold to India.

Pinctada margaritifera is known to occur and has probably delivered some of the larger pearls that can be found in the superb pieces of jewelry produced at the courts of the Mogul rulers during the 17th and 18th centuries. River pearls were found in the whole country during the 1960s and 1970s. In 1960, pink pearls from Bangladesh, which was then still Pakistan, were exhibited at the World Fair in New York.

Although there are well-known sites of *Pteria penguin* in all these areas, there have only been occasional discoveries of individual pearls. They can reach a size of up to 13 mm and are usually of baroque shape. The colors resemble the strongly iridescent nacre of the shells, the prominent colors of which are light pink, bluish-pink and dark pink, but also golden-metallic and grayish to black; hence the popular names "golden winged oyster" or "black winged oyster". The thinness of the shells has meant, however, that they were of no use to the MOP industry in the 19th century.

1.4. INDIAN/PACIFIC OCEAN

1.4.1. General

Pearl oysters are found between the Indian and Pacific Ocean in a widespread area that extends from the Andaman Sea to Torres Strait and Melanesia, where the Solomon Islands form the eastern border. In the north, the area includes the Celebes and Sulu Sea up to the Philippines and the South China Sea, from where it spreads as far as the Gulf of Tongking along the coast of China and Vietnam (Fig. 1.11).

Within the widespread geographical areas there are certain places with particularly rich pearl grounds and these have represented the center of the so-called pearling fleets in the past. The name was used to describe fleets equipped with diving teams that set off in search of shells of the large *Pinctada maxima* (see Section 9.3.1, Figs. 9.6 and 9.7). The pearls, although large and beautiful, played little role during the high time of the MOP industry, while the thick, heavy shells were highly sought after. They were divided into "gold-lipped" and "silver-lipped" shells; the former with a more or less broad, golden-colored rim and the latter with a distinctly white or silvery-white rim.

The pronounced European interest in the large shells of *P. maxima* began when MOP manufacturers were confronted with a rising demand from the newly emerging Bourgeoisie classes of the 19th century. The rapid industrial progress led to the development of machinery that helped to speed up production of buttons and the many beautiful small decorative utility and jewelry items. The demand for MOP objects appeared boundless as prosperity spread throughout society. Germany had small MOP industries in Adorf in Saxony and in Upper Franconia, which initially used the local freshwater shells. In Austria, Vienna was the center while Paris assumed this role in France. The first button factories opened in the mid-west of the United States at the end of the 19th century as a direct result of the so-called "pearl rush". They used the shells of local freshwater mussels. The resources were soon insufficient and the American MOP industry began to rely on the apparently endless supplies fished by the pearling fleets. New York became a leading import center. A large button-making industry also developed in Japan.

FIGURE 1.11   Southeast Asia.

Natural pearls from *P. maxima* emerged for the first time in noticeable, though still small quantities, on the European market during the late 19th century. In the centuries before, the rather sporadic finds had probably all gone to China and India. Size is the first evident characteristic of these pearls as baroques can reach 20 mm and more in length. They were never classified as oriental pearls and the natural pearl trade still today describes them as Australian pearls. More precisely, the name is used for silvery-white pearls from the Australian coast, while yellow and golden pearls are often called Philippine, Indonesian or Burmese pearls. The direct location of the find, if known, is added to the description. It is, however, impossible to specify a definite origin once the pearls are on the market.

Natural pearls from *P. maxima* have always been rare, even during the high days of the 19th century's pearl trade. Reports from divers indicated that not one pearl was found in 100t of shells. The shells of *P. margaritifera* and Akoya pearl oysters,

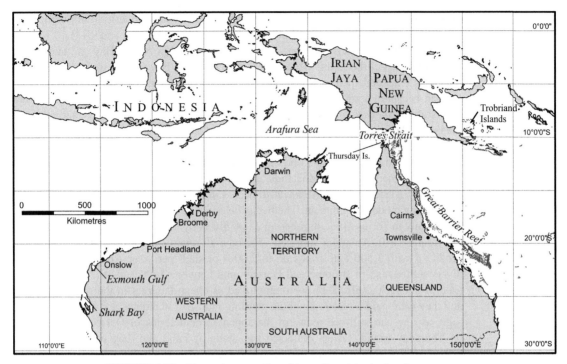

FIGURE 1.12   The northern coastline of Australia and the Arafura Sea.

occurring in the areas described above, were fished to a limited extent, the latter ones being named "Australian Lingah" or "Bastard Shell". There was not one place where Akoya oysters yielded pearls to a similar extent as in the Persian Gulf or the Strait of Manaar.

1.4.2. Australia

The most extensive grounds of *Pinctada maxima* are found along the Australian coast to the north of the Tropic of Capricorn for a length of nearly 3,500 km from the North West Cape (Exmouth Gulf) in the west to south of Cape York (Cooktown, north of Cairns) in the east (Fig. 1.12). The silver-lip is predominant while the gold-lip is confined to passages between islands on the northwestern coast. *P. margaritifera* occurs only in small populations and shark Bay on the Western Australian coast has *P. albina* (Lamarck, 1819).

The Australian MOP industry started abruptly in Shark Bay when 3t of *P. albina* shells were shipped to London in 1850. The tiny pearls <3mm in size were marketed in India and China. Already in 1861, the search for the giant, plate-like shells of *P. maxima* began in Exmouth Gulf. At first, the shells were picked by hand during low tide, followed by local divers who went out on small boats to areas further away. Twenty years later, a real fleet had emerged. When modern diving equipment was introduced the harvests were increased even higher (Figs. 9.9 and 9.11).

When further grounds of *P. maxima* were discovered in Nickol Bay and later along the whole coast of the Eighty Mile Beach up to Roebuck Bay just off Broome, the pearling fleets extended their activities and Broome became the roaring center (Fig. 9.7). Darwin became an additional center when the northern coast in the area of the Timor and Arafura Sea was developed. News of the new grounds spread quickly and attracted many people from different backgrounds to the then still wild North. From the 1870s onwards, pearling fleets also traveled to the northeast to Torres Strait where the pearl grounds stretch from the upper end of the Great Barrier Reef to New Guinea (Fig. 1.12). There are numerous reefs and islands, the most prominent of them are Thursday Island and Friday Island. *P. margaritifera* also occurs in the area. The first Japanese arrived in 1885, using modern equipment. The fleets, belonging to Japanese ship owners, settled at first mainly in Torres Strait, numbering 2,000 divers by the year 1900.

Around 1910, Australia supplied 75% of high-quality MOP around the world and produced with 3,500 employees up to 2,000 t of shell per year. After 1900, Japanese divers settled also in Broome and Darwin, and between 1900 and 1935, they fished a total of about 70,000 t. They were employed as cheap laborers in their home country and they were paid a piece rate in the pearling industry. Accidents and death rates were high, as one can still see today in the Japanese cemeteries in Darwin, Broome and on Thursday Island. Divers collected the shells by hand and they were immediately opened on board of the ships and searched for pearls, which the divers nearly always kept for themselves. The ship owners more or less silently tolerated this as, compared with the vast tonnes of MOP, the pearls presented only a small part of the profit and good divers were rare. The shells were usually dried, sorted and cleaned directly on board and then packed immediately into cases holding about 100 kg (Fig. 9.6).

The industry came to an end with the 1920s and was practically non-existent with the outbreak of World War II in 1939. Since the early 1960s, Australia has produced cultured pearls that are considered to be among the finest of the world (see Section 9.3.4.2).

The pearling fleets also extended their search to the surroundings of Papua New Guinea and into the Solomon Sea. It is interesting to note that the Trobriand Islands, belonging today to Papua New Guinea, had rich banks of Akoya pearl oysters that the locals called "lapi shell". They produced small, predominantly yellowish, pearls that mainly Russian dealers bought during the 19th century. They sold them in their own country, together with their local river pearls.

1.5. SOUTHEAST ASIA

1.5.1. Myanmar (Burma)

Myanmar has rich banks of *Pinctada maxima* in the Mergui Archipelago in the Andaman Sea (Fig. 1.11). They are considered as the grounds geographically farthest to the west, with the hundreds of islands and coral reefs offering good ecological conditions. The grounds continue down the western coast of Thailand to Malaysia and from there to the Strait of Malacca between Malaysia and Sumatra. In the Mergui Archipelago, pearl fishing was a century-old tradition for the local Salongs or Salons, maritime nomads who had always been acquainted with the products of the sea. Shells and pearls, and probably also the meat of the pearl oysters, were an early trade item for them, exchanging them with passing ships from India and China. The age-old traditions changed with the arrival of the pearling fleets.

The "Mergui Shells", as they were soon known, began to play a role on the world market and were shipped to London and Hamburg via Singapore. The pearls, no more than occasional finds, soon acquired a reputation due to their fine quality. They were succeeded in the 1960s by cultured pearls of equally fine quality.

1.5.2. The Philippines

The rather shallow waters with their coral reefs between the numerous islands of the Philippines provide good conditions for both *Pinctada maxima* and *P. margaritifera*. Especially rich grounds of *P. maxima* are located in the Sulu Sea between the Philippines and the Indonesian Kalimantan Island (Borneo) (Fig. 1.11). The Sulu Archipelago was under the rule of the Islamic Sultanate of Sulu, and its inhabitants lived an independent life as maritime nomads. Being talented divers, they engaged in sporadic pearl fishing, relishing the meat of pearl oysters that they called "muchia". The pearls did not really seem to have played an economic role as they were worn by children.

Chinese trading ships crossing the waters to the Philippines appear to have been the first foreigners who showed an interest in both the pearls and the shells, and it can be assumed that they traded pearls and shells for porcelain hundreds if not thousand years ago. The Chinese were followed by Arabian and Indian dealers, who came with the Islamic missionaries. In the 19th century, the Chinese dominated once more, taking probably all pearls of fine quality to China where they found a suitable use on the head-dresses or earrings worn by Mandarins. However, the most beautiful pearls remained in the family of the Sultan of Sulu.

The Sulu Archipelago became a popular destination for the pearling fleets, at first owned by Chinese who introduced diving equipment. Later on, Australian and Japanese fleets, constantly on the search for new fishing grounds, also came. The shells were exported from Manila, earning themselves the name "Manila Shell", which became something like a quality name.

The Philippines remained an important source of MOP for the United States until after World War II. Within the country, MOP is still used locally, but the golden age of elaborate processing is over. An example of the latter is the fine ornaments that still adorn the verandas of early Spanish houses in Manila. The Philippines have had a flourishing cultured pearl industry for the last 30 years (see Section 9.3.6).

1.5.3. Indonesia

Pinctada maxima is found on reefs between nearly all the islands of Indonesia. Pearls and MOP from the Malaysian Archipelago have been known in Europe since the times of the Portuguese seafarers. The islanders, who dived only in shallow waters and collected the shells by hand, exchanged their products with passing merchant ships.

The local inhabitants did not, however, discover their own rich pearl grounds until pearling fleets from the Torres Strait and the northwest of Australia arrived in the late 19th century.

The Aru Islands (Kep Aru) within the Arafura Sea proved to be very rich fishing grounds. The city of Dobo on Wokam Island in the north became a center for the pearl trade, and the large, silvery-white "Dobo pearls" were soon known on the European market. Dobo is still a trading place for natural pearls (Fig. 1.11).

The pearling fleets of the 19th century skillfully ignored any bans issued by the Dutch colonial government by entering into agreements with the local chiefs. The Dutch showed little interest in their MOP resources, while the international market considered shells from Indonesia as valuable as those from Australia. "Makassar Shells", "Banda Shells", "Ceram Shells" and "Penang Shells" soon earned themselves a reputation on the world market. Indonesia has been a producer of South Sea cultured pearls since the 1960s (see Section 9.3.5; Table 9.2).

1.5.4. China

Three species of *Pinctada* have been found producing pearls along the southern and southeastern coast of China (Fig. 1.11). *Pinctada maxima* occurs in the Gulf of Tongking, particularly on the coasts of Hainan Island from where some of the large, so-called Mandarin pearls, of the past came. These pearls were also imported from the Mergui and the Malayan Archipelago. *P. margaritifera* is found along the whole southern coast into the Strait of Taiwan. The grounds were probably reduced by overfishing as early as 400–500 years ago. Black pearls have been known in China for centuries and were considered a symbol of wisdom.

The Akoya pearl oyster has yielded large quantities of pearls for more than 2,000 years. Natural pearls have come from along the whole Chinese coast from the border with Vietnam to the coasts of Fujian province. Some areas excelled insofar as they have regularly produced pearls of good quality. In the Gulf of Tongking, especially, the pearls from Hepu Bay have been famous in the past. When the last dowager empress, Cixi, died in 1908, she was reported to have held a necklace with pink "Hepu Pearls" in her hands and that the necklace was subsequently placed into her grave. Other significant areas are Hainan Island, Leizhou Peninsula in Guangdong Province and the Strait of Taiwan between Fujian and Taiwan Island (Fig. 1.11).

Pearls are mentioned in old scripts dating from the third millennium BC. Until the 19th century, large pearls were used to decorate caps and dresses worn by mandarins. The pearls were drilled with two holes, like a button, so that they could be sewn on. The double drilling reduced the value of a pearl, at least from the point of view of the Western market. Today's market is still familiar with the so-called "Mandarin pearls" and also calls them "Chinese pearls".

Pearl fishing started near Hepu and Beihai in Guangxi Province as early as during the Han dynasty (206–220 AD) and it is estimated that 2,000 to 4,000 t of pearl shells have come from the area during the nearly 2,000 year since then. In all dynasties, the emperors had the exclusive rights and all pearls were first offered to them. During the Ming dynasty (1368–1644 AD), pearls were highly appreciated. The empress Tzu Hsi employed approximately 100,000 pearl fishers in about 1526. When rakes and dredges were used in later centuries, this led to irreparable destruction of the pearl grounds.

Only sporadic fishing took place after World War II. The end came with the Cultural Revolution and with the first cultured pearl farms that were built in the middle of the 1960s.

The processing of MOP into artistic carvings is an old Chinese tradition. Examples of these are gambling chips that were produced primarily between 1736 and 1796 during the rule of Emperor Ch'ing Lung with meticulous engravings. These depict scenes of Chinese life taken from a total of 34,000 parables that the emperor himself wrote in an attempt to show his subjects a more spiritual way of life according to the teachings of Confucius.

1.5.5. Vietnam

Vietnam borders on to the Gulf of Tongking in the northeast; *Pinctada maxima*, *P. margaritifera* and the Akoya pearl oyster are found there in the same way as along the Chinese coast. Little is known about natural marine pearls in Vietnam today.

Orange-yellow pearls from the large marine gastropod *Melo melo* (Lightfoot, 1786) have become known from Vietnam. They represent some of the most valuable pearls available on the market. A Vietnamese legend speaks of the tears of a princess who had lost a war against an enemy, her own husband. Molluscs collected her tears in the eastern Sea and transformed them into colorful pearls. There have been pilot projects for cultured pearls since the 1990s and an Akoya cultured pearl industry, even if small, has been successfully instigated (see Section 9.2.4).

1.6. PACIFIC OCEAN

1.6.1. Japan

The Japanese people did not seem to have appreciated pearls until the rise of the Chinese influence from the 6th century onwards. Systematic pearl fisheries started only in the late 19th century, after the country opened up during the Meiji restoration. They concentrated on the southeastern coast where the Akoya pearl oyster ("akoya gai") is especially found. The fishing areas included Ago Bay on Ise Peninsula in Mie Prefecture where in the early 1890s Mikimoto started the production of cultured pearls (Fig. 1.13). The natural pearls, among them the so-called "Ise pearls" from Ago Bay, were sold via Tokyo and Shanghai to the international pearl market and tonnes of small seed pearls sold to China were processed into pearl medicine. The shells, which were not very significant in the MOP industry, became known in Europe as "Japanese Lingah".

The shells of *Pinctada margaritifera*, occurring to the south of Kyushu Island in the islands of the Ryukyu group, were fished for the 19th century MOP industry.

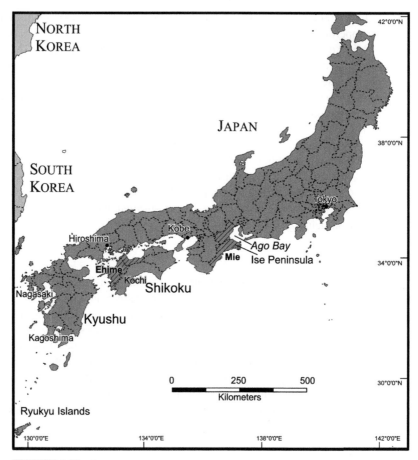

FIGURE 1.13   Japan.

The species was used by Mikimoto when he started cultured pearl production in the Ryukyu Islands. The Ryukyu Islands have notable grounds of *Pteria penguin*, the "winged oyster" or "butterfly shell". The Japanese name is "Mabé gai"; the word "Mabé" most likely stems from an old dialect that is spoken in the Ryukyu Islands. Another Japanese name, "Eboshi gai", draws a similarity between the shell's shape and the ornaments that Shinto priests wear on their heads. The species has been used successfully for the production of blister pearls since the 1950s (see Section 9.6.2).

1.6.2. Oceania

The island world of Oceania in the Pacific Ocean (Fig. 1.14), divided into Melanesia, Micronesia and Polynesia, is the largest geographical area for pearls from *Pinctada margaritifera*. The market did not discover black pearls until the 19th century

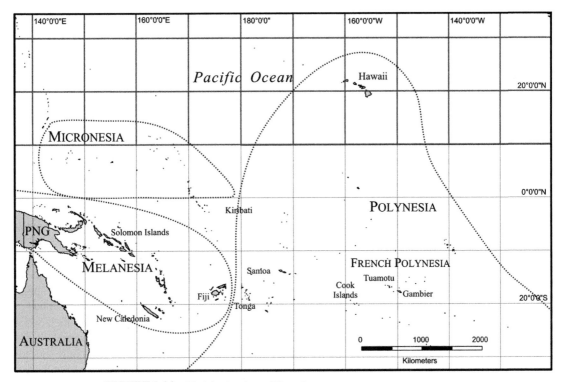

FIGURE 1.14 The island nations of Oceania.

(Lintilhac, 1985). The pearls come in a wide range of black and gray hues that may pass into dark blue, dark green and dark brown.

Although Catherine the Great added a necklace with 30 black pearls to the Russian crown jewels in the 18th century, there was only reluctant interest in Europe. Black pearls from Mexico appeared shortly after 1800, but interest started only when the French empress Eugenie had a necklace with black pearls made for her. When it was sold after the Franco–Prussian War by Christie's for £4,000, prices for black pearls began to increase. By the year 1900, they had multiplied 25 times, when compared with the year 1800. In 1908, G.F. Kunz mentions a drop-shaped black pearl of 49 grains (12.25 carats) in weight that was sold for US$40,000.

Today's natural pearl market still speaks of Pacific pearls to describe colored pearls from the vast area of the Pacific Ocean. French Polynesia and the Cook Islands have remained the most important sources, and the term "Tahiti pearl" is exclusively reserved for pearls from French Polynesia. The natural pearls are still considered absolute rarities that can reach high prices, although current prices no longer reach the exorbitant heights of about 100 years ago.

Pacific pearls were probably known in China from early times while Europeans came to know them only after the conquest of the Pacific Islands by Portuguese and

Spanish seafarers in the 16th century. The Polynesians themselves used the black shells for making tools, hooks and harpoon heads, and also for personal decoration and jewelry. Captain Cook brought a priest's mourning dress back to England from his first visit to Tahiti in 1769. It had a mask and breastplate made of whole, polished black shells from *P. margaritifera.*

The first regular shipments of black MOP went to Sydney between 1811 and 1814, but organized fishing did not start until 1827 when European traders sent many large ships from Chile. In 1827/1828 the Englishman Hugh Cuming, a self-made man and passionate shell collector (after which the French Polynesian subspecies, *P. margaritifera cumingii*, received its name) went with his own ship around the islands of the Tuamotu Archipelago and set up pearl fishing on Marutea South atoll by bringing divers from other atolls for help. During a period of 48 days they gathered 28,000 pearl oysters. They were of little value and Cuming soon abandoned his plans.

Soon ships from all nations headed for the islands, and the black shells became known as "Sydney Shell" (due to their export via Sydney), "Gambier Shell" or "Auckland Shell" (when exported via Auckland). Buttons made of black MOP were called "smoked pearl" in the trade and became fashionable from 1880 onwards. The local Polynesian divers were able to remain underwater for 60–90 seconds and simple wooden diving goggles were used later. The French colonial government issued a ban on diving suits in 1892 as the inhabitants of many atolls made their living from shell diving (Fig. 1.15). Export figures rose rapidly between 1900 and 1930, leading to the depletion of stocks in many atoll lagoons. Pearls were originally a by-product, found only sporadically, and the Polynesians themselves disregarded pearls. They also did not know how to drill them. When black pearls came into fashion in the second half of the 19th century, middlemen and agents of well-known Paris companies resided permanently in Papeete so that they could make an immediate offer once a spectacular pearl had been found.

The pearl grounds of New Caledonia were already overfished by 1910. Many Pacific islands suffered a similar fate, especially Hawaii. There was intensive fishing

FIGURE 1.15   Pearl divers on an island in the Tuamoto Archipelago, around 1900. (*Source*: Kunz and Stevenson, 1908.)

for a short period on Pearl and Hermes Reef between 1927 and 1930 for factories in New York and San Francisco. The US Bureau of Fisheries sent Dr. Paul Galtsoff to Hawaii in 1927 in order to implement a re-population program with *P. margaritifera* on the larger islands.

Over the past 40 years, a successful cultured pearl industry has emerged in the Pacific area in French Polynesia and the Cook Islands (see Section 9.4). Other Pacific nations are on their way. The so-called "Pipi pearls" have to be mentioned here. They come from *P. maculata* (Gould, 1850), which occurs throughout the Pacific Ocean. The tiny golden pearls, rarely larger than 5 mm, called "poe pipi" in Maori language, are mainly found in the Cook Islands.

1.7. CENTRAL AMERICA

The natural pearl market calls pearls from Central America "American pearls", but also uses more specific names like Mexican pearls, Venezuelan pearls, Panama pearls or La Paz pearls, etc. The old description "occidental pearls" goes back to the 16th century, but it was no longer used after 1930 and is, contrary to "oriental pearls", outdated today.

American pearls seem to always show a certain resemblance to each other, even when they come from different species. They occur in different shapes and sizes. Pearls from *Pinctada mazatlanica* reach the largest of up to 20 mm while the average size is less than 10 mm. Pearls from *Pteria sterna*, which usually produces better yields, reach a maximal size of 13 mm. Pearls from Akoya pearl oysters, coming from Colombia and Venezuela, rarely reach more than 7 mm, with an average size of only 3–4 mm. In general, shapes are small and irregular and there are many tiny white seed pearls.

American pearls occur in different colors, often of intensive and even dark hues: yellowish, dark brown and black, silver and gray. In the early 20th century, Venezuela was considered the main source of valuable golden pearls.

The pearls became nearly forgotten in the decades following World War II but since around 1990 a few specimens of the "Baja Pearls" as they are called in the United States, are seen again on the market.

1.7.1. Western Atlantic Zone

Akoya pearl oysters originally occurred as far north as Virginia on the North American coast and south to northern Brazil in South America (Fig. 1.16). The shells of the Atlantic Pearl Shell, used to distinguish them from the similar or identical species in the Indian Ocean, were used to a limited extent in the European MOP industry. They were very thin and more colorful than those from the Indian Ocean, even bronze colored and black, and they became known as "Venezuela Lingahs".

Pteria colymbus (Röding, 1798) occurs together with Akoya oyster and represents, with its thin, bluish shells, the western Atlantic species of the genus *Pteria*. *Pteria longisquamosa* (Dunker, 1852) also occurs. It is another, even rarer Atlantic species, but is prevalent on the coasts of Colombia and Venezuela. Neither species is significant

FIGURE 1.16   The countries of central and South America.

for pearl formation, but the shells have a rather strong iridescence and were occasionally used by the European MOP industry.

Columbus set foot on the North American continent during his third voyage on August 5, 1498. Ten days later, he discovered the Margarita and Cubagua islands about 20 km off the coast (Fig. 1.16). The first Indian women who crossed his path wore numerous pearls, and he proceeded immediately to exchange the pearls for colorful pieces of broken glass and earthenware. He collected about 1.5 kg of valuable pearls, the most beautiful of which probably later became a part of the legendary Venezuelan necklace owned by Queen Isabella of Spain.

Margarita Island, named after the pearls found along the coasts, was the most important site for pearls in Venezuela until the 1920s. The first pearl boom was already over in 1543, due to reckless plundering, and fishing came to an end in the 17th century. The pearl grounds stretched over a length of 800 km from Paria Peninsula to Coro, including a number of small islands to the south of Margarita Island. After a break of nearly 200 years the pearl industry started again in 1823, when the Congress of Great Colombia granted the monopoly for pearl fishing to Bridge & Rundell, an English company. Soon the threat of overfishing arose once more and the government banned dredging. The last golden age of Venezuelan pearls lasted from around 1900 until 1930, when a veritable pearling fleet with hundreds of ships set off from Cumana and Carupano. Diving suits were introduced in 1909. Diving concentrated on shells for the MOP industry, and the pearls were a by-product. There were enough pearls, however, for European pearl traders to be traveling to Venezuela again. When the government commissioned a scientific survey in the early 1990s it was found that the pearl grounds were still suffering from the Spanish ravages of the early 16th century. There have been small pilot projects for pearl culture.

In Colombia, the pearl grounds concentrated on the coasts of Guajira Peninsula, between Riohacha and Cabo de la Vela, and also into the Gulf of Venezuela. Cartagena was the center of the pearl trade, where the Spanish ships anchored in order to collect the tribute of pearls for the Spanish crown. Like Venezuela, the pearl grounds fell into neglect and the early 19th century brought better times. Around 1900, about 20 million pearl oysters were still fished annually. The divers were local Indians. For several years, there were a number of pilot projects for pearl cultures in Colombia.

Pearl fishing also took place in the Gulf of Darien in the border region between Colombia and Panama, extending to the Panamanian coast. It is fair to assume today that the pearls were marketed together with pearls coming from Panama's Pacific coast.

1.7.2. Eastern Pacific Zone

1.7.2.1. Panama

The Eastern Pacific Zone features *Pinctada mazatlanica* (Hanley, 1855). The zone runs along the eastern coast from southern California to Peru and includes the Galapagos Islands (Fig. 1.16). Pearls from *P. mazatlanica*, the Panamian or Calafia pearl oyster, only ever came in significant quantities from the Gulf of California in Mexico and the coasts of Panama and Costa Rica.

The pearl grounds of Panama are within the Gulf of Panama on the Archipelago de las Perlas, which consists of 43 islands, including Isla del Rey. The most famous pearl in the world, the Peregrina, came from one of the small islands. Vasco Nunez de Balboa obtained it from a slave to whom he gave his freedom. The Peregrina weighs 203.84 grains (50.96 carats) and has a drop shape. Balboa presented the pearl to the Spanish king Ferdinand V and, in 1513, it became part of the Spanish crown jewels. When the daughter of Philipp IV, Maria Theresa, married the French king Louis XIV in 1660, she brought the pearl with her to France. In 1814, Napoleon's brother Joseph Bonaparte took the pearl on his way back to France and gave it to Hortense de Beauharnais a few years later. After 1837, the pearl passed on to her son, Emperor Napoleon III. In 1873, after he was forced to leave France, he sold it to the Duke of Abercorn. In 1969, Richard Burton bought it at an auction of the Parke Bernett Galleries in New York, he paid US$37,000. The pearl, later set by Cartier in the drop pendant of a lavish necklace, still belongs to Elizabeth Taylor.

The Spanish town of Seville held a monopoly of pearls from Panama from the 16th century onwards. Wealthy Spanish merchants in Panama conducted the pearl diving by leaving their slaves to do the work. Pearls still came regularly on the market during the 19th century, and the European MOP industry discovered shells from Panama around 1850. Soon several hundred divers worked in the Gulf of Panama, fishing for the rather thick shells with their dark rim. The "Panama Shell" or "Bullock Shell" were less valuable than Australian shells.

Pearls, at first rather a by-product of the fisheries, became fashionable on the European market around 1900, due to the popularity of Pacific pearls that they resembled. The local inhabitants did their own fishing and used the shells to decorate houses and graves, and wore the pearls as talisman. They disregarded black pearls as omens of bad luck and even went so far as to burn them.

Costa Rica, Nicaragua and Honduras (Fig. 1.16) had their own fisheries for MOP in the late 19th century. Diving suits were introduced in 1906, but soon afterwards the expeditions came to an end. The Indian inhabitants of Ecuador and Peru owned enormous quantities of pearls when the Spanish came, pearls were even quite common. The local pearl grounds, about which little information is available, ceased to exist during the 17th century.

1.7.2.2. Mexico

In Mexico, the Gulf of California has been the center of pearl fishing. *Pinctada mazatlanica* occurs in the Gulf, accompanied by substantial quantities of *Pteria sterna*. The so-called "western winged oyster" has a gray to violet nacre and displays a strong iridescence. While the shells were usually considered too thin for use in the MOP industry, the pearls have reached the same importance as those from *Pinctada mazatlanica*. Today's Mexican cultured pearl production is based on *Pteria sterna* (see Section 9.6.6.2). The pearl riches of the Gulf of California were discovered when Hernando Cortez came in 1535 to what is now called Baja California. The Spanish were deeply impressed with the new pearl riches and pearl fishing was declared a right of the crown in 1586. Adventurers from Europe came in the 17th century, which became the "century of the pearl seekers", but from the 1680s onwards the pearl grounds fell under the control of the Jesuit missionaries of the Catholic church. The pearl grounds were left to themselves until 1740, leading to rich harvesting seasons that were mainly supervised by a certain Manuel de Ocio who, due to his wealth acquired by pearl fishing, developed an independent economic structure for the area. Pearl fishing came to a temporary stop in the early 19th century until the shells of *Pinctada mazatlanica* were discovered by the European MOP industry in 1836. The export of shells rose continuously from 1840 onwards and eventually became more important than the export of pearls. La Paz became the center for the pearling fleets that employed Indian divers and entered a period of prosperity that was to last until the 1920s. In 1874, new compressed air diving equipment was introduced and a new era began in La Paz. The exclusive right for pearl fishing was issued to five foreign companies between 1884 and 1912 and the Mexican pearl grounds suffered from reckless exploitation (see Section 9.5.1).

In 1903, Gaston Vives from an influential La Paz family established the Compania Criadora de Conchy y Perla California for running a large hatchery facility on Esperito Santo Island, in the Bay of San Gabriel, by collecting and rearing larvae of *P. mazatlanica* (see Section 9.5.2; Figs. 9.22 and 9.23). Between 1909 and 1914, Vives harvested between 200,000 and 500,000 pearl oysters annually for ca. 900 t of shells. Gaston Vives himself sold the pearls directly to the large European jewelry houses. The MOP industry came to a stop with the Mexican revolution, and Vives was especially personally harmed as the revolutionary troops destroyed all his property including the hatchery that had aimed at restoring and maintaining the pearl grounds.

The years following the revolution saw a decline, and fishing was totally banned in 1940 after there was a disease in 1938 that wiped out stocks of pearl oysters. The 1960s saw renewed interest from the government in developing re-population strategies and in eventually developing a cultured pearl industry. In 1997, a research team from the University of Guaymas produced the first commercial Mexican cultured pearls and has kept on with production since (see Section 9.6.3; Kiefert *et al.*, 2004).

1.8. SUMMARY

The beauty of pearls and MOP has been appreciated from very early times and they have been used as decorations since antiquity. Pearls have also been traditionally used as symbols of wealth and power for royalty and aristocrats. Records in the Bible, Koran, and ancient Greek, Roman and Chinese writings show appreciation of their value and special qualities. Throughout the Middle Ages pearls were used lavishly and luxuriously in Europe, the Middle East and India. Pearls were also given religious significance and used to decorate cathedrals and Christian artifacts. The European pearl trade reached its final high period after 1850, and lasted until the end of the 1920s. During this period the middle class became increasingly wealthy and was able to purchase modest pearl jewelry.

Pearl oysters occur around the world in tropical and subtropical regions. There are or have been many fishing industries for MOP and pearls around the world, and they often have long and complex histories. Many countries with a long history of pearling and which were major producers now contribute little to world markets, which are dominated by cultured pearls. Other countries, such as Australia and French Polynesia, which have relatively brief histories of pearling, are now significant contributors. The scene changed dramatically in the 1900s with development of a technique for producing cultured pearls. The cultured pearl industries that subsequently developed are described in Chapter 9.

References

Kiefert, L., McLaurin-Moreno, D., Arizmendi, E., Hanni, H.A., Elen, S., 2004. Cultured pearls from the Gulf of California, Mexico. *Gems Gemmol.* 40, 26–38.

Kunz, G.F., Stevenson, C.H., 1908. *The Book of the Pearl: the history, art, science and industry of the queen of gems.* The Century Company, New York. 548 pp.

Lintilhac, J.P., 1985. *Black Pearls of Tahiti.* Royal Tahitian Peark Book, Papeete. 109 pp.

Strack, E., 2006. *Pearls.* Ruhle-Diebener-Verlag GmbH & Co, Stuttgart. 707 pp.

CHAPTER 2

Taxonomy and Phylogeny

Katsuhiko T. Wada and Ilya Tëmkin

2.1. INTRODUCTION

The term "pearl oysters" has traditionally been applied to bivalves of the genera *Pinctada* Röding, 1798, and *Pteria* Scopoli, 1777, included in the family Pteriidae Gray, 1847, that is believed to have originated in the Triassic approximately 230 million years ago (Hertlein and Cox, 1969; Skelton and Benton, 1993). Pteriidae are distributed predominantly in shallow waters of the tropical and subtropical continental shelf regions around the globe, and are particularly abundant in the Indo-Pacific. Geographic ranges of two species, *Pteria hirundo* (Linnaeus, 1758) in the eastern Atlantic and members of the *Pinctada fucata/martensii/radiata/imbricata* species complex[1] in the western Pacific, however, extend into temperate regions (Ranson, 1961; Hayes, 1972). As throughout their history, pteriids continue to occupy epifaunal habitats forming byssal attachments, either individually or in large clusters, to diverse substrata in coral reef, hard and soft bottom environments. Pteriids have planktotrophic larvae and substratum selection occurs prior to metamorphosis. The ecological constraints of epifaunal life habits are largely due to requirements for physical stabilization and defence against predation.

Members of Pteriidae are characterized by a laterally compressed and obliquely ovate shell with a straight hinge line, deep byssal notch, and frequently elongated, posteriorly projecting extensions on the hinge (posterior auricles). Extensive individual

[1]See Section 2.3.1.13.

variation in shell shape and color (largely influenced by environmental conditions), and a limited set of discrete, easily diagnosable morphological characters precluded establishment of an unambiguous taxonomy. The identification of many pearl oyster species by shell characters is difficult, particularly at the early stages of development, because shells are similar in shape and color, and typically thin, fragile, and easily damaged when dried. Many problems of pteriid systematics have recently been addressed using new molecular and anatomical data, but these efforts have not yet been translated into a new taxonomic framework. Given the present state of flux of pteriid systematics and taxonomy, this chapter is not intended to propose a new taxonomic system but to provide a review of the current state of pteriid systematics and phylogenetics, and to establish comprehensive redescriptions of commercially important species of the family in light of recent findings.

2.2. GENERA

In addition to the pearl oyster genera *Pteria* and *Pinctada*, the family Pteriidae includes genera *Electroma* Stoliczka, 1871, and *Pterelectroma* Iredale, 1939 (Newell, 1969; Boss, 1982; Butler, 1998). The taxonomic status of monotypic *Pterelectroma*, containing a single enigmatic species, *P. zebra* (Reeve, 1857), is uncertain. The currently accepted taxonomic framework for pearl oysters is presented below:

Phylum Mollusca
    Class Bivalvia Linnaeus, 1758
        Subclass Pteriomorphia Beurlen, 1944
            Order Pterioida Newell, 1965
                Suborder Pteriina Newell, 1965
                    Superfamily Pterioidea Gray, 1847 [1820]
                        Family Pteriidae Gray, 1847 [1820]
                            Genus *Electroma* Stoliczka, 1871 (U. Cretaceous – Recent)
                            Genus *Pinctada* Röding, 1798 (Miocene – Recent)
                            Genus *Pteria* Scopoli, 1777 (Triassic – Recent)
                            Genus *Pterelectroma* Iredale, 1939 (Recent)

2.2.1. Outline of taxonomy at the generic level

Previous taxonomic works on Pteriidae focused either on individual species (Mikkelsen *et al.*, 2004), regional revisions (Prashad and Bhaduri, 1934; Hynd, 1955; Rao, 1968; Matthews, 1969; Hayes, 1972; Rao & Rao, 1974; Wang, 1978; Bosch *et al.*, 1995; Hayami, 2000; Hwang and Okutani, 2003; Matsukuma, 2004), or discussed pteriid taxa in the context of large-scale faunistic studies (Abbott, 1974; Habe, 1977, 1981; Oliver, 1992; Lamprell and Healy, 1998; Higo *et al.*, 1999). The only worldwide revision of the genus *Pinctada* (Jameson, 1901) was based exclusively on shell characters and its taxonomic value was greatly compromised by not being consistent with the guidelines of the Code of Zoological Nomenclature (International Commission on Zoological Nomenclature, 1999). Subsequent significant publications (Hynd, 1955; Ranson 1961) provided extensive synonymies for the majority of species of *Pinctada* but considered few type specimens and did not include justification for the taxonomic decisions. The genera *Pteria* and *Electroma* have never been revised [notwithstanding the extensive synonymies compiled by Fischer-Piette (1980, 1982)]. A worldwide, family-level revision has not been attempted since the largely outdated and exclusively shell-based monographs by Reeve (1857) and Dunker (1872–1880).

Recent phylogenetic analyses aimed at resolving higher-level relationships within the subclass Pteriomorphia or, more narrowly, within the superfamily Pterioidea based on molecular (Steiner and Hammer, 2000; Giribet and Distel, 2003; Matsumoto, 2003; Tëmkin, 2004) and morphological (Tëmkin, 2004, 2006a) data indicated that Pteriidae is not a natural (monophyletic) group. To circumvent the issue of dealing with problematic phylogenetic affinities, it would suffice to treat the genera *Pteria* and *Pinctada* independently in the context of the superfamily Pterioidea (see Section 2.4).

2.2.2. Generic descriptions

The genera of Pteriidae have traditionally been defined by shell shape (Hertlein and Cox, 1969; Hayes, 1972; Habe, 1977; Oliver, 1992; Lamprell and Healy, 1998). In addition to gross differences in shell shape, the genera *Pteria* and *Pinctada* are distinguished by the pattern of hinge teeth and the shape of the posterior adductor muscle scars. Several soft anatomical characters, most notably the pattern of intestinal coiling and the relationship of the ventricle and the intestine, have also been proposed as genus-level diagnostic characters (Hayes, 1972; Mikkelsen *et al.*, 2004). The following brief descriptions of these two genera are not intended to be exhaustive, but they aim to provide sufficient morphological details to allow a definitive identification of the genera in the field or in the laboratory.

Genus *Pteria*: The shell is strongly to moderately obliquely-ovate in outline. The hinge line is straight and extended anteriorly past the umbo into subtriangular to subquadrate anterior auricles and posteriorly into characteristically long and deeply sinuated posterior auricles. The exterior shell surface sculpture consists of commarginal periostracal and, in some species, calcitic scales that typically abrade with age. The hinge teeth are represented by an anterior subumbonal tooth in the left valve and a posterior submarginal oblique ridge in the right valve with complementary sockets in respective opposite valves. The adductor muscle is oval-shaped with the distal extremities of the posterior pedo-byssal retractor muscle adjacent or proximal to its anterior border. The byssus is composed of fused byssal threads that emerge from the base of the byssal groove ventral to the short foot. The inner surfaces of the mantle lobes are not fused to the sides of the visceral mass. The mantle margins are divided into four folds. The ascending arm of the intestine passes to the right side over the straight, posteroventrally extending descending arm. The rectum passes through the ventricle of the heart and terminates with a membranous anal funnel, subtriangular in outline. The ctenidial ocelli ("cephalic eyes") are absent in some species but present in both ctenidia in other species.

Genus *Pinctada*: The shell is anteriorly oblique, inequilateral, laterally compressed, and subcircular to quadrate in outline. The hinge line is straight and extended anteriorly past the umbo into subtriangular anterior auricles and posteriorly into weakly developed and slightly sinuated posterior auricles (obsolete in adults in some species). The exterior shell surface sculpture consists of commarginal prismatic scales that typically abrade with age. When present, hinge teeth are represented by an anterior subumbonal tooth in the right valve and posterior submarginal oblique ridge in the left valve with complementary sockets in respective opposite valves. The periostracum is thin and typically abraded in adults. The adductor muscle is crescent-shaped with the distal extremities of the posterior pedo-byssal retractor muscle inserted in the concavity on its anterior border. The anterior pedo-byssal retractor muscles are asymmetrical. The byssus is composed of discrete byssal threads and emerges from the base of the byssal groove ventral to the short foot. The inner surfaces of the mantle lobes are fused to the sides of the visceral mass. The mantle margins are divided into three folds. The ascending arm of the intestine passes from left to right side over the straight, posteroventrally descending arm. The rectum passes dorsal to the heart and terminates with a membranous anal funnel, lanceolate in outline. The ctenidial ocelli are absent or present in the left ctenidium only.

2.2.3. Approximate number of species and distribution

Although the exact number of pteriid species cannot be established at present, new results suggest that previously reported numbers of pteriid species – in the order of 50 (Mikkelsen *et al.*, 2004) – are likely to be overestimated (Yu and Chu, 2005a, b). Within the genus *Pteria* approximately 20 species have been named from the Indo-Pacific region, whereas only four species have been recognized from the Atlantic coasts of the Americas. As for the genus *Pinctada*, about 14 scientific names are applied to species from the Indo-Pacific region, including Baja California, and only two species have been recognized from the Atlantic coasts of North and South America. Most species of Pteriidae are tropical and subtropical, but the distributions of some species extend to higher latitudes. Among the Pacific species, *Pinctada fucata* has been recorded from Iwate Prefecture in Japan (as *P. radiata* in Higo *et al.*, 1999) and in Victoria, Australia (Hynd, 1955). *Pteria hirundo* recorded from southern England may be the northmost record for an Atlantic species while records from the southern Atlantic included Rio de la Plate between Uruguay and Argentina for the same species (probably in error) (Hayes, 1972).

There are several reasons for such unrestrained proliferation of pteriid names. Early authors typically considered one or few specimens in establishing new species and failed to appreciate the extent of intraspecific variation. Moreover, they were often unaware of other contemporary works describing the same taxa. With rare exceptions, the delineation of pteriid species was based exclusively on gross shell morphology, primarily shell shape and color; the two features that are notoriously variable, developmentally plastic, and greatly influenced by the environment. In addition, pteriids display a rather limited suite of discrete shell characters that are typically used in classification of other bivalve groups: they lack prominent and differentiated hinge teeth, elaborate shell sculpture, and significantly variable arrangement of muscle scars. Consequently, minor differences in color, shape (particularly in the degree of obliquity and the extent of the posterior auricle), and size (such as seen between juvenile and adult forms) had frequently been used as criteria for establishing new names. Even though most subsequent authors had access to much greater samples, they rarely employed adequate methodology to examine variation by population genetic, morphometric, or any other quantitative methods. Moreover, many species names have been created based on presumed endemism of local populations, even despite the absence of morphological differences. It is significant that none of the revisionary studies to date have provided a comprehensive survey of the extant type material, thus making the revisionary taxonomy of Pteriidae a matter of opinion based on interpretation of secondary sources.

2.2.4. Habitat and ecology

Most species of Pteriidae inhabit shallow littoral and sublittoral zones of continental shelf regions. Some species are, however, found on sandy bottoms of about 80 m depth. An exceptionally deep record of 589 m was reported for *Pteria hirundo hirundo* from Portugal (Hayes, 1972).

Pearl oysters attach to various substrata by the byssus either individually or forming gregarious clumps. Nevertheless, they are capable, particularly during early developmental stages, of moving short distances using the foot upon dislodging the byssus and prior to secreting a new one. Large populations of some *Pinctada* species have been observed to form banks in shallow coasts [Hynd (1955) for *P. albina sugillata* in Australia; Rao and Rao (1974) for Akoya pearl oysters in India]. The preferred habitats for species of *Pinctada* are hard substrata (often in crevices), such as rocky, gravel, and shelly bottoms, live and dead coral, but sometimes individuals are found on sandy bottoms (Ranson, 1961; Rao and Rao, 1974; Butler, 1998; Hayami, 2000). Species of *Pteria* are predominantly epizoic, preferentially attaching to stems of gorgonians (Hayes, 1972; Habe, 1977; Lamprell and Healy, 1998; Hayami, 2000; Hwang and Okutani, 2003). Species of *Electroma* are commonly found on hard corals, seagrasses, and algae (Tantanasiriwong, 1979; Ludbrook and Gowlett-Holmes, 1989; Butler, 1998). *Pinctada longisquamosa* (Dunker, 1852) and *Pterelectroma zebra* (Reeve, 1857) are atypical in their choice of substrata, typically attaching to seagrasses (Mikkelsen *et al.*, 2004) and thyroid *Lytocarpus* (Odhner, 1917; Prashad, 1932), respectively.

A number of studies have been conducted on settlement preferences in both natural and cultured populations of pearl oysters (Allen, 1906; Alagarswami *et al.*, 1987; Victor and Velayudhan, 1987; Friedman and Bell, 1996; Zhao *et al.*, 2003).

2.3. COMMERCIAL SPECIES

Several species of *Pteria* and *Pinctada* have received much attention as a commercial source of pearls and mother-of-pearl (nacre) over the centuries (see reviews by Wada, 1991; Gervis and Sims, 1992; Donkin, 1998; Landman *et al.*, 2001; Hamel and Mercier, 2003; Chapters 1 and 9) and, more recently, as a model system for bone regeneration studies (Lopez *et al.*, 1995; Westbroek and Marin, 1998; Currey *et al.*, 2000; Pereira *et al.*, 2001; Mouriès *et al.*, 2002; Rousseau *et al.*, 2003). Consequently, much of the research on pearl oysters has focused on major aquaculture species. This section provides descriptions for five commercially important pteriid species (Fig. 2.1)

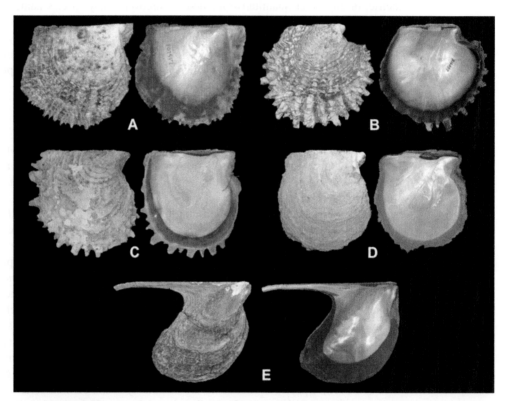

FIGURE 2.1   Commercial species of pearl oysters. Size is expressed as shell length (measured as the longest distance between the extremities of the anterior and posterior shell margins parallel to the straight hinge line). A: specimen of *P. fucata/martensii/radiata/imbricata* species complex (Nada, Wakayama Prefecture, Japan; AMNH 307417, 54.76 mm); B: *Pinctada margaritifera* (Kontiki Lagoon, Vanua Levu Island, Fiji; AMNH 250418, 97.88 mm); C: *Pinctada mazatlanica* (Bahia de Bacochibampo, Guaymas, Mexico; AMNH 311788, 87.82 mm); D: *Pinctada maxima* (Vansittart Bay, Northern Territory, Australia; AMNH 232675, 146.16 mm); E: *Pteria penguin* (Amami Islands, Kagoshima Prefecture, Japan; AMNH 304802, 165.09 mm). (See Color Plate 5)

based primarily on a synthesis of results from a large number of previously published studies and additional observations by Tëmkin (2006a). Despite the fact that there is still some disagreement on the identity of some of the cultured pearl oysters, the results of recent phylogenetic and descriptive studies provide sufficient grounds for establishing morphological diagnoses and stabilizing most of their names. Problematic issues and additional comments are discussed in the *Remarks* sections that follow each species description. The most extensive description is provided for the Akoya pearl oyster, the longest established species in pearl culture, and the rest of the descriptions are made in reference to it. A complimentary description of the anatomy of the black-lip pearl oyster, *Pinctada margaritifera*, is provided in Chapter 3. The generalized shell. morphology of *Pinctada* is shown in Figure 2.2 and an overview of soft anatomy of *Pinctada* and *Pteria* are shown in Figure 2.3.

2.3.1. *Pinctada fucata/martensii/radiata/imbricata* species complex, the Akoya pearl oyster

2.3.1.1. Shell morphology

The shell morphology and anatomy of specimens from Sri Lanka have been thoroughly studied for over a century (Thurston, 1894; Herdman, 1903, 1904; Hornell, 1922, 1951; Rao and Rao, 1974; Velayudhan and Gandhi, 1987). Although no comparable morphological treatments have been published on the Japanese populations of the species [an exception being an anatomical atlas by Shiino (1952) and a review by Wada (1991)], an enormous amount of morphological detail is scattered throughout the vast pearl culture literature.

The shell (Fig. 2.1A) is prosocline, inequilateral, typically rounded to anteriorly obliquely ovate in outline. The umbones are subterminal, located posterior to the anterior auricles, and slightly projected beyond the straight hinge, with the umbo in the left valve elevated slightly more dorsally than the umbo in the right valve. The left valve is more convex than the right valve. The growth gradient is markedly retrocrescent in early stages, gradually becoming infracrescent later in ontogeny. Length typically exceeds height in juveniles but is nearly equal to it in adults. The outline of the inner nacreous layer roughly follows the outline of the shell margin investing the auricles. The prismatic margin is widest along the ventral border and of approximately the same width in both valves; it is flexible and capable of bending during closure to create an airtight seal when the shell is closed. The shell is variable in thickness, gradually thinning towards the margins. The anterior subtriangular auricles are continuous with the straight hinge line. The posterior auricles are typically short, broadly rounded, and sinuated. The byssal notch is narrow and slit-like. The shell exterior is concordant in color between valves and most frequently is brown-purple, green, red, yellow, or white, with a varying number of distally widening rays or radially arranged blotches of purple or red-brown color radiating from the umbonal region. The nacre is iridescent, with white metallic luster and sometimes with a yellow, silver, gold, or pink tint. The commarginal sculpture of the exterior surface consists of projecting imbricating scales (lamellae) that overlap with neighboring rows to produce a radial pattern. The scales are typically small, fluted, tapering, and concave in cross section, with finely rounded ends, projecting from shell surface or slightly raised lamina at an oblique angle. Typically, the surfaces of the scales bear thin, transverse, dark brown, and evenly spaced lines. The scales are generally abraded in adults, resulting in a smooth appearance of the exterior surface. The muscle attachment sites (muscle scars) lie entirely within the nacreous border. The posterior adductor muscle scar is most conspicuous, large, crescent-shaped, and placed slightly posterior to the center; it occupies from one-third to one-half of the diameter of the shell. The pallial line is discontinuous, consisting of rounded or irregularly shaped scars marking insertions of radial pallial muscles, stretching along a curve from the umbonal region to the anteroventral extremity of the posterior adductor muscle scar. There are several elongated scars in the area between the dorsal tip of the posterior

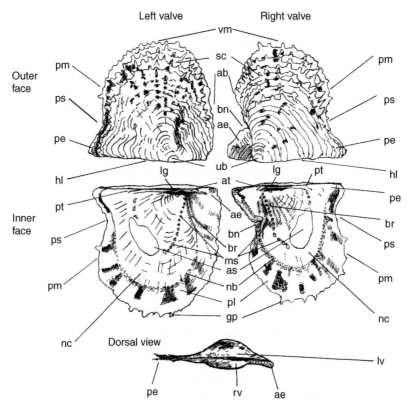

FIGURE 2.2 Overview of shell morphology showing diagrammatic views of the exterior and interior surfaces of both valves of a generalized individual of *Pinctada*. Abbreviations: ab: anterior margin (border); ae: anterior ear (auricle); as: adductor muscle scar; at: anterior teeth; bn: byssal notch; br: byssal ridge; gp: growth process (commarginal calcitic scale); hl: hinge line; lg: ligament; lv: left valve; ms: pallial muscle scar; nb: nacreous border (edge of nacre); nc: nacre (mother-of-pearl layer); pe: posterior ear (posterior process or auricle); pl: non-nacreous (prismatic) layer (prismatic margin); pm: posterior margin (border); ps: posterior sinus; pt: posterior teeth; rv: right valve; sc: scales (commarginal calcitic scale); ub: umbo; vm: ventral margin.

adductor muscle scar and the hinge line just posterior to the ligament. The hinge teeth are well developed in young individuals, consisting of an anterior subumbonal tooth in the right valve and a corresponding socket in the left valve, and a posterior oblique submarginal ridge in the left valve and a corresponding socket in the right valve. The dentition becomes less distinct in adults. The ligament is alivincular, amphidetic, and dorsal submarginal. The fibrous component of the ligament is deposited in a shallow, subtriangular, monoserial resilifer, slightly concave in cross section and weakly rounded along the ventral surface. The resilifer is directed posteriorly from the umbo, greatly widening towards the ventral margin of the ligament area, so that the functional part of the fibrous ligamental layer is situated approximately at the center along the hinge axis. Shell microstructure is nacro-prismatic and has been extensively studied (reviewed by Carter, 1990). The periostracum (studied by Kawakami and Yasuzumi, 1964 and Wada, 1968) is very thin and typically abraded in adults (Guenther et al., 2006).

Embryonic and larval shells were studied by several authors (Ota, 1957; Tanaka, 1958; Alagarswami et al., 1983; Wada, 1991; Fujimura et al., 1995). The prodissoconch I (embryonic shell growth stage) is D shaped and is approximately $70 \mu m$ in length and $60 \mu m$ in height. The prodissoconch II (pelagic larval growth stage) is subtriangular in outline, opisthogyrous (coiling slightly posteriorly), with regularly spaced commarginal growth lines, and approximately $215 \mu m$ in length and $200 \mu m$ in height.

2.3.1.2. Mantle structure

The visceral mass occupies a relatively small area (compared to the size of the shell) and is produced posteroventrally toward the anteroventral surface of the posterior adductor muscle. The mantle lobes occupy most of the area between the valves extending from the hinge line around the circumference of the nacreous border of the shell. The mantle lobes are differentiated into four areas of distinct appearance: the mantle isthmus, the homogeneous central zone, the contractile pallial or distal zone, and the pigmented marginal zone (see Fig. 3.4). The margins of the mantle lobes are fused dorsally immediately under the hinge forming the mantle isthmus; the remaining mantle margins are free. The mantle isthmus is non-pigmented; it extends anteriorly past the anterior part of the visceral mass, forming a hood concealing the lips, the mouth, and the dorsal extremities of the labial palps. The inner surfaces of the left and right mantle lobes are connected by a short supramyal septum stretching from the dorsalmost surface of the posterior adductor muscle to the posterior surface of the visceral mass above the point of emergence of the terminal part of the intestine. The homogeneous central zone of the mantle is non-pigmented (effectively translucent). The inner surfaces of the central zone of the mantle lobes attach to the muscles, and are fused to the sides of the visceral mass, the outer surface of the pericardium, and the kidneys. The pallial zone is demarcated from the central zone by the sinuses of the radial pallial retractor muscles and contains muscle bundles, nerves, and blood vessels. The marginal zone (mantle edge) is divided into three folds with the periostracal groove separating the outer fold from the middle and inner folds (see Section 3.3; Fig. 3.14). Transverse and longitudinal muscle bundles underlie the outer epithelium and inner epithelium, respectively. See Ojima (1952) and Hayasi (1938) for characterization of cell types found in the epithelial tissue. The mantle edge is typically deeply pigmented in the form of alternate, roughly regularly spaced, blotches of black and gray colors, and shades of yellow, brown, and orange. It is divided into three folds: the inner and middle folds with alternating simple tentacles (typically several in a row) and larger branched tentacles, and the outer fold devoid of tentacles.

2.3.1.3. Musculature

The anterior adductor muscle is lost in early ontogeny (the monomyarian condition). The posterior adductor muscle is large, bean-shaped in cross section, with slightly tapering rounded dorsal end and widely rounded ventral end (Fig. 2.3). It is placed slightly posterior to the visceral mass, stretching transversely between the valves.

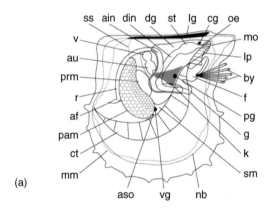

(a)

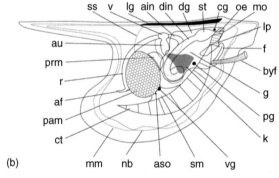

(b)

FIGURE 2.3 Overview of soft anatomy showing diagrammatic views from the right side with the right valve and mantle removed with anterior to the right. A: *Pinctada*; B: *Pteria*. Abbreviations: af: anal funnel; ain: ascending intestine; aso: abdominal sense organ; au: auricles; by: byssus; byf: byssal thread fusion; cg: cerebral ganglia; ct: ctenidia; dg: digestive gland; din: descending intestine; f: foot; g: gonad; k: kidney; lg: ligament; lp: labial palp; mm: mantle margin; mo: mouth; nb: border of nacreous layer; oe: esophagus; pam: posterior adductor muscle; pg: pedal ganglia; prm: posterior pedo-byssal retractor muscle; r: rectum; sm: suspensory membrane; ss: supramyal septum; st: stomach; v: ventricle; vg: visceral ganglia.

It is comprised of nearly equal semi-translucent anterior and opaque posterior lobes, corresponding to "quick" and "catch" muscles, respectively. The posterior pedo-byssal retractors are symmetrical muscle bundles lying in the horizontal plane of the body. Their proximal extremities are rounded in cross section and attached to the posterior end of the byssal gland, from where they diverge and slightly flatten (resulting in an oval cross section), widening distally to be inserted in the concavity of the anterior surface of the posterior adductor muscle. A pair of fused anterior pedo-byssal retractor muscles is present on each side of the visceral mass extending from the byssal root to their attachment point on the inner surface of the shell in the umbonal arch. The left fused anterior pedo-byssal retractor is stronger than the right one and passes anterior to the latter. The symmetrical accessory pedo-byssal muscles branch off the fused anterior pedo-byssal retractors and attach to the shell interior slightly postero ventral to the attachments of the fused anterior pedo-byssal retractors. Numerous radial pallial retractor muscles are present on the entire periphery of the pallial mantle zone. The radial pallial retractor muscle bundles branch and converge with neighboring ones, and fuse proximally forming attachments of the pallial line to inner nacreous shell layer.

The foot emerges from the anterior surface of the visceral mass. It is noticeably twisted to the left with the tip pointing dorsally (and slightly leftward). It is muscular, tongue shaped, oval in cross section, and slightly tapering distally. The dorsal surface of the foot is typically covered with dark brown to black spots. The byssal groove runs on its ventral side and proximally forms a pit through which the byssus protrudes. The byssal gland is located at the antero ventral part of the visceral mass that lodges a common root of a bundle of discrete, stout, flattened, bronze-green byssal fibers with adhesive tips at their distal extremities. Locomotion and the process of byssal attachment were discussed by Kelaart (1858) and Herdman (1903).

2.3.1.4. Labial palps

Paired symmetrical labial palps are located at the anterodorsal part of the visceral mass under the hood formed by the fusion of the mantle lobes at the anterior most region of the mantle isthmus. They project dorsoventrally on either side of the visceral mass from their connection to the lips to the level of the base of the foot. Each palp consists of outer and inner elongated subtriangular folds attached to the visceral mass along the longer side, wider at the base, and narrowing dorsally. The outer and inner flaps are continuous with the upper and lower lips, respectively. The inner surfaces of the outer and inner folds of the palps are grooved. Association of the labial palps and ctenidia is of Category III (Stasek, 1963), characterized by the anterior filament of the inner demibranch not inserted into the distal oral groove.

2.3.1.5. Ctenidia (gills)

Typically colorless ctenidia are large, symmetrical, plicate, broadly sickle-shaped, encircling the visceral mass and the posterior adductor muscle ventrally. The inner and outer demibranchs are subequal, widest in the middle, each with a marginal food groove. The dorsal edges of the descending lamellae within each ctenidium are fused forming gill axes. The gill axes are fused to the visceral mass and are connected to the adductor muscle by a thin and translucent suspensory membrane. The anatomy of the ctenidial axes, containing longitudinal ctenidial retractor muscles, and longitudinal afferent and efferent vessels, was described by Atkins (1943). The dorsal edges of the outer demibranchs are connected to the inner surfaces of the mantle lobes by ciliary junctions. The dorsal edges of the inner demibranchs are connected to each other by ciliary junctions, but a definite narrow band of tissue fusion is present along the ventral edge of the contact area between the demibranchs. The filamental organization of the ctenidia is heterorhabdic with large principal filaments (U-shaped in cross section). The interlamellar junctions form continuous septa that stretch between the opposing aboral surfaces of the principal filaments. The adjacent filaments are in a serial tissue fusion along the ventral edges of the demibranch margins, dorsalmost edges along ctenidial axes, and the free margins of the ascending lamellae. Elsewhere the filaments are joined by ciliated junctions and, occasionally, by tissue fusion. Thus, the association of the filaments is intermediate between eleutherorhabdic and synaptorhabdic grades. The filaments are equipped with the frontal, microlaterofrontal, and lateral cilial tracts. Overall, gill morphology corresponds to Type B (1b), characterized by frontal currents moving dorsally in plical grooves and ventrally on crests (Atkins, 1937).

2.3.1.6. Feeding and digestive system

The food, feeding, digestive system and digestive processes of pearl oysters are reviewed in Chapter 4. The following is a description of the processes in Akoya pearl oysters. Akoya pearl oysters are typical species in this aspect of their biology and this description must be considered within the broader context of Chapter 4.

The produced, smooth lips conceal the slit-like, dorsoventrally compressed mouth. The short esophagus opens into the anterodorsal part of the stomach that lies within the dorsal part of the visceral mass. The stomach is enveloped by the digestive gland and covered superficially by the gonadal tissue. The ovoid stomach is placed asymmetrically on the left side of the visceral mass. The overall stomach morphology corresponds to the Type III of Purchon (1957). The internal morphology, histology, and particle sorting were studied by Herdman (1904), Purchon (1957), Nakazima (1958), and Kuwatani (1965). Gross morphology was visualized by the corrosion casting method by Handa and Yamamoto (2003). The descending arm of the intestine (cojoined with the style sac) extends ventrally from the posteroventral stomach wall into a small intestinal pouch twisting over itself from the left to right side. The ascending arm stretches to the posterodorsal extremity of the visceral mass running parallel to and on the right side of the descending arm until it exits from the posterodorsal part of the visceral mass. The terminal part of the intestine (the rectum) extends straight to the dorsal surface of the posterior adductor muscle passing above the ventricle and perforating the supramyal septum in its course. The rectum descends along the midline of the posterior surface of the adductor muscle and terminates just before reaching its posteroventral extremity. The intestine terminates in a posteriorly oriented membranous process (anal funnel) surrounding the anus. The anal funnel is lanceolate in outline, with a narrow tapering tip, facing ventrally, perpendicular to the posterior surface of the adductor muscle, and contains the anal opening at its base. The surface of the anal funnel is translucent. The digestive gland is extensive and occupies most of the visceral mass. The feeding behavior, physiology of digestion, and food were studied by Herdman (1903, 1904) and Chellam (1987).

2.3.1.7. Circulatory system

The circulatory system consists of a heart, blood vessels, and lacunae distributed irregularly throughout the body. The heart occupies nearly the entire pericardial cavity and consists of a dark-colored median ventricle situated dorsal to two lateral symmetrical, pale yellow-white auricles. The dorsal extremity of the ventricle is adjacent to the ventral wall of the rectum but is not perforated by it. A detailed account of the heart was provided by Suzuki (1985). Major blood vessels include the anterior and posterior aortae, left and right pallial arteries, and the paired branchial afferent and efferent vessels. The anterior aorta is directed anteriorly, passing dorsal to the midline over the left side of the intestine, and branching into the visceral mass and, more anteriorly, dividing into two pallial arteries along the mantle edge. The posterior aorta passes backward along the right side of the intestine, branching above the anus into the interior of the adductor muscle. The branchial afferent and efferent vessels pass longitudinally through the ctenidial axes. The anterior aorta passes over the left side of the intestine, running medially on its dorsal surface, producing numerous branches upon its entry into the visceral mass, and subdividing anteriorly into the left and right pallial arteries projected into the mantle lobes. The posterior aorta follows the right side of the intestine and branches into the interior of the adductor muscle above the anus. Blood, carried by the ramifications of the arteries, passes into lacunae (irregular spaces between tissues) and drains into larger venous sinuses connecting directly to the heart or indirectly by intermediation of the ctenidia. The blood is colorless; its cellular composition was analyzed by Kawamoto and Nakanishi (1957).

2.3.1.8. Excretory system

The excretory system consists of the pericardial wall and cavity, paired symmetrical nephridia (kidneys), and pericardial glands projecting from the walls of the auricles. A detailed account of the excretory system was given by Suzuki (1985). The pericardium is placed within the space between the posterodorsal part of the visceral mass and the anterodorsal part of the posterior adductor muscle. The anterior wall of the pericardium is fused to the visceral mass, its floor is formed by the transverse branch connecting the right and left nephridia; the posterior and lateral walls of the pericardium are free of fusion to adjacent parts. The pericardium is penetrated by the rectum in its dorsal part. The renopericardial canals extend from the pericardium on either side of the visceral mass and connects the pericardium to the kidneys via a narrow slit. The nephridia are roughly triangular in outline, consisting of two large pouch-like sacs lying on either side of posterior portion of visceral mass and connected to the exterior by minute pores. The posterior portions of the nephridia are connected to the posterior adductor muscle via the ctenidial suspensory membrane; its outer walls are fused to the body wall. The small and oval external renal aperture is equipped with a sphincter muscle and is situated immediately ventral to the genital aperture within an inconspicuous renogenital vestibule, which is placed at the junction of the inner plate of the inner gill with the visceral mass approximately midway between the ventral border of the latter and the base of the foot.

2.3.1.9. Nervous system

The nervous system conforms to the general bivalve body plan, with three pair of ganglia: (1) the cerebropleural ganglia surrounding the esophagus; (2) the fused pedal ganglia at the base of the foot; and (3) the visceral ganglia at the anteroventral side of the adductor muscle. The cerebropleural ganglia connect to the visceral ganglia by the cerebrovisceral connectives and to the pedal ganglia by the cerebropedal connectives. The anterior common pallial nerves stretch from the anterior surface of the cerebral ganglia and give off two major nerves penetrating the mantle. The cerebropleural ganglia innervate the labial palps and the otocysts. The pedal ganglia innervate the foot and the byssal gland by three sequentially branching principal nerves: (1) the superior pedal nerve passing along the dorsal part of the foot to its tip; (2) the inferior pedal nerve supplying the byssal groove; and (3) the byssal nerve passing to the byssal gland. The visceral ganglia give off (1) the ctenidial and (2) the posterior common pallial nerves from its anterior and posterior parts, respectively, and (3) the two nerves passing posteriorly along the surface of the posterior adductor muscle to the abdominal sense organs situated slightly ventral to the anus. The posterior common pallial nerves pass posteriorly and outward, and branch into the external, median, and internal pallial nerves. The external pallial nerve runs along the mantle margins; the median pallial nerve runs through the central part of the mantle and the inner pallial nerve passes along the line of the insertions of the radial pallial retractor muscles (See Fig. 3.14). The three corresponding anterior and posterior branches of the common pallial nerves join to form a single complex network, the pallial plexus. The ctenidial nerves proceed through the suspensory membranes and pass through the gill axes to the posterior extremities of the ctenidia along the afferent vessels. The abdominal sense organs are pigmented, laterally compressed, ridge like, and situated at the postero-ventral surface of the posterior adductor muscle, slightly ventral to the anus with their axes lying transversely to the anteroposterior axis. The right abdominal sense organ is larger than the left one and is situated slightly dorsal to the latter. The otocyst contains numerous otoconia and is positioned on the anterior side of the visceral mass. The osphradium is bounded by a well-marked projection close to the visceral ganglia. The tentacles of the mantle edge contain the projections of the nerves and are extremely sensitive to touch. Photoreceptors are diffused through the mantle, being concentrated at the pigmented areas of the mantle edge and foot. The postlarval ctenidial ocelli are absent.

2.3.1.10. Reproductive system

While the following is a consideration of reproduction of Akoya pearl oysters, there is a comprehensive review of reproduction in *Pinctada* and *Pteria* species in Chapter 5.

The reproductive system and reproductive biology were discussed by Herdman (1903, 1904), Chellam (1987), Wada (1991), and reviewed by Ranson (1961). The environmental effects on gametogenesis, spawning, and maturation were investigated by Cahn (1949), Uemoto (1958), and Chellam (1987). The reproductive system is dioecious but sexual change was observed between different spawning seasons, and protandry was detected in many cases. The gonads are paired, forming a thick envelope covering the digestive gland and the intestine but not extending into the foot. Male and female gonads are indistinguishable at low magnification; both are creamy yellow and consisting of branched tubuli with many caeca (alveoli). The mature gametes pass into tubules that converge into two vessels (one on each side of the visceral mass) that open immediately dorsal to the renal apertures. An account of larval development was provided by Herdman (1903) and Fujimura *et al.* (1995) (see also Section 5.7).

2.3.1.11. Distribution

Members of the *P. fucata/martensii/radiata/imbricata* species complex are distributed throughout the equatorial zone between the Tropic of Cancer and the Tropic of Capricorn of the Indo-Pacific and western Atlantic regions (Fig. 2.4). They have been reported from India, East Africa, Sri Lanka, Southeast Asia, Korea, southern China, Papua New Guinea, Hawaii, Palau, Australia, the Red Sea, the Persian Gulf, the Mediterranean region, the Caribbean region, and the Gulf of Mexico (Prashad and Bhaduri, 1934; Hornell, 1951; Hynd, 1955; Ranson, 1961; Work, 1969; Hayes, 1972; Kay, 1979; Tantanasiriwong, 1979; Bosch and Bosch, 1982; Roberts *et al.*, 1982; Wada, 1991; Gervis and Sims, 1992; Oliver, 1992; Bernard *et al.*, 1993; Shirai, 1994; Lai, 1998; Lamprell and Healy, 1998; Hwang and Okutani, 2003). In Japan, the geographic range was reported by Kuroda (1960), Kuroda *et al.* (1972), Habe (1981), Inaba (1982), Shirai (1994), Higo *et al.* (1999), Hayami (2000), and Matsukuma (2004).

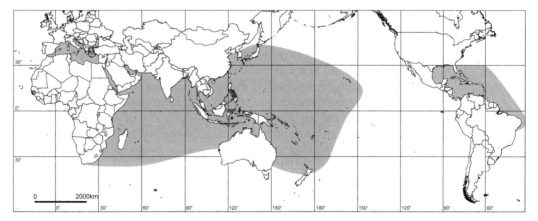

FIGURE 2.4 Approximate distribution of the Akoya pearl oyster; pearl oysters within the *Pinctada fucata/martensii/radiata/imbricata* species complex (see Section 2.3.1.13). Details of exact sites of distribution are provided in the text.

2.3.1.12. Habitat and ecology

Members of the *P. fucata/martensii/radiata/imbricata* species complex occur in the shallow littoral zone, often inhabiting small closed inlets or bays, enriched with freshwater or terrestrial enriched water. They attach by byssus to a variety of substrata on rocky, gravel, and, more rarely, sandy bottoms, particularly in crevices, occurring either sparingly or gregariously. Among recorded substrata are seagrasses and clumps of dead shells (Ranson, 1961; Abbott, 1974; Mahadevan and Nagappan, 1974; Tantanasiriwong, 1979; Butler, 1998; Lamprell and Healy, 1998). The animals typically recline on the right valve with the plane of commissure at approximately 45° to the attachment surface (Stanley, 1970). They occur in shallow waters to a depth of approximately 20 m (Hynd, 1955; Hayami, 2000).

2.3.1.13. Remarks

The taxonomic status of the Akoya pearl oyster remains unsettled because of substantial morphological variation within and among populations, local geographical isolation of some populations, transport by humans, hybridization, and erratic taxonomic practice. Traditionally, three distinct species were recognized to correspond to three biogeographic realms: *Pinctada imbricata* Röding, 1798 in the western Atlantic region, *Pinctada radiata* (Leach, 1814) in the eastern Indian Ocean and the Red Sea regions, and *Pinctada fucata* (Gould, 1850) in the Indo-Pacific region. The Japanese populations have frequently been recognized to be distinct either at a species level as *Pinctada martensii* (Dunker, 1872) (Hayami, 2000) or subspecies *Pinctada fucata martensii* (Kuroda et al., 1972; Habe, 1977, 1981; Matsukuma, 2004). The distinctions were based either on continuous characters or on polymorphic and ontogenetic variations, despite the fact that most authors expressed doubts of the utility of these characters.

Recent genetic analyses (Colgan and Ponder, 2002; Masaoka and Kobayashi, 2002, 2003, 2005a, 2006b; Atsumi et al., 2004; Yu et al., 2006) based on a variety of genetic (allozymes, partial nuclear 18S and 28S rDNA, internal transcribed spacers ITS1 and ITS2, partial mitochondrial 16S rDNA) have shown that the populations from Australia, South East Asia, and Japan form a highly supported monophyletic group and are most likely conspecific (Fig. 2.5). Moreover, morphometric studies did not provide means for separating the three populations (Colgan and Ponder, 2002). In addition, mating experiments have corroborated conspecificity of the South East Asian and Japanese populations (Atsumi et al., 2004).

Regarding the Persian Gulf populations, Beaurnent and Khamdan (1994) in their electrophoretic and morphometric study of pearl oyster specimens from the Natural History Museum, London, indicated that the Japanese shells clearly fall within the range of specimens from the Arabian Gulf. Masaoka and Kobayashi (2006a), however, discriminated the Persian Gulf specimens from the Japanese specimens in a phylogenetic analysis based on the hypervariable intergenic spacer of nuclear ribosomal RNA and mitochondrial 16S ribosomal RNA gene regions.

Molecular studies showed that the Atlantic populations are differentiated from the Indo-Pacific populations by phylogenetic analysis of partial hypervariable intergenic

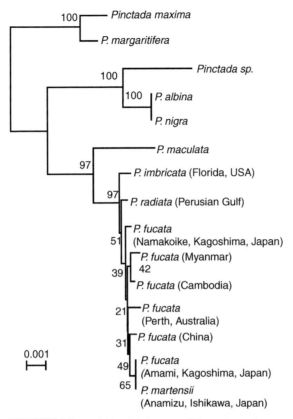

FIGURE 2.5 Neighbor-jointing tree of *Pinctada* species constructed with pairwise distances calculated following the application of Kimura's two parameter correlation for multiple substitutions in 28S rRNA gene region. The tree was produced using MEGA 2. The numbers above the branches are the percentage of 1000 bootstrap replicates. From Masaoka and Kobayashi (2005d).

spacer of nuclear ribosomal RNA genes (IGS) and the mitochondrial 16S rRNA gene regions, the loci commonly used for intraspecific (population-level) analyses (Masaoka and Kobayashi, 2003, 2004a, b, 2005a–d) (Fig. 2.5).

Comparative karyotype analysis did not reveal any difference between the Atlantic (Caribbean) and the Pacific (Japanese) populations (Wada, 1978; Komaru and Wada, 1985; Wada and Komaru, 1985). To date, no discrete morphological characters were reported to be diagnostic of populations from the three biogeographical regions.

Taken together, these studies suggest that the Akoya pearl oyster is a cosmopolitan, globally distributed species, characterized by substantial intraspecific variation largely due to clinal genetic differentiation and morphological plasticity. Considering disjunct distribution of some populations, incomplete sampling of morphological and genetic data, and the lack of sound taxonomic revision, more research is required to establish whether *Pinctada imbricata*, *P. radiata*, *P. fucata and P. martensii* comprise a single species [as anticipated by Shirai (1994) although without providing justification for his claim]. Given the paramount importance of members of this species complex in pearl culture, proper understanding of phylogenetic affinities among different populations of this species complex has important implications for the genetic programs in the pearl industry (Wada, 1982, 1984, 1994, 1997, 1999; Wada *et al.*, 1995).

2.3.2. *Pinctada margaritifera* (Linnaeus, 1758), the Black-lip pearl oyster

2.3.2.1. Shell morphology

Shell morphology has been described in many studies (Prashad and Bhaduri, 1934; Hynd, 1955; Ranson, 1961; Rao and Rao, 1974). The shell (Fig. 2.1B) is large, prosocline, laterally compressed, typically rounded in outline, with the left valve more convex than the right valve. The hinge line is straight and extended into the anterior and posterior auricles. The umbones are subterminal and located posterior to the anterior auricles. Shell height is equal to or slightly exceeds length. Shell growth is retrocrescent in early ontogeny, becoming infracrescent in adults, and markedly procrescent in some individuals. The outline of the inner nacreous layer roughly follows the outline of the shell margin investing the auricles. The prismatic margin is widest along the ventral border and of approximately the same width in both valves; it is flexible and capable of bending during closure to create an airtight seal when the shell is closed. Average dimensions are approximately 120 mm in height and 100 mm in length. The largest recorded specimen is 300 mm (probably referring to the widest dimension; Lintilhac, 1985). The anterior subtriangular auricles are continuous with the straight hinge line and greatly reduced in large individuals. When present, the posterior auricles are very short, weakly and broadly sinuated, often continuous with the posterior margin. The byssal notch is narrow, slit-like, increasing transversely with size and gaping widely in some large individuals. The color of the exterior surface of the shell is typically dark green (bronze), brown, or black, with a radial pattern of well-defined discrete, and distally widening white rays or partially continuous white blotches. The umbones often lack a color pattern, turning opaque white as the shell abrades with age. Juveniles are generally of much lighter color and do not show distinct white rays. The interior prismatic margin is typically black with no or few irregular lighter blotches. The interior nacreous surface is strongly iridescent, silvery, with pink, red, or green tint. Frequently the nacre has dark smoky (black) color, with a hint of red and green iridescence, and a distinct yellow band along the nacro-prismatic border. The commarginal sculpture consists of prismatic scales projecting from the densely spaced commarginal growth lines. The scales are long, straight, often curved distally in either direction, with rounded or tapering distal ends; in cross section they are flat at the base but concave, convex, or twisted distally. The scales are arranged either discretely or project from continuous calcitic lamina; they overlap forming radial rows extending at a low angle (nearly horizontal) relative to the valve surface. Scales are typically opaque white proximally and bronze or black distally, sometimes with parallel transverse lines.

   Muscle attachment sites are located within the nacreous border. The large adductor muscle scar is crescent-shaped, with a slightly tapering rounded dorsal end, and a widely rounded ventral end. The suboval posterior pedo-byssal retractor scar is inset within the adductor scar, producing a fused three-lobed scar located subcentrally along the dorsoventral axis, slightly posterior to the anteroposterior axis. A variable number of small discontinuous pallial muscle scars extends from a dorsal extremity of the posterior adductor muscle scar dorsally to the area posterior to the resilifer, and from the anteroventral extremity of posterior adductor muscle scar along a curve leading to the scar of the fused anterior pedo-byssal retractor muscles. The fused anterior pedo-byssal retractor muscle scars are small, distinct, round, and placed within the umbonal arch; the attachment scars of the accessory pedo-byssal retractor muscles are placed slightly ventral and posterior to them. The ligament is alivincular, amphidetic, and dorsal submarginal. The fibrous component of the ligament is deposited in a shallow, subtriangular resilifer, slightly concave in cross section and weakly rounded along its ventral surface. The resilifer is directed posteriorly from the umbo, greatly widening towards the ventral margin of the ligamental area. There are no hinge teeth at any stage of development. The nacro-prismatic microstructure was reviewed by Carter (1990).

   Larval shells were studied by Alagarswami *et al.* (1989), Doroudi and Southgate (2003) and Paugam *et al.* (2006). The prodissoconch I (D-shaped larval shell growth stage) is approximately 80 µm long and 60 µm high. The prodissoconch II is subtriangular, opisthogyrous, with regularly spaced commarginal growth lines, approximately 140 µm long and 130 µm high. A large body of literature discusses shell growth rates and factors that affect it, particularly in cultured stocks (Nicholls, 1931; Sagara and Takemura, 1960; Service de la Pêche, 1970; Nasr, 1984; Sims, 1990, 1993; Gervis and Sims, 1992; see Section 5.7).

2.3.2.2. Soft anatomy

The soft anatomy was described in detail by Fougerouse-Tsing and Herbaut (1994). See Chapter 3 of this text. To a large extent, it resembles that of the Akoya pearl oyster. The mantle anatomy, histology, histochemistry, and ultrastructure were described by Jabbour-Zahab *et al.* (1992, 2003). The margins of the inner and middle folds bear a row of alternating large and small, digiform, tapering tentacles. Both surfaces of the inner fold and the inner surface of the middle fold are densely covered with black-brown, and, sometimes, orange blotches. The foot is also pigmented with dark brown to black speckles. The anatomy of the labial palps was discussed by Thiele (1886) and of the ctenidia – by Ridewood (1903) and Atkins (1938, 1943). The ctenidia are either black or bright orange-red. The typhlosole of the intestinal tract is orange and the sheath covering the external part of the intestine is either light brown-black or orange (as seen in preserved specimens). The anal funnel is large, flat, wide, with a narrow tapering tip, and the surface either solid gray or translucent-white with a median dark brown-black stripe on the dorsal surface. The abdominal sense organs are translucent ridges and extend laterally with their distal extremities attached to the inner surfaces of mantle lobes (Thiele, 1889). No ctenidial ocelli are present.

The reproductive system is protandric hermaphrodite. A detailed histology and ultrastructure of the gonadal tissue at different stages of maturation were described by Tranter (1958) and Thielley and Grizel (2003). Embryonic and larval development were was discussed by Doroudi and Southgate (2003), and the reproductive biology was reviewed by Ranson (1961). (See also Chapter 5).

2.3.2.3. Distribution

Pinctada margaritifera is a widely distributed Indo-Pacific species (Fig. 2.6). Its distribution range includes the Red Sea, the Arabian Sea, the Persian Gulf, India, Sri Lanka, Southeast Asia, southern Japan, Australia, New Caledonia, Polynesia, Micronesia, New Guinea, Hawaii, Cocos-Keeling Islands, and Madagascar (Smith, 1891; Saville-Kent, 1893; Jameson, 1901; Odhner, 1919; Boutan, 1925; Prashad, 1932; Hornell, 1951; Hynd, 1955; Macpherson and Gabriel, 1962; Maes, 1967; Tantanasiriwong, 1979; Bosch and Bosch, 1982; Springsteen and Leobrera, 1986; Lintilhac and Durand, 1987; Bernard *et al.*, 1993; Hayami, 2000; Hylleberg and Kilburn, 2002). There are recurring recent records of living individuals found in the western Atlantic (Florida), where it might have been introduced by commercial overseas shipping (Burch, 1995a), but there is no evidence to date that the species has established there (Chesler, 1994; Carlton, 1996; Camp *et al.*, 1998).

2.3.2.4. Habitat and ecology

Individuals of *Pinctada margaritifera* are most abundant in littoral and sublittoral zones of reef environments (Odhner, 1919; Bullivant, 1962; Maes, 1967; Tantanasiriwong, 1979). Occasionally, some individuals are found in deeper waters up to a depth of 66 m (Lamprell and Healy, 1998). The animals attach by byssus to hard substrata, such as

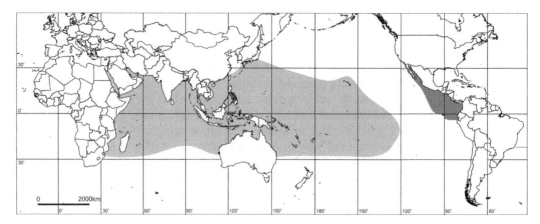

FIGURE 2.6 Approximate distribution of the Black-lip pearl oyster, *Pinctada margaritifera* (pale gray), and the Panamanian pearl oyster, *P. mazatlanica* (dark gray). Details of exact sites of distribution are provided in the text.

rocks, gravel, live and dead coral (Butler, 1998; Hayami, 2000), and occur either singly or in large clusters (Hynd, 1955; Bullivant, 1962; Bosch and Bosch, 1982). They are used as an indicator for ecological monitoring of heavy metals (Klumpp and Burdon-Jones, 1982; Sarver *et al.*, 2004).

2.3.2.5. Remarks

Pinctada margaritifera is the type species of the genus *Pinctada* (designated by Iredale, 1915) and is the oldest valid pearl oyster name. There is little disagreement on the identity of the species in the present, but the intraspecific variation found among and within populations of this widely distributed species have raised a possibility that it might in fact represent more than a single species. As noted by Allan (1959), shape and size differences exist between different populations of *P. margaritifera*, however no meristic data were provided to establish their significance. Several aspects of soft anatomy, namely the color of the ctenidia, intestinal sheath, and mantle edge, are concordant within the individual but appear to be polymorphic, being either black or orange-red within the same population (Shirai, 1994; Hwang and Okutani, 2003). This led certain authors to recognize these variants as subspecies (Pouvreau *et al.*, 2000). Consistent with morphological observations, the population genetic studies have revealed a geographical differentiation of *P. margaritifera* populations based on allozyme and DNA sequence data (Blanc and Durand, 1989; Arnaud-Haond *et al.*, 2002, 2003, 2004; see Section 12.2.2). These studies showed a low degree of divergence, typical for an intraspecific level of variation, and demonstrated that gene flow exists among different populations. Representatives of French Polynesian and Hawaiian populations have sometimes been referred to as separate species or subspecies, *P. cumingii* and *P. galtsoffi*, respectively.

2.3.3. *Pinctada maxima* (Jameson, 1901), the Silver- or Gold-lip pearl oyster

2.3.3.1. Shell morphology

General descriptions of the shell were given by Jameson (1901), Hynd (1955), Takemura and Okutani (1958), and Ranson (1961). This is the largest species of the genus: the largest recorded specimen is 250 mm high and 282 mm long (Torres Straits; Iredale, 1939). The shell (Fig. 2.1D) is prosocline, typically rounded in outline, with length typically equalling height. The left valve is more convex; the right valve nearly flat to weakly convex. The convexity decreases with age. The prismatic margin is wider in the right valve. The anterior auricles are very small, subtriangular or irregular in shape, sometimes nearly obsolete in large specimens. The byssal notch is small, narrow, slit-like, vertically elongated, and nearly obsolete in large individuals (exceeding approximately 150 mm in height). The posterior auricles are absent or short, and typically indistinct from the valves; they are rarely very weakly and broadly sinuated. In young individuals (approximately 50 mm in height) the posterior auricles are well developed. The exterior color is usually pale yellow-gray, sometimes with valves partially green or ochre, and, occasionally, with dark green or purple-brown (always darker than the ground color), distally broadening radial rays. Juveniles (>120 mm in height) display a range of colors (green, purple-black, yellow, creamy white, gray, brown, sometimes with zigzag pattern of purple-brown) that typically disappear with age but sometimes are retained in the umbonal region. The exterior shell margins and scales have commarginal (transverse for scales) alternating dark and light bands. The inner prismatic margin is of the same ground color as the exterior. The interior nacreous surface is thick, lustrous, iridescent, silvery, rarely showing any dark tints, sometimes with a rich golden rim that develops on the periphery of the nacreous layer. The sculpture of the exterior surface of the shell consists of commarginal prismatic scales which are most prominent in young individuals and typically abraded in large individuals. The scales are long, irregularly clavate, twisted, and sparsely distributed, forming radial rows; they are curved outward at their proximal ends, and downward and inward at the distal ends. They typically narrow toward the rounded or blunt end, some widening distally with blunt or irregular distal margins; in cross section they are flat, concave, convex, or irregular. The scales at the shell margins of opposing valves frequently interdigitate. The scales are typically of the same color as the exterior.

The posterior adductor muscle scar is large, wide, very slightly curving, with the dorsal extremity narrower than the ventral extremity, adjacent to the smaller, narrow, oval shaped, and anteriorly positioned posterior pedo-byssal retractor muscle scar. The dorsal lobe of the posterior adductor muscle scar is substantially larger and broader in proportion compared to that of *Pinctada margaritifera* due to the difference in the relationship of the posterior adductor muscle and the posterior pedo-byssal retractor muscles (Odhner, 1917). There are no hinge teeth present in large individuals, whereas hinge teeth are occasionally observed in young specimens. Embryonic and larval shells were studied by Tanaka and Kumeta (1981), Rose and Baker (1989) and Zhao *et al.* (2003). The prodissoconch I is approximately 57–80 μm long; the prodissoconch II is opisthogyrous, subtriangular, and is 234 μm long. The microstructure is nacroprismatic (Kobayashi, 1971). The periostracum is typically abraded in adults.

2.3.3.2. Soft anatomy

The general soft anatomy was previously described by Takemura and Kafuku (1957). No discrete or meristic differences in anatomical characters differentiating *P. maxima* from *P. margaritifera* were found with the exception of byssus sloughing in adult individuals exceeding 150 mm in length and the unique shape of the anal funnel [illustrated by Hynd (1955: Fig. 2A)]. Feeding behaviour and reproduction of *P. maxima* are reviewed in Chapters 4 and 5, respectively.

2.3.3.3. Distribution

Pinctada maxima is distributed in the central Indo-Pacific region from Myanmar to the Solomon Islands, including Southeast Asia, the Philippines, South China Sea, Australia,

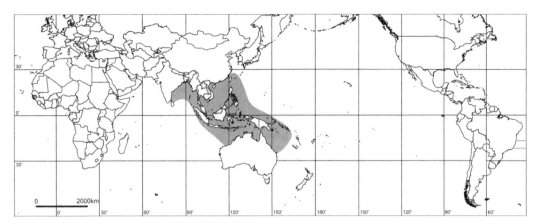

FIGURE 2.7 Approximate distribution of the Silver- or Gold-lip pearl oyster, *Pinctada maxima*. Details of exact sites of distribution are provided in the text.

Papua New Guinea, Indonesia, Polynesia, Micronesia, and southern Japan (Hynd, 1955; Takemura and Okutani, 1958; Ranson, 1961; Wu, 1980; Habe, 1981; Springsteen and Leobrera, 1986; Lintilhac and Durand, 1987; Gervis and Sims, 1992; Bernard *et al.*, 1993; Shirai, 1994; Burch, 1995a, b; Hu and Tao, 1995; Higo *et al.*, 1999; Hayami, 2000; Swennen *et al.*, 2001). The range extends north to Hainan Island off the coast of China to 25°S on the west coast of Australia and 16° on the east coast (Hynd, 1955) (Fig. 2.7).

2.3.3.4. Habitat and ecology

Individuals of *Pinctada maxima* inhabit shelly, rocky (Butler, 1998), gravel, sandy bottoms (Tantanasiriwong, 1979; Shirai, 1994), and reef environments (Hwang and Okutani, 2003). In early stages of growth, they are attached by byssus to hard substrata, but once the byssus is lost they remain on the bottom unattached (Allan, 1959). The individuals are typically found in shallow waters of littoral and sublittoral zones occasionally reaching the maximal recorded depths of 100–120 m (Ranson, 1961; Shirai, 1994). Predators and parasites were discussed by Allan (1959), Wolf and Sprague (1978), and Dix (1973), and reviewed by Butler (1998); they are also covered in detail in Chapter 12.

2.3.3.5. Remarks

In addition to being one of the principal species in pearl culture in the Indo-Pacific region (Gervis and Sims, 1992; Shirai, 1994; see Section 9.3), the species has been used extensively as a model for studying biomineralization and osteogenic properties of proteins contained in the nacre (Bédouet *et al.*, 2001). Aspects of the population genetics of *P. maxima* are discussed in Section 12.2.1.

2.3.4. *Pinctada mazatlanica* (Hanley, 1855), the Panamanian pearl oyster

2.3.4.1. Shell morphology

Hanley's (1856) original description adequately captured the most important aspects of shell morphology that appears to be morphologically intermediate between *Pinctada margaritifera* and *Pinctada maxima*. The shell morphology was subsequently re-described by Jameson (1901), Ranson (1961), and Hayes (1972). The shell is large, thick, prosocline, typically rounded in outline, with length typically equal to height. The shell (Fig. 2.1C) is nearly equivalve with the left valve slightly more convex than the right valve. The width of the prismatic margins is approximately the same in both valves. The anterior auricles are very small (sometimes nearly obsolete) and of irregular shape. When present, the posterior auricles are very weakly and broadly sinuated. The byssal notch varies in shape but is typically elongated vertically. The exterior surfaces of the valves are concordant in color, usually light to dark brown-green, with variously expressed (typically faint and indistinct) darker brown, slightly distally broadening radial rays or radially arrayed blotches of the same color. The inner prismatic margin is of the same color as the exterior surface but often of a lighter shade. The nacre is thick, iridescent but somewhat cloudy, silvery, with a purple tint along the nacreous border. The sculpture of the exterior surface consists of commarginal calcitic scales projecting at low angles from the shell surface or slightly raised lamina. The scales are large, long, wide, flat or fluted, with tapering, blunt, or rounded ends, typically curving downward and concave in cross section. The color of the scales corresponds to the exterior color of shell surface. The scales along the margin typically project beyond the shell margins and interdigitate with those of the opposite valve when the shell is closed.

The posterior adductor muscle scar is large, wide and bean-shaped, with a narrow rounded dorsal extremity and a broadly rounded ventral extremity. It is adjacent to a smaller, oval-shaped, anteriorly positioned posterior pedo-byssal retractor muscle scar. No hinge teeth are present at any stage of development. The Prodissoconch II is opisthogyrous, subtriangular, and measures $276 \mu m$ in length and $270 \mu m$ in height (Ranson, 1961). The maximum shell height is $180 mm$. Shell microstructure of this species was studied by Grégoire (1960) and Carter (1990). Early ontogeny of the shell was discussed by Bernard (1898). The periostracum is thin and abraded in most individuals.

2.3.4.2. Soft anatomy

The general anatomy resembles that of *Pinctada margaritifera*. The anal funnel has a dark longitudinal stripe on its dorsal surface as found in some, but not all, specimens of *P. margaritifera*. The reproductive system and reproduction of *P. mazatlanica* are described in Chapter 5.

2.3.4.3. Distribution

Pinctada mazatlanica is distributed along the western coasts of the Americas extending from the Gulf of California to northern Peru, including reported occurrences from

Baja California, Baja California Sur, Golfo de California, Isla del Coco, and west coasts of Mexico, Panama, Costa Rica, and Ecuador (Hayes, 1972; Abbott, 1974; Moore, 1983; Sanchez, 1990; Shirai, 1994; Saucedo and Montefore, 1997; Arnaud-Haond *et al.*, 2001) (Fig. 2.6).

2.3.4.4. Habitat and ecology

Individuals of *Pinctada mazatlanica* attach by byssus to hard substrata among rocks (Moore, 1983; Shirai, 1994) in shallow offshore water at 5–27 m depth (Hayes, 1972). They have been reported to form large colonies on rocks of the wading zone to depth of about 22 m (reaching maximal depth of 36 m) in Panama (MacKenzie, 1999).

2.3.4.5. Remarks

This species is sometimes referred to as a subspecies of *Pinctada margaritifera* (i.e., *P. margaritifera mazatlanica*) in literature on aquaculture (e.g., MacKenzie, 1999).

2.3.5. *Pteria penguin* (Röding, 1798), the Winged pearl oyster

2.3.5.1. Shell morphology

The shell is large, thick, strongly prosocline, with the angle between the hinge and normal axes increasing with size (shell becoming less oblique with age) reaching approximately 40–45 degrees (Fig. 2.1). The shell is inequivalve with the left valve more convex than the right valve. The posterior shell margin (underlying the posterior auricle) is straight to weakly concave. Shell length exceeds height in smaller specimens but is equal to or shorter than height in large specimens due to a shift in the growth gradient from retrocrescent in juveniles to infracrescent in adults. The prismatic margin is substantially wider in the right valve. The anterior auricles are subtriangular in outline and variable in size. The byssal notch, located immediately ventral to the anterior auricles, is either broadly rounded or vertically elongated and narrow. The posterior auricles are typically long, especially in young specimens, and narrow, gradually tapering, with rounded or tapering ends, and deeply sinuated. In some individuals, however, the posterior auricles are very short. The exterior surfaces of the valves are concordant in color and are usually dark brown to black with variants to lighter shades of brown or green (often co-occurring with light brown), with or without numerous, thin, faint radial rays of a slightly lighter shade than the background color. In lighter-colored specimens darker colored commarginal stripes are occasionally present. The inner prismatic margin is solid black or light brown. The nacre is iridescent, silvery, often with a strong pink tint, typically more pronounced along the nacreous border. Exterior surface sculpture is absent; the shell surface is smooth. The periostracum is light brown and often persists to the adult stage forming small, straight or curving downward, narrow, fluted (to nearly tubular in young individuals), tapering scales projecting from the shell surface or raised lamina at a low angle or nearly parallel relative to the shell surface. The posterior adductor muscle scar is oval, adjacent to the anteriorly positioned, smaller, rounded, posterior pedo-byssal retractor muscle scar. The hinge teeth are represented by the weakly to nearly obsolete subumbonal anterior tooth in the left valve and a submarginal posterior ridge in the right with complementary sockets in the opposing valves. In young individuals, however, the hinge teeth are strongly developed. Beer (1999) described general embryonic and larval development of *Pt. penguin*. The maximum hinge length is 250 mm (Oliver, 1992). Shell microstructure was studied by Grégoire (1960) and Wada (1961), and was reviewed by Carter (1990).

2.3.5.2. Habitat and ecology

Individuals of *Pteria penguin* attached to a variety of substrata, typically hydroids, within 25 m depth (Mahadevan and Nagappan, 1974; Oliver, 1992; Butler, 1998; Hayami, 2000).

2.3.5.3. Distribution

The distribution of *Pteria penguin* extends from the Red Sea and the Arabian Gulf throughout the tropical eastern Indo-Pacific to southern Japan (Okinawa to Honshu) and includes Southeast Asia, the Philippines, Queensland round to northwest Australia, Thursday Island, southern China, and Taiwan (Prashad, 1932; Allan, 1959; Habe, 1977, 1981; Abbott and Dance, 1982; Reid and Brand, 1985; Springsteen and Leobrera, 1986; Oliver, 1992; Bernard *et al.*, 1993; Shirai, 1994; Burch, 1995a; Lamprell and Healy, 1998; Higo *et al.*, 1999; Hayami, 2000; Swennen et al., 2001; Hwang and Okutani, 2003) (Fig. 2.8).

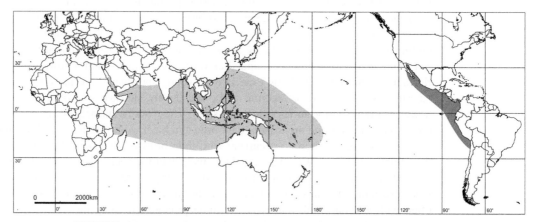

FIGURE 2.8 Approximate distribution of the winged pearl oysters, *Pteria penguin* (pale gray) and *Pteria sterna* (dark gray). Details of exact sites of distribution are provided in the text.

2.3.5.4. Remarks

Pteria penguin is a principal source of half-pearls (mabé pearls) (Landman *et al.*, 2001) (see Section 9.6). *Pteria sterna*, which has a more limited distribution (Fig. 2.8) is also used for commercial mabé and round pearl production in Mexico (see Section 9.6.6.2).

2.4. PHYLOGENY

2.4.1. Higher-level phylogenetic relationships

Pearl oysters belong to the superfamily Pterioidea in the bivalve subclass Pteriomorphia. In addition to Pteriidae (which contains the pearl oyster genera *Pteria* and *Pinctada*), the Pterioidea includes families Isognomonidae, Malleidae, and Pulvinitidae, traditionally distinguished by shell shape and ligament structure. All phylogenetic studies agree that the crown-group Pterioidea is a natural (monophyletic) taxon, but the question of its most closely related group (sister taxon) is still debated favouring Ostreidae, Pinnidae, or both (reviewed by Tëmkin, 2006a). The evolutionary relationships of the Recent Pterioidea inferred from phylogenetic analyses of molecular (Steiner and Hammer, 2000; Giribet and Distel, 2003; Matsumoto, 2003), morphological (Tëmkin, 2006a), and combined data (Tëmkin, 2004) support monophyly for the superfamily but also show that three of the families – including the Pteriidae – are polyphyletic. The only presumably monophyletic pterioid family is the monotypic Pulvinitidae (reviewed by Tëmkin, 2006b).

The taxonomic composition of Pteriidae has not been stable. In the analysis of 18S rDNA data, species of *Pteria* branch out sequentially (not forming a monophyletic group) basal to the rest of Pterioidea, whereas species of *Pinctada* and its sister clade, composed of species of pteriid *Electroma* and malleid *Vulsella*, were most derived (Steiner and Hammer, 2000). A simultaneous analysis of several loci had also recovered a derived *Electroma/Vulsella* clade but failed to resolve the relationships among the rest of the pteriid genera resulting in a polytomy (Giribet and Distel, 2003). An analysis of morphological data alone has confirmed the monophyly of all pteriid genera and produced a sister-group relationship of *Pteria* and *Pinctada* species that, in turn, are a sister group to a clade composed of species of *Electroma*, *Vulsella*, and *Crenatula* (an isognomonid) (Tëmkin, 2006a). The results of the latter study most closely approximate the currently accepted taxonomic composition of the Pteriidae. It is noteworthy, however, that the relationship of *Pteria* and *Pinctada* was not supported by non-homoplastic synapomorphies (shared derived characters) and had low support values. The relationships among the pterioid genera remain uncertain.

2.4.2. Species-level phylogenetic relationships

There are no studies focusing exclusively on the species-level relationships within the genus *Pteria*. Prior studies addressing higher-level relationship within the

Pteriomorphia or Pterioidea were limited in sampling to provide insights into a phylogenetic structure of the genus. On the other hand, there have been a number of recent phylogenetic analyses of the genus *Pinctada* focusing on the relationship among commercial species.

Although not based on an explicit phylogenetic framework, Jameson (1901) proposed a subdivision of *Pinctada* based on the presence or absence of hinge teeth into two groups a was subsequently followed by Hynd (1955) and Ranson (1961). Among species used in the cultured pearl industry, only species of the *P. fucata/martensii/radiate/imbricata* complex belong to the former category; *P. maxima, P. margaritifera,* and *P. mazatlanica* fall into the latter group. Based on the frequent occurrence of small and pointed scales on the exterior surface sculpture, Hynd (1955) suggested that the toothed forms were primitive and the toothless forms were derived. Despite the fact that later studies showed the presence of hinge teeth in *P. maxima,* recent molecular studies (Masaoka and Kobayashi, 2003, 2004a, b, 2005a–d, 2006a, b; He *et al.,* 2005; Yu *et al.,* 2006) confirmed that the two groups form distinct clades (Fig. 2.5). However, contrary to the phylogenetic conclusions reached based on morphology (Hynd, 1955), the analysis of base composition in 18S rDNA and 28S rDNA sequences implied that *P. maxima* and *P. margaritifera* are more primitive (Kobayashi and Masaoka, 2001, 2002, 2005d; He *et al.,* 2005; Yu and Chu, 2005a, b; Yu *et al.,* 2005a, b). Masaoka and Kobayashi (2002, 2005d, 2006b) considered *P. maxima* to be the most primitive *Pinctada* species.

Masaoka and Kobayshi (2005d, 2006b) proposed an evolutionary scenario according to which *P. maxima* first emerged from an ancestral species possibly in the Indonesian region and subsequently diversified producing *P. albina/P. nigra,* and *P. maculata* which dispersed throughout the tropical to subtropical Indo-Pacific. Subsequently, *P. imbricata* (or, the *P. fucata/martensi/radiata/imbricata* species complex) evolved and dispersed widely from the tropics to cooler regions northward (Japan, Persia, and North American western Atlantic) and southward (Southern Australia, South Africa, South American western Atlantic). Concurrently, *P. margaritifera* evolved from the *P. maxima* lineage, an event likewise associated with invading cooler subtropical environments.

2.4.3. The fossil record

The genus *Pteria* did not evolve until the Triassic (Hertlein and Cox, 1969), presumably from the extinct family Bakevelliidae, a paraphyletic stem group that gave rise (among living taxa) to toothed oysters (pterioid genus *Isognomon*) and common oysters (the superfamily Ostreoidea) (Tëmkin, 2006a). The relationships among the extinct species of *Pteria* and their living descendants are not well understood (Fig. 2.9).

The fossil record of the genus *Pinctada* is much more recent: the earliest records date back to Miocene of Western Europe and Caucasus (Hertlein and Cox, 1969). Tamura (1961) had established the subgenus *Pinctada* (*Eopinctada*) to accommodate a single species, *P. (E.) matsumotoi,* from the Middle Cretaceous of Japan. Despite superficial similarity to the Miocene and Recent species of *Pinctada* (and *Pteria*), there is no substantial morphological evidence to consider the Mesozoic species to be ancestral or closely allied to the Recent *Pinctada* without reservation. Moreover, this would imply a considerable stratigraphic gap in the history of *Pinctada* and major geographic

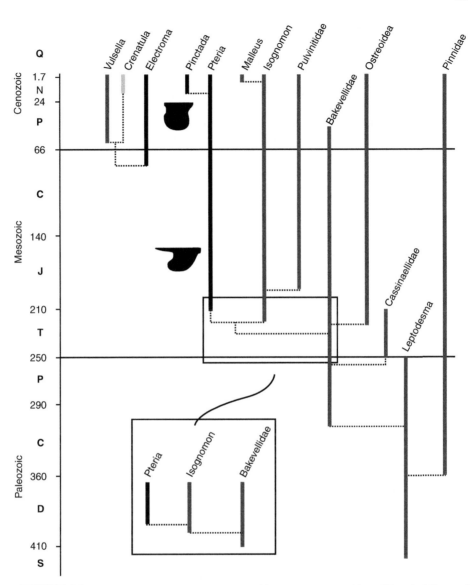

FIGURE 2.9 Phylogeny of the Pterioidea. Sratigraphic ranges represented by solid vertical lines and relationships indicated by horizontal dotted lines. The numbers on the left show time in million years before present. The inset shows alternative phylogeny for the Bakevelliidae, *Pteria*, and *Isognomon* lineages. Modified from Tëmkin (2006a).

disjunction because the Miocene fossils have been reported from Western Europe, Caucasus, and Turkmenistan (Bagdasaryan, 1983; Caretto *et al.*, 1989; Nevesskaya, 1993). Among extant and commercially important pearl oysters, *P. mazatlanica* is reported from the Pliocene of Baja California Sur (Moore, 1983).

2.5. SUMMARY

The taxonomy of pearl oyster species is confused and there are probably considerably fewer species than the ca. 50 that have been described. This is particularly because many species descriptions were based on small numbers of specimens and exclusively on shells, without recognising that the shells of pearl oyster species are highly variable, developmentally plastic and greatly influenced by environmental conditions. There have been a number of taxonomic studies of pearl oysters, but these have concentrated on individual species or regional revisions or they have considered pteriid taxa within the context of large-scale faunistic studies. Other studies of genera and the Pteriidae have been compromised by factors such as limited access to type specimens.

A contemporary taxonomic problem of considerable importance is the status of a species-complex, *Pinctada fucata/martensii/radiata/imbricata*. It seems that this complex may constitute a cosmopolitan, globally-distributed species, characterised by substantial intraspecific variation over its range. Taxonomic resolution of the species-complex is progressing through morphological and genetic studies, but it is currently unresolved and the species-complex is referred to here and elsewhere in this book as the Akoya pearl oyster: Akoya pearls being the widely-used name for pearls derived from members of this species-complex in Japan. Given the paramount important of its members in the global pearl culture industry, a good understanding of the phylogenetic affinities among different populations of Akoya pearl oysters has important implications for genetic programs within the pearl industry.

Very detailed descriptions of shell morphology and soft anatomy of the Akoya pearl oyster are given in this chapter as a reference point for other morphological studies of pearl oysters. Details of distribution and habitat are also provided. Similar descriptions, but in less detail, are given for four other pearl oyster species of major commercial importance: *Pinctada margaritifera* (Blacklip pearl oyster), *P. maxima* (Silverlip or Goldlip pearl oyster), *P. mazatlanica* (Panamanian or Calafia pearl oyster), *Pteria penguin* (Winged pearl oyster).

Despite considerable advances in resolving the broader systematics of the Pteriidae, further research is required to provide a more complete view of the phylogeny of *Pteria* and *Pinctada* in the context of better-resolved relationships within the superfamily Pterioidea. Furthermore, an important task for the future will be to integrate analyses of extant species with the fossil record, biogeographic, and palaeoenvironmental evidence to provide a comprehensive view of the evolution of pearl oysters.

References

Abbott, T.R., 1974. *American Seashells: The Marine Mollusca of the Atlantic and Pacific Coasts of North America*. Van Nostrand Reinhold Company, New York, 541 pp.

Abbott, T., Dance, S.P., 1982. *Compendium of Seashells: A Color Guide to More Than 4200 of the World's Marine Shells*. E.P. Dutton Inc., New York, 411 pp.

Alagarswami, K., Dharmaraj, S., Velayudhan, T.S., Chellam, A., Victor, A.C.C., Gandhi, A.D., 1983. Larval rearing and production of spat of pearl oyster *Pinctada fucata* (Gould). *Aquaculture* 34, 287–301.

Alagarswami, K., Dharmaraj, S., Velayudhan, T.S., Chellam, A., 1987. Hatchery technology for pearl oyster production. In: Alagarswami, K. (Ed.), *Pearl Culture*. Central Marine Fisheries Research Institute (CMFRI) Bulletin, 39, pp. 62–71.

Alagarswami, K., Dharmaraj, S., Chellam, A., Velayudhan, T.S., 1989. Larval and juvenile rearing of black-lip pearl oyster, *Pinctada margaritifera* (Linnaeus). *Aquaculture* 76, 43–56.

Allan, J., 1959. *Australian Shells with Related Animals Living in the Sea, in Freshwater and on the Land.* Charles T. Branford Company, Boston, MA, 487 pp.

Allen, A.W., 1906. Some notes on the life history of *Margaritifera panasesae. J. Linn. Soc. (Zool.)* 29, 410–413.

Arnaud-Haond, S., Monteforte, M., Galtier, N., Blanc, F., Bonhomme, F., 2001. Population structure and genetic variablity of pearl oyster *Pinctada mazatlanica* along Pacific coasts from Mexico to Panama. *Conserv. Genet.* l, 299–307.

Arnaud-Haond, S., Boudry, P., Saulnier, D., Seaman, T., Vonau, V., Bonhomme, F., Goyard, E., 2002. New anonymous nuclear DNA markers for the pearl oyster *Pinctada margaritifera* and other *Pinctada* species. *Mol. Ecol. Notes* 2, 220–222.

Arnaud-Haond, S., Monteforte, M., Blanc, F., Bonhomme, F., 2003. Evidence for male-biased effective sex ratio and recent step-by-step colonization in the bivalve *Pincatada mazatlanica. J. Evol. Biol.* 16, 790–796.

Arnaud-Haond, S., Vonal, V., Bonhomme, F., Boudry, P., Blanc, F., Pror, J., Seaman, T., Goyard, E., 2004. Spatio-temporal variation in the genetic composition of wild populations of pearl oyster (*Pinctada margaritifera cumingii*) in French Polynesia following 10 years of juvenile translocation. *Mol. Ecol.* 13, 2001–2007.

Atkins, D., 1937. On the ciliary mechanisms and interrelationships of lamellibranchs. Part 3: Types of lamellibranch gills and their food currents. *Q. J. Microsc-. Sci.* 79, 375–421.

Atkins, D., 1938. On the ciliary mechanisms and interrelationships of lamellibranchs. Part 7: Latero-frontal cilia of the gill filaments and their phylogenetic value. *Q. J. Microsc. Sci.* 80, 345–436.

Atkins, D., 1943. On the ciliary mechanisms and interrelationships of lamellibranchs. Part 8: Notes on gill musculature in the Microciliobranchia. *Q. J. Microsc. Sci.* 84, 187–256.

Atsumi, T., Komaru, A., Okamoto, C., 2004. Genetic relationship among the Japanese pearl oyster *Pinctada fucata martensii* and foreign pearl oysters. *Fish. Genet. Breed. Sci.* 33, 135–142. (in Chinese,with English Abstract)

Bagdasaryan, K.G., 1983. On the palaeobiology of east-Paratethian Miocene Pteriidae and Isognomonidae. *Bull. Acad. Sci. Georgian SSR* 110, 193–196.

Beaurnent, A.R., Khamdan, S.A.A., 1994. Electrophoretic and morphometric characters in population differetiation of the pearl oyster, *Pinctada radiata* (Leach), from around Bahrain. *J. Moll. Stud.* 57, 433–441.

Bédouet, L., Schuller, M.H., Marin, F., Milet, C., Lopez, E., Giraud, M., 2001. Soluble proteins of the nacre of the giant oyster *Pinctada maxima* and of the abalone *Haliotis tuberculata*: extraction and partial analysis of nacre proteins. *Comp-. Biochem. Physiol. B* 128, 389–400.

Beer, A., 1999. Larval culture, spat collection and juvenile growth of the winged pearl oyster, *Pteria penguin*. World Aquaculture '99. In: *The Annual International Conference and Exposition of the World Aquaculture Society*, 26th April to 2nd May 1999, Sydney, Australia. Book of Abstracts, p. 63.

Bernard, F., 1898. Recherches ontogéniques et morphologiques sur la coquille des Lamellibranches. Ann. Sci. Nat. (Zoologie) Ser. 8. 8, 1–208.

Bernard, F.R., Cai, Y.Y., Morton, B., 1993. *Catalogue of the Living Marine Bivalve Molluscs of China*. Hong Kong University Press, Hong Kong,. 146 pp.

Beurlen, K., 1944. Beiträge zur Stammesgeschichte der Muscheln. *München Akad. Sitzungsberichte* 11, 113–131.

Blanc, F., Durand, P., 1989. Genetic variability in natural bivalves populations: the case of the black-lipped pearl oyster *Pinctada margaritifera. La Mer* 27, 125–126.

Bosch, D., Bosch, E., 1982. *Seashells of Oman*. Longman Group Limited, London, 212 pp.

Bosch, D.T., Dance, S.P., Moolenbeek, R.G., Oliver, P.G., 1995. *Seashell of Eastern Arabia* 29. Motivate Publishing, Dubai, 296 pp.

Boss, K.J., 1984. Mollusca. In: Parker, S.P. (Ed.), *Synopsis and Classification of Living Organisms.* McGraw-Hill, New York, pp. 945–1166.

Boutan, L., 1925. *La perle: Étude générale de la perle histoire de la méléagrine et des mollusques product-eurs de perles.* G. Doin, Paris, 421 pp.

Bullivant, J.S., 1962. Direct observation of spawning in the blacklip pearl shell oyster (*Pinctada margaritif-era*) and the thorny oyster (*Spondylus* sp.). *Nature* 193, 700–701.

Burch, B.L., 1995a. Pearly shells-Part 5: Transport by man of commercial pearl shells and their hitch-hikers. *Hawaiian Shell News 43 (New Series 426)* 3, 6.

Burch, B.L., 1995b. Pearly shells-Part 9: Pearl oyster taxonomy and its problems. *Hawaiian Shell News 43 (New Series 430)*, 10–12.

Butler, A.J., 1998. Order Pterioida. In: Beesley, P.L., Ross, G.J.B., Wells, A. (Eds.), *Mollusca: The Southern Synthesis. Fauna of Australia.* CSIRO Publishing, Melbourne, pp. 261–267.

Cahn, A.C., 1949. *Pearl Culture in Japan. Natural Resources Secion Section Report 122. Gen.* Supreme Commandant for the Allied Forces, Tokyo, 91 pp.

Camp, D.K., Lyons, W.G., Perkins, T.H., 1998. *Checklists of Selected Shallow-Water Marine Invertebrates of Florida.* Florida Department of Environmental Protection, Florida Marine Research Institute, St. Petersburg, Florida, 238 pp.

Caretto, P.G., Durand, P., Blanc, F., 1989. Apport de l'analyse biométrique à l'étude des relations phylo-génétiques de la nacre fossile *Pteria margaritifera studeri* (Mayer) (Mollusque, bivalve, Pteriidae). *Atti. Soc. Ital. Sci. Nat. Mus. Civ. Stor. Nat. Milano* 130, 205–216.

Carlton, J.T., 1996. Marine bioinvasions: the alteration of marine ecosystems by nonindigenous species. *Oceanography* 9, 36–43.

Carter, J.G., 1990. Evolutionary significance of shell microstructure in the Palaeotaxodonta, Pteriomorphia and Isofilibranchia (Bivalvia: Mollusca). In: Carter, J.G. (Ed.), *Skeletal Biomineralization: Patterns, Processes and Evolutionary trends.* Vol. I. Van Nostrand Reinhold, New York, pp. 135–296.

Chellam, A., 1987. Biology of pearl oyster *Pinctada fucata* (Gould). In: Alagarswami, K. (Ed.), *Pearl Culture.* Central Marine Fisheries Research Institute (CMFRI) Bulletin 39, pp. 13–20.

Chesler, J., 1994. Not just bilge water. *Am. Conchologist* 22, 13.

Colgan, D.J., Ponder, W.F., 2002. Genetic discrimination of morphologically similar, sympatric species of pearl oysters (Mollusca: Bivalve: *Pinctada*). in eastern Australia. *Mar. Freshwat. Res.* 53, 697–709.

Currey, J.D., Zioupos, P., Davies, P., Casinos, A., 2000. Mechanical properties of nacre and highly mineral-ized bone, *Proc. R. Soc. Lond. B Biol. Sci.* 268, 107–111.

Dix, T.G., 1973. Histology of the mantle and pearl sac of the pearl oyster *Pinctada maxima* (Lamellibranchia). *J. Mal. Soc. Aust.* 2, 365–375.

Donkin, R.A., 1998. *Beyond Price. Pearls and Pearl-Fishing: Origins to the Age of Discoveries.* American Philosophical Society, Philadelphia, 448 pp.

Doroudi, M.S., Southgate, P.C., 2003. Embryonic and larval development of *Pinctada margaritifera* (L.). *Moll. Res.* 23, 101–107.

Dunker, W., 1872–1880, Die Gattung *Avicula* in Abbildungen nach der Natur mit Beschreibungen. pp. 1–84, in: h. c. Küster, (ed.), *Systematisches Conchylien-Cabinet von Martini und Chemnitz.* Vol. 7, part 3. Bauer und Raspe, Nürnberg. [Pp. 1–56, pl. 1–18, 1872; 57–68, pl. 19–24, 1879; pp. 69–84, pl. 25–27, (1880).

Fischer-Piette, E., 1980. Revision des Aviculidae (*Pinctada* excl.) IV: *Electroma. Boll. Malac.* 16, 397–406.

Fischer-Piette, E., 1982. Revision des Aviculidae (*Pinctada* excl.). V: *Pteria. Boll. Malac.* 18, 151–174.

Fougerouse-Tsing, A., Herbaut, C., 1994. *Atlas Anatomique de l'Huître Perlière: Pinctada margaritif-era.* Établissement pour la Valorisation des Activités Aquacoles et Maritimes (EVAAM), Centre Universitaire de Polynésie Française, 68 pp.

Friedman, K.J., Bell, J.D., 1996. Effects of different substrata and protective mesh bags on collection of spat of the pearl oysters, *Pinctada margaritifera* (Linnaeus, 1758) and *Pinctada maculata* (Gould, 1850). *J. Shellfish. Res.* 15, 525–541.

Fujimura, T., Wada, K., Iwaki, T., 1995. Development and morphology of the pearl oyster larvae, *Pinctada fucata. Venus* 54, 25–48.

Gervis, M.H., Sims, N.A., 1992. The *biology* and *culture* of *pearl oysters* (Bivalvia: Pteriidae). ICLARM Studies and Reviews 21, pp. 49.

Giribet, G., Distel, D.L., 2003. Bivalve phylogeny and molecular data. In: Lydeard, C., Lindberg, D.R. (Eds.), *Molecular Systematics and Phylogeography of Mollusks*. Smithsonian Books, Washington and London, pp. 45–90.

Gould, A.A., 1850. Shells collected by the United States Exploring Expedition under the command of Charles Wilkes. *Proc. Boston Soc. of Nat. Hist.* 3, 309–312.

Gray, J.E., 1847. List of the genera of Recent Mollusca, their synonyma and types, *Proc. Zool. Soc. Lond.*, 15, 129–219.

Grégoire, C., 1960. Further studies on structure of the organic components in mother-of-pearl, especially in pelecypods. Part 1. *Bull. Inst. R. Sci. Nat. Belg.* 36, 1–22.

Guenther, J., Southgate, P.C., de Nys, R., 2006. The effect of age and shell size on accumulation of fouling organisms on the Akoya pearl oyster. *Pinctada fucata* (Bivalvia, Pteriidae). *Aquaculture*. 253, 366–373.

Habe, T., 1977. *Systematics of Mollusca in Japan. Bivalvia and Scaphopoda*. Zukan-no-Hokuryukan, Tokyo, 372 pp. (In Japanese)

Habe, T., 1981. Bivalvia. In: Koyama, Y., Yamamoto, T., Toki, Y., Minato, H. (Eds.), *A Catalogue of Mollusks of Wakayama Prefecture, the Province of Kii*. Publications of the Seto Marine Biological Laboratory-Special Publication Series. 7(1), 25–223.

Hamel, J.-F., Mercier, A., 2003. Hosting a living treasure: pearl oysters. *Aquarium* 26, 108–118.

Handa, T., Yamamoto, K., 2003. Corrosion casting of the digestive diverticula of the pearl oyster, *Pinctada fucata martensi* (Mollusca: Bivalvia). *J. Shellfish Res.* 22, 777–779.

Hanley, S., 1842–1856. *An Illustrated and Descriptive Catalogue of Recent Bivalve Shells*. Williams and Norgate, London, xviii+392 pp.

Hayami, I., 2000. Family Pteriidae Order Pterioida. In: Okutani, T. (Ed.), *Marine Mollusks in Japan*. Tokai University Press, Tokyo, Japan, pp. 879–883. (In Japanese with English species, descriptions)

Hayasi, K., 1938. On the so-colled bottle-shaped glands of *Pinctada martensii*, with special reference to Tanaka's theory of pearl formation. *Annot. Zool. Japon* 17, 105–106.

Hayes, H.L., 1972. *The Recent Pteriidae (Mollusca) of the Western Atlantic and Eastern Pacific Oceans*. Unpublished Ph.D. Dissertation, Georgetown University, Washington D.C., 202 pp., 14pls. (Dissertation Abstract International *33B*(8): 4039).

He, M., Huang, L., Shi, J., Jiang, Y., 2005. Variability of ribosomal DNA ITS-2 and its utility in detecting genetic relatedness of pearl oyster. *Mar. Biotechnol.* 15, 40–45.

Herdman, W.A., 1903. Observations and experiments on the life-history and habits of the pearl oyster. In: Herdman, W.A. (Ed.), *Report to the Government of Ceylon on the Pearl Oyster Fisheries of the Gulf of Manaar*. Part I. The Royal Society, London, pp. 125–146.

Herdman, W.A., 1904. Anatomy of the pearl oyster (*Margaritifera vulgaris*, Schum.). In: Herdman, W.A. (Ed.), *Report to the Government of Ceylon on the Pearl Oyster Fisheries of the Gulf of Manaar*. Part 2. The Royal Society, London, pp. 37–69.

Hertlein, L.G., Cox, L.R., 1969. Family Pteriidae Gray, 1847 (1820). In: Cox, L.R., Newell, N.D., Boyd, D.W., Branson, C.C., Casey, R., Chavan, A., Coogan, A.H., Dechaseaux, C., Fleming, C.A., Haas, F., Hertlein, L.G., Kauffman, E.G., Keen, A.M., LaRocque, A., McAlester, A.L., Moore, R.C., Nuttall, C.P., Perkins, B.F., Puri, H.S., Smith, L.A., Soot-Ryen, T., Stenzel, H.B., Trueman, E.R., Turner, R.D., Weir, J. (Eds.), *Treatise on Invertebrate Paleontology. Part N. Mollusca 6: Bivalvia*. Vol. 1. Geological Society of America and University of Kansas, Lawrence, Kansas, pp. N302–N306.

Higo, S., Callomon, P., Goto, Y., 1999. *Catalogue and Bibliography of the Marine Shell-Bearing Mollusca of Japan*. Elle Science Publications, Osaka, Japan, 749 pp.

Hornell, J., 1922. The common molluscs of South India. *Madras Fish. Bull.* 13, 97–215.

Hornell, J., 1951. *Indian Molluscs*. Bombay Natural History Society, Bombay, India, 96 pp.

Hylleberg, J., Kilburn, R.N., 2002. Annotated inventory of molluscs from the Gulf of Mannar and vicinity. Zoogeography and Inventory of Marine Molluscs Encountered in Southern India. Eleventh International Congress & Workshop conducted at the Suganthi Devadason Marine Research Institute (SDMRI) Tuticorin, Tamil Nadu, India. Phuket Marine Biological Center Special Publication 26, 19–79.

Hu, C.H., Tao, H.J., 1995. *Shells of Taiwan Illustrated in Color*. National Museum Taiwan, Taiwan, iii+483 pp.

Hwang, J-J., Okutani, T., 2003. Taxonomy and distribution of the genera *Pteria* and *Pinctada* (Bivalvia: Pteriidae) in Taiwan. *J. Fish. Soc. Taiwan* 30, 199–216.

Hynd, J.S., 1955. A revision of the Australian pearl-shells, genus *Pinctada* (Lamellibranchia). *Aust. J. Mar. Freshwat. Freshwater Res.* 6, 98–137.

Inaba, A., 1982. *Molluscan Fauna of the Seto Inland Sea*. Hiroshima Shell Club, Hiroshima, 181 pp. (In Japanese)

International Commission on Zoological Nomenclature (ICZN), 1999. *International code of zoological nomenclature 5 Code international de nomenclature zoologique; adopted by the International Union of Biological Sciences*. 4th ed. London: International Trust for Zoological Nomenclature, c/o Natural History Museum, London, xxix+ 306 pp.

Iredale, T., 1915. Some more misused molluscan generic names. *Proc. Malac. Soc. Lond.*, *11*, 291–306.

Iredale, T., 1939. Mollusca. Part I. Scientific Reports of the In: Great Barrier Reef Expedition 1928–1929, 5, pp. 209–425.

Jabbour-Zahab, R., Chagot, D., Blanc, F., Grizel, H., 1992. Mantle histology, histochemistry and ultrastructure of the pearl oyster *Pinctada margaritifera* (L.). *Aquat. Living Resour.* 5, 287–298.

Jabbour-Zahab, R., Blanc, F., Grizel, H., 2003. The mantle. In: Grizel, H. (Ed.), *Atlas d'Histologie et de Cytologie des Mollusques Bivalves Marins: An Atlas of Histology and Cytology of Marine Bivalve Molluscs*. Ifremer, Plouzané, France, pp. 11–34.

Jameson, H.L., 1901. On the identity and distribution of the mother-of-pearl oysters; with a revision of the subjenus *Margaritifera. Proc. Gen. Meetings Sci. Bus. Zoo. Soc. Lond. 1*, 372–394.

Kawakami, I.K., Yasuzumi, G., 1964. Electron microscope studies on the mantle of the pearl oyster *Pinctada martensii* Dunker. Preliminary report. The fine structure of the periostracum fixed with permanganate. *J. Electr. Microsc.* 13, 119–123.

Kawamoto, N.Y., Nakanishi, S., 1957. Microscopical observation on the blood corpuscles of pearl-oyster *Pinctada martensii* (Dunker). *Bull. Natl. Pearl Res. Lab.* 3, 201–206.

Kay, E.A., 1979. *Hawain Marine Seashells*. Bernice P. Bishop Museum, Honolulu. 653 pp.

Kelaart, E.F., 1858. Introductory report on the natural history of the pearl oyster (*Meleagrina margaritifera*, Lam.) of Ceylon. *Proc. R. Phys. Soc. Edinburgh 1*, pp. 399–405.

Klumpp, D.W., Burdon-Jones, C., 1982. Investigations of the potential of bivalve molluscs as indicators of heavy metal levels in tropical marine waters. *Aust. J. Mar. Freshwat. Res.* 33, 285–300.

Kobayashi, I., 1971. Internal shell microstructure of recent bivalvian molluscs *Sci. Rep. Niigata Univ. Ser. E: Geol. and Mineral.*, *2*, 27–50.

Kobayashi, M., Masaoka, T., 2001. Estimation of the phylogenetic relationships of pearl oyster species based on mitochondrial ribosomal RNA gene sequence data. *DNA Polymorphism* 9, 90–94. (in Japanese)

Komaru, A., Wada, K.T., 1985. Karyotype of the Japanese pearl oyster, *Pinctada fucata martensi*, observed in the trochophore larvae. Yoshoku Kenkyusho Kenkyu Hokoku (*Bull. Natl. Res. Inst. Aquacult.*) 7, 105–107.

Kuroda, T., 1960. *A Catalogue of Molluscan Fauna of the Okinawa Islands*. Ryukyu University, 106 pp.

Kuroda, T., Habe, T., Oyama, K., 1972. *The Seashells of Sagami Bay*. Maruzen, Tokyo. 19-741 (In Japanese) + 489 (In English) + 51pp., 121 pls.

Kuwatani, Y., 1965. On the anatomy and function of stomach of Japanese pearl oyster, *Pinctada martensii* (Dunker). *Nippon Suisan Gakkaishi (Bull. Jap. Soc. Sci. Fish.)* 31, 174–186. (in Japanese, with English abstract).

Lai, K.Y., 1998. *Shell (2)*. Recreation Press Co. Ltd., 199 pp. (In Chinese)

Lamprell, K., Healy, J., 1998. *Bivalves of Australia*, Vol. 2. Backhuys Publishers, Leiden, 288 pp.

Landman, N.H., Mikkelsen, P.M., Bieler, R., Bronson, B., 2001. *Pearls: A Natural History*. Harry N. Abrams, Inc., New York, 232 pp.

Leach, W.E., 1814. *Zoological Miscellany; Being Descriptions of New, or Interesting Animals*. Vol. 1. E. Nodder & Son, London, 144 pp.

Linnaeus, C., 1758. *Systema naturæ Per regna tria naturæ, Secundum classes, ordines, genera, species, Cum characteribus, differentiis, synonymis, locis*. Vol. 1. Impensis Direct, Laurentii Salvii, Holmiæ [Stockholm], 824 pp.

Lintilhac, J.-P., 1985. *Les perles noires de Tahiti*. Pacific Editions, Tahiti, 88 pp.

Lintilhac, J.-P., Durand, A., 1987. *Black pearls of Tahiti*. Royal Tahitian Pearl Book, Tapeete, Tahiti, 109 [+4] pp.

Lopez, E., Berland, S., LeFaou, A., 1995. Nacre, osteogenic and osteoinductive properties. *Bull. Inst. Océanogr.* 14, 49–57.

Ludbrook, N.H., Gowlett-Holmes, K.L., 1989. Chitons, gastropods and bivalves. In: Shepherd, S.A., Thomas, I.M. (Eds.), *Marine Invertebrates of Southern Australia. Part II.* South Australian Government Printing division, Adelaide, pp. 504–724.

MacKenzie, C.L., 1999. A history of the pearl oyster fishery in the archipelago de las Perlas, Panama. *Mar. Fish. Rev.* 61, 58–65.

Macpherson, J.H., Gabriel, C.J., 1962. *Marine Molluscs of Victoria*. Melbourne University Press, Melbourne, 475 pp.

Maes, V.O., 1967. The littoral marine mollusks of Cocos-Keeling Islands (Indian Ocean). *Proc. Acad. Nat. Sci. of Phila.*, *119*, pp. 93–217.

Mahadevan, S., Nagappan, N.K., 1974. The commercial molluscs of India. In: Nair, R.V., Rao, K.S. (Eds.), *Ecology of Pearl Oyster and Chank Beds*. Central Marine Fisheries Research Institute, India, Cochin, pp. 106–121.

Masaoka, T., Kobayashi, T., 2002. Phylogenetic relationship in pearl oysters (Genus: *Pinctada*) based on rRNA sequences. *DNA Polymorphism* 10 100-104. (In Japanese)

Masaoka, T., Kobayashi, T., 2003. Estimation of phylogenetic relationships in pearl oysters (Genus: *Pinctada*) based on 28SrRNA and ITS sequences. *DNA Polymorphism* 11, 76–81. (In Japanese)

Masaoka, T., Kobayashi, T., 2004a. Polymerase chain reaction-based species identification of pearl oyster using nuclear ribosomal DNA internal transcribed regions. *Fish. Genet. Breed. Sci.* 33, 101–105.

Masaoka, T., Kobayashi, T., 2004b. DNA extraction method for organisms contain mucopolysaccharides. *DNA Polymorphism* 12, 82–86. (In Japanese)

Masaoka, T., Kobayashi, T., 2005a. Natural hybridization between *Pinctada fucata* and *Pinctada maculata* inferred from internal transcribed spacer regions of nuclear ribosomal RNA genes. *Fish. Sci.* 71, 829–836.

Masaoka, T., Kobayashi, T., 2005b. Species identification of *Pinctada imbricata* using intergenic spacer of nuclear ribosomal RNA genes and mitochondrial 16S ribosomal RNA gene regions. *Fish. Sci.* 71, 837–846.

Masaoka, T., Kobayashi, T., 2005c. Development of new DNA markers of pearl oyster (*Pincatada martensii* and *P. fucata*) using ISSR (Inter-Simple Sequence Repeat) – PCR methods. *DNA Polymorhpism* 13, 145–150. (In Japanese)

Masaoka, T., Kobayashi, T., 2005d. Estimation of phylogenetic relationships in pearl oysters (Mollusks: Bivalve: *Pinctada*) used for pearl production based on rRNA gene sequences. *DNA Polymorphism.* 13, 151–162. (In Japanese)

Masaoka, T., Kobayashi, T., 2006a. Species identification of *Pinctada radiata* using intergeneric spacer of nuclear ribosomal RNA genes and mitochondrial 16S ribosomal RNA gene regions. *Fish. Genet. Breed. Sci.* 35, 49–59.

Masaoka, T., Kobayashi, T., 2006b. Review-phylogeny and identification of *Pinctada* secies: Introduction and application of molecular genetic approach. *Fish. Genetic. Breed. Sci.* 36, 1–14. (In Japanese, with English abstract)

Matsukuma, A., 2004. Family Pteriidae, Order Pterioida. In: Okutani, T. (Ed.), *Encyclopedia of Shellfish.* Sekaibunkasha, Tokyo, Japan, pp. 288. (In Japanese)

Matsumoto, M., 2003. Phylogenetic analysis of the subclass Pteriomorphia (Bivalvia) from mtDNA COI sequences. *Mol. Phylogenet Evol.* 27, 429–440.

Matthews, H.R., 1969. Notas sobre a família Pteriidae no nordeste Brasileiro (Mollusca: Pelecypoda). *Arq. Cienc. Mar. 9*, 178–180.

Mikkelsen, P.M., Těmkin, I., Bieler, R., Lyons, W.G., 2004. *Pinctada lonisquamosa* (Dunker, 1852) (Bivalvia: Pteriidae), an unrecognized pearl oyster in the western Atlantic. *Malocologia* 46, 473–501.

Moore, E.J., 1983. Tertiary marine pelecypods of California and Baja California: Nuculidae through Malleidae. *Geological Survey Professional Paper* 1228-A, A1–A108.

Mouriès, L.P., Almeida, M.-J., Milet, C., Berland, S., Lopez, E., 2002. Bioactivity of nacre water-soluble organic matrix from the bivalve mollusk *Pinctada maxima* in three mammalian cell types: fibroblasts, bone marrow stromal cells and osteoblasts. *Comp. Biochem. Physiol. B* 132, 217–229.

Nakazima, M., 1958. On the differentiation of stomach of Pelecypoda (I). *Venus* 20, 197–207.

Nasr, D.H., 1984. Feeding and growth of the pearl oyster, *Pinctada margaritifera* in Dongonab Bay, Sudan. *Red Sea Hydrobiologia* 110, 241–246.

Nevesskaya, L.A., 1993. Opredelitel miotenovih dvustvorcatiyh molluskov Yugo-Zapadnoy Yevrasii. Russkaya Akademia Nauk, Trudiyh Paleontologiceskovo Instituta, Tom. 247, 412 pp. (in Russian)

Newell, N.D., 1965. Classification of the Bivalvia. *Am. Mus. Novit.* 2206, 1–25.

Newell, N.D., 1969. Classification of Bivalvia. In: Cox, L.R., Newell, N.D., Boyd, D.W., Branson, C.C., Casey, R., Chavan, A., Coogan, A.H., Dechaseaux, C., Fleming, C.A., Haas, F., Hertlein, L.G., Kauffman, E.G., Keen, A.M., LaRocque, A., McAlester, A.L., Moore, R.C., Nuttall, C.P., Perkins, B.F., Puri, H.S., Smith, L.A., Soot-Ryen, T., Stenzel, H.B., Trueman, E.R., Turner, R.D., Weir, J. (Eds.), *Treatise on Invertebrate Paleontology, Part N. Mollusca 6: Bivalvia.* Volume 1. Geological Society of America and University of Kansas, Lawrence, Kansas, pp. N205–N224.

Nicholls, A.G., 1931. On the breeding and growth rate of the black lip pearl oyster (*Pinctada margaritifera*). *Rept. Gt. Barrier Reef Comm.* 2, 26–31.

Odhner, N.H., 1917. Results of Dr. E. Mjöbergs Swedish scientific expeditions to Australia 1910–1913. XVII: Mollusca. Kungl. *Svenska Vetenskapsakademiens Handlingar* 52, 1–115.

Odhner, N.H., 1919. Contribution à la faune malacologique de Madagascar. Arkiv för Zoologi Utgifvet af K. *Svenska Vetenskapsakademiens* 12, 1–52.

Ojima, Y., 1952. Histological studies of the mantle of pearl oyster (*Pinctada martensii* Dunker). *Cytologia* 17, 134–143.

Oliver, P.G., 1992. *Bivalved Seashells of the Red Sea.* Verlage Christa. Hemmen and National Museum of Wales, Wiesbaden and Cardiff, 330 pp.

Ota, S., 1957. Notes on the identification of free swimming larvae of pearl oyster (*Pinctada martensii*). *Kokoritsu Shinju Kenkyusho Hokoku (Bull. Natl. Pearl Res. Lab.) 2, 128–132.* (in Japanese).

Paugam, A., D'Ollone, C., Cochard, J.-C., Garen, P., Le Penne, M., 2006. The limits of morphometric features for the identification of black-lip pearl oyster (*Pinctada margaritifera*) larvae. *J. Shellfish Res.* 25, 959–967.

Pereira, L., Milet, C., Almeida, M.J., Rousseau, M., Robichon, F., Lopez, E., 2001. Biological effect of the water-soluble matrix extract from the nacre of the bivalve mollusc *Pinctada maxima* on vertebrate cell proliferation and differentiation. *Bone* 28, S249–S249.

Pouvreau, S., Bacher, C., Heral, M., 2000. Ecophysiological model of growth and reproduction of the black pearl oyster, *Pinctada margaritifera*: potential applications for pearl farming in French Polynesia. *Aquaculture* 186, 117–144.

Prashad, B., 1932. *The Lamellibranchia of the Siboga Expedition. Systematic Part 2: Pelecypoda (exclusive of the Pectinidae).* Siboga-Expeditie. Vol. 53c. E. J. Brill, Leiden, 353 pp.

Prashad, B., Bhaduri, J.L., 1934. The pearl oysters of Indian waters. *Rec. Indian Mus. (J. Indian Zool.)* 35, 167–174.

Purchon, R.D., 1957. The stomach in the Filibranchia and Pseudolamellibranchia. *Proc. Zool. Soc. Lond.*, 129, 27–60.

Ranson, G., 1961. Les espèces d'huîtres perlières du genre *Pinctada* (biologie de quelques-unes d'entre elles). *Institute Royal des Sciences Naturelles de Belgique, Mémoires, 2e série, fasc.* 67, 1–95. 42 pls.

Rao, K.V., 1968. Pearl oysters of the Indian region. *Proceedings of the Symposium on Mollusca held at Cochin from January 12–16.* Marine Biological Association of India, Mandapam Camp, pp 1017–1028.

Rao, K.V., Rao, S.K., 1974. Pearl oysters. In: Nair, R.V., Rao, S.K. (Eds.), *The Commercial Molluscs of India.* ICAR Bulletin of the Central Marine Fisheries Research Institute 25, 84–105.

Reeve, L.A., 1857. Monograph of the genus Avicula. In: Reeve, L.A., *Conchologia Iconica, or Illustrations of the Shells of Molluscous Animals.* Vol. 10. Lovell Reeve, London. [20 pp.], 18 pls.

Reid, R.G.B., Brand, D.G., 1985. Some unusual renal characteristics of the bivalves *Pteria penguin*, *Spondylus nicobaricus* and *Spondylus barbatus*. In: Morton, B., Dudgeon, D. (Eds.), *The Malacofauna of Hong Kong and Southern China. II Proceedings of the Second International Workshop on the Malacofauna of Hong Kong and Southern China, Hong Kong, 6–24 April 1983*. Vol. 2. Hong Kong University Press, Hong Kong, pp. 513–518.

Ridewood, W.G., 1903. On the structure of the gills of the Lamellibranchia. *Philos. Trans. R. Soc. Lond. Biol. Sci.* 195, 147–284.

Roberts, D., Soemodihardjo, S., Kastoro, W., 1982. *Shallow Water Marine Molluscs of North-West Java*. Lembaga Oseanologi Nasional, Lembaga Limu Pengetahuan Indonesia, Jakarta, 143 pp.

Röding, P.F., 1798. *Museum Boltenianum sive Catalogus cimeliorum e tribus regnis naturæ quæ olim collegerat Fon. Fried Bolten, M.D.p.d. per XL annus Proto physicus Hmburgensis. Pars Secunda continens Conchylia sive Testacea univalvia, bivalvia & multivalvia*. J.C. Trappii, Hamburgi [Hamburg], viii + 199 pp.

Rose, R. A., Baker, S. B. *1989*. Research and development of hatchery and nursery culture for the pearl oyster *Pinctada maxima*. FIRTA Project 87/82. Final Report September 1989.

Rose, R.A., Dybdahl, R., Harders, S., 1990. Reproductive cycle of the Western Australian silver lip pearl oyster *Pinctada maxima* (Jameson) (Mollusca: Pteriidae). *J. Shellfish Res.* 9, 261–272.

Rousseau, M., Pereira-Mouriès, L., Almeida, M.-J., Milet, C., Lopez, E., 2003. The water-soluble matrix fraction from the nacre of *Pinctada maxima* produces earlier mineralization of MC3T3-E1 mouse pre-osteoblasts. *Comp. Biochem. Physiol. B* 137, 1–7.

Sagara, J., Takemura, Y., 1960. Studies on the age determination of the silver-lip pearl oyster, *Pinctada maxima* (Jameson). *Bull. Tokai Reg. Fish. Res. Lab.* 27, 7–14.

Sanchez, E.M., 1990. Catalogo de bivalves marinos del Ecuador. *Bol. Sci. y Tech. (Instituto Nacional de Pesca, Ecuador)* 10, 1–136.

Saucedo, P., Montefore, M., 1997. Breeding cycle of pearl oysters *Pinctada mazatlanica* and *Pteria sterna* (Bivalvia: Pteriidae) at Bahia de La Paz, Baja California sur, Mexico. *J. Shellfish Res.* 16, 103–110.

Sarver, D.J., Ellis, A., Sims, N.A., Wise, D., 2004. Using pearl oysters as heavy metal monitors in tropical waters. *J. Shellfish Res.* 23, 310. (Abstract only)

Saville-Kent, W., 1893. *The Great Barrier Reef of Australia; its Products and Potentialities*. W. H. Allen, London, 387 pp.

Scopoli, G.A.I.A., 1777. *Introductio ad historiam naturalem sistens genera lapidum, plantarum et animalium: hactenus detecta, caracteribus essentialibus donata, in tribus divisa, subinde ad leges naturae*. Wolfgangum Gerle, Pragae [Prague], 506 pp.

Service de la Pêche, 1970. *Etude sur l'Industrie Nacrière en Polynésie Française*. Service de la Pêche, Tahiti, Polynésie Française, 34 pp.

Shiino, S.M., 1952. Anatomy of *Pteria (Pinctada) martensii* (Dunker), mother-of-pearl mussel. Special Publication of the Fisheries Experimental Station of Mie Prefecture, [i] p, 12 pls.

Shirai, S., 1994. *Pearls and Pearl Oysters of the World*. Marine Planning Co., Okinawa, 108 pp. (In Japanese, with English comments)

Sims, N.A., 1990. *A Select Bibliography on the Biology, Ecology and Culture of Pearl Oysters*. South Pacific Commission, Noumea, New Caledonia, 24 pp.

Sims, N.A., 1993. Size, age and growth of the black-lip pearl oyster, *Pinctada margaritifera* (L.) (Bivalvia: Pteriidae). *J. Shellfish Res.* 12, 223–228.

Skelton, P.W., Benton, M.J., 1993. Mollusca: Rostroconchia, Scaphopoda and Bivalvia. In: Benton, M.J. (Ed.), The Fossil Record 2. Chapman & Hall, London, pp. 237–263.

Smith, E.A., 1891. On a collection of marine shells from Aden, with some remarks upon the relationship of the molluscan fauna of the Red Sea and the Mediterranean. *Proc. Zool. Soc. Lond.*, 390–436.

Springsteen, F.J., Leobrera, F.M., 1986. *Shells of the Philippines*. Carfel Seashell Museum, Manila, Philippines, 377 pp.

Stanley, S.M., 1970. *Relation of Shell Form to Life Habits of the Bivalvia (Mollusca)*. The Geological Society of America, Boulder, Colorado, 296 pp.

Stasek, C.R., 1963. Synopsis and discussion of the association of ctenidia and labial palps in the bivalved Mollusca. *The Veliger* 6, 91–97.

Steiner, G., Hammer, S., 2000. Molecular phylogeny of the Bivalvia inferred from 18S rDNA sequences with particular reference to the Pteriomorphia. In: Harper, E.M., Taylor, J.D., Crame, J.A. (Eds.), *The Evolutionary Biology of the Bivalvia, Proceedings of "Biology & Evolution of the Bivalvia," an international symposium organized by the Malacological Society of London, 14–17 September 1999, Cambridge, UK*. Special Publications, 177. Geological Society, London, pp. 11–29.

Stoliczka, F., 1871. *Cretaceous fauna of Southern India, 3: The Pelecypoda, with a review of all known genera of this class, fossil and recent. Memoirs of the Geological Survey of India.* Palæontologia Indica Series 6, 1–537.

Suzuki, T., 1985. Comparative anatomy of the excretory organs and heart of bivalves-subclasses Pteriomorphia, Palaeoheterodonta and Heterodonta. *Bull. Natl. Res. Inst. Aquaculture* 7, 59–82.

Swennen, C., Moolenbeek, R.G., Ruttanadakul, N., Hobbelink, H., Dekker, H., Hajisamae, S., 2001. *The Molluscs of the Southern Gulf of Thailand.* Biodiversity Research and Training Program, Bangkok, Thailand, 210 pp.

Takemura, Y., Kafuku, T., 1957. Anatomy of the silver-lip pearl oyster *Pinctada maxima* (Jameson). *Bull. Tokai Reg. Fish. Res. Lab.* 16, 39–40 1–23 (in Japanese with English Abstract)

Takemura, Y., Okutani, R.T., 1958. On the identification of species of *Pinctada* found attached to *Pinctada maxima* (Jameson) in the Arafra Sea. *Bull. Tokai Reg. Fish. Res. Lab.* 20, 47–60.

Tamura, M., 1961. New subgenus of pearl oyster, *Eopinctada*, from the Cretaceous Mifune Group in Kumamoto Prefecture, Japan, *Trans. Proc. Palaeontol. Soc. Japan* (New Series), 44, 147–151.

Tanaka, Y., 1958. Identification of larva of *Pinctada martensii. Venus* 19, 215–218.

Tanaka, Y., Kumeta, M., 1981. Successful artificial breeding of the silver-lip pearl oyster, *Pinctada maxima* (Jameson). *Yoshoku Kenkyusho Kenkyu Hokoku (Bull. Natl. Res. Inst. Aquacult.)* 2, 21–28 (in Japanese, with English Abstract)

Tantanasiriwong, R., 1979. A checklist of marine bivalves from Phuket Island, adjacent mainland and off-shore islands, western peninsular Thailand. *Phuket Mar. Biol. Center, Res. Bull.* 27, 1–15.

Tёmkin, I., 2004. A new system for Pterioidea (Mollusca: Bivalvia). In: Wells, F.E., (Ed.), *Molluscan megadiversity: Sea, Land and Freshwater. World Congress of Malacology*, July 11–16, Western Australian Museum, Perth, Western Australia, p. 145.

Tёmkin, I., 2006a. Morphological perspective on classification and evolution of Recent Pterioidea (Mollusca: Bivalvia). *Zool. J. Linn. Soc.* 148, 253–312.

Tёmkin, I., 2006b. Anatomy, shell morphology, and microstructure of the living fossil *Pulvinites exempla* (Hedley, 1914) (Bivalvia: Pulvinitidae). *Zool. J. Linn. Soc.* 148, 523–552.

Thiele, J., 1886. Die Mundlappen der Lamellibranchiaten. *Z. Wiss. Zool.* 44, 239–272.

Thiele, J., 1889. Die abdominalen Sinnerorgane der Lamellibranchier. *Z. Wiss. Zool.* 48, 47–59.

Thielley, M., Grizel, H., 2003. Reproductive organs and gametogenesis. In: Grizel, H. (Ed.), *Atlas d'Histologie et de Cytologie des Mollusques Bivalves Marins: An Atlas of Histology and Cytology of Marine Bivalve Molluscs.* Ifremer, Plouzané, France, pp. 169–200.

Thurston, E. (1894). Notes on the pearl and chank fisheries of the Gulf of Manaar. Madras Government Museum, Bulletin 1, 116 pp.

Tranter, D.J., 1958. Reproduction in Australian pearl oysters (Lamellibranchia). IV. *Pinctada margaritifera* (Linnaeus). *Aust. J. Marine Freshwat. Res.* 9, 509–525.

Uemoto, H., 1958. Studies on the gonad of the pearl oyster *Pinctada martensii* (Dunker): II. Histological observation with regard to both seasonal variation and the change during the course of the artificial spawning *Kokoritsu Shinju Kenkyusho Hokoku (Bull. Natl. Pearl Res. Lab.)* 4, 287–307.

Velayudhan, T.S., Gandhi, A.D., 1987. Morphology and anatomy of Indian pearl oyster. In: Alagarswami, K. (Ed.), *Pearl Culture.* Central Marine Fisheries Research Institute (CMFRI) Bulletin 39, 4–12.

Victor, A.C.C., Velayudhan, T.S., 1987. Ecology of pearl culture grounds. In: Alagarswami, K. (Ed.), *Pearl Culture.* Central Marine Fisheries Research Institute (CMFRI) Bulletin 39, 78–86.

Wada, K., 1961. On the relationship between shell growth and crystal arrangement of the nacre in some Pelecypoda. *Venus* 21, 204–211.

Wada, K., 1968. An electron microscope study of the formation of the periostracum of *Pinctada martensii. J. Electron. Microsc.* 17, 275.

Wada, K.T., 1978. Chromosome karyotypes of these bivalves: the oyster *Isognomon alatus* and *Pinctada imbricata*, and the bay scallop, *Argopecten irradians irradians*. *Biol. Bull. (Woods Hole)* 155, 225–245.

Wada, K.T., 1982. Inter- and itraspecific electrophoretic variation in three species of the pearl oysters from the Nansei Island of Japan. *Bull. Natl. Res. Inst. Aquaculture* 3, 1–10.

Wada, K.T., 1984. Breeding study of the Japanese pearl oyster, *Pinctada fucata martensii*. *Bull. Natl. Res. Inst. Aquaculture* 6, 79–157. (in Japanese with English Abstract)

Wada, K.T., Komaru, A., 1985. Karyotypes in five species of the Pteriidae (Bivalvia: Pteriomorphia). *Venus* 44, 183–192.

Wada, K.T., 1991. The Japanese pearl oyster. *Pinctada fucata* (Gould) (Family Pteridae). In: Menzel, W. (Ed.), *Estuarine and Marine Bivalve Mollusk Culture*. CRC Press, Boca Raton, FL, pp. 245–260.

Wada, K.T., 1994. Genetics of pearl oyster in relation to aquaculture. *JARQ (Japn. Intnatl. Res. Cntr. Agi. Sci., Min. Agri., Forest. Fish., Tsukuba, Japan)* 28, 276–282.

Wada, K.T., 1997. Recent research on genetic improvement of aquacultured species with reference to quantitative or population genetics in Japan. *Bull. Natl. Res. Inst. Aquaculture* Suppl.3, 25–28.

Wada, K.T., 1999. Aquaculture genetics of bivalve molluscs: a review. In: Xu, H., Colwell, R.R., (Eds.), *Proceedings of International Symposium on Progress Prospective Marine Biotechnology* (ISPPMB '98). China Ocean Press, Beijing, China, pp. 52–67.

Wada, K.T., Komaru, A., Ichimura, Y., Kurosaki, H., 1995. Spawning peak occurs during winter in the Japanese subtropical population of the pearl oyster, *Pinctada fucata fucata* (Gould, 1850). *Aquaculture* 133, 207–214.

Wada, R., Wada, S., 1939. Sexes in the silver-lip pearl oyster. *Kagaku-Nanyo* (*Science of the South Seas*) 2, 40–43.

Wang, Z., 1978. A study of the Chienese of the family Pteriidae (Mollusca). *Studia Marina Sinica* 14, 101–115 (in Chinese).

Westbroek, P., Marin, F., 1998. A marriage of bone and nacre. *Nature* 392, 861–862.

Wolf, P.H., Sprague, V., 1978. An unidentified protistan parasite of the pearl oyster, *Pinctada maxima*, in tropical Australia. *J. Inverteb. Pathol.* 31, 262–263.

Work, R.C., 1969. Systematics, ecology, and distribution of the molluscs of Los Roques, Venezuela. *Bull. Mar. Sci.* 19, 614–711.

Wu, W., 1980. The list of Taiwan bivalve fauna. *Quat. J. Taiwan Mus.* 33, 55–208.

Yu, D.H., Chu, K.H., 2005a. Phylogenetics of the common pearl oysters in the genus *Pinctada*: evidence from nrDNA ITS sequence. *Biodiversity Sci.* 13, 315–323. (In Chinese, with English summary)

Yu, D.H., Chu, K.H., 2005b. Species identity and phylogenetic relationship of the oysters in *Pinctada*, based on ITS sequence analysis. *Biochem. Syst. Ecol.* 34, 240–250.

Yu, D.H., Li, Y., Wu, K., 2005a. Analysis on nucleotide sequence variation of rDNA ITS2 in *Pinctada fucata* from China, Japan and Australia. *South China Fish. Sci.* 1, 1–6. (In Chinese, with English abstract)

Yu, D.H., Li, Y., Wu, K., 2005b. Analysis on nucleotide sequence variation of rDNA ITS2 in *Pinctada fucata* from China, Japan and Australia. *South China Fish. Sci.* 1, 1–6. (In Chinese, with English abstract)

Yu, D.H., Jia, X., Chu, K.H., 2006. Common pearl oysters in China, Japan, and Australia are conspecific: Evidence from ITS sequences and AFLP. *Fish. Sci.* 72, 1183–1190.

Zhao, B., Zhang, S., Qian, P.Y., 2003. Larval settlement of the silver- or goldlip pearl oyster *Pinctada maxima* (Jameson) in response to natural biofilms and chemical cues. *Aquaculture* 220, 883–901.

CHAPTER 3

Soft Tissue Anatomy, Shell Structure and Biomineralization

Angelique Fougerouse, Marthe Rousseau and John S. Lucas

3.1. INTRODUCTION

The first detailed anatomical observations on a pearl oyster were made by Herdman (1904) on "*Margaritifera vulgaris*" (=Akoya pearl oyster[1]) from Sri Lanka. There have been a number of subsequent anatomical studies on this "species" (Velayudhan and Gandhi, 1987; Chellam *et al.*, 1991), and on other species including *Pinctada maxima* (Takemura and Kafuku, 1957) and *P. margaritifera* (Seurat, 1905; Hervé, 1934; Ranson, 1952; Fougerouse-Tsing and Herbaut, 1994). The structure of the pearl oyster shell has also received considerable research attention, particularly the structure and growth of nacre (e.g., Wada, 1972; Erben and Watabe, 1974) which, more recently, has been investigated for potential biomedical roles (e.g., Lopez *et al.*, 1992; Duplat *et al.*, 2007). This chapter includes a comprehensively illustrated description of the soft anatomy of *P. margaritifera* as a general guide to these structures in other species. The reader is also referred to Section 2.3.1 which provides comparative anatomical details for the Akoya pearl oyster. The chapter goes on to describe the structure of the shells and nacre of pearl oysters and the mechanisms by which nacre is formed.

---

[1]Throughout this text the term 'Akoya' is used for pearl oysters in the currently unresolved complex of *Pinctada fucata-martensii-radiata-imbricata* (see Section 2.3.1.13).

3.2. ANATOMY OF SOFT TISSUES OF *PINCTADA MARGARITIFERA*

3.2.1. General anatomy

In bivalve molluscs, the hinge is the dorsal side of the animal and the area of widest shell valve opening is ventral. The pearl oyster's typical orientation is to sit hinge down, attached to the substrate with the byssal threads, with the ventral side uppermost. The mouth is situated anteriorly. The organs, in a visceral mass, are enclosed within the left and right lobes of the mantle (Fig. 3.1). After removal of one valve, the large adductor muscle is obvious in a posterior-ventral position. It is separate from the visceral mass, except dorsally where a portion of free rectum binds it to the digestive mass (Figs. 3.2 and 3.3).

The foot and the byssus are in the anterior region, ventral to the mouth and surrounded by the labial palps (Figs. 3.2 and 3.3). The whitish visceral mass is located in the dorsal region, enclosing a large part of the digestive system and gonad. The central region, between the adductor muscle and the foot, contains the heart (one light dorsal ventricle and two dark auricles) (Fig. 3.2), the byssal/pedal retractor muscles (Fig. 3.3) and the brown excretory system running along the anterior part of the branchial axis (Fig. 3.3).

The "pearl pocket" (the region where the pearl nucleus is placed during the nucleation process, see Section 8.9.4) is located ventrally. It contains the intestinal loop and connective tissue (Fig. 3.3), and the gonad when the animal is sexually mature. The large pigmented gills (ctenidia) are positioned anteriorly, between the labial palps.

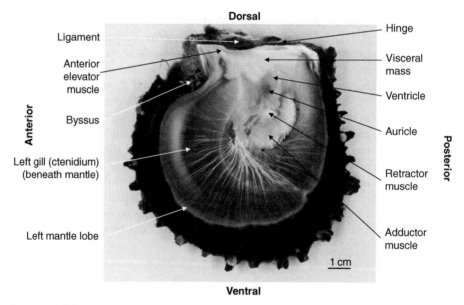

FIGURE 3.1   Lateral view of *Pinctada margaritifera* after removal of the left valve (from Fougerouse-Tsing and Herbaut, 1994). (See Color Plate 6)

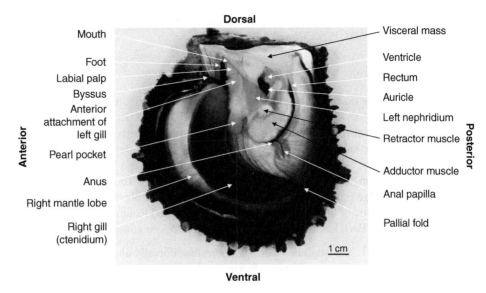

FIGURE 3.2 Lateral view of *Pinctada margaritifera* after removal of the left valve, left mantle lobe, left ctenidium and pericardium (from Fougerouse-Tsing and Herbaut, 1994). (See Color Plate 7)

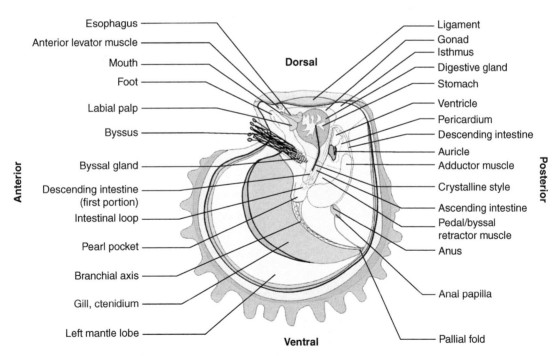

FIGURE 3.3 Diagram of lateral view of *Pinctada margaritifera* (from Fougerouse-Tsing and Herbaut, 1994).

They run ventrally and end posterior-ventrally at the pallial fold, in front of the anal papilla (Figs. 3.2 and 3.3).

3.2.2. The mantle

The mantle is the most external organ, lining the insides of the shell valves and enclosing all other soft tissues and organs (Fig. 3.1). Its fundamental function is to secrete the shell valves and ensure their growth. It consists of two large lobes, with each lining one valve of the shell. These two lobes are separated anteriorly, ventrally and posteriorly. Fused to the visceral mass and the adductor muscle, they join together dorsally along the hinge to form the isthmus (Fig. 3.4). The black pigmentation of the external edge of the mantle and the outer nacre of the shell gives *P. margaritifera* its common name of black-lip pearl oyster (Salomon and Roudnitska, 1986).

Each mantle lobe may be divided into four zones (See also Fig. 3.14):

1. the isthmus, where the lobes fuse dorsally along the hinge;
2. the central area, adhering to the visceral mass and adductor muscle;
3. a distal or pallial area, which is very contractile and showing many fan-shaped radial muscular bundles that are visible to the naked eye;
4. a free marginal area that is thick and pigmented, and fringed with short tentacles.

This division corresponds to that described in *P. maxima* (Dix, 1973). Some authors, however, considered that the isthmus was not distinct from the central zone in the Akoya

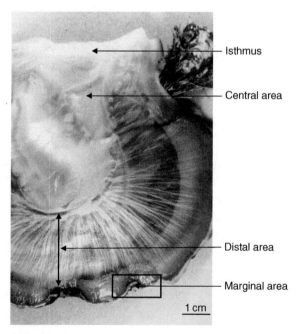

FIGURE 3.4   The four areas of a mantle lobe of *Pinctada margaritifera* (from Fougerouse-Tsing and Herbaut, 1994).

pearl oyster [=*Pinctada martensii* (Ojima, 1952), =*Margaritifera vulgaris* (Herdman, 1904), =*Pinctada fucata* (Chellam *et al.*, 1991)] nor that the central zone was distinctly separated from the pallial zone in *P. margaritifera* (Grizel and Chagot, 1991).

3.2.3. The gills (ctenidia)

There are two symmetrical, flat, crescent-shaped and filamentous gills (Figs. 3.2 and 3.3), characteristic of bivalves with lamellibranch gills. The gills function in filter feeding and respiration (see Chapters 4 and 6). The feeding function is via their connection to the labial palps to form the pallial organ (see Section 4.2.1). Each gill has a dorsal main axis that is vascular and muscular. It has two long branchial sheets (hemictenidia) that arise from the main axis to form a narrow inverted V-shape in transverse section. Each branchial sheet is folded back on itself to create a descending and an ascending lamellae. Thus, each gill is composed of four elongate lamellae and is W-shaped in transverse section (Fig. 3.5). The anterior insertion point of each gill, left and right, is located laterally between the labial palps of the same side (Fig. 3.2). Each gill is fused anteriorly with the visceral mass by its branchial axis and then it is fused with the adductor muscle in the anterior-ventral region of the pearl oyster. The gills then fuse via their axes and pass round the ventral zone of the adductor muscle. The fused gills end at the pallial fold facing the anal papilla, in a posterior-ventral position.

Each gill is attached to the mantle lobe and gill is attached to gill, where this occurs, by long interlocking cilia. The indentation of the epithelium in these joining areas also strengthens the cohesion of the whole.

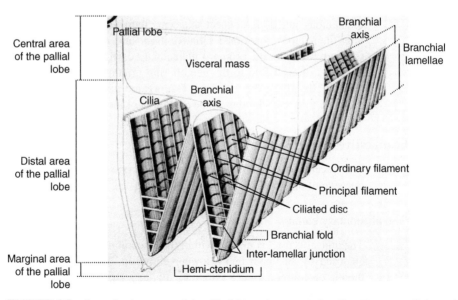

FIGURE 3.5 Generalized structure of the gill of *Pinctada margaritifera* (from Fougerouse-Tsing and Herbaut, 1994).

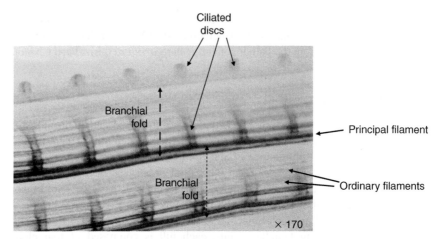

FIGURE 3.6   Branchial filaments of *Pinctada margaritifera* (from Fougerouse-Tsing and Herbaut, 1994).

Each branchial sheet and its two lamellae consist of a series of parallel filaments along its length. These are the basic functional elements of the gills. The branchial sheet is pleated along its axis and made up of two types of filaments: the ordinary filaments and the principal filaments (Figs. 3.5 and 3.6). The ordinary filaments are more numerous and occur over the folds; whereas the principal filaments, which are larger, occur in the hollow of the each fold (Fig. 3.6). The cohesion and support of the gills is accomplished by two types of connections: the inter-lamellar junctions and the inter-filamentous junctions (Herdman, 1904). The inter-lamellar junctions connect the descending and ascending components of each principal filament while the inter-filamentous junctions connect adjacent filaments (Fig. 3.5). There are two kinds of inter-filamentous junctions: ciliated discs and organic junctions (Herdman, 1904). Ciliated discs join the ordinary filaments to each other, as well as the first and the last filaments of each fold to the neighboring principal filaments by long thick and inter-locking cilia (Figs. 3.5 and 3.6). The organic junctions join the ordinary filaments of the same fold to each other in the axial area of the gills.

3.2.4. The digestive system

The visceral mass is obvious after removing one shell valve, mantle lobe and gill from one side of the pearl oyster (Fig. 3.2).

3.2.4.1. The alimentary canal

The digestive tract is mainly within the visceral mass (Fig. 3.2). It begins at the mouth, enclosed by an upper and a lower yellow lip. Thick and smooth at the level of the mouth, each lip continues symmetrically as a strip of thinner tissue up to the insertion point of the gills, these strips spread out in triangular flat laminae to form the labial palps, characterized by an internal surface covered with regular grooves, parallel to

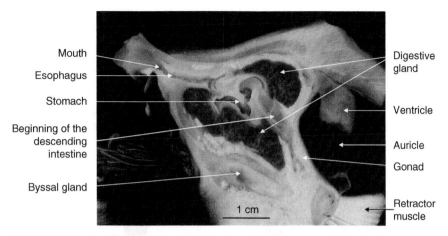

Mouth

Esophagus

Stomach

Beginning of the
descending
intestine

Byssal gland

Digestive
gland

Ventricle

Auricle

Gonad

Retractor
muscle

1 cm

FIGURE 3.7   Sagittal section of the visceral mass of *Pinctada margaritifera*, showing the first parts of the digestive tract included in the visceral mass (from Fougerouse-Tsing and Herbaut, 1994).

each other and presenting a curve. These yellowish labial palps receive the fine organic and inorganic particles from the filter-feeding processes of the gills and may discharge some of the filtered material as pseudofeces (see Section 4.2).

Dissection of the visceral mass is necessary to observe the digestive tract (Fig. 3.7). The mouth leads into the esophagus which opens into the stomach. The stomach, almost totally covered by the digestive gland, has many chambers (Figs. 3.3 and 3.7) (see Section 4.2.6). However, when the pearl oyster is not sexually mature, the stomach can be seen through transparent tissues. Of significant size, it is mainly located in the left side of the visceral mass. The intestine follows on from the stomach and may be divided into three distinct zones:

1. the anterior descending intestine;
2. the ascending intestine;
3. the rectum or posterior descending intestine.

The initially descending intestine goes down in the ventral area between the two retractor muscles and describes a loop with posterior convexity (Fig. 3.8). It contains a long translucent crystalline rod, the crystalline style, which opens into the gastric cavity at its anterior end (Fig. 3.3). The junction between the initially descending intestine and the ascending intestine is characterized by an intestinal loop in the posterior region. This can be observed in the upper area of the "pearl pocket" (Fig. 3.8). The intestinal loop is oriented to the left side of the pearl oyster. In this loop, the ascending intestine passes around the initially descending intestine to the left, then abruptly passes up to the dorsal region, near the posterior wall of the visceral mass, parallel to the heart axis. The ascending intestine is characterized by the presence of a typhlosole: an infolding of the internal wall of the intestine. The orange-colored typhlosole occurs all along this section of the intestine. It decreases in size toward the dorsal region of the ascending intestine.

The rectum is located outside the visceral mass (Figs. 3.2. and 3.9). The ascending intestine emerges from the visceral mass above the heart and covers the ventricle before

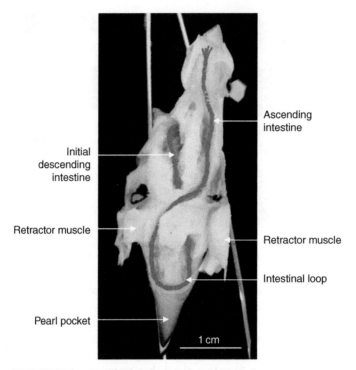

FIGURE 3.8   The descending intestine and the ascending intestine of *Pinctada margaritifera* are marked in gray (from Fougerouse-Tsing and Herbaut, 1994).

descending as the rectum along the convex side of the adductor muscle. The pigmented rectum ends at the anus with a leaf-shaped anal papilla, located in the posterior-ventral region against the adductor muscle (Figs. 3.2 and 3.9). The rectum also has a typhlosole along its length, although smaller than in the ascending intestine, and it decreases at the anal end.

3.2.4.2. The digestive gland

The dark brown digestive gland is essentially located in the dorsal region of the visceral mass, above the retractor muscles (Fig. 3.7). It is near the esophagus and forms a bulky mass around the stomach. In the posterior region of the visceral mass, the digestive gland is contiguous with parts of descending and ascending intestines, except when the gonad occupies most of the visceral mass during sexual maturity.

3.2.5. The reproductive system

As in the Akoya pearl oyster [*Pinctada albina* (Tranter, 1958b) and *P. fucata* (Tranter, 1958d)], *P. margaritifera* is a protandrous hermaphrodite (Tranter, 1958c; Thielley, 1993) (see Section 5.3.1). Young sexually mature pearl oysters are male, whereas large older animals are usually female. Some females may revert to being male under some

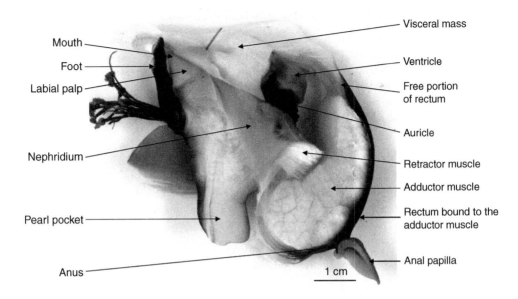

FIGURE 3.9 The visceral mass, adductor muscle and rectum of *Pinctada margaritifera* (from Fougerouse-Tsing and Herbaut, 1994). (See Color Plate 8)

environmental conditions. However, bisexual animals are uncommon, suggesting that sexual inversion may be a rapid process (Tranter, 1958b; Thielley, 1993). In *P. margaritifera*, as observed by Ranson (1952) and Thielley (1989), it is impossible to determine the sex of the animal from the external appearance of the gonad. The gonads in both sexes are whitish. They form a diffuse tissue at the periphery of the visceral mass that is indistinguishable as paired organs (Figs. 3.3 and 3.7). The gonad develops in the dorsal regions of the visceral mass, surrounding the esophagus and stomach, and totally enclosing the digestive gland, and the descending and ascending intestines. Gonad penetrates the tissue between the two retractor muscles, spreads ventrally and invades the "pearl pocket" at sexual maturity.

A similar lack of sexual dimorphism in gonad color has been found in the Akoya pearl oyster (Herdman, 1904; Tranter, 1958d). However, sexual dimorphism in terms of gonad appearance was described in *P. albina* by Tranter (1958a). Whether the gonads are testes or ovaries can only be determined in *P. margaritifera* by making an incision in the gonad wall and observing the gametes that flow out; "milt" is a dense white suspension of spermatozoa, while ova have a granular appearance. The nature of these gametes should be confirmed by microscopic examination.

Tranter (1958a) described a urogenital papilla in *P. albina*. Herdman (1904) observed a urogenital orifice in the Akoya pearl oyster where two distinctive ducts, a urinary and a more dorsal genital duct, joined. The gametes in *P. margaritifera* are released into fine gonoducts within the gonadal tissue, near the descending intestine. They pass into larger and symmetrical gonoducts between the digestive gland and gonad and, when they reach the renopericardial cavity, are ejected through the buttonhole-like urogenital orifices, located under the anterior area of the gills.

3.2.6. The muscular, locomotor and attachment systems

The principal muscular elements are the adductor muscle, the byssal/pedal retractor muscles and the elevator muscles. Except in the very early stages after settlement, pearl oysters remain fixed to the substrate by the byssus and movement is essentially limited to activity of the foot. In the early stages after settlement, however, pearl oyster spat may actively detach their byssal threads from the substrate, move with their foot and reattach at another position (Gervis and Sims, 1992; Taylor *et al.*, 1997).

3.2.6.1. The adductor muscle

In the Pteriidae, the opening and closing of the shell valves are controlled by the largest and most important muscle in their body, the adductor muscle. It is crescent-shaped, with a constant thickness and an enlarged ventral area. The adductor muscle has two regions (Fig. 3.9):

1. a narrow whitish posterior strip, of constant thickness;
2. a larger anterior muscular mass, semi-translucid and yellowish.

3.2.6.2. The byssal/pedal retractor muscles

The pearl oyster has two symmetrical retractor muscles projecting laterally from the visceral mass. These muscles form two large white oval masses adhering to the adductor muscle and located in its ventral concavity. They are V-shaped and surround the byssal organ (Fig. 3.10).

3.2.6.3. The levator muscles

There are two pairs of levators located dorsally behind the mouth. The posterior pair are smaller and more difficult to observe than the anterior pair. The insertion zone of

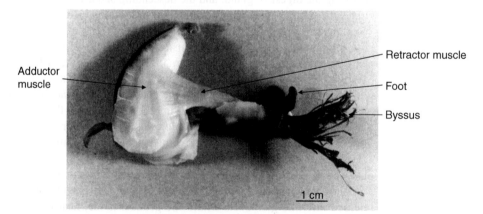

FIGURE 3.10   The adductor and the retractor muscles of *Pinctada margaritifera* (from Fougerouse-Tsing and Herbaut, 1994).

the anterior pair is conspicuous in the inner face of the valves and forms a hollow in the nacreous layer in the anterior-dorsal region (Fig. 3.3).

3.2.6.4. The foot

The foot is an oblong tongue-shaped organ that is oval in transverse section. It is located in the anterior part of the animal, between the mouth and the byssus. It is pigmented and has a ventral longitudinal groove, the pedal groove (Figs. 3.3, 3.10 and 3.11). This very turgescent organ is not used in burrowing, as in some bivalve molluscs, but it has tactile and locomotor roles, at least in the early stages (Ranson, 1952).

3.2.6.5. The byssal gland

This gland is a double lobed compact mass, located in the central part of the retractor muscles (Fig. 3.11). It consists of two bundles of grooves, more or less radiating, which converge and end externally to form the byssal threads. These greenish grooves are initially clear. They darken gradually until they are deep bronze-green where the byssus is formed. The distal end of these byssal threads flatten as a disc, which anchors the pearl oyster to the substrate.

3.2.7. The "pearl pocket"

Located under the retractors, between the concave edge of the gills and the ventral part of the adductor muscle, is a transparent turgescent organ known as the "pearl pocket"

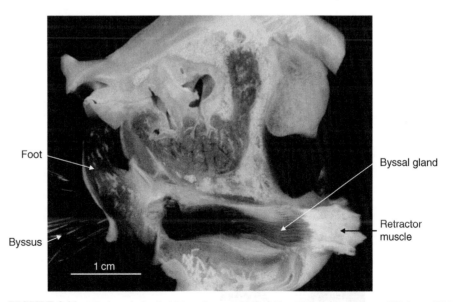

FIGURE 3.11   The byssal gland of *Pinctada margaritifera* (from Fougerouse-Tsing and Herbaut, 1994).

(Figs. 3.2, 3.3 and 3.9). This receives the nucleus and piece of mantle tissue that are necessary in the formation of the pearl culture (see Section 8.9.4). It is a region of tissue of variable size that is flattened laterally, with a slender anterior region. The "pearl pocket" contains the intestinal loop in its posterior region at the level of its attachment to the visceral mass (Fig. 3.3). It may be pigmented. It is invaded by the gonad when the animal is sexually mature and is usually identified as gonad.

3.2.8. The circulatory system

The circulatory system consists of a heart, and system of lacunae (cavities) and blood vessels, which assure the circulation of the hemolymph throughout the pearl oyster. After one of the shells is removed, the heart can be seen through the transparency of the mantle lobe (Fig. 3.1). It is located within a pericardial cavity in the posterior region of the visceral mass, and covered by a thin pericardial membrane. It is limited dorsally by the small portion of free rectum, located between the visceral mass and the adductor muscle, and limited ventrally by the retractor muscles (Fig. 3.9). It consists of a ventricle and two auricles, which are all triangular (Fig. 3.12). The yellowish ventricle is median and dorsally located in comparison with the dark lateral symmetrical auricles. Each of the auricles communicates with the ventricle at its dorsal end. They join ventrally at their base and are in close connection with the excretory system.

Two large blood vessels, the anterior aorta and the posterior aorta, are visible to the naked eye. They emerge from the dorsal end of the ventricle and the former plunges into the anterior left side of the visceral mass, while the latter goes posteriorly and connects with the rectum. Each aorta subdivides into smaller vessels which irrigate the various organs of the animal.

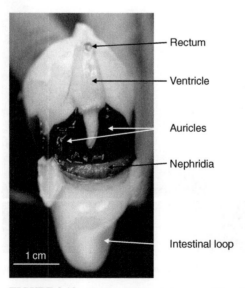

FIGURE 3.12   The heart of *Pinctada margaritifera* (from Fougerouse-Tsing and Herbaut, 1994).

3.2.9. The excretory system

The excretory system consists of a pair of light brown nephridia, on each side of the visceral mass (Figs. 3.2 and 3.9), and numerous small pericardial glands. They join each other ventrally, beneath the auricles (Fig. 3.12). The excretory system is covered by a fine epidermis of mesodermal origin, forming the renopericardial cavity. A distal branch of the nephridium follows the branchial axis, while a proximal branch goes to the heart. The excretory products are released through the buttonhole-like urogenital orifices that are located under the anterior area of the gills.

3.2.10. The nervous system

Like all bivalve molluscs, the nervous system of pearl oysters is not very apparent to macroscopic observation.

It consists of three pairs of ganglia:

1. the cerebral ganglia, located at the sides of the esophagus (Fig. 3.13);
2. the pedal ganglia joined together as a single mass at the base of the foot;
3. the visceral ganglia or parieto-splanchnic ganglia, lying upon the anteroventral surface of the adductor muscle.

Commissures connect homologous ganglia. Thus, the supra-esophageal commissure connects the cerebral ganglia (Fig. 3.13) and there is a large commissure between the visceral ganglia. There are further connectives to non-homologous ganglia and other nervous tissues connect ganglia to organs.

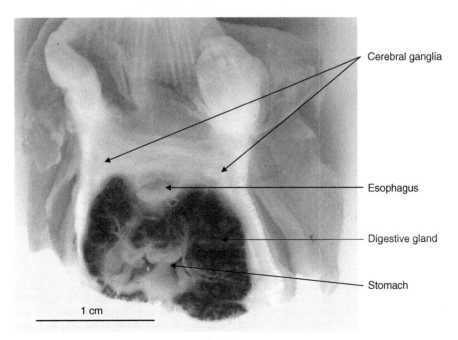

FIGURE 3.13 The cerebral ganglia of *Pinctada margaritifera* (from Fougerouse-Tsing and Herbaut, 1994).

3.3. SHELL STRUCTURE AND MINERALIZATION

The morphological characteristics of shells of the pearl oyster species are comprehensively described at generic level in Section 2.2.2 and for each of the major commercial species in Section 2.3. There is considerable variation in the shell morphology of each species and this has led to considerable confusion in the taxonomy of the Family, based on shell morphology (see Section 2.3). There is, however, a fundamental anatomical structure of pearl oyster shells that is common to all species and which is characteristic of bivalves and other molluscs. The shell consists of about 95% calcium carbonate as calcite and aragonite, and there are three layers (Fig. 3.14):

1. an outer periostracum, largely composed of proteins;
2. a medial ostracum or prismatic layer, composed of calcite crystals in an organic matrix;
3. an inner hypostracum or nacreous layer (mother-of-pearl, MOP), composed of aragonite crystals in an organic matrix.

The shell grows concentrically from the umbo. Growth along the shell edge is, however, not necessarily uniform, it may occur as projecting growth processes or "fingers" (Fig. 3.1), which are subsequently filled in between. Growth rates of the shell may change seasonally or in response to other environmental variation, leaving concentric lines in the periostracum.

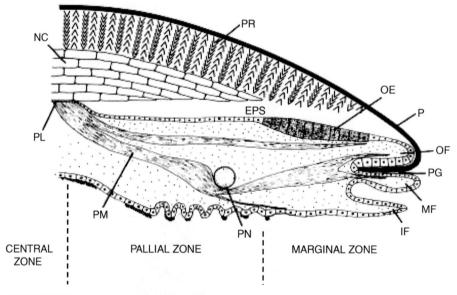

FIGURE 3.14 Diagrammatic cross-section through the mantle and growing edge of a bivalve shell. The layers of periostracum, prismatic calcite (ostracum), and aragonitic nacre (hypostracum) are underlain by the extrapallial space and mantle, EPS, extrapallial space; IF, inner fold of mantle; MF, middle fold of mantle; NC, nacreous shell layer; OE, outer epithelium of the mantle; OF, outer fold of the mantle; P, periostracum; PG, periostracal groove; PL, pallial line; PM, pallial muscle; PN, pallial nerve; PR, prismatic shell layer (from Wilt *et al.*, 2003). (used with permission)

The mantle, lining the shell to the extent of its outer shell edge, secretes the growing edges of the shell. The edge of the mantle can be divided into three zones: the outermost marginal zone, a medial pallial zone and a central zone, which extends inward across the surface of the mantle (Chellam *et al.*, 1991; Garcia-Gasca *et al.*, 1994) (Fig. 3.14). The marginal zone has three folds: the inner, the middle, and the outer fold which lays against the inner surface of the shell valve (Fig. 3.14). The middle and the outer folds are separated by the periostracal groove. The surfaces of the mantle are complex and regions of specialized epithelial cells secrete mucus and the three layers of the shell (Dix, 1973: Jabbour-Zahab *et al.*, 1992; Garcia-Gasca *et al.*, 1994; Du *et al.*, 2002). The epithelial secretions of the various zones and folds in the mantle of *Pinctada mazatlanica* are shown in Table 3.1. There is particular interest in the nacreous layer of pearl oysters and most of the following section will be devoted to this layer.

3.3.1. Periostracum

The periostracal layer is secreted by the epithelial cells of the periostracal gland in the periostracal groove, which is situated between the middle and outer folds of the marginal zone of the mantle (Fig. 3.14, Table 3.1). It does not usually increase in thickness after secretion (Gervis and Sims, 1992) and remains as a relatively thin layer of coarse horny material that coats the outer surfaces of the shells.

The periostracum is secreted as a soluble precursor, the periostracin (Waite *et al.*, 1979), which is then cross-linked by o-diphenols and tyrosinase (for phenoloxidases) to form the insoluble periostracum, consisting of quinone-tanned proteins (known as

TABLE 3.1

Epithelial secretions of zones and folds in the mantle of *Pinctada mazatlanica*. (Modified from Garcia-Gasca *et al.*, 1994.)

Mantle structure	Secretion
Marginal zone	
Inner fold	
Inner side	Mucus
Outer side	Mucus
Middle fold	
Inner side	Mucus
Outer side	Periostracum
Outer fold	
Inner side	Periostracum
Outer side	Prismatic layer
Pallial zone	
Inner side	Mucus
Outer side	Nacre
Central zone	
Inner side	–
Outer side	Nacre

conchiolin) (Waite *et al.*, 1983; Saleudin *et al.*, 1983). It is among the strongest and most chemically inert structures in the animal kingdom (Zhang *et al.*, 2006). It provides mechanical protection for the shell and protection from fouling (see Section 15.3.1). However, over time, it decreases in thickness and shell coverage through erosion, affecting its anti-fouling properties (Guenther *et al.*, 2006). Another essential function of the periostracal layer is to facilitate shell mineralization. The mantle epithelium is only in contact with the inner shell surface at the periostracal groove and the periostracum isolates the extrapallial space from the external environment, seawater (Fig. 3.14). This permits the formation of super-saturation conditions in the extrapallial fluid: the conditions required for calcium deposition.

Little research has been undertaken specifically on the periostracum of pearl oysters, apart from its role in fouling reduction. A recent study has shown that a novel tyrosinase, a copper-containing phenoloxidase that share similarities with cephalopod tyronases, plays an important role in periostracum formation (Zhang *et al.*, 2006).

3.3.2. Prismatic layer

As the periostracum is secreted from the inner side of the outer mantle fold, the prismatic layer, or ostracum, is secreted beneath it from the outer side of the outer mantle fold (Fig. 3.14, Table 3.1). Like the periostracal layer, it is secreted by one epithelial region of the outer mantle and not subsequently enhanced, unless there is damage to the shell, e.g., by shell borers. The prismatic layer consists of relatively large polygonal prisms of calcite, which are arranged in a lattice-like format with long axes perpendicular to the shell surface (Gervis and Sims, 1992).

As with the periostracal layer, there are few studies of the prismatic layer specifically in pearl oysters, but the processes are similar to those that have been observed in other bivalves. The prisms form in the extrapallial fluid, between the periostracum and the mantle (see above). Electron-dense lamellae serve as the internal boundary of the future prisms in *Pinctada radiata* (Nakahara and Bevelander, 1971). Fragments of the lamellae become detached and migrate in the extrapallial spaces until bounded externally by the periostracum. The lamellar fragments form organic frameworks in which crystal initiation and growth occur. Stout interprismatic walls develop and mature elongated prisms form by the addition of crystallites. Each crystallite is enclosed by a thin organic envelope. Thus, there are two main organic components of the prismatic layer: the interprismatic walls separating the prisms, and the thin intraprismatic envelops enclosing each component crystallite of a prism (Nakahara *et al.*, 1980). These authors investigated the fine structure and amino acid composition of five bivalves, including Akoya pearl oysters and *P. margaritifera*. They found that the intraprismatic envelopes of these species contained unusually high levels (23–80%) of the amino acid aspartic acid and suggested that the aspartic acid-rich envelope may be essential for the initiation and growth of calcite crystals.

3.3.3. Nacre

The unique properties of nacre (MOP) are due to a small proportion (1–5% by weight) of an organic matrix which includes proteins, glycoproteins, chitin and lipids. This results in a combination of high mechanical strength, similar to ceramics, with elasticity

Currey (2005). Nacre consists of flat polygonal tablets of crystalline calcium carbonate with extracrystalline and intracrystalline organic networks. In this respect, the basic structure of nacre in units is like that of the prismatic shell layer. The nacre layer, or hypostracum, however, differs fundamentally from the outer two shell layers in that, although like them, its initial point of secretion is at the distal edge of the mantle (Fig. 3.14, Table 3.1), it continues to be secreted and thickened by the entire outer surface of the mantle (Fig. 3.4) throughout the pearl oyster's life (Gervis and Sims, 1992).

The structure of mollusc nacre has been well studied over a considerable period (e.g., Schmidt, 1924; Wada, 1961; Bevelander and Nakahara, 1969; Mutvei, 1969, 1970, 1979; Wise, 1970; Grégoire, 1972; Erben and Watabe, 1974). Studies have often used electron microscopy; for example Grégoire et al. (1955) used this technique to show the inter-laminar sheet of organic matter. Very recently, the sophisticated method of cryo-transmission electron microscopy (TEM) was employed to cross-check the fine structure usually observed by dehydration and staining techniques with the same matrix but hydrated and vitrified at low temperature (Levi-Kalisman et al., 2001). The composite mineral/organic structure was also studied, particularly the role of the organic matrix using a biochemical approach (Weiner and Traub, 1980, 1984; Mann, 1989; Weiner and Addadi, 1991; Mann et al., 1993; Belcher et al., 1996; Falini et al., 1996; Feng et al., 1999; Thompson et al., 2000). It has resulted in different models proposed by Schäffer et al. (1997) and Levi-Kalisman et al. (2001).

One of the recent aspects of the biochemical study of pearl oyster nacre is its potential use as a natural bioceramic for bone regeneration (Lopez et al., 1992; Rousseau et al., 2003, 2004, 2005a; Duplat et al., 2007).

3.3.3.1. Nanostructure of the aragonite tablets

Very early, Schmidt (1924) demonstrated that there was an intracrystalline organic matrix present within the aragonite tablets of mollusc nacre. This was confirmed by Watabe (1965), Mutvei (1970, 1979), and Weiner and Traub (1984). Mutvei showed that the matrix was on a micrometer scale.

The image in Figure 3.15 was obtained by atomic force microscopy (AFM) measuring the phase difference between the excitation signal and the cantilever response on the polished surface of a nacre tablet from P. maxima (Rousseau et al., 2005a, c). It shows mineral nanograins surrounded by an organic phase. The organic matrix is organized in the form of "foam" with very thin walls and closed cells. The mineral nanograins are thus encapsulated inside the organic framework (vesicle). The mean nanograin size was 44 ± 23 nm (Rousseau et al., 2005a). The aragonite tablet appears as a single crystal in electron diffraction (not shown here; see Liu et al., 1992, for example). The tablet is thus made up of a coherent aggregation of nanograins and Figure 3.15 shows that the organic framework is interconnected within the tablet. There is no evidence that the mineral nanograins are interconnected.

3.3.3.2. Continuity of the organic framework in the tablet

Figure 3.16 is a darkfield TEM image of tablet layers of P. maxima (see Rousseau et al., 2005a, c). Only the organic matrix is in contrast (part of it only). Firstly, the

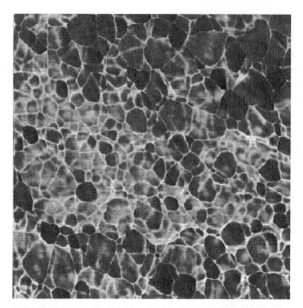

FIGURE 3.15   Atomic force microscopy image in phase contrast of the polished surface of nacre tablets of *Pinctada maimaa* showing nanograin crystallites (dark) within a foam-like organic matrix (1×1 μm^2) (from Rousseau *et al.*, 2005a).

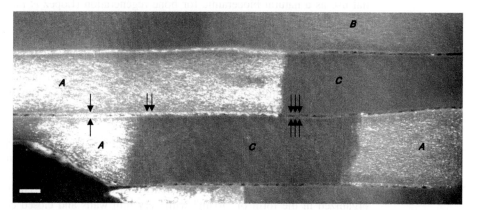

FIGURE 3.16   Darkfield TEM image of tablets in the nacre of *Pinctada maxima* showing the crystalline structure of the organic matrix (bar is 100 nm). See text for explanation of lettering and arrows (from Rousseau *et al.*, 2005a).

highest contrast comes from the organic matrix of the inter-laminar region between tablets. Secondly, and most importantly, this technique shows the intracrystalline organic matrix in a different way to AFM. The intracrystalline matrix of each aragonite tablet diffracts in the form of a fine speckled texture, with a similar texture for each tablet. The variation of the contrast from one tablet to the other in Figure 3.16 suggests that the intracrystalline organic matrix is a "single crystal" network.

- If the organic matrix is oriented under the Bragg angle[2] it brightens all over the tablet (e.g., tablets A in Fig. 3.16).
- If it is just slightly misoriented, only a faint contrast appears all over the tablet B. This is related to a Bragg tolerance phenomenon classically observed in the case of small coherent domains such as these.
- In other tablets C the intracrystalline organic network is extinguished because it is not selected in this direction or not at the Bragg angle.

The distortions appearing in one tablet A, may be real or related to sampling artifacts. The relationships between the intracrystalline and inter-laminar matrix are notable in this darkfield TEM (Fig. 3.16). The inter-laminar matrix is bright, in high contrast, where the intracrystalline matrix is bright. If the neighboring tablets are bright, both sides of the inter-laminar sandwich are bright (single arrows), but only one side is bright if the neighbor is extinguished (double arrow). When the two neighbors are extinguished, the inter-laminar matrix is also extinguished or faintly in contrast (triple arrow). The median layer of the inter-laminar sandwich is formed by a porous layer appearing as a succession of bridges in cross-section (Rousseau *et al.*, 2005b).

3.3.3.3. Biomineralization and nacre secretion

The active role of proteins in biomineralization is fundamental (Belcher *et al.*, 1996; Falini *et al.*, 1996; Feng *et al.*, 1999); it represents a source of inspiration for future nanotechnology with a bottom-up approach. The "bricks and mortar" ordering of nacre has already inspired the toughening of ceramic materials by co-processing rigid ceramic, such as silicon carbide, and compliant interlayers such as boron nitride (Naslain *et al.*, 1998; Smith, 1998), and there are many other examples (Simkiss and Wilbur, 1989).

Rousseau *et al.* (2005b) demonstrated that the front of nacre biomineralization in *Pinctada margaritifera* has a unique feature: a "garden terrace" or "stairs-like" front. Figure 3.17b shows the developing nacre layers on the epithelial tissue of the mantle when examined by scanning electron microscopy (SEM) in the direction of the arrows indicated in Figure 3.17a. This image shows that nacre growth does not occur layer by layer; all the layers grow at the same time at their stair-like front. Each step is the front of a growing layer which is spatially offset (above and below) from its neighbors, with a significant distance from one step to another (Fig. 3.17b). This offset is responsible for the dynamics and the hierarchy of the nacre biomineralization processes. Each layer is secreted in the form of extrapallial fluid: a liquid "precursor" which forms a film and then undergoes progressive crystallization to become mature nacre.

Thus, sheet-nacre growth is not a cycled mechanism row-by-row as in columnar nacre (Wise, 1970; Nakahara *et al.*, 1982). This evidence of simultaneous fronts of growing layers agrees with cross-sections obtained by equivalent techniques of other

---

[2]The angle between the crystal plane and the diffracted beam for constructive interference.

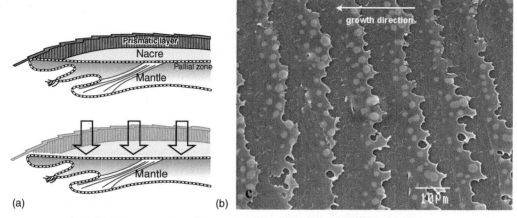

FIGURE 3.17 Zone of nacre formation in *Pinctada margaritifera*. (a) Diagram of the direction and position of the SEM scan of the mantle side of nacre formation shown in (b). (b) Incipient nacre layers with round tablet nuclei (light gray) in film from extrapallial fluid. There are faint imprints in a layer below of more mature tablets that were removed with the shell in the preparation process (from Rousseau *et al.*, 2005b).

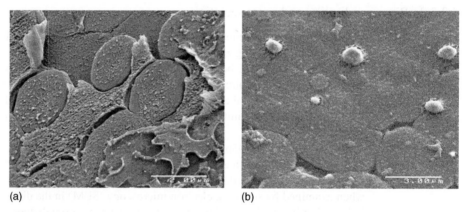

FIGURE 3.18 SEM views of the nacre of *Pinctada margaritifera* on the shell side of nacre formation (cf. Fig. 3.17a). (a) Incipient nacre showing the circular tablet nuclei in a film from the extrapallial fluid. (b) Mature aragonite tablets remaining after the surface was treated with alcohol, removing most of the film and developing nuclei (from Rousseau *et al.*, 2005b).

researchers (Wada, 1961; Towe and Hamilton, 1968; Wada, 1972; Nakahara, 1991). It also agrees with the steps of development observed on the shell side of the developing nacre (Fig. 3.18). The nacre tablets shown in Figure 3.18 are at about stage 3 of the five stages of mineralization proposed in the model shown in Figure 3.19. The study of Rousseau *et al.* (2005b) has furnished support for the "compartment theory" described by Bevelander and Nakahara (1969). The compartments are "open" (Fig. 3.18), i.e. the systematic offset observed from one row to the other is an efficient way to organize interactions such as gradients of matter, such as water, pH variation, nucleation launching, and transmission of the crystallographic information by heteroepitaxy, etc.

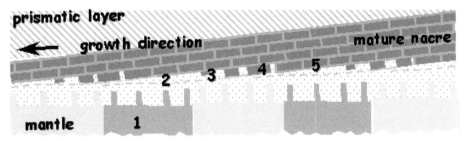

FIGURE 3.19 Model for the "stairs-like" mineralization front in the sheet nacre. It shows the simultaneous multilayered growth of shell: (1) discharging epithelial cells of mantle, (2) extrapallial fluid organized as a film, (3) nucleation of tablet, (4) incipient nacre developing by self-ordering of the film and (5) mature nacre.

An important point is the progressive mineralization in a continuous process from fluid to biomineral.

The nacre layered structure is developed by a sequence of events (Fig. 3.19):

1. The discharge of a mixture of organic and ionic species by some specialized cells of the mantle.
2. Each layer of this fluid forms a film at the interface between the mantle and the previous layer already formed (shaping of the film as a compartment).
3. This film further undergoes a self-ordering process until its complete mineralization. It starts with nucleation. The location and signal for nucleation comes from the underlying row of nacre: nucleation starts at the moment when the intracrystalline matrix forms in the preceding layer.
4. Next is the growth of nuclei in all directions until the total thickness of the film is reached. Then the tablets grow in the form of cylinders.
5. Finally the tablets come into contact which each other. The layer is completed when the triple points are mineralized. This results in a Voronoi-type 2D arrangement.

Some dynamic aspects of nacre growth are that the offset from one step to another is under control. There is a distance of 10–15 μm from one front to the next at the adult stage near the edge. A distance of 20–30 μm is thus necessary for the complete mineralization involving the following steps: (i) secretion, (ii) shaping of the film, (iii) nucleation, (iv) growth and (v) polygonization as a completely mature nacre layer, on the same step level. A growth rate of 3.4 ± 1.4 layers per day in the case of the pearl of *P. margaritifera* was measured by Caseiro (1995).

Nucleation of new tablets has to stay well controlled in order to be relatively constant in density, ca. 11 nuclei per 100 μm^2. This makes one nucleus ca. 10 μm^2, which is consistent with the mean size of the tablets in *P. margaritifera* and Akoya pearl oysters (Wada, 1972). Each aragonite tablet starts from a single center (Wada, 1972; Rousseau *et al.*, 2005b). The growth first evolves freely at the same rate in all directions. As a result the nuclei rapidly acquire a cylindrical shape. Then, when making contact with neighboring tablets they become polygonal (Fig. 3.20). In Figure 3.20a, nucleation follows a hexagonal array, thus Voronoi cells turn into perfect hexagons. In Figure 3.20b

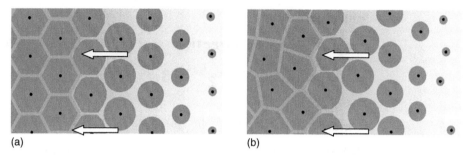

FIGURE 3.20   Model of the front of mineralization within the film of the extrapallial fluid. (a) A Voronoi diagram for a two-dimensional lattice of developing tablets. (b) A more realistic case with the same amount of tablets but less well ordered (From Rousseau *et al.*, 2005b).

the same number of nuclei was used but with a small deviation from the regular lattice. As a result, the standard deviation of the tablets size increases and the shape turns into more realistic polygons sometimes with five or seven (or more sides). Rousseau *et al.* (2005b) hypothesized about why the tablets first grow with a spherical shape and then turn polygonal when contacting each other. The presence of nanograins encapsulated in an organic vesicle suggests an aggregation-like control of the extension of the tablet by the organic template and not directly by the atomic forces.

3.4. SUMMARY

This chapter provides a highly illustrated description of the various components of the soft anatomy of the Blacklip pearl oyster, *Pinctada margaritifera*. The shell structure of pearl oysters, which is fundamentally similar to other bivalves, is also described. It consists of three layers. The outermost layer, the periostracum, is largely composed of quinone-tanned proteins (known as conchiolin). It provides mechanical and fouling protection for the shell. It tends, however, to decrease in thickness and shell coverage through erosion. The middle layer, the ostracum or prismatic layer, consists of relatively large polygonal prisms of calcite, which are arranged in a lattice-like format with long axes perpendicular to the shell surface. There are organic interprismatic walls separating the prisms and thin intraprismatic envelops enclose each component crystallite of the prism. The inner layer, hypostracum, is composed of nacre (MOP). Nacre consists of flat polygonal tablets of aragonite with complex extracrystalline and intracrystalline organic matrices in a "bricks and mortar" format. Nacre is laid down in layers in a complex sequence of events involving progressive mineralization within the extrapallial fluid. A Voronoi model is proposed to describe the two-dimensional lattice of developing tablets.

The three shell layers are secreted by different sections of the distal mantle edge, but the nacreous layer differs fundamentally from the other layers. While the nacreous layer is initially secreted at the distal edge of the mantle, like the other layers, it continues to be secreted and thickened over the entire outer surface of the mantle lobes. This creates the thick MOP lining inside of the pearl oyster shell.

Understanding the anatomy of pearl oysters provides a basis for assessing their physiological status and well being. This relates to important aspects of the culture and husbandry of pearl oysters such as reproduction and disease. However, arguably the most interesting part of their anatomy is the mantle, which is responsible for secreting the MOP that lines pearl oyster shells. While the process of nacre secretion has been tapped for cultured pearl production (see Chapter 8), our understanding of the physiological, biochemical and molecular mechanisms of nacre secretion is incomplete. The particular and unique characteristics of nacre have inspired research with applications in fields such as ceramics and biomedical science; the outcomes of such research may provide significant benefits to mankind.

References

Belcher, A.M., Wu, X.H., Christensen, R.J., Hansma, P.K., Stucky, G.D., Morse, D.E., 1996. Control of crystal phase switching and orientation by soluble mollusc-shell proteins. *Nature* 381, 56–58.

Bevelander, G., Nakahara, H., 1969. An electron microscope study of the formation of nacreous layer in the shell of certain bivalve molluscs. *Calc. Tiss. Res.* 3, 84–92.

Caseiro, J., 1995. Evolution de l'épaisseur des dépôts de matériaux organiques et aragonitiques durant la croissance des perles de *Pinctada margaritifera*. *C.R. Acad. Sci. Paris* 321, 9–16.

Chellam, A., Victor, A.C.C., Dharmaraj, S., Velayudhan, T.S., Satyanaryana Rao, K., 1991. *Pearl Oyster Farming and Pearl Culture. Training Manual 8, Regional Seafarming Development and Demonstration Project (RAS/90/002)* Central Marine Fisheries Research Institute (CMFRI), Tuticorin, India. 87 pp.

Currey, J.D. 2005. Hierarchies in biomineral structures. *Science* 309, 253–254.

Dix, T.G., 1973. Histology of the mantle and pearl sac of the pearl oyster *Pinctada maxima* (Lamellibranchia). *J. Malacol. Soc. Aust.* 2, 365–375.

Du, X., Li, G., Liu, Z., Ye, F., Wang, R., 2002. Genetic diversity of two wild populations in *Pinctada martensii. Zhongguo Shuichan Kexue (J. Fish. Sci. China)* 9, 100–105. (in Chinese with English Abstract).

Duplat, D., Chabadel, A., Gallet, M., Berland, S., Bedouet, L., Rousseau, M., Kamel, S., Milet, C., Jurdic, P., Brazier, M., Lopez, E., 2007. The *in vitro* osteoclastic degradation of nacre. *Biomaterials* 28, 2155–2162.

Erben, H.K., Watabe, N., 1974. Crystal formation and growth in bivalve nacre. *Nature* 248, 128–130.

Falini, G., Albeck, S., Weiner, S., Addadi, L., 1996. Control of aragonite or calcite polymorphism by mollusk shell macromolecules. *Science* 271, 67–69.

Feng, Q.L., Li, H.B., Cui, F.Z., Li, H.D., 1999. Crystal orientation domains found in the single lamina in nacre of the *Mytilus edulis* shell. *J. Mater. Sci. Lett.* 18, 1547–1549.

Fougerouse-Tsing, A., Herbaut, C., 1994. *Atlas anatomique de l'huître perlière Pinctada margaritifera.* Etablissement pour la Valorisation des Activités Aquacoles et Maritimes (EVAAM) Centre Universitaire de Polynésie Française, Tahiti. 68 pp.

Garcia-Gasca, A., Ochoa-Baez, R.I., Betancourt, M., 1994. Microscopic anatomy of the mantle of the pearl oyster *Pinctada mazatlanica* (Hanley, 1856). *J. Shellfish Res.* 13, 85–91.

Gervis, M.H., Sims, N.A., 1992. The Biology and Culture of Pearl Oysters (Bivalvia: Pteriidae). *ICLARM Stud. Rev.* 21. 49 pp.

Grégoire, C., 1972. Structure of the molluscan shell. *Chemical Zoology* Vol. 7. Academic Press, New York. pp. 45–102.

Grégoire, C., Duchateau, G., Florkin, M., 1955. La trame protidique des nacres et perles. *Ann. Océano* 31, 1–36.

Grizel, H. and Chagot, D., 1991. Etude des lésions du manteau. *Rapport Final Action Cordet 88/210. Pinctada margaritifera, le Manteau, ses Altérations en Polynésie*, pp. 123–140.

Guenther, J., Southgate, P.C., de Nys, R., 2006. The effect of age and shell size on accumulation of fouling organisms on the Akoya pearl oyster *Pinctada fucata* (Gould). *Aquaculture* 253, 366–373.

Herdman, W.A., 1904. Anatomy of the pearl oyster (*Margaritifera vulgaris*, Schum). In: Herman, W.A. (Ed.), *Report to the Government of Ceylon on the Pearl Oyster Fisheries of the Gulf of Manaar*. The Royal Society, London, pp. 37–76.

Hervé, F., 1934. *Extrait du Bulletin de l'Agence Générale des Colonies*, 297 et 298, 149 pp.

Jabbour-Zahab, R., Chagot, D., Blanc, F., Grizel, H., 1992. Mantle histology, histochemistry and ultrastructure of the pearl oyster *Pinctada margaritifera* (L.). *Aquat. Living Resour.* 5, 287–298.

Levi-Kalisman, Y., Falini, G., Addadi, L., Weiner, S., 2001. Structure of the nacreous organic matrix of a bivalve mollusk shell examined in the hydrated state using Cryo-TEM. *J. Struct. Biol.* 135, 8–17.

Liu, J., Sarikaya, M., Aksay, I.A., 1992. A hierarchically structured model composite: A TEM study of the hard tissue of red abalone. *Mater. Res. Soc. Symp. Proc.* 255, 9–17.

Lopez, E., Vidal, B., Berland, S., Camprasse, S., Camprasse, G., Silve, C., 1992. Demonstration of the capacity of nacre to induce bone formation by human osteoblasts maintained *in vitro*. *Tiss. Cell* 24, 667–679.

Mann, S., 1989. Chemical and biological perspectives. In: Mann, S., Williams, R.J.P. (Eds.), *Biomineralization*. Wiley-VCH, Weinheim, Germany, pp. 146–148.

Mann, S., Archibald, D.D., Didymus, J.M., Douglas, T., Heywood, B.R., Meldum, T.C., Reeves, N.J., 1993. Crystallization at inorganic–organic interfaces: Biominerals and biominetics synthesis. *Science* 261, 1286–1292.

Mutvei, H., 1969. On the micro and ultrastructure of conchiolin in the nacreous layer of some recent and fossil molluscs. *Stockholm Centr. Geol.* 20, 1–17.

Mutvei, H., 1970. Ultrastructure of the mineral and organic components of molluscan nacreous layers. *Biomineral. Res. Rep.* 2, 48–72.

Mutvei, H., 1979. On the internal structures of the nacreous tablets in molluscan shells. *Scan. Electr. Micros.* 2, 457–462.

Nakahara, H., 1991. Nacre formation in bivalve and gastropod mollusks. In: Suga, S., Nakahara, H.(Eds.), *Mechanisms and Phylogeny of Mineralization in Biological Systems*. Springer-Verlag, Tokyo, pp. 343–350.

Nakahara, H., Bevelander, G., 1971. The formation and growth of the prismatic layer of *Pinctada radiata*. *Calc. Tiss. Res.* 7, 31–45.

Nakahara, H., Kakei, M., Bevelander, G., 1980. Fine-structure and amino-acid composition of the organic envelope in the prismatic layer of some bivalve shells. *Venus (Jap. J. Malacol.)* 39, 167–177. (in Japanese with English Abstract).

Nakahara, H., Bevelander, G., Kakei, M., 1982. Electron microscopic and amino acid studies on the outer and inner shell layers of *Haliotis rufescens*. *Venus (Jap. J. Malacol.)* 41, 33–46. (in Japanese with English Abstract).

Naslain, R., Pailler, R., Bourrat, X., Heurtevent, F., 1998. Mimicking the layered structure of natural shells as a design approach to fiber–matrix interface in CMCs. In: Girelli-Visconti, J. (Ed.), *Proceedings of the ECCM-8 Conference*, Vol. 4. Woodhead Publications, Cambridge, UK, pp. 191–199.

Ojima, Y., 1952. Histological studies of the mantle of pearl oyster (*Pinctada martensii*, Dunker). *Cytologia* 17, 134–143.

Ranson, G., 1952. *Préliminaires. Un Rapport sur l'huître Perlière dans les Etablissements Français de l'Océanie*. Muséum d'Histoire Naturelle, Paris. 76pp.

Rousseau, M., Pereira-Mouries, L., Almeida, M.J., Milet, C., Lopez, E., 2003. The water-soluble matrix fraction from the nacre of *Pinctada maxima* produces earlier mineralization of MC3T3-E1 mouse pre-osteoblasts. *Comp. Biochem. Physiol. B, Biochem. Mol. Biol.* 135, 1–7.

Rousseau, M., Lopez, E., Coute, A., Mascarel, G., Smith, D.C., Naslain, R., Bourrat, X., 2004. Multi-scale structure and growth of nacre: A new model for bioceramics. *Bioceramics* 16, 1009–1012.

Rousseau, M., Lopez, E., Stempfle, P., Brendle, M., Franke, L., Guette, A., Naslain, R., Bourrat, X., 2005a. Multiscale structure of sheet nacre. *Biomaterials* 26, 6254–6262.

Rousseau, M., Lopez, E., Couté, A., Mascarel, G., Smith, D.C., Naslain, R., Bourrat, X., 2005b. Sheet nacre growth mechanism: A Voronoi model. *J. Struct. Biol.* 149, 149–157.

Rousseau, M., Bourrat, X., Stempfle, P., Brendle, M., Lopez, E., 2005c. Multi-scale structure of the *Pinctada* mother of pearl: Demonstration of a continuous and oriented organic framework in a natural ceramic. *Bioceramics* 17, 705–708.

Salomon, P., Roudnitska, M., 1986. *La Magie de la perle noire Tahiti Perles*, Times Ed., Singapore. 225 pp.

Saleuddin, A.S.M., Peit, H.P., 1983. Shell formation. In: Saleuddin, A.S.M., Wilbur, K.M. (Eds.), *The Mollusca, Physiology, Vol. 4,* Academic Press, New York, pp. 235–287.

Schäffer, T.E., Ionescu-Zanetti, C., Proksch, R., Fritz, M., Walters, D.A., Almqvist, N., Zaremba, C.M., Belcher, A.M., Smith, B.L., Stucky, G.D., Morse, D.E., Hansma, P.K., 1997. Does abalone nacre form by heteroepitaxial nucleation or by growth through mineral bridges? *Chem. Mater.* 9, 1731–1740.

Schmidt, N.J., 1924. *Die Bausteine des Tierkörpers in Polarisierten Lichte.* Cohen, Bonn.

Seurat, L.G., 1905. Observations anatomiques et biologiques sur l'huître perlière (*Margaritifera margaritifera* var. Cumingi Reeve) des lagons des Tuamotu et des Gambier. *Journal Officiel des Etablissements Français de l'Océanie (Papeete, Tahiti)*, 210–215.

Simkiss, K., Wilbur, K.M., 1989. *Biomineralization.* Academic Press, New York. 337 pp.

Smith, B.L., 1998. Studying shells: A growth industry. *Chem. Indust.* 16, 649–653.

Strack, E., 2006. *Pearls,* Rühle-Diebener-Verlag GmbH & Co., KG, Germany. 706 pp.

Takemura, Y., Kafuku, T., 1957. Anatomy of the silver-lip pearl oyster *Pincatda maxima* (Jameson). *Bull. Tokai Fish. Res. Inst.* 16, 1–40.

Taylor, J.J., Rose, R.A., Southgate, P.C., 1997. Inducing detachment of silver-lip pearl oysters (*Pinctada maxima,* Jameson) spat from collectors. *Aquaculture* 159, 11–17.

Thielley, M., 1989. *Etude histologique et cytochimique de la gamétogenèse chez la nacre Pinctada margaritifera (L.) var.* cumingii *(Jameson).* Diplôme d'Etudes Approfondies Université Française du Pacifique, Tahiti, 22 pp.

Thielley, M., 1993. *Etude cytologique de la gamétogénèse, du sex-ratio et du cycle de reproduction chez l'huître perlière, Pinctada margaritifera (L.) var.* cumingii *(Jameson), (mollusques bivalves). Comparaison avec le cycle de* Pinctada maculata *(Gould).* Thèse de doctorat, Université Française du Pacifique, Tahiti, 233 pp.

Thompson, J.B., Poloczi, J.T., Kindt, J.H., Michenfelder, M., Smith, B.L., Stucky, G., Morse, D.E., Hansma, P.K.,2000. Direct observation of the transition from calcite to aragonite growth as induced by abalone shell proteins. *Biophys. J.* 79, 3307–3312.

Towe, K.M., Hamilton, G.H., 1968. Ultrastructure and inferred calcification of the mature and developing nacre in bivalve mollusks. *Calc. Tiss. Res.* 1, 306–318.

Tranter, D.J., 1958a. Reproduction in Australian pearl oysters (Lamellibranchia). I. *Pinctada albina* (Lamarck): Primary gonad development. *Aust. J. Mar. Freshwat. Res.* 9, 135–143.

Tranter, D.J., 1958b. Reproduction in Australian pearl oysters (Lamellibranchia). III. *Pinctada albina* (Lamarck): Breeding season and sexuality. *Aust. J. Mar. Freshwat. Res.* 9, 191–216.

Tranter, D.J., 1958c. Reproduction in Australian pearl oysters (Lamellibranchia). IV: *Pinctada margaritifera* (Linnaeus). *Aust. J. Mar. Freshwat. Res.* 9, 509–525.

Tranter, D.J., 1958d. Reproduction in Australian pearl oysters (Lamellibranchia). V. *Pinctada fucata* (Gould). *Aust. J. Mar. Freshwat. Res.* 10, 45–66.

Velayudhan, T.S., Gandhi, A.D., 1987. Morphology and anatomy of Indian pearl oyster. *Bull. Cent. Mar. Fish. Res. Inst.* 39, 4–12.

Wada, K., 1961. Crystal growth of molluscan shells. *Kokoritsu Shinju Kenkyusho Hokoku (Bull. Natl. Pearl Res. Lab.)* 7, 703–828. (in Japanese with English summary)

Wada, K., 1972. Nucleation and growth of aragonite crystals in the nacre of some bivalve molluscs. *Biomineralization* 6, 141–159.

Waite, J.H., 1983. Quinone-tanned scleroproteins. In: Saleuddin, A.S.M., Wilbur, K.M. (Eds.), *The Mollusca, Physiology, Vol. 4,* Academic Press, New York, pp. 467–504.

Waite, J.H., Saleuddin, A.S.M., Andersen, S.O., 1979. Periostracin – a soluble precursor of sclerotized periostracum in *Mytilus edulis L., J. Comp. Physiol. B* 130, 301–307.

Watabe, N., 1965. Studies on shell formation, XI: Crystal–matrix relationships in the inner layers of mollusk shells. *J. Ultrastruct. Res.* 12, 351–370.

Weiner, S., Addadi, L., 1991. Acidic molecules of mineralized tissues: The controllers of crystal formation. *TIBS* 16, 252–256.

Weiner, S., Traub, W., 1980. X-diffraction study of the insoluble organic matrix of mollusc shells. *Eur. Biochem. Soc.* 111, 311–316.

Weiner, S., Traub, W., 1984. Macromolecules in mollusc shells and their functions in biomineralization. *Phil. Trans. R. Soc. Lond. B* 304, 425–434.

Wilt, F.H., Killian, C.E., Livingstone, B.T., 2003. Development of calcareous skeletal elements in invertebrates. *Differentiation* 71, 237–250.

Wise, S.W., 1970. Microarchitecture and mode of formation of nacre (mother-of-pearl) in pelecypods, gastropods, and cephalopods. *Eclog. Geol. Helvet.* 63, 775–797.

Zhang, C., Xie, L.P., Huang, J., Chen, L., Zhang, R.Q., 2006. A novel putative tyrosinase involved in periostracum formation from the pearl oyster (*Pinctada fucata*). *Biochem. Biophys. Res. Commun.* 342, 632–639.

CHAPTER 4

Feeding and Metabolism

John S. Lucas

4.1. FOOD SOURCES

Like most other bivalve molluscs, pearl oysters are filter or suspension feeders throughout the free-living stages of their lifecycle. They filter fine suspended particles, seston,
from the water around them. This suspended particulate matter (SPM) is very hetero-
geneous in composition and varies enormously in particle size and abundance over the
various habitats of pearl oysters. There may be a large component of particulate inorganic matter (PIM) that is often composed of resuspended particles from the benthic
sediment. Particulate organic matter (POM) is made up of a wide range of living organisms together with non-living material such as feces, decomposing cells and mucus.
The living organisms may be divided according to size.

1. Picoplankton (<0.4–2 μm diameter) includes bacteria, cyanobacteria (blue-
   green algae) and picophytoplankton (the smallest unicellular algae).
2. Nanoplankton (2–20 μm) and microplankton (20–200 μm) include most
   phytoplankton species, i.e. unicellular autotrophic algae such as diatoms and
   chlorophyte flagellates, as well as various heterotrophic organisms, such as
   protozoans, some dinoflagellates and other heterotrophic flagellates.
3. Mesoplankton (200–2,000 μm) consists mainly of zooplankton, such as eggs,
   larval stages and copepods.

There are particles of organic detritus and inorganic matter of all sizes.

There are often significant amounts of dissolved organic matter (DOM) in seawater and this is another potential source of nutrients for pearl oysters, although there are no data for this. The large surfaces of exposed epithelium within their mantle cavity and water currents through it facilitate exchanges of inorganic and organic molecules between the soft tissues and ambient water. For example, Siebers and Winkler (1984) found that there was 29–66% net absorption of various amino acids from water flowing through the mantle cavity of the mussel *Mytilus edulis*. They suggested that high biomasses of mussels would successfully compete with ambient bacteria for the available DOM and play an important role in recycling DOM. This may also apply to pearl oysters at high culture densities.

There have been studies of seston in two Japanese bays where there is major farming of Akoya pearl oysters.[1] Mean SPM and mean phytopigment concentrations (chlorophyll *a* + phaeopigment) near a pearl farm in Ohmura Bay, western Kyushu are shown in Table 4.1. SPM was sorted into size fractions, 1.2–10, 10–50, 50–100, 100–300 and >300 μm, and the dry weight and phytopigment content of each fraction were measured in monthly samples (Numaguchi, 1994a). Dry weights and phytopigment contents of the finest particle fraction (1.2–10 μm) were always greater than those of any other size fraction. Mean dry weight of the 1.2–10 μm fraction was 42% of the total SPM dry weight (range 27–61%). Mean phytopigment content of the 1.2–10 μm fraction was 55% of the total SPM content (range 31–77%). Considering that it takes many more of these finest particles to make an equivalent mass of larger particles, these finest particles were much more numerically abundant than indicated by the percent dry weight data. Since pearl oysters filter these particles, Numaguchi (1994a) considered that these pico- and nano-phytoplankton are very important components of the seston for pearl oyster culture.

Tomaru *et al.* (2002) monitored the seasonal changes in seston where Akoya are cultured in Uchiumi Bay, the Inland Sea, from June 1997 to February 1999 (Table 4.1). Mean percent values for particulate organic carbon (POC) of picoplankton, nanoplankton and seston >20 μm were 40, 33 and 27%, respectively, over the study period. There were similar orderings of mean percent values for particulate organic nitrogen (PON), phosphorus (POP) and chlorophyll *a*. Erratic blooms of various diatoms, up to ca. 600 cells/mL, occurred during the period and the blooms were most common during spring and summer, March–August.

There are data for a very different environment to these coastal Japanese bays from seston studies in the oligotrophic waters of atolls lagoons of the Tuamotu Archipelago, French Polynesia (FrP), where *Pinctada margaritifera* is farmed (Table 4.1). Data for Takapoto atoll lagoon from 19 studies over 24 years are summarized by Pouvreau *et al.* (2000a). Total POC values over the period ranged from 115 to 249 μg/L, but were mostly in the order of 180 μg/L. Picoplankton was the dominant component of the seston with picodetritus being a major component. Most of the phytoplankton biomass

---

[1]Throughout this text the term 'Akoya' is used for pearl oysters in the currently unresolved complex of *Pinctada fucata-martensii-radiata-imbricata* (see Section 2.3.1.13). All these names, especially the first two, have been used in studies that are included in this chapter. The source of the studied Akoya is specified where the source is not from the major pearl farming regions of Japan.

TABLE 4.1

Parameters of the seston at some pearl oyster localities. Values in brackets are ranges, other values are means.

Species	Locality (period of study)	SPM (mg/L)	POC (μg/L)	chl a + (μg/L)	PIM (mg/L)	Major components of seston	Source
Akoya	Ohmura Bay, western Kyushu (May to December 1989)	3.4 (2.1–6.3)	NA	2.6 (1.6–5.2)	NA	Picoplankton + nanoplankton =42% dwt = 55% chl a +	Numaguchi (1994a)
	Uchiumi Bay, Inland Sea (June 1997 to February 1999)	NA	100–250[a]	1–3[a]	NA	Picoplankton = 40% POC Nanoplankton = 33% POC	Tomaru et al. (2002)
Pinctada margaritifera	Takapoto lagoon, FrP (1974 to 1998)	(0.4–1.6)	ca. 180[a] (115–249)	0.2–0.3[a] (≤0.8)	(0.2–0.9)	Picoplankton (mainly picodetritus) Nanoplankton	Pouvreau et al. (2000a) Loret et al. (2000a) and Delesalle et al. (2001)
	Orpheus Island GBR lagoon (May 1996 to June 1997)	1.4 (1.2–1.7)	ca. 280 (200–350)	NA	ca. 0.8	NA	Yukihira et al. (1999; 2006)
Pinctada maxima	Coastal bay GBR lagoon (May 1996 to June 1997)	11.1 (2–60)	ca. 1,050 (≤4,000)	NA	ca. 9	Nanoplankton	Yukihira et al. (1999; 2006)

chl a +: chlorophyll a + phytopigments; GBR: Great Barrier Reef; FrP: French Polynesia; NA: not available; PIM: particulate inorganic matter; POC: particulate organic carbon; SPM: suspended particulate matter.
[a]Most values are within this range or in this vicinity.

was picophytoplankton (Delesalle *et al.*, 2001). The larger seston usually consisted of very low levels of nanophytoplankton, protozoans and zooplankton: typical POC values for each of these categories were 1–9 μg/L (Pouvreau *et al.*, 2000a). Detritus usually made up the bulk of POC over all sizes, and the POC of autotrophs was usually a very minor component. Total suspended particular matter (SPM), both POC and PIM, increased with wind velocity. A notable feature of these data for Takapoto lagoon is the degree of consistency of POC values both spatially and temporally, and this is also shown by chlorophyll *a* and primary production levels (Delesalle *et al.*, 2001). However, the composition of the phytoplankton community was not stable. In 1974, the major nanophytoplankton groups were diatoms (>60 species), dinoflagellates (35 species) and coccolithophorids, with dinoflagellates being most numerically abundant (Sournia and Ricard, 1976). Studies in 1996–1997 found communities dominated by picocyanobacteria and nanoplankton, which were largely chlorophytes, prymnesio-phytes and heterotrophic dinoflagellates (Loret *et al.*, 2000a; Delesalle *et al.*, 2001).

Yukihira *et al.* (1999, 2006) considered contrasting coastal inshore and offshore environments in the Great Barrier Reef (GBR) lagoon where *Pinctada maxima* and *P. margaritifera* occur (Table 4.1). A silty coastal bay, typical of a *P. maxima* habitat, was characterized by large and erratic fluctuations in SPM, especially PIM. The fluctuat-ing SPM levels were strongly correlated with wind velocity, as found in Takapoto lagoon by Pouvreau *et al.* (2000a), and it is likely that the increases in SPM were due to resus-pension from the benthos. POM in the coastal bay fluctuated erratically with a mean of 2.1 mg/L over the year. Most particles were in the 4–15 μm diameter range (Table 4.1). In contrast to the coastal bay, SPM and POM remained within narrow limits near a fring-ing reef of Orpheus Island, offshore in the GBR lagoon, where *P. margaritifera* occurs (Yukihira *et al.*, 2006). POM values for the GBR study are presented as POC in Table 4.1 using the approximate conversion POC = POM × 0.5 (Pouvreau *et al.*, 2000a).

Comparisons of seston abundance and composition at the five localities presented in Table 4.1 are limited by incomplete data sets for four localities. The consistent pat-tern is of pico- and nanoplankton or nanoplankton dominating the seston. There are some characteristics of the habitats of each species. The two semi-enclosed localities of Akoya are characterized by high phytopigment levels and presumably high primary production within the seston. The two localities of *P. margaritifera* are physiographi-cally quite different: an atoll lagoon in FrP versus an island fringing reef of the inner GBR lagoon. Furthermore, Takapoto lagoon has the unusual feature of seston domi-nated by picodetritus and correspondingly low phytopigment. However, the lagoon's low levels of SPM, POC and PIM overlap with corresponding values at Orpheus Island. The coastal locality of *P. maxima* with very high and fluctuating seston levels was quite distinct.

4.2. FEEDING MECHANISMS

Gosling (2003) and Ward and Shumway (2004) present comprehensive reviews of feeding and digestion in bivalves. Their reviews only contain several brief references to these processes in pearl oysters, but much is of general application to pearl oyster species.

As described, bivalves encounter a wide range of particles as they feed by filtering the seston content of water flowing through their mantle cavity (Fig. 4.1). There are many factors that affect their ability to capture these particles and the nutritional value of the particles, i.e. whether they are organic or inorganic, their abundance, size, shape, surface characteristics, density, digestibility and biochemical composition. It would be very inefficient for bivalves to act as passive sieves in capturing particles, ingesting all particles and indiscriminately processing them through their complex digestion sys-tem. Thus, some bivalves have developed mechanisms for preferential capture of par-ticles, controlling capture of particle, and pre-ingestive and post-ingestive sorting of particles (Gosling, 2003; Ward and Shumway, 2004).

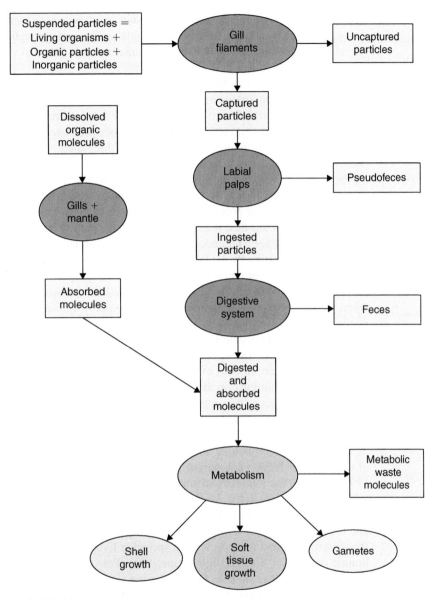

FIGURE 4.1 Sequence of suspended particle capture and processing in pearl oysters.

4.2.1. Pallial organ

The pallial organ, consisting of gills and labial palps, is the pre-ingestive structure of bivalves. It is responsible for capture and sorting food particles before ingestion.

Feeding and respiration (see Section 4.3.1) occur as currents of water are drawn into the anterior and ventral regions of the mantle cavity, and through the gill filaments (Figs. 4.1 and 4.2). Pearl oysters have eulamellibranch gills and the following brief description relates to this kind of gill structure.

Water currents through the gill filaments are maintained by beating of lateral cilia on the inter-filament spaces and particles in suspension are captured at least in part by these cilia. Often there are laterofrontal cirri at the entrances to the inter-filament spaces. The laterofrontal cirri are double rows of specialized cilia that form a mesh-work, facilitating the capture of particles and pushing them onto frontal cilia. Trapped particles are bound in a mucous layer of frontal cilia on the gill filaments and moved by these cilia to ventral food grooves on the filaments. The particles and mucus are formed into mucous strings in these ciliated ventral food grooves and passed to labial palps at the mouth. The opposing palp lamellae fluidize the mucus to release the bound particles. Particles are then ingested as a mucous slurry (Ward and Shumway, 2004). The mechanism is not well understood in bivalves with eulamellibranch gills, but there is often particle sorting by the labial palps, e.g. according to nutritional value and size, and material that is not ingested is released from the palps in mucus-bound bundles as pseudofeces. Pseudofeces are also formed at the food groove and labial palp junction when the rate of particle capture exceeds the processing capacity of the labial palps or the ingestion rate (Fig. 4.1). These pseudofeces are carried to the posterior of the mantle cavity and pass gently out of the mantle cavity in the vicinity of the excurrent region (Fig. 4.2). The pseudofeces are mucus permeated and fragile masses compared with feces, which are ribbons of compact pellets. The water that has flowed in through the gills forms excurrents on the inside of the gills and flows vigorously to the poste-rior, collecting feces as the anus releases them. The feces are carried further from the body than the pseudofeces, apart from emerging from the pearl oyster at a different position (Fig. 4.2).

Details of gill morphology are described in Section 3.2.3. There are several features of the gills of FrP *Pinctada margaritifera* that affect their functioning (Pouvreau *et al.*, 1999a). Firstly, the gill size and surface area is relatively very large. Gill surface area (GA) for 1 g dry soft tissue weight (dwt) is 3,502 mm^2. Equivalent values for six other bivalves that inhabit low-turbidity environments ranged from 285 to 2,740 mm^2. Only one bivalve, *Geukensia demissa*, has been reported with a GA relatively greater than *P. margaritifera*, and this species is a specialist living high in the intertidal zone with limited periods for filter feeding (Franz, 1993; Pouvreau *et al.*, 1999a). Secondly, the gill microstructure is notable for not having laterofrontal cirri on the gill filaments (Pouvreau *et al.*, 1999a). As indicated above, these cirri supplement the frontal and lateral cilia in their role of capturing suspended particles as they pass between the gill filaments. Their effect is to improve the capture efficiency of small particles, espe-cially picoplankton. A grouping of bivalve species according to whether they have well developed, reduced or no laterofrontal cirri has shown that their capture efficiencies of 1 μm particles are 30–70%, 0–40% and 0–20%, respectively (Pouvreau *et al.*, 1999a). A similar tabulation of comparative capture efficiency was presented by Ward and Shumway (2004) and noted by Gosling (2003).

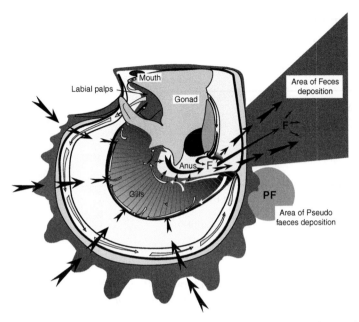

FIGURE 4.2 Lateral diagram of a pearl oyster showing passage of water through the mantle cavity and gills (black arrows), capture of particles on gills, passage of particles to labial palps and mouth (white arrows), and production of pseudofeces (long grey arrows) and feces (from Pouvreau *et al.*, 2000b) (with permission).

4.2.2. Clearance rate and particle capture

Clearance rate (CR) is the rate at which particles are captured by the gill filament cilia from the flow of water through the mantle cavity. When all or almost all the particles in the incurrent water are retained on the gill filaments, CR becomes the same as "pumping rate" (PR), the rate of water flow through the mantle cavity. CR is usually measured by what are termed "indirect" methods. These methods involve measuring the rate of removal of particles by a bivalve in a container with a fixed volume or a fixed flow through of water, e.g. in *P. maxima* and *P. margaritifera* by Yukihira *et al.* (1998a). "Direct" methods of measuring CR and PR involve measuring particle concentrations of water samples taken from the incurrent and excurrent flows, e.g. Pouvreau *et al.* (1999a) and Yukihira *et al.* (1999), and by measuring the rate of excurrent flow, e.g. Numaguchi (1994b) and Sukhotin *et al.* (2003). Capture efficiency of the various sized particles in the incurrent water is determined by measuring the size-frequency structure of particles before and after filtering through the gill filaments.

CRs for four species of pearl oysters are shown in Table 4.2. With the exception of Hawkins *et al.* (1998), data for *P. margaritifera* and *P. maxima* show very high rates: in the range of 11–22 L/h. In comparable data for six other bivalves, the CR values for 1 g dwt were 0.08–8.71 L/h (Pouvreau *et al.*, 1999a). They point out that the high CR values in FrP *P. margaritifera* result from a combination of large gill surface area (GA)

TABLE 4.2

Clearance rates of 1 g dry soft tissue weight (dwt) *Pinctada* species feeding *in situ* or on natural particles.

Species	Clearance rate (L/oyster/h)	Source
Akoya (Bermuda)	5.2	Ward and MacDonald (1996)
Akoya	14.7[a]	Sato *et al.* (1964)
P. margaritifera (GBR)	14.6[b]	Yukihira *et al.* (1999)
P. margaritifera (FrP)	22	Pouvreau *et al.* (2000b)
P. margaritifera (Malaysia)	5.5	Hawkins *et al.* (1998)
P. maxima	10.7[b]	Yukihira *et al.* (1999)

FrP: French Polynesia; GBR: Great Barrier Reef.
[a]Not 1 g dwt, 72 mm shell height.
[b]Calculated from (Yukihira *et al.*, 1998a; 1999).

and high PR per unit GA (PR/GA ratio). Bivalve PR/GA ratios are usually within the range 0.4–4.1 mL/mm^2/h, but in FrP *P. margaritifera* it is 5.1 mL/mm^2/h.

Interestingly, the CRs for FrP *P. margaritifera* when feeding on cultured algae and on natural particles were about twice the equivalent values for GBR *P. margaritifera*.

P. margaritifera and *P. maxima* may grow to more than 200 mm shell height (SH) and CR increases exponentially with size as an exponent <1 of dwt. CRs for these species feeding on the cultured microalga *Isochrysis* aff. *galbana* Tahitian isolate (T-Iso) in laboratory studies, were related to dwt as:

$$CR = 10.73 \times dwt^{0.617} \, L/h \quad P. \, maxima \, (GBR) \, (Yukihira \, et \, al., \, 1998a)$$

$$CR = 12.34 \times dwt^{0.604} \, L/h \quad P. \, margaritifera \, (GBR) \, (Yukihira \, et \, al., \, 1998a)$$

$$CR = 25.88 \times dwt^{0.573} \, L/h \quad P. \, margaritifera \, (FrP) \, (Pouvreau \, et \, al., \, 1999a)$$

There are only minor differences between maximum CR estimates from these laboratory studies and those for oysters feeding *in situ* or in seawater with natural particles (Yukihira *et al.*, 1998b, 1999; Pouvreau *et al.*, 1999a, 2000b). Thus, large >10 g oysters of the two species from the GBR have PRs of >50 L/h and similarly large FrP *P. margaritifera* have PRs of about 100 L/h. There may be temporal variations in PRs, perhaps diurnal rhythms. However, Li *et al.* (1997) did not find significant diurnal rhythms in the PRs of Akoya (South China Sea) juveniles and Pouvreau *et al.* (2000b) observed that *P. margaritifera* were generally feeding throughout the day and night. If these large oyster sustain these PRs for long periods, they will filter thousands or even >10,000 L each week. Large numbers of cultured oysters in lagoonal pearl farms could potentially have significant effects on the lower trophic levels through their clearance

TABLE 4.3

Seston particle diameters corresponding to 50% and 90% capture efficiency by *Pinctada* species. These data are estimates from published figures.

Species	Particle diameters (μm)		Source
	50% capture	90+% capture	
Akoya (Bermuda)	<2.5[a]	8–35	Ward and MacDonald (1996)
P. margaritifera (FrP)	2	3–7 + [b]	Pouvreau *et al.* (1999a)
P. margaritifera (GBR)	<1.5[c]	3–10 + [b]	Yukihira *et al.* (1999)
P. maxima	2	4–10 + [b]	Yukihira *et al.* (1999)

FrP: French Polynesia; GBR: Great Barrier Reef.
[a]About 75% at 2.5 μm, the smallest particle size tested.
[b]Particle size was not tested to the upper level where capture efficiency declines.
[c]About 75% at 1.5 μm, the smallest particle size tested.

of phytoplankton, but this was not the case in well-farmed Takapoto lagoon. Farmed *P. margaritifera* and their associated "pest" *P. maculata* were estimated to consume only 0.24% of the gross primary production by phytoplankton within the lagoon (Niquil *et al.*, 2001). The current extent of farming appears to be comfortably within the capacity of the lagoon.

Akoya (Bermuda) have a substantially lower CR (Table 4.2). This is not a function of size as the data are standardized to 1 g dwt size, although there is probably an element of temperature effect as the data for this sub-tropical species were obtained at the relatively low 22°C. There are no 1 g dwt size data for the other Akoya data, from Japan (Table 4.2), but it is notable from these data that the relatively small pearl oyster may filter >1,000 L per week. Three and four years old Akoya have maximum CRs of 350 and 550 L/day, respectively (Uyeno and Inouye, 1961; Sato *et al.*, 1964; Numaguchi, 1996). Unlike *P. margaritifera* farming in atoll lagoons, high density farming of this species in bays in Japan impacts the lower trophic levels of the environment (Uyeno and Inouye, 1961).

There are intra- and interspecific differences in the particle capture efficiency versus particle diameter (Table 4.3). However, they are characterized by the ability to capture 2 μm particles, the upper range of picoplankton, with at least 50% efficiency. Thus, they can feed to some extent on the abundant picoplankton. Furthermore, GBR *P. margaritifera* captured 1.5 μm diameter particles with 75% efficiency and will undoubtedly capture 1 μm particles with moderate efficiency, putting this species among those bivalves with highest retention of these finest particles (Pouvreau *et al.*, 1999a). It does not appear to belong within the group of bivalves without laterofrontal cirri shown in the tables of Pouvreau *et al.* (1999a) and Ward and Shumway (2004) (Section 4.2.1).

There is a downward curving relationship between capture efficiency and diminishing particle size in Akoya (Bermuda) (Ward and MacDonald, 1996). In FrP and GBR

P. margaritifera and *P. maxima* there is a plateau of high capture efficiency with a critical lower point of particle size at which there is an abrupt decline in capture efficiency
(Pouvreau *et al.*, 1999a; Yukihira *et al.*, 1999) (Fig. 4.3). Hawkins *et al.* (1998) found
much lower capture efficiencies for *P. margaritifera* in the silty conditions, 6–40 mg/L,
of the Merbok mangrove system, Malaysia. Values ranged from about 30% in the
3–10 μm particles size range, down to about 15% for 16 μm particles. These data were,
however, within the context of four other bivalve species, a mussel and three rock oyster species, none of which was found to have capture efficiencies >40%. The capture
efficiency of *P. margaritifera* over the particle size range tested was comparable to the
other four species.

Tomaru *et al.* (2002) considered that seston particles <2 μm were too small to be
captured by adult Akoya and thus of no nutritious value (Tomaru, personal communication). This was despite picoplankton being the dominant fraction of the bay where the
pearl oysters are cultured. They recommended only monitoring the composition of seston >2 μm to evaluate the food available for these oysters. It has also been considered
that *P. margaritifera* cannot feed on particles <2 μm (Loret *et al.*, 2000b). However, the
data on capture efficiency in Table 4.3 show that 2 μm diameter is not a cut-off point.
There is particle capture, be it inefficient, below this size and this may well be the case
for Akoya.

As described above, the capture of seston particles by bivalves is often more than
purely a sieving process, determined by gill morphology. Thus, pearl oysters have
mechanisms to control the rate of particle capture and possibly preferentially capture
some particles.

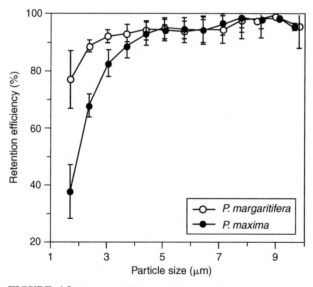

FIGURE 4.3 Capture efficiency versus particle size in *Pinctada margaritifera* and *P. maxima* from
the Great Barrier Reef (from Yukihira *et al.*, 1999) (with permission). Concentration of suspended
particles = 2 mg/L.

CRs of GBR *P. margaritifera* and *P. maxima* feeding on the microalgae T-Iso and *D. tertiolecta* at concentrations ca. 0–10 mg dwt/L declined exponentially with increasing cell concentration (Yukihira *et al.*, 1998b). CRs (L/h) were related to food concentration (FC, mg dwt/L) of T-Iso as:

$$CR = 56.67 \times 10^{-0.127FC} \quad \textit{P. margaritifera}$$

$$CR = 47.0 \times 10^{-0.102FC} \quad \textit{P. maxima}$$

Similarly, CR declined exponentially with increasing particle concentrations when these two species were feeding on natural particles (Fig. 4.4) (Table 4.4).

Bivalve gill filaments are innervated by branchial nerves from visceral ganglia and the filament cilia are under nervous control (Gosling, 2003). CR is controlled by altering the angle at which the cilia beat or by adjusting the apertures of the ostia (perforations through which the water enters eulamellibranch gill filaments). Presumably the declining CR with increasing SPM reflects this nervous control. This pattern is not universal in bivalves. For instance, it is quite different to *Mytilus edulis*. The mussel ceased filtering at very low cell densities of a diatom, then maintained maximum CR over a range of cell densities before CR declined progressively at very high cell densities (Forster-Smith, 1975).

FrP *P. margaritifera* show strongly preferential ingestion of some phytoplankters compared to their abundance in the seston (Loret *et al.*, 2000a) (Section 4.2.3). It was not established where on the pallial organ this pre-ingestion sorting was occurring: on the gills or labial palps or both. Very high concentrations of some phytoplankters in the

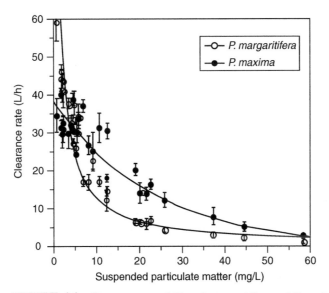

FIGURE 4.4 Clearance rates *of Pinctada margaritifera* and *P. maxima* from the Great Barrier Reef at various concentrations of natural suspended particles (from Yukihira *et al.*, 1999) (with permission). Temperature = 23–28°C.

TABLE 4.4
Clearance rate (CR), pseudofeces production rate (rejection rate, RR), ingestion rate (IR), absorption efficiency (AEff), absorption rate (AbsR), defecation rate (FecR) and daily solids production of Great Barrier Reef (GBR) _Pinctada margaritifera_ and _P. maxima_ at various suspended particle matter (SPM) and particulate organic matter (POM) levels at 23–28°C. All weights are dry soft tissue weight (dwt). The data are standardised to 10 g dwt oysters and calculated from (Yukihira _et al._, 1999).

Species	SPM (mg/L)	POM (mg/L)	CR (L/h)	RR (mg/h)	IR (mg/h)	AEff (%)	AbsR (mg/h)	FecR (mg/h)	Total solids released (g/day)
P. margaritifera	2	0.7	55	3	71	52	37	34	0.9
	5	1.1	24	15	144	38	55	89	1.8
	10	1.9	13	31	124	30	37	87	2.3
	25	4.0	6	55	69	22	15	54	2.4
	50	7.7	3	65	38	17	6	32	2.3
P. maxima	2	0.7	35	0	71[a]	59	42	29	0.1
	5	1.1	31	7	149	53	79	70	1.9
	10	1.9	25	28	281	45	126	155	4.4
	25	4.0	13	60	249	27	67	182	5.8
	50	7.7	4	44	162	12	19	143	4.5

[a]Estimated from Fig. 10 of Yukihira _et al._ (1999).

digestive system relative to their abundance in the seston suggest that at least selective capture was occurring at the gills.

4.2.3. Labial palps

Captured particles are transported from the gills to the labial palps where there is usually major pre-ingestive sorting (Figs. 4.1 and 4.2). The POM/PIM ratio is increased by differentially rejecting PIM and large particles that cannot be easily ingested or digested are ejected in the pseudofeces. The palps may also reject captured particles that reach the mouth region at a rate beyond the handling capacity of the digestive system.

When Akoya adults were fed charcoal particles 2.5–52.5 μm in length and 2.5–20 μm in width there were relatively more small particles in the esophagus than in the suspension (Kuwatani, 1965a). Particles up to 12.5 μm in length and 5 μm in width were relatively more abundant in the esophagus compared to the suspension. Virtually no particles >30 μm length and >17.5 μm width were ingested. Kuwatani (1965a) did not record the occurrence of pseudofeces and their particle size content, but there was pre-ingestive sorting on the basis of size.

There are conflicting data on the functioning of labial palps in pearl oysters in improving the organic fraction of ingested particles. Studies have shown increased pseudofeces production with increasing SPM, but no relative increase in the PIM content of the pseudofeces, which would correspondingly enhance the ingested organic fraction. However, there have been quite different results from two studies of *P. margaritifera* in an atoll lagoon.

As sediment levels and SPM increased, Akoya (Bermuda) decreased their CR and increased their rate of pseudofeces production (MacDonald and Ward, 1994). However, these oysters indiscriminately rejected both POM and PIM in the pseudofeces: rejecting particles without sorting. Hawkins *et al.* (1998) found insignificant particle sorting in *P. margaritifera* feeding on natural particles. Although the CR of Malaysian *P. margaritifera* was comparable to four nearby bivalve species, its organic ingestion rate, 2.0 mg/h, was substantially less than similar rates for 1 g dwt individuals of the other species (Hawkins *et al.*, 1998). The organic ingestion rates of the five species were correlated with their ability to increase the organic fraction by separating POM and PIM before ingestion. *P. margaritifera* was lowest, having no measured ability. Similarly, Yukihira *et al.* (1999) found no evidence of sorting before ingestion. There was no significant difference between the POM/PIM ratio of the natural SPM particles and the oyster's pseudofeces. Percent organic content of the pseudofeces declined exponentially from ca. 35% to ca. 13% as SPM increased from 1 to 100 mg/L (Yukihira *et al.*, 1999). This conformed to the decreasing proportion of POM in the environmental SPM (cf. Table 4.4). *P. maxima* similarly showed no evidence of sorting before ingestion (Yukihira *et al.*, 1999).

The gut contents of Akoya pearl oysters in the Hansando and Chungmu coastal areas of South Korea were predominately small diatoms, such as *Skeletonema costatum*, and *Chaetoceros*, *Nitzschia* and *Thalassiosira* species. These were the dominant phytoplankters in these coastal environments and from this Chang *et al.* (1988) considered that there was a lack of selectivity in feeding. Lack of selectivity may be further supported by the pearl oysters ingesting *Nitzschia* species, which are of poor nutritional value for pearl oysters (Numaguchi, 2000).

In contrast to these studies, *P. margaritifera* showed POM selectivity in the very low seston levels (<0.6–1.2 mg/L) in Takapoto lagoon, FrP. Their pseudofeces were white, being largely $CaCO_3$ and showing preferential rejection of mineral-rich particles before ingestion (Pouvreau *et al.*, 2000b). PIM content of the pseudofeces was 80% at 0.4 mg L SPM, increasing to 90% at 1.25 mg/L. Equivalent values for the PIM component of the atoll lagoon water were 35% and 50%, respectively. This divergence from other studies may reflect the very low seston levels and the nature of the SPM in an atoll lagoon or a morphological difference (see below).

Even further than enhancing the organic fraction at ingestion, Loret *et al.* (2000a) found preferential ingestion of some phytoplankters, largely *Leucocryptos* spp., by FrP *P. margaritifera*. The oysters' proximal guts contained from 9 to 80 times the relative amount of cryptophyte pigment compared to the adjacent water column. Dinoflagellates were also selectively ingested, with the proximal guts containing up to 22 times the relative amount of their pigment than the adjacent water. Particle selection of cryptophytes was not based on size. Their diameter, 5–10 μm, was within the ranges of two microalgae groups, chlorophyte and prymnesiophyte species, which were more abundant in the seston (Loret *et al.*, 2000a). There was apparent strong selection against picocyanobacteria, as indicated by their relatively low occurrence in the proximal gut, but this must be an artifact of inefficient capture of particles $<1\,\mu m$ diameter (see Section 4.2.2). The study did not establish whether the strong selectivity was at the gills or palps. The preferred microalgae may have been selected by epicellular or intracellular chemical cues (Ward and Shumway, 2004).

There are substantial differences between data for FrP, GBR and Malaysian *P. margaritifera* in CR, particle capture efficiency and particle selectivity, suggesting genetic differences. However, some of the differences may reflect plasticity in these pearl oysters' filtering and sorting structures, as shown in some other bivalves. For example, nearby populations of the Pacific oyster, *Crassostrea gigas*, had substantially different gill/body, palp/body and gill/palp ratios, and probably differences between length and spacing of the laterofrontal cirri of the gill filaments (Barille *et al.*, 2000). Gill area was 25% larger in low-turbidity conditions and palp area was 60% larger in high-turbidity conditions. Gill cirri tended to be longer and more evenly spaced in oysters at high turbidity. Seasonal variations in relative gill and palp sizes, and variations in these organs in response to laboratory conditions have also been shown (Honkoop *et al.*, 2003; Drent *et al.*, 2004). It is quite conceivable that *P. margaritifera* and other pearl oysters show some plasticity in these vital pre-ingestive functions.

4.2.4. Pseudofeces

This section deals with rate of pseudofeces production (rejection rate, RR, mg/h), while recognizing the inconsistent evidence for POM versus PIM sorting in pseudofeces production.

CRs (L/h) of *P. maxima* and *P. margaritifera* decline exponentially with increasing natural SPM over 1–60 mg/L (Fig. 4.4). However, they did not compensate adequately and particle filtration rate (FR, mg/L) ($FR = CR \times SPM$) still increased (Yukihira *et al.*, 1999). As the FR increased, the rate of pseudofeces production (RR) increased in a linear fashion:

$$RR = -13.2 + 0.182FR\ mg/h \ \text{for}\ P.\ maxima$$

$$RR = -29.8 + 0.457FR\ mg/h \ \text{for}\ P.\ margaritifera$$

Hawkins *et al.* (1998) also measured RR in *P. margaritifera* and found a linear relationship between RR and FR:

$$RR = -1.1(\pm 4.3) + 0.811(\pm 0.14)FR\ mg/h$$

The threshold for pseudofeces production for both species from the GBR was about 3 mg/L SPM (Yukihira *et al.*, 1999). Pseudofeces production by *P. margaritifera* peaked at 40% of filtered particles when SPM was about 30 mg/L (Table 4.4). Pseudofeces production by *P. maxima* peaked at 20% of filtered particles when SPM was about 30 mg/L. Of these two species, *P. maxima* is the one that is adapted to silty environments and two adaptations compared to *P. margaritifera* are being able to ingest larger volumes of food particles including higher PIM levels.

The threshold for pseudofeces production of *P. margaritifera in situ* in Takapoto lagoon, FrP, was substantially lower (Pouvreau *et al.*, 2000b). It was slightly more than 0.4 mg/L SPM, within the context of an SPM range of 0.4–1.5 mg/L. Pouvreau *et al.* noted that it is very low compared to thresholds for other bivalves and *P. margaritifera* and *P. maxima* from the GBR are more typical.

4.2.5. Ingestion

Pearl oysters mainly ingest particles in the 2–20 µm range (e.g., Kuwatani, 1965a), after whatever pre-ingestive passive and active selective processes are occurring (Fig. 4.1). The pre-ingestive processes are fallible: there are records of pearl oyster having much larger particles, up to 300 µm, in their guts. These large particles include small live zooplankters such as protozoans, cladocerans, copepods and rotifers, and invertebrate larvae and eggs (Gervis and Sims, 1992; Li *et al.*, 1999). Chellam (1983) deliberately fed trochophore larvae to adult Akoya (India) and they were ingested and passed through the oysters' guts. The trochophores emerged live in the feces and developed on to straight-hinge veligers. It seems that while mesoplankton-sized organisms may be ingested, they are poorly, if at all, digested. Pouvreau *et al.* (2000a) considered that these mesoplanktonic organisms are unlikely to be a significant source of food.

4.2.6. Digestion

Post-ingestive sorting follows the pre-ingestive processes to further enhance the availability of appropriate particles for digestion and absorption (Fig. 4.1). Pearl oysters have a complex stomach and digestive system (see Section 3.2.4). The complexity is comprehensively described for Akoya by Kuwatani (1965b). The asymmetric stomach is divided into anterior and posterior compartments, with the esophagus entering the posterior compartment. There are ciliary areas and tracts that move particles in complex patterns that sort digested and fine particles from undigested and coarse particles, transporting the digested and fine particles to five digestive diverticula. A crystalline style rotates against the adjacent gastric shield, releasing enzymes as it is abraded. Kuwatani (1965b) categorized the functions of the stomach as sorting, digestion, absorption and ejection via a series of processes that involve seven successive regions of the stomach. There is initial extracellular digestion through enzymes released by the crystalline style. Intracellular digestion and absorption then occur in the digestive diverticula. Undigested particles are ejected into the intestinal grooves from the various food sorting areas and tracts.

Kuwatani (1965a) studied the functioning of the digestive tract of Akoya using charcoal particles. Most of the smaller particles were coated with mucus after ingestion, the mucus was presumed to come from the crystalline style. The particles were transported into the digestive tubules where cells phagocytosed particles <5 µm. This selective phagocytosis resulted in 95% of the particles in the digestive tubules being <5 µm 3 hours after ingestion. Subsequently, the phagocytic cells that contained these indigestible particles underwent fragmentation and released the particles, which passed to the intestine and rectum. The processes were quite rapid. Large charcoal particles were moved to the intestinal grooves and appeared in the oyster's rectum almost immediately after ingestion. Feces consisted of large particles until about 3 hours after feeding (Kuwatani, 1965a) (Fig. 4.5). Similarly, when Akoya were fed microalgae, *Pavlov lutheri*, the amount of phaeophytin in the digestive diverticula increased rapidly (Numaguchi, 1985; 1994a). It was most abundant after 3 hours and then decreased gradually. When *P. margaritifera* and *P. maxima* were feeding on natural particles in seawater, it took 42 minutes and 56 minutes, respectively, from ingestion to initial release of feces, at 23–28°C (Yukihira *et al.*, 1999). Their feces contained a high proportion of PIM, so these times reflect the initial rapid passage of indigestible particles.

There is no information on the digestion of the detritus, but there have been studies of digestion of phytoplankton in the field and microalgae in laboratory conditions. Some studies have shown that pearl oysters cannot or have limited ability to digest some components of the phytoplankton that they consume.

Chlorophytes, especially *Chlamydomonas* species, were relatively abundant in the atoll lagoon environment of *P. margaritifera* (Loret *et al.*, 2000a). This was reflected in the large numbers of chlorophyte cells in the oysters' proximal guts. However, there were large numbers of chlorophyte cells in the distal gut and living motile cells were passed out in the feces (Loret *et al.*, 2000a). Similarly, from the presence of intact autotrophic dinoflagellates in the feces, it appears that these too were poorly digested. High Performance Liquid Chromatography (HPLC) analysis of the break down of plant

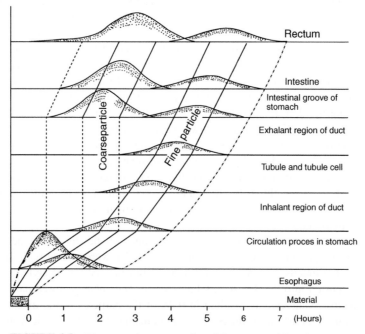

FIGURE 4.5 Diagrammatic representation of the passage of fine charcoal particles through the alimentary canal of an Akoya pearl oysters (from Kuwatani, 1965a) (with permission). Temperature = 21–22°C.

pigments through the gut suggested that the digestive efficiency of *P. margaritifera* in this environment was low. This is a surprising result considering that these oysters live in oligotrophic conditions (Loret *et al.*, 2000a).

4.2.7. Absorption

Absorption efficiency (AEff) is determined by comparing the ratio of organic matter in the feces to the ratio in the injected particles, from a concept developed by (Conover, 1966). The indigestible inorganic material is used as a "marker" for the organic material and its degree of absorption.

Feeding on a uniform diet of cultured microalgae (T-Iso) at a concentration of 5,000 cells/mL (=0.5 mg dwt/L), there was no relationship between AEff and body size in *Pinctada margaritifera* and *P. maxima* over the size range 0.5–9 g dwt (Yukihira *et al.*, 1998a). Despite this very uniform diet, AEff varied from about 20–95% in the data for both species. Mondal (2006) also found no relation between AEff and oyster size in Akoya and it ranged from 42 to 57%. On the other hand, there were strong inverse relationships between food concentration and AEff for *P. margaritifera* and *P. maxima* feeding on T-Iso and *Dunaliella primolecta* (a green flagellate). Mean AEff was about 70% at very low concentrations of T-Iso (<0.5 mg/L). It declined to about 6% in *P. margaritifera* and 15% in *P. maxima* at 10 mg/L T-Iso.

The highly significant inverse relationships are described by:

$$AEff = 67.39 \times 10^{-0.108FC} \text{ for } P. \text{ margaritifera}$$

$$AEff = 70.20 \times 10^{-0.072FC} \text{ for } P. \text{ maxima}$$

where FC is food concentration of T-Iso in mg/L

The strongly inverse relationships between AEff and FC on monospecific microalgae diets were reflected in the oysters feeding on SPM from ambient seawater (Yukihira *et al.*, 1999). AEffs for both species were about 65% in <0.5 mg/L dwt of SPM. They declined to 30% in *P. margaritifera* and 45% in *P. maxima* at 10 mg/L SPM (Table 4.4), which is interesting because these values are higher than at 10 mg/L of monospecific microalgae diets. There was <2 mg/L POM in the SPM at this level (Table 4.4), so the organic content of 10 mg/L SPM was much lower than in 10 mg/L unicellular algae. Not unexpectedly, it appears that AEff reflects the amount of POM ingested rather than total particulate matter ingested. There was >8 mg/L POM at 60 mg/L SPM and AEffs at this level declined to 10–15% (Table 4.4). AEff declined to this level at equivalent POM concentrations of unicellular microalgae (Yukihira *et al.*, 1998b). Hawkins *et al.* (1998) found results for *P. margaritifera*. They used natural SPM from seawater and related AEff to the proportion of POM in the SPM. There was a direct relationship and Hawkins *et al.* (1998) suggested that AEff was related to *proportion* of POM. AEff decreased as the proportion of POM decreased at the progressively higher SPM levels, but the *absolute concentration* of POM (mg/L) increased with SPM (Hawkins *et al.*, 1998). There was an inverse relationship between AEff and POM concentration, as found in other studies using cultured algae and natural POM.

4.2.8. Feces

Defecation is discharge of undigested materials that were never part of the body's metabolic processes (and may include undigested living organisms). It is distinct from excretion: the discharge of waste molecules that were part of the metabolic processes of the animal (Fig. 4.1, Section 4.3.2).

Yukihira *et al.* (1999) collected feces from pearl oysters that were feeding on natural particles to measure their organic content and obtain AEffs. They measured ingestion rates but did not consider defecation rates. However, defecation rates can be determined from ingestion rates and AEff values (Table 4.4). Variations in defecation rate resulted from a combination of the increase and then decline of ingestion rate with increasing POM and SPM levels, and the decline of AEff as POM and SPM increase. The major determinant was ingestion rate as changes in AEffs were more conservative. Defecation rate peaked at about 90 mg/h for *P. margaritifera* in the 5–10 mg/L SPM range. It peaked at about twice this rate in *P. maxima* at about 25 mg/L SPM (Table 4.4). The highest defecation rates were at or near the optimum ingestion rates and highest absorption rates.

P. margaritifera and *P. maxima* feeding cultured microalgae also showed increasing rates of defecation and then decline with increasing POM levels (Yukihira *et al.*, 1998b). There were the same trends of increasing ingestion rate and decreasing AEff as POM increased, resulting in progressively more feces production, until ingestion rate began to decline. There were peaks of 39 mg/h for 10 dwt *P. margaritifera* and 32 mg/h for *P. maxima* at about 4 mg/L T-Iso.

In the field, it is notable that 10 g dwt (ca. 150 mm SH) GBR *P. margaritifera* may produce a quarter of their dry weight in solid wastes each day as pseudofeces and feces at high levels of SPM (Table 4.4). *P. maxima* of 10 g dwt (ca. 180 mm SH) may produce a half of their dwt in solid wastes each day. These values relate to specific environmental conditions where this study was conducted, but they are indicative of potentially high levels of biodeposition at moderate SPM levels. These are the highest values estimated for large specimens of the two largest pearl oyster species and, as with other physiological parameters, pseudofeces production and defecation rate are strongly correlated with size (Yukihira *et al.*, 1998a; Pouvreau *et al.*, 2000b). However, they are exponentially correlated with size, with an exponent <1, and thus feces production and defecation are relatively greater in smaller oysters of these species (Yukihira *et al.*, 1998a).

In studying feces production in *P. margaritifera* in Takapoto atoll lagoon, Pouvreau *et al.* (2000b) began with the premise that fecal production reaches a plateau when the digestive system becomes full and reaches its capacity to process food. This was shown in *Crassostrea gigas* at SPM levels >90 mg/L (Barille *et al.*, 1997), but may not be relevant to the SPM levels of pearl oysters' environments. Following the plateau model for feces production, Pouvreau *et al.* (2000b) derived the relationship:

$$F = F_{max-1} \times W^{0.49} \times (1 - e^{-0.06SPM})$$

where F is feces production (mg/h), W is dry tissue weight (g)

There was a strong relationship between body size and feces production, with values ranging from 5 to 40 mg/h, according to size. The amount of feces production was never more than 20 mg/h g dwt and this was considered to be the plateau level. SPM had little effect, but this was in an environment where SPM levels never exceeded 2 mg/L. SPM values are far beyond 2 mg/L in the data for GBR *P. margaritifera* and *P. maxima* in Table 4.4 and there is no evidence of plateaus of feces production. As described, there is a peak and decline of feces production as SPM increases.

4.3. METABOLISM AND SCOPE FOR GROWTH

Little is known of the multitude of biochemical processes that comprise metabolism in pearl oysters and, for that matter, other bivalve molluscs. Most information on metabolism is for the processes that are readily measured: feeding, digestion and absorption, respiration, excretion, and changes in the proximate levels of protein, lipid and carbohydrate in the whole body or in various organs in relation to nutritional status and gametogenesis.

Studies of biochemical processes and component organic compounds have focused on the secretion and composition of the shell structures (see Section 3.3.3). There have been a number of studies of variations in the proximate levels of protein, lipid and carbohydrate through gametogenic cycles (see Section 5.5) and several studies have considered the influence of diet on the biochemical and energy composition of pearl oysters.

The topics of feeding, digestion and absorption are treated in Sections 4.1 and 4.2. The data on absorption are only in terms of general organic content, and hence energetic value, determined from the difference between the relative organic content of ingested particles and relative organic content of the feces. There are no observations on absorption of specific food components. In their modeling of carbon fluxes in *Pinctada margaritifera*, Pouvreau *et al.* (2000a) assumed that the absorption of protein, lipid and carbohydrate was unbiased, reflecting their proportions in the oyster's diet (0.5, 0.25 and 0.25, respectively).

4.3.1. Respiration

The energy-yielding oxidative processes of metabolism are usually aerobic in pearl oysters, although anaerobic metabolism may occur in some circumstances. Oxygen for aerobic metabolism is extracted from the water flowing through the mantle cavity and CO_2 is discharged. The gill filaments contain hollow tubes through which hemolymph flows and deoxygenated hemolymph thereby comes into close proximity with the seawater flowing over the filaments. There is reciprocal exchange of O_2 and CO_2 as they diffuse passively across the fine filament membranes according to the gradients between seawater and hemolymph. There is also some gas exchange between seawater and hemolymph across the large permeable surfaces of the mantle.

Table 4.5 shows data on the respiration rates of a number of pearl oyster species. The relationships between respiration rate (μL O_2/h) and SH (mm) of *Pinctada albina sugillata* over SH ranges 10.0–66.3 mm (from an pearl bank) and 22.5–57.4 mm (from a shore) were linear (Dharmaraj, 1983):

$$RR = 0.573 + 0.029 \times SH \text{ (shore oysters)}$$

However, respiration rate is not usually directly related to a linear body measurement, since it is related to metabolizing mass of tissue. In *P. margaritifera* and *P. maxima* there were exponential relationships between respiration rate ($\mu L\, O_2$/h) and dwt (g) (Robert *et al.*, 1998; Yukihira *et al.*, 1998a):

$$RR = 0.84 \times dwt^{0.72} \quad \text{\textit{P. margaritifera} (FrP)}$$

$$RR = 1.039 \times dwt^{0.642} \quad \text{\textit{P. margaritifera} (GBR)}$$

$$RR = 0.857 \times dwt^{0.561} \quad \text{\textit{P. maxima}}$$

See also Table 4.7, which shows the respiration rates of three sizes of *P. maxima* as energy units, J/h (see Section 4.3.3). It shows that respiration consumes a considerable percentage of the absorbed energy (AE), with the percentage decreasing with size.

4.3.2. Excretion

Excretion is usually a relatively minor component of pearl oyster metabolism, often <5% of the AE (Fig. 4.1, Table 4.6) (Mondal, 2006). Its major components are waste

TABLE 4.5

Respiration rates (RR) of *Pinctada* species. These values are derived from equations and figures in the sources.

Species	dwt (g)	Temperature (°C)	RR (μL/g/h)	Source
Akoya	4.0–4.3	21–23	5,000 (?)	Mori (1948)
Akoya	1	28	990[a]	Itoh (1976)
	3		936[a]	
Akoya (India)	ca. 2.3	ca. 28	580	Dharmaraj (1983)
	ca. 4.4	ca. 28	410	
P. albina sugillata	ca. 3.9	ca. 28	300	Dharmaraj (1983)
P. margaritifera (GBR)	1	28 ± 1	1,040	Yukihira *et al.* (1998a)
	10		290	
P. margaritifera (FrP)	1	28–30	840	Robert *et al.* (1998)
	10		440	
P. maxima	1	28 ± 1	860	Yukihira *et al.* (1998a)
	10		310	

dwt: dry soft tissue weight; GBR: Great Barrier Reef; FrP: French Polynesia.
[a]Converting 1 μg atom of oxygen to 12.37 μL of gaseous O_2 at 28°C.

nitrogen molecules from protein catabolism, mainly NH_3/NH_4^+, with some amino acids, urea and other minor components. Wu *et al.* (1997) also found substantial amounts of NO_3–N and NO_2–N in excretions of Akoya, which is unusual for bivalves.

Estimates of nitrogen excretion in some *Pinctada* species are of the same order, except for those reported by Saucedo *et al.* (2004) (Table 4.6). There was an approximately linear relationship between excretion rate (ER) and body size in Akoya feeding on natural particles. Excreted NH_4–N was proportional to (wet tissue weight)$^{0.941}$ (Itoh, 1976). ER was proportional to dwt$^{0.78}$ in *P. margaritifera* in FrP (Robert *et al.*, 1998). There were linear relationships between excretion and body size in GBR *P. margaritifera* and *P. maxima* feeding on uniform concentrations of cultured algae (Yukihira *et al.*, 1998a):

$$ER (\mu g/h) = 81.37 \times 0.642 \text{ dwt (g)} \quad \text{for } P. \text{ margaritifera}$$

$$ER (\mu g/h) = 72.83 \times 0.789 \text{ dwt (g)} \quad \text{for } P. \text{ maxima}$$

ER rates were very variable, however, when these two species were feeding on natural particles (Yukihira *et al.*, 1999). In these short-term studies with healthy oysters, there were no consistent changes in the rate of protein catabolism under near starvation and starvation conditions (low and negative scope for growths, SFGs). This is not surprising in the short term, but ER will probably increase as rates of protein catabolism increase during prolonged conditions of negative SFG (see Section 6.2.2).

Linked with nitrogen excretion, there were losses of carbon as amino acids and urea (Méro, 1996; Pouvreau *et al.*, 1998, 2000a). These organic molecules accounted for less than 10% of total nitrogen excreted and increased with dwt according to:

$$ER (\text{mg carbon/day}) = 0.02 \times \text{dwt}^{0.78} (g)$$

TABLE 4.6
Rates of nitrogen excretion in *Pinctada* species.

Species	dwt (g)	Temperature (°C)	Nitrogen excretion (µg/g h)	Percent of AE	Source
Akoya	1	28	57[a]	NA	Itoh (1976)
	3		51[a]	NA	
P. mazatlanica	NA	23–33	0.44–1.6[b]	3.5–7.5	Saucedo *et al.* (2004)
P. margaritifera and	0.1	28 ± 1	80	2–4	Yukihira *et al.* (1998a)
P. maxima (GBR)	1		70	3	
	10		19	3–5	
P. margaritifera (FrP)	1	28–30	29[b]	ca. 2	Pouvreau *et al.* (1998),
	5		20[b]	ca. 2	Robert *et al.* (1998)

AE: absorbed energy; dwt: dry soft tissue weight; GBR: Great Barrier Reef; FrP: French Polynesia.
[a]Converting 1 µg atom of NH_3–N to 14 µg Nitrogen.
[b]Converting 1 µmole of NH_3/NH_4–N to 17.5 µg Nitrogen.

For a 1 g dwt oyster, this amounts to excreting 0.02 mg of carbon/day, representing <0.1% of carbon intake per day (calculated from Pouvreau et al., 2000a).

4.3.3. Scope for growth

The concept of a balance between energy uptake and energy expenditure, with any positive balance being available for expenditure on growth and reproduction, has general application for animals. However, it has been particularly used in bivalves:

$$SFG\ (J/h) = AE - (RE + EE)\ \text{e.g., Widdows (1985)}$$

where SFG, scope for growth is surplus energy available for growth and reproduction; AE is absorbed energy, which is the difference between ingested and fecal energy; RE is respired energy or energy cost of metabolism; and EE is excreted energy.

During aerobic respiration the rates of oxygen consumption and carbon dioxide release directly reflect the animal's metabolic rate and oxygen consumption is usually measured. Energy production (J) per unit oxygen consumption depends on which kinds of molecules are oxidized as the source of energy. Yukihira et al. (1998a) used a conversion factor 1 mL O_2 = 20.33 J, following Crisp (1971).

Yukihira et al. (1998b) determined the SFG of *P. margaritifera* and *P. maxima* using a standard microalgae diet, 5,000 cells/mL of T-Iso. This cell concentration was below the level that pseudofeces were produced, so the rate of particle ingestion was equal to particle CR. Table 4.7 shows the SFG values for three sizes of *P. maxima* derived from the three components, AE, RE and EE. Similar data for *P. margaritifera* are also presented by Yukihira et al. (1998a). As described above, AEff was the only component parameter of SFG that was unrelated to size and a mean value was used. Also as described above, the parameters CR, RE and EE per unit body mass were exponentially correlated with size, with exponents being less than 1. The fact that CR declined exponentially with size and AEff was independent of size result in AE per unit body mass declining exponentially with size.

RE decreased as a percentage of AE as pearl oyster size increased and, since RE was the major component of energy expenditure, SFG increased with pearl oyster size as a percentage of AE, i.e. from 61 to 71% over the size range 0.1 to 10 g dwt (Table 4.7). However, because CR, and hence AE, only increased by about 20-fold over this 100-fold increase in dwt, there was only a 23-fold increase in SFG corresponding to the 100-fold increase in body size (Table 4.7). The energy surplus decreased from 8.8% of body energy per day in 0.1 g dwt individuals to 1.5% of body energy per day in 10 g dwt individuals in *P. maxima* (Table 4.7). Similarly, the energy surplus in *P. margaritifera* decreased from 5.3% to 1.8% of body energy per day over this size range (Yukihira et al., 1998a). This pattern of relative SFG being inversely related to size underlies higher relative growth rates in small pearl oysters compared with larger oysters in similar environmental conditions. Furthermore, SFG is divided between gametogenesis, and

TABLE 4.7

Scope for Growth of three sizes of *Pinctada maxima* feeding on *Isochrysis* aff. *galbana* Tahitian at 5,000 cells m/L (Yukihira *et al.*, 1998a) (with permission). Italicised numbers in brackets represent the % of total AE.

Parameter	Body size (dwt, g)		
	0.1	1	10
CR (L/h)	2.8	11.5	47.1
AEff (%)	58	58	58
AE (J/h)	13.2	58.9	263.1
RE (J/h)	4.8 (*36*)	17.4 (*29*)	63.4 (*23*)
EE (J/h)	0.3 (*2*)	1.8 (*3*)	12.2 (*5*)
SFG = AE − (RE + EE) (J/h)	8.1 (*61*)	39.7 (*67*)	187.5 (*71*)
SFG as percentage per day increase in body energy	8.8	3.8	1.5

AEff: absorption efficiency; AE: absorbed energy; CR: clearance rate; dwt: dry soft tissue wt; EE: excreted energy; RE: respired energy; SFG: scope for growth.

growth of somatic tissue and shell in larger pearl oysters, which further reduces their relative growth rates.

Pouvreau *et al.* (2000a) developed a sophisticated estimation of SFG for *P. margaritifera*. They considered carbon fluxes instead of energy fluxes and developed a model that incorporated pearl oyster size and variability in environmental factors in Takapoto lagoon, FrP. The SFG model was based on the standard concept of a surplus resulting from the balance of uptake versus expenditure, but allometric functions based on PIM, POC and dwt were developed for each of the SFG components. SFG, as net caron gained, was then partitioned between somatic tissue, gonad and shell growth, based on the asymptotic increase in reproductive expenditure with age. The excellent model was validated through providing realistic predictions of growth of somatic tissue, gonad and shell for pearl oysters from 1 to 4 years in the Takapoto lagoon (Pouvreau *et al.*, 2000a).

In environmental conditions where the pearl oyster cannot obtain sufficient food particles for AE to exceed RE + EE, e.g. in low POM or indigestible SPM conditions, SFG will be negative and the animal loses body weight and may ultimately die (see Section 6.2.2). There were brief periods of negative SFG for *P. margaritifera* in an atoll lagoon (Pouvreau *et al.*, 2000a). Yukihira *et al.* (1999) found negative SFG values for *P. margaritifera* and *P. maxima*, especially the former, feeding on natural particles composed of very high PIM levels.

There is disagreement over the energetics of pearl oysters. Hawkins *et al.* (1998) considered them to have poor rates of energy uptake, based on the rate of absorption of POM in *P. margaritifera* relative to four other tropical bivalves. This explained for them the poor growth rates of pearl oysters. However, Yukihira *et al.* (1998a) and Pouvreau *et al.* (1999b) considered pearl oysters to be among the most energetic bivalves. In CR data for 27 bivalves from all major groups, standardized to 1 g dwt, the 11.5 L/h rates of *P. margaritifera* and *P. maxima* from the GBR exceeded all other tropical and temperate bivalves (Yukihira *et al.*, 1998a; Table 5). The CR value of 25.9 L/h for FrP *P. margaritifera*, found by Pouvreau *et al.* (1999b), is twice as high. The only bivalve approximating even the CR rates of the GBR pearl oysters was a scallop, *Argopecten irradians* (Bricelj and Shumway, 1991). In standardized respiration data for 25 bivalves, similar rates to *P. margaritifera* and *P. maxima*, 1.04 and 0.86 mL O_2/h, respectively, have only been recorded in the giant clam, *Tridacna gigas*, and *A. irradians* (Bricelj *et al.*, 1987; Klumpp and Griffiths, 1994). In standardized SFG data for 14 bivalves, similar or greater levels to *P. margaritifera* and *P. maxima*, 36–40 J/h, have only been recorded in *T. gigas* (Klumpp and Griffiths, 1994), and the mussels *Mytilus edulis*, under certain conditions (Tedengren *et al.*, 1990; Okumus and Stirling, 1994), and *M. galloprovincialis* (Labarta *et al.*, 1997). It must be noted that SFG is very dependent on environmental conditions, especially food, as is evident in the data for *M. edulis* where recorded SFG values range from 7.3 to 72.6 J/h (Yukihira *et al.*, 1998a, Table 5). The influences of environmental factors on SFG of *P. margaritifera* and *P. maxima*, and other pearl oyster species, are considered in Chapter 6.

4.4. SUMMARY

Most research in this field has been conducted on *P. margaritifera*, *P. maxima* and Akoya pearl oysters. Insofar as these three are representative, and it is likely that they are, pearl oysters are typical bivalve molluscs in the general morphology and functioning of their feeding, ingestion, digestion, defecation and excretion. There are, however, some notable features. *P. margaritifera* has relatively large gill sizes and surface areas, which result in exceptional PRs through the mantle cavity. *P. maxima* also has high PRs and these two large species may filter thousands of litres per week. Akoya do not match these PRs, even on a per unit weight basis. The three species are able to capture SPM down to 2 μm with at least 50% efficiency, but *P. margaritifera* is notably able to capture particles down to 1.5 μm, i.e. bacteria and the smallest microalgae. Some results show that, unlike many bivalves, the species are unable to sort organic and inorganic particles at the palps and discharge unsorted pseudofeces in high SPM conditions. In terms of responses to various levels of SPM, there are interspecific differences. The feeding mechanisms of *P. maxima* and *P. margaritifera* reflect adaptation to their favored environments: high SPM, including silt, and low SPM, respectively.

There are virtually no data on the metabolism of pearl oysters *per se*, but there are good data on energy metabolism. Energy metabolism is summarized here using the equation: Scope for growth (SFG) = Absorbed energy (AE) − Respired energy (RE) − Excreted energy (EE). SFG is energy available for both growth and reproduction. AE and hence SFG are strongly influenced by SPM, including silt levels: there are levels of SPM above which intake and AEff decline. AE and RE levels per unit body weight decline exponentially with size, while EE is a very minor component. The outcome is that, while SFG increases with body size, it decreases relative to size. Thus, smaller pearl oysters grow at a relatively faster rate than larger oyster. The overall "picture" that emerges for *P. margaritifera* and *P. maxima* is of highly energetic bivalves for which their "competitors" in the bivalve energetics stakes are amongst other fast-growing commercial bivalves.

References

Barille, L., Prou, J., Heral, M., Razet, D., 1997. Effects of high natural seston concentrations on the feeding, selection, and absorption of the oyster *Crassostrea gigas* (Thunberg). *J. Exp. Mar. Biol. Ecol.* 212, 149–172.

Barille, L., Haure, J., Cognie, B., Leroy, A., 2000. Variations in pallial organs and eulatero-frontal cirri in response to high particulate matter concentrations in the oyster *Crassostrea gigas*. *Can. J. Fish. Aquat. Sci.* 57, 837–843.

Bricelj, V.M., Shumway, S.E., 1991. Physiology: Energy acquisition and utilizaiton. In: Shumway, S.E. (Eds.), *Scallops: Biology, Ecology and Aquaculture. Developments in Aquaculture and Fisheries Science*, Vol. 21. Elsevier, Amsterdam, pp. 305–346.

Bricelj, V.M., Epp, J., Malouf, R.E., 1987. Comparative physiology of young and old cohorts of bay scallop *Argopecten irradians irradians* (Lamarck): Mortality, growth, and oxygen consumption. *J. Exp. Mar. Biol. Ecol.* 112, 73–91.

Chang, M., Hong, J.S., Huh, H.T., 1988. Environmental conditions in the pearl oyster culture grounds and food organisms of *Pinctada fucata martensii* (Dunker) (Bivalvia, Pterioida). *Ocean Res.* 10, 67–77. (in Korean, with English Abstract)

Chellam, A., 1983. Study on the stomach contents of pearl oyster *Pinctada fucata* (Gould) with reference to the inclusion of bivalve eggs and larvae, *Symposium on Coastal Aquaculture*, Cochin, January 12–14, 1980. Marine Biological Association of India, Cochin, India, pp. 604–607.

Conover, R.J., 1966. Assimilation of organic matter by zooplankton. *Limnol. Oceanogr.* 11, 338–345.

Crisp, D.J., 1971. Energy flow measurements. In: Holme, N.A., McIntyre, A.D. (Eds.), *Methods for the Study of Marine Benthos, International Biological Progress 16*. Blackwell, Oxford, pp. 197–323.

Delesalle, B., Sakka, A., Legendre, L., Pages, J., Charpy, L., Loret, P., 2001. The phytoplankton of Takapoto Atoll (Tuamotu Archipelago, French Polynesia): Time and space variability of biomass, primary production and composition over 24 years. *Aquat. Living Resour.* 14, 175–182.

Dharmaraj, S., 1983. Oxygen consumption in pearl oyster *Pinctada fucata* (Gould) and *Pinctada sugillata* (Reeve). *Proc. Symp. Coast. Aquac.* 2, 627–632.

Drent, J., Luttikhuizen, P.C., Piersma, T., 2004. Morphological dynamics in the foraging apparatus of a deposit feeding marine bivalve: Phenotypic plasticity and heritable effects. *Funct. Ecol.* 18, 349–356.

Forster-Smith, R.L., 1975. The effect of concentration of suspension on filtration rates and pseudofaecal production of *Mytilus edulis* L., *Cerastoderma edule* (L.) and *Venerupis pullastra* (Montagu). *J. Mar. Biol. Assoc. UK* 17, 1–22.

Franz, D.R., 1993. Allometry of shell and body weight in relation to shore level in the intertidal bivalve *Geukensia demissa* (Bivalvia: Mytilidae). *J. Exp. Mar. Biol. Ecol.* 174, 193–207.

Gervis, M.H., Sims, N.A., 1992. *The Biology and Culture of Pearl Oysters (Bivalvia: Pteriidae)*, International Center for Living Aquatic Resources Management, Manila, 49 pp.

Gosling, E., 2003. *Bivalve Molluscs: Biology, Ecology and Cultivation*, Fisheries News Books, Blackwell Publishing, Oxford, 443 pp.

Hawkins, A.J.S., Smith, R.F.M., Tan, S.H., Yasin, Z.B., 1998. Suspension-feeding behaviour in tropical bivalve molluscs: *Perna viridis, Crassostrea belcheri, Crassostrea iradelei, Saccostrea cucculata* and *Pinctada margarifera*. *Mar. Ecol. Prog. Ser.* 166, 173–185.

Honkoop, P.J.C., Bayne, B.L., Drent, J., 2003. Flexibility of size of gills and palps in the Sydney rock oyster *Saccostrea glomerata* (Gould, 1850) and the Pacific oyster *Crassostrea gigas* (Thunberg, 1793). *J. Exp. Mar. Biol. Ecol.* 282, 113–133.

Itoh, K., 1976. Relation of oxygen consumption and ammonia nitrogen excreted to body size and to water temperature in the adult pearl oyster *Pinctada fucata* (Gould). *Kokoritsu Shinju Kenkyusho Hokoku (Bull. Natl. Pearl Res. Lab.)* 20, 2254–2275. (in Japanese, with English summary)

Klumpp, D.W., Griffiths, C.L., 1994. Contributions of phototrophic and heterotrophic nutrition to the metabolic and growth requirements of four species of giant clam (Tridacnidae). *Mar. Ecol. Prog. Ser.* 115, 103–115.

Kuwatani, Y., 1965a. A study on feeding mechanism of Japanese pearl oyster, *Pinctada martensii* (Dunker), with special reference to passage of charcoal particles in the digestive system. *Nippon Suisan Gakkaishi (Bull. Jap. Soc. Sci. Fish.)* 31, 789–798. (in Japanese, with English abstract)

Kuwatani, Y., 1965b. On the anatomy and function of the stomach of Japanese pearl oyster, *Pinctada martensii* (Dunker). *Nippon Suisan Gakkaishi (Bull. Jap. Soc. Sci. Fish.)* 31, 174–186. (in Japanese, with English abstract)

Labarta, U., Fernandez-Reiriz, M.J., Babarro, J.M.F., 1997. Differences in physiological energetics between intertidal and raft cultivated mussels *Mytilus galloprovincialis*. *Mar. Ecol. Prog. Ser.* 152, 167–173.

Li, H., He, H., Jin, Q., 1999. Trophic niches and potential yields of three economic bivalves in Daya Bay. *Tropic Oceanol. (Redai Haiyang, Guangzhou)* 18, 53–60. (in Chinese, with English Abstract)

Li, H., Jin, Q., Guo, C., Lian, J., 1997. Study on the feeding rate and feeding rhythm of larvae and juveniles of pearl oyster, *Pinctada martensii*. *Tropic Oceanol (Redai Haiyang, Guangzhou)* 16, 41–48. (in Chinese, with English Abstract)

Loret, P., Pastoureaud, A., Bacher, C., Delesalle, B., 2000a. Phytoplankton composition and selective feeding of the pearl oyster *Pinctada margaritifera* in the Takapoto lagoon (Tuamotu Archipelago, French Polynesia): *In situ* study using optical microscopy and HPLC pigment analysis. *Mar. Ecol. Prog. Ser.* 199, 55–67.

Loret, P., Le Gall, S., Dupuy, C., Blanchot, J., Pastoureaud, A., Delesalle, B., Caisey, X., Jonquieres, G., 2000b. Heterotrophic protists as a trophic link between picocyanobacteria and the pearl oyster *Pinctada margaritifera* in the Takapoto lagoon (Tuamotu Archipelago, French Polynesia). *Aquat. Microb. Ecol.* 22, 215–226.

MacDonald, B.A., Ward, J.E., 1994. Variation in feeding behavior of two subtropical bivalves in response to acute increases in sediment load. *J. Shellfish Res.* 13, 289–290.

Méro, D., 1996. *Contribution á l'étude de l'excrétion azotée de l'huître perliére Pinctada margaritifera.* Cergy Pontoise, France, 45 pp. (with English Abstract)

Mondal, S.K., 2006. Effects of temperature and body size on food utilisation in the marine pearl oyster *Pinctada fucata* (Bivalvia: Pteriidae). *Indian J. Mar. Sci.* 35, 43–49.

Niquil, N., Pouvreau, S., Sakka, A., Legendre, L., Addessi, L., Le Borgne, R., Charpy, L., Delesalle, B., 2001. Trophic web and carrying capacity in a pearl oyster farming lagoon (Takapoto, French Polynesia). *Aquat. Living Resour.* 14, 165–174.

Numaguchi, K., 1985. Studies on phaeophytin contained in the digestive diverticula of pearl oyster, *Pinctada fucata martensii*, as index of feeding of phytoplankton. *Bull. Natl. Res. Inst. Aquacult. (Japan)* 7, 91–96. (in Japanese, with English Abstract)

Numaguchi, K., 1994a. Fine particles of the suspended solids in the pearl farm. *Suisan Kogaku (Fish. Engin.)* 30, 181–184. (in Japanese with English Abstract)

Numaguchi, K., 1994b. Effect of water temperature on the filtration rate of Japanese pearl oyster, *Pinctada fucata martensii*. *Suisanzoshoku (J. Jap. Aquacult. Soc.)* 42, 1–6. (in Japanese with English Abstract)

Numaguchi, K., 1996. A review on the feeding ecology and food environment of the Japanese pearl oyster, *Pinctada fucata martensii*. *Bull. Natl. Res. Inst. Fish. Sci.* 8, 123–138. (in Japanese, with English Abstract)

Numaguchi, K., 2000. Evaluation of five microalgal species for the growth of early spat of the Japanese pearl oyster *Pinctada fucata martensii*. *J. Shellfish Res.* 19, 153–157.

Okumus, I., Stirling, H.P., 1994. Physiological energetics of cultivated mussel (*Mytilus edulis*) populations in two Scottish west coast sea lochs. *Mar. Biol. (Berlin)* 119, 125–131.

Pouvreau, S., Bodoy, A., Buestel, D., 1998. Détermination du bilan énergétique chez l'huître perlière, *Pinctada margaritifera*, et premier modèle écophysiologique de croissance dans le lagon d'atoll de Takapoto. Rapport RI DRV 98-01. IFREMER, Tahiti, French Polynesia, pp. 69.

Pouvreau, S., Jonquieres, G., Buestel, D., 1999a. Filtration by the pearl oyster, *Pinctada margaritifera*, under conditions of low seston load and small particle size in a tropical lagoon habitat. *Aquaculture* 176, 295–314.

Pouvreau, S., Tiapari, J., Gangnery, A., Garnier, M., Lagarde, F., Robert, S., Jonquieres, G., Teissier, H., Prou, J., Bennett, A., Caisey, X., Haumani, G., Buestel, D., Bodoy, A., 1999b. Energy budget of the black-lip pearl oyster, *Pinctada margaritifera*, in the lagoon of Takapoto, World Aquaculture '99. *The Annual International Conference and exposition of the World Aquaculture Society*, April 26 to May 2, 1999 Sydney, Australia. World Aquaculture Society, Sydney, p. 609.

Pouvreau, S., Bacher, C., Heral, M., 2000a. Ecophysiological model of growth and reproduction of the black pearl oyster, *Pinctada margaritifera*: potential applications for pearl farming in French Polynesia. *Aquaculture* 186, 117–144.

Pouvreau, S., Bodoy, A., Buestel, D., 2000b. *In situ* suspension feeding behaviour of the pearl oyster, *Pinctada margaritifera*: combined effects of body size and weather-related seston composition. *Aquaculture* 181, 91–113.

Robert, S., Pouvreau, S., Tiapari, J., Bennett, A., Caisey, X., Jonquieres, G., Teissier, H., Mero, D., Goulletquer, P., Haumani, G., Buestel, D., Bodoy, A., 1998. Estimation of oxygen consumption and nitrogenous excretion of the black-lip pearl oyster, *Pinctada margaritifera*, in relation to its body size. *Aquaculture Europe 98*, October 7–10, 1998, Bordeaux, France (Abstract only).

Sato, T., Matsumoto, S., Horiguchi, Y., Tsugii, T., 1964. Filtering and feeding rate of the pearl oyster *Pteria (Pinctada) martensii* Dunker determined by crude silicate as an indicator. *Nippon Suisan Gakkaishi (Bull. Jap. Soc. Sci. Fish.)* 30, 717–722. (in Japanese with English summary)

Saucedo, P.E., Ocampo, L., Monteforte, M., Bervera, H., 2004. Effect of temperature on oxygen consumption and ammonia excretion in the Calafia mother-of-pearl oyster, *Pinctada mazatlanica* (Hanley, 1856). *Aquaculture* 229, 377–387.

Siebers, D., Winkler, A., 1984. Amino-acid uptake by mussels, *Mytilus edulis*, from natural sea water in a flow-through system. *Helgol. Meeresunt.* 38, 189–199.

Sournia, A., Ricard, M., 1976. Données sur l'hydrologie et la productivité du lagon d'un atoll fermé (Takapoto, Iles Tuamotu). *Vie Milieu* 26, 243–279. (English Summary)

Sukhotin, A.A., Lajus, D.L., Lesin, P.A., 2003. Influence of age and size on pumping activity and stress resistance in the marine bivalve *Mytilus edulis* L. *J. Exp. Mar. Biol. Ecol.* 284, 129–144.

Tedengren, M., Andre, C., Johannesson, K., Kautsky, N., 1990. Genotypic and phenotypic differences between Baltic and North Sea populations of *Mytilus edulis* evaluated through reciprocal transplantations. III: Physiology. *Mar. Ecol. Prog. Ser.* 59, 221–227.

Tomaru, Y., Udaka, N., Kawabata, Z., Nakano, S., 2002. Seasonal change of seston size distribution and phytoplankton composition in bivalve pearl oyster *Pinctada fucata martensii* culture farm. *Hydrobiologia* 481, 181–185.

Uyeno, F., Inouye, H., 1961. Relationships between basic production of foods and oceanographical conditions of seawater in pearl farms, with special reference to overcrowded culture and food chains around pearl oysters. *Kokoritsu Shinju Kenkyusho Hokoku (Bull. Natl. Pearl Res. Lab.)* 7, 829–864.

Ward, J.E., MacDonald, B.A., 1996. Pre-ingestive feeding behaviors of two sub-tropical bivalves (*Pinctada imbricata* and *Arca zebra*): Responses to an acute increase in suspended sediment concentration. *Bull. Mar. Sci.* 59, 417–432.

Ward, J.E., Shumway, S.E., 2004. Separating the grain from the chaff: particle selection in suspension- and deposit-feeding bivalves. *J. Exp. Mar. Biol. Ecol.* 300, 83–103.

Widdows, J., 1985. Physiological procedures. Chapter 7 In: Bayne, B.L., Brown, D.A., Burns, K., Dixon, D.R., Ivanovici, A., Livingstone, D.R., Lowe, D.M., Moore, M.N., Stebbing, A.R.D., Widdows, J. (Eds.), *The Effects of Stress and Pollution on Marine Animals*. Praeger Press, New York, pp. 161–178.

Wu, M.C., Gao, H., Ding, M., Xie, Y., Chang, S., Qian, P., Wu, C., 1997. The growth promoting effect of metabolites of *Pinctada martensii* on *Kappaphycus alvarezii*. *Oceanol. Limnol. Sinica* 28, 453–457. (in Chinese with English Summary)

Yukihira, H., Klumpp, D.W., Lucas, J.S., 1998a. Effects of body size on suspension feeding and energy budgets of the pearl oysters *Pinctada margaritifera* and *P. maxima*. *Mar. Ecol. Prog. Ser.* 170, 119–130.

Yukihira, H., Klumpp, D.W., Lucas, J.S., 1998b. Comparative effects of microalgal species and food concentration on suspension feeding and energy budgets of the pearl oysters *Pinctada margaritifera* and *P. maxima* (Bivalvia: Pteriidae). *Mar. Ecol. Prog. Ser.* 171, 71–84.

Yukihira, H., Klumpp, D.W., Lucas, J.S., 1999. Feeding adaptations of the pearl oysters Pinctada *margaritifera* and *P. maxima* to variations in natural particulates. *Mar. Ecol. Prog. Ser.* 182, 161–173.

Yukihira, H., Lucas, J.S., Klumpp, D.W., 2006. The pearl oysters, *Pinctada maxima* and *P. margaritifera*, respond in different ways to culture in dissimilar environments. *Aquaculture* 252, 208–224.

CHAPTER 5

Reproduction, Development and Growth

Pedro E. Saucedo and Paul C. Southgate

5.1. INTRODUCTION

There have been several major revisions of the reproductive biology and physiology of commercially important families within the class Bivalvia, including Ostreidae, Mytilidae and Pectinidae (Bayne, 1976; Seed, 1976; Andrews, 1979; Giese and Pearse, 1979; Sastry, 1979; Mackie, 1984; Román et al., 2002; Barber and Blake, 2006). For the Pteriidae, Gervis and Sims (1992) presented a compilation of the biology, ecology and culture of pearl oysters, but provided only basic information relating to reproduction, and included only members of the genus *Pinctada*.

In the Pteriidae, the basic reproductive cycles of the major commercial species have been studied since the 1900s using smears and histological or histochemical examination of the gonad (Table 5.1). We know from these studies that most aspects of the reproductive cycle are common to all members of the genus *Pinctada*, with temporal differences attributable primarily to latitudinal and habitat variations among species, and among populations of the same species. However, many aspects of the reproductive biology and factors affecting reproductive effort and success are poorly understood for the Pteriidae despite the importance of these issues to the pearling industry.

Hatchery culture methods for the major commercial species within the Pteriidae are well established (see Section 7.4) and this has allowed greater understanding of development and growth of these species. For hatchery-based pearl operations and those that rely on wild spat collection, the growth rate of juvenile oysters determines the

TABLE 5.1

Historical review of studies of the reproductive cycles of pearl oysters.

Species	Locality	Method	Summary of reproductive events	Reference
Pinctada fucata	India, Sri Lanka	NS SM	Maturity reached during 1st year; continuous breeding over the year with 2 peaks: May–July and November–January.	Herdam and Hornell (1906)
	India, Sri Lanka	NS	Sexual maturity seen by 1st year; continuous spawning throughout the year with 2 peaks: April–May and September–October.	Hornell (1922)
	Australia	MS-3 HA VI	Maturity attained within 6 months; breeding from December through May with a major spawning peak in April–May and a minor peak in January–February; protandric and protogynic sex change observed, together with 1 bisexual condition.	Tranter (1959)
	India	MS-1 BA	Continuous breeding with 2 spawning peaks (March and November); glycogen and lipids rise before spawning and decline after it; proteins increase only during June–September.	Desai *et al.* (1979)
	India	MS-2 HA	Maturity reached within 7–8 months; multiple spawnings during the year; females dominate at size of first maturation and larger sizes.	Chellam (1987)
	Iran	MS-1 HA	Marked protandrous behavior, bisexuality uncommon; bimodal cycle with peaks in April–July and September–December; spawning occurs at 28°C and over.	Behzadi *et al.* (1997)
	Iran	MS-1 HA	Active breeding from spring through summer with 2 peaks: June–July and November–December; spawning occurs between 25–26°C.	Jamili *et al.* (1999)
Pinctada fucata martensii	Japan	MS-1 HA	Full ripeness within 1st year; continuous breeding over the year with 1 peak in summer; low tendency for bisexuality.	Ojima and Maeki (1955)
	Japan	MS-1 HA	Continuous breeding throughout the year with a major peak in winter (21°C) and a minor peak in summer (28°C); few cases of bisexuality observed.	Wada *et al.* (1995)
	Korea	MS-1 HA	Breeding season extends from April through August, with 1 spawning peak in June–July when temperature rises to 21°C.	Choi and Chang (2003)
Pinctada albina	Australia	MS-3 HA	Description of primary male and female gonad development; color differences of male and female gonads established; full ripeness within 4–5 months.	Tranter (1958a)
	Australia	FS-1 MS-3 HA	Sexual maturity occurs within 4–5 months; continuous breeding over the year with 1 spawning peak in autumn.	Tranter (1958b)
	Australia	FS-1 MS-3 HA	Continuous breeding over the year with 2 spawning peaks, one major in April–May and a minor in November; some cases of bisexuality; sex change discussed.	Tranter (1958c)

(Continues)

TABLE 5.1 (*Continued*)

Species	Locality	Method	Summary of reproductive events	Reference
Pinctada albina sugillata	Australia	MS-2 HA	Sexual maturity seen by 1st year (41 mm); continuous breeding cycle with peaks in October, March, January, and April; male predominated below 71 mm; few cases of hermaphroditism, reproductive condition improves at 17–18°C.	O'Connor (2002)
Pinctada margaritifera	Red Sea	NS HA	Sexual maturity seen by 2nd year; breeding season extends from March to September with 1 spawning peak in October.	Crossland (1957)
	Australia	MS-3 HA	Maturity reached until 2nd year; continuous breeding over the year with 2 peaks: July–August and November–December; some cases of protandry and protogyny observed.	Tranter (1958d)
	French Polynesia	MS-1 HA UA	Ultrastructural study of spermatogenesis; analysis of stem cell differentiation and sexual cells development; no data of breeding season or provided.	Thielley *et al.* (1993a)
	French Polynesia	MS-1 VI HA, HiA	Cytological description of spermatogenesis and oogenesis; marked protandry tendency; sex reversal analyzed; no data of breeding season provided.	Thielley *et al.* (1993b)
	French Polynesia	BS-1 HA	Maturity seen by 1st year (40 mm); breeding season from November through May and 2 spawning peaks in oysters <100 mm and 5 peaks in oysters >100 mm.	Pouvreau *et al.* (2000b)
	Australia	MS-1 HA	Continuous breeding over the year with a sampling peak from August to February; spawning occurs between 26°C and 28°C sex ratio skewed towards maleness.	Acosta-Salmón and Southgate (2005)
Pinctada maxima	NW Australia	NS HA	Sexual maturity attained by the end of 1st year; marked protandrous tendency; spawning occurs all year round, with a peak in September–November between 27°C and 29°C.	Wada (1953)
	Australia	FS-6 HA VI	Maturity occurs by the end of 1st year; bisexuality was uncommon; breeding season from September through April with multiple spawnings, a major peak in October and a minor peak in April.	Rose *et al.* (1990)
Pinctada mazatlanica	Mexico to Panama Mexico	MS-1 HA	The species spawns once a year in summer (July–August) Breeding season from April to September with a minor spawning in May and a major peak in September; spawning temperature of 28–29°C.	Sevilla (1969)
	Mexico	MS-1 HA	Spawning occurs throughout the year with 3 peaks: June–July, October, and April. Sequential sex change seen; spawning temperature varied from 28°C to 33°C.	García-Domínguez *et al.* (1996)
	Mexico	MS-1 HA	Maturity seen by 1st year (35 mm); breeding season from April–August with a peak in September; sex change occurs after 100 mm; spawning occurs at 28–29°C	Saucedo and Monteforte (1997)
	Costa Rica	MS-1 HA	Sexual maturity reached by 1st year (45 mm); multi-spawning behavior with 2 clear peaks: October and June; spawning is triggered at 25–26°C.	Solano-López *et al.* (1997)

(*Continues*)

TABLE 5.1 *(Continued)*

Species	Locality	Method	Summary of reproductive events	Reference
	Mexico	MS-1 HA BA	Active breeding from March to August with a major peak in April and a minor in August; massive spawning in summer (September); spawning temperature of 28–29°C.	Saucedo *et al.* (2002a)
	Mexico	FS-1 HiA	Breeding season from March to August with a major peak in April and minor peak in August; protandrous and protogynic sex change found; bisexuality is discussed.	Saucedo *et al.* (2002b)
	Mexico Mexico	HA SS-1 HA BA	Sampling over 18 months; longer reproductive cycle during El Niño; spawning occurred from July to November during El Niño and from August to October during La Niña. Intense breeding from April to September with a spawning peak in summer; population spawning at 29.5°C.	García-Cuellar *et al.* (2004) Vite-García (2005)
	Mexico	SS-1 HiA DA	Oocyte growth and quality assessed; lipid and protein indices measured; data show 2 breeding peaks over the year (April and August) and 1 spawning peak (August).	Gómez-Robles *et al.* (2005)
Pinctada imbricata	Australia	MS-2 HA	Breeding season from spring to autumn with 2 peaks: a major in November–December and a minor in March–April; male sex predominated below 71 mm shell height.	O'Connor and Lawler (2004)
Pteria penguin	Thailand	MS-1 HA	Breeding season throughout the year with 2 peaks for males (March–June and December–January) and 1 peak for females (June–July).	Arjarasirikoon *et al.* (2004)
Pteria sterna	Mexico	MS-1 HA	Continuous breeding from October through May with multiple spawnings and 2 clear peaks: December–January and April–May.	Arizmendi-Castillo (1996)
	Mexico	BS-1 HA	Maturity reached by 1st year; females mature sooner than males; multiple spawnings with peaks in December, February and August.	Hernández-Díaz and Bückle-Ramírez (1996)
	Mexico	MS-1 HA	Multiple spawnings over the year with a major peak in winter (November–February); sexual maturity attained by first year (~25 mm).	Saucedo and Monteforte (1997)
	Mexico	SS-1 HA BA	Breeding season from February through October with 2 peaks in January and April; energy for gametogenesis comes from the digestive gland (firstly) and adductor muscle (secondly); only proteins from both tissues are used and carbohydrates are stored.	Vite-García (2005)

NS: not specified; MS-1: monthly samplings for 1 year; MS-2: monthly samplings for 2 years; MS-3: monthly samplings for 3 years; FS-1: fortnight samplings for 1 year; FS-6: fortnight samplings for 6 years; BS-1: bimonthly samplings for 1 year; SS-1: seasonal samplings for 1 year; VI: visual inspection of gonads; SM: smears; HA: histological analysis of gonads; HiA: histochemical analysis; DA: digital image analysis; BA: biochemical analysis; UA: ultrastructural analysis.

time required for them to reach a size suitable for pearl production. This may have an important bearing on the economic viability of a pearl farm and has provided impetus for research into the factors affecting growth of pearl oysters (e.g. Sims, 1994; Pouvreau and Prasil, 2001; Yukihira *et al.*, 2006). This Chapter presents an overview of reproduction, development and growth in pearl oysters.

5.2. FUNCTIONAL ANATOMY OF THE REPRODUCTIVE SYSTEM AND ACCESSORY TISSUES

5.2.1. Gonad tissue

Like most bivalves, with the exception of the Pectinidae, pearl oysters lack a discrete gonad. Instead, gonad tissue arises from the mesoderm and develops gradually and diffusely around the digestive gland, intermingled with other organs of the visceral mass (see Section 3.2.5; Fig. 3.7). Differentiation and development of gonad tissue is a complex process occurring by means of seasonal soma–germline interactions. Macroscopically the gonad is formed by two massive and relatively symmetric lobules (one corresponding to each valve) that grow toward the dorsal region of the visceral mass as gametogenesis proceeds (Figs. 3.3 and 3.7), thus constituting a massive paired gland when specimens ripen (Tranter, 1958a).

When observed using a light microscope in transverse section (Fig. 5.1a), gonad tissue appears delineated from the central region by the digestive gland and from the body wall by a layer of neutral and acid mucopolysaccharides and reticular collagen fibers, as well as longitudinal and radial muscular fibers arranged in round compact pockets. When stained with Alcian blue (PAS), eosinophilic cells of the mucopolysaccharide layer show two clearly contrasting fractions; the neutral inactive (with a pink-magenta color) and the acid active (with an intense blue color). A thick matrix of interstitial connective tissue, seen as a complex vascular system running among the acini, provide support, communication and substrate for the differentiation of gonad tissue and other kinds of somatic nourishing storage cells, namely the vesicular connective tissue (VCT) cells and auxiliary (follicle) cells, whose morphology and biochemical composition varies (see Section 5.2.3). The size and diffusion of the inter-connective tissue matrix shows an inverse relationship with the progress of gametogenesis and the area covered by VCT and auxiliary cells.

The structure of gonad tissue is simple, as it is formed by a series of small, compact granular bags (follicles, or more correctly, acini) that enlarge and transform into a complex network of branched and stratified tubules as gametogenesis proceeds (Tranter, 1958a). The acini represent the structural units of gonad tissue. They are sacs of connective tissue containing numerous gametes that are distinguished as testis or ovary (Figs. 5.1b, 5.2b and 5.3a) (Beninger and Le Pennec, 2006). Follicles consist of an oocyte surrounded by layers of auxiliary cells. Gametes, either oocytes or spermatozoa, develop within these units by permanent mitosis of oogonial or spermatogonial stem cells, which produce cloned daughter cells. Stem cells (mother cells or germinal cells) (Figs. 5.3a, b) are aggregations of small oval-shaped basophilic cells that have three major properties: (1) capacity to divide and renew themselves for long periods; (2) lack of specialized, tissue-specific structures; and (3) potential to remain as stem cells or give rise to specialized cell types. Thus, stem cells are totipotent, multipotent or pluripotent cells. In gonochoristic molluscs, differentiation of stem cells into oogonia or spermatogonia is controlled by a neurosecretion, the androgenic factor. If this factor is present, the oyster becomes male, but if absent, the sex is female (Sastry, 1979). Additional information about neuroendocrine regulation of gametogenesis is provided in Section 5.6.1. After stem cell differentiation, spermatogonia or oogonia

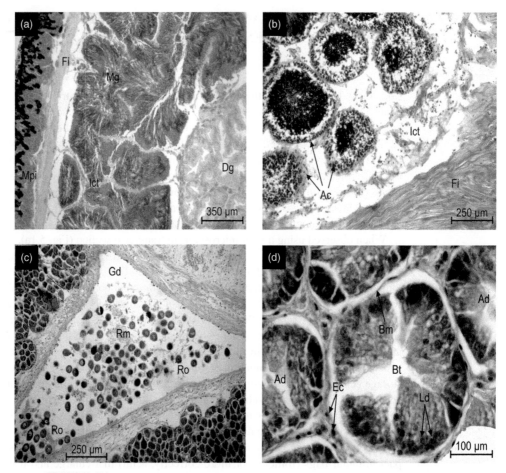

FIGURE 5.1 Photomicrographs of transversal sections of *Pinctada mazatlanica* reproductive system and accessory tissues. (a) Visceral mass showing the neutral (clear) and acid (dark) mucopolysaccharide layer (Mpl), collagen fibers (Fi), inter-connective tissue (Ict), male gonad (Mg), and digestive gland (Dg); (b) Male gonad showing small, round-shaped acini (Ac) starting to develop within the Ict matrix; (c) Female gonad showing a secondary gonoduct (Gd) filled with residual oocytes (Ro) and large amounts of resorptive material (Rm); (d) Digestive gland with adenomeres (Ad) revealing the basal membrane (Bm) and lipid droplets (Ld) among eosinophilic cells (Ec) near the central blind tubules (Bt).

develop into ripe spermatozoa and oocytes that are later discharged to the seawater through branched tubules that open into primary and secondary gonoducts of larger size as the gonad reaches the near ripe stage (Fig. 5.1c).

5.2.2. Digestive gland

Merged with gonad tissue as part of the visceral mass, the digestive gland is formed by a large number of blind-end granular tubules that are connected to the stomach by branched conduits. When observed microscopically in transverse section (Fig. 5.1d), tubules are composed of structural units, the adenomeres which have an external layer

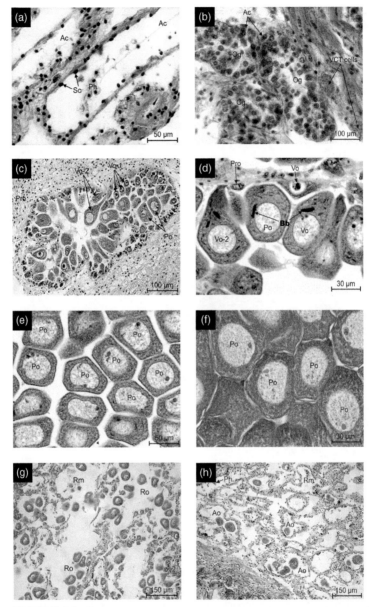

FIGURE 5.2 Photomicrographs of female gonads, showing the sexual stages of oogenesis in *Pinctada mazatlanica*. (a) Inactive stage, with empty and collapsed acini (Ac) containing stem cells (Sc), some phago-cytes (Ph), and no signs of gamete development; (b) Multiplicative phase evidencing the presence of oogonia (Og) proliferating within acini, and among them, large amounts of VCT cells; (c) Early development, domi-nated by previtellogenic oocytes (Pro), while few vitellogenic oocytes (Vo) and postvitellogenic oocytes (Po) start to differentiate; (d) Mid development with previtellogenic, vitellogenic, and postvitellogenic oocytes (Pro, Vo, Po) in relative equal proportion; the Balbiany body (Bb) can be easily observed; (e) Late develop-ment, showing free postvitellogenic oocytes (Po) filling acini, but still leaving a wide interoocyte space; (f) Ripe stage, with free, larger postvitellogenic oocytes (Po) causing the interoocyte space to reduce; (g) Partial spawning, presenting many scattered residual oocytes (Ro) and resorptive material (Rm); (h) Spent stage, showing empty acini with atresic oocytes (Ao), resorptive material (Rm), and phagocytes (Ph).

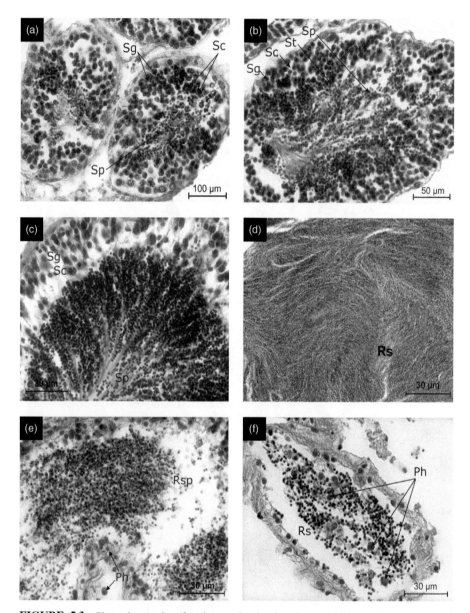

FIGURE 5.3 Photomicrographs of male gonads, showing the sexual stages of spermatogenesis in *Pinctada mazatlanica*. (a) Early development, with many stem cells (Sc) and spermatogonia (Sg) rapidly differentiating into primary and secondary spermatocytes and spermatozoa (Sp); (b) Mid development, characterized by the presence of all type of cellular stages, such as stem cells (Sc), spermatogonia (Sg), spermatocytes (Sc), and spermatozoa (Sp); (c) Late development, in which stem cells (Sc) and spermatogonia (Sg) decrease in number while ripe, tailed spermatozoa (Sp) increase in the lumen; (d) Ripe stage, evidencing a dense volume of ripe spermatozoa (Rs) packing the acini; (e) Partial spawning, showing distended but empty acini, residual spermatozoa (Rsp), and some phagocytes (Ph) starting to appear; (f) Spent stage, with collapsed acini and evident signs of residual spermatozoa (Rs) lysed by phagocytes (Ph).

of epithelial cells, with round basophilic nuclei, resting on a thin, acidophilic basal membrane. A thin layer of inter-connective tissue surrounds and supports the adenomeres. The digestive gland has dual function: (1) it controls the extracellular digestion process of the oyster and (2) it stores energy reserves (lipid and glycogen), together with the adductor muscle, to sustain gametogenesis (Saucedo *et al.*, 2002b). The latter role is described in detail in Section 5.5.2.

5.2.3. Non-germinal nourishing cells

Two types of non-germinal cells that participate in the process of nourishing the developing gametes have been described for the Pteriidae; (1) vesicular connective tissue (VCT) cells and (2) auxiliary cells (Saucedo *et al.,* 2002b). VCT cells consist of a dense matrix of small, spherical or oval pleomorphic somatic cells contained within a larger envelope (Fig. 5.4a). They store mainly glycogen, but also lipid droplets and they constitute an important store of nutrients. Present in both the gonad and digestive gland, VCT cells form an integral part of the connective tissue network in which development and morphogenesis of male and female gametes takes place (Figs. 5.4b and 5.4c). These eosinophilic cells are widely distributed among adenomeres and acini and usually surround wide excretory conduits. It is likely that these conduits connect with each other via the interconnective tissue and may be important in communication between the gonad and the digestive gland in two ways: (1) transportation to the digestive gland of nutrients recycled from recently lysed gametes and their storage as VCT cell reserves (Fig. 5.4b) and (2) transportation of nutrients to the gonad from VCT cells in the digestive gland to sustain gametogenesis (Saucedo *et al.*, 2002b). So far, VCT cells have only been observed in *P. mazatlanica* (Saucedo *et al.*, 2002b) while *Pteria* species (at least *Pt. sterna*) probably lack VCT and adipogranular nourishing cells (Vite-García and Saucedo, 2008).

Auxiliary cells are specialized intra-gonad cells observed only in female acini and always in intimate contact with developing oocytes to which they attach by desmosome-like gap junctions. This contact, first observed with the early developing oogonia or oocytes, probably remains until oocytes are surrounded by the vitelline envelope (Dorange and Le Pennec, 1989) (Figs. 5.4d). Auxiliary cells, which appear to participate in the formation of the vitelline envelope, range in size from 20 to 30 μm and show enormous plasticity in morphology according to the stage of gametogenesis and their position relative to growing oocytes. Thus, they may be observed basally near the stalk region, eccentrically or laterally to the oocyte (Fig. 5.5). Auxiliary cells serve a nutritive function for young oocytes undergoing rapid development, especially when synthesizing the lipid and protein cortical granules of the ooplasm during vitellogenesis. These cells gradually disappear in ripe gametes or, if still present, are detached from the oocytes.

5.3. SEX CHARACTERISTICS

5.3.1. Sex differentiation and sex reversal

Pearl oysters are typically protandrous hermaphrodites. They develop first as males and retain this condition for one or several reproductive cycles until changing sex. The two

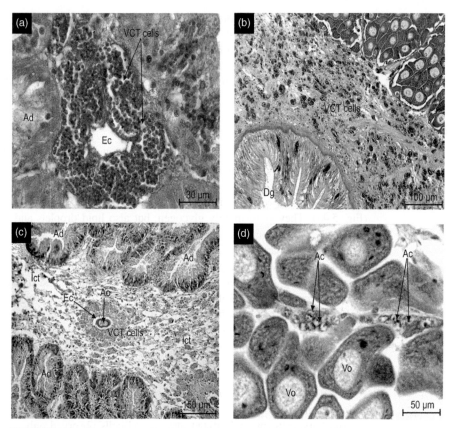

FIGURE 5.4 Photomicrographs of transversal sections of *Pinctada mazatlanica* non-germinal nourishing cells. (a) Vesicular connective tissue (VCT) cells distributed around adenomeres (Ad) in the digestive gland and surrounding an excretory conduit (Ec); (b) VCT cells spread between the female gonad (Fg) and the digestive gland (Dg); (c) VCT cells in the digestive gland showing an atretic oocyte within an excretory conduit and suggesting communication with the gonad through the Ict matrix; (d) Female gonad revealing vitellogenic oocytes (Vo) nourished by groups of auxiliary cells (Ac).

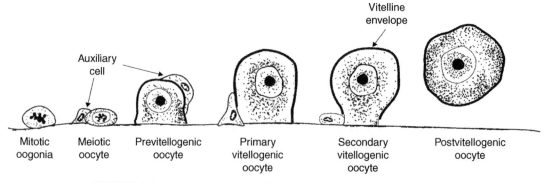

FIGURE 5.5 Functional anatomy of auxiliary cells nourishing during oocyte development process.

sexual phases overlap in the same acinus only temporarily, while the few residual sper-matozoa associated with the new oogonial development undergo progressive regres-sion and lysis (Fig. 5.6a). Protogyny, in which the sex of first maturation is female, is uncommon among marine bivalves. Tranter (1959) and Saucedo *et al.* (2002b), how-ever, reported some cases of natural protogyny in *P. margaritifera* and *P. mazatlanica*, respectively (Fig. 5.6b).

Pearl oysters are rhythmical hermaphrodites, having more than one sex reversal during their lifetime. This phenomenon has been observed in *P. mazatlanica* (Galtsoff, 1950; Saucedo and Monteforte, 1997; Saucedo *et al.*, 2002b; Vite-García and Saucedo, 2008; Figs. 5.6a and 5.6b), *P. maxima* (Wada, 1953), *P. albina* (Tranter, 1958c), *P. margaritifera* (Tranter, 1958d; Thielley *et al.*, 1993b) and *P. fucata* (Ojima and Maeki, 1955; Tranter, 1959; Chellam, 1987; Wada *et al.*, 1995). Rhythmical sex reversal occurs repeatedly, either annually or at shorter intervals. Its expression is the

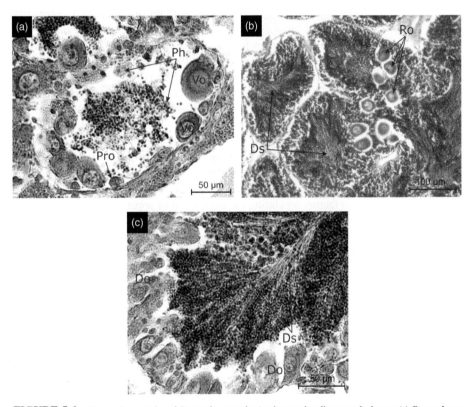

FIGURE 5.6   Photomicrographs of *Pinctada mazatlanica* hermaphrodite sexual phases. (a) Protandrous hermaphrodite phase with residual spermatozoa destroyed by phagocytes (Ph) and substituted by newly developed previtellogenic oocytes (Pro) and vitellogenic oocytes (Vo); (b) Rhythmical hermaphrodite phase showing a female–male sex reversal, with residual oocytes (Ro) and developing spermatozoa (Ds); (c) Apparent functional bisexual phase evidencing developing spermatozoa (Ds) and developing oocytes (Do) in equal proportion.

result of a complex interaction of endogenous factors and varying environmental conditions. In wild populations, for example, sex reversal has been attributed to several factors, from premature death of male oysters affecting the sex ratio (Wada, 1953), physiological damage and nutritive stress (Pouvreau *et al.*, 2000a; Yukihira *et al.*, 2000; Acosta-Salmón and Southgate, 2005), excessive fouling (Taylor *et al.*, 1997b), and partial spawning events associated with new re-development of the gonad (Tranter, 1958b, 1958c; Pouvreau *et al.*, 2000b; Saucedo *et al.*, 2002a, 2002b). Tranter (1958c) attributed sex reversal to one of two causes: (1) "a weak hereditary sex-determining mechanism" or (2) "the germ cell rudiments responsive of the food reserve levels in the body tissue". Although no data were provided by Tranter in support of these hypotheses, he argued that "maleness" usually occurred when energy reserves within the animal's tissues were low, while "femaleness" was favored when such reserves were high. Conversely, under culture conditions, sex change has been related to prolonged overcrowding within culture devices (Saucedo *et al.*, 2002b; Acosta-Salmón and Southgate, 2005) and this is not incompatible with the argument of Tranter (1958c).

Although not reported as a normal condition, simultaneous hermaphroditism, has also been observed in low frequency in some pearl oyster species, including Akoya[1] pearl oysters (Ojima and Maeki, 1955; Tranter, 1959), *P. albina* (Tranter, 1958c, d) and *P. mazatlanica* (Saucedo *et al.*, 2002b) (Fig. 5.6c). In contrast, simultaneous hermaphroditism, was considered to be uncommon in Akoya from India (Chellam, 1987) and Iran (Behzadi *et al.*, 1997), *P. maxima* from Australia (Rose *et al.*, 1990), *P. margaritifera* from French Polynesia (Thielley *et al.*, 1993b) and Australia (Acosta-Salmón and Southgate, 2005) and in *Pt. sterna* from Mexico (Arizmendi-Castillo, 1996; Hernández-Díaz and Bückle-Ramírez, 1996). So far, the simultaneous hermaphroditic condition has not been shown to be functional in pearl oysters, and overlapping of male and female conditions may be interpreted as a simple transition phase during sex reversal. Since factors regulating the expression of apparent simultaneous hermaphroditic phases in pearl oysters are likely to be the same as those mentioned above for sex reversal, then the eco-physiological significance of protandry in wild versus cultured pearl oysters requires further study.

5.3.2. Sex ratio in wild and cultured oysters

Pearl oyster populations have a sex ratio approaching 1:1 with increasing age (Wada, 1953; Tranter, 1958c; Gervis and Sims, 1992). Wild *P. maxima* in north Western Australia were confirmed as protandrous hermaphrodites by Hart and Joll (2006) who noted that a sex ratio of 1:1 was not achieved until females had reached a dorso-ventral measurement (DVM) (maximal shell length, see Section 5.7.4) of 170 mm. Under culture conditions, however, sexually undifferentiated oysters may predominate over those with active developing gonads, and males of all sizes and ages out-number females.

[1]Throughout this text the term 'Akoya' is used for pearl oysters in the currently unresolved complex of *Pinctada fucata–martensii–radiata–imbricata* (see Section 2.3.1.13).

While the first condition implies retention of the vegetative phase (non-differentiation of stem cells) indefinitely, the second condition suggests some difficulty in developing the energetically expensive female gonad. Evidently, both cases may be taken as indicators of ambient stress and/or unsuitable culture conditions. In natural populations, for example food availability, water temperature, acute water temperature changes (e.g. those observed during El Niño or La Niña events), genetics and neuroendocrine influences are recognized as the most important cues controlling sexual differentiation of germ cells in the direction of maleness or femaleness in bivalves (Coe, 1943). Stocking density (overcrowding) may also play a key role in determining greater expression of maleness. Examples of this have been documented for *P. maxima* (Taylor, 1999), *P. margaritifera* (Thielley *et al.*, 1993b; Acosta-Salmón and Southgate, 2005) and *P. mazatlanica* (Saucedo and Monteforte, 1997; Vite-García and Saucedo, 2008) under culture conditions. The adaptive value of protandry under such conditions may be related to either the consequent size-independent nature of male reproductive success versus the size-dependent nature of reproductive success in females (Wright, 1988), or the difficulty for overcrowded, stressed oysters to allocate the energy required to develop oocytes instead of energetically cheaper spermatozoa (Saucedo *et al.*, 2002b).

5.4. REPRODUCTIVE, GAMETOGENIC AND BREEDING CYCLES

5.4.1. General

The overall pattern of reproduction in pearl oysters is synchronous, with the gonad undergoing a series of continuous and sequential events that end with a simultaneous breeding period within a population. This cyclical process consists of three general phases that usually occur as part of an annual reproductive cycle:

1. *Vegetative phase*: Typified by individuals lacking gametes and displaying no trace of sexual activity (Fig. 5.2a). This phase usually prevails during periods of high food supply and unsuitable water temperatures.
2. *Cellular differentiation and proliferation phase*: When environmental conditions improve, undifferentiated stem cells develop into spermatogonia or oogonia that multiply within acini and give rise to viable gametes (Figs. 5.2b to 5.2g; 5.3a to 5.3e). These gametes are later spawned.
3. *Recovery phase*: Consisting of the regression of the gonad remaining after spawning to recycle nutrients from residual gametes and ensure the start of a new cycle (Figs. 5.2h; Fig. 5.3f).

5.4.2. Timing of reproductive events

Bivalve molluscs are known to show considerable variation in the timing of reproductive cycles. Differences in the expression of ripening and spawning events within a species occur because critical water temperatures are attained at different latitudinal ranges and at different times (Giese, 1959; Giese and Pearse, 1974).

Among the Pteriidae, there are slight to marked differences in expression of ripening and spawning peaks, which result from the wide or narrow latitudinal distribution of a species (Gervis and Sims, 1992). Of all members of the family, *P. margaritifera* and *P. mazatlanica* are among the species that show the widest latitudinal range. *P. margaritifera* inhabits the Indian and Pacific Oceans, from the east coast of Africa to the west coast of the Americas, as well as the eastern Mediterranean Sea, covering a broad belt from 30°N to 28°S (Yukihira *et al.*, 2000; Fig. 2.6). Consequently, reports of spawning seasons vary considerably from region to region and range from only once a year (March through September) in the Red Sea (Crossland, 1957), twice a year in Australia (July–August and November; Tranter, 1958d) and French Polynesia (Thielley *et al.*, 1993a, b) to five times in French Polynesia in oysters of >100 mm shell height (Pouvreau *et al.*, 2000b). In contrast, *P. mazatlanica* is found along the eastern Pacific coasts, from the Gulf of California (~28°N) to Peru (~20°S) (Keen, 1970; Fig. 2.6). Populations inhabiting the tropical coast of Costa Rica spawn multiple times during the year (Solano-López *et al.*, 1997), while those from Mexico can spawn once in summer (Sevilla, 1969; Saucedo and Monteforte, 1997; Saucedo *et al.*, 2002a, b; Gómez-Robles *et al.*, 2005; Vite-García and Saucedo, 2008), twice (García-Cuellar *et al.*, 2004), or even three times a year (García-Domínguez *et al.*, 1996).

The timing of reproductive seasons among the other pearl oyster species also varies markedly between species and between populations of the same species. Species that have been reported to spawn twice a year include Akoya pearl oysters from India, Iran and eastern Australia (Herdam and Hornell, 1906; Hornell, 1922; Tranter, 1959; Desai *et al.*, 1979; Chellam, 1987; Behzadi *et al.*, 1997; Jamili *et al.*, 1999; O'Connor and Lawler, 2004) and *P. albina* from Australia (Tranter, 1958b, c). Those that spawn only once a year include Akoya pearl oysters from Japan (Ojima and Maeki, 1955; Wada *et al.*, 1995; Choi and Chang, 2003) and Australian *P. maxima* (Rose *et al.*, 1990). For *P. albina sugillata* also from Australia, O'Connor (2002) observed at least three spawning peaks during the year. For species of *Pteria*, Arjarasirikoon *et al.* (2004) found multiple spawning peaks for *Pt. penguin* in Thailand while *Pt. sterna* behaves as a multiple spawner in Mexico (Arizmendi-Castillo, 1996; Hernández-Díaz and Bückle-Ramírez, 1996; Saucedo and Monteforte, 1997; Hernández-Olalde, 2003; Vite-García and Saucedo, 2008). For more detailed information on the timing of breeding and spawning seasons among the Pteriidae see Table 5.1.

5.4.3. Inter-annual variations

Breeding seasons also vary from year to year within the same geographic area. These inter-annual variations result mostly from changing global temperatures, either when normal conditions prevail or when abnormal oceanographic events occur, such as El Niño or La Niña (Lluch-Belda *et al.*, 2003; García-Cuellar *et al.*, 2004). Inter-annual variations evidently determine inter- and intra-specific variations in somatic growth and reproduction in marine bivalves (Newell *et al*, 1982; MacDonald *et al.*, 1987).

If water temperature remains relatively high after a partial spawning, and tissue energy reserves maintain high levels, a population may skip the recovery (resting) phase and begin a new, short-term reproductive cycle. The strategy involves recycling of nutrients obtained from unspawned residual gametes and their rapid incorporation into newly formed gametes. This has been documented in wild *P. albina* (Tranter, 1959) and *P. margaritifera* (Pouvreau *et al.*, 2000b) and cultured *P. mazatlanica* (Saucedo *et al.*, 2002a, b).

5.4.4. Oogenesis and spermatogenesis

Oogenesis and spermatogenesis in pearl oysters proceed relatively synchronously within the population. They involve identical nuclear processes (karyokinesis) of proliferation and differentiation of stem cells. There are, however, major differences in the cytoplasmic processes (cytokinesis) that need to be taken into account when comparing (Rappaport, 1991).

5.4.4.1. Difference in cell size

Unlike spermatozoa that shed their cytoplasm (=ooplasm) as they grow, oocytes continuously incorporate nutrients from exogenous (diet) and endogenous (reserve tissues and cells) sources to provide nutrients for the zygote (Pipe, 1987; De Gaulejac *et al.*, 1995). Vitellogenesis is very energy demanding and strongly sensitive to changes in external factors and selective pressures. It ends when the oocyte is fully ripe and requires no further nourishment when it detaches from the acinus basal membrane in preparation for spawning. Vitellogenesis forms the yolk, which is synthesized from a lipoprotein, vitelline, present in the hemolymph. It provides the embryo with the nutrients and energy required to complete its development until the resulting larva is able to nourish itself from the surrounding environment (Gallager and Mann 1986; see Section 5.7.3).

5.4.4.2. Difference in quantity and quality of cell products

The final product of oogenesis is an oocyte that is genetically diploid. Instead, spermatogenesis produces four haploid spermatozoa. Such a disparity in gametogenesis occurs because the primary oocyte ceases development in Prophase I of meiosis (diplotene in particular) and remains arrested in this stage while growing and fattening during the rest of the vitellogenesis (Raven, 1966; De Gaulejac *et al.*, 1995; Krantic and Rivailler, 1996). It is not until the germinal vesicle (=nuclear membrane) breaks that the oocyte restart meiosis to reduce its genotype to a haploid state through release of the polar bodies (see Section 5.7.1, Fig. 5.12). Germinal vesicle breakdown (GVBD) in bivalves may occur within the acini before spawning, inside the gonoducts during spawning, or in the mantle cavity after spawning (Lora-Vilchis *et al.*, 2003). In most cases the compound responsible for GVBD and spawning is serotonin (Guerrier *et al.*, 1993; Kyozuka *et al.*, 1997; Martinez *et al.*, 2000; Lora-Vilchis *et al.*, 2003). So far, no study of GVBD and meiosis resumption has been conducted for any species of pearl oyster.

5.4.5. Oogenesis (vitellogenesis)

Cytokinetic changes in the oocyte during vitellogenesis have been well studied for bivalve molluscs. Dorange and Le Pennec (1989) divided oogenesis into three general stages.

1. *Pre-meiotic phase*: Involves a rapid differentiation and proliferation of stem cells to form primary oogonia that later enlarge and give rise to secondary oogonia (Fig. 5.2b). This is the pre-meiotic phase because all divisions occurring within acini are mitotic.
2. *Previtellogenic phase*: Secondary oogonia filling acini give rise to the first primary oocytes that suspend development in Prophase I of meiosis (Fig. 5.2c). These oocytes are small and immature.
3. *Vitellogenic phase*: Oocytes still arrested in diplotene and being thus diploid, start growing and developing by incorporating nutrients from endogenous and exogenous sources to form the yolk (Figs. 5.2d to 5.2f).

Of these stages, gamete development is the most complex and has the longest duration. Atresia of oocytes also requires special consideration. This process represents the deterioration and disintegration of the cellular membrane and most of the major cellular constituents of the oocyte, and thereby its breakdown (Figs. 5.2h). It was first described in the Pectinidae (Tang, 1941). Tranter (1958b, c) also made general observations of this process in the Pteriidae, but did not use the term atresia.

This process begins in the cytoplasm by the action of hydrolytic enzymes distributed throughout the entire oocyte, finally resulting in the bursting of the vitelline envelope and release of lysed contents into the acinus lumen (residual material). Although it is common to relate atresia with the post-spawning event, it can also occur during oogenesis as a result of abnormal or stressful conditions, such as variations in water temperature, salinity, photoperiod, food availability, etc. (Orton, 1920; Sastry, 1968; Bayne, 1976; Lowe *et al.*, 1982). Atresia has also been suggested to act as a "buffer" in which oocyte energy supply is re-allocated for gametogenesis when food supply and stored reserves are insufficient (Paulet and Boucher, 1991). This situation may result in a period of continued oogenesis with no external transfer of nutrients to the gonads. When this happens, a partial spawning, where small parts of the oocyte cytoplasm are shed, may produce a mixture of full, empty and atretic acini. On this basis, atresia may also act as a "synchronizer" of new ovarian development, where high concentrations of hemocytes surrounding gonoducts after a partial spawning are commonly observed (Duinker, 2002). In the pearl oyster, *Pt. sterna*, there are groups of phagocytes, tentatively named macro-phagocytes, which destroy residual gametes and re-distribute nutrients among storage tissues (Saucedo and Monteforte, 1997).

5.4.6. Spermatogenesis

Spermatogenesis in bivalve molluscs consists of three phases: (1) spermatocytogenesis, (2) meiosis and (3) spermiogenesis (Maxwell, 1983) (Fig. 5.3).

1. *Spermatocytogenesis* is the mitotic phase involving differentiation of stem cells into spermatogonia and primary spermatocytes. These cells have spherical or oval nuclei and are anchored to the acini membrane.

2. *Meiosis* occurs when primary spermatocytes go through the first meiotic division to give rise to secondary spermatocytes. When these cells quickly complete the second meiotic division, they produce four spermatids, which are still immature.
3. *Spermiogenesis* involves differentiation of spermatids to form ripe, elongated spermatozoa. No further mitosis or meiosis occurs. During spermiogenesis, the acrosome and flagella forms and excessive cytoplasm is separated and agglomerated into Sertoli cells. Spermatozoa are released into the acini lumen.

5.4.7. Classification of reproductive stage in pearl oysters

Tranter (1958b, d) originally used eight stages to follow the progress of oogenesis and spermatogenesis in *P. albina* and *P. margaritifera*, respectively. Of these, five stages involved the development of the gonad *per se* and three its regression. All stages included letters and numbers. Briefly for oogenesis, stages were: Fd_1 (mostly oogonias), Fd_2 (immature oocytes with nuclei), Fd_3 (growing oocytes with larger nuclei), Fd_4 (free and attached oocytes in equal number), Fd_5 (mainly ripe free oocytes), Fr_1 (spawning and beginning of oocyte lysis), Fr_2 (active regression of residual oocytes), Fr_3 (empty gonad or starting to develop again). Spermatogenesis was organized in a similar way by the author (Md_1, Md_2, Md_3, Md_4, Md5, Mr_1, Mr_2 and Mr_3). Subsequent authors have proposed modified versions of Tranters' pioneering work, including minor or major changes with regard to the species studied (e.g. Rose *et al.*, 1990; Thielley *et al.*, 1993b). Wada (1953) employed a scheme of seven stages to track changes in gonad development of Akoya pearl oysters as follows: early growth, late growth, maturation, spawning, late spawning, early follicle and follicle stage. Arjarasirikoon *et al.* (2004) utilized a complex classification format of eleven stages for spermatogenesis (spermatogonium stage, primary spermatocyte, leptotene spermatocyte, zygotene spermatocyte, pachytene spermatocyte, diplotene spermatocyte, metaphase spermatocyte, secondary spermatocyte, spermatid I, spermatid II and spermatozoa) and six stages for oogenesis (oogonium state, oocyte I, oocyte II, oocyte III, oocyte IV, oocyte V) for the winged pearl oyster, *Pt. penguin*.

Given the variety of classification methods used to describe reproductive cycles in pearl oysters since Tranter's proposal, recent years have seen development of more practical and consistent classification schemes integrating all distinctive cytological changes during male and female gonad development. While not universal, a typical scheme should include at least five general stages:

1. resting, also called inactive, indeterminate, or undifferentiated;
2. active development, also named gametogenesis or growing stage;
3. ripeness or near ripe stage, also tagged as maturation stage;
4. partial spawning;
5. spent or post-spawning stage.

The stage involving gamete development is the most conspicuous and may be divided into sub-stages, such as early and late development, or early, mid and late development. For didactic purposes, a step-by-step sequence for spermatogenesis and oogenesis in pearl oysters using a large scheme is summarized in Table 5.2.

TABLE 5.2

Stages of oogenesis and spermatogenesis observed for many species of marine bivalves, including pearl oysters.

Stage	Ovary	Testis
Resting	Also called inactive, undifferentiated or stage 0. There is no trace of gonad development and specimens are unable to be sexed. Acini are empty and thin, occasionally containing some granulocytes and phagocytes remaining from the previous cycle. The vascular inter-connective tissue is distended and greatly diffused (Fig. 5.2a).	
Early development	Also called multiplicative phase or stage 1, marks the onset of oogenesis. Acini are filled with small (8–10 μm diameter) round oogonias (Fig. 5.2b) that later differentiate into few larger (12–14 μm) previtellogenic oocytes, recognized by their dark-purple color (Fig. 5.2c). They are immature cells, lack yolk, and are still connected to the acinus wall by the stalk region, in which some auxiliary cells can be observed (Fig. 5.4d).	Acini are small and compact. Spermatogonia (8–9 μm) proliferate and give rise to several layers of smaller primary (4–5 μm) and secondary spermatocytes (2–3 μm) expanding toward the lumen (Fig. 5.3a). These cells are the dominant stage. Few pockets of spermatids and spermatozoa are also observed. The proportion of inter-connective tissue and VCT cells is large.
Mid development	Acini grow and stratify. The inter-connective tissue and VCT cells decrease. All developmental stages can be seen within acini (Figs. 5.2d). Vitellogenic oocytes, recognized by the growth of their ooplasm, are the dominant cell type. Because of their size differences, they can be divided into primary (15–25 μm) and secondary (28–35 μm). These oocytes are still connected to the acini wall, have a peduncle or pear shape, a large nucleus with 1–2 nucleoli, and sometimes 1–2 small basophilic structures called *Balbiani body*. This vitelline body, described by Guraya (1975), may correspond to the "yolk nuclei" reported by Tranter (1959) and Rose *et al.* (1990) for pearl oysters. It likely serves an endogen supply of lipids for nourishing oocytes.	Acini enlarge and stratify, while the inter-connective tissue matrix and VCT cells reduce. Starting from the outer layers to the center, all developmental stages are present: spermatogonia, abundant spermatocytes, smaller spermatids (1.5–1.8 μm), and few ripe spermatozoa showing their acidophilic tails as pink lines radiating from the center of the lumen (Fig. 5.3b).
Late development	Acini growth and stratification continues. The inter-connective tissue and VCT cells have almost disappeared. Auxiliary cells decrease in number, or if still present, they appear separated from oocytes. The inter-oocyte space reduces as the frequency of ripe oocytes increases (Fig. 5.2e). These oocytes, the dominant stage now, are free in the lumen, have a polyhedral shape, and increase to 36–45 μm. A chromatin mass in the nucleoplasm is dense and amorphous. The nucleolus is dense, small, and usually in marginal position. The Balbiani body may still be present.	Acini keeps growing and expanding, while the amount of inter-connective tissue and nourishing cells diminish. Spermatogonia and spermatocytes have been reduced in thickness to a few layers of cells at the periphery, as acini are packed now with a dense, dark-blue band of ripe spermatozoa several cells deep and measuring 1–1.3 μm diameter (Fig. 5.3c).
Ripeness	Also called stage 2, marks the end of vitellogenesis. Acini are strongly stratified and packed with free, ripe oocytes 46–48 μm diameter (Fig. 5.2f). VCT cells have almost disappeared. The nucleus now occupies a large area in middle of the oocyte. Many oocytes begin to enter the germinal vesicle stage, indicating the resumption of meiosis and the readiness for spawning.	Acini are enlarged and distinguishing the boundaries between them is difficult. The dominant stage now is the spermatozoa, which strongly pack acini. Stem cells and spermatogonia are restricted to a thin layer at the periphery of the acini (Fig. 5.3d). Only a small amount of interstitial connective tissue is developed at this stage.

(Continues)

TABLE 5.2 (*Continued*)

Stage	Ovary	Testis
Partial spawning	*Stage 3*: Gametes are expelled into the ocean. Acini walls look broken and empty, but still distended (Fig. 5.2g). Much residual material is noticeable. After the germinal vesicle breakdown (Fig. 5.6a), some types of phagocytes and granulocytes appear in the space between the free residual oocytes, which now look rounded. The nucleolus and chromatin disaggregate. Because spawning is never complete, new gonad replenishment may occur when environmental conditions are still suitable.	After gamete release, many residual spermatozoa are observed scattered within acini, with the first signals of phagocytic activity starting. A gap between acinus walls and the mass of residual spermatozoa appears (Fig. 5.3e).
Spent	*Post-spawning or stage 4*: Acini are collapsed and empty, with clear signs of phagocytic activity, oocyte lysis, and presence of resorptive material (Fig. 5.2h). Residual oocytes, especially the smaller ones may also undergo atresia (Fig. 5.6b). The inter-connective tissue matrix and VCT cells develop again.	Gamete resorption starts. Empty acini show no evidence of gonad development. There is a rapid proliferation of different kinds of hemocytes (e.g. phagocytes, granulocytes, and amoebocytes) surrounding and destroying residual spermatozoa (Fig. 5.3f).

5.5. REPRODUCTIVE PHYSIOLOGY

5.5.1. Seasonal acquisition and allocation of energy during gametogenesis

The annual cyclical periodicity of reproduction in marine invertebrates was identified by Giese (1959). The author claimed that temporal changes in the energy content of somatic tissues occur in individuals subjected to a universal principle of seasonality. Within a particular set of latitudinal and habitat features, seasonality influences each species to develop adaptive strategies to deal with the available energy in the environment (Giese and Pearse, 1979). Acquired energy has to be partitioned and allocated by an individual for three main purposes: (1) general metabolism; (2) somatic growth; and (3) reproduction. This is expressed in the formula:

$$SFG = AE - (RE - EE)$$

where AE is acquired energy, RE is the energy required for general metabolism and EE is excreted energy. SFG (scope for growth) is the net energy gain that is available for somatic growth and reproduction (see Section 4.3.3). The factors affecting the regulatory mechanisms by which energy is partitioned to these biological processes are discussed in Section 5.6.

In the majority of bivalves, reproductive cycles are coupled with seasonal changes of storage and depletion of biochemical compounds within somatic and germinal tissues (Epp *et al.*, 1988; Barber and Blake, 2006). Reproductive seasonality, in biochemical terms, has been well documented for commercially important species, particularly scallops (e.g. Sastry and Blake, 1971; Taylor and Venn, 1979; Lodeiros *et al.*, 2001; Racotta *et al.*, 2003) and a number of similar studies have been conducted with pearl oysters (Desai *et al.*, 1979; Saucedo *et al.*, 2002a; Vite-García, 2005; Vite-García and Saucedo, 2008, Cáceres-Puig, 2007). These studies show that nutrients (glycogen, lipid and protein) may be stored in somatic tissues when food is abundant and, later, during times of food limitation or of reduced metabolic rate, used to support gametogenesis (Epp *et al.*, 1988).

Gametogenesis in bivalves represents a period of particularly high energy demand, when the costs associated with cellular maintenance, somatic and gamete production are met by exogenous food supply, endogenous reserves, or a combination of both (Epp *et al.*, 1988; Barber and Blake, 2006). Several authors have discussed the relative contribution of food intake versus energy reserves to satisfy metabolic demands of growth and reproduction in marine bivalves. However, because these processes are highly dependent on a complex interaction of exogenous and endogenous factors, no clear pattern has yet been established. For example, energy reserves are required for initiating and sustaining gametogenesis in some scallops (Comely, 1974; Taylor and Venn, 1979), while food intake is necessary for gonad growth in other species (Ansell, 1974; Barber and Blake, 1983; Brokordt and Guderley, 2004) because energy reserves from the muscle, digestive gland and mantle are inadequate. Some authors have even stressed that the immediate use of carbohydrates, proteins and lipids obtained from ingested food is an advantage for saving energy needed for gametogenesis, when compared to that provided by catabolic degradation and *de novo* synthesis of these compounds using ATP (Brokordt and Guderley, 2004). An intermediate situation occurs in scallops (Sastry, 1979; Barber and Blake, 1981; Robinson *et al.*, 1981), and the winged pearl oyster *Pt. sterna* (Vite-García and Saucedo, 2008), where gametogenesis is fueled by a combination of stored reserves and food supply.

Additional examples are those of species having two periods of gonad growth sustained from stored reserves in winter and food supply in summer; these include *P. mazatlanica* from Bahía de La Paz, Mexico (Saucedo *et al.*, 2002a; Vite-García, 2005). In this case, the geographic area (Bahía de La Paz) is a transition ecotone between cooler temperate and warmer tropical provinces. Consequently, expression of a spring reproduction peak relying on high phytoplankton levels during winter, usually ends with regression of gonad tissue rather than with a massive spawning. This occurs mainly because water temperature during spring is insufficient to trigger spawning (<24°C). Despite this, spring peaks can be more conspicuous than summer peaks in terms of higher frequency and size of postvitellogenic oocytes, higher concentrations of protein and lipids in the gonad (Fig. 5.7), and lower levels of proteins in the adductor muscle (Saucedo *et al.*, 2002a; Vite-García, 2005) (Fig. 5.8).

5.5.2. Participating tissues

The major tissues that participate in storage of energy reserves (gonad, digestive gland, adductor muscle and mantle) are used differently over time through clear seasonal cycles of energy storage, mobilization and depletion.

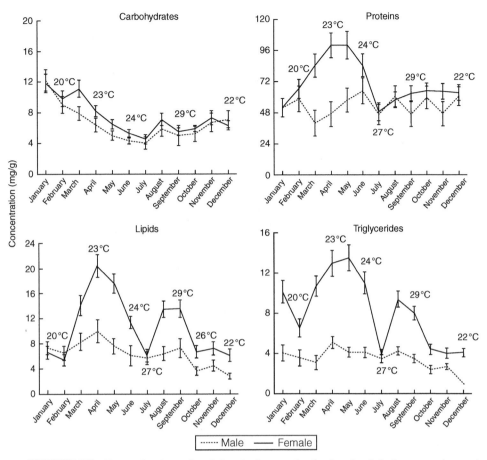

FIGURE 5.7 Temporal and sexual variations in the mean levels of total carbohydrates, proteins, total lipids, and triglycerides of *Pinctada mazatlanica* gonad tissue. Bars denote standard deviation.

5.5.2.1. Gonad

The gonad is not generally considered to be a somatic storage compartment *per se*, but rather the germinal-target site where energy from other tissues is directed. In practice, however, some accessory cell compartments located around acini function as energy reservoirs that participate in developing gonad tissue. In *P. mazatlanica*, for example, carbohydrate (glycogen) may be stored in non-germinal VCT cells scattered throughout the inter-connective tissue matrix of the gonad and digestive gland (Figs. 5.4a to 5.4c). Accumulation of glycogen in these cells does not occur during gonad development, but during the previous winter season as a conservative strategy to ensure gonad maturation. A similar pattern has also been reported for Akoya pearl oysters (Desai *et al.,* 1979). Thus, when water temperature is adequate for the onset of gametogenesis,

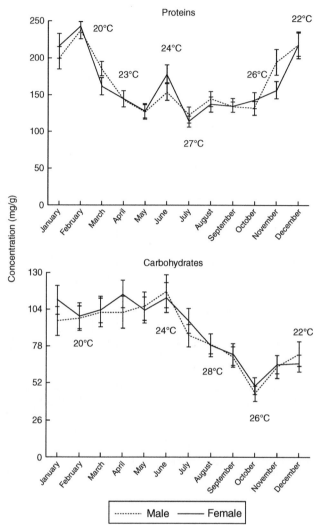

FIGURE 5.8 Temporal and sexual variations in the mean levels of total carbohydrates and proteins of *Pinctada mazatlanica* adductor muscle. Bars denote standard deviation.

P. mazatlanica follows one of two metabolic pathways for using glycogen reserves: (1) their constant use and depletion from VCT cells present in the gonad until spawning takes place (Fig. 5.7a); or (2) a complementary uptake of glycogen from ingested food. In both cases, the nutrients are used as an immediate source of energy for build-up of gametes via their conversion into lipids. At completion of the cycle, lipid levels in female gonads are depleted with spawning and rise again when oysters are nearly ripe. This scenario agrees with observations by Desai *et al.* (1979) for Akoya pearl oysters and Vite-García and Saucedo (2008) for *Pt. sterna*.

5.5.2.2. Adductor muscle

The adductor muscle is the most important energy storage site in many marine bivalves (see recent reviews by Barber and Blake, 2006; Chantler, 2006). Although the adductor muscle accumulates mostly proteins and carbohydrates, these metabolites vary in their relative importance between species. *P. mazatlanica* and *Pt. sterna* use predominantly protein reserves to sustain gametogenesis (Saucedo *et al.*, 2002a; Vite-García, 2005; Cáceres-Puig, 2007; Vite-García and Saucedo, 2008). *P. mazatlanica*, for example, stores very high protein levels (~250 mg/g) in adductor muscle at the beginning and end of the annual cycle (winter) as a conservative strategy to ensure gonad development (Fig. 5.8). Thereafter, proteins are rapidly consumed during the first stages of gametogenesis until spring ripening peaks (Fig. 5.8). This event is accompanied by a significant loss of muscle weight to less than half its maximum value and, in consequence, by substantial growth of gonad tissue and increases in levels of its component proteins, total lipids and triglycerides (Fig. 5.7). The scenario with *Pt. sterna* for storage and allocation of protein and carbohydrate reserves is different, as only adductor muscle protein reserves are mobilized to the gonad during the main breeding season (January through April), with muscle carbohydrates being stored during the same period (Fig. 5.9). Unlike other pearl oyster species, whose reproductive peaks are in summer, *Pt. sterna* reproduces primarily in winter and spring.

5.5.2.3. Digestive gland

While some authors report an inverse relationship between substrates stored in the digestive gland (lipids, carbohydrates, proteins) and gametogenesis in bivalves (Sastry, 1970), others attribute a secondary role in gonad development for the digestive gland (Barber and Blake, 1983). Additionally, a number of studies found no significant change in the condition of the digestive gland during the ongoing reproductive cycle of bivalves (Epp *et al.*, 1988; Lodeiros *et al.*, 2001; Racotta *et al.*, 2003). In *P. mazatlanica*, the exact role of the digestive gland during gametogenesis is unclear (Saucedo *et al.*, 2002a; Vite-García, 2005).

In this species, carbohydrates are gradually stored in the digestive gland despite gonad development, suggesting that this metabolite is not used for the reproductive cycle underway and is accumulated to cover subsequent less demanding energy requirements (Fig. 5.10). This senario differs from the pattern observed by Desai *et al.* (1979) for Akoya pearl oysters, where the glycogen content of the digestive gland was high during early gametogenesis and declined during and after spawning. Protein levels in the digestive gland are high during the resting period, when water temperature is unsuitable, and declines as gametogenesis proceeds (Fig. 5.10). These substrates are controlled independently of muscle proteins and muscle carbohydrates. Lipids and triglycerides are stored in the digestive gland for short periods during the spring and summer ripening peaks and are immediately mobilized to the gonad, suggesting again a conservative strategy in this species regarding the use of carbohydrate and lipid reserves during gonad growth. These results agree with findings of Barber and Blake (1981),

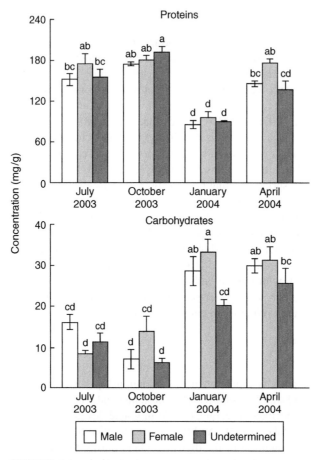

FIGURE 5.9 Temporal and sexual variations in the mean levels of total carbohydrates and proteins of *Pteria sterna* adductor muscle. Bars denote standard error.

Racotta *et al.* (1998), Román *et al.* (2002), Arellano-Martínez *et al.* (2004) for pectinids that the digestive gland acts as a short-term storage and transfer site of lipids and carbohydrates to meet reproductive demands.

In *P. sterna*, however, a significant proportion of the energy channeled to activate and sustain gametogenesis originates from reserves stored in the digestive gland (~35%), with a lesser amount (~28%) from the adductor muscle (Vite-García and Saucedo, 2008). In this species, carbohydrates are slowly depleted during most of the year, indicating relatively low but constant use during gametogenesis (Fig. 5.11). In contrast, protein content decreases from ~200 mg/g in July (resting period) to less than 90 mg/g during the main spawning season in January (Fig. 5.11). Thus *P. mazatlanica* and *Pt. sterna* show opposing metabolic pathways with regard to the use of

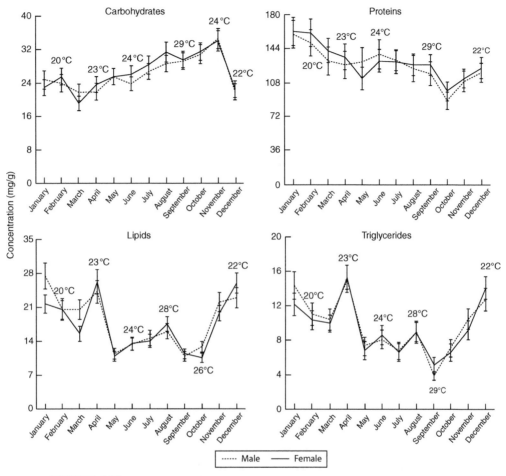

FIGURE 5.10 Temporal and sexual variations in the mean levels of total carbohydrates, proteins, total lipids, and triglycerides of *Pinctada mazatlanica* digestive gland. Bars denote standard deviation.

carbohydrate and protein reserves during gametogenesis; gametogenesis relies primarily on energy obtained from the digestive gland in *Pt. Sterna* and from adductor muscle in *P. mazatlanica*.

5.5.2.4. Mantle tissue

Mantle tissue has a minimal role in the reproductive process of marine bivalves. The only exception is in the Mytilidae, where there are two related physiological functions: (1) the seasonal accumulation of metabolic reserves for gametogenesis and (2) the occurrence of gametogenesis *per se* within its connective tissue matrix (Gabbott, 1975; Bayne *et al.*, 1982; Lowe *et al.*, 1982; Gabbott and Peek, 1991). In other species, the role of mantle

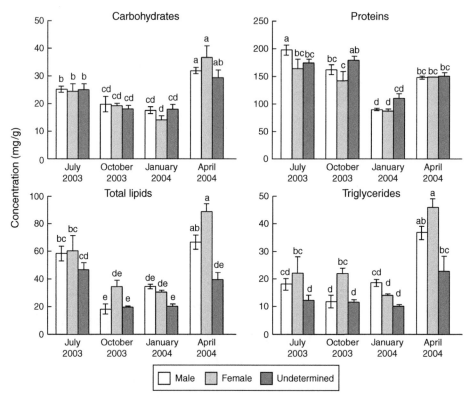

FIGURE 5.11 Temporal and sexual variations in the mean levels of total carbohydrates, proteins, total lipids, and triglycerides of *Pteria sterna* digestive gland. Bars denote standard error.

tissue for storage of nutrients used for reproduction is secondary or negligible. This is true for *P. mazatlanica* (Saucedo *et al.*, 2002a) and *Pt. sterna* (Caceras-Puig, 2007; Vite-García and Saucedo, 2008) where the amounts of glycogen, protein, lipid and triglyceride reserves within mantle tissues are significantly lower than those present in the gonad, digestive gland and adductor muscle. None of these metabolites (with the exception of glycogen that varied from 12 mg/g during the resting period in spring to 3 mg/g during spawning in winter) shows significant seasonal variations during gametogenesis.

5.5.3. Reproductive strategies

Following the pioneering work of Bayne (1976), it has been further demonstrated that conservative and opportunistic strategies are not discrete, since many species adopt a combination of both to ensure somatic and reproductive growth phases (Racotta *et al.*, 1998; Kang *et al.*, 2000; Luna-González *et al.*, 2000; Saucedo *et al.*, 2002a; Racotta *et al.*, 2003; Arellano-Martínez *et al.*, 2004; Vite-García, 2005; Vite-García and Saucedo, 2008). This mixed strategy occurs mostly in response to changes in water temperature and food supply at a specific locality, which in turn depend on other environmental variables that influence gross primary productivity. Therefore, site-dependent variations have been observed for many species. Similar shifts between conservative and opportunistic strategies have also been observed in species inhabiting regionally related localities. This is true for *Pt. sterna* from Bahía de La Paz, Mexico (Cáceres-Puig, 2007; Vite-García and Saucedo, 2008). This species use available food rather than tissue reserves to maintain metabolism and sustain gonad development when primary productivity is high; later, when productivity drops, muscle reserves (either protein or glycogen) are mobilized to the gonad. Conversely, in subtropical *P. mazatlanica* from Bahía de La Paz, the continuity of gonad growth relies mostly on the large amount of nutrients stored in various tissues and cell compartments (e.g. adductor muscle, digestive gland, VCT cells, auxiliary cells, Balbiani body) during winter, and on ingestion of food following winter. Similar information on the strategies used by other species of pearl oyster is lacking.

5.5.4. Bioenergetics of reproduction

The theoretical model of partitioning and allocation of energy hypothesizes that because bivalve molluscs can only ingest and assimilate limited amounts of energy from the environment, somatic growth and reproduction are generally separated in time to maximize reproductive output (MacDonald and Thompson, 1985; Duinker, 2002). Evidently, the different ways by which energy is allocated to growth and reproduction is species dependent (Epp *et al.*, 1988; Ramírez-Llodra, 2002).

Total production (=SFG, see Section 5.5.1) is the amount of energy expended for somatic tissue growth and gamete output. Reproductive effort (RE), the amount of energy allocated specifically to reproduction, is usually expressed as a proportion of total production in the following formula:

$$RE = P_r/(P_r + P_g + P_s)$$

where P_g refers to somatic production or growth, P_r is gamete production and P_s is production of shell. Because of the variety of egg production patterns found among bivalves, a wide range of methodologies has been developed to quantify fecundity. These include direct egg counts in brooding species, spawning induction in live individuals and smear or histological studies of preserved samples (Ramírez-Llodra, 2002). Fecundity is generally correlated with other life-history traits, such as egg size, female size and age, age at first reproduction, reproductive effort, etc. Although reproductive effort and fecundity in bivalves usually decline with age, the timing varies from a combination of exogenous and endogenous cues (MacDonald and Thompson, 1985; Gosling, 2003). It has also been shown that reproductive output in some bivalve species varies not only between populations separated by a few kilometers, but also between consecutive years in a given population (MacDonald and Thompson, 1985).

The concept of separate periods of somatic and reproductive growth is applicable to many bivalve species, particularly those displaying synchronous cyclic resting periods (typical of subtropical and temperate areas). In these species, there are moments when energy allocated to the gonads is stopped and preferentially channeled into growth of somatic tissues (Paulet and Boucher, 1991; Duinker, 2002). Under these conditions, the lack of oocyte development during resting periods might be a strategy to stop oogenesis when the amount of substrate available within ovaries is below a threshold level. It appears, however, that oogenesis never stops completely in many species during the resting period, as it is normal to observe ovaries filled with developing and residual oocytes at the same time. Situations like this may reflect physiological and nutritive stress (e.g. oocyte atresia), but also indicate that momentary activity of new, short-term re-developmental processes are occurring within the gonad. It has been suggested that tissue composition could remain constant as the individual increases in size until attaining sexual maturity (Lodeiros *et al.*, 2001). When this happens, and somatic growth slows to benefit investment in gonad development, chemical composition of somatic tissues can change according to the requirements of reproduction. These short-term shifts in tissue composition indicate partitioning of energy reserves to meet the demands of periods devoted primarily to either reproduction or somatic growth.

The strategies used by pearl oysters to partitioned and allocated energy for growth and reproduction are not yet clear. In *P. mazatlanica* and *Pt. sterna*, for example, allocation of assimilated energy appears to vary during growth, but certainly there are shifts between the different phases of the reproductive cycle (resting, active development, ripeness, spawning and spent), and also between sexes in each species (Vite-García, 2005; Vite-García and Saucedo, 2008, Cáceres-Puig, 2007; Tables 5.3 and 5.4). The build-up of the gonad is supported by energy taken, in decreasing order, from adductor muscle and digestive gland (in *P. mazatlanica*) and from the digestive gland and adductor muscle (in *Pt. sterna*) (Saucedo *et al.*, 2002a; Vite-García, 2005; Vite-García and Saucedo, 2008, Cáceres-Puig, 2007). In *P. mazatlanica*, $\sim 21\,kJ/g$ are provided by the adductor muscle and $\sim 16\,kJ/g$ are provided by the digestive gland. In *Pt. sterna*, $\sim 17\,kJ/g$ are obtained from the digestive gland and $\sim 13\,kJ/g$ from the adductor muscle. In both species, the adductor muscle and digestive gland contribute around two-thirds of the energy needed to meet reproductive demands. In terms of energy per gram tissue, proteins and carbohydrates represent, respectively, the substrates with the highest and lowest caloric values (Tables 5.3 and 5.4). In both species, all somatic tissues (excepting mantle tissue) normally lose around $3\,kJ/g$ of mass during gonad maturation, resulting in a gain of $\sim 10\,kJ/g$ for testis maturation and $\sim 40\,kJ/g$ for ovary maturation. These values are similar to those reported for other bivalves (e.g. Brokordt and Guderley, 2004). Evidently, these caloric differences are more noticeable during the active development and ripeness stages, in which lipids in females more than doubled their caloric values relative to lipids in males and undifferentiated animals.

5.6. FACTORS INFLUENCING REPRODUCTION

5.6.1. Endogenous factors

5.6.1.1. Neuroendocrine control of reproduction

Very little is known about the neuroendocrine regulation of reproduction in bivalves. Several researchers have proposed that the nervous system is linked to gametogenic cycles by means of dynamic feedback controls that operate every time the animal

TABLE 5.3

Changes in mean energy content (kJ/g) of germinal and somatic tissues of *Pteria sterna* in relation to stage of gonad developmental and sex.

| | Gonad tissue | | | | | | Digestive gland | | | | | | Mantle tissue | | | | | | Muscle | | | |
| | Carbohydrates | | Proteins | | Total lipids | | Carbohydrates | | Proteins | | Total lipids | | Carbohydrates | | Proteins | | Total lipids | | Carbohydrates | | Proteins | |
	♂	♀	♂	♀	♂	♀	♂	♀	♂	♀	♂	♀	♂	♀	♂	♀	♂	♀	♂	♀	♂	♀
UND	0.7		3.5		1.3		1.4		12.3		4.0		0.6		6.8		1.1		0.8		15.8	
DEV	0.8	0.8	8.5	7.0	1.9	3.3	1.4	1.1	9.9	9.7	5.7	4.8	0.6	0.7	8.1	7.7	1.2	1.2	1.6	1.3	11	12
RIP	0.8	0.9	10.7	11.8	1.9	4.7	1.7	1.5	9.3	8.0	6.8	6.7	1.0	0.6	9.6	7.8	1.2	1.3	3.1	1.9	10.2	11.0
SPW	0.7	Nf	5.3	Nf	2.1	Nf	1.4	Nf	11.3	Nf	6.2	Nf	0.3	Nf	5.6	Nf	1.2	Nf	2.0	Nf	8.9	Nf
SPE	0.9	0.5	6.1	5.5	1.8	1.7	2.0	1.0	10.3	8.4	8.7	4.2	1.2	0.2	9.4	5.3	1.3	1.2	1.8	1.6	13.7	10.7

UND: undifferentiated; DEV: active development; RIP: ripeness; SPW: spawning; SPE: spent; Nf: no females.

TABLE 5.4

Changes in mean energy content (kJ/g) of germinal and somatic tissues of *Pinctada mazatlanica* in relation to stage of gonad developmental and sex.

| | Gonad tissue | | | | | | Digestive gland | | | | | | Mantle tissue | | | | | | Muscle | | | |
| | Carbohydrates | | Proteins | | Total lipids | | Carbohydrates | | Proteins | | Total lipids | | Carbohydrates | | Proteins | | Total lipids | | Carbohydrates | | Proteins | |
	♂	♀	♂	♀	♂	♀	♂	♀	♂	♀	♂	♀	♂	♀	♂	♀	♂	♀	♂	♀	♂	♀
UND	0.6		5.6		1.2		1.6		9.8		3.6		0.6		6.8		1.1		3.9		14.1	
DEV	0.7	1.1	8.5	10.9	2.1	2.8	1.7	1.8	10.9	11.3	3.7	4.1	0.7	0.5	8.8	8.9	1.7	1.5	4.4	4.8	12.3	14.3
RIP	0.8	0.8	10.5	12.1	2.1	6.4	1.9	1.6	9.8	9.8	3.5	4.1	0.6	0.4	8.2	6.0	2.1	1.9	5.5	5.3	11.6	10.9
SPW	0.5	Nf	6.3	Nf	1.0	Nf	1.4	Nf	10.0	Nf	3.2	Nf	0.9	Nf	7.3	Nf	1.4	Nf	3.9	Nf	11.5	Nf
SPE	0.6	Nf	6.7	Nf	1.3	Nf	1.5	Nf	10.3	Nf	3.3	Nf	0.7	Nf	7.9	Nf	1.2	Nf	4.0	Nf	14.8	Nf

UND: undifferentiated; DEV: active development; RIP: ripeness; SPW: spawning; SPE: spent; Nf: no females.

reaches a certain physiological (reproductive) condition (Sastry, 1979; Sastry and Blake, 1971). A cue (e.g. changes in water temperature, photoperiod, food supply, etc.) received by the animal stimulates a signal and release of a neurosecretion which is transported through the hemolymph to the target tissue (e.g. gonad, digestive gland, mantle, adductor muscle). The signal and neurosecretion usually operate through dynamic "response thresholds" that act as on/off switches to govern all cyclic events associated with reproduction (e.g. stem cells differentiation, gonad development or regression, spawning induction, etc.). Neurosecretions control a wide variety of reproductive processes, such as glycogen metabolism, lipogenesis, storage and mobilization of energy reserves, synthesis of enzymes and biocompounds, etc. (Sastry and Blake, 1971; Bayne, 1976; Sastry, 1979; Mackie, 1984). However, while some information on the roles of neurosecretions (e.g. serotonin) in reproduction of bivalves is available (Matsutani and Nomura, 1986), nothing is known of their roles or mediation in pearl oysters.

5.6.1.2. Genetic Control of reproduction

The influence of genotype in facilitating gametogenesis in bivalves is evident, regardless of changes in environmental factors (Martínez and Mettifogo, 1998; Gosling, 2003). Field and laboratory-based studies showed that transplanted edible oyster stocks maintained the timing of their gametogenic cycle for at least six generations at a new locality; this differed from the timing of the native population (Loosanoff and Davis, 1963; Loosanoff, 1969). However, studies related to genetics of pearl oysters are scarce and they focus primarily on selective breeding and have no relation to the reproductive process (see Chapter 12).

5.6.2. Exogenous factors

5.6.2.1. Water temperature

Lubet and Aloui (1987) considered that all bivalve species have a particular "temperature window" where gametogenesis proceeds optimally and where events such as gonad growth and spawning are either triggered or stopped. Other authors (e.g. Shpigel *et al.*, 1992) emphasized that a water temperature threshold has to be reached to initiate gametogenesis, and after that, continuation of the cycle depends on both water temperature and the level of stored nutrient reserves. Based on these and previous arguments, it is likely that the critical water temperatures required for reproduction of pearl oysters vary considerably between species, between locations and also from year to year (Table 5.1). For example, optimum spawning temperatures range from 20–21°C for temperate Akoya pearl oysters in northern Japan (Ojima and Maeki, 1955) and Korea (Choi and Chang, 2003), to 22–23°C in southern Japan (Wada *et al.*, 1995) and to 27–29°C in the Persian Gulf (Behzadi *et al.*, 1997). For *Pt. sterna*, which is a multi-spawning species that breeds in winter inside the Gulf of California, there are reports of a single spawning season occurring either between 19°C and 22°C near Guaymas, Sonora (Arizmendi-Castillo, 1996) or between 23°C and 25°C in Laguna Ojo de Liebre, in the state of Baja California Sur (Hernández-Olalde, 2003). Conversely, there were two spawning periods for this species in the area of Bahía de La Paz: one from January to February (20°C to 22°C) and other from April to May (23°C to 25°C) (Saucedo and Monteforte, 1997). At northern latitudes near Ensenada, Baja California, there were three spawning peaks over the year: December (18.5°C), February (14.2°C) and August (27.8°C) (Hernández-Díaz and Bückle-Ramírez, 1996).

5.6.2.2. Food availability

Gametogenesis could not proceed beyond a certain point in the scallop, *Argopecten irradians*, if food was absent and water temperature was below a critical point (Sastry, 1968). Low water temperature, <15°C, could be inhibitory in well-fed scallops that had already begun gametogenesis, yet increases in water temperature, without sufficient food, resulted in resorption rather than proliferation of gametes in the tropical mussel, *Perna perna* (Vélez and Epifanio, 1981). Thompson *et al.* (1996) found that under adverse temperatures and food conditions, nutrient reserves in the oyster, *Crassostrea virginica,* were used for metabolism rather than gametogenesis. Adequate food supply as well as appropriate water temperature are required for reproductive maturation to be reached in bivalves (Utting, 1993; Section 7.4.1).

Where bivalves are produced by hatchery culture, the ability to produce juveniles outside their normal reproductive season is desirable. This allows extended hatchery production and flexibility in the timing of production. Manipulation of water temperature and food availability can be used to "condition" bivalve broodstock and is commonly applied in commercial bivalve culture (Gosling, 2003). This approach was used by Saucedo *et al.* (2000) to condition *P. mazatlanica* broodstock prior to hatchery production; however, there is limited information relating to reliable broodstock conditioning protocols for pearl oysters and this area has received relatively little research attention (see Section 7.4.1).

5.7. DEVELOPMENT AND GROWTH

5.7.1. Embryological and early larval development

Spawning in pearl oysters is usually triggered by a change in the oysters's environment, such as a rise or fall in water temperature or change in salinity (Gervis and Sims, 1992), and similar changes are used to induce spawning under culture conditions (see Section 7.4.2). Pearl oysters also spawn in response to water-borne gametes from other individuals (Fig. 14.2). Spawning results from muscular contractions (Tranter, 1958a) and oocytes are activated immediately prior to spawning in the follicle (Tranter, 1958d). Sperm and eggs are released into the water where fertilization takes place. Extrusion of the first and second polar body occurred within 5 minutes and 15–20 minutes of insemination, respectively, in Akoya pearl oysters (Wada *et al.*, 1989), while first polar body extrusion in *P. margaritifera* was recorded after 24 minutes (Doroudi

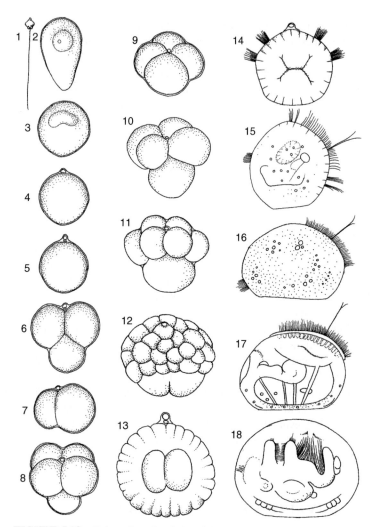

FIGURE 5.12 Embryonic and early larval development of *Pinctada maxima* showing: 1, sperm; 2, egg; 4, fertilized egg with first polar body; 14, development of the prototroch, a band of cilia characteristic of the trochophore larval stage 5 hours after fertilization; 15, late trochophore with apical flagellum and pre- and post-oral tufts of cilia; 17, D-stage veliger, now with a shell and a ciliated bilobed velum protruding from the ventral margin of the shell. Muscles used to retact the velum into the shell are also visible. Redrawn from Wada (1942).

and Southgate, 2003). The pattern of cleavage and embryological development has been reported for *Pinctada maxima* (Wada, 1942; Wada, 1953; Rose and Baker, 1994), *P. margaritifera* (Doroudi and Southgate, 2003) and Akoya pearl oysters (Alagarswami *et al.*, 1983a; Chellam *et al.*, 1991), and is shown for *P. maxima* in Fig. 5.12.

The timing of each of the major developmental stages during embryological and larval development of *P. maxima*, *P. margaritifera* and *P. fucata* is shown in Table 5.5. Rose and Baker (1994) reported the gastrula stage of *P. maxima* to appear within

TABLE 5.5

Embryonic and larval development of some important pearl oyster species (Doroudi and Southgate, 2003).

	Pinctada fucata				*Pinctada maxima*				*Pinctada margaritifera*			
Temperature (°C)	24.3–29.8		25.0–29.5		27–31		26–28		26–30		27–29	
Stage:	S	T	S	T	S	T	S	T	S	T	S	T
Egg	47	–	–	–	59	–	–	–	45	–	49.7	–
Polar body	–	–	–	–	–	–	–	–	–	–	59.9	24 min
Four blastomeres	–	–	60	1 h	–	–	–	–	–	–	60	2 h
Gastrula	–	–	75	5 h	–	–	–	–	–	–	69.6	5 h
Trochophore	–	–	75	8 h	–	–	–	–	–	–	70.4	8–12 h
D-shape veliger	67	20 h	85	24 h	–	–	75	20 h	75	24 h	79.7	24 h
Early umbo	100	–	110	8 d	96	5 d	115	10 d	110	9 d	109.9	8 d
Umbo	135	11 d	140	14 d	125	12 d	–	–	140	12 d	140.7	12 d
Eye-spot	210	15 d	235	21 d	–	–	–	–	210	16 d	230.8	22 d
Authors*	1		2		3		4		5		6	

Mean size (S) of egg or larval diameter or shell length (in µm); time (T) is shown in minutes (min), hours (h) or days (d).
The sources of the data are as follows: 1: Alagarswami *et al.* (1983); 2: Rose and Baker (1994); 3: Minaur (1969); 4: Tanaka and Kumeta (1981); 5: Alagarswami *et al.* (1989); 6: Doroudi and Southgate (2003).

5 hours of fertilization and the trochophore within 8 hours; similar timing for these developmental stages was reported for *P. margaritifera* by Doroudi and Southgate (2003) (Table 5.5). Development of *P. mazatlanica* was reported to be slow by comparison, reaching the blastula, gastrula and trochophore stages 4–5 hours, 6–12 hours and 18–24 hours after fertilization, respectively (Martinez-Fernandez *et al.*, 2003). Chellam *et al.* (1991) reported that *P. fucata* reached the blastula and gastrula stages within 5 and 7 hours of fertilization, respectively, but did not report timing of the trochophore stage.

The trochophore stage measures around 75 µm in *P. maxima* and 70 µm in *P. margaritifera* (Table 5.5). Trochophores are broader at the anterior end which has an apical flagellum providing propulsion (Fig. 5.12). Thay have pre-oral and post-oral tufts of cilia which assist with movement. Dorsal ectodermal cells begin to secrete the embryonic shell or prodissoconch I, which is fully formed when the larvae reach the veliger stage (Fig. 5.12). Early veliger larvae have a characteristic D-shaped shell giving rise to the common names of "D-stage" or "D-shaped" veliger (Figs. 5.12 and 5.13a). This stage is reached around 20–24 hours after fertilization by the major culture species (Tanaka and Kumeta, 1981; Alagarswami *et al.*, 1983b; Alagarswami *et al.*, 1989; Rose and Baker, 1994; Doroudi and Southgate, 2003) (Table 5.5). Veligers possess a bilobed, ciliated organ, the velum, which provides propulsion and a means to capture food particles (Fig. 5.12). The antero-posterior axis of the shell of newly formed

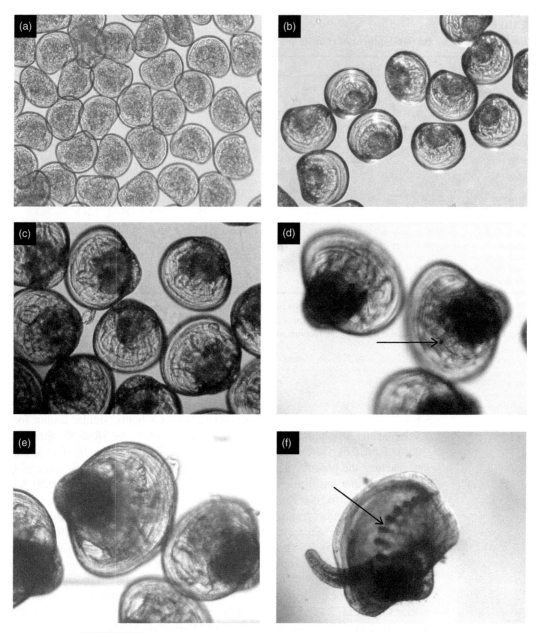

FIGURE 5.13   Stages in the development of *Pinctada maxima* larvae: (a) D-stage veliger larvae 19 hours after fertilization; (b) 3 day old veliger larvae; (c) 8 day old veliger larvae with prominent umbo development; (d) 12 day old "eyed" larvae (arrow indicating position of eye-spot); (e) 14 day old pediveliger larvae (larger specimen showing partially extended foot); (f) 18 day old plantigrade showing extended foot and developing gill (arrowed) visible through shell. (Photos: Jens Knauer).

D-stage veligers measure 75–85 μm in *P. maxima*, 75–80 μm in *P. margaritifera*, around 67 μm in Akoya pearl oysters (Table 5.5) and ca. 81 μm in *Pt. penguin* (Beer, 1999).

5.7.2. Larval development and growth

Larval development in pearl oysters requires from 16 to 30 days and is influenced by factors such as water temperature, nutrition and the availability of an appropriate settlement substrate (Gervis and Sims, 1992). D-stage veligers develop preliminary growth rings in the shell 1–2 days after fertilization (Fig. 5.13b) and begin to develop opposing umbos which arise dorsally above the hinge axis. The umbo stage (Fig. 5.13c) is reached 11–14 days after fertilization, depending on species, when larvae measure around 125–140 μm along the antero-posterior axis (Table 5.5). As larvae continue to grow, the umbos become increasingly prominent (Fig. 5.13d) as does the larval foot (Fig. 5.13e). Larvae develop a pigmented spot, commonly called the "eye-spot", on either side of the base of the foot primordium (Fig. 5.13d) and ctenidial ridges begin to develop. In *P. margaritifera* and *P. fucata*, the eye-spots are darkly pigmented (brown/black) and they develop in larvae with an antero-posterior measurement (APM) of ca. 210–230 μm (Alagarswami et al., 1983b, 1989; Doroudi and Southgate, 2003). The eye-spots of *P. maxima* larvae are red in color. They first appeared as faint 5–7 μm diameter spots in larvae with an APM of 205 μm, and attain a diameter of 10 μm in larvae with an APM of around 230 μm (Rose and Baker, 1994). The presence of an eye-spot indicates competence to metamorphose (Rose and Baker, 1994) and that settlement is imminent. The well developed foot is used during short periods on the substrate as the larvae enter the transitional stage between swimming and crawling. Eye-spot larvae are seen around 15–22 days after fertilization and have an APM of 210–235 μm depending on the species (Table 5.5).

Larvae with a well developed foot, known as pediveligers, select an appropriate substrate for settlement. The velum is lost and gills develop during the transition between the velum- and gill-based feeding systems (Fig. 5.13f). This is a period when bivalve larvae must draw on endogenous energy reserves accumulated during larval development (Holland and Spencer, 1973; Holland, 1978; see Section 5.7.3). Byssal threads secreted by the byssal gland are used to attach plantigrade post-larvae to the substrate. However, they retain the ability to move away from unfavorable conditions by ejecting byssus, crawling to a more favorable position and secreting new byssal attachments (Taylor et al., 1997a). Shell growth of newly settled pearl oysters is rapid in the form of the thin transparent dissoconch (adult shell) produced around the ventral and lateral shell margins (Fig. 5.14a). This process extends the hinge line laterally and the young juveniles, or spat, begin to resemble adult oysters in shape. Newly settled spat of *P. maxima* averaged an APM of 349 μm and a dorso-ventral measurement (DVM) of 289 μm at 28–35 days old (Rose and Baker, 1994) whereas Akoya pearl oyster spat had an APM of around 300 μm at 24 days of age (Alagarswami et al., 1983b). Rose and Baker (1994) noted that the transparent dissoconch of *P. maxima* spat (Fig. 5.14b) gradually became opaque with increasing mantle pigmentation.

Around 24 days after settlement, *P. maxima* spat begin to develop growth processes on the left (upper) shell valve and, by 39 days after settlement, spat with an average

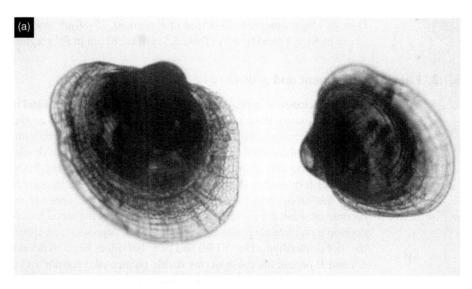

FIGURE 5.14 Young spat of (a) *Pinctada margaritifera* showing development of dissoconch and (b) 26 day old *Pinctada maxima* spat showing well developed gills and pigmented mantle margin with projections (Photo: Jens Knauer).

shell length of 6 mm had 4 or 5 growth processes on both shell valves (Rose and Baker, 1994). Growth processes arise initially from radial ridges running from the umbo to the shell margin and are broader distally than proximally. Similarly shaped growth processes are found on the shells of *P. margaritifera* while those of the Akoya pearl oyster differ by tapering to a fine point (Rose and Baker, 1994; Section 2.3.1.1).

Growth rates of pearl oyster larvae are influenced by their surrounding environment, in particular food availability (Doroudi *et al.*, 1999a; Doroudi and Southgate,

2002), food quality (Martinez-Fernandez *et al.*, 2006) and water quality (Doroudi *et al.*, 1999b)(see Section 7.4.6). Tanaka *et al.* (1970) reported a growth rate of 5 μm per day for *P. margaritifera* during the first 7 days of larval rearing, whereas Doroudi *et al.* (1999a) reported a mean daily growth rate for the same species of 7.2 μm over the 22 day period up to the eye-spot stage. *P. margaritifera*, *P. fucata* and *P. maxima* reach settlement at approximately the same size (230–266 μm) and age 20–23 days from fertilization (Rose and Baker, 1994; Doroudi and Southgate, 2003).

5.7.3. Larval energetics

Pearl oysters show similar patterns of embryological and larval development to that of other bivalves. The first feeding D-stage veliger develops within approximately 20–24 hours of fertilization. Development up to this point in temperate bivalves is generally fueled by maternally-derived lipid reserves (Gallager and Mann, 1986; Gallager *et al.*, 1986). D-stage veliger larvae actively feed and begin to accumulate energy reserves that will fuel metamorphosis. Development in bivalves incorporates a transitional phase from the velum-based feeding system of larvae to a gill-based feeding system in newly metamorphosed spat. During this process, the velum is resorbed or shed (Little, 1998) as the gills develop (Figs. 5.13f and 5.14b). Until the gills are fully formed the larvae are unable to feed (Cole, 1938) and further development and metamorphosis relies on endogenous energy reserves accumulated during larval life (Holland and Spencer, 1973; Holland, 1978). In the larvae of the majority of bivalves so far examined, lipid, specifically neutral lipid, is accumulated during larval development (Holland and Spencer, 1973; Whyte *et al.*, 1987) and is utilized as an energy reserve during metamorphosis until the gill-based feeding system is established (Holland, 1978). Success at metamorphosis is thought to be closely correlated with the level of lipid reserves accumulated during larval development (Holland and Spencer, 1973). Similar biochemical and energetic trends have been found in the tissues of *P. margaritifera* larvae during development. Strugnell and Southgate (2003) reported that depletion of lipid reserves between day 1 and day 4 contributed 56% of the total energy utilized by larvae during this period; protein also contributed substantially (40%) to this total. Lipid subsequently accumulated in *P. margaritifera* larvae between day 18 and day 21 immediately prior to settlement. It contributed almost 70% of the total energy gain per larva during this period (Strugnell and Southgate, 2003), suggesting that lipid is the primary energy reserve utilized by *P. margaritifera* during metamorphosis.

5.7.4. Growth measurement

Growth in bivalves is commonly determined by measuring changes in shell dimension(s) over time. While it would be more accurate to determine growth as a function of tissue mass, this is impossible to do by non-destructive means and, on this basis, shell measurement is the most common method of determining growth rate in pearl oysters (Gervis and Sims, 1992; Rose and Baker, 1994; Southgate and Beer, 1997).

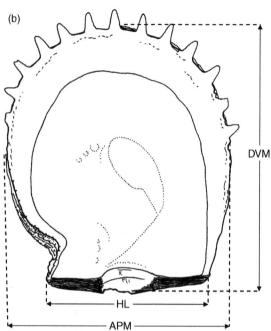

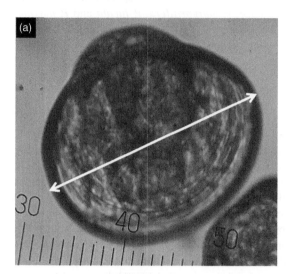

FIGURE 5.15 Measurement of (a) larval pearl oyster shell with arrow indicating antero-posterior measurement (APM) and (b) adult pearl oyster shell, showing: APM; DVM: dorso-ventral measurement; and HL: hinge length.

The longest shell axis of pearl oyster larvae is the antero-posterior axis or APM (Fig. 5.15a). However, in juveniles and adults the longest axis becomes the dorso-ventral axis and the dorso-ventral measurement (DVM) extends perpendicular from the hinge line to the distal shell margin (Hynd, 1955; Fig. 5.15b). Measurement of the dorso-ventral axis of pearl oysters can be complicated by the growth processes (fingers) that protrude from the growing edge of the shell (Fig. 5.15b); they are fragile and may vary in length. DVM is therefore measured to the base of the growth processes (Fig. 5.15b).

DVM is commonly used for comparing growth rates of juvenile and adult pearl oysters (Gervis and Sims, 1992) and to determine legal or operable sizes for them (Hynd, 1955; Alagarswami and Chellam, 1977). DVM is often referred to as "shell height" and APM as "shell length" (e.g. Taylor *et al.*, 1997d).

Other measures which have been used in studies with pearl oysters include: hinge length (HL), the distance along the hinge line between the tips of the anterior and posterior auricles (Fig. 5.15b); heel depth, the thickness of the valve at the hinge line; hinge width, the gape between the dorsal edge of each hinge line; and thickness or shell width, which is the maximum distance between the outer surfaces of the two shell valves when closed. The latter is influential in the size of nucleus that can be implanted for pearl production (see Section 8.9).

5.7.5. Growth rates

Growth rates of bivalves can be determined as either absolute or relative changes (Gosling, 2003). True growth rates determine the change in size over a specified period or "the velocity of growth" (Malouf and Bricelj, 1989). Relative growth rate plots growth increment per unit body mass per unit time; it is typically rapid in young animals and slows with age. Absolute growth rate, which determines the rate of increase in size or mass per unit time, is commonly used to describe growth of cultured bivalves and is particularly valuable when comparing between treatments within the same study. It is influenced by the size and age of bivalves as well as seasonal variations in food, water temperature and locations. The relationship between size and age is represented by a sinusoidal curve which reflects a period of rapid growth (the exponential phase of growth), followed by an inflection and a more gentle slope towards asymptotic length (L_∞) (Southgate and Lucas, 2003). Pearl oysters, for example show a rapid initial increase in shell size to near maximum size subsequent to increase in shell thickness (Gervis and Sims, 1992).

5.7.6. Modeling growth of pearl oysters

Site-related variation in growth has been reported for *Pinctada maxima* (Saville-Kent, 1893; Gervis and Sims, 1992), *P. margaritifera* (Friedman and Southgate, 1999), *Pt. penguin* (Smitasiri *et al.*, 1994) and Akoya pearl oysters (Alagarswami, 1970; Nalluchinnappan *et al.*, 1982). It is important from a commercial perspective to be able to accurately determine and compare the growth rates of pearl oysters at different sites, and to be able to estimate their growth potential. However, measurement of size alone provides insufficient information to fully describe and compare growth. A combination of size, the time to reach a particular size (i.e. growth rate) and the maximum size an animal can theoretically attain, better describes growth performance of a species under given culture conditions (Vakily, 1992). This is generally achieved by fitting mathematical models to length or weight data.

The model that has found widest application in fisheries science is the von Bertalanffy Growth Function (VBGF). It is preferred over other models as it is biologically interpretable, can be used for comparative growth studies and its parameters are relatively easy to determine (Vakily, 1992). Such models can be used to estimate important growth parameters such as asymptotic length (L_∞), maximum length (L_{max}) and the growth coefficient (K), which is often used to compare growth rates between populations and species. The VBGF has been widely used to estimate growth parameters of pearl oysters, including *P. margaritifera* (Nasr, 1984; Pouvreau *et al.*, 2000c; Pouvreau and Prasil, 2001; Yukihira *et al.*, 2006), *P. imbricata* (Prajneshu and Venugopalan, 1999; Urban, 2000; Urban, 2002; Marcano *et al.*, 2005), *P. mazatlanica* (Saucedo and Monteforte, 1997) and *P. maxima* (Yukihira *et al.*, 2006; Lee *et al.*, 2008) (Table 5.6).

In much of the research with bivalves, age and length-at-age are estimated (e.g. Quayle and Newkirk, 1989; Pouvreau and Prasil, 2001), however, with the advent of hatchery production, the absolute age of an individual is known and the parameters of growth models may be estimated with greater accuracy. In a recent study with hatchery

TABLE 5.6
von Bertalanffy growth parameters reported for pearl oysters at various locations.

Species	Location	Growth parameters		Author(s)
		L_∞ (mm)	K (per year)	
Pinctada margaritifera	Cook Islands (wild)	183	0.260	Sims (1994)
	Cook Islands (cultured)	310–157	0.254–0.528	Sims (1994)
	French Polynesia	161	0.46	Pouvreau *et al.* (2000a)
	French Polynesia	147–186.5	0.42–0.58	Pouvreau and Prasil (2001)
	Queensland, Australia	136–157	0.54–0.58	Yukihira *et al.* (2006)
	Kenya	127.2	0.3	Mavuti *et al.* (2006)
	Red Sea	153	1.52	Nasr (1984)
Pinctada maxima	Indonesia	168.38	0.930	Lee *et al.* (2008)
	Vietnam	260	0.4816	Tuyen and Tuan (2000)
	Queensland, Australia	205–229	0.39–0.41	Yukihira *et al.* (2006)
	Western Australia	194–210	0.72–0.79	Hart and Joll (2006)
	Thailand*	168	0.85	Traithong *et al.* (1996)
Pinctada imbricata	Venezuela	85.15	1.42	Marcano *et al.* (2005)
	Colombia (Caribbean)	84	0.939	Urban (2000)
	Colombia (Caribbean)	65.7	1.767	Urban (2002)
Pinctada fucata	India	79.31	0.075 (per month)	Chellam (1998)
Pinctada mazatlanica	Mexico	110	0.45	Saucedo *et al.* (1998)
Pteria sterna	Mexico	100	0.69	Saucedo *et al.* (1998)
Pteria colymbus	Venezuela	71		Lodeiros *et al.* (1999)

L_∞: asymptotic length, K: growth coefficient.
*Values from Yukihira *et al.* (2006) by calculation from original data given by authors (source not accessed).

cultured *P. maxima* in Indonesia, Lee *et al.* (2008) determined the suitability of five growth models (Gompertz, Richards, Logistic, Special VBGF and General VBGF) to length-at-age data. The criteria used to determine the best fit model were a low mean residual sum of squares (MRSS), high coefficient of determination (r^2) and low deviation of the asymptotic length (L_∞) from the maximum length (L_{max}). The Special VBGF yielded the best fit ($L_\infty = 168.38$ mm; $K = 0.930$/year; $t_0 = 0.126$; MRSS $= 208.64$; $r^2 = 0.802$; Deviation of L_∞ from $L_{max} = 37.52$ mm) (Table 5.7; Fig. 5.16).

Most pearl farms are able to choose from a variety of potential farm sites within the lease area available to them. These sites vary in their ability to support pearl oyster growth because of differences in water quality parameters and food availability which may also vary on a temporal basis. Pearl farms may therefore use different sites to culture pearl oysters at different times of the year or to culture pearl oysters of different ages. The parameters generated by growth models such as the VBGF allow comparison between sites and identification of those supporting greater growth rates. In bivalve culture generally, sites are chosen to maximize growth rate and reduce the time required for the culture species to reach market size. However, this is not an exclusive requirement in pearl oyster culture because while it is desirable to maximize the growth rate

TABLE 5.7

Growth parameters of models fitted to growth data for *Pinctada maxima* aged 0.58–4.83 years ($n = 8010$).

Model	Formula	L_∞ (mm)	K (per year)	t_0	B (per year)	MRSS	r^2	Dev
Special VBGF	$L_\infty[1-e^{-k(t-t_0)}]$	168.38	0.930	0.126		208.64	0.802	37.52
General VBGF	$L_\infty[1-e^{-k(t-t_0)}]^b$	172.86	0.651	0.414	0.514	207.23	0.803	33.04
Gompertz	$L_\infty e[-e^{(-k(t-t_0))}]$	165.85	1.232	0.565		212.20	0.798	40.07
Richards	$L_\infty[1-be^{(-k(t-t_0))}]^{1/b}$	165.85	0.0001	1.232	0.00006	212.23	0.798	40.05
Logistic	$L_\infty[1 + e^{-k(t-t_0)}]^{-1}$	172.43	0.200	0.748		240.79	0.771	33.47

L_∞: asymptotic length, K: growth coefficient, t_0: theoretical age when length = 0, MRSS: mean residual sum of squares, $B_{Richards}$: growth parameter[*], $B_{GenVBGF}$: surface factor[*], Dev: deviation of L_∞ from L_{max}[*]. [*]As defined by Urban (2002) (*source*: Lee *et al.*, 2008).

of juveniles to a size where they can be used for pearl production, slower growth rates are required for conditioning of oysters prior to pearl seeding (see Section 8.9.2) and to improve pearl quality immediately prior to pearl harvest (see Section 8.9.7). Lee *et al.* (2008), for example, identified sites with high K and L_∞ values as being preferred for culturing juvenile *P. maxima* and sites with low K values as being preferred for weakening *P. maxima* prior to pearl seeding. Sites with low K values would also be preferred for optimizing pearl quality as it is generally acknowledged that slower nacre deposition results in higher nacre quality. Shell growth and the rate of deposition of nacre onto the pearl nucleus are strongly correlated (Coeroli and Mizuno, 1985).

5.7.7. Other growth parameters

The growth index "phi prime" (ϕ') is calculated as:

$$\phi' = \log_{10}K + 2 \times \log_{10}L_\infty$$

where K is the growth constant and L_∞ is asymptotic DVM calculated using the VBGF (Pauly and Munro, 1984). It has been used in a number of studies for comparing growth rates of oysters between sites or those held under different conditions (Gervis and Sims, 1992; Friedman and Southgate, 1999; Pouvreau and Prasil, 2001; Yukihira *et al.*, 2006). Another growth index, $T_{(120)}$ or $T_{(100)}$, determines the time required for oysters to reach a size suitable for pearl production (e.g., 120 mm DVM for *P. maxima* and 100 mm DVM for *P. margaritifera*; Gervis and Sims, 1992) and has clear practical value. It has been used to compare culture sites and culture conditions (Gervis and Sims, 1992; Pouvreau and Prasil, 2001; Yukihira *et al.*, 2006). For example, *P. margaritifera* held in deep water (35 m) had ϕ' values of 3.67–3.77 and $T_{(120)}$ between 6.5 and 6.8 years, while those at a depth of 15–17 m showed much better growth with a ϕ' value of 4.02 and $T_{(120)}$ of 2.7 years (Sims, 1990).

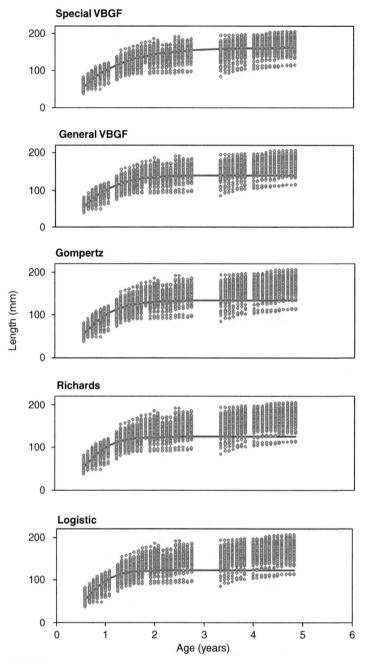

FIGURE 5.16 Changes in shell length (dorso-ventral measurement) of *Pinctada maxima* cultured in West Papua, Indonesia, plotted against age (years) and fitted with various growth models; Special VBGF, General VBGF, Gompertz, Richards and Logistic. (Lee *et al.*, 2008).

5.7.8. Shell dimensions and morphometrics

A number of studies have reported on the relationships between various shell dimensions, or between shell dimensions and weight, in pearl oysters (e.g. Galtsoff, 1931; Alagarswami and Chellam, 1977; Yoo *et al.*, 1986; Gaytan-Mondragon *et al.*, 1993; Southgate and Beer, 1997; Saucedo *et al.*, 1998). Such data can be used in assessing the effects of different growth conditions and to describe ontogenetic changes (Gervis and Sims, 1992). *Pinctada* spp. generally undergo changes in shell shape with age. For example, Alagarswami and Chellam (1977) found that young (<36 mm) Akoya pearl oysters changed from a sub-quadrate shape to an oblong shape. Similarly, *P. mazatlanica* was reported to undergo isometric growth up to a size of around 100 mm when shape began to change from being "symmetrical" to "oblong" (Saucedo *et al.*, 1998). Strong variation in shell morphology was reported for *Pt. sterna*, even in oysters of the same size and age, with no clear pattern of isometric or allometric growth in young or adult oysters (Saucedo *et al.*, 1998). Morphometric relationships may also be influenced by culture conditions. The ratio of shell height to shell length (DVM:APM) of *P. maxima* juveniles is influenced by stocking density and increased from 0.85 for juveniles held at a density of 1.3/100 cm², to 0.93 for those at a density of 20/100 cm² (Taylor *et al.*, 1997d).

5.7.9. Scope for growth

As outlined above in Section 5.5.1, SFG is a measure of the balance between energy uptake and energy expenditure in an organism, with energy in excess to that required for respiration, metabolism and excretion, available for growth and reproduction. SFG has been applied widely to bivalves, including pearl oysters (e.g. Yukihira *et al.*, 1998) and a detailed discussion of this is provided in Section 4.3.3.

5.7.10. Factors affecting growth

The minimum desirable DVM for pearl seeding or nucleus implantation is 50 mm, ≥100 mm and ≥120 mm for Akoya pearl oysters, *P. margaritifera* and *P. maxima*, respectively (Gervis and Sims, 1992). Clearly it is desirable to minimize the time required for oysters to reach this size. Rapid growth is also desirable after pearl seeding because growth rate is correlated to the rate of nacre deposition (Coeroli and Mizuno, 1985) and larger pearls are generally more valuable (see Section 8.10). Growth rates of pearl oysters are influenced by many interacting endogenous (e.g. genetic potential and neurohormonal expression) and exogenous (e.g. environmental and culture conditions) factors (Pouvreau and Prasil, 2001). An understanding of these factors has important commercial implications for the cultured pearl industry.

5.7.10.1. Major environmental parameters

Correlation between pearl oyster growth and food availability has been shown in field studies (e.g. Yamaguchi and Hasuo, 1977; Numaguchi, 1994a; Pouvreau *et al.*, 2000a) and in laboratory studies with juveniles and larvae (e.g. del Rio-Portilla *et al.*, 1992; Doroudi *et al.*, 1999a; Doroudi and Southgate, 2002). Food availability itself is influenced by other environmental parameters such water temperature, depth and nutrient load and has been shown to vary seasonally at pearl culture sites (Numaguchi, 1994b; Tomaru *et al.*, 2002). It is also influenced by culture conditions, in particular, characteristics of culture units such as mesh size and the degree fouling (both of which influence water flow) and stocking density which influences the food available to individual oysters (see below). The influences of different types of culture units and stocking density on pearl oyster growth are discussed in detail in Section 7.6.

Food particles available to pearl oysters must be of an appropriate size and have favorable morphology to allow ingestion and digestion. Rose (1990), for example, reported that *P. maxima* larvae were unable to digest *Nannochloropsis oculata* and that small larvae had difficulty ingesting cells of the diatom *Chaetoceros gracilis*. A similar study with *P. margaritifera* larvae showed preferential ingestion of flagellates over diatoms (Doroudi *et al.*, 2003). *Nitzschia closterium* is a poor food for Akoya pearl oysters when fed alone (Numaguchi, 2000) possibly because of inedibility. Indeed, oyster mortalities and loss of condition have been reported under field culture conditions in Japan, when *Nitzschia* spp. predominate the natural phytoplankton population (Fukushima, 1970, 1972; Tomaru *et al.*, 2001; Tomaru *et al.*, 2002).

Growth rates of pearl oysters are strongly influenced by the nutritional composition of ingested food particles (Taylor *et al.*, 1997c; Southgate *et al.*, 1998; Martinez-Fernandez *et al.*, 2006; Martinez-Fernandez and Southgate, 2007; see Section 7.4.6.4). Unfortunately, we know little of the nutritional requirements of pearl oysters although some inferences can be made from our knowledge of the general nutritional requirements of bivalves (Knauer and Southgate, 1999). Growth rates of *P. margaritifera* larvae were, however, correlated with the levels of certain nutrients in micro-algae fed to them (Martinez-Fernandez *et al.*, 2006; see Section 7.4.6.4); of particular importance were dietary carbohydrate and highly unsaturated fatty acids. Under hatchery or laboratory conditions, where pearl oysters are fed micro-algal diets composed of a small number of species (e.g. Rose and Baker, 1994; Southgate and Beer, 1997; see Section 7.4.6.4), the nutritional composition of those species is an important consideration. Under field conditions, however, many species of micro-algae and other particulate organic matter (POM) are available to pearl oysters and nutrient deficiency is improbable.

Water temperature has an acute effect on the metabolic rate and feeding activity of pearl oysters (e.g. Numaguchi and Tanaka, 1986; Yukihira *et al.*, 2000) which in turn influences growth rate (Pouvreau and Prasil, 2001). A positive relationship between growth rate and water temperature has been reported in a number of studies with pearl oysters (Table 6.3). The effects of water temperature on metabolism and growth of pearl oysters are detailed in Sections 6.3 and the reader is referred to this section for more detail. Water temperature is an important influence on the thickness of nacre layers (Chellam, 1987) and their rate of deposition (Hollyer, 1984). It is generally acknowledged that thinner layers of nacre produce better quality nacre with greater luster (Matsui, 1958) and pearls are commonly harvested in the cooler months of the year because of this (see Section 8.9.7).

Numaguchi and Tanaka (1986) reported better growth rates of Akoya pearl oyster spat at lower than ambient salinities of 26.5‰ and 18.9‰, when compared to other salinities tested over a 25 day period (15.1‰, 30.3‰, 34.1‰ and ambient salinity of 37.9‰). The Japanese Akoya industry prefers to culture oysters in areas with freshwater influx as it improves pearl quality (Gervis and Sims, 1992). *P. maxima* juveniles can also tolerate a broad range of salinities (25–45‰) without a significant effect on survival (Taylor *et al.*, 2004). However, growth rate is influenced by salinity and, like Akoya pearl oysters (Numaguchi and Tanaka, 1986), growth rate was improved at a reduced salinity (30‰) when compared to ambient salinity (34‰) (Taylor *et al.*, 2004). The effects of salinity on physiological functions and growth of pearl oysters are discussed in detail in Section 6.4.3 and the reader referred to this section for more detail. Salinity ranges favorable to good growth and survival of various life stages of pearl oysters are shown in Table 6.6.

Co-culture of Akoya pearl oysters with the red seaweed *Kappaphycus alvarezii* in China was shown to improve growth rates of both pearl oysters and seaweed at a water temperature of 20°C (Qian *et al.*, 1996). Seaweed growth was positively influenced by nitrogenous metabolites produced by the pearl oysters and oyster growth was presumed to have benefited from improved water quality (Qian *et al.*, 1996). In subsequent research, pearls with thicker and more homogeneous nacre coverage were produced by Akoya oysters that were co-cultured with *K. alvarezii* (Wu *et al.*, 2003). Moderation of water quality, particularly pH, O_2 and CO_2, resulting from seaweed metabolism was assumed to provide a more suitable environment for nacre deposition (Wu *et al.*, 2003).

The effects of other environmental parameters including dissolved oxygen, currents, depth, aerial exposure, turbidity, pollution and toxins, on the physiological processes and growth of pearl oysters are discussed in detail in Chapter 6.

5.7.10.2. Other factors

Many studies have presented data showing the effects of different culture sites, and different positions within a site, on growth rates of pearl oysters. These differences relate primarily to differences in water temperature, flow rate and food availability. Sims (1994) reported on the growth rates of *P. margaritifera* held under different environmental and culture conditions in the Cook Islands. Oysters on chaplets (see Section 7.5.4.1) were either suspended from a longlines or from platforms fixed by legs to the sea bed. He reported L_∞, K, ϕ' and $T_{(120)}$ values of 241.9, 0.322, 4.24 and 1.4, respectively, for platform cultured oysters and 309.7, 0.254, 4.39 and 1.2, respectively, for longline cultured oysters at the same site. Sims (1994) also reported considerable differences in these parameters between sites regardless of culture method. Wild oysters at a depth of 35 m recorded L_∞, K, ϕ' and $T_{(120)}$ values of 130.7, 0.346, 3.77 and 6.5, respectively, on sand and 170.7, 0.162, 3.67 and 6.8, respectively, on rock; both showed considerably slower growth than wild oysters in shallower water (15–17 m), which had L_∞, K, ϕ' and $T_{(120)}$ values of 151.1, 0.459, 4.02 and 2.7, respectively (Sims, 1994).

The growth rates of *P. margaritifera* held on longlines at a depth of 7 m at nine sites in French Polynesia were described by Pouvreau and Prasil (2001). They reported L_∞

to range from 147 to 186.5 between sites with corresponding K values of between 0.42 and 0.58. The growth performance index, ϕ', was lower when cultured in atoll lagoons ($\phi' = 4.09$) than at islands or in the open ocean ($\phi' = 4.16$). $T_{(100)}$ was also greater at atoll sites ($T_{(100)} = 25$ months) than at islands sites or in the open ocean ($T_{(100)} = 22$ months). The authors pointed out that atoll lagoon waters in French Polynesia are characterized by low levels of POM compared to island lagoons, but that open ocean sites are known to have lower levels of POM than lagoons (Charpy *et al.*, 1997). Clearly differences in POM could not alone explain their results. Pouvreau and Prasil (2001) hypothesized that the high pumping rate of *P. margaritifera*, water flow, and the rate of food renewal at the study sites, were more likely to explain growth differences between sites.

Yukihira *et al.* (2006) assessed growth of cultured *P. maxima* and *P. margaritifera* at two dissimilar sites; a mainland bay with relatively high suspended particulate matter (SPM; range, 2–60 mg/L, mean, 11.1 mg/L) and high POM (mean 2.1 mg/L), and an off-shore coral reef site with SPM and POM levels of 1.4 mg/L and 0.56 mg/L, respectively. Growth rates of *P. maxima* did not differ significantly between sites but the growth of *P. margaritifera*, which typically lives in oligotrophic waters with low turbidity, was adversely affected by high levels of SPM at the mainland site (Yukihira *et al.*, 2006).

The major predators of pearl oysters are described in Section 11.8 and, while predation is normally associated with mortality of pearl oysters, the action of some predators is more evident by its impact on growth rates. "Fish grazing" was described as a cause of non-nacreous shell loss in *P. margaritifera* juveniles in the Cook Islands (Sims, 1994). Similar damage was caused to *P. margaritifera* juveniles in Australia by leatherjackets, *Paramonacanthus japonicus*, which lived within oyster culture units and trimmed the growing margin of the shell (Southgate and Beer, 1997). The activities of these fish reduced the mean DVM, APM, HL and wet weight of the oysters (Southgate and Beer, 1997).

5.8. SUMMARY

Pearl oysters are typical marine bivalves in many features of their reproductive biology, embryological and larval development, and growth. They are protandrous hermaphrodites with various sex reversals during their lifetime in response to complex interactions of endogenous and environmental factors. Thus, overcrowded oysters may be predominantly males because spermatogenesis is energetically cheaper than oogenesis. Pearl oysters have diffuse gonadial tissue, which is composed of small granular bags, acini. The acini contain stem cells, which may develop into oocytes or spermatocytes, the gametogenic processes being largely supported by energy and metabolites from the digestive gland and adductor muscle. The overall pattern of reproduction in pearl oyster populations is synchronous, with male and females undergoing sequential processes that lead to a simultaneous breeding period. This is a cyclical process with three general phases: vegetative phase, during which the oysters display no evidence of sexual activity; cellular differentiation phase, during which the proliferation and differentiation of stem cells lead to mature gametes and spawning; recovery phase, during which there is regression of the gonad after spawning and recovery of nutrients from unspawned gametes for recycling. Pearl oyster populations may spawn once or multiple times during a year. While regular patterns have been identified for species in some regions, there are inter-, intra-specific and inter-annual variations in frequency and timing of spawning. Some of this variation is genetic, but much is due to differences in important environmental conditions such as water temperature and food availability.

Spawning in pearl oysters is usually triggered by a change in the environment or presence of water-borne gametes. Early development follows the typical marine bivalve pattern of trochophore, D-stage veliger, umbo stage, eye-spot stage, pediveliger, metamorphosis and newly settled spat. It takes in the order of 3–4 weeks. Development rates are particularly influenced by food availability and stored lipid is probably the primary energy reserve used during metamorphosis.

The relationship between size and age in juvenile and adult pearl oysters is represented by a sinusoidal curve, which reflects a period of exponential growth followed by an inflection and gentle slope towards asymptotic length (L_∞). The Special von Bertalanffy Growth Function (VBGF) best models this size–age relationship. Growth of juvenile and adult pearl oysters is, like other aspects of their biology, strongly influenced by environmental factors. Even the method of culture, e.g. longline or rack, will affect their growth rates in a similar environment.

Our understanding of reproduction in pearl oysters is still rather rudimentary and relates primarily to reproductive seasonality. Greater knowledge of the energetic and nutritional basis of gamete development in pearl oysters will improve gamete quality and hatchery production, and drive research to develop appropriate methods for out-of-season conditioning. Furthermore, management of growth rates requires better understanding of the biotic and abiotic factors involved, particularly in terms of the sites and methods used for pearl oyster culture. Commercial pressures will insure continuing research in this field.

References

Acosta-Salmón, H., Southgate, P.C., 2005. Histological changes in the gonad of the blacklip pearl oyster (*Pinctada margaritifera* Linnaeus, 1758) during the reproductive season in north Queensland, Australia. *Moll. Res.* 25, 71–74.

Alagarswami, K., 1970. Pearl culture in Japan. Lessons for India. *Proceedings of the symposium on molluska*, Part III, Marine Biological Association, India, pp. 975–998.

Alagarswami, K., Chellam, A., 1977. Change of form and dimensional relationship in the pearl oyster *Pinctada fucata* from the Gulf of Mannar. *Indian J. Fish.* 24, 1–14.

Alagarswami, K., Dharmaraj, S., Velayudhan, T.S., Chellam, A., Victor, A.C.C., 1983a. Embryonic and early larval development of pearl oyster *Pinctada fucata* (Gould). Symp. On Coastal Aquaculture, 12–18 Januatu 1980. Part 2: Molluscan Culture. *Symp. Mar. Biol. Assoc. India* 6, 598–603.

Alagarswami, K., Dharmaraj, S., Velayudhan, T.S., Chellam, A., Victor, A.C.C., Gandhi, A.D., 1983b. Larval rearing and production of spat of pearl oyster *Pinctada fucata* (Gould). *Aquaculture* 34, 287–301.

Alagarswami, K., Dharmaraj, S., Chellam, A., Velayudhan, T.S., 1989. Larval and juvenile rearing of blacklip pearl oyster, *Pinctada margaritifera* (Linnaeus). *Aquaculture* 76, 43–56.

Andrews, J.D., 1979. Pelecypoda: Ostreidae. In: Giese, C., Pearse, J.S. (Eds.), *Reproduction of Marine Invertebrates*. Academic Press, New York, pp. 293–341.

Ansell, A.D., 1974. Seasonal changes in biochemical composition of the bivalve *Chlamys septemradiata* from the Clyde Sea area. *Mar. Biol.* 25, 85–99.

Arellano-Martínez, M., Racotta, I.S., Ceballos-Vázquez, B.P., Elorduy-Garay, J.F., 2004. Biochemical composition, reproductive activity and food availability of the lion's paw *Nodipecten subnodosus* in the Laguna Ojo de Liebre, Baja California Sur, Mexico. *J. Shellfish Res.* 23, 15–23.

Arizmendi-Castillo, E., 1996. Ciclo reproductivo de las ostras perleras *Pinctada mazatlanica* (Hanley, 1856) Y *Pteria sterna* (Gould, 1851) (Pteriidae), en el área de Guaymas, Sonora, México. M.Sc. Thesis, Inst. Tecn. Est. Sup. Monterrey, Nuevo León, Mexico.

Arjarasirikoon, U., Kruatrachue, M., Sretarugsa, P., Chitramvong, Y., Jantataeme, S., 2004. Gametogenic processes in the pearl oyster, *Pteria penguin* (Roding, 1798) (Bivalvia, Mollusca). *J. Shellfish Res.* 23, 403–410.

Barber, B.J., Blake, N.J., 1981. Energy storage and utilization in relation to gametogenesis in *Argopecten irradians concentricus* (Say). *J. Exp. Mar. Biol. Ecol.* 52, 121–134.

Barber, B.J., Blake, N.J., 1983. Growth and reproduction of the bay scallop, *Argopecten irradians* (Lamarck) at its Southern distributional limit. *J. Exp. Mar. Biol. Ecol.* 66, 247–256.

Barber, B.J., Blake, N.J., 2006. Reproductive physiology. In: Shumway, S.E., Parsons, G.J. (Eds.), *Scallops: Biology, Ecology and Aquaculture*. Amsterdam, The Netherlands, pp. 357–416. Elsevier, 2nd Edition

Bayne, B.L., 1976. Aspects of reproduction in bivalve mollusks. In: Wiley, M. (Ed.), *Estuarine Processes*, Vol.1. Academic Press, London, UK, pp. 432–448.

Bayne, B.L., Bubel, A., Gabbott, P.A., Livingstone, D.R., Lowe, D.M., Moore, M.N., 1982. Glycogen utilization and gametogenesis in *Mytilus edulis* L. *Mar. Biol. Lett.* 3, 89–105.

Beer, A., 1999. Larval culture, spat collection and juvenile growth of the winged pearl oyster, *Pteria penguin*. World Aquaculture '99. The Annual International Conference and Exposition of the World Aquaculture Society 26th April to 2nd May 1999, Sydney, Australia. Book of Abstracts, p, 63.

Behzadi, S., Parivar, K., Roustaian, P., 1997. Gonad cycle of pearl oyster, *Pinctada fucata* (Gould) in Northeast Persian Gulf, Iran. *J. Shellfish Res.* 16, 129–135.

Beninger, P.G., Le Pennec, M., 2006. Structure and function in scallops. In: Shumway, S.E., Parsons, G.J. (Eds.), *Scallops: Biology, Ecology and Aquaculture*. Amsterdam, The Netherlands, pp. 123–228. Elsevier, 2nd Edition

Brokordt, K.B., Guderley, H.E., 2004. Energetic requirements during gonad maturation and spawning in scallops: Sex differences in *Chlamys islandica* (Müller, 1776). *J. Shellfish Res.* 23, 25–32.

Cáceres-Puig, J.I., 2007. Dinámica del esfuerzo reproductivo de *Pteria sterna* (Gould, 1851) en la Bahía de La Paz, B.C.S., Mexico. B.Sc Thesis, Centro de Investigaciones Biológicas del Noroeste. La Paz, B.C.S., Mexico.

Chantler, P.D., 2006. Scallop adductor muscles: Structure and function. In: Shumway, S.E., Parsons, G.J. (Eds.), *Scallops: Biology, Ecology and Aquaculture*. The Netherlands, Amsterdam, pp. 229–316. Elsevier, 2nd Edition

Charpy, L., Dufour, P., Garcia, N., 1997. Particulate organic matter in sixteen Tuamotu Atoll lagoons (French Polynesia). *Mar. Ecol. Prog. Ser.* 151, 55–65.

Chellam, A., 1987. *Biology of pearl oyster*. Central Marine Fisheries Research Institute (CMFRI), Cochin, India. Bull. 39: Pearl culture, pp. 13–21.

Chellam, A., Victor, A.C.C., Dharmaraj, S., Velayudhan, T.S., Satyanaryara Rao, T.S., 1991. *Pearl oyster farming and pearl culture. Training Manual 8*. Central Marine Fisheries Research Institute, Tuticorin, India. pp. 104.

Choi, Y.H., Chang, Y.J., 2003. Gametogenic cycle of the transplanted-cultured pearl oyster, *Pinctada fucata martensii* (Bivalvia: Pteriidae) in Korea. *Aquaculture* 220, 781–790.

Coe, W.R., 1943. Sexual differentiation in mollusks. *I. Pelecypods. Quart. Rev. Biol.* 18, 154–164.

Coeroli, M., Mizuno, K., 1985. Study of different factors having an influence upon the pearl production of the black-lip pearl oyster. *Proc. 5th Int. Coral Reef Symp.*, Tahiti. pp. 551–556.

Cole, H.A., 1938. The fate of the larval organs in the metamorphosis of *Ostrea edulis*. *J. Mar. Biol. Assoc. UK* 22, 469–484.

Comely, C.A., 1974. Seasonal variation in the flesh weights and biochemical content of the scallop *Pecten maximus* (L) in the Clyde Sea Area. *J. Cons. Int. Explor. Mer.* 35, 281–295.

Crossland, C., 1957. The cultivation of the mother-of-pearl oyster in the Red Sea. *Aus. J. Mar. Freshwater Res.* 8, 111–130.

De Gaulejac, B., Henry, M., Vicente, N., 1995. An ultrastructural study of gametogenesis of the marine bivalve *Pinna nobilis* (Linnaeus 1758). *I. Oogenesis. J. Mollusc. Stud.* 61, 375–392.

del Rio-Portilla, M.A., Re-Araujo, A.D., Voltolina, D., 1992. Growth of the pearl oyster *Pteria sterna* under different thermic and feeding conditions. *Mar. Ecol. Prog. Ser.* 89, 221–227.

Desai, K., Hirani, G., Nimavat, D., 1979. Studies on the pearl oyster *Pinctada fucata* (Gould): Seasonal biochemical changes. *Ind. J. Mar. Sci.* 8, 49–50.

Dorange, G., Le Pennec, M., 1989. Ultrastructural study of oogenesis and oocytic degeneration in *Pecten maximus* from the Bay of Saint Brieuc. *Mar. Biol.* 103, 339–348.

Doroudi, M., Southgate, P.C., 2002. The influence of algae ration and larval density on growth and survival of blacklip pearl oyster *Pinctada margaritifera* (L.) larvae. *Aquacult. Res.* 209, 621–626.

Doroudi, M., Southgate, P.C., 2003. Embryonic and larval development of *Pinctada margaritifera* (Linnaeus, 1758). *Mollus. Res.* 23, 101–107.

Doroudi, M., Southgate, P.C., Mayer, R., 1999a. Growth and survival of blacklip pearl oyster (*Pinctada margaritifera* L.) larvae fed different densities of microalgae. *Aquacult. Inter.* 7, 179–187.

Doroudi, M., Southgate, P.C., Mayer, R., 1999b. The combined effects of temperature and salinity on embryos and larvae of the black-lip pearl oyster, *Pinctada margaritifera* (L.). *Aquacult. Res.* 30, 1–7.

Doroudi, M., Southgate, P.C., Lucas, J.S., 2003. Variation in clearance and ingestion rates by larvae of black-lip pearl oyster (*Pinctada margaritifera* L.) feeding on various micro-algae. *Aquacult. Nutrit.* 9, 11–16.

Duinker, A., 2002. Processes related to reproduction in great scallops (*Pecten maximus* L.) from Western Norway. D.Sc. Thesis, University of Bergen, Norway.

Epp, J., Bricelj, V.M., Malouf, R.E., 1988. Seasonal partitioning and utilization of energy reserves in two age classes of the bay scallop *Argopecten irradians irradians* (Lamarck). *J. Exp. Mar. Biol. Ecol.* 121, 113–136.

Friedman, K.J., Southgate, P.C., 1999. Growout of blacklip pearl oysters, *Pinctada margaritifera* collected as wild spat in the Solomon Islands. *J. Shellfish Res.* 18, 159–167.

Fukushima, K., 1970. Mass mortalities of the Japanese oyster *Pinctada fucata martensii* in Matoya Bay and Gokasho Bay. *Shinjugijutsu Kenkyukai Kaiho* 9, 1–21. (in Japanese)

Fukushima, K., 1972. The health of *Pinctada fucata martensii* and food organism environment. *Shinjugijutsu Kenkyukai Kaiho* 10, 15–31. (in Japanese)

Gabbott, P.A., 1975. Storage cycles in marine bivalve molluscs: A hypothesis concerning the relationship between glycogen metabolism and gametogenesis. In: H. Barnes. (Ed.), Proceedings of 9th European Marine Biology Symposium, Aberdeen University Press, Scotland, pp. 191–211.

Gabbott, P., Peek, K., 1991. Cellular biochemistry of the mantle tissue of the mussel *Mytilus edulis* L. *Aquaculture* 94, 165–176.

Gallager, S.M., Mann, R., 1986. Growth and survival of larvae of *Mercenaria mercenaria* (L.) and *Crassostrea gigas* (Gmelin) relative to broodstock conditioning and lipid content of eggs. *Aquaculture* 56, 81–103.

Gallager, S.M., Mann, R., Sasaki, G.C., 1986. Lipid as an index of growth and viability in three species of bivalve larvae. *Aquaculture* 56, 81–103.

Galtsoff, P.S., 1931. The weight–length relationship of the shells of the Hawaiian pearl oyster *Pinctada* sp. *Am. Natural.* 65, 423–433.

Galtsoff, P.S., 1950. The Pearl Oyster Resources of Panama. US Dept. Interior, Fish and Wild Life Service, Spec. Sci. Rep. Fish. No. 28.

García-Domínguez, F., Ceballos-Vázauez, B.P., Tripa-Quezada, A., 1996. Spawning cycle of pearl oyster, *Pinctada mazatlanica* (Hanley, 1856), (Pteriidae) at Isla Espíritu Santo, B.C.S. *México J. Shellfish Res.* 15, 297–303.

García-Cuellar, J.A., García-Domínguez, F., Lluch-Belda, D., Hernández-Vázquez, S., 2004. El Niño and La Niña effects on reproductive cycle of the pearl oyster *Pinctada mazatlanica* (Hanley, 1856) (Pteriidae) at Isla Espíritu Santo in the Gulf of California. *J. Shellfish Res.* 23, 113–120.

Gaytan-Mondragon, I., Caceras-Martinez, C., Tobias-Sanchez, M., 1993. Growth of the pearl oysters *Pinctada mazatlanica* and *Pteria sterna* in different culture structures at La Paz Bay, Baja California Sur, Mexico. *J. World Aquacult. Soc.* 24, 541–546.

Gervis, M.N., Sims, N.A., 1992. The Biology and Culture of Pearl Oysters (Bivalvia: Pteriidae). *ICLARM Stud. Rev.* 21, 49.

Giese, A.C., 1959. Reproductive cycles of some west coast invertebrates. In: Withrow, R.B. (Ed.), *Photoperiodism and Related Phenomena in Plants and Animals*. Americna Associatin of Advanced Science, New York, pp. 625–638.

Giese, A.C., Pearse, J.S., 1974. Introduction: General principles. In: Giese, A.C., Pearse, J.S. (Eds.), *Reproduction of Marine Invertebrates*, Vol. 1. Academic Press, New York, pp. 1–49.

Giese, A.C., Pearse, J.S., 1979. Molluscs: Pelecypods and lesser classes. In: Giese, A.C., Pearse, J.S. (Eds.), *Reproduction of Marine Invertebrates*, Vol.5. Academic Press, New York, pp. 112–292.

Gómez-Robles, E., Rodríguez-Jaramillo, C., Saucedo, P.E., 2005. Digital image analysis of lipid and protein histochemical markers for measuring oocyte development and quality in pearl oyster *Pinctada mazatlanica* (Hanley, 1856). *J. Shellfish Res.* 24, 1197–1202.

Gosling, E., 2003. *Bivalve Molluscs: Biology, Ecology, and Culture*. Fishing New Books, Blackwell Science, New York. 443 pp.

Guerrier, P., Leclerc-David, C., Moreau, M., 1993. Evidence for the involvement of internal calcium stores during serotonin-induced meiosis reinitiation in oocytes of the bivalve mollusc *Ruditapes philippinarum*. *Dev. Biol.* 159, 474–484.

Guraya, S.S., 1975. Balbiani's vitelline body in the oocytes of vitellogenic and nonvitellogenic females of the domestic fowl: A correlative cytological and histochemical study. *Acta Morphol. Acad. Sci. Hung.* 23, 251–261.

Hart, A.M., Joll, L., 2006. Growth, mortality, recruitment, and sex ratio in wild stocks of the silver-lipped pearl oyster *Pinctada maxima* (Jameson) (Mollusca: Pteriidae) in Western Australia. *J. Shellfish Res.* 25, 201–210.

Herdam, W.A., Hornell, J., 1906. General summary and recommendations: Reproduction and life history of the pearl oyster. *Rep. Pearl. Fish. Manaar* 5, 114–118.

Hernández-Díaz, A., Bückle-Ramírez, L.F., 1996. Gonad cycle of *Pteria sterna* (Gould, 1851) (Mollusca, Bivalvia) in Baja California, Mexico. *Cien. Mar.* 22, 495–509.

Hernández-Olalde, L., 2003. Ciclo reproductivo de la ostra perlera *Pteria sterna* (Mollusca: Bivalvia) en la Laguna Ojo de Liebre, Baja California Sur. M.Sc. Thesis, Centro Interdisciplinario de Ciencias Marinas (CICIMAR), La Paz, Baja California Sur, México.

Holland, D.L., 1978. Lipid reserves and energy metabolism in the larvae of benthic marine invertebrates. In: Malins, D.C., Sargent, J.R. (Eds.), *Biochemical and Biophysical Perspectives in Marine Biology*. London Academic Press, pp. 85–123.

Holland, D.L., Spencer, B.E., 1973. Biochemical changes in fed and starved oyster, *Ostrea edulis* L., during larval development, metamorphosis and early spat growth. *J. Mar. Biol. Assoc. UK* 53, 287–298.

Hollyer, J., 1984. Pearls – jewels of the sea. *Infofish Marketing Digest* 5, 32–34.

Hornell, J., 1922. The Indian pearl fishery of the Gulf of Mannar and Palk Bay. *Madras Fish. Bull.* 16, 1–188.

Hynd, J.S., 1955. A revision of the Australian pearl-shells genus *Pinctada* (Lamellibranchia). *Aust. J. Mar. Freshwat. Res.* 6, 98–135.

Jamili, S., Amini, G., Oryan, S., 1999. Gonad changes and serum steroid levels during the annual reproductive cycle of the pearl oyster *Pinctada fucata* Gould. *Iran J. Fish. Sci.* 1, 67–75.

Kang, C.K., Park, M.S., Lee, P.Y., Choi, W.J., Lee, W.C., 2000. Seasonal variations in condition, reproductive activity, and biochemical composition of the Pacific oyster, *Crassostrea gigas* (Thunberg) in suspended culture in two coastal bays of Korea. *J. Shellfish Res.* 19, 771–778.

Keen, M., 1970. *Seashells of the Tropical West America*. Stanford University Press, Stanford, California. 1068 pp.

Knauer, J., Southgate, P.C., 1999. A review of the nutritional requirements of bivalves and the development of alternative and artificial diets for bivalve aquaculture. *Rev. Fish. Sci.* 7, 241–281.

Krantic, S., Rivailler, P., 1996. Meiosis reinitiation in molluscan oocytes: A model to study the transduction of extracellular signals. *Invertebr. Reprod. Dev.* 30, 55–69.

Kyozuka, K., Deguchi, R., Yoshida, N., Yamashita, M., 1997. Change in intracellular Ca_2^+ is not involved in serotonin-induced meiosis reinitiation form the first prophase in oocytes of the marine bivalve *Crassostrea gigas*. *Dev. Biol.* 182, 33–41.

Lee, A.M., Williams, A., Southgate, P.C., 2008. Modelling and comparison of growth of the silverlip pearl oyster *Pinctada maxima* (Jameson, Mollusca: Pteriidae) cultured in west Papua, Indonesia. *Marine Freshwater Res.* 59, 22–31.

Little, C., 1998. Mollus. Life Histor. In: Beesley, P.L., Ross, G.J.B., Wells, A. (Eds.), *Mollusca: The Southern Synthesis. Fauna of Australia*, Vol.5. CSIRO Publishing, Melbourne, pp. 23–29. Part B viii 565–1234 pp.

Lodeiros, C.J., Rengel, J.J., Himmelman, J.H., 1999. Growth of *Pteria colymbus* (Roding, 1798) in suspended culture in Golfo de Cariaco, Venezuela. *J. Shellfish Res.* 18, 155–158.

Lodeiros, C., Rengel, J., Guderley, H.E., Nusetti, O., Himmelman, J.M., 2001. Biochemical composition and energy allocation in the tropical scallop *Lyropecten* (*Nodipecten*) *nodosus* during the months leading up to and following the development of gonads. *Aquaculture* 199, 63–72.

Loosanoff, V.L., 1969. Maturation of gonad of oysters, *Crassostrea virginica*, of different geographical areas subjected to relatively low temperatures. *The Veliger* 11, 153–163.

Loosanoff, V.L., Davis, H.C., 1963. Temperature requirements for maturation of gonads of northern oysters. *Biol. Bull.* 103, 80–96.

Lora-Vilchis, M.C., Saucedo, P.E., Rodríguez-Jaramillo, C., Maeda-Martínez, A.N., 2003. Histological characterization of the spawning process in the catarina scallop, *Argopecten ventricosus* (Sowerby II, 1842) induced by thermal shock and serotonin injection. *Inv. Repro. Dev.* 44, 79–88.

Lowe, D.M., Moore, M.N., Bayne, B.L., 1982. Aspects of gametogenesis in the marine mussel *Mytilus edulis*. *J. Mar. Biol. Assoc. UK* 62, 133–145.

Lubet, P., Aloui, N., 1987. Limites letales thermmiques et action de la temperature sur les gametogenesis et l'activité neurosecretrice chez la moule (*Mytilus edulis* et *M. galloprovincialis*, Moolusque Bivalve). *Haliotis* 16, 309–316.

Luna-González, A., Cáceres-Martínez, C., Zuñiga-Pacheco, C., López-López, S., Ceballos-Vázquez, B.P., 2000. Reproductive cycle of *Argopecten ventricosus* (Sowerby II, 1842) (Bivalvia: Pectinidae) in the Rada of Puerto de Pichilingue, B.C.S., Mexico and its relation to temperature, salinity, and quality of food. *J. Shellfish Res.* 19, 107–112.

Lluch-Belda, D., Lluch-Cota, D.B., Lluch-Cota, S.E., 2003. Scales of interannual variability in the California Current System: Associated physical mechanisms and likely ecological impacts. *Calif. Coop. Ocean. Fish. Inv. Rep. (CalCOFI)* 44, 76–85.

Mackie, G.L., 1984. Bivalves. In: Wilburg, K.M. (Ed.), *The Mollusca, Vol. 7: Reproduction*. Academic Press, New York, pp. 351–417.

MacDonald, B.A., Thompson, R.J., 1985. Influence of temperature and food availability on the ecological energetics of the giant scallop *Placopecten magellanicus*. II. Reproductive output and total production. *Mar. Ecol. Prog. Ser.* 25, 295–303.

MacDonald, B.A., Thompson, R.J., Bayne, B.L., 1987. Influence of temperature and food availability on the ecological energetics of the giant scallop *Placopecten magellanicus* IV. Reproductive effort, value and cost. *Oecologia* 72, 550–556.

Malouf, R.E., Bricelj, M., 1989. Comparative Biology of Clams: Environmental tolerances of feeding and growth. In: Manzi, J.J., Castagna, M. (Eds.), *Clam Mariculture in North America*. Elsevier, Amsterdam, pp. 23–74.

Marcano, J.S., Larez, A., Alio, J.J., Sanabria, H., Prito, A., 2005. Growth and mortality of *Pinctada imbricata* (Mollusca: Pteriidae) in Guamachito, Araya Peninsula, Sucre State, Venezuela. *Ciencias Marinas* 31, 387–397.

Martínez, G., Mettifogo, L., 1998. Adductor muscle and mobilization of trophic material for gametogenesis of the scallop *Argopecten purpuratus* Lamarck. *J. Shellfish Res.* 17, 113–116.

Martinez, G., Olivares, A.Z., Metiffogo, L., 2000. In vitro effects of monoamines and prostaglandins on meiosis reinitiation and oocyte release in *Argopecten purpuratus* Lamarck. *Inv. Repro. Dev.* 38, 61–69.

Martinez-Fernandez, E., Acosta-Salmón, H., Rangel-Davalos, C., Olivera, A., Ruiz-Rubio, H., Romo-Pinera, A.K., 2003. Spawning and larval culture of the pearl oyster *Pinctada mazatlanica* in the laboratory. *World Aquacult.* 34, 36–39.

Martinez-Fernandez, E., Southgate, P.C., 2007. Use of tropical microalgae as food for larvae of the black-lip pearl oyster *Pinctada margaritifera*. *Aquaculture* 263, 220–226.

Martinez-Fernandez, E., Acosta-Salmón, H., Southgate, P.C., 2006. The nutritional value of seven species of tropical microalgae for black-lip pearl oyster (*Pinctada margaritifera*, L.) larvae. *Aquaculture* 257, 491–503.

Matsui, Y., 1958. Aspects of environment of pearl-culture grounds and problems of hybridization in the genus *Pinctada*. In: Buzzati-Traverso, (Eds.), *Perspectives in Marine Biology*. University of California Press, Berkley, pp. 519–531.

Matsutani, T., Nomura, T., 1986. Serotonin-like immunoreactivity in the central nervous system and gonad of scallop, *Patinopecten yessoensis*. *Cell. Tiss. Res.* 224, 515–517.

Maxwell, W.L., 1983. Mollusca. In: Adiyodi, K.G., Adiyodi, R.G. (Eds.), *Reproductive Biology of Invertebrates, Vol. II: Spermatogenesis and Sperm Function*. John Wiley & Sons, New York, pp. 275–320.

Minaur, J., 1969. Experiments on the artificial rearing of the larvae of *Pinctada maxima* (Jameson) (Lamellibranchia). *Aust. J. Mar. Freshwat. Res.* 20, 175–187.

Nalluchinnappan, I., Dev, D.S., Irulandy, M., Jeyabaskaran, Y., 1982. Growth of the pearl oyster, *Pinctada fucata*, in caged culture at Kundugal Channel, Gulf of Mannar India. *Indian J. Marine Sci.* 11, 193–194.

Nasr, D.H., 1984. Feeding and growth of the pearl oyster, *Pinctada margaritifera* (L.) in Dongonab Bay, Red Sea. *Hydrobiologia* 110, 241–245.

Newell, R.I.E., Hilbish, T.J., Koehn, R.K., Newell, C.J., 1982. Temporal variation in the reproductive cycle of *Mytilus edulis* (L.) (Bivalvia: Mytilidae) from localities on the east coast of the United States. *Biol. Bull.* 162, 299–310.

Numaguchi, K., 1994a. Growth and physiological condition of the Japanese pearl oyster, *Pinctada fucata martensii* (Dunker, 1850) in Ohmura Bay, Japan. *J. Shellfish Res.* 13, 93–99.

Numaguchi, K., 1994b. Fine particles of the suspended solids in the pearl farm. *Suisan Kogaku (Fish. Engin.)* 30, 181–184. (in Japanese with English Abstract)

Numaguchi, K., 2000. Evaluation of five microalgal species for the growth of early spat of the Japanese pearl oyster *Pinctada fucata martensii*. *J. Shellfish Res.* 19, 153–157.

Numaguchi, K., Tanaka, Y., 1986. Effects of salinity on mortality and growth of spat of the pearl oyster, *Pinctada fucata martensii*. *Bull. Natl. Res. Inst. Aquacult.* 9, 41–44.

O'Connor, W.A., 2002. Latitudinal variation in reproductive behavior in the pearl oyster, *Pinctada albina sugillata*. *Aquaculture* 209, 333–345.

O'Connor, W.A., Lawler, N.F., 2004. Reproductive condition of the pearl oyster, *Pinctada imbricata*, Roding, in Port Stephens, New South Wales, Australia. *Aquat. Res.* 35, 385–396.

Ojima, Y., Maeki, K., 1955. Some cytological account of the maturation of the gonad in the pearl oyster *Pinctada martensii* Dunker. *Bull. Inst. Sci. Res. Pearls* 2, 13–25. (in Japanese)

Orton, J.H., 1920. Sea temperatures, breeding and distribution in marine animals. *J. Mar. Biol. Assoc. UK* 12, 339–366.

Paulet, Y.M., Boucher, J., 1991. Is reproduction mainly regulated by temperature or photoperiod in *Pecten maximus*? *Invertebr Reprod. Dev.* 19, 61–70.

Pauly, D., Munro, J.L., 1984. Once more on the comparison of growth in fish and invertebrates. *Fishbyte* 2, 21.

Pipe, R.K., 1987. Oogenesis in the marine mussel *Mytilus edulis*: An ultrastructural study. *Mar. Biol.* 95, 405–414.

Pouvreau, S., Bodoy, A., Buestel, D., 2000a. *In situ* suspension feeding behaviour of the pearl oyster, *Pinctada margaritifera*: Combined effects of body size and weather-related seston composition. *Aquaculture* 181, 91–113.

Pouvreau, S., Gangnery, A., Tiapari, J., Lagarde, F., Garnier, M., Bodoy, A., 2000b. Gametogenic cycle and reproductive effort of the tropical blacklip pearl oyster, *Pinctada margaritifera* (Bivalvia: Pteriidae), cultivated in Takapoto Atoll (French Polynesia). *Aquat. Living Resour.* 13, 37–48.

Pouvreau, S., Tiapari, J., Gangnery, A., Lagarde, F., Garnier, M., Teissier, H., Haumani, G., Buestel, D., Bodoy, A., 2000c. Growth of the black-lip pearl oyster, *Pinctada margaritifera*, in suspended culture under hydrobiological conditions of Takapoto lagoon (French Polynesia). *Aquaculture* 184, 133–154.

Pouvreau, S., Prasil, V., 2001. Growth of the blacklip pearl oyster, *Pinctada margaritifera* at nine culture sites in French Polynesia: Synthesis of several sampling designs conducted between 1994 and 1999. *Aquat. Living Res.* 145, 155–163.

Prajneshu, Venugopalan, R., 1999. von Bertalanffy growth model in a random environment. *Can. J. Fish. Aquat. Sci.* 56, 1026–1030.

Qian, P.Y., Wu, C.Y., Wu, M., Xie, Y.K., 1996. Integrated cultivation of the red algae *Kappaphycus alvarezii* and the pearl oyster *Pinctada martensi. Aquaculture* 147, 21–35.

Quayle, D.B., Newkirk, G.F., 1989. *Farming bivalve molluscs: Methods for study and development.* World Aquaculture Society, Baton Rouge, LA.

Racotta, I.S., Ramírez, J.L., Avila, S., Ibarra, A.M., 1998. Biochemical composition of gonad and muscle in the catarina scallop, *Argopecten ventricosus*, after reproductive conditioning under two feeding systems. *Aquaculture* 163, 111–122.

Racotta, I.S., Ramírez, J.L., Ibarra, A.M., Rodríguez-Jaramillo, C., Carreño, D., Palacios, E., 2003. Growth and gametogenesis in the lion-paw scallop *Nodipecten* (*Lyropecten*) *subnodosus. Aquaculture* 217, 335–349.

Ramírez-Llodra, E., 2002. Fecundity and life-history strategies in marine invertebrates. *Adv. Mar. Biol.* 43, 87–170.

Rappaport, R., 1991. Cytokinesis. In: Kinne, K.H. (Ed.), *Oogenesis, Spermatogenesis and Reproduction.* Karger Editorial, Dortmund, Germany, pp. 1–36.

Raven, C.P., 1966. Maturation and fertilization. In: Raven, C.P. (Ed.), *Morphogenesis: The Analysis of Molluscan Development.* Pergamon Press, Oxford, pp. 226–254.

Robinson, W.E., Wehling, W.E., Morse, M.P., Leod, G.C., 1981. Seasonal changes in soft-body component indices and energy reserves in the Atlantic deep-sea scallop, *Placopecten magellanicus. Fish. Bull.* 79, 449–458.

Román, G., Martínez, G., García, O., Freites, L., 2002. Reproducción. In: Maeda-Martínez, A.N. (Ed.), *Los Moluscos Pectinidos de Iberoamérica: Ciencia y Auicultura.* Ed. Limusa, Mexico, pp. 27–60.

Rose, R.A., 1990. A manual for the artificial propagation of the silverlip or goldlip pearl oyster, *Pinctada maxima* (Jameson) from Western Australia. Fisheries Department, Perth. 41 pp.

Rose, R.A., Baker, S.B., 1994. Larval and spat culture of the western Australian silver-or goldlip pearl oyster, *Pinctada maxima* Jameson (Mollusca): Pteriidae. *Aquaculture* 12, 35–50.

Rose, R.A., Dybdahl, R., Harders, S., 1990. Reproductive cycle of the Western Australian silver-lip pearl oyster, *Pinctada maxima* (Jameson) (Mollusca: Pteriidae). *J. Shellfish Res.* 9, 261–272.

Sastry, A.N., 1968. Relationships among food, temperature and gonad development of the bay scallop, *Aequipecten irradians concentricus* Say, reared in the laboratory. *Bull. Mar. Sci.* 15, 417–434.

Sastry, A.N., 1970. Reproductive physiological variation in latitudinally separated populations of the bay scallop, *Argopecten irradians* Lamarck. *Biol. Bull.* 138, 56–65.

Sastry, A.N., 1979. Pelecypoda (excluding Ostreidae). In: Giese, A.C., Pearse, J.S. (Eds.), *Reproduction of Marine Invertebrates.* Academic Press, New York, pp. 113–292.

Sastry, A.N., Blake, N.J., 1971. Regulation of gonad development in the bay scallop, *Aequipecten irradians* Lamarck. *Biol. Bull.* 140, 274–282.

Saucedo, P.E., Monteforte, M., 1997. Breeding cycle of pearl oysters *Pinctada mazatlanica* (Hanley 1856) and *Pteria sterna* (Gould 1851) at Bahía de La Paz, Baja California Sur, Mexico. *J. Shellfish Res.* 16, 103–110.

Saucedo, P., Monteforte, M., Blanc, F., 1998. Changes in shell dimensions of pearl oysters, *Pinctada mazatlanica* (Hanley 1856) and *Pteria sterna* (Gould 1851), during growth as criteria for Mabe pearl implants. *Aquacult. Res.* 29, 801–814.

Saucedo, P.E., Rodríguez-Jaramillo, C., Aldana-Avilés, C., Monsalvo-Spencer, P., Reynoso-Granados, T., Villarreal, H., Monteforte, M., 2000. Gonadic conditioning of the Calafia mother-of-pearl oyster, *Pinctada mazatlanica* (Hanley 1856) under two temperature regimes. *Aquaculture* 195, 103–119.

Saucedo, P.E., Racotta, I.S., Villarreal, H., Monteforte, M., 2002a. Seasonal changes in the histological and biochemical profile of the gonad, digestive gland, and muscle of the Calafia mother-of-pearl oyster *Pinctada mazatlanica* (Hanley, 1856). *J. Shellfish Res.* 21, 127–135.

Saucedo, P.E., Rodríguez-Jaramillo, C., Monteforte, M., 2002b. Microscopic anatomy of gonad tissue and specialized storage cells associated with oogenesis and spermatogenesis in the Calafia mother-of-pearl oyster, *Pinctada mazatlanica* (Hanley, 1856). *J. Shellfish Res.* 21, 145–155.

Saville-Kent, W., 1893. Pearl and pearl shell fisheries. In: Allen, W.H. (Ed.), *The Great Barrier Reef of Australia: Its products and potentialities.* London, pp. 204–224.

Seed, R., 1976. Ecology. In: Bayne, B.L. (Ed.), *Marine Mussels: Their Ecology and Physiology.* Cambridge University Press, Cambridge, pp. 13–65.

Sevilla, M.L., 1969. Contribución al conocimiento de la madreperla *Pinctada mazatlanica* (Hanley, 1845). *Rev. Soc. Mex. Hist. Nat.* 30, 223–262.

Shpigel, M., Barber, B.J., Mann, R., 1992. Effects of elevated temperature on growth, gametogenesis, physiology, and biochemical composition in diploid and triploid Pacific oysters, *Crassostrea gigas* Thunberg. *J. Exp. Mar. Biol. Ecol.* 161, 15–25.

Sims, N.A., 1990. *The Black-Lip Pearl Oyster, Pinctada margaritifera, in the Cook Islands.* Master's thesis, University of New South Wales, Australia 109 p.

Sims, N.A., 1994. Growth of wild and cultured Black-Lip Pearl Oysters, *Pinctada margaritifera* (L.) Pteriidae, Bivalvia, in the Cook Islands. *Aquaculture* 122, 181–191.

Smitasiri, R., Kajitwiwat, J., Tantichodok, P., 1994. *Growth of a Winged Pearl Oyster, Pteria penguin Suspended at Different Depths.* Phuket Marine Biology Center, Special Publication, 13, 213–216.

Solano-López, Y., Cabrera-Peña, J., Palacios, J.A., Cruz, R.A., 1997. Madurez sexual, índice de condición y rendimiento de *Pinctada mazatlanica* (Pterioida: Pteriidae), Golfo de Nicoya, Costa Rica. *Rev. Biol. Trop.* 45, 1049–1054.

Southgate, P.C., Beer, A.C., 1997. Hatchery and early nursery culture of the blacklip pearl oyster (*Pinctada margaritifera* L.). *J. Shellfish Res.* 16, 561–567.

Southgate, P.C., Beer, A.C., 2000. Growth of blacklip pearl oyster (*Pinctada margaritifera* L.) juveniles using different nusery culture techniques. *Aquaculture* 187, 97–104.

Southgate, P.C., Lucas, J.S., 2003. Reproduction, life cycles and growth. In: Lucas, J.S., Southgate, P.C. (Eds.), *Aquaculture: Farming Aquatic Animals and Plants.* Blackwell Science, Oxford, pp. 111–122.

Southgate, P.C., Beer, A.C., Duncan, P.F., Tamburri, R., 1998. Assessment of the nutritional value of three species of tropical microalgae, dried *Tetraselmis* and a yeast-based diet for larvae of the blacklip pearl oyster, *Pinctada margaritifera* (L.). *Aquaculture* 162, 247–257.

Strugnell, J.M., Southgate, P.C., 2003. Changes in tissue composition during larval development of the blacklip pearl oyster, *Pinctada margaritifera* (L.). *Mollus. Res.* 23, 179–183.

Tanaka, Y., Kumeta, M., 1981. Successful artificial breeding of the silverlip pearl oyster, *Pinctada maxima* (Jameson). *Bull. Natural Res. Instit., Aquacult.* 2, 21–28.

Tanaka, Y., Inoha, S., Kakazu, K., 1970. Studies on seed production of blacklip pearl oyster *Pinctada margaritifera* in Okinawa. I. Spawning induction by thermal stimulation. *Bull. Tokai Reg. Fish. Res. Lab.* 63, 75–78.

Tang, S.P., 1941. The breeding of the scallop (*Pecten maximus* L.) with a note on the growth rate. *Proc. Liverpool Biol. Soc.* 54, 9–28.

Taylor, A.C., Venn, T.J., 1979. Seasonal variation in weight and biochemical composition of the tissues of the queen scallop, *Chlamys opercularis*, from the Clyde Sea area. *J. Mar. Biol. Assoc. UK* 59, 605–621.

Taylor, J.J., 1999. Spat production, nursery rearing and grow-out of the silver-lip pearl oyster, *Pinctada maxima* (Jameson). Ph.D. thesis, James Cook University of North Queensland, Australia.

Taylor, J.J., Rose, R.A., Southgate, P.C., 1997a. Inducing detachment of silver-lip pearl oyster (*Pinctada maxima*, Jameson) spat from collectors. *Aquaculture* 159, 11–17.

Taylor, J.J., Southgate, P.C., Rose, R.A., 1997b. Fouling animals and their effect on the growth of silver-lip pearl oysters, *Pinctada maxima* (Jameson) in suspended culture. *Aquaculture* 153, 31–40.

Taylor, J.J., Southgate, P.C., Wing, M.S., Rose, R.A., 1997c. The nutritional value of five species of microalgae for spat of the silver-lip pearl oyster, *Pinctada maxima* (Jameson) (Mollusca:Pteriidae). *Asian Fish. Sci.* 10, 1–8.

Taylor, J.J., Rose, R.A., Southgate, P.C., Taylor, C.E., 1997d. Effects of stocking density on growth and survival of early juvenile silverlip pearl oysters *Pinctada maxima* (Jameson), held in suspended nursery culture. *Aquaculture* 153, 41–49.

Taylor, J.J., Southgate, P.C., Rose, R.A., 2004. Effects of salinity on growth and survival of silver-lip pearl oyster, *Pinctada maxima* (Jameson) spat. *J. Shellfish Res.* 23, 375–378.

Thielley, M., Weppe, M., Herbaut, C., 1993a. Ultrastructural study of gametogenesis in the French Polynesian Black pearl oyster *Pinctada margaritifera* (Mollusca, Bivalvia). I) Spermatogenesis. *J. Shellfish Res.* 12, 41–47.

Thielley, M., Cabral, P., Weppe, M., Herbaut, C., 1993b. *Etude histologique et cytochimique de la gametogenese chez la nacre Pinctada margaritifera (L) var. cumingii (Jameson)*. Mem. DEA. Univ. Française du Pacifique, French Polynesia.

Thompson, R.J., Newell, R., Kennedy, V., Mann, R, 1996. Reproductive processes and early development. In: Kennedy, V., Newell, R., Eble, A. (Eds.), *The Eastern Oyster Crassostrea virginica*. Maryland Sea Grant Books, Maryland, pp. 335–370.

Tomaru, Y., Kawabata, Z., Nakano, S.I., 2001. Mass mortality of the Japanese pearl oyster *Pinctada fucata martensii* in relation to water temperature, chlorophyll and phytoplankton composition. *Dis. Aquat. Org.* 44, 61–68.

Tomaru, Y., Udaka, N., Kawabata, Z., Nakano, S., 2002. Seasonal change of seston size distribution and phytoplankton composition in bivalve pearl oyster *Pinctada fucata martensii* culture farm. *Hydrobiologia* 481, 181–185.

Traithong, T., Poomtong, T., Sookchuay, C., 1996. Growth of the hatchery-produced juvenile pearl oyster *Pinctada maxima* (Jameson) in the Gulf of Thailand. *Proceedings of Sixth Workshop of the Tropical Marine Mollusc Programme (TMMP)* at Centre of Advanced Study in Marine Biology, Annamalai University, India. Phuket marine Biological Centre Special Publication, Vol. 16, pp. 327.

Tranter, D.J., 1958a. Reproduction in Australian pearl oysters (Lamellibranchia). I. *Pinctada albina* (Lamarck): Primary gonad development. *Aust. J. Mar. Freshwat. Res.* 9, 135–143.

Tranter, D.J., 1958b. Reproduction in Australian pearl oysters (Lamellibranchia). II. *Pinctada albina* (Lamarck): Gametogenesis. *Aust. J. Mar. Freshwat. Res.* 9, 144–158.

Tranter, D.J., 1958c. Reproduction in Australian pearl oysters (Lamellibranchia). III. *Pinctada albina* (Lamark): Breeding season and sexuality. *Austr. J. Mar. Freshwater Res.* 9, 191–216.

Tranter, D.J., 1958d. Reproduction in Australian pearl oysters (Lamellibranchia). IV. *Pinctada margaritifera* (L.). *Austr. J. Mar. Freshwat. Res.* 9, 509–523.

Tranter, D.J., 1959. Reproduction in Australian pearl oysters (Lamellibranchia). V. *Pinctada fucata* (Gould). *Aust. J. Mar. Freshwat. Res.* 10, 45–66.

Urban, H.J., 2000. Culture potential of the pearl oyster (*Pinctada imbricata*) from the Caribbean II. Spat collection and growth and mortality in culture systems. *Aquaculture* 189, 375–388.

Urban, H.J., 2002. Modelling growth of different developmental stages in bivalves. *Marine Ecology Progress Series* 238, 109–114.

Utting, S., 1993. Procedures for maintenance and hatchery conditioning of bivalve broodstock. *World Aquaculture* 24, 78–82.

Vakily, J.M., 1992. Determination and comparison of bivalve growth with emphasis on Thailand and other tropical areas. *ICLARM Tech. Rep.* 36, 25.

Vélez, A., Epifanio, C.E., 1981. Effects of temperature and food ration on gametogenesis and growth in the tropical mussel *Perna perna* (L.). *Aquaculture* 22, 21–26.

Vite-García, M.N., 2005. Variaciones estacionales en el uso de reservas energéticas entre los tejidos relacionados con la gametogenesis en las ostras perleras *Pteria sterna* (Gould, 1851) y *Pinctada mazatlanica* (Hanley, 1856). M.Sc. Thesis, Centro de Investigaciones Biológicas del Noroeste, La Paz, B.C.S., Mexico.

Vite-García, M.N., Saucedo, P.E., 2008. Energy storage and allocation during reproduction of Pacific winged pearl oyster *Pteria sterna* (Gould, 1851) at Bahía de La Paz, Baja California Sur, Mexico. *J. Shellfish Res.* 27, 375–383.

Wada, S., 1942. Artificial insemination and development in *Pinctada maxima* (Jameson). *Kagaku-Nanyo (Science of South Seas)* 4(3), 26–32.

Wada, S., 1953. Life history. In: Wada, S. (Ed.), *Biology and Fisheries of the Silver-lip Pearl Oyster*. Okada Publ, Japan, pp. 13–35.

Wada, K.T., Komaru, A., Uchimura, Y., 1989. Triploid production in the Japanese pearl oyster, *Pinctada fucata martensii*. *Aquaculture* 76, 11–19.

Wada, K.T., Komaru, A., Uchimura, Y., Kurosaki, H., 1995. Spawning occurs during winter in the Japanese subtropical population of the pearl oyster, *Pinctada fucata fucata* (Gould, 1850). *Aquaculture* 133, 207–214.

Whyte, J.N.C., Bourne, N., Hodgson, C.A., 1987. Assessment of biochemical composition and energy reserves in larvae of the scallop *Patinopecten yessoensis*. *J. Exp. Mar. Biol. Ecol.* 113, 113–124.

Wright, W.G., 1988. Sex change in the Mollusca. *Trends Ecol. Evo.* 3, 137–140.

Wu, M., Mak, S.K.K., Zhang, X., Qiian, P.Y., 2003. The effect of co-cultivation on the pearl yield of *Pinctada martensii* (Dunker). *Aquaculture* 221, 347–356.

Yamaguchi, K., Hasuo, M., 1977. Relation between activity of pearl oyster and seasonal changes of environmental factors in culture ground. *Kokoritsu Shinju Kenkyusho Hokoku (Bull. Natl. Pearl Res. Lab.)* 21, 2315–2324. (in Japanese)

Yoo, S.K., Chang, Y.J., Lim, H.S., 1986. Growth comparison of pearl oyster, *Pinctada fucata* between the two culturing areas. *Bull. Korea Fish. Soc.* 19, 593–598.

Yukihira, H., Klumpp, D.W., Lucas, J.S., 1998. Comparative effects of microalgal species and food concentration on suspension feeding and energy budgets of the pearl oysters *Pinctada margaritifera* and *P. maxima* (Bivalvia: Pteriidae). *Mar. Ecol. Prog. Ser.* 171, 71–84.

Yukihira, H., Lucas, J.S., Klumpp, D.W., 2000. Comparative effects of temperature on suspension feeding and energy budgets of the pearl oysters *Pinctada margaritifera* and *P. maxima*. *Mar. Ecol. Pro. Ser.* 195, 179–188.

Yukihira, H., Lucas, J.S., Klumpp, D.W., 2006. The pearl oysters, *Pinctada maxima* and *P. margaritifera*, respond in different ways to culture in dissimilar environments. *Aquaculture* 252, 208–224.

CHAPTER 6

Environmental Influences

John S. Lucas

6.1. INTRODUCTION

Pearl oysters in their natural environment experience the simultaneous effects of a wide array of environmental factors. Some of these factors, such as wavelength of ambient light may never have a major influence, although it seems that blue light promotes nacre secretion on cultured pearls. Other factors, both biological and physical may continually influence the pearl oyster, e.g. predation, food and water temperature, while others may be important at particular times, e.g. salinity, disease and aerial exposure. Furthermore, the effects of environmental factors on a pearl oyster differ according to the developmental stage and the pearl oyster's physiological condition, e.g. post-spawning and after pearl nucleation.

Major environmental factors are treated in this Chapter and they are largely treated individually to clarify their particular influences. The data are typically from experimental studies where only one environmental factor is manipulated and others kept as constant as possible. Examples of compounded effects of environmental factors on pearl oysters in a particular physiological condition come from postulated causes of the disastrous mass mortalities of Akoya pearl oysters[1] in farms in Japanese bays since the 1960s (Itoh and Muzamoto, 1978; Numaguchi, 1994a; Tomaru *et al.*, 2001).

---

[1]Throughout this text the term 'Akoya' is used for pearl oysters in the currently unresolved complex of *Pinctada fucata/martensii/radiata/imbricata* (see Section 2.3.1.13). The first two names have been particularly used in the Japanese studies described in this chapter. If the Akoya is shown to be one very widely distributed species, there may be adaptive genetic and acclimatisation variation over its range. For this reason, the source of the studied Akoya is specified where the source is not from the major pearl farming regions of Japan.

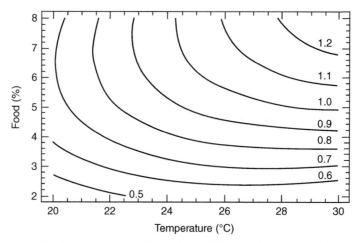

FIGURE 6.1. Response surface analysis of the synergistic effects of food ration and temperature on growth rates of juvenile *Pteria sterna*. Food ration is percentage dry weight per oyster day. Growth response contours, 0.5–1.2 mm/week, are for the shell axis of maximum growth. From del Rio-Portilla *et al.* (1992).

These field observations of pearl oyster environments are only indicative, but it is possible to study the synergistic effects of two or more environmental factors through experimentation. This involves a multifactorial design that generates response surface contours of the parameters tested (Figs. 6.1 and 6.2). These demonstrate the distinct synergistic effects of only two environmental factors.

Some environmental factors that strongly influence pearl oysters are not treated in this Chapter, being treated in other chapters in the context of pearl farming. These are competitors, diseases and parasites (Chapter 11), and biofouling (Chapter 14).

6.2. FOOD

Like other bivalves, pearl oysters filter feed upon a variety of living and dead organic particles, and probably absorb organic molecules across their body surfaces (see Section 4.1). Knauer and Southgate (1999) provide a comprehensive review of the current knowledge of biochemistry and nutritional requirements of bivalves, e.g. total protein, essential amino acids, carbohydrates, total lipid, essential fatty acids, sterols, minerals, trace elements and vitamins. The data are limited, unevenly focused and sometimes inconsistent. There have been some studies on pearl oysters that have focused on biochemical changes during maturation and spawning (see Section 5.5), and during larval development (Strugnell and Southgate, 2003). There are very limited data on the general biochemistry and nutritional requirements of pearl oysters, i.e. lipids and fatty acids (Teshima *et al.*, 1987; Saito, 2004) and carbohydrates (Shinomiya

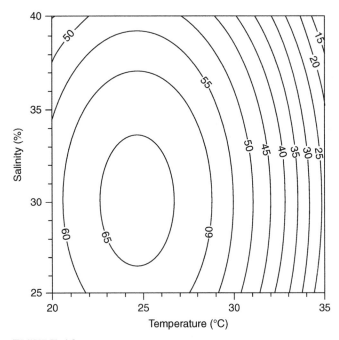

FIGURE 6.2. Response surface analysis of the synergistic effects of temperature and salinity on percentage survival to day 6 of the early development of *Pinctada margaritifera*. The contours are percentage survival. From Doroudi *et al.* (1999).

et al., 1999). However, a number of studies have determined the nutritional value of micro-algae for pearl oysters and elucidated information on their nutritional requirements from the results (e.g. Taylor *et al.*, 1997; Martinez-Fernandez *et al.*, 2006). The focus has been on larvae to achieve optimum growth and survival rates through this critical and technically demanding phase of mariculture (see Section 7.4.6.4). There is also the advantage of rapid experimental results obtained by studying the early stages that are developing rapidly. Similar effects of food quality and quantity occur during later stages of the pearl oyster lifecycle, but their effects on growth and survival take much longer to become evident. Most long-term studies are field based, where it is not possible to control the food quality and quantity, only to take appropriate measurements on a regular basis and then correlate oyster growth and survival with these data, within the context of variability of other environmental parameters.

6.2.1. Quality

Small diatoms are the predominate food of Akoya pearl oysters in some Japanese and Korean pearl farming areas, including species of *Bacteriastrum*, *Chaetoceros*, *Leptocylindrus*, *Melosira*, *Nitzschia*, *Rhizosolenia*, *Skeletonema*, *Thalassionema* and *Thalassiosira* (Fukushima, 1970; Ohwada and Uemoto, 1985; Chang *et al.*, 1988). These diatoms undergo rather abrupt and independent blooms in Uchiumi Bay, Japan (Tomaru *et al.*, 2002a; Hashimoto and Nakano, 2003). However, pearl oyster condition tends to decline and mortalities occur when there are long periods of predominantly *Nitzschia* species in the phytoplankton (Fukushima, 1970, 1972; Tomaru *et al.*, 2001; Tomaru *et al.*, 2002a). *N. closterium* was shown to be poor food for Akoya as a sole diet, possibly due to inedibility (Numaguchi, 2000). Thus, Tomaru *et al.*, (2001, 2002a) hypothesized that the poor condition of pearl oysters during the blooms of the inedible *Nitzschia* species led to vulnerability to infections and subsequent mass mortality in Uchiumi Bay. *Nitzschia* species, however, may not be so food limiting in the field. Akoya ingested *N. closterium* at a similar rate to four other similar-sized microalgae and absorbed it with ca. 40% efficiency, i.e. digestion and absorption were not problems (Numaguchi, 2002). While *Nitzschia* species may be nutritionally deficient as sole diets, they will not be the sole algae ingested in the field, even during a bloom.

A comparison of *Isochrysis* aff. *galbana* Tahitian (T-Iso) and *Dunaliella primolecta* as food micro-algae for *P. margaritifera* and *P. maxima* found that the former was more nutritious (Yukihira *et al.*, 1998). In energy components calculated for both diets for both pearl oyster species, T-Iso was always the superior food in absorbed energy (AbsE) and scope for growth (SFG) (see Section 4.3). Clearance rates (CR) and absorption efficiencies (AEff) were similar for each pearl oyster species feeding on the two diets over the range 0.1–10 mg dwt micro-algae per liter. It seems that T-Iso was the superior food because its energetic value was 20.27 J/mg, compared to that of *D. primolecta*, 15.06 J/mg, rather than because of differences in pre- and post-ingestion processes.

The particulate inorganic component (PIM) of the seston tends to increase with the concentration of suspended particle matter (SPM) (Hawkins *et al.*, 1998; Yukihira *et al.*, 1999), affecting the quality of the SPM for pearl oysters as the proportion of particulate organic matter (POM) declines. *P. margaritifera* and *P. maxima* typically inhabit different environments, the former in low SPM environments and the latter in high SPM and high PIM environments (Gervis and Sims, 1992; Yukihira *et al.*, 1999). Food quality, in terms of the proportion of POM affects both species profoundly, but to different extents, reflecting their environment-related ability to filter feed over a range of SPM levels. Over proportions of POM 0.16–0.28 in natural SPM, the peak AbsE of *P. margaritifera* was at the highest POM level tested (i.e. lowest SPM), and the peak for *P. maxima* was at 0.2 (Yukihira *et al.*, 1999). Maximum SFGs for these two species corresponded to the maximum POM values.

Some phytoplankters, typically dinoflagellates and cyanobacteria, have been implicated in pearl oysters mortality. Blooms, known as "red tides" have been a problem for pearl farms in Japan since they began pearl culture (see Section 11.6.2.1).

6.2.2. Quantity

Quantity of food is also a major factor in the physiological condition, metabolic function, growth and survival of pearl oysters. There is not, however, a simple relationship between these parameters and food particle concentration, i.e. more food leading to better condition, metabolic function, growth and survival. This kind of relationship may occur at low particle concentrations. From the nature of filter feeding, however, it is possible for the pre-ingestion and post-ingestion mechanisms to become overloaded (see Section 4.2). Milke and Ward (2003) found that high particle concentrations markedly increased the times for pre-ingestive processing in two bivalves. Overloading the digestive system reduces AEff and this may occur to the extent that absolute rate of absorption declines. SFG will become negative under these conditions.

When feeding on micro-algae over the concentration range 6–120×10^3 cells/mL, specific growth rates of *P. maxima* spat increased steadily with cell concentration from 0 to 50×10^3 cells/mL (Mills, 2000). Mills noted that these spat grew moderately well even at 6×10^3 cells/mL density, but they were "skinny" spat. Organic content of these spat was about 9% compared with 13% in spat at the higher cell concentrations. The efficiency of converting organic material from captured food cells into animal tissue declined from 40% to <10% with increasing cell concentration. Loss of conversion efficiency probably occurred during pre-ingestion, i.e. more pseudofeces were produced, and post-ingestion, i.e. poor digestion and AEff.

When *Pinctada margaritifera* and *P. maxima* fed on T-Iso at cell concentrations from 1 to 10 mg/L (ca. 10–100×10^3 cells/mL), SFG of *P. margaritifera* peaked at 1–2 mg/L and declined to zero by 6 mg/L (Table 6.1) (Yukihira *et al.*, 1998). SFG of *P. maxima* peaked at 2–3 mg/L and declined to zero by 9 mg/L. There were similar but slightly lower cell concentrations for the equivalent SFG values when feeding on a larger flagellate, *Dunaliella primolecta* (Table 6.1). Major factors in these patterns of SFG versus cell concentration were declining CR and AEff, and increasing pseudofeces production with increasing cell concentration (see Sections 4.2.2, 4.2.4 and 4.2.7).

Again, when *P. margaritifera* and *P. maxima* fed over a natural SPM range of 3–40 mg/L, SFG peaked and then declined to zero (Table 6.1) (Yukihira *et al.*, 1999). As for the unicellular algae, the peak and zero particle levels were higher for *P. maxima* than *P. margaritifera*. These levels for peak and zero values of SPG when feeding on natural SPM are substantially higher than those for feeding on unicellular diets. This is not surprising considering that the bulk of SPM is inorganic. However, when POM levels within SPM are considered, the POM levels for optimal and zero SFG are lower than the concentrations for pure micro-algae diets (Table 6.1). Thus, the pearl oysters obtained proportionally more of the available organic matter in the diets containing inorganic matter. This has been found in other bivalves and is possibly due to inorganic particles facilitating the mechanical processes of digestion (Knauer and Southgate, 1999).

Longer experimental studies have shown the effects of food quantity on growth. Groups of small *Pteria sterna* were reared for 15 weeks on three "rations" of a micro-alga, *Chaetoceros* species, at three temperatures (del Rio-Portilla *et al.*, 1992). The algae rations were based on feeding dry weight (dwt) of alga as a percentage of pearl oyster dry soft tissue dwt. Growth rate increased with food quantity over most of the temperature range, although at the lowest algae ration, 2%, growth was very slow at all temperatures and was obviously very food limited. Growth rate was greatest at the highest algae ration, 8%, combined with the highest temperature, 30°C, but these were not optimum conditions of either factor (del Rio-Portilla *et al.*, 1992). Higher levels of food, in particular, are required for optimum growth.

Several field studies have shown a correlation between growth and food quantity. Yamaguchi and Hasuo (1977) found that the ratio dry weight/wet weight of soft tissue (a measure of pearl oyster condition) was positively correlated with chlorophyll a (chl a) levels in a pearl culture region over 3 years. There was also a positive correlation between the growth rate of Akoya and phytopigment concentration, 2–6 µg/L, in Ago Bay (1967–1969) and Ohmura Bay (1984–1985), Japan (Numaguchi, 1994a). Chl a levels influenced rate of shell growth of Akoya in Yusu and Uchima Bays, Japan (Tomaru et al., 2002b). Shell growth rate was positively correlated with food abundance when water temperature was >20°C, but, interestingly, the correlation was only with shell thickness, not other shell dimensions.

Concentrations of SPM in the Takapoto atoll lagoon are always low, usually 0.6–1.3 mg/L, and never in excess of the feeding capacity of P. margaritifera (Pouvreau et al., 2000) (cf. values in Table 6.1). Growth rates of somatic tissue, shell and gonad weights follow variations in POM throughout the year. Pouvreau et al. (2000) modeled the effects of food quantity on P. margaritifera growth in Takapoto lagoon. The model predicted that there would be 20% less growth if POM levels were 20% lower over the course of a year. The result of a 30% lower POM level over the course of a year was a 50% reduction in growth. The SFG model was validated from field measurements of pearl oyster growth in the field. Furthermore, an extended period of supplementing Takapoto lagoon water with micro-algae at 20,000 cells/mL found that P. margaritifera on the enhanced diet had higher metabolic rate (MR) and improved body condition than pearl oysters on the natural levels of phytoplankton (Cuzon et al., 2004).

In another case of SFG validation and field testing, the effects of food quantity and quality on growth rate, survival and other parameters were tested for P. margaritifera and P. maxima at two Great Barrier Reef (GBR) lagoon sites (Yukihira et al., 2006). One site was characterized by high and very variable levels of SPM (mean 11.1 mg/L, range 2–60 mg/L; mean POM 2.1 mg/L), while the other was characterized by low and

TABLE 6.1

Concentrations of two unicellular microalgae and natural suspended particulate matter at which there is optimal Scope for Growth (SFG), and high levels at which SFG declines to zero for *Pinctada margaritifera* and *P. maxima*.

	Food levels (mg/L)					
	T-Iso		D. primolecta		natural SPM	
Species	Optimum SFG	SFG = 0	Optimum SFG	SFG = 0	Optimum SFG	SFG = 0
P. margaritifera	1–2	6	1	4	3 (0.8)[a]	15 (2.6)[a]
P. maxima	2–3	9	2	7	7–10 (1.7)[a]	40 (6.2)[a]

Data from Yukihira et al. (1998; 1999).

T-Iso: *Isochrysis cf. galbana* Tahitian; *D. primolecta*: *Dunaliella primolecta*; POM: particulate organic matter; SPM: natural suspended particulate matter.

[a]POM content in the natural SPM.

stable levels of SPM (mean 2.1 mg/L; mean POM 0.56 mg/L). Growth rates, condition index (CI) and survival of *P. maxima* were similar at both sites. Growth rate, CI, survival, asymptotic size (shell height, SH_∞) and other parameters of *P. margaritifera* were all lower at the high SPM site. As described in Section 6.2.1, *P. margaritifera* typically inhabits low SPM environments, e.g. Takapoto above, and SFG studies have shown optima at low particle concentrations. Thus, SFG predictions of poor performance at the high SPM site were validated.

Akoya pearl oysters endured long periods of starvation with progressive loss of soft tissues (Numaguchi, 1995a; 1995b). Mortality at 23–29°C finally reached 95% after 115 days of starvation. Initial response to starvation was the glycogen content of the adductor muscle decreasing markedly during the first week. Then, as the oysters' tissues were consumed in catabolism, there was more than 70% loss of dwt and condition index (CI) decreased from about 13.7 to <4. Numaguchi (1995b) considered these to be critical levels for the survival of this pearl oyster. Temperature has a strong positive influence on MR (see below) and thus on catabolic rate during starvation. Rates of tissue and reserve energy loss increased with temperature during starvation (Numaguchi, 1995a).

As a final note in this section, Ohwada and Uemoto (1985) gave an estimate of 100 g dwt POM/oyster/year required for good growth and reproduction of Akoya. They indicated that POM consisting of diatoms is preferable, but did not specify oyster size.

6.3. TEMPERATURE

6.3.1. Metabolic rate

Like other aquatic invertebrates and most fish, the body temperature of pearl oysters is variable (poikilothermy), closely approximately their immediate environmental temperature and essentially derived from it (ectothermy). Pearl oysters are rarely exposed to lethal temperatures in their environments, but, within their normal ranges, temperature has a profound effect on MR and related functions.

Studies of the influence of temperature on metabolic and related physiological processes of bivalves have shown that there is usually an optimum temperature or narrow temperature range for each species at which there is maximum MR, growth rate and survival (Yukihira *et al.*, 2000). Maximum MR is reflected in many of the processes of the body occurring at near their maximum rates, e.g. rates of respiration, clearance, absorption, excretion, tissue growth, shell secretion and gametogenesis. The pattern of MR reaching a maximum with increasing temperature and then declining at higher temperatures partly results from two opposing factors. Biochemical reaction rates, the bases of MR, increase with temperature, but rates of enzyme denaturation increase with temperature.

A common pattern of the influence of temperature on MR and related functions is shown in Figure 6.3. CR, a measure of MR, for these two *Pinctada* species is maximal near 28°C. (Measurements were made at 4–5°C intervals and it is not possible to be precise from these data.) CRs declined from 28 to 23°C and then showed a stronger

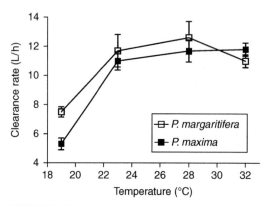

FIGURE 6.3. Effect of temperature on the clearance rates of *Pinctada margaritifera* and *P. maxima*. From Yukihira *et al.* (2000).

rate of decline below 23°C. As temperature becomes much lower than optimum there is a progressively greater decline in MR with each °C decrease. From 28 to 32°C, CR is unchanged in *P. maxima*, but it declines in *P. margaritifera* (Fig. 6.3). As temperature increases above optimum, MR initially declines slowly and then abruptly with each °C increase as it approaches sub-lethal temperatures. *P. margaritifera* has reached the declining MR phase within the 28–32°C range, while *P. maxima* is not adversely affected. The MR of *P. maxima* is more affected at low temperatures than *P. margaritifera*. The two species show similar patterns of MR versus temperature, but the MR–temperature relationship of *P. maxima* is shifted to higher temperatures compared with *P. margaritifera* (Yukihira *et al.*, 2000).

Various MR-related functions of Akoya have been tested over a wide range of temperatures. They have usually shown the typical pattern of very low rates at low temperature, increasing rates with increasing temperature and an optimum and then decline at high temperature. The term "hibernation" has been used for the oyster's condition at the lowest temperatures (Ohwada and Uemoto, 1985). The valves are completely closed in seawater below 7°C during winter (Kobayashi and Tobata, 1949b). Gill cilia do not beat at 7–8°C (Kobayashi and Matsui, 1953; Wada, 1991) and CR is virtually zero at 8–10°C (Miyauti, 1962; Numaguchi, 1994b). The crystalline style is absent during the period when temperature is below 10°C, corresponding with the period of starvation (Ōta, 1958). Oxygen consumption and ammonia excretion virtually cease in this temperature range, reflecting minimal MR (Seki, 1972). From 7–13°C there are occasional slight valve movements and these increase in strength and frequency up to about 15°C. Above 15°C there are normal valve movements associated with regular filter feeding. Rate of cilia beating on the gills increases almost linearly with temperature (Yamamoto *et al.*, 1999; Yamamoto, 2000). CR was <20L/oyster/day at 10–13°C, then rose sharply from 19°C to a peak at 25–28°C. In this optimum temperature range, CR was about 270L/oyster/day (Numaguchi, 1994b). The crystalline style is re-established after winter when the water temperature is about 15°C. Its length is then strongly correlated with

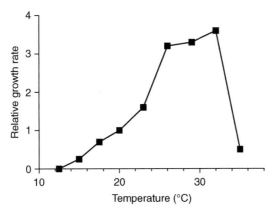

FIGURE 6.4. Relative growth rates of Akoya spat at various temperatures over 24 days. From Numaguchi and Tanaka (1986b).

temperature during the warmer period of the year (Ōta, 1958). Similarly, oxygen consumption, NH_3–N excretion and heart rate increase with temperature (Uemoto, 1968; Seki, 1972; Itoh, 1976; Numaguchi and Tanaka, 1986b). There were some differences, however, at the upper temperature level. As Yukihira *et al.,* (2000) found with CR of *P. margaritifera*, CR of Akoya declined precipitously from 28 to 33°C (Numaguchi, 1994b). In contrast, heart rate increased up to 93 beats/minutes at 32°C, the highest temperature tested (Numaguchi and Tanaka, 1986b). Pumping rate (PR) and ciliary beating rate peaked and decline precipitously at 32°C and 34°C, respectively, in 28°C acclimatized oysters (Yamamoto, 2000). Some of the variation is due to the duration of exposure to high temperatures in the different techniques, i.e. it takes longer to measure CR at each temperature than heart rate. Reflecting these rates of various functions versus ambient temperature in Akoya, the growth rates of spat over a 24 day period show rapid growth up to a plateau about 29–32°C and then a precipitous decline to 35°C (Fig. 6.4). There was highest survival within the range 17.5–29°C and total mortality at 12.5 and 35°C (Numaguchi and Tanaka, 1986b).

Pearl secretion ceases at <12°C and cultured pearls may lose weight at low temperatures in Akoya (Wada, 1969a). This reflects the strong influence of temperature on nacre secretion rate in pearl production, mediated through its effects on the host oyster's metabolism (Wada, 1972).

Temperature coefficient, Q_{10}, is a means of describing the relative effects of temperature changes on MR or MR-related parameters over 10°C intervals (Yukihira *et al.,* 2000).

$$Q_{10} = (R_2/R_1)\exp[10/(t_2 - t_1)]$$

where R_1 and R_2 are rates at temperature t_1 and t_2, respectively.

Q_{10} values are high where there are marked changes in MR with each °C temperature change, i.e. they tend to be high (>3) in the lower part of the tolerated temperature regime where MR decreases markedly with decreasing temperature, as described above. Q_{10} values are moderate (1–2) near optimum temperature and then <1 as MR declines with increasing temperature above the optimum. Q_{10} values in Table 6.2 show this pattern, with two exceptions (Uemoto, 1968; Saucedo *et al.*, 2004). Uemoto (1968) found little variation in Q_{10} except for a high value in the 27–30°C range and interpreted 27°C as the upper limit of optimal temperature. The extraordinarily high Q_{10} value for the 10–14°C interval in Akoya results from the extremely low CR value at 10°C (Miyauti, 1962). Q_{10} values for each species tend to be lower for long term, e.g. growth rates, than short-term studies, e.g. respiration rates, at the upper end of

TABLE 6.2
Q_{10} values for metabolic rate-related functions in Pinctada species.

Species	Function	Q_{10} value Temperature range					Source
Akoya	Respiration rate	6.4 *10–14°C*	4.0 *14–18°C*	3.0 *18–22°C*	2.5 *22–26°C*	2.2 *26–28°C*	Itoh (1976)
	Respiration rate	2.7 *13–15 °C*	1.9 *15–21 °C*	2.2 *21–27 °C*	4.4 *27–30 °C*	2.7 *30–33 °C*	Uemoto (1968)
	Excretion rate	7.7 *10–14°C*	4.5 *14–18°C*	3.4 *18–22°C*	2.7 *22–26°C*	2.4 *26–28°C*	Itoh (1976)
	Clearance rate[a]	20 *16–19°C*	14 *19–22°C*	2 *22–25°C*	0.8 *25–28°C*	0.02 *28–31°C*	Numaguchi (1994a)
	Clearance rate[b]	2,175 *10–14°C*	11 *14–18°C*	3.1 *18–22°C*	1.1 *22–26°C*	0.4 *26–30°C*	Miyauti (1962)
	Growth rate[c]	85 *15–17.5°C*	6.0 *17.5–23°C*	4.8 *23–29°C*	1.5 *29–32°C*	3×10^{-4} *32–35°C*	Numaguchi and Tanaka (1986b)
P. maxima	Clearance rate	6.2 *19–23°C*		1.1 *23–28°C*		1.0 *28–32°C*	Yukihira *et al.* (2000)
	Scope for Growth	15 *19–23°C*		1.5 *23–28°C*		0.9 *28–32°C*	Yukihira *et al.* (2000)
	Growth rate[d]	4.9 *20–23°C*		2.0 *23–26°C*	1.2 *26–29°C*	0.4 *29–32°C*	Mills (2000)
P. margaritifera	Clearance rate	3.4 *19–23°C*		1.2 *23–28°C*		0.7 *28–32°C*	Yukihira *et al.* (2000)
	Oxygen consumption	4.7 *15–24°C*		2.1 *25–30°C*	2.3 *30–33°C*	0.9 *33–35°C*	Sugiyama and Tomori, (1988)
	Scope for Growth	2.7 *19–23°C*		1.8 *23–28°C*		0.3 *28–32°C*	Yukihira *et al.* (2000)
P. mazatlanica	Respiration rate	1.5 *18–23°C*		2.3 *23–28°C*		1.2 *28–33°C*	Saucedo *et al.* (2004)

[a]Calculated from Figure 6.2.
[b]Calculated from Table 6.1.
[c]Calculated from Table 6.2.
[d]Calculated from Figure 6.3.

the temperature range (Table 6.2). The longer-term results are most likely to be better reflections of tolerances in the field.

In an indirect approach to measuring the relationship between MR and temperature, Numaguchi (1995a) determined the effects of three temperatures on the catabolic losses and CI of unfed Akoya. Dry meat weight and CI gradually decreased over 60 days of starvation with losses being lowest at 15°C and highest at 28°C. MR was supported by catabolism of the oysters' soft tissues and was positively related to temperature.

6.3.2. Growth rate

As already indicated, growth and development rate are linked to MR, which is strongly influenced by temperature. Temperature's influence on growth rate is one of the most obvious environmental effects and it is often of considerable importance for the pearl farmer.

There have been many studies showing the positive relationship between temperature and growth rate in all stages of the pearl oyster lifecycle and some are shown in Table 6.3. However, some studies have shown negative relationships between growth rate or SFG and temperature near the upper limits of a species (Pandya, 1976; Yukihira et al., 2000; Pouvreau and Prasil, 2001). These reflect declining MR or declining availability of energy for growth at temperatures above optimum.

Tissue growth may cease or tissue is lost at low temperatures when there is insufficient energy for tissue growth, i.e. SFG is near zero or negative. One to 3-year-old Akoya lost tissue during winter months in Japan (Kobayashi and Tobata, 1949a). The oysters went into declining body weight when water temperature dropped below about

TABLE 6.3
Range of positive temperature affects on growth and development rate of *Pinctada* and *Pteria* species.

Species	Stage	Temperature range of positive effects (°C)	Temperature range tested (°C)	Source
Akoya	Embryo	18–28		Komaru et al. (1990)
	Spat	20–28	7.5–35	Numaguchi and Tanaka (1986b)
	Adult	14–30		Kuwatani et al. (1974)
Akoya (India)	Larva	24–29		Alagarswami et al. (1983)
	Adult	19–29	19 to >29	Pandya (1976)
P. maxima	Spat	20–29	20–35	Mills (2000)
	Adult (SFG)	19–32		Yukihira et al.(2000)
P. margaritifera	Adult (SFG)	19–28	19–32	Yukihira et al. (2000)
Pteria sterna	Spat/juvenile	20–30	20–30	del Rio-Portilla et al. (1992)
	Juvenile	15–27	15–27	Araya-Nunes et al. (1991)
	Larvae	21–28	21–28	Araya-Nunes et al. (1991)

13°C. These authors estimated that the oysters spent at least 5 months (ca. 150 days) of the year in "hibernation" and subsequent recovery of tissue that was lost during winter months. Similarly, on the basis of apparent lack of feeding and other activity at low winter temperatures, Kim (1969) estimated ceased growth for periods of 115, 130 and 150 days at three locations on the South Korean coast, reflecting the respective periods of very low water temperatures at these locations.

6.3.3. Survival

As well as the pervasive influence of temperature on growth, MR, and other physiological functions over a range of environmental temperatures, there are upper and lower temperatures where the animal is severely stressed, and upper and lower lethal temperatures.

Sub-lethal, i.e. where the oyster persists indefinitely under physiological stress, and lethal temperatures, where the oyster will die if the conditions persist, have been investigated in short-term laboratory studies of various life stages of pearl oysters (Table 6.4). For, instance, 3 mm hinge length Akoya juveniles survived well from 17.5–29°C (Numaguchi and Tanaka, 1986b). Most were in sufficiently good condition to attach to the walls of the rearing tanks with byssal threads. While some juveniles survived at 12.5° and 32°C, these temperatures were clearly marginal. There was no survival at 7.5 and 35°C (Table 6.4). In adult Akoya the lower lethal temperature results from the cessation of feeding and hence starvation. The upper lethal temperature is near 35°C, when gill cilia movements and heart beatings cease.

Different stages of a lifecycle may have different temperature tolerances and sometimes tolerance increases during development. There is, however, no clear evidence of this phenomenon in the data for pearl oysters (Table 6.4).

Although lethal temperatures may sometimes be reached quickly during aquaculture procedures, such as pearl oysters being left too long exposed to the sun, it is very rare for bivalves in their natural environment to be exposed to extreme temperatures such that they die quite abruptly. Mortality occurs as a function of time of exposure and the severity of the temperature extreme. There is a time–temperature function, such that it takes longer exposure for mortality at marginally lethal temperatures than it does at more extreme lethal temperatures (Kennedy et al., 1974; Mihursky et al., 1974). Furthermore, because mortality at lethal temperatures results from metabolic dysfunction, the consequences of metabolic dysfunction develop more slowly at low temperatures than at high temperatures. The time factor in the time–temperature lethal stress is longer at low than at high temperatures. Lethal effects of high and low temperatures are also confounded by other factors associated with them, especially at high temperatures (Wada, 1991). Low levels of dissolved oxygen, lowered salinity during summer rainfall, toxic algal blooms and high levels of metabolic wastes may be associated with high water temperatures. Stress in the physiological condition of the oysters, e.g. post-spawning, will also increase vulnerability to extreme temperatures.

Low water temperature was considered a factor in the mortality of farmed pearl oysters, *P. maxima*, in northern Australia. Mortality apparently resulted from the additional

TABLE 6.4

Temperature ranges of *Pinctada* and *Pteria* species. These temperatures are approximations as the studies were conducted using temperature intervals of 2–5°C and most values are from short-term studies. Long-term studies may reveal more conservative upper and lower lethal temperatures. Sub-lethal and lethal temperatures (see text for definitions).

Species	Stage	Temperature (°C)					Source
		Lower lethal	Lower sub-lethal	Moderate	Upper sub-lethal	Upper lethal	
Akoya	Embryo			23[a]			Komaru *et al.* (1990)
	8 cell embryo					29–30[b]	Okamura and Nakamura, 1988
	Larva					35 +[b]	Okamura and Nakamura (1988)
	Larva	15		17–29[c]	32	35	Wada (1991)
	Spat					35 +[b]	Okamura and Nakamura (1988)
	Juvenile	7.5	12.5	17.5–29	32	35	Numaguchi and Tanaka (1986b)
	Adult	7–8		13–25		35	Wada (1991)
	Adult				32	35	Watanabe, (1988)
	Adult				>30		Numaguchi (1994a)
	Adult	<10					Je *et al.* (1988)
Akoya (India)	Juvenile			17–35			Rao (2001)
Akoya (Eastern Australia)	Embryo	14		18–26			O'Connor and Lawler (2004)
	Juvenile			14–26			O'Connor and Lawler (2004)
P. margaritifera	Embryo			20–30		35	Doroudi *et al.* (1999)
	Larva			21–26.5		35	Doroudi *et al.* (1999)
	Juvenile			22–31.5[d]			Nasr (1982)
	Adult			23–28	>28		Yukihira *et al.* (2000)
	Adult			26– < 30			Pouvreau and Prasil (2001)
	Adult					35	Sugiyama and Tomori (1988)
P. maxima	Spat	17.7[e]	<20	23–32[e]		35	Mills (2000)
	Adult		<19	23–32			Yukihira *et al.* (2000)
P. mazatlanica	Adult			18–28	33		Saucedo *et al.*, (2004)
Pteria sterna	Larva	18		21–28			Araya-Nunes *et al.* (1995)

[a]Lower rates of abnormal chromosome behavior than at 18 and 28°C.
[b]50% mortality after 24 hours.
[c]Estimated.
[d]Temperature range in field nursery.
[e]In long-term tank culture with low mortality.

stress of being handled during the winter months and movements to farms were subsequently made during the warmer months (Pass *et al.*, 1987). Low temperature reduced survival of *P. maxima* near the southern end of its distribution along the GBR (18°S). Its survival and growth rate were found to be more adversely affected by the lower winter temperatures at an inshore site compared to an offshore site (Yukihira *et al.*, 2006). High temperatures together with other factors, organic pollution, post-spawning stress, starvation, dinoflagellate blooms, have been implicated, but not conclusively demonstrated, in the mass mortalities in pearl farms over recent decades in Japan (Tomaru *et al.*, 2001).

There is an interesting effect of seasonally increasing temperature on survival in the progress of the "red adductor muscle disease", which has been responsible for heavy mortality of Akoya (Morizane *et al.*, 2001; Nagai *et al.*, 2007) (see Section 9.2.6).

Wada (1991) noted that there were few data on the effects on pearl oysters of various environmental parameters combined with high water temperatures. This would equally apply to low temperatures.

6.3.4. Geographic range and acclimatization

Optimum temperature and lethal temperature ranges are not necessarily absolute for each species or population. There may be genetic differences between populations of a species that influence their temperature tolerances. Populations of pearl oysters at higher latitudes may have a lower optimum temperature and lower lethal temperatures than populations at lower latitudes. For instance, *P. margaritifera* from French Polynesia (FrP) and the GBR are above their optimum temperatures and apparently entering their zone of temperature stress at >30°C (Table 6.4). However, this species has been cultured on the Sudanese coast of the Red Sea where water temperatures range up to 34°C (Reed, 1966).

Pearl oysters at one locality may thermally acclimatize through the seasons of the year. A degree of compensatory non-genetic change in the MR–temperature relationship has been described in some bivalves (Gosling, 2003). The result is that the temperature optimum is lower in winter and higher in summer, with the compensatory advantages of this acclimatization being that MR and growth are not as low in winter and there is greater tolerance of high temperatures in summer. There is evidence of acclimatization in pearl oysters, but not consistently or of a compensatory level. Akoya showed thermal acclimatization through the year at a site in Japan (Yamamoto, 2000). Oysters collected at 12°C (February), 20°C (May) and 27°C (September) were progressively exposed to higher temperatures over a short period and found to have maximum oxygen consumptions at 30°C, 30°C and 36°C, respectively, and maximum PRs at 17°C, 25°C and 32°C, respectively. This is clearly temperature acclimation, but it is hardly compensatory with the oysters at 12°C having very low rates. Numaguchi (1994a) followed the CR–temperature relationship of Akoya through the seasons, finding seasonal variations in the relationship that were greater than predicted and contrary to compensatory acclimation. Itoh (1976) measured rates of oxygen consumption and N excretion of Akoya at ambient seawater temperature over long periods in 3 years.

There were large seasonal variations in these physiological parameters and no suggestion of varying rate–temperature relationships. Saucedo *et al.* (2004) considered that *P. mazatlanica* showed temperature compensation in its rates of oxygen consumption and N excretion, based on moderate Q_{10} values. They did not, however, test whether the rate–temperature relationship changed with seasonal variations in water temperature.

6.4. SALINITY

6.4.1. Osmoconforming

As in other bivalves, water constitutes a high proportion of the soft tissues of pearl oysters. Values for Akoya range from 70% in hepatopancreas tissue to 85% in mantle tissue (Dharmaraj *et al.*, 1987). Bivalves are osmoconformers, i.e. their intracellular fluid (ICF) and extracellular fluid (ECF) (=hemolymph) are approximately isosmotic with the water in their mantle cavity and conform to its variations within tolerated limits. This is the case for Akoya (Funakoshi *et al.*, 1985). Similarly, the inorganic ion composition of their body fluids is very similar to that of the water mantle cavity. There is usually some ion regulation across the large surfaces of exposed epithelium, at least for Na^+ and K^+ ions. During periods of salinity stress, such as extremes or rapid changes, it is possible for some bivalves to hold the valves tightly closed for two days or more (Funakoshi *et al.*, 1985). They buffer the rate of osmotic and ionic changes in the mantle cavity water and thence in the body fluids where rapid changes may be disruptive. However, if the oyster is to tolerate long periods of tightly closed valves, it requires physiological adaptations to the low dissolved oxygen (DO) levels that develop in the mantle cavity water and its tissues.

Buffering against rapid salinity changes was demonstrated when specimens of Akoya from the Gulf of Mannar, India, were abruptly transferred from seawater, 34‰ salinity, into salinities ranging from 14 to 58‰, i.e. abrupt changes in salinity up to about 20‰ above and below seawater (Alagarswami and Victor, 1976). Control oysters in seawater opened immediately and had the longest total period of open valves and most valve movements during the first day after transfer (Table 6.5). The pumping rate (PR) is linked to the degree and duration that the valves and mantle lobes are open and showing low stress (Yamamoto, 2000). The period to first opening increased, while total periods of open valves and number of valve movements were inversely related to the degree of salinity change during the first day. The oysters were buffering salinity changes between their mantle cavity fluid and the ambient water, with those exposed to the greatest gradients making least exposure to the new salinity. Oysters at extreme salinities gaped for most of the time on the second day after transfer and, subsequently, many died (Table 6.5).

Even greater salinity extremes may be used in the management of cultured pearl oysters, as they are used for other bivalves (MacNair and Smith, 1999). Pearl oysters can tolerate extremely lethal salinities for short periods through tight closure of their valves, although this kind of exposure is not part of their normal experience. Their tight valve closure may be used in culture to eliminate soft-bodied pest species on their

TABLE 6.5

Periods to first valve opening, duration of valve openings on the first day and subsequent mortality in groups of Akoya (India) transferred abruptly from seawater to a wide range of test salinities.

	Test salinities (‰)								
	14	15	17	19–29	34	38–45	52	55	58
Period to first valve opening (hours)	22	19	19	<1	0	<1	?	?	5
Valve openings on 1st day	Nil	Brief	Brief	Moderate to long	Long	Long to moderate	Brief	Brief	Brief
Subsequent mortality (%)	100	50	0	0	0	0	67	100	100

Data from Alagarswami and Victor (1976).

external surfaces (see Section 15.5). To kill mud worm (*Polydora ciliata*), Akoya are immersed in freshwater for 15 minutes to induce tight valve closure, and then they are immersed in 22% brine for 20 minutes. This kills the *Polydora* without harming the oyster (Waki and Yamaguchi, 1964). A similar technique rids Akoya (New South Wales, Australia) of a stylochid flatworm predator (O'Connor and Newman, 2001).

6.4.2. Tolerance

Pearl oysters are recognized as marine bivalves, but many species occur in coastal waters adjacent to large landmasses in at least part of their geographic distributions. There is potential lowering of salinity during periods of rainfall and runoff from rivers and estuaries in these coastal waters. This is particularly the case in semi-enclosed water masses, such as embayments, where surface waters may be particularly affected. Salinity declined from >34‰ in winter 1997 to <28‰ in summer where Akoya are farmed in Gokasho Bay, central Japan (Abo and Toda, 2001) and from 36‰ to 25‰ over these seasons at another location (Seki, 1972). Cahn (1949) also reported haloclines with a 1–2 m surface layer of low salinity water overlying seawater salinity in Japanese pearl farming areas. There was high mortality during periods of low surface salinity. Alternatively, there may be strongly hypersaline conditions where water masses have limited connection to the open ocean and little input of freshwater, e.g. up to 60‰ salinity in the Arabian Gulf where Akoya occur (Al-Sayed *et al.*, 1997). Species diversity decreases in the highest salinity regions of the Gulf and Akoya are notably one of the marine organisms that are abundant in these conditions. Even *P. margaritifera*, which is typically associated with coral reefs and reef lagoons (Yukihira, 1998), is abundant where salinity ranges over 26–37‰ through the seasons in Gazi Bay, Kenya (Kimani and Mavuti, 2002).

Some species of pearl oyster must therefore be quite euryhaline, tolerating a considerable range of salinities. Confirming this, data on salinity tolerances of embryo, larvae, spat and adult stages of the lifecycle of some *Pinctada* species are shown in Table 6.6. These data, despite their inadequacy, show many species tolerating at least

TABLE 6.6
Favorable salinity ranges of high survival and growth of lifecycle stages of *Pinctada* species. These are not precise salinity levels as the tests were at least 2‰ salinity or more intervals.

Species	Stage	Favorable range (‰ salinity)	Range tested (‰ salinity)	Means of establishing this	Source
Akoya (eastern Australia)	Embryo[a]	29–35	11–35	high survival in rearing	O'Connor and Lawler (2004)
	Juvenile	29–35	11–35	No mortality over 7 days[b]	O'Connor and Lawler (2004)
Akoya	Larva	11–29	5–29	Limited mortality during development	Ōta (1957)
	Spat	15–38	7.5–38	High survival over 25 days[a]	Numaguchi and Tanaka (1986a)
	Spat	19–38	7.5–38	Good growth over 25 days[b]	Numaguchi and Tanaka (1986a)
	Adult	18–30	10–30	No mortality over 48 hours[b]	Katada (1959)
	Adult[c]	>18–30	10–30	No mortality over 48 hours[b]	Katada (1959)
	Adult	>18	Field observation	Normal growth	Ohwada and Uemoto (1985)
	Adult	26–35	0–70	No mortality over 48 hours[b]	Funakoshi *et al.*, (1985)
	adult	>25	22–32	optimum feeding rate	Ōta (1961)
Akoya (India)	Adult	17–45	14–58	High survival over ≥2 days[b]	Alagarswami and Victor (1976)
P. margaritifera	Embryo[a]	<25–39	25–40	High survival in rearing	Doroudi *et al.* (1999)
	Larva	<25–32	25–40	High survival in rearing	Doroudi *et al.* (1999)
P. maxima	Spat	25–45	25–45	High survival over 20 days	Taylor *et al.* (2004)
	Spat	30–34	25–45	Optimum growth over 20 days	Taylor *et al.* (2004)

[a]Embryo to early larval stage.
[b]Abrupt exposure.
[c]Nucleated 2 months before test, not tested between 18 and 30‰.

a 15‰ salinity range, including early developmental stages, although they may not grow optimally over the full extent of this range. Favorable ranges may not be well defined. Akoya larvae show a progressive inverse relationship between level of mortality and salinity rather than abrupt change from near total survival to near total mortality in lowered salinity conditions (Ōta, 1957). Duration of exposure is also important for these larvae. The broad salinity range for larval development of *P. margaritifera* might be expected in inshore and embayment pearl oysters, such as Akoya, but it is surprising for a species that tends to occur in oceanic environments. Akoya from eastern Australia are exceptional with no survival through embryonic development at 26‰ salinity and below, and limited salinity tolerance by juveniles (Table 6.6). Limited salinity tolerance through early development will affect recruitment and subsequent life stages, no matter how salinity tolerant the later stages may be. The comparatively limited salinity tolerance of these oysters is surprising, since they are found in environments in eastern Australia that are subject to low salinities. The specimens for this study of salinity tolerance were from the mouth of an estuary (O'Connor and Lawler, 2004).

Table 6.6 presents favorable (non-lethal) salinities, but favorable salinity ranges are bordered by stressful salinities, and then upper and lower lethal salinity ranges, e.g. Table 6.5. Numaguchi and Tanaka (1986a) determined mortality and other parameters for Akoya oyster spat exposed to salinities down to 7.5‰. Spat were transferred abruptly from seawater into the test salinities and degree of initial stress was shown by rate of byssal reattachment. All spat at 26.5–38‰ reattached within a day. No spat attached below 15‰. There was 95–100% survival over 25 days in 15–38‰ salinity and no survival at 7.5 and 11‰. There was optimum growth at 26.5‰, good growth over 19–38‰ and slightly poorer growth at 15‰ salinity. The spat at this early stage of development were surprisingly euryhaline. However, Numaguchi and Tanaka (1986a) suggested that the optimum salinity for Akoya oyster spat is about 26.5‰ and that they should be cultured at salinities greater than 23‰.

Salinities for good survival of adult Akoya are ca. >18‰ (Table 6.6) The reported value of at least >26‰ salinity required for 100% survival over 48 hours is quite inconsistent with other data for juveniles and adults (Funakoshi *et al.*, 1985). Nucleate Akoya are less tolerant of low salinity than normal pearl oysters 2 months after the nucleus is inserted Table 6.6. There was 37% and 70% mortality of normal and nucleated pearl oysters, respectively, after 36 hours in 9.5‰ salinity (Katada, 1959). The nucleated oysters showed increased sensitivity to stress even 2 months after nucleation.

Dholakia *et al.* (1997) studied the limits of salinity tolerance and tolerance of abrupt salinity changes for Akoya from the Gulfs of Kutch and Mannar. These Gulfs are located on the northwest and southeast coasts of India, respectively, with the latter one characterized by less silty conditions and greater development of coral reefs. Limits of salinity tolerance were tested by progressively raising or lowering the salinity by 2‰ each day to allow the oysters time to acclimatize. The salinities for total mortality of pearl oysters from the two regions were 12–70‰ (Kutch) and 15–67‰ (Mannar). There was progressive daily mortality as the extreme salinities were approached and the oysters would not have survived for any extended periods near the salinity extremes. It is not possible to make any definitive statement about levels of long-term survival, but the observations establish that Akoya from these regions are quite euryhaline. This was reinforced when the oysters were found to tolerate abrupt salinity increases and decreases of about 18‰ (Dholakia *et al.*, 1997).

As well as the observations of Cahn (1949) there appears to be one study where salinity is clearly implicated in mortalities in the field. Surface salinity abruptly declined from 30‰ to 16‰ in Ago Bay, Japan, in May 1976 (Itoh and Muzamoto, 1978). There was an abrupt increase in mortality of 40–50% among the cultured

Akoya at the surface, but not at lower depths where salinity was quite stable. The usual high rates of mortality occurring in these pearl oysters after spawning are attributed to a combination of environmental effects on the weakened post-spawning oysters. However, these oysters had died before spawning during the period of unusually low salinity.

6.4.3. Effects of salinity on physiological functions and growth

Salinity affects the physiological functions and hence growth rates in a wide range of osmoconforming marine invertebrates, including bivalves. However, there are limited data on the mechanisms by which salinity affects the MR and other physiological functions in bivalves and no data for pearl oysters.

Heart rate varies with salinity in Akoya oyster spat, demonstrating the effect of salinity on the MR of heart muscle cells via changes in the osmolality and ion content of the hemolymph (Numaguchi and Tanaka, 1986a). The pattern of heart rate versus salinity is rather like that of MR-related parameters versus temperature in reaching an optimum and then declining. Heart rate steadily increased from 38 beats/minutes at 11‰ salinity to 90 beats/minutes at 26.5‰. Heart rate then declined to 72 beats/minutes at 38‰, which was seawater concentration. It is interesting that peak heart rate occurred at about 70% of seawater concentration.

Similarly, feeding rate versus salinity in Akoya (Gulf of Mannar) showed progressively increasing CR with increasing salinity from a very low rate at 14‰ (Alagarswami and Victor, 1976). CR peaked at seawater salinity, 34‰, and then declined at salinities greater than seawater. CR was about half maximum rate at 57‰ salinity. Abo and Toda (2001) found that the relative CR of Akoya also increased with salinity from no filtering at <15‰ to a maximum at 30–35‰. The increase in relative CR with increasing salinity followed a sigmoidal pattern to the maximum level. These authors did not follow CR at salinities above seawater. Ōta and Fukushima (1961) measured feces production as an indirect measure of feeding rate. They found that the relative amount of feces declined abruptly below about 25‰. All these studies showed that CR was reduced by salinity where the latter deviated substantially from seawater levels.

PR arises from the rate of beating of the cilia on the gills and the degree of expansion of the valves and mantle lobes (Yamamoto, 2000). Ciliary movements on the gills of Akoya were abnormal at 13‰ salinity and ceased at 9.5‰ (Kobayashi and Matsui, 1953). These levels fit well with the value (above) for the cessation of CR (Abo and Toda, 2001). There is a further factor inhibiting feeding at this salinity level in that the crystalline style disappears at ca. 14‰ and lower salinities (Wada, 1969b). This is probably due to the cessation of secretions by the style sac epithelium, reflecting general metabolic stress.

Jeyabaskarn et al. (1983) suggested that high salinities within the range 29–34‰ reduced growth of Akoya in farms in the Gulf of Manmar. However, fouling increased at times of high salinity and temperature was relative low at times of high salinity, so the influence of salinity cannot be separated from at least two other factors that influence growth.

There have been two short-term studies of the influence of salinity on growth and development rates of pearl oyster spat under controlled conditions (Table 6.6). The longest study was for 25 days considering the effects of a wide salinity range on the growth rates of Akoya oyster spat (Numaguchi and Tanaka, 1986a). Two of these brief studies involved abrupt transfers into the test salinities. The exceptions were development of embryonic and early larval stages of two *Pinctada* species during their culture (Doroudi *et al.*, 1999; O'Connor and Lawler, 2004). They showed growth over broad salinity ranges, although salinity had a significant effect on growth rate with poorest growth and survival at the extremes of the ranges tested (Fig. 6.2).

Even brief exposure to stressful salinities can affect subsequent growth over a substantial period. Groups of Akoya were exposed to low salinities for 36 and 48 hours and there was 6–88% mortality at the lower levels (Katada, 1959). The groups of oyster were subsequently returned to the field and measured after 3 months. All groups where there had been any mortality during salinity stress, and only those groups, showed significantly reduced shell and wet weight growth compared with controls. Mean growth was as low as 4.38 g over 3 months for briefly stressed oysters compared with 9.75 g for controls.

Akoya live at high salinities in the Arabian Gulf and it appears that hypersalinity adversely affects asymptotic size (Al-Sayed *et al.*, 1997). The mean SH of randomly collected oysters from the north and east coasts of Bahrain (40–42‰ salinity) was twice those of randomly collected oysters from the west coast of Bahrain (50–60‰ salinity). Poly Acrylamide Gel Electrophoresis analysis did not suggest any genetic basis for the size differences. While there must be environmental differences other than salinity, these data point to hypersaline stress as being a significant factor in the size differences between populations.

6.5. OXYGEN

Pearl oysters, like other bivalves, obtain oxygen from the ambient water as it passes through the mantle cavity. The gill filaments and mantle have large surfaces lined with fine permeable epithelia where the circulating hemolymph is in close proximity with the water flowing through the mantle cavity. Total oxygen consumption, e.g. μL/h, increases with pearl oyster size, i.e. mass of metabolizing tissue. However, oxygen consumption per unit body mass, e.g. μL O_2/g dwt/h, reflecting MR per unit body mass (J/g dwt/h, Table 4.5), declines. This follows from a ubiquitous pattern of MR per unit body mass being inversely related to size (Gillooly *et al.*, 2001). Respiration rates per unit body mass will be higher again than the values in Table 4.5 in spat and small juveniles. Thus, oxygen availability will be more crucial for these earliest stages.

6.5.1. Uptake versus dissolved oxygen

Oxygen diffuses passively from the higher partial pressure of oxygen (PO_2) in mantle cavity water to the lower PO_2 levels in the hemolymph that result from oxygen consumption for aerobic metabolism. Rate of oxygen uptake depends on the PO_2 gradient between the ambient water and hemolymph, and keeping the gills and mantle well irrigated with high PO_2 water is important.

There is an approximate relationship between the oxygen dissolved per unit volume of water (DO) and PO_2. Temperature, salinity and pH are factors that influence the DO: PO_2 relationship. However, at the relatively stable salinities and pH of seawater, and within moderate temperatures there is a good relationship. Many publications present DO instead of PO_2, especially in relation to the uptake of oxygen for respiration, and DO and PO_2 will both be used as in the original publications.

Under moderate to high DO conditions, Akoya remove <10% of DO from the water flowing through the mantle cavity, even at high MRs (Fig. 6.7)(Yamamoto *et al.*, 1999; Yamamoto, 2000). It is evident that oxygen is generously available under normal conditions. DO levels can, however, decline to a point where there is insufficient PO_2 gradient to maintain sufficient diffusion rates. Thus, Akoya respire maximally at DO levels from 4.5 mL O_2/L down to the quite low values of 1–1.5 mL O_2/L found by Mori (1948) and 1.5 mL O_2/L found by Miyauti (1968b). Oxygen consumption then declines precipitously to zero at DO levels below 1 mL O_2/L (Fig. 6.5). Yamamoto *et al.* (1999) similarly found maximum levels of oxygen consumption over a broad range of PO_2 levels at 14, 20 and 27°C and then precipitous declines at low PO_2 levels of ca. 15 and 30 mm Hg (2 and 4 kPa) (Fig. 6.6). Furthermore, these experimental observations are supported by field observations that low DO levels alone are not critical for Akoya (Ohwada and Uemoto, 1985). They tolerate substantial periods of low DO unless the level drops below 1 mL O_2/L.

Akoya are not, however, oxyregulators in the sense of controlling ventilation to compensate for variations in ambient DO level. Gill ventilation rate (=PR, see Section 4.2.2) and the related ciliary beating rate on the gill surfaces hardly change over a wide range of DO levels (Yamamoto *et al.*, 1999). Unlike the influence of suspended particle concentration on rate of water passage through the gills (Yukihira *et al.*, 1999),

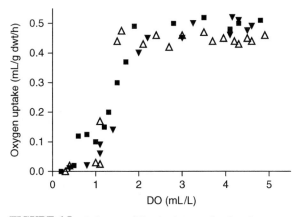

FIGURE 6.5.   Influence of dissolved oxygen level on the oxygen consumption rate of Akoya pearl oysters at ca. 23°C. The different symbols are for three oysters. The oysters were measured from high to progressively lower DO levels. Modified from Miyauti (1968b).

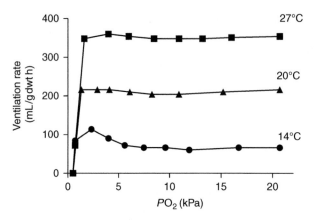

FIGURE 6.6. Influence of oxygen tension (PO_2) on the ventilation rate of Akoya pearl oysters at 14°C, 20°C and 27°C. Modified from Yamamoto *et al.* (1999).

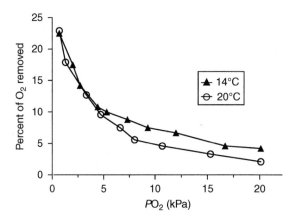

FIGURE 6.7. Influence of oxygen tension (PO_2) on the percentage uptake of oxygen from the ambient water by Akoya pearl oysters at 20°C. Modified from Yamamoto *et al.* (1999).

there is no evidence of a mechanism to maximize the PO_2 gradient at low DO levels by increasing ventilation of the gills (Yamamoto *et al.*, 1999). Only ca. 25% of the available DO is extracted from the ambient water, even at the lowest DO levels (Fig. 6.7). The increasing percent extraction is not related to increased ventilation of the gills and must result from the increasing PO_2 gradient between the ambient water and hemolymph.

Dharmaraj (1983) obtained similar results for respiration at low DO levels without directly measuring oxygen consumption. Dharmaraj found that Akoya (India) showed increasing valve gape and number of valve movements as DO level declined from 3.5 to 0.5 mL O_2/L. The valves then closed and there were minimal valve movements below 0.5 mL/L. Reduction in gape with retraction of the mantle edges is correlated with reduced ventilation rate (Jorgensen, 1996). The precipitous decline to no ventilation at near zero DO levels in the data of Yamamoto *et al.* (1999), presumably reflects valve closure.

The above were short-term studies in the laboratory. One field study by (Miyauti, 1968a) showed a chronic phenomenon where fouling substantially reduced oxygen consumption in farmed pearl oysters. Reduction in oxygen consumption was directly related to the level of fouling. This is due to obstruction of the pearl oyster's feeding and respiratory current. The fouling organisms also have high levels of oxygen consumption and it is quite possible that, in combination with reducing current flow, they create localized environments of low DO. Just one component of the oysters' fouling organisms, *Mytilus edulis*, had 13 times greater rate of oxygen consumption than the fouled oysters (Miyauti, 1968a). The pearl oysters' oxygen consumption increased about two-fold immediately after shell cleaning and further increased to about three-fold, 2 weeks after shell cleaning (Miyauti, 1968a).

In another field study, Miyauti (1965) investigated the oxygen consumption of Akoya kept in "spawning baskets" to induce spawning. Spawning baskets reduce exchange with the adjacent water and ratios of DO in a basket to DO in ambient water ranged from 78% to 92%. However, DO level inside the basket never declined below 3 mL/L and, as anticipated from the laboratory studies, respiratory rate was unaffected. Oysters in the spawning baskets initially gaped more, had fewer valve movements and had lost their diurnal rhythm, but this was due to removal of the byssus (Miyauti, 1965). They showed essentially normal valve movements after 3 weeks.

6.5.2. Tolerance of low DO and anaerobic metabolism

Akoya and *Pinctada sugillata* from the Gulf of Mannar were found to be surprisingly tolerant of anoxic conditions (Dharmaraj, 1983; Dharmaraj *et al.*, 1987). Akoya and *P. sugillata* pearl oysters began dying after 19 and 24–27 hours, respectively, during zero DO conditions. Anaerobic metabolism in pearl oysters has not been studied, but their survival for such periods in anoxic conditions provides strong evidence for it.

Anaerobic metabolism probably occurs when the pearl oysters close their valves for extended periods to buffer abrupt environmental changes, such as aerial exposure (see Section 6.7.2) and lowered salinity from influx of freshwater, or in toxic conditions. When rapid salinity changes induced prolonged valve closure in an estuarine bivalve, *Scrobicularia plana,* there was anaerobic metabolism and accumulation of acid metabolites in the tissues within an hour of continuous valve closure (Trueman and Akberali, 1981).

There were mass mortalities of Akoya during a dense dinoflagellate bloom when the pearl oysters were unable to maintain their PR for an extended period (Matsuyama *et al.*, 1995). The pearl oysters may have gone beyond their capacity for anaerobic metabolism despite the high DO levels in the environment.

Low DO may be a problem in some environments where pearl oysters are cultured at high densities. DO levels may be substantially reduced due to anoxic fecal accumulation in the substrate, pollution, stratified and stagnant water, and high water temperature.

Low DO may be one of several factors causing stress and mortality in these environ-ments. This has been postulated for mortalities of pearl oyster in farms within the inner region of Matushima Bay, Japan (Kan-no *et al.*, 1965).

6.6. CURRENTS

As for other sedentary marine filter feeders, water currents bring suspended food parti-cles to pearl oysters, maintain high DO and disperse their wastes. During reproduction they disperse the pearl oysters' planktonic gametes and larvae (Kailola *et al.*, 1993; Sims, 1993; Friedman and Bell, 1999; Condie *et al.*, 2006). Where low water currents prevail there may be food limitation, low DO stress and accumulation of fecal material on the substrate, which may become anoxic.

There has not been a rigorous study of the impacts of current velocity on pearl oys-ters in the field. A major factor is that water currents at a location are usually irregular: varying with depth, tides, wind and regional factors.

Early field observations of *P. maxima* on the GBR and *P. margaritifera* in Hawaii suggested that strong currents promoted growth rate (Saville-Kent, 1890; 1893; Galtsoff, 1933). Strong currents were found to increase the rate of nacre secretion on pearl nuclei in Akoya, but pearl quality was lower (Kafuku and Ikenoue, 1983).

Low current velocities may be intrinsic to an area or they may be caused on a pearl oyster farm by high-density culture and fouling of culture structures causing resistance to current flow. In either case, there is a combination of depleted food levels, lowered DO and reduced currents for the down-stream oysters within the culture structures. Mass mortalities of *P. margaritifera* during 1985–1986 in Takapoto lagoon, FrP, in the absence of detected pathogens, were attributed to depletion of natural food particles by the high density of pearl oysters in culture (Vacelet *et al.*, 1996; Charpy *et al.*, 1997). A comprehensive ecophysiological model of *P. margaritifera* in this atoll lagoon predicted that POM depletion and reduced growth would occur at current velocities <1 cm/s (Pouvreau *et al.*, 2000). Since currents vary between <1 and 10 cm/s in the lagoon (Rougerie, 1979), it is conceivable that growth depletion will occur due to POM depletion under dense culture conditions, although what constitutes dense culture con-ditions in these long-line farms has not been determined (Pouvreau *et al.*, 2000).

P. maxima spat were found to grow more rapidly in 3 mm mesh compared to 1.5 and 0.75 mm meshes, although there were similar rates of survival (Taylor *et al.*, 1998). The largest mesh size had intrinsically better current flow and hence food avail-ability, but the current flow difference between the largest and two smaller meshes was increased as the smaller meshes had greater fouling.

On the other hand, strong currents are not entirely positive. In areas of fine sedi-mentary substrates, they increase the amount of suspended inorganic particles and may hamper the feeding processes of some species. *P. maxima* is tolerant of high levels of suspended inorganic particles, and favors high current and high suspended matter envi-ronments. These are not favorable environments for *P. margaritifera* (Yukihira *et al.*, 2006).

There has been one brief study of Akoya in basket enclosures in the field, which showed that rate of feces production, i.e. feeding, was related to current (Ōta, 1961). In laboratory experiments, Miyauti and Irie (1965a; 1965b) measured the effects of water current velocity on valve activity and shell-edge regeneration of Akoya over 0.5–16 cm/s current velocity. Regular shell activity, interpreted as normal feeding, and rate of shell regeneration increased with current velocity up to 12 cm/s (ca. 0.4 km/hour) and then declined. Shell-edge regeneration still occurred at 0.5 cm/s, but it was only 20% of the rate at 12 cm/s. Ohwada and Uemoto (1985) describe how MR of this species increases with current velocity up to ca. 15 cm/s and is quite disrupted above 20 cm/s. These are similar to Miyauti and Irie's findings, but no source is given. Growth and CR are reduced in scallops above current velocities of 10–15 cm/s. There was a similar result to that of Akoya despite the considerable morphological and habitat differences between these bivalves (MacDonald *et al.*, 2006).

Considering the effects of water currents on planktonic larvae of pearl oysters, Condie *et al.* (2006) modeled the transport of *P. maxima* larvae off Australia's North West Shelf. The large tidally driven currents in this region will transport larvae back and forth for long distances across the shelf, while lower frequency currents affect their net transport. Despite the large distances traveled by the modeled larvae, most would complete planktonic development less than 30 km from the point of release (Condie *et al.*, 2006). Thus, there should be substantial recruitment into local parent populations about 4 weeks after the main spawning events.

6.7. DEPTH AND AERIAL EXPOSURE

6.7.1. Depth

Some pearl oysters occur over a wide range of depths, e.g. Gervis and Sims (1992) reported that *P. maxima*, *P. margaritifera* and Akoya range from the intertidal zone down to 80, 40 and 30, respectively.

Depth *per se*, i.e. hydrostatic pressure, has not been shown to influence pearl oysters; other environmental associated with depth have significant influences. These factors are often present as gradients. They included light, temperature, salinity, DO, currents, turbidity and biological factors such as levels of fouling and parasitic infection. Pearl oysters in the intertidal zone experience some conditions that are peculiar to that habitat and these are treated in Section 6.7.2.

There were major differences between mortality rates of Akoya at different depths in Ago Bay, Japan, in the years 1974–1977 (Fig. 6.8). Although the general mortality rates varied from year to year, the mortality versus depth pattern for 2- and 3-year-old pearl oysters over these years was fairly consistent. There was highest mortality at the surface or 2 m depth. Temperature, dissolved oxygen (DO), salinity, and infection rates by the shell boring sponge *Polydora ciliata* were environmental factors measured during this period (Itoh and Muzamoto, 1978). There was little or no temperature stratification over much of the year. The 10-year mean winter minimum was ca. 11°C at

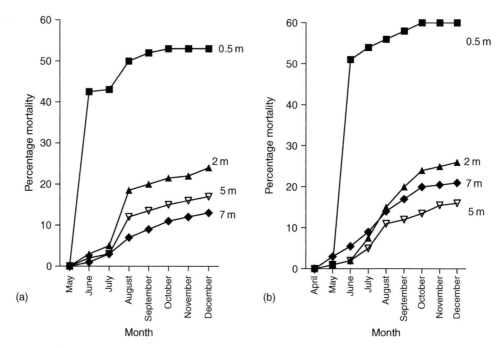

FIGURE 6.8. Progressive mortality of (a) 2- and (b) 3-year-old Akoya pearl oysters at various depths in the pearl culture grounds of Ago Bay, Japan, during 1976. Modified from Itoh and Muzamoto (1978).

all depths, and the summer maxima were ca. 28.5°C at the surface and 25°C at 7 m depth. There were transient and erratic variations of approximately ±2°C from the 10-year mean values at all depths over the years 1974–1977. DO levels largely fluctuated between 3.5 and 6 mL/L, tending to be lowest at 7 m depth. DO plunged to 1 mL/L at 7 m depth in August 1977. Rainfall in the Ago Bay area was erratic during 1974–77, but there was a strong pattern of abrupt salinity declines in the surface water during the spring–summer period. Salinity was usually ca. 31‰ at the surface, but it plunged to 19‰ in 1974, 16‰ in 1976 and 23‰ in 1977. Salinity variations were more conservative at other depths, including 2 m. These low salinities persisted for several weeks. Shell infection rates by *Polydora* tended to increase with depth, with 100% for 2-year-old oysters at 7 m in 1977.

The high rates of mortality occurred during the mid-year spawning season and Itoh and Muzamoto (1978), like some other Japanese biologists, attributed the mortality to a combination of environmental effects on the weakened post-spawning oysters. The pearl oysters at shallow depths often experience higher temperatures and lower salinities and this may be the basis for their higher mortality. The 16‰ salinity in May 1976, pre-spawning, was associated with abrupt increases to 40–50% mortality, implicating low salinity on this occasion. Effects of the DO level near 1 mL/L DO level at 7 m in August 1977 were not evident in distinctly increased mortality. That post-spawning pearl oysters are susceptible to environmental stresses is indirectly support by data for 1-year-old pearl oysters in Ago Bay, 1977 (Itoh and Muzamoto, 1978). These

TABLE 6.7
Settlement of spat versus depth for Pinctada and Pteria species from studies of settlement of test substrates.

Species	Location	Depths tested (m)	Depths of greatest settlement (m)	Source
Akoya	Uchiumi Bay, Japan	1–30	1–10	Tomaru et al. (1999)
	Golfo de Cariaco, Venezuela	5–8, 18–21	5–8	Lodeiros et al. (2001)
P. margaritifera	Orpheus Isl., GBR	2, 6	No difference	Beer and Southgate (2000)
P. mazatlanica	Baja de la Paz, Mexico	0–6	No difference	Monteforte and Garcia-Gasca (1994)
Pt. colymbus	Golfo de Cariaco, Venezuela	6–8, 19–21	6–8	Marquez et al. (2000)
Pt. penguin	Orpheus Is., GBR	2, 6	6	Beer and Southgate (2000)
Pt. sterna	Baja de la Paz, Mexico	0–15	0–5	Monteforte et al. (1995)

pearl oysters suffered no more than 5% mortality from May to December, while 2- and 3-year-old oysters experienced at least 20% mortality.

Depth is a very important factor in the settlement of pearl oyster pediveligers. Levels of spatfall on substrates suspended at various depths are shown in Table 6.7. Pearl oyster veligers are considered to be photopositive (Nayar et al., 1987). Table 6.7 shows studies that found higher levels of settlement at shallow depths at various locations. This variety of locations represents a wide range of conditions. The only common depth-related feature is light intensity. The settling pediveligers apparently show positive phototaxis at the time of selecting depth of substrates, even if they choose concealing substrates at that depth. The study of spatfall versus depth by Tomaru et al. (1999) is a good example of the phenomenon. They measured levels of spatfall of Akoya over the depth range 1–30 m in Uchiumi Bay, Japan, and found high densities over 1–10 m depth on collectors at 1 m intervals through this range, with highest abundance at 6 m depth. Few spat were found on collectors at 15, 20 and 30 m depth. The water column was quite well mixed for much of the test period and there was no correlation between spatfall and either water temperature or chlorophyll level. They considered that some unmeasured factor was influencing settlement depth. There was "plenty of light" in the upper levels of the water column and no evidence of negative phototaxis (Tomaru et al., 1999).

Hydrostatic pressure has been shown to influence the swimming behavior of some bivalve larvae (Cragg and Gruffydd, 1975). There are, however, no data on whether this phenomenon occurs or does not occur in pearl oyster larvae.

6.7.2. Aerial exposure

While it is uncommon to find commercial pearl oyster species in the intertidal zone over much of their distributions, this reflects fishing pressure rather than their natural depth distribution. The depth distribution of most pearl oyster species extends up to the intertidal zone (Gervis and Sims, 1992).

Periods of intertidal exposure may require bivalves to keep their valves closed tightly to maintain water in the mantle cavity and prevent the soft tissues from drying. During exposure, without the usual water current through the gills, oxygen uptake may virtually cease and the bivalve may switch to anaerobic metabolism. An alternative strategy is periodic gaping of the valves to aerate the gills and mantle cavity, through diffusion of oxygen from the air, while tolerating some progressive desiccation. Many bivalves employ the latter compromise strategy during exposure to air.

Periods of aerial exposure of pearl oysters inhabiting the intertidal zone would usually be a matter of unit hours. However, their tolerance of longer periods is of commercial interest and it was found to be quite substantial in one species. Groups of Akoya (47–60 mm SH) from the Gulf of Kutch, India, tolerated up to 21 hours of exposure to air without mortality on re-immersion (Dharmaraj, 1983; Dharmaraj *et al.*, 1987). There was an initial 6 hours period of tight valve closure during aerial exposure and some intermittent valve moments. After 9 hours there was partial gaping, and after 12 hours the valves gaped widely and the mantle was withdrawn from the outer shell region. Valve movements ceased after 18 hours exposure. Oysters exposed for >6 hours took up to 3 hours to resume normal valve movements on re-immersion. Where the mantle was withdrawn (≥12 hours exposure), it took 3–5 hours to resume its normal position on re-immersion, although the gills resumed function before the mantle recovered. There tended to be a period of high oxygen consumption after re-immersion when it is likely that the pearl oysters were compensating for or recovering from a period of anaerobic metabolism. Dharmaraj (1983) reported that batches of 25 specimens each of Akoya and *P. sugillata* were transported for 43 hours under a covering of wet sacking bags, with intermittent immersion in seawater "for a while". There were no mortalities during this period.

6.8. TURBIDITY

Turbidity, which reflects the density of suspended particulate matter (SPM) (or total suspended solids, TSS), has a major influence on the filtering and pre-ingestive sorting processes in pearl oysters (see Section 4.2.2). Although the density of POM may increase with SPM, the density of POM tends to decline relative to PIM (Hawkins *et al.*, 1998; Yukihira *et al.*, 1999). Progressively increasing levels of SPM have been shown to initially increase the intake of POM and then reduce the intake of POM in *P. margaritifera* and *P. maxima* (Table 4.4) (Yukihira *et al.*, 1999). This is because high levels of SPM and PIM interfere with the filtering and pre-ingestive processes in these species. This is a widespread phenomenon in bivalves and other filter feeders, and is very likely to be the case in other pearl oysters species.

As most pearl oysters occur close to the coast they must have some tolerance of turbidity and probably there are interspecific differences according to environment. Table 4.4 shows the different tolerances of *P. margaritifera* and *P. maxima*, the former species tending to occur in much less turbid water than the latter. Furthermore, the different tolerances of turbidity were reflected in relative growth and survival rates of the two species at high and low turbidity sites in the GBR over a year (Yukihira *et al.*, 2006).

Kripa *et al.* (2007) measured a number of hydrological parameters where nucleated Akoya oyster were being cultured. They found that pearl oyster mortality was only correlated with turbidity, among the various parameters measured, and that there was higher mortality with higher turbidity. Several other authors have suggested that turbidity was involved in pearl oyster mortalities, but only on the basis that they coincided temporally or spatially. Farah (1991) attributed the sudden mass mortality of *P. margaritifera* in the Red Sea, north of Port Sudan, from March to May 1969, to exceptionally high turbidity during this period. Other hydrological parameters were being routinely measured and there were no other obviously exceptional factors that may have contributed to the mortality. Similarly, Kan-no *et al.* (1965) found that the highest rates of mortality of Akoya in Matsushima Bay, Japan, were generally in the areas of high water temperature, low salinity, high turbidity and high content of sulfide in benthic sediment.

6.9. POLLUTION AND TOXINS

6.9.1. Pollution

Pollution affecting the environment quality of farmed pearl oysters comes from two main sources. One is exogenous, the other is potentially intrinsic to high-density farming.

Exogenous pollution includes sewage, chemical discharges, oil spills (Samhan *et al.*, 1979) and even discharge from adjacent finfish culture (Ohwada and Uemoto, 1985). Like many bivalves, pearl oysters accumulate heavy metals and hydrocarbon pollutants and have been suggested as bio-indicators in monitoring (Samhan *et al.*, 1979; Klumpp and Burdon-Jones, 1982; Al-Madfa *et al.*, 1998).

Gifford *et al.* (2005) exposed Akoya to various levels of heavy metals, lead and zinc, and aliphatic hydrocarbons, hexadecane and octacosane, for 2 months. There were significant reductions in oyster growth at high concentrations of zinc, while high concentrations of lead completely blocked shell growth. Aliphatic hydrocarbons had no deleterious effect on oyster growth, while moderate levels of octacosane resulted in significant increases in shell growth. Gifford *et al.* (2005) suggested that Akoya are relatively tolerant of some pollutants.

Cumulative mortalities of Akoya exposed to oxidized oils in suspension (fish feed oil) or emulsion (methyl linoleate) were ca. 40% after 8–9 weeks, compared with negligible mortality in Akoya exposed to unoxidized oils (Sugishita *et al.*, 2005). Pathological changes observed in experimental oysters were characterized by blistering and necrosis of cells in various organs and were identical to those observed in diseased oysters from natural mass mortalities. Oxidized oil also caused conspicuous blistering and necrosis in the epithelial cells of the digestive organ during *in vitro* exposure.

Organic pollution caused by suspended solids containing lipid peroxides or emulsion of oxidized oils can chronically damages tissues of cultured oysters and may cause mass mortalities (Sugishita *et al.*, 2005).

Exposure of pearl oysters to crude oil and its complex components clearly impact adversely on pearl oysters in the field. Mass mortalities have been associated with oil spills. There were heavy losses of *P. maxima* following the grounding of the oil tanker *Ocean Grandeur* in March 1969 in Torres Strait, Australia. In the following 1970 season, there were mortalities of up to 80–100% amongst the pearl oysters transported to farms, with over 50% of non-fished stock dying. Extreme mantle collapse was the major sign. Surviving oysters were in weak condition, had depressed growth rates, developed a double-backed shell abnormality and showed abnormal nacre deposition on seeded and half pearls. In the years following the grounding, no juveniles were observed on adjacent grounds. The oil spill and the use of a non-biodegradable detergent were considered significant causes of the mortalities by some observers (Yamashita, 1986), but other investigations suggested an uncharacterized infectious agent as the cause. Long-term ecological damage following the event was also recorded, with declining recruitment and ultimate disappearance of pearl oysters, sea grass and other molluscs with the end result that the grounds resembled an undersea desert by 1979 (Yamashita, 1986).

In Florida there was a mass mortality of subtidal Akoya as part of a severe adverse ecological impact following the 1975 Florida Keys oil spill. Losses in pearl oysters were attributed to a toxic water soluble component of the oil (Chan and Monney, 1977).

Farming pearl oysters may itself be a source of environmental pollution (see Section 14.4.1). High densities of pearl oysters may pollute the adjacent environment, especially through accumulation of organic substances on the substrate below the suspended oysters. This pollution affects the pearl oysters in turn. There have been progressive declines in productivity of oyster farms during repeated culture at high densities in Japanese bays (Ohwada and Uemoto, 1985). Progressive degradation of the substrate occurred to the extent that it was responsible for critical reductions in DO and even release of hydrogen sulfide during stagnant periods in summer.

Substrate degradation not only results from the oysters' feces and pseudofeces. Cleaning fouling organisms from equipment used to culture pearl oysters is another potentially significant source of organic material being deposited (Ohwada and Uemoto, 1985). The total amount of fouling matter removed from 100 cages of farmed Akoya (50 oysters/cage) was estimated as 394 kg/year.

6.9.2. Toxic phytoplankton

Massive phytoplankton blooms, called "red" tides, have been a major problem for the Akoya pearl industry in Japan for a long time (Section 10.6.2.1). Since the 1960s they have become more frequent. For example, the number of red tides per year in the Gokasho and Ago Bay area increased from four in 1973 to 25 in 1980 (Ohwada and Uemoto, 1985). Red tides in Ohmura Bay were caused by blooms of the dinoflagellates *Heterosigma* species in August and *Prorocentrum* species in October 1985 (Numaguchi,

1994a). Mean phytopigment levels rose 10-fold during the red tides, oysters' filtration rates declined sharply and this was reflected in low amounts of algae in their digestive glands. There have been red tides of another dinoflagellate, *Heterocapsa circularisquama*, at 14 locations on the western Japanese coast since 1988 (Matsuyama *et al.*, 1995; Matsuyama *et al.*, 2000). Large-scale blooms of *H. circularisquama* occurred in Ago Bay during 1992 with cell densities up to 87,000 cells/mL. Concurrently, there were mass mortalities of Akoya and other commercial bivalves. The history, etiology, epidemiology and pathology of the toxic phytoplankters are treated in detail in Section 10.6.2.1. It is possible, however, that some of the deleterious effects of dinoflagellate and other phytoplankton blooms on pearl oysters are through interfering with their feeding and respiration at "choking" levels. Alternatively, they may not be nutritious while being so abundant that they limit the intake of nutritious algae. There are some suggestions of this in the literature:

- Filtering rates of Akoya declined sharply during blooms of the red tide dinoflagellates, *Heterosigma* sp. and *Prorocentrum* sp., although there was no immediate mortality Numaguchi (1994a).
- Tomaru *et al.* (2001) found that the cell density of *Nitzschia* spp., considered to be inedible micro-algae for Akoya, increased in Uchiumi Bay before and during a mass mortality event in 1998. They suggested that the oysters, weakened by starvation during the dominance of inedible micro-algae, contracted an infectious disease that caused the mortality.
- Negri *et al.* (2004) suggested that the cyanobacterium *Trichodesmium erythraeum*, while not toxic, contributed to *P. maxima* mortalities by limiting intake of nutritious algae during blooms.

6.9.3. Other toxins

There have been two laboratory studies of the toxicity of specific compounds for Akoya. Kuwatani *et al.* (1970) studied the effects of NH_4^+ on respiration, filtration and other physiological functions on Akoya. Respiration increased, and byssal and crystalline style secretion were disrupted at $>100\,\mu$g-at N/L. All physiological functions were disrupted and mucus production increased at $>1,000\,\mu$g-at N/L. There was mortality at $1,000–3,000\,\mu$g-at N/L depending on Nitrogen concentration, duration of exposure and temperature, e.g. 1,000 and 2,000 μg-at N/L were lethal at 26–28°C after 72 hours and 48 hours, respectively; 2,000 μg-at N/L was lethal after 72 hours at 20–22°C.

Takayanagi *et al.* (2000) similarly found that formaldehyde concentration, duration of exposure and temperature affected mortality in Akoya. Median lethal concentration (LC50) declined progressively as duration of exposure increased, with the LC50 for 1-year oysters being 7.7 mg/L after 96 hours at 20°C and 5.3 mg/L after 96 hours at 25°C. The LC50s were higher for 2-years oysters. The pearl oysters lowered the ambient formaldehyde concentration as they absorbed it.

6.10. MODELING ENVIRONMENTAL INFLUENCES

There are comprehensive environmental data for some bivalves and measured effects of various parameters on their energetics (SFG). From these it is possible to develop energetics models that quite accurately predicted growth in the complex environmental conditions of the field, e.g. Grant (1996), Barillé *et al.*(1997), Scholten and Smaal (1998).

Pouvreau et al. (2000) developed a bioenergetic model for *Pinctada margaritifera* based on C/day to simulate growth and reproduction in this lagoon in Takapoto atoll lagoon, FrP (Fig. 6.9). The model was fundamentally a very sophisticated

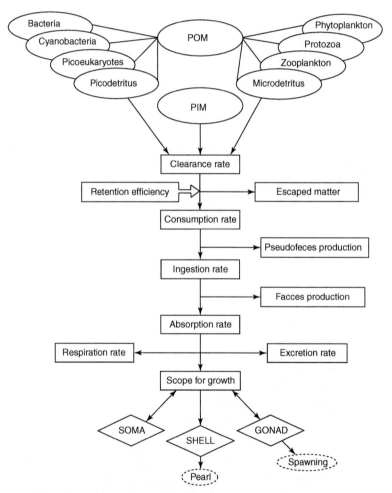

FIGURE 6.9.   Scheme of a growth model developed by Pouvreau *et al.* (2000) for *Pinctada margaritifera* in Takapoto atoll lagoon, FrP, based on C fluxes. PIM: particulate inorganic matter; POM: particulate organic matter.

scope for growth (SFG) model of the balance of energy gain and loss (see Section 4.3.3).

$$SFG = AE - (RE + EE)$$

where SFG is surplus energy available for growth and reproduction; AE is absorbed energy, which is the difference between ingested and fecal energy; RE is respired energy or energy cost, EE is excreted energy.

Pouvreau *et al.* (2000) had comprehensive data on hydrology, and size range and components of POM for Takapoto atoll lagoon over long periods. They also had comprehensive quantitative data on physiological function rates and coefficients obtained from eight studies, enabling them to calculate allometric functions for rates of food acquisition versus POM and PIM, pseudofeces and feces production, oxygen consumption and excretion, resource allocation to growth and reproduction, gamete release, and shell and cultured pearl formation. The test of the model was how well it predicted the amount of surplus energy (SFG) according to oyster size, and its allocation to shell and soft tissue growth, including gonad development and spawning events. The model successfully simulated growth and reproduction of 1- to 4-year-old pearl oysters in the lagoon according to the PIM and POM levels over the yearlong study. The model succeeded partly due to the stability of the hydrological and SPM conditions throughout the year in Takapoto lagoon. Temperature was not incorporated in the model. Its narrow range of annual variation, 26 to 30°C in the lagoon, was not found to have any significant effect on the physiology of *P. margaritifera* in these FrP energetics studies (Pouvreau *et al.*, 2000). They noted, however, that during El Niño years the water temperature may rise to 31.5°C in Takapoto lagoon and that it varies between 22 and 32°C in FrP atolls at various latitudes. They concluded that temperature must be taken into account in broader studies, a conclusion supported by temperature effects on energetics outside the 23–29°C range found by Yukihira *et al.* (2000).

Pouvreau *et al.* (2000) considered the implications of their results for the cultured pearl oyster industry in Takapoto lagoon. They calculated that the current population of farmed pearl oysters in the lagoon has an insignificant role in consuming the primary production of planktonic organisms compared to wild stocks of other bivalves. Niquil *et al.* (2001) came to the same conclusion "oyster farming in this lagoon is thus very far from being food limited". However, local stocking of pearl oysters at high densities in low currents will limit growth.

Abo and Toda (2001) also developed a bioenergetics model was for Akoya, based on a sophisticated SFG model, to evaluate the effects of farming density on pearl oyster growth. One component of the model described the influences of environmental parameters, temperature, salinity and food density, on physiological parameters of the pearl oyster, i.e. filtration rate, assimilation efficiency, respiration rate, and somatic and reproductive growth. The other component determined food density, chl a, from the balance of phytoplankton growth, pearl oysters' filter feeding and water exchange. Abo and Toda (2001) applied the model to Gokasho Bay, central Japan, simulating the effects of pearl oyster density on the food density and growth rates of oysters in the bay. Levels of chl a and oyster growth rates were inversely related to farming density, especially chl a in summer. However, the decrease in growth rate with even a seven-fold increase over current stocking density was not as dramatic as might be expected. Abo and Toda (2001) considered that the current farming density in Gokasho Bay is appropriate to phytoplankton levels in the bay. However, they observed that impact of primary production on growth is not the only factor to consider in farming density. There is potential for benthic degradation due to deposition of organic matter from densely cultured oysters in Gokasho Bay (see Section 14.4.1).

Yukihira (1998) acquired much information on the influence of various environmental parameters on the SFG of *Pinctada maxima* and *P. margaritifera*. This led to a study of growth and survival of the two species at contrasting sites in the GBR lagoon (Yukihira *et al.*, 2006). A major objective was to compare growth rates at the two sites with growth rates predicted from the SFG studies (Yukihira, 1998). The predicted rates were derived from measured environmental parameters, temperature, SPM, PIM and POM, at the two sites. Unlike the studies outlined above, however, there was very poor correlation between the predicted and actual growth rates over the 14-month study. There were insufficient hydrological and other field data for an adequate predictive model of growth. For instance, SPM, of great importance for feeding and nutrition, varied erratically over short periods and could not be accurately measured at sufficiently short intervals.

6.11. PEARL QUALITY

Studies of environmental influences on pearl oysters *per se* are justified in their own right and they extend our general knowledge of bivalve molluscs. However, pearl oysters are commercial bivalves and studies of environment influences have been driven to a considerable extent by the need to determine optimum field conditions for their culture. Furthermore, culturing healthy oysters is not an end in itself. The ultimate goal is to produce a high frequency of high quality pearls. One simple approach to studying the effects of salinity on pearl quality was to abruptly immerse nucleated oysters into various salinities for periods of 12 to 48 hours and then determine pearl quality at harvest (Katada, 1959). Twelve-hour exposure to salinities ≤19‰ resulted in pearls with discolored nacre and 24 hours exposure resulted in pearls with thin nacre at harvest 4 months later. All the nucleated oysters survived these periods of exposure, but mortality began after 36 hours at 11‰ and 13‰ salinities. As short exposures to low salinities adversely affected growth in nucleated oysters over subsequent months (Katada, 1959), they had long-term effects on nacre secretion in the pearl sac. Salinity must be at least 21‰ for normal pearl production (Ohwada and Uemoto, 1985) and at least 25–27‰ if pearl quality is to be unaffected (Ōta and Fukushima, 1961).

Exposure to air for up to an hour during shell cleaning results in transient disruption of nacre secretion on the current pearl surface (Ōta and Shimizu, 1961). There is not a strong pattern; however there tends to be loss of pearl quality between oysters exposed for 60 minutes versus 10 minutes. This tends to be more during the months of July to October, than during the months of May and June.

Horiguchi and Maegawa (1978) considered the effects of dredging on shell regeneration, mortality and pearl quality of operated Akoya at four sites at various distances from dredging in Ago Bay, Japan. Presumably, degree of turbidity was a factor here, but it was not measured. Mortalities ranged from 0% to 7.2% and nucleus rejection rates from 4.5% to 14.9%, and were not significantly greater at the two stations closest to the dredging. There was no consistency in average pearl quality versus distance from dredging. However, one of the stations closest to the dredging produced the greatest value of pearls by a substantial margin.

Pearl quality consistently differs at various localities and it is very likely that environmental parameters, with genetics and technology, are responsible for this variation. There were a number of early studies by Japanese scientists seeking to relate pearl quality to oceanographical parameters at the localities of various pearl farms, e.g. Sawada and Tange (1959), Hasuo *et al.* (1962) and Horiguchi and Maegawa (1978). Most of these studies are published in Japanese in journals such as *Kokoritsu Shinju Kenkyusho Hokoku (Bull. Natl. Pearl Res. Lab.)* and not readily accessible to western biologists, although some are considered in this chapter. Despite considerable research, the studies were generally unable to quantify any clear relationship between environmental parameters and pearl quality. It is generally considered that, in Akoya, whiter pearls and pearls that are more lustrous are obtained more frequently from farms in enclosed waters that are influenced by terrestrial runoff and have strong temperature fluctuations, especially low winter temperatures. Poorer quality pearls are obtained from farms that are adjacent to open ocean and influenced by ocean currents. Calcium metabolism in the pearl sac epithelium is influenced by environmental temperature and by the physiological condition of the oyster and these affect pearl quality (Wada, 1972).

6.12. SUMMARY

Pearl oysters in their natural environments experience the simultaneous effects of a wide range of environmental factors. Rigorous data on simultaneous effects, however, are very limited and this Chapter generally considers the most important factors affecting pearl oysters as individual factors.

Food, almost inevitably, is a major environmental factor. Pearl oysters filter feed on suspended particulate matter (SPM), consisting mainly of bacteria, micro-algae, suspended organic matter and inorganic particles. Particle size, density, composition and digestibility affect the nutritional value of the SPM. Optimum densities of micro-algae are in the $10–100 \times 10^3$ cells/mL range. As in all poikilotherms, ambient temperature profoundly influences pearl oysters through its effects on MR and related processes, such as respiration and feeding rates. There is an optimum temperature, typically ca. 28°C, or temperature range for maximum MR. Below this, MR usually declines steeply and may virtually cease in low temperatures. MR also declines at near-lethal high temperatures. Growth rates reflect MRs and thus show a similar pattern in relation to temperature. Pearl oysters are osmoconformers. While unable to regulate their body fluid osmolality, they use sustained valve closure to buffer rapid salinity changes. Some pearl oysters occur in coastal habitats where salinity is lowered by terrestrial run-off. Thus, it is not surprising that at least some species are quite euryhaline. MR-related functions, however, decline in low salinities and hypersaline conditions. Pearl oysters are also oxyconformers in that they make no response to variation in DO levels. Their oxygen consumption remains at maximum rate down to quite low DO levels, 1–1.5 mL O_2/L, then declines precipitously below this. Unlike PR responses to various SPM levels, they show no response to low DO by increasing water flow through the gills and mantle cavity.

Water currents are very important in bringing food and oxygen to pearl oysters and carrying away their wastes; however, strong currents may be deleterious by increasing suspended inorganic matter, interfering with filtering and pre-ingestive processes. Depth *per se* does not influence pearl oysters, but physical and biological factors may vary with depth and result in depth-linked variation in mortality and growth rates

Two bioenergetic models have been developed for pearl oysters. One model aimed to predict growth and reproduction of *P. margaritifera* in an atoll lagoon and the other aimed to determine the effects of farming density on growth of Akoya. The models required comprehensive data on environmental parameters and the influence of these parameters on the pearl oysters. Both predicted that current farming densities were well within the capacity of the farming environment.

The ultimate aspect of environment influences on pearl oysters, i.e. environmental influences on pearl quality, is a major aspect that requires more research. This is not an easy topic to research, judging by the limited results of Japanese research over a number of years. With more data on the genetics of pearl oysters, however, it should be possible to better distinguish environmental influences from genetic influences.

References

Abo, K., Toda, S., 2001. Evaluation model of farming density of Japanese pearl oyster, *Pinctada fucata martensii*, based on physiology and food environment. *Suisan Kiyo Kenkyu (Bull. Jap. Soc. Fish. Oceanogr.)* 65, 135–144. (in Japanese, with English abstract)

Alagarswami, K., Victor, A.C.C., 1976. Salinity tolerance and rate of filtration of the pearl oysters, *Pinctada fucata. J. Mar. Biol. Ass. India* 18, 149–158.

Alagarswami, K., Dharmaraj, S., Velayudhan, A., Chellam, A., Victor, A.C.C., Gandhi, A.D., 1983. Larval rearing and production of spat of pearl oyster *Pinctada fucata* (Gould). *Aquaculture* 34, 287–302.

Al-Madfa, H., Abdel-Moati, M.A.R., Al-Gimaly, F.H., 1998. Pinctada radiata (pearl oyster): A bioindicator for metal pollution monitoring in the Qatari waters (Arabian Gulf). *Bull. Environ. Contam. Toxicol.* 60, 245–251.

Al-Sayed, H., El-Din, A.K.G., Saleh, K.M., 1997. Shell morphometrics and some biochemical aspects of the pearl oyster *Pinctada radiata* (Leach 1814) in relation to different salinity levels around Bahrain. *Arab Gulf J. Sci. Res.* 15, 767–782.

Araya-Nunes, O., Ganning, B., Bueckle-Ramirez, F., 1991. Gonad maturity, induction of spawning, larval breeding, and growth in the American pearl-oyster (*Pteria sterna*, Gould). *Calif. Fish Game* 77, 181–193.

Araya-Nunes, O., Ganning, B., Bueckle-Ramirez, F., 1995. Embryonic development, larval culture, and settling of American pearl-oyster (*Pteria sterna*, Gould) spat. *Calif. Fish Game.* 81, 10–21.

Barillé, L., Heral, M., Barillé-Boyer, A.-L., 1997. Modélisation de l'écophysiologie de l'huître *Crassostrea gigas* dans un environment estuarien. *Aquat. Living Resour.* 10, 31–48. (with English abstract)

Beer, A.C., Southgate, P.C., 2000. Collection of pearl oyster (family Pteriidae) spat at Orpheus Island, Great Barrier Reef (Australia). *J. Shellfish Res.* 19, 821–826.

Cahn, A.R., 1949. Pearl culture in Japan. *Fish. Leafl. US Fish Wildl. Serv.* 357, 1–91.

Chan, E.I., Monney, N.T., 1977. Oil pollution and tropical littoral communities: Biological effects of the 1975 Florida Keys oil spill. In: *Proceedings 1977 Oil Spill Conference (Prevention, Behaviour, Control, Cleanup)*. American Petroleum Inst., Washington DC.

Chang, M., Hong, J.S., Huh, H.T., 1988. Environmental conditions in the pearl oyster culture grounds and food organisms of *Pinctada fucata martensii* (Dunker) (Bivalvia, Pterioida). *Ocean Res.* 10, 67–77. (in Korean, with English abstract)

Charpy, L., Dufour, P., Garcia, N., 1997. Particulate organic matter in sixteen Tuamotu Atoll lagoons (French Polynesia). *Mar. Ecol. Prog. Ser.* 151, 55–65.

Condie, S.A., Mansbridge, J.V., Hart, A.M., Andrewartha, J.R., 2006. Transport and recruitment of silver-lip pearl oyster larvae on Australia's North West Shelf. *J. Shellfish Res.* 25, 179–185.

Cragg, S.M., Gruffydd, L.D., 1975. The swimming behaviour and the pressure responses of the veliconcha larvae of *Ostrea edulis* L. *Ninth European Marine Biology Symposium, Oban, 2 Oct 1974*, Aberdeen University Press. Aberdeen, UK. pp. 43–57.

Cuzon, G., Soyez, C., LeMoullac, G., 2004. Metabolic rate of *Pinctada margaritifera* during gametogenesis. Abstracts of Technical Papers Presented at the 96th Annual Meeting of the National Shell Fisheries Association, March 1–5, 2004, Honolulu, Hawaii, USA. *J. Shellfish Res.* 23, 286.

del Rio-Portilla, M.A., Re-Araujo, A.D., Voltolina, D., 1992. Growth of the pearl oyster *Pteria sterna* under different thermic and feeding conditions. *Mar. Ecol. Prog. Ser.* 89, 221–227.

Dharmaraj, S., 1983. Oxygen consumption in pearl oyster *Pinctada fucata* (Gould) and *Pinctada sugillata* (Reeve). *Proc. Symp. Coast. Aquac.* 2, 627–632.

Dharmaraj, S., Kandasami, D., Alagarswami, K., 1987. Some aspects of physiology of pearl oyster. *Central Mar. Fish. Res. Inst. Bull. (Cochin)* 39, 21–28.

Dholakia, A.D., Bhola, D.V., Patel, P.H., Makwana, D.M., 1997. Salinity tolerance limit for survival of pearl oyster, *Pinctada fucata* (Gould) and edible oyster, *Crassostrea gryphoides* (Schlotheim). *Fish. Chimes* 17, 17–18.

Doroudi, M.S., Southgate, P.C., Mayer, R.J., 1999. The combined effects of temperature and salinity on embryos and larvae of the black-lip pearl oyster, *Pinctada margaritifera* (L.). *Aquac. Res.* 30, 271–277.

Farah, O.M., 1991. Water characteristics of Dongonab Bay, Sudanese Red Sea. *Bull. Natl. Inst. Oceano. Fish.* 16, 81–87.

Friedman, K.J., Bell, J.D., 1999. Variation in abundance of blacklip pearl oyster (*Pinctada margaritifera* Linne.) spat from inshore and offshore reefs in Solomon Islands. *Aquaculture* 178, 273–291.

Fukushima, K., 1970. Mass mortalities of the Japanese oyster *Pinctada fucata martensii* in Matoya Bay and Gokasho Bay. *Shinjugijutsu Kenkyukai Kaiho* 9, 1–21. (in Japanese)

Fukushima, K., 1972. The health of *Pinctada fucata martensii* and food organism environment. *Shinjugijutsu Kenkyukai Kaiho* 10, 15–31. (in Japanese)

Funakoshi, S., Suzaki, T., Wada, K., 1985. Salinity tolerances of marine bivalves. In: Sindermann, C.J. (Ed.), *Environmental Quality and Aquaculture Systems. Proceedings of the 13th US-Japan Meeting on Aquaculture*, Mie, Japan, October 24–25, 1984. NOAA Technical Report NMFS 69. US Department of Commerce, National Oceanic and Atmospheric Administration, National Marine Fisheries Service, Sringfield, VA, USA, pp. 15–18.

Galtsoff, P.S., 1933. Pearl and Hermes reef, Hawaii: hydrographical and biological observations. *Bull. Bishop Mus. Honolulu* 107, 3–49. (cited by Gervis and Sims, 1992)

Gervis, M.H., Sims, N.A., 1992. *The Biology and Culture of Pearl Oysters (Bivalvia: Pteriidae)*. International Center for Living Aquatic Resources Management, Manila. 49 pp.

Gifford, S., Dunstan, H., O'Connor, W., Macfarlane, G.R., 2005. Quantification of *in situ* nutrient and heavy metal remediation by a small pearl oyster farm at Port Stephens, Australia. *Mar. Poll. Bull.* 50, 417–422.

Gillooly, J.F., Brown, J.H., West, G.B., Savage, V.M., Charnov, E.L., 2001. Effects of size and temperature on metabolic rate. *Science* 293, 2248–2251.

Gosling, E., 2003. *Bivalve Molluscs: Biology, Ecology and Cultivation.* Fisheries News Books. Blackwell Publishing, Oxford, UK. 443 pp.

Grant, J., 1996. The relationship of bioenergetics and the environment to the field growth of cultured bivalves. *J. Exp. Mar. Biol. Ecol.* 200, 239–256.

Hashimoto, T., Nakano, S., 2003. Effect of nutrient limitation on abundance and growth of phytoplankton in a Japanese pearl farm. *Mar. Ecol. Prog. Ser.* 258, 43–50.

Hasuo, M., Sakaguchi, S., Yamaguchi, K., Murakami, E., 1962. Studies on the quality of the pearl and growth rate of the pearl oyster, *Pinctada martensii*, which is cultured at culture ground in Nagasaki Prefecture. *Kokoritsu Shinju Kenkyusho Hokoku (Bull. Natl. Pearl Res. Lab.)* 8, 920–947. (in Japanese)

Hawkins, A.J.S., Smith, R.F.M., Tan, S.H., Yasin, Z.B., 1998. Suspension-feeding behaviour in tropical bivalve molluscs: *Perna viridis, Crassostrea belcheri, Crassostrea iradelei, Saccostrea cucculata* and *Pinctada margarifera. Mar. Ecol. Prog. Ser.* 166, 173–185.

Horiguchi, Y., Maegawa, M., 1978. Studies on marine environments and maricultural resources in the waters around Zaga-Shima (Ago Bay). 3: Effects of dredging works upon growth and mortality of operated pearl oyster and quality of pearl. *Report of the Fisheries Research Laboratory*, Mie University. Shima No, 1, 31–38 (in Japanese).

Itoh, K., 1976. Relation of oxygen consumption and ammonia nitrogen excreted to body size and to water temperature in the adult pearl oyster *Pinctada fucata* (Gould). *Kokoritsu Shinju Kenkyusho Hokoku (Bull. Natl. Pearl Res. Lab.)* 20, 2254–2275. (in Japanese, with English summary)

Itoh, K., Muzamoto, S., 1978. Mortality of the pearl oyster *Pinctada fucata* in pearl cultured ground Ago Bay, Japan. *Kokoritsu Shinju Kenkyusho Hokoku (Bull. Natl. Pearl Res. Lab.)* 22, 2383–2404. (in Japanese, with English summary)

Je, J.G., Hong, J.S., Yi, S.K., 1988. A study on the fouling organisms in the pearl oyster culture grounds in the southern coast of Korea. *Ocean Res.* 10, 85–105. (in Korean).

Jeyabaskarn, Y., Dev, D.S., Nalluchinnappan, I., Radhakrishnan, N., 1983. On the growth of the pearl oyster *Pinctada fucata* (Gould) under farm conditions at Tuticorin, Gulf of Mannar. *Symp. Ser. Mar. Biol. Assoc. India* 6, 587–589.

Jorgensen, C.B., 1996. Bivalve filter feeding revisited. *Mar. Ecol. Prog. Ser.* 142, 287–302.

Kafuku, T., Ikenoue, H., 1983. Pearl oyster *(Pinctada fucata). Modern Methods in Aquaculture in Japan.* Elsevier Scientific Publishing, Amsterdam. pp. 161–171.

Kailola, P.J., Williams, M.J., Stewart, P.C., Reichelt, R.E., Mcnee, A., Grieve, C., 1993. *Australian Fisheries Resources. Pearl Oysters.* Bureau of Resource Sciences, Department of Primary Industries and Energy, and the Fisheries Research and Development Corporation, Canberra Australia. pp. 87–89.

Kan-no, H., Sasaki, M., Sakurai, Y., Watanabe, T., Suzuki, K., 1965. Studies on the mass mortality of the oyster in Matsushima Bay I. General Aspects of the mass mortality of the oyster in Matsushima Bay and its environmental conditions. *Bull. Tohoku Reg. Fish. Res. Lab.* 25, 1–26. (in Japanese, with English Abstract)

Katada, S., 1959. The influence of low salinity seawater on death and growth of the pearl oyster (*Pinctada martensii*) and quality of cultured pearls. *Kokoritsu Shinju Kenkyusho Hokoku (Bull. Natl. Pearl Res. Lab.)* 5, 489–493. (in Japanese)

Kennedy, V.S., Roosenberg, W.H., Zion, H.H., Castagna, M., 1974. Temperature–time relationships for survival of embryos and larvae of *Mulinia lateralis* (Mollusca: Bivalvia). *Mar. Biol. (Berl.)* 24, 137–145.

Kim, I.O., 1969. Study on the culture pearl oyster. I. On the wintering and growth of pearl oyster, *Pinctada martensii. Bull. Fish. Res. Dev. Agency* Pusan 4, 109–118. (in Korean)

Kimani, E.N., Mavuti, K.M., 2002. Abundance and population structure of the blacklip pearl oyster, *Pinctada margaritifera* L. 1758 (Bivalvia: Pteriidae), in coastal Kenya. *West. Indian Ocean J. Mar. Sci.* 1, 169–179.

Klumpp, D.W., Burdon-Jones, C., 1982. Investigations of the potential of bivalve molluscs as indicators of heavy metal levels in tropical marine waters. *Aust. J. Mar. Freshw. Res.* 33, 285–300.

Knauer, J., Southgate, P.C., 1999. A review of the nutritional requirements of bivalves and the development of alternative and artificial diets for bivalve aquaculture. *Rev. Fish. Sci.* 7, 241–280.

Kobayashi, H., Matsui, J., 1953. Studies on the resisting power of the pearl-oyster (*Pinctada martensii*) against the change in environment. *I.* On ciliary movement of the gill. *Norinsho Suisan Koshusho Kenkyu Hokoku (J. Shimonoseki Coll. Fish.)* 3, 123–131. (in Japanese)

Kobayashi, S., Tobata, M., 1949a. Studies on culture of pearl. II. Activity of pearl oyster in winter. *Nippon Suisan Gakkaishi (Bull. Jap. Soc. Fish. Sci.)* 14, 196–199. (in Japanese, with English abstract)

Kobayashi, S., Tobata, M., 1949b. Studies on culture of pearl. III. Activity of pearl-oyster in winter. *Nippon Suisan Gakkaishi (Bull. Jap. Soc. Fish. Sci.)* 14, 200–202. (in Japanese, with English abstract)

Komaru, A., Matsuda, H., Yamakawa, T., Wada, K.T., 1990. Meiosis and fertilization of the Japanese pearl oyster eggs at different temperature observed with a fluorescence microscope. *Nippon Suisan Gakkaishi (Bull. Jap. Soc. Fish. Sci.)* 56, 425–430.

Kripa, V., Mohamed, K.S., Appukuttan, K.K., Velayudhan, T.S., 2007. Production of Akoya pearls from the Southwest coast of India. *Aquaculture* 262, 347–354.

Kuwatani, Y., Nishii, T., Wada, K., 1970. Effect of ammonia concentration on physiological conditions of Japanese pearl oyster. *Kokoritsu Shinju Kenkyusho Hokoku (Bull. Natl. Pearl Res. Lab.)* 15, 1874–1899. (in Japanese, with English summary)

Kuwatani, Y., Nishii, T., Wada, K.T., 1974. Growth and maturation of Japanese pearl oysters reared in tanks in winter. *Kokoritsu Shinju Kenkyusho Hokoku (Bull. Natl. Pearl Res. Lab.)* 18, 2118–2129. (in Japanese with English abstract)

Lodeiros, C., Pico, D., Nunez, M., Narvaez, N., De Donato, M., Graziani, C., 2001. Experimental culture of the pearl oyster *Pinctada imbricata* in the Golfo de Cariaco, Venezuela. *Aquaculture 2001: Book of Abstracts.* World Aquaculture Society, Baton Rouge, USA. p. 383.

MacDonald, B.A., Bricelj, M., Shumway, S.E., 2006. Physiology: energy acquisition and utilisation. Chapter 7. In: Shumway, S.E., Parsons, G.J. (Eds.), *Scallops: Biology, Ecology and Aquaculture.* Amsterdam, Elsevier. pp. 417–492.

MacNair, N., Smith, M., 1999. Investigations into treatments to control fouling organisms affecting oyster production. *J. Shellfish Res.* 18, 331.

Marquez, B., Lodeiros, C., Jimenez, M., Himmelman, J.H., 2000. Disponibilidad de juveniles por captacion natural de la ostra *Pteria colymbus* (Bivalvia: Pteriidae) en el Golfo de Cariaco, Venezuela. *Rev. Biol. Trop.* 1, 151–158. (with English abstract)

Martinez-Fernandez, E., Acosta-Salmón, H., Southgate, P.C., 2006. The nutritional value of seven species of tropical microalgae for black-lip pearl oyster (*Pinctada margaritifera*, L) larvae. *Aquaculture* 257, 491–503.

Matsuyama, Y., Nagai, K., Mizuguchi, T., Fujiwara, M., Ishimura, M., Yamaguchi, M., Uchida, T., Honjo, T., 1995. Ecological features and mass mortality of pearl oysters during red tides of *Heterocapsa* sp. in Ago Bay in 1992. *Nippon Suisan Gakkaishi (Bull. Jap. Soc. Fish. Sci.)* 61, 35–41.

Matsuyama, Y., Uchida, T., Honjo, T., 2000. Impact of harmful dinoflagellate *Heterocapsa circularisquama* on shellfish aquaculture in Japan. *J. Shellfish Res.* 19, 636.

Mihursky, J.A., Kennedy, V.S., McErlean, A.J., Roosenburg, W.H., Gatz, A.J., Castagna, M., O'Connor, S.G., O'Connor, J.M., Gibson, C.I., Zion, H.H., Amende, L., 1974. The thermal requirements and tolerances of key estuarine organisms, Maryland Water Resources Research Center Technical Report Series no. 26.

Milke, L.M., Ward, J.E., 2003. Influence of diet on pre-ingestive particle processing in bivalves. II: Residence time in the pallial cavity and handling time on the labial palps. *J. Exp. Mar. Biol. Ecol.* 293, 151–172.

Mills, D., 2000. Combined effects of temperature and algal concentration on survival, growth and feeding physiology of *Pinctada maxima* (Jameson) spat. *J. Shellfish Res.* 19, 159–166.

Miyauti, T., 1962. Filtering rates in the Japanese pearl oyster. II: Effects of temperature and specific gravity of seawater on the filtering rates. *Suisanzoshoku (J. Jap. Aquacult. Soc.)* 10, 7–13. (in Japanese)

Miyauti, T., 1965. Studies on artificial spawning in pearl culture. I. On shell movement and oxygen consumption of the Japanese pearl oyster, *Pteria (Pinctada) martensii* (Dunker) inside the spawning-basket and dissolved oxygen of seawater inside and outside the spawning-basket. *Seiri Seitai* 13, 81–87. (in Japanese, with English summary)

Miyauti, T., 1968a. Studies on the effect of shell cleaning in pearl culture-III. The influence of fouling organisms upon the oxygen consumption in the Japanese pearl oysters. *Jpn. J. Ecol. (Otsu)* 18, 40–43. (in Japanese, with English abstract)

Miyauti, T., 1968b. Studies on the vitality of the pearl oyster, *Pteria (Pinctada) martensii* (Dunker) under abnormal conditions. I: Oxygen uptake and shell movement in sea water of low oxygen content. *The Veliger* 10, 342–348.

Miyauti, T., Irie, H., 1965a. Relation between pearl oysters (*Pteria martensii*) and the current-velocity of the environmental waters. *Bull. Fish. Nagasaki Univ.* 19, 56–64. (in Japanese, with English summary)

Miyauti, T., Irie, H., 1965b. Supplemental report on the relation between pearl oysters (*Pteria martensii*) and current velocity of environmental water. Effects of the velocity of the environmental waters on the motion of shell opening and shutting and shell regeneration. *Bull. Fish. Nagasaki Univ.* 20, 22–28. (in Japanese, with English summary, the senior author is shown as Miyauchi in the English title)

Monteforte, M., Garcia-Gasca, A., 1994. Spat collection studies on pearl oysters *Pinctada mazatlanica* and *Pteria sterna* (Bivalvia, Pteriidae) in Bahia de La Paz, South Baja California, Mexico. *Hydrobiologia* 291, 21–34.

Monteforte, M., Kappelman-Pina, E., Lopez-Espinosa, B., 1995. Spatfall of pearl oyster, *Pteria sterna* (Gould), on experimental collectors at Bahia de La Paz, South Baja California, Mexico. *Aquac. Res.* 26, 497–511.

Mori, S., 1948. On the respiration of *Pinctada martensii* (Dunker) in the sea water of low oxygen tension (A preliminary note). *Kairuigaku Zasshi* 15, 1–4. (in Japanese)

Morizane, T., Takimoto, S., Nishikawa, S., Matsuyama, N., Tyohno, K., Uemura, S., Fujita, Y., Yamashita, H., Kawakami, H., Koizumi, Y., Uchimura, Y., Ichikawa, M., 2001. Mass mortalities of Japanese pearl oyster in Uwa Sea, Ehime in 1997–1999. *Fish Pathol.* 36, 207–216.

Nagai, K., Go, J., Segawa, S., Honjo, T., 2007. A measure to prevent relapse of reddening adductor disease in pearl oysters (*Pinctada fucata martensii*) by low-water-temperature culture management in wintering fisheries. *Aquaculture* 262, 192–201.

Nasr, D.H., 1982. Observations on the mortality of the pearl oyster, *Pinctada margaritifera*, in Dongonab Bay, Red Sea. *Aquaculture* 28, 271–281.

Nayar, K.N., Mahadevan, S., Alagarswami, K., 1987. Ecology of pearl oyster beds. *Bull. Cent. Mar. Fish. Res. Inst.* 39, 29–36.

Negri, A.P., Bunter, O., Jones, B., Llewellyn, L., 2004. Effects of the bloom-forming alga *Trichodesmium erythraeum* on the pearl oyster *Pinctada maxima*. *Aquaculture* 232, 91–102.

Niquil, N., Pouvreau, S., Sakka, A., Legendre, L., Addessi, L., Le Borgne, R., Charpy, L., Delesalle, B., 2001. Trophic web and carrying capacity in a pearl oyster farming lagoon (Takapoto, French Polynesia). *Aquat. Living Resour.* 14, 165–174.

Numaguchi, K., 1994a. Growth and physiological condition of the Japanese pearl oyster, *Pinctada fucata martensii* (Dunker, 1850) in Ohmura Bay, Japan. *J. Shellfish Res.* 13, 93–99.

Numaguchi, K., 1994b. Effect of water temperature on the filtration rate of Japanese pearl oyster, *Pinctada fucata martensii*. *Suisanzoshoku (J. Jap. Aquacult. Soc.)* 42, 1–6. (in Japanese, with English abstract)

Numaguchi, K., 1995a. Effects of water temperature on catabolic losses of meat and condition index of unfed pearl oyster *Pinctada fucata martensii*. *Nippon Suisan Gakkaishi (Bull. Natl Res. Inst. Fish. Sci.)* 61, 735–738.

Numaguchi, K., 1995b. Influences of unfed condition on the mortality of pearl oyster *Pinctada fucata martensii*. *Fish. Sci. (Tokyo)* 61, 739–742.

Numaguchi, K., 2000. Evaluation of five microalgal species for the growth of early spat of the Japanese pearl oyster *Pinctada fucata martensii*. *J. Shellfish Res.* 19, 153–157.

Numaguchi, K., 2002. Comparison of the ingestion rate, digestion rate and phyto-pigment content in the digestive diverticula of Japanese pearl oyster, *Pinctada fucata martensii*, juveniles grazing six different micro-algae. *Nippon Suisan Gakkaishi (Bull. Jap. Soc. Fish. Sci.)* 68, 534–537. (in Japanese, with English abstract)

Numaguchi, K., Tanaka, Y., 1986a. Effects of salinity on mortality and growth of the spat of the pearl oyster, *Pinctada fucata martensii*. *Yoshoku Kenkyusho Kenkyu Hokoku (Bull. Natl. Res. Inst. Aquacult.)* 9, 41–44. (in Japanese)

Numaguchi, K., Tanaka, Y., 1986b. Effects of temperature on mortality and growth of the spat of the pearl oyster, *Pinctada fucata martensii*. *Yoshoku Kenkyusho Kenkyu Hokoku (Bull. Natl. Res. Inst. Aquacult.)* 9, 35–39. (in Japanese, with English abstract)

O'Connor, W.A., Lawler, N.F., 2004. Salinity and temperature tolerance of embryos and juveniles of the pearl oyster, *Pinctada imbricata* Roding. *Aquaculture* 229, 493–506.

O'Connor, W.A., Newman, L.J., 2001. Halotolerance of the oyster predator, *Imogine mcgrathi*, a stylochid flatworm from Port Stephens, New South Wales, Australia. *Hydrobiologia* 459, 157–163.

Ohwada, K., Uemoto, H., 1985. Environmental conditions in pearl oyster culture grounds in Japan. In: Sindermann, C.J. (Ed.), *Environmental Quality and Aquaculture Systems. Proceedings of the 13th US Japan Meeting on Aquaculture*, Mie, Japan, October 24–25, 1984. NOAA Technical Report NMFS 69. US Department of Commerce, National Oceanic and Atmospheric Administration, National Marine Fisheries Service, Springfield, VA, USA, pp. 45–50.

Okamura, T., Nakamura, Y., 1988. Thermal tolerance of the eggs, planktonic larvae and settled juveniles of the pearl oyster, *Pinctada fucata martensii* (Dunker). *Rep. Mar. Ecol. Res. Inst.* No.88203, 1–32. (in Japanese, with English abstract)

Ōta, S., 1957. Physio-ecological studies of the free swimming larva of the pearl oyster *Pinctada martensii* (Dunker). I: On the resistability of the larva for low salinity sea water. *Kokoritsu Shinju Kenkyusho Hokoku (Bull. Natl. Pearl Res. Lab.)* 3, 207–212. (in Japanese)

Ōta, S., 1958. Studies of the feeding habits of *Pinctada martensii* (Dunker). I: Seasonal variation of the length of the crystalline style. *Kokoritsu Shinju Kenkyusho Hokoku (Bull. Natl. Pearl Res. Lab.)* 4, 315–317. (in Japanese with English summary)

Ōta, S., 1961. Studies on the feeding habits of *Pinctada martensii*. VII. Differences of the amount of feces due to the current in culture ground (a preliminary report). *Kokoritsu Shinju Kenkyusho Hokoku (Bull. Natl. Pearl Res. Lab.)* 6, 572–575. (in Japanese)

Ōta, S., Fukushima, Y., 1961. Studies on the feeding habits of *Pinctada martensii*. VI: The effect of diluted seawater on the amount of faeces of the pearl oyster. *Kokoritsu Shinju Kenkyusho Hokoku (Bull. Natl. Pearl Res. Lab.)* 6, 567–572. (in Japanese with English summary)

Ōta, S., Shimizu, S., 1961. Influence on the quality of pearls caused by the exposure to the air for the shell cleaning of the pearl oyster. *Kokoritsu Shinju Kenkyusho Hokoku (Bull. Natl. Pearl Res. Lab.)* 6, 688–694. (in Japanese)

Pandya, J.A., 1976. Influence of temperature on growth ring formation in the pearl oyster, *Pinctada fucata* (Gould), of the Gulf of Kutch. *Indian J. Mar. Sci.* 5, 249–251.

Pass, D.A., Dybdahl, R., Mannion, M.M., 1987. Investigations into the causes of mortality of the pearl oyster, *Pinctada maxima* (Jamson), in Western Australia. *Aquaculture* 65, 149–169.

Pouvreau, S., Prasil, V., 2001. Growth of the black-lip pearl oyster, *Pinctada margaritifera*, at nine culture sites of French Polynesia: synthesis of several sampling designs conducted between 1994 and 1999. *Aquat. Living Resour.* 14, 155–163.

Pouvreau, S., Bacher, C., Heral, M., 2000. Ecophysiological model of growth and reproduction of the black pearl oyster, *Pinctada margaritifera*: Potential applications for pearl farming in French Polynesia. *Aquaculture* 186, 117–144.

Rao, G.S., 2001. Ecofriendly onshore marine pearl culture – An overview. In: Menon, N.G., Pillai, P.P. (Eds.), *Perspectives in Mariculture*. The Marine Biological Association of India, Cochin, India. 259–264.

Reed, W., 1966. Cultivation of the black-lip pearl oyster *Pinctada margaritifera* (L.). *J. Conchol.* 26, 26–32.

Rougerie, F., 1979. L'environnement de l'atoll de Takapoto-Tuamotu. Caractéristiques générales du milieu liquide lagunaire de l'atoll de Takapoto. *J. Soc. Océanistes* 33, 35–45.

Saito, H., 2004. Lipid and FA composition of the pearl oyster *Pinctada fucata martensii*: Influence of season and maturation. *Lipids* 39, 997–1005.

Samhan, O., Anderlini, V., Al-Harmi, L., 1979. The pearl oyster, *Pinctada margaritifera*, as an indicator of oil pollution in Kuwait. *Annual research report*, Kuwait Institute for Scientific Research, Safat, pp. 81–82

Saucedo, P.E., Ocampo, L., Monteforte, M., Bervera, H., 2004. Effect of temperature on oxygen consumption and ammonia excretion in the Calafia mother-of-pearl oyster, *Pinctada mazatlanica* (Hanley, 1856). *Aquaculture* 229, 377–387.

Saville-Kent, W., 1890. On the experimental cultivation of the mother-of-pearl shell *Meleagrina margaritifera* in Queensland. *Report Australian Ass. Adv. Sci.* Vol. 2, 541–548.

Saville-Kent, W., 1893. *Pearl and pearl-shell fisheries. The Great Barrier Reef of Australia; its products and potentialities*. W. H. Allen & Co, London. 1075–1078.

Sawada, Y., Tange, M., 1959. The oceanographical studies on the pearl culture ground. II: On the comparison with the qualities of the pearls cultured at four stations in Ago Bay, and the oceanographical observations for them. *Kokoritsu Shinju Kenkyusho Hokoku (Bull. Natl. Pearl Res. Lab.)* 5, 459–480. (in Japanese)

Scholten, H., Smaal, A.C., 1998. Responses of *Mytilus edulis* L. to varying food concentrations: Testing EMMY, an ecophysiological model. *J. Exp. Mar. Biol. Ecol.* 219, 217–239.

Seki, M., 1972. Studies on natural factors affecting the growth and pearl quality of Japanese pearl oyster *Pinctada fucata* in culture environments. *Bull. Mie Pref. Fish. Res.* 1, 32–149. (in Japanese, with English summary)

Shinomiya, Y., Iwanaga, S., Kohno, K., Yamaguchi, T., 1999. Seasonal variations in carbohydrate-metabolizing enzyme activity and body composition of Japanese pearl oyster *Pinctada fucata martensii* in pearl culture. *Nippon Suisan Gakkaishi (Bull. Jap. Soc. Fish. Sci.)* 65, 294–299.

Sims, N.A., 1993. Pearl oysters. In: Wright, A., Hill, L. (Eds.), *Nearshore Marine Resources of the South Pacific*. FFA/ICOD. 409–430.

Strugnell, J.M., Southgate, P.C., 2003. Changes in tissue composition during larval development of the blacklip pearl oyster *Pinctada margaritifera* (L.). *Molluscan Res.* 23, 179–183.

Sugishita, Y., Hirano, M., Tsutsumi, K., Mobin, S.M.A., Kanai, K., Yoshikoshi, K., 2005. Effects of exogenous lipid peroxides on mortality and tissue alterations in Japanese pearl oysters *Pinctada fucata martensii*. *J. Aquat. Anim. Health* 17, 233–243.

Sugiyama, A., Tomori, A., 1988. Oxygen consumption of black-lip pearl oyster. *Suisanzoshoku (J. Jap. Aquacult. Soc.)* 36, 121–125. (in Japanese)

Takayanagi, K., Sakami, T., Shiraishi, M., Yokoyama, H., 2000. Acute toxicity of formaldehyde to the pearl oyster *Pinctada fucata martensii*. *Water Res.* 93–98.

Taylor, J.J., Southgate, P.C., Wing, M.S., Rose, R.A., 1997. The nutritional value of five species of microalgae for spat of the silver-lip pearl oyster, *Pinctada maxima* (Jameson) (Mollusca: Pteriidae). *Asian Fish. Sci.* 10, 1–8.

Taylor, J.J., Southgate, P.C., Rose, R.A., 1998. Effects of mesh covers on the growth and survival of silver-lip pearl oyster (*Pinctada maxima*, Jameson) spat. *Aquaculture* 162, 243–248.

Taylor, J.J., Southgate, P.C., Rose, R.A., 2004. Effects of salinity on growth and survival of silver-lip pearl oyster, *Pinctada maxima*, spat. *J. Shellfish Res.* 23, 375–377.

Teshima, S., Kanazawa, A., Shimamoto, R., 1987. Effects of algal diets on the sterol and fatty acid compositions of the pearl oyster *Pinctada fucata*. *Nippon Suisan Gakkaishi (Bull. Jap. Soc. Sci. Fish.)* 53, 1663–1667.

Tomaru, Y., Kawabata, Z.i., Nakagawa, K.i., Nakano, S.i., 1999. The vertical distribution of pearl oyster *Pinctada fucata martensii* spat in Uchiumi Bay. *Fish. Sci. (Tokyo)* 65, 358–361.

Tomaru, Y., Kawabata, Z., Nakano, S.I., 2001. Mass mortality of the Japanese pearl oyster *Pinctada fucata martensii* in relation to water temperature, chlorophyll a and phytoplankton composition. *Dis. Aquat. Org.* 44, 61–68.

Tomaru, Y., Udaka, N., Kawabata, Z., Nakano, S., 2002a. Seasonal change of seston size distribution and phytoplankton composition in bivalve pearl oyster *Pinctada fucata martensii* culture farm. *Hydrobiologia* 481, 181–185.

Tomaru, Y., Kumatabara, Y., Kawabata, Z., Nakano, S., 2002b. Effect of water temperature and chlorophyll abundance on shell growth of the Japanese pearl oyster, *Pinctada fucata martensii*, in suspended culture at different depths and sites. *Aquacult. Res.* 33, 109–116.

Trueman, E.R., Akberali, H.B., 1981. Responses of an estuarine bivalve, *Scrobicularia plana* (Tellinacea) to stress. *Malacologia* 21, 1–2.

Uemoto, H., 1968. Relationship between oxygen consumption by the pearl oyster and its environmental temperature. *Kokoritsu Shinju Kenkyusho Hokoku (Bull. Natl. Pearl Res. Lab.)* 13, 1617–1623. (in Japanese, with English summary)

Vacelet, E., Arnoux, A., Thomassin, B., 1996. Particulate material as an indicator of pearl-oyster excess in the Takapoto Lagoon (Tuamotu, French Polynesia). *Aquaculture* 144, 133–148.

Wada, K., 1969a. Relations between growth of pearls and water temperature. *Suisan Kaiyo Kenkyu (Bull. Jap. Soc. Fish. Oceanogr.)*, Spec. No. pp. 299–301 (in Japanese, with English abstract)

Wada, K., 1969b. Studies on the crystalline style of the Japanese pearl oyster-I. Effect of diluted sea water on the crystalline style. *Nippon Suisan Gakkaishi (Bull. Jap. Soc. Sci. Fish.)* 35, 133–140.

Wada, K., 1972. Relations between calcium metabolism of pearl sac and pearl quality. *Kokoritsu Shinju Kenkyusho Hokoku (Bull. Natl. Pearl Res. Lab.)* 16, 1949–2027. (in Japanese, with English summary)

Wada, K.T., 1991. The pearl oyster, *Pinctada fucata* (Gould) (Family Pteriidae). In: Menzel, W. (Ed.), *Estuarine and Marine Bivalve Mollusk Culture*. CRC Press, Boston.pp. 245–260.

Waki, S., Yamaguchi, K., 1964. Extermination of mud worm, penetrating into the shell of the pearl oyster, by brine treatment. *Kaiho (Natl. Fed. Pearl Cult. Co-op. Assoc.)* 64, 15–25. (in Japanese)

Watanabe, Y., 1988. Thermal tolerance of juveniles, young and adults of the pearl oyster, *Pinctada fucata martensii* (Dunker). *Rep. Mar. Ecol. Res. Inst. No. 88203* 123, 33–71. (in Japanese, with English abstract)

Yamaguchi, K., Hasuo, M., 1977. Relation between activity of pearl oyster and seasonal changes of environmental factors in culture ground. *Kokoritsu Shinju Kenkyusho Hokoku (Bull. Natl. Pearl Res. Lab.)* 21, 2315–2324. (in Japanese)

Yamamoto, K., 2000. Effects of water temperature on respiration in the pearl oyster *Pinctada fucata martensii*. *Suisanzoshoku (Bull. Jap. Soc. Aquacult.)* 48, 47–52. (in Japanese, with English summary)

Yamamoto, K., Adachi, S., Koube, H., 1999. Effects of hypoxia on respiration in the pearl oyster, *Pinctada fucata martensii*. *Suisanzoshoku (Bull. Jap. Soc. Aquacult.)* 47, 539–544. (in Japanese, with English summary)

Yamashita, S., 1986. The Torres Strait pearling industry. In: Haines, A.K., Williams, G.C., Coates, C. (Eds.), *Torres Strait Fisheries Seminar*. Australian Government Publishing Service, Canberra. pp. 118–119.

Yukihira, H., 1998. *Feeding, energy budgets and nutritional ecology of Pinctada margaritifera (Linnaeus) and P. maxima* Jameson, PhD thesis, School of Biological Sciences. James Cook University of North Queensland, Townsville, Australia. pp. xxiii, 177.

Yukihira, H., Klumpp, D.W., Lucas, J.S., 1998. Comparative effects of microalgal species and food concentration on suspension feeding and energy budgets of the pearl oysters *Pinctada margaritifera* and *P. maxima* (Bivalvia: Pteriidae). *Mar. Ecol. Prog. Ser.* 171, 71–84.

Yukihira, H., Klumpp, D.W., Lucas, J.S., 1999. Feeding adaptations of the pearl oysters *Pinctada margaritifera* and *P. maxima* to variations in natural particulates. *Mar. Ecol. Prog. Ser.* 182, 161–173.

Yukihira, H., Lucas, J.S., Klumpp, D.W., 2000. Comparative effects of temperature on suspension feeding and energy budgets of the pearl oysters *Pinctada margaritifera* and *P. maxima*. *Mar. Ecol. Prog. Ser.* 195, 179–188.

Yukihira, H., Lucas, J.S., Klumpp, D.W., 2006. The pearl oysters, *Pinctada maxima* and *P. margaritifera*, respond in different ways to culture in dissimilar environments. *Aquaculture* 252, 208–224.

CHAPTER 7

Pearl Oyster Culture

Paul C. Southgate

7.1. INTRODUCTION

The major pearl culture industries in Japan, Australia and French Polynesia rely traditionally on oysters collected from the wild. They are collected as adults, or as juveniles which are on-grown to a size suitable for pearl production. However, hatchery production has become increasingly important to the industry over the past 25–30 years with resulting changes to the general culture techniques used for pearl oysters. Hatchery propagation of pearl oysters allows selection for commercially important traits such as nacre color and growth rate. It also provides a basis for techniques such as gamete cryopreservation and induction of triploidy, which are well established in other bivalve culture industries but are yet to be fully assessed for their potential in pearl oyster culture. This chapter covers the culture methods used for pearl oysters during all stages of culture, from hatchery production to the husbandry of oysters "seeded" for pearl production.

7.2. SOURCES OF CULTURE STOCK: COLLECTION OF ADULTS

Given that it takes around 2 years for most pearl oyster species to reach a size suitable for pearl production, there are obvious advantages in obtaining adult stock from the wild and using them directly for pearl production. However, overfishing can lead to depletion of pearl oyster stocks and this may seriously affect the sustainability of an industry. Examples of this were seen in the Cook Islands and Kiribati where over-collection reduced stocks of adult *Pinctada margaritifera* to below self-sustaining levels (Sims, 1993a). If properly managed, however, an industry based on wild collected adults is sustainable, and perhaps the best example of this is the Australian pearling industry in north-western Australia. Collection of adult *P. maxima* by divers is strictly controlled through the use of a quota system, which determines the number of oysters that may be taken from the wild and their minimum size (Wells and Jernakoff, 2006; see Section 9.3); divers primarily collect oysters between 3- and 6-years old in the 120–170 mm size range (Hart and Joll, 2006).

An excellent description of the methods used for collection of *P. maxima* in Western Australia was provided by Wells and Jernakoff (2006) and further information is provided in Section 9.3.2. Collecting boats are equipped with booms which extend from either side of the boat and are fitted with weighted ropes that hang vertically to a height of 1–1.5 m from the bottom. Each rope is fitted with a collecting bag to hold oysters. Divers operate within depths of <20 m and place collected oysters in a neck bag before transferring them to the larger collecting bag. Divers generally collect around 250 oysters per day and each boat may have six divers. After cleaning, oyster are placed into metal-framed pocket or panel nets (see Section 7.5.4) and held in tanks with a recirculating water supply during transport to a holding area where the nets are attached to lines on the sea bottom. Oysters may remain in the holding area for around 2 months to allow recovery before being used for pearl production.

7.3. SOURCES OF CULTURE STOCK: SPAT COLLECTION

Perhaps the best example of an industry based on juveniles collected from the wild is the "black" pearl industry in French Polynesia (see Section 9.4). In eastern Polynesia, the atolls are generally small and enclosed, and their lagoons are poorly flushed by oceanic water. Furthermore, they have good natural stocks of pearl oysters producing large numbers of larvae that are retained within the atoll lagoons. Pearl oyster spat may be collected by deploying an appropriate substrate or "spat collector" into lagoon waters when larvae are abundant and approaching settlement (see Section 5.7.2). Spat collectors may be composed of various materials, but more commonly used are natural substrates such as tree branches, oyster shells and pieces of coral as well as plastic materials such as shade-cloth and frayed rope (Table 7.1). As well as being suitable as a substrate for spat collection, the materials used in spat collectors should ideally be cheap, durable and locally available (Vakily, 1989).

Recent years have seen greater use of light-weight spat collectors made from plastic mesh (e.g. shade-cloth or shade-mesh) or ropes to form a three-dimensional structure offering a large surface area and protective spaces for newly recruited spat (Fig. 7.1). These include the "flower type" collector, where a strip of plastic sheet or shade-mesh is folded concertina-style and tied in the middle (Gervis and Sims, 1992), and the "accordion-style" collector, where approximately 25 m lengths of 8–12 cm wide strips of shade-mesh are threaded accordion-fashion and compressed onto a 1 m length of rope (Haws and Ellis, 2000).

TABLE 7.1
Some materials used for spat collection of pearl oysters.

Location	Species	Material	Author(s)
Japan	*P. martensii*	Cedar sprigs Mollusc shells Old fishing nets	Shirai (1970)
India	*P. fucata*	Oyster baskets Nylon mesh Nylon frills	Victor *et al.* (1987) Nayar *et al.* (1978) Achari (1980)
Cook Islands	*P. margaritifera*	Hyzex film *Pemphis acidula* sprigs Plastic rope	Passfield (1989)
French Polynesia	*P. margaritifera*	Netron tube Shade mesh *Pemphis acidula* sprigs	Coeroli *et al.* (1984)
Sudan	*P. margaritifera*	Wooden boards Split bamboo	Crossland (1957) Rahma and Newkirk (1987)
Mexico	*P. mazatlanica* *Pteria sterna*	Nylon gill net in mesh bags	Monteforte and Garcia-Gasca (1994)
Australia	*Pinctada* spp. *Pteria* spp.	Shade cloth in mesh bags	Beer and Southgate (2000)
Solomon Islands	*P. margaritifera*	Shademesh sheet/strips Plastic sheet/strips	Friedman *et al.* (1998)

Source: Adapted from Gervis and Sims (1992).

7.3.1. Factors influencing spat collection success

Spat collectors are generally suspended from floating structures such as longlines or rafts (see Section 7.5). As well as the type of material used to construct the collectors, spat collection success is influenced by factors such as location, depth and season, as well as local currents and hydrodynamics (Coeroli *et al.*, 1984; Monteforte and García-Gasca, 1994; Knuckey, 1995; Friedman *et al.*, 1998). The timing of spat collector deployment is critical and must coincide with the availability of metamorphically competent larvae of the target species. Beer and Southgate (2000), for example, showed that 76% of *Pteria penguin* spat collected over a 12-month trial occurred between March and May in northern Australia. Poor timing of spat collector deployment can result in significant "by-catch" of unwanted species (Crossland, 1957).

The period of immersion or "soak" time for which spat collectors are deployed may also vary. In French Polynesia and the Cook Islands spat collectors are generally deployed for around 6 months and juveniles are harvested when they attain a dorsoventral shell height of approximately 50–65 mm (Coeroli *et al.*, 1984; Sims, 1993a; Pouvreau and Prasil, 2001). In northern Australia, Beer and Southgate (2000) deployed spat collectors composed of shade-cloth loosely compressed within mesh bags for

FIGURE 7.1 *Pinctada margaritifera* juveniles attached to plastic net collector in French Polynesia (Photo: Sandra Langy).

4-week durations to collect pearl oyster spat. While increasing immersion time provides the opportunity to harvest larger pearl oysters from spat collectors, there is some risk associated with the inevitable increase in fouling and predation, particularly if the collector material is house within a protective mesh bag. Research in the Solomon Islands showed that when spat collectors within mesh bags were deployed for 6 months the mortality of pearl oyster recruits was high (Friedman *et al.*, 1998). The authors noted large numbers of predators including crabs and *Cymatium* gastropods found within the collectors. The larvae of predators also recruit to spat collectors and may grow rapidly.

Ideally, spat collectors should be deployed for periods that do not allow the juveniles of predators to become large enough to consume pearl oyster spat.

7.4. SOURCES OF CULTURE STOCK: HATCHERY PRODUCTION

7.4.1. Broodstock

A critical factor in the success of hatchery production of pearl oysters is selection of appropriate broodstock. Clearly, broodstock should be selected on the basis of desirable qualities in the offspring. In pearl oysters these include rapid growth rate, aspects of shell morphometrics and, perhaps most importantly, nacre coloration. The major implications of broodstock selection are outlined in Section 12.3.4.4 and consideration must be given to the potential for loss of genetic diversity and inbreeding depression. Given that pearl oysters are protandrous hermaphrodites (see Section 5.3.1) the size of the broodstock selected is also important to ensure an adequate supply of both male and female oysters. Rose (1990) recommended that individuals with prominent growth processes on their shells (see Section 5.7.4), indicating a healthy actively growing individual, should be chosen as broodstock. Stunted and small oysters, as well as very large oysters that are more difficult to spawn in the hatchery, should be avoided (Rose, 1990). Broodstock must be thoroughly cleaned prior to spawning induction to remove and reduce bacterial contamination. It is not uncommon for broodstock oysters to be immersed in fresh water for 1–2 minutes following cleaning.

Pearl oyster broodstock are easily inspected to determine their sex and reproductive condition (Rose, 1990). The various stages of reproductive maturity in pearl oysters have been described for the major culture species (Tranter, 1958; Rose, 1990; Saucedo *et al.*, 2001) and are outlined in detail in Section 5.4.7. The major stages of gonad maturation in *Pinctada maxima* are shown in Table 7.2 and a more detailed scheme for pearl oysters is shown in Table 5.2.

Increasing water temperature and food availability are the major cues for reproductive maturation in pearl oysters (see Section 5.6.2). Despite this knowledge, however, conditioning of pearl oyster broodstock to bring about reproductive maturation outside normal breeding season, has received little research attention compared to other bivalves of commercial importance. This may reflect that two of the major culture species, *P. maxima* and *P. margaritifera*, are tropical with extended spawning periods (Tranter, 1958; Rose, 1990; Acosta-Salmon and Southgate, 2005) which reduces the need for out of season conditioning. It may further reflect the relatively large size of pearl oysters and logistical difficulties in providing the large volumes of micro-algae that would be needed for adequate conditioning to occur. Certainly, reproductive conditioning of the larger species of pearl oyster (e.g. *P. maxima* and *P. margaritifera*) has proved problematic (Rose and Baker, 1989) and hatcheries generally rely on broodstock that mature under "natural" oceanic conditions.

TABLE 7.2

Stages of gonad maturation in *Pinctada maxima*.

Stage	Description
0	Gonad tissue flaccid or invisible, sex indeterminate.
1	Gonad visible but proliferation to gut loop is slight and proximal gonad appears granular and difficult to sex by color (male white and female yellow).
2	Sex easily determined by color. Tissue has proliferated distally along lateral walls of gut loop and appears semi confluent (at this stage spawning could occur, but gametes are usually immature or non-viable.
3	Gonad ripe and bulging, gonad tissue extends over the surface of stomach, gut loop and liver. Gonad appears confluent and when pierced, gametes run freely and profusely.

Source: Rose (1990).

Pinctada maxima broodstock require a minimum of 5 weeks to progress from stage 0 to spawning ripe condition at stage 3 (Table 7.2) when held on a surface longline (Rose, 1990). Akoya[1] pearl oysters show gonad development within a water temperature range of 18–24°C and conditioning to reproductive maturation outside normal breeding season can be achieved by increasing water temperature from ambient to 20–24°C for 3 weeks (Hayashi and Seko, 1986). Akoya broodstock can be maintained within a temperature range of 25–28°C and fed a mixed micro-algal diet supplemented with corn flour to maintain spawning condition (Alagarswami *et al.*, 1987). Individuals of the same species with maturing gonads were successfully brought to spawning condition when held at 25–28°C and fed a mixed algal diet (4 L/oyster/day) supplemented with corn flour (30 mg/oyster/day) for 45 days (Chellam *et al.*, 1991). *P. mazatlanica* was successfully conditioned when held at elevated water temperatures and supplied with cultured micro-algae (Saucedo *et al.*, 2001); a gradual increase in water temperature, from 20°C to 29°C over a 2-month period was more effective than direct exposure of oysters to water temperatures of either 20°C, 24°C or 28°C.

7.4.2. Spawning induction

Exposure to elevated, or to fluctuating water temperature, is the most common means of spawning induction in pearl oysters. Indeed ripe pearl oysters will often spawn in response to being removed from field culture conditions to the hatchery. Thermal stimulation has been reported as a successful means of spawning induction in all major commercial pearl oyster species (Alagarswami *et al.*, 1983a, b; Alagarswami *et al.*, 1989; Chellam *et al.*, 1991; Rose and Baker, 1994; Southgate and Beer, 1997). The

[1]Throughout this text the term 'Akoya' is used for pearl oysters in the currently unresolved complex of *Pinctada fucata-martensii-radiata-imbricata* (see Section 2.3.1.13.)

emperatures used and the range over which they may fluctuate will vary according to the target species. For example, Akoya pearl oysters from temperate Australian waters experience a natural water temperature range of approximately 14–26°C and they spawn at temperatures above 18°C (O'Connor and Lawler, 2004). For tropical species though, the level by which water temperature may be increased is limited by higher ambient water temperature. For spawning induction of *P. maxima*, for example, Rose (1990) proposed an increase from 2°C to 5°C above ambient, with a maximum of 33–34°C. Spawning in Akoya pearl oysters from tropical waters in India has been induced by raising water temperature from 28.5°C to 35°C (Alagarswami *et al.*, 1983a). It is common practice to alternate exposure of pearl oyster broodstock between elevated water temperature and ambient water temperature until spawning occurs (Tanaka *et al.*, 1970). Broodstock may be held at a water temperature of around 5–6°C above ambient for 30 min before being transferred back to ambient water temperature and so on (Gervis and Sims, 1992).

To increase the water temperature range to which tropical pearl oyster species can be exposed for spawning induction, Southgate and Beer (1997) used "cold conditioning" where *P. margaritifera* broodstock were held overnight in tanks containing aerated water at 20°C; this is 5–6°C below the ambient water temperature, but within the normal temperature range experienced by *P. margaritifera*. The following day, oysters were placed directly into another tank containing water heated to 30–31°C to induce spawning.

Chemical stimulation of spawning is also commonly used for pearl oysters. A widely used method is based on the observation that pearl oysters, like other bivalves, readily spawn in response waterborne gametes released by other individuals in the same body of water (Lucas, 2003). Male pearl oysters usually spawn first and the presence of sperm in the water stimulates the females to spawn (Alagarswami *et al.*, 1983a). It is common practice in pearl oyster hatcheries to stimulate spawning by introducing a suspension of sperm obtained from a sacrificed male into the spawning tank. Shallow cuts are made into the gonad of the male and sperm may be washed from the incisions using filtered seawater or can be removed using a pipette. Filtered seawater is added to the collected sperm and the resulting sperm suspension is added directly to the spawning tank.

Changes to the chemical composition of seawater have also been shown to induce spawning of pearl oysters. Rose (1990) reported a spawning response in *P. maxima* when exposed to a weak solution of hydrogen peroxide (final concentration of 0.006%) following adjustment of the holding water to pH 9 using ammonium hydroxide (NH_4OH), sodium hydroxide or tris buffer. Hydrogen peroxide combined with either normal or alkaline seawater (pH 9.1) also resulted in spawning of Akoya pearl oysters (Chellam *et al.*, 1991). Direct injection of 0.2 mL of 0.1 N NH_4OH solution into the adductor muscle of Akoya oysters has also been reported to stimulate a high (48%) spawning response (Chellam *et al.*, 1991).

Broodstock are normally induced to spawn in a communal spawning tank or shallow spawning table (Figs. 7.2 and 14.2). However, once spawning commences oysters are removed and placed individually into aquaria where they complete spawning (Fig. 7.3). This is an important step, because reproductive maturation is not completely synchronous and not all oysters will produce good quality eggs or sperm. Inspection of

FIGURE 7.2 *Pinctada maxima* in a shallow communal spawning tank. (Photo: Clare Flanagan).

FIGURE 7.3 Female *Pinctada maxima* spawning following removal from communal spawning tank (see Fig. 7.2). Note that this individual has been marked as a female following inspection to determine sex and reproductive condition. Spawned eggs (E) are negatively buoyant. (Photo: Clare Flanagan). (See Color Plate 9)

gametes from individual pearl oysters allows selection of good quality gametes for use in the hatchery. Sperm should be active while eggs should be spherical and negatively buoyant (Rose, 1990).

7.4.2.1. "Strip" spawning

Hatcheries producing Akoya pearl oysters in Japan may utilize gametes that are "stripped" from the gonads of broodstock. This technique involves washing gametes from shallow cuts made into the gonad following removal of one of the shell valves. Although destructive, this allows the nacre quality of the broodstock shells to be observed. Because pearl oyster eggs activate in the follicle immediately before a spawning (Tranter, 1958), stripping does not result in viable eggs unless they are treated. Stripped gametes are usually treated with dilute ammonium solution to improve the fertilization rate (Kuwatani, 1965). This technique has also been used with *Pinctada maxima* and *P. margaritifera* (Tanaka *et al.*, 1970; Tanaka and Kumeta, 1981; Rose, 1990). Tanaka and Kumeta (1981) reported that eggs stripped from *P. maxima* and treated with 0.001 N NH_4OH for 40 min produced larvae that grew well and metamorphosed. Rose (1990) added a suspension of stripped eggs from *P. maxima* to sufficient 0.1 N NH_4OH to produce a 0.5% ammonium solution of NH_4OH. After stirring, the egg suspension was left for 40–50 min until the germinal vesicle in the majority of eggs was no longer visible and the eggs had become spherical. Sperm was then added to the treated eggs for 1–2 min before the fertilized eggs were washed and resuspended in 1 μm filter sea water (Rose, 1990). Ito (1998) used ammonia solution to mature ova stripped from "spent" *P. maxima* gonads prior to fertilization and reported good rates of larval growth and survival.

Although gamete stripping is commonly used in bivalve propagation (Quayle and Newkirk, 1989), it is destructive and is generally used with lower value species such as edible oysters. Given the relatively high value of pearl oysters and their potential to produce a high value product, this technique is not widely used in pearl oyster hatcheries and gametes resulting from thermal induction are generally acknowledged to be of higher quality (Tanaka *et al.*, 1970; Tanaka and Kumeta, 1981; Rose 1990).

7.4.3. Fertilization of eggs

It is common practice in pearl oyster hatcheries to combine the eggs from a number of individuals prior to fertilization. The same is done for sperm. Eggs and sperm from each oyster can be inspected microscopically and those of better quality can be identified and combined. While no special preparation is required for sperm, it is common for eggs to be passed through a nylon sieve mesh to remove debris (e.g. feces, etc.) before fertilization. Eggs may also be washed by holding them on a submerged fine mesh sieve and washing with clean 1 μm filtered seawater (Rose and Baker, 1994).

It is important that too much sperm is not used during fertilization of bivalve eggs. Excessive sperm may lead to polyspermy that can affect survival and development of embryos and larvae (Lucas, 2003), and can impact on water quality when it decays. To avoid these potential problems, it is common in bivalve hatcheries to add mixed sperm suspension to the combined egg suspension initially at a low level (e.g. 1 mL of sperm suspension per litre of egg suspension). Additional sperm can be added if necessary following microscopic examination to determine the number of available sperm per egg. Fertilization is indicated by the presence of the first polar body resulting from meiotic division of the egg nucleus (Fig. 5.12).

7.4.4. Cryopreservation of gametes

Cryopreservation of gametes is practiced widely in the aquaculture industry and appropriate protocols have been developed for a number of rock oyster species (Bougrier and
Rabenomanana, 1986; Yankson and Moyse, 1991). Despite this, cryopreservation of
gametes has only recently been investigated for pearl oysters (Lyons *et al.*, 2005; Acosta-
Salmon *et al.*, 2007). Recent commercial interest in developing specific breeding lines
for pearl oysters, to improve traits such as pearl quality (see Chapter 12), brings about
a requirement to be able to store gametes from genetically superior oysters. For example, commercially important aspects of pearl quality, such as color, are directly related
to the donor mantle tissue used during the pearl seeding process (Acosta-Salmon *et al.*,
2004). Cryopreservation of the gametes from superior donor oysters and their subsequent
use in breeding programs, would allow more efficient use of genes from such animals
and allow general improvement to the overall quality of donors within a culture population. Factors influencing the success of cryopreservation include the cryoprotectant
agents (CPAs) used, equilibrium time and cooling rate. Lyons *et al.* (2005) determined
the influence of CPAs on the motility of spermatozoa from *Pinctada margaritifera* and
reported that sperm preserved with 1 M trehalose and 5% dimethyl sulfoxide retained a
total sperm motility of 86% while sperm preserved with propylene glycol showed poor
motility. In a follow-up study Acosta-Salmon *et al.* (2007) reported high motility of spermatozoa following cryopreservation using 0.45 M trehalose with 0, 5% or 12% dimethyl
suloxide and frozen at a cooling rate of 2.1 to 5.2°C/min.

7.4.5. Egg incubation

Following fertilization, pearl oyster eggs are generally incubated in tanks containing
gently aerated and filtered (usually to 1 µm) seawater (Rose and Baker, 1994; Southgate
and Beer, 1997). Given the potentially high mortality of eggs and embryos during
incubation, it is not uncommon for an antibiotic to be added to egg incubation tanks.
A broad-spectrum antibiotic, such as streptomycin-sulfate, may be used during egg incubation at a concentration of 10 mg/L (Southgate and Beer, 1997; Beer, 1999).

Egg density can have a significant effect on the survival of pearl oyster embryos. In
general, survival to the larval stage is inversely proportional to egg density. For example, Southgate *et al.* (1998a) determined the effect of egg density on survival of fertilized eggs of *Pinctada maxima* and *P. margaritifera* through to the veliger larval stage,
when incubated at densities of 10, 20, 30, 40, 50 and 100/mL and 10, 20, 30, 50, 100 and
150/mL, respectively. Highest survival was found at the lowest egg density (10/mL) for
both species, while lowest survival was found at a density of 100/mL for both species.
Rose (1990) recommended that egg density should not exceed 30/mL for *P. maxima*.

Information on the effects of other culture parameters on development and survival
of eggs and embryos of pearl oysters is scarce. Doroudi *et al.* (1999a) determined the
effects of varying water temperature (20–35°C) and salinity (25–40‰) combinations on
survival of *P. margaritifera* embryos. Development of embryos was significantly influenced by water temperature and salinity with optimal development of eggs occurring

within the ranges of 25–30°C and <24–39‰, respectively. There was no development of embryos at 20°C or 35°C. In a similar study, embryos of Akoya pearl oysters incubated at water temperatures in the range of 14–26°C and salinities in the range of 11–35‰ showed greatest survival within a temperature range of 18–26°C and a salinity range of 29–35‰ (O'Connor and Lawler, 2004).

Embryonic and early larval development of pearl oysters has been described for a number of species (Wada, 1942; Minaur, 1969; Tanaka and Kumeta, 1981; Alagarswami *et al.*, 1983b; Alagarswami *et al.*, 1989; Rose and Baker, 1994; Doroudi and Southgate, 2003) and is discussed in detail in Section 5.7 (Fig. 5.12). The first motile stage is the trochophore which is ciliated but lacks a shell and is unable to feed (Fig. 5.12). The shelled veliger stage is reached 20–24 h after fertilization and early veliger larvae have a characteristic "D"-shaped shell giving rise to the common names of "D-stage" or "D-shaped" veliger (Figs. 5.12 and 5.13(a)).

Incubation tanks are generally maintained for around 24 h to allow larval development through to the veliger stage. Now with a shell, the larvae are more robust and can be drained from incubation tanks and retained on an appropriately sized sieve mesh to separate them from undeveloped eggs and moribund larvae (Rose and Baker, 1994; Southgate and Beer, 1997; Fig. 7.4). Larvae can then be washed and transferred to larval culture tanks.

7.4.6. Larval culture

Pearl oyster larvae are generally cultured in 1 μm filtered seawater. They must also be provided with cultured micro-algae as a food source which necessitates regular

FIGURE 7.4 Drain-down of egg incubation and larval culture tanks in a pearl oyster hatchery. Water drains from the tank through an appropriately sized sieve mesh which retains the larvae. Larvae are resuspended into clean sea water, counted and placed back into clean culture tanks.

replacement of culture water to maintain water quality. A number of factors influence the success of hatchery culture of pearl oysters.

7.4.6.1. Water quality parameters

Information on the effects of water quality parameters on survival, growth rate and development of pearl oyster larvae is limited. However, Doroudi *et al.* (1999a) reported maximum survival of 6-day-old *Pinctada margaritifera* larvae to occur within a salinity range of 26.5–33.5‰ and within a water temperature range of 22.5–26.5°C. They noted that for each degree (°C) increase above 29°C, survival decreased by approximately 5% to a level of around 20% at 35°C. For older larvae (15 days), maximal survival occurred at a salinity of <31‰ and a water temperature of <25°C, however, growth of larvae within this temperature range was very slow (Doroudi *et al.,* 1999a). The greatest growth rates of 6-day-old *P. margaritifera* larvae occurred within a salinity range of 29–33.5‰ and a water temperature range of 26–29°C, while the greatest growth rate of 15-day-old larvae occurred at 25–36‰ and 24.5–30°C, respectively (Doroudi *et al.,* 1999a). Similar information for the larvae of other pearl oyster species is limited, however, Rose (1990) recommended that larvae of *P. maxima* should be reared within a temperature range of 27–30°C.

Whether or not to provide pearl oyster larval culture tanks with aeration is debatable based on available literature. Rose (1990), for example, recommended that *P. maxima* larvae should be cultured in tanks without aeration until just before settlement. Similarly, no aeration was provided to culture tanks containing Akoya pearl oyster larvae by Alagarswami *et al.* (1983b) as it is thought to negatively affect growth and spatfall of this species (Chellam *et al.*, 1991). In contrast, the larvae of *P. margaritifera* and *P. mazatlanica* are generally supplied with gentle aeration throughout the larval culture period (Southgate and Beer, 1997; Martínez-Fernández *et al.*, 2003).

7.4.6.2. Water exchange

Pearl oyster hatcheries replace the water in larval culture tanks periodically during larval culture. During this process the larvae are generally retained on a submerged mesh sieve as the tank is drained (Fig. 7.4). The tank is subsequently cleaned and refilled with clean filtered seawater before larvae are placed back into it. This provides a good opportunity for hatchery staff to estimate the numbers of larvae in each culture tank and calculate survival. Larvae are washed from the sieve on which they were caught into a vessel of known volume (e.g. 20 L). Replicate samples are removed from the vessel and the numbers of larvae within the samples is determined using a microscope. The total number of larvae can be estimated on the basis of these counts. Samples of larvae may also be preserved at each water change for subsequent measurement or examination.

The frequency at which water in larval culture tanks is replaced varies. Rose and Baker (1994), for example, varied water change frequency according to the density of *Pinctada maxima* larvae; only one change on day 10 for larvae stocked at a density of

1/mL, but 4–5 water changes (one every 5–6 days) for larvae stocked at densities of 5–8/mL. It is more common, however, for water used to culture pearl oyster larvae, to be replaced every 2–3 days (Alagarswami *et al.,* 1989; Southgate and Beer, 1997; Beer, 1999; Martínez-Fernández *et al.,* 2003).

Flow-through systems have also been used to culture pearl oyster larvae, where water flows through larval culture tanks on a continuous or intermittent basis (Southgate and Beer, 1997; Southgate and Ito, 1998; Martínez-Fernández *et al.*, 2003). Larvae are retained in the tank by fine mesh through which the water leaves the tank. The assumed advantages of flow-through systems relate to the potential for improved water quality and reduced manpower associated with tank maintenance. However, Southgate and Beer (1997) showed that a flow-through system used to culture *P. margaritifera* larvae did not result in improved growth and survival of larvae when compared to those held in culture tanks with bi-daily water change, despite improved water quality.

A number of potential problems were identified when using flow-through culture systems for young pearl oyster larvae. These relate primarily to the fine mesh required to retain larvae within the tanks and the ease with which it can become clogged, providing potential for tanks to overflow. Furthermore, the flushing of dead larvae and uneaten food and feces from the tank bottom is difficult to achieve using a flow-through system and these items must be removed from culture tanks by manual siphoning (Southgate and Beer, 1997; Southgate and Ito, 1998). However, the chance of mesh blockage is reduced when a large mesh size is used to retain older larvae and a number of pearl oyster hatcheries use static culture system for the first part of larval development and a flow-through system for culturing larger larvae.

7.4.6.3. Grading

The regular grading or culling of larvae is necessary to remove slow growing and moribund larvae. Grading reduces variation in size between larvae and improves the synchrony with which larvae reach settlement; it also enhances post-settlement survival (Alagarswami *et al.*, 1987). Rose and Baker (1994) suggested that culling slower developing *Pinctada maxima* larvae reduced the period over which settlement occurred from about 7 days to 3 days. Grading of larvae is easily achieved using different sizes of mesh sieves; it is usually conducted during the tank draining procedure. Rose (1990) recommended that the larvae of *P. maxima* could be graded on day 14 by passing them through a 100 μm and 60 μm sieve mesh. Larvae caught on the 100 μm mesh are retained and placed back into larval culture tanks, while those caught on the 60 μm sieve mesh are discarded. Hayashi and Seko (1986) graded Akoya larvae at 8, 13 and 18 days of age.

7.4.6.4. Larval nutrition

The nutritional composition of the diet provided to pearl oyster larvae during hatchery culture has a major influence on their growth and survival (Southgate *et al.*, 1998b; Martínez-Fernández *et al.*, 2006). The larvae of bivalve molluscs generally exhaust maternally-derived energy reserves over the first few days of development and must

FIGURE 7.5 Micro-algae culture laboratory associated with a pearl oyster hatchery.

then build up energy reserves during the remainder of larval life to fuel metamorphosis (Mann and Gallager, 1985; Gallager *et al.*, 1986; Whyte *et al.*, 1987; see Section 5.7.3). This pattern of energy utilization and assimilation is also seen in pearl oyster larvae (Strugnell and Southgate, 2003).

Pearl oyster larvae are fed cultured micro-algae which must provide energy and the specific nutrients required for tissue synthesis and accumulation of energy reserve in growing larvae. Micro-algae are normally cultured on-site at the hatchery (Fig. 7.5). This process is labor and resource demanding (Southgate, 2003) and makes a significant contribution to hatchery running costs (Brown and Jeffrey, 1992; Coutteau *et al.*, 1994). The reader is referred to standard texts for further information relating to micro-algae culture methods.

Micro-algae vary in their nutrient content and some species better suit the nutritional requirements of pearl oyster larvae than others. It is generally accepted that a mixture of micro-algae species provides a better balance of nutrients for bivalve larvae than a single species (Webb and Chu, 1982) and, on this basis, pearl oyster larvae are usually fed a mixture of at least two species of micro-algae. Species from the genera *Isochrysis*, *Pavlova* and *Chaetoceros* are commonly used as a food for bivalve larvae, including pearl oysters, although a wide variety of micro-algae species have been used successfully with pearl oyster larvae (Table 7.3).

Martínez-Fernández *et al.* (2006) determined the nutritional value of seven species of micro-algae for larvae of *Pinctada margaritifera*. Members of the genus *Pavlova* were shown to have particularly high nutritional value for both D-stage and later stage larvae with the diatom *Chaetoceros* spp. being of high nutritional value for later stage larvae only. Larval growth rates were shown to be correlated with the levels of a number of key nutrients within the micro-algae. Growth of D-stage *Pinctada margaritifera* larvae was significantly correlated with protein, carbohydrate, lipid, saturated fatty acid (SFA), and highly unsaturated fatty acid (HUFA) contents of micro-algae. Of the SFA and HUFA, levels of myristic acid (14:0) and palmitic acid (16:0), and docosahexaenoic

TABLE 7.3

Some of the species of micro-algae assessed as a food for pearl oyster larvae and their reported nutritional values.

Species	Micro-algae	Nutritional value	Author(s)
P. margaritifera	*Pavlova pinguis* *Pavlova salina* *Pavlova* sp. (PS 50) *Isochrysis* (T-Iso) *Chaetoceros muelleri* *Chaetoceros* sp. *Skeletonema* sp. *Micromonas pusilla*	*Pavlova salina* and *Pavlova* sp. supported greatest growth of D-stage and umbone larvae. *Chaetoceros* sp. supported poor growth of D-stage larvae. Larval growth correlated with carbohydrate, lipid, protein and (HUFA) content of micro-algae.	Martinez-Fernandez *et al.* (2006)
P. margaritifera	*Pavlova salina* *Pavlova* sp. (PS 50) *Isochrysis* (T-Iso) *Micromonas pusilla* *Chaetoceros muelleri* *Chaetoceros* sp. *Skeletonema* sp. fed in binary and ternary combinations.	*Pavlova* sp. and *Pavlova* sp./*Chaetoceros muelleri* recommended for D-stage and umbone larvae, respectively.	Martínez-Fernández and Southgate (2007)
P. margaritifera	*Pavlova salina* *Isochrysis* (T-Iso) *Chaetoceros simplex*	*Isochrysis* (T-Iso) supported greatest growth rate and survival	Southgate *et al.* (1998b)
P. margaritifera	*Chaetoceros muelleri* *Chaetoceros simplex* *Isochrysis* (T-Iso) *Pavlova salina* *Pavlova lutheri*	Clearance rate (CR) and ingestion rate (IR) lower with diatoms than flagellates. CR and IR of *Pavlova salina* were about five-times higher than those for *Chaetoceros* spp.	Doroudi *et al.* (2003)
P. maxima	*Isochrysis* (T-Iso) *Chaetoceros calcitrans* *Chaetoceros gracilis* *Nannochloropsis oculata*	*Nannochloropsis oculata* not easily digested. Spines of *Chaetoceros gracilis* impeded ingestion by small larvae.	Rose and Baker (1994)
P. maxima	*Isochrysis* (T-Iso) *Chaetoceros calcitrans* *Chaetoceros gracilis* *Nannochloropsis oculata* *Tetraselmis suecica*	*Isochrysis* (T-Iso), *Chaetoceros calcitrans* and *Chaetoceros gracilis* were most nutritious.	Rose (1990)
P. maxima	*Brachiomonas submarina* *Chlorococcum* sp. *Dicrateria* sp. *Dunaliella tertiolecta* *Isochrysis galbana* *M. (Pavlova) lutheri* *Nannochloris* sp. *Nitzschia closterium* *Pyramimonas grossi* *Chlorella* sp.* Isolate (green flagellate)*	*Isochrysis galbana* and *Monochrysis lutheri* were the most nutritious species. At the larval rearing temperature of 30°C both species lost motility. The greater suitability of tropical micro-algae species was suggested.	Minaur (1969)

(Continues)

TABLE 7.3   (*Continued*)

Species	Micro-algae	Nutritional value	Author(s)
P. fucata	*Isochrysis galbana* *Pavlova* sp. *Chromulina* sp. *Dicrateria* sp.	High. Good. Good. Good.	Chellam *et al.* (1991)
P. mazatlanica	*Isochrysis* (T-Iso) and *P. lutheri* (1:1) mixture	Supported production of spat.	Martinez-Fernandez *et al.* (2003)
Pteria sterna	*Phaeodactylum tricornutum* *Chaetoceros calcitrans* *Chaetoceros muelleri* *Thalassiosira weisflogii* *Dunaliella salina* *Nannochloris* sp. *Tetraselmis tetrahele* *Tetraselmis suecica* *Isochrysis* (T-Iso) *Pavlova lutheri*	Only *Nannochloris* sp. *Isochrysis* (T-Iso) and *Pavlova lutheri* were ingested. Only *Isochrysis* and *P. lutheri* were digested.	Martínez-Fernández *et al.* (2004)

*Local isolates.

acid (DHA), respectively, were particularly important (Martínez-Fernández *et al.*, 2006). The growth on umbone *P. margaritifera* larvae was significantly correlated with carbohydrate, and HUFA contents of micro-algae.

As well as nutrient content, micro-algae fed to pearl oysters larvae must have appropriate morphological characteristics and be suited to the culture conditions of the larvae. For example, Rose and Baker (1994) reported that the chlorophyte, *Nannochloropsis oculata*, was unable to be digested by *P. maxima* larvae because of its thick cell wall. The same authors reported that spines on the diatom, *Chaetoceros gracilis*, impeded ingestion by small *P. maxima* larvae. A number of authors have pointed out potential problems when using micro-algae of temperate origin as a food for tropical pearl oyster larvae cultured at relatively high water temperatures (Minaur, 1969; Tanaka and Inoha, 1970). While tropical species of micro-algae are likely to be better suited to this role (Minaur, 1969; Tanaka *et al.*, 1970) it is only in recent years that a good variety of these species have become available to pearl oyster hatcheries (Martínez-Fernández *et al.*, 2006). Selection of micro-algae used as a food source for pearl oyster larvae should take into account the origin of the micro-algae and its suitability under specific larval culture conditions.

7.4.6.5. Food ration

Successful hatchery culture of pearl oysters relies on providing an appropriate ration to larvae that supports high growth rates without compromising water quality or causing undue increase in hatchery running costs. Feeding schedules for the larvae and early post-larvae of a number of pearl oyster species are shown in Table 7.4; however,

TABLE 7.4
Feeding schedule (cells/mL/day) used for larvae and early post larvae of some pearl oyster species. Values in parentheses are mean antero-posterior shell sizes of the larvae.

Age (days)	Species			
	P. maxima[a]	*P. margaritifera*[b]	*P. fucata*[c]	*P. mazatlanica*[d]
1	5,000 (80)	1,000 (82)	5,000 (67.5)	10,000
2	6,000	1,000	5,000	10,000 (80)
3	7,000	1,000	5,000	10,000
4	8,000	1,000	5,000	10,000
5	10,000 (98)	2,000	5,000	10,000
6	12,000	2,000	5,000	10,000
7	14,000	2,000	5,000	10,000
8	18,000	4,000	5,000	10,000
9	20,000	8,000	5,000	10,000
10	25,000 (115)	8,000	5,000 (135)	10,000
11	28,000	8,000 (138)	10,000	10,000
12	30,000	8,000	10,000	10,000 (139)
13	32,000	10,000	10,000	10,000
14	34,000	10,000	10,000	10,000
15	38,000 (144)	12,000	10,000 (180)	10,000
16	40,000	12,000	10,000	10,000
17	42,000	12,000	10,000	10,000
18	44,000 (165)	12,000	10,000 (200)	10,000
19	48,000	12,000	15,000	15,000
20	50,000 (185)	10,000 (214)	15,000 (220)	15,000 (193)
21	50,000	10,000	50,000*	15,000
22	50,000	10,000	50,000*	15,000
23	50,000	10,000	50,000*	15,000
24	50,000	10,000	50,000* (300)	15,000 (223)
25	50,000 (237)	10,000	50,000*	15,000
26	50,000	12,000	50,000*	15,000
27	50,000	12,000	50,000*	30,000
28	50,000	18,000	50,000*	30,000
29	50,000	18,000	50,000*	30,000
30	50,000 (281)	25,000	50,000*	30,000
Author(s):	Rose (1990)	Southgate and Beer (1997)	Chellam *et al.* (1991)	Martinez-Fernandez *et al.* (2003)

*50,000 cells/day/spat.
[a]Diet composed of *Isochrysis* (T-Iso), *Chaetoceros gracilis* and *Nannochloropsis oculata* (~5–10%).
[b]Diet composed of *Isochrysis* (T-Iso), *Pavlova salina* and *Chaetoceros simplex*.
[c]Diet composed of *I. galbana* only to 15 days after settlement when *Chaetoceros* is added to the diet.
[d]Diet composed of 1:1 mixture of *Isochrysis* (T-Iso) and *Pavlova lutheri*.

research to determine optimal rations of micro-algae for pearl oyster larvae of different ages is scarce. Research with *Pinctada margaritifera* larvae assessed six different rations (1, 2, 5, 10, 20 and 30 $\times$ 10^3 cells/mL) of a 1:1 mixture of *Isochrysis* (T-Iso) and *Pavlova salina* (Doroudi *et al.*, 1999b). Growth rates of D-stage larvae fed micro-algae at rations greater than 5 $\times$ 10^3 cells/mL were significantly greater than those of larvae fed lower rations. Greatest growth rate was shown by larvae fed a ration of 20 $\times$ 10^3 cells/mL, while maximal survival was shown by larvae receiving 10 $\times$ 10^3 cells/mL (Doroudi *et al.*, 1999b). Doroudi *et al.* (1999b) stated that optimal algal ration is likely to vary according to larval age and also noted the likely relationship between algal density and larval density.

In follow-up research, Doroudi and Southgate (2000) assessed the influence of algal ration and larval density on growth and survival of *P. margaritifera* larvae. One-day-old D-stage larvae were cultured at initial densities of 1, 2 and 5/mL and fed algal rations of 0, 2.5, 5 or 12.5 $\times$ 10^3 cells/mL, while in a second experiment, 13-day-old larvae were cultured at the same densities but fed algal rations of 0, 7.5, 15 and 37.5 $\times$ 10^3 cells/mL. D-stage larvae showed high survival when cultured at the density of 3/mL and fed a ration of 4.5–11.5 $\times$ 10^3 cells/mL; greatest growth rate was also shown by larvae fed within this range. Older larvae showed the greatest growth rate within an algal ration range of 15–32 $\times$ 10^3 cells/mL; however, greatest survival occurred at an algal ration of <2.5 $\times$ 10^3 cells/mL which supported poor growth rates. Doroudi and Southgate (2000) recommended that *P. margaritifera* larvae should be fed at an algal ration of around 8 $\times$ 10^3 cells/mL and cultured at an initial density of 3 larvae/mL up to 8 days of age, while older (13–20 day) larvae should be fed an algal ration of around 25 $\times$ 10^3 cells/mL and cultured at a density of <2 larvae/mL.

Given the relatively high cost and technical requirements of micro-algae culture (Southgate, 2003) some effort has been made over recent years to determine the nutritional value of alternatives to live micro-algae for pearl oyster larvae. Southgate *et al.* (1998b) reported that commercially available dried *Tetraselmis* could be used to replace 50% of a standard diet of live micro-algae composed of an equal mixture of *Isochrysis* (T-Iso), *Pavlova salina* and *Chaetoceros simplex* when fed to *Pinctada margaritifera* larvae. Indeed, a diet composed of a 1:1 mixture of dried *Tetraselmis* and the ternary algal diet supported greater growth and survival than the ternary algal diet alone. The potential for dried *Tetraselmis* to replace live micro-algae as a food for *P. margaritifera* larvae was further investigated by Doroudi *et al.* (2002) who showed that a 25% substitution of live micro-algae with dried *Tetraselmis* is possible without affecting larval growth rates, when larvae are less than 150 μm in size (the live algae were an equal mixture of T-Iso and *Pavlova salina*). However, dried *Tetraselmis* was of high nutritional value for older *Pinctada margaritifera* larvae, substituting up to 75% of the live micro-algal diet without affecting larval growth.

A commercially available yeast-based diet has also been assessed for its nutritional value for 15-day-old umbone larvae of *P. margaritifera*. Although larvae fed the yeast-based diet showed similar survival to those fed live micro-algae, their growth rate was comparitively poor (Southgate *et al.*, 1998b).

7.4.6.6. Stocking density

The density at which pearl oyster larvae are cultured has a significant impact on food availability and may influence water quality. Chellam *et al.* (1991) noted that density plays a significant role in the growth of pearl oyster larvae and at higher densities growth and spatfall are reduced. A range of initial stocking densities have been reported for pearl oyster larvae such as 1–2/mL for *Pinctada margaritifera* (Alagarswami *et al.*, 1989; Southgate and Beer, 1997), 3/mL for *P. mazatlanica* (Martínez-Fernández *et al.*, 2003) and 1–8/mL for *P. maxima* (Rose and Baker, 1994). More specific information for *P. maxima* was provided by Rose (1990) who recommended that initial larval stocking density should be no more than 5/mL and this should be reduced to 2/mL by day 10 and to 1/mL by day 14. It was further recommended that larval stocking density should be adjusted at water change (Rose, 1990).

As outlined in Section 7.4.6.5, larval stocking density is an important factor influencing the optimal micro-algal ration fed to pearl oyster larvae, and there is clear interaction between these factors (Doroudi and Southgate, 2000). On the basis of research that assessed the combined effects of larval density and algal ration, Doroudi and Southgate (2000) recommended an initial stocking density for *P. margaritifera* larvae of 3/mL and an algal ration of around 8×10^3 cells/mL; older (13–20 day) larvae should be cultured at a density of <2 larvae/mL and fed an ration of around 25×10^3 cells/mL.

7.4.6.7. Settlement

Generalized larval development for *Pinctada* spp. is shown in Fig. 7.6 and is described in detail in Section 5.7.2 and Table 5.5. The proportion of *P. maxima* larvae at each stage of development over time is shown in Table 7.5. Larvae of the three major commercial pearl oyster species (*P. maxima, P. margaritifera* and the Akoya pearl oyster) settle and metamorphose at approximately the same age (20–23 days after fertilization) and size (230–266 μm) (Alagarswami *et al.*, 1989; Rose and Baker, 1994) although this may vary according to culture conditions.

Once the larvae have developed a prominent eye-spot (see Section 5.7.2; Fig. 5.12(d)), they are competent to metamorphose. Pearl oyster larvae are generally retained on an appropriately sized sieve mesh to separate larger metamorphically competent larvae from their less developed siblings. *P. maxima* larvae, for example, are collected on a 140 μm mesh sieve (with a diagonal pore size of 198 μm) when the larvae measure 200 μm or greater (Rose, 1990). Similarly, *P. margaritifera* larvae are passed through a 150 μm sieve; larvae that are retained on the sieve are transferred to settlement tanks, those that pass through the sieve are returned to larval culture tanks for further development (Southgate and Beer, 1997). This process may be repeated a number of times.

Settlement and metamorphosis in pearl oysters is asynchronous and occurs over a number of days. Rose and Baker (1994) reported *P. maxima* pediveligers to begin metamorphosis into post-larvae on day 24 (Table 7.5). By day 25 metamorphosing post-larvae or "plantigrades" accounted for 22% of the population with an antero-posterior

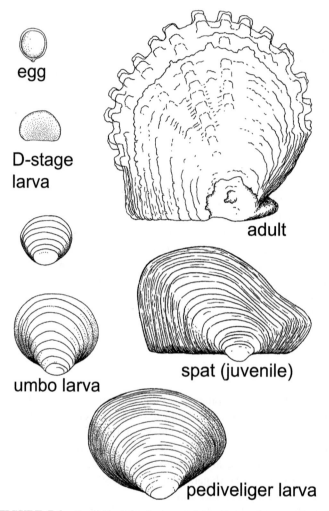

FIGURE 7.6 Generalized development scheme for pearl oysters from fertilized egg through larval development to adult.

measurement (APM) of 230–330 μm. Rudimentary gill filaments were visible by day 25. By 28 days after fertilization, 5% of the cultured population were still umbonal veligers, 60% were pediveligers, 14% were plantigrades and 23% were newly settled spat (Table 7.5). Newly settled spat averaged 349 μm APM and by this stage they were bysally attached to the substratum. Gill filaments could be very clearly seen (Fig. 5.12(f)) and increase in shell mass was evident as a transparent dissoconch (Fig. 5.13(b)). Settlement was completed by 35 days after fertilization.

Settlement tanks for pearl oysters generally contain an appropriate settlement substrate such as plastic slats within pocket nets (Taylor *et al.*, 1998a) or shade-cloth

TABLE 7.5

Proportions of *Pinctada maxima* larvae at various stages of development during hatchery culture.

Day	Stage of development						
	D-stage veliger	Early umbone	Umbonal veliger	Pediveliger	Eyespot	Plantigrade	Settled spat
1	100%	–	–	–	–	–	–
9	20–30%	70–80%	–	–	–	–	–
10			Present	–	–	–	–
12–14				Present	–	–	–
16	0%				–	–	–
17	–			<5%	–	–	–
19	–		85–90%		–	–	–
20	–				Present	–	–
22–24	–			70%		Present	–
25	–					22%	–
28	–		5%	60%		14%	23%
35	–	–	–	–	–	–	100%

Source: Rose and Baker (1994).

housed in mesh bags similar to spat collector (Southgate and Beer, 1997), that are hung into settlement tanks (Fig. 7.7). Settlement tanks are generally aerated more vigorously than larval culture tanks to facilitate larval settlement onto collectors, not onto the tank bottom. Aeration must also be adequate enough to provide water circulation through spat collectors. Settlement of larvae onto the collectors normally occurs within a day or two when a flow-through water system can be established for settlement tanks (Southgate and Beer, 1997; Martínez-Fernández *et al.*, 2003). For *P. maxima*, the water in settlement tanks was filtered to 10 μm and was exchanged every 3 days once settlement had occurred (Rose, 1990).

Settlement of pearl oyster larvae on the sides and bottom of settlement tanks will occur in the absence of spat collectors. Spat can then be removed from tank surfaces with a thin piece of plastic film or by gentle brushing and will readily reattach to other substrates. Using this method, the density of spat on substrates transferred from the hatchery can be controlled (Taylor *et al.*, 1997b, e). However, spat mortality can result from high settlement to the tank bottom where feces and uneaten micro-algae also accumulate. Care should be taken not to overfeed. In the absence of spat collectors, the number of larvae placed into settlement tanks needs greater control to ensure sufficient surface area for settlement. Rose (1990) recommended 0.5×10^6–10×10^6 pediveligers per 1,000 L for *P. maxima*.

FIGURE 7.7   Spat collectors suspended into settlement tanks to collect *Pinctada maxima* larvae. Collectors are composed of powder-coated metal frame with polypropylene rope stretched horizontally between the frame (see Fig. 7.8). (Photo: Joseph Taylor, Atlas Pearls.)

Very little research has been conducted into factors influencing the success of the settlement process. Saucedo *et al.* (2005) determined the effectiveness of different types of settlement substrates for recruitment of *P. matzalantica* spat. Spat collectors were composed of an outer bag and inner substrate made of either onion bag, mosquito net, fishing net or shade-cloth. Recruitment of spat was influenced by collector material with fishing net and shade-cloth yielding the highest and lowest densities of spat, respectively. *P. matzalantica* spat also preferentially settled to collectors located in the middle of the water column (60–90 cm tank depth) and, when provided with collectors of different colors, showed preference for red and green substrates. Recruitment of *P. maxima* spat was greater on glass or plastic plates than on monofilament nylon and shade-cloth (Rose and Baker, 1994). In a later study, *P. maxima* larvae were reported to preferentially settle onto polypropylene rope and a combination of rope and PVC slat collectors when compared to monofilament nylon and PVC slat collectors alone (Taylor *et al.*, 1998a). The PVC slats used by Taylor *et al.* (1998a) had a concave and convex surface and *P. maxima* spat showed greater recruitment to the concave surface of the PVC slat and to slats positioned horizontally within the water column. PVC slats or polypropylene rope positioned horizontally within frames are now commonly used within the pearl oyster industry as a substrate for larval settlement (Fig. 7.8).

The presence of an epifloral bio-film has been shown to improve recruitment of bivalve spat to artificial substrates (Weiner *et al.*, 1989; O'Foighil *et al.*, 1990). This phenomenon was also shown to improve recruitment of *P. maxima* spat to PVC slat collectors (Taylor *et al.*, 1998a). It is widely acknowledged that such bio-films provide specific chemical cues that can induce settlement and metamorphosis of bivalve

FIGURE 7.8 Spat collectors with *Pinctada maxima* spat attached. (Photo: Joseph Taylor, Atlas Pearls.)

larvae. It has further been suggested that a diatomaceous bio-film may provide nutrition to bivalve larvae by means of "deposit feeding" (O'Foighil *et al.*, 1990). However, Taylor *et al.* (1998a) reported no significant difference in mean shell length between *P. maxima* spat settled onto conditioned PVC slats and those on non-conditioned slats. This indicates that any potential nutritional benefit associated with the epifloral film may be lost as the spat become competent filter feeders. Rose and Baker (1994) suggested that inappropriate conditioning of spat collectors may provide insufficient stimulus for settlement of *P. maxima* larvae resulting in extension of the settlement period.

Doroudi and Southgate (2002) showed that metamorphically competent larvae of *P. margaritifera* could be induced to settle in the presence of chemicals such as γ-aminobutyric acid (GABA). This chemically mediated settlement occurred in higher frequency in the presence of an appropriate substrate. In the presence of GABA the rate of settlement of *P. margaritifera* larvae increased from 14% in treatments with plastic substrate only to 25% in treatments with the plastic substrate and GABA (Doroudi and Southgate, 2002). This result implies that to optimize settlement of pearl oysters, both appropriate chemical and physical cues (i.e. substrate) are required.

The level of illumination in settlement tanks has been shown to influence recruitment patterns of bivalve larvae which show light avoidance behavior (Ritchie and Menzel, 1969). There is some evidence that illumination also influences settlement patterns of pearl oyster larvae. Alagarswami *et al.* (1987), for example, reported improved settlement of Akoya pearl oyster larvae in dark-colored tanks compared to lighter ones and Chellam *et al.* (1991) noted higher spatfall in black tanks compared to white and blue tanks. Rose and Baker (1994) speculated that light colored semi-transparent rearing tanks may have inhibited settlement of *P. maxima* larvae.

The culture density of metamorphically competent pearl oyster larvae has been shown to influence subsequent growth and survival. Taylor *et al.* (1998c) reported reduced growth of *Pinctada maxima* spat, with increasing density of larvae at settlement, resulting from increased competition for space and food. A setting density of no more than 1 larva/mL was recommended for *P. maxima* (Taylor *et al.*, 1998c).

Prior to settlement, the larvae of edible oysters may be separated from their culture water and collected into dense aggregations which, with appropriate conditions of moisture and reduced temperature, can be transported to distant hatcheries or farms where settlement takes place (Lucas, 2003). Successful cold storage transport and remote settlement of eyed *P. maxima* larvae has been undertaken in Indonesia, where larvae packed in 45 µm mesh and damp toweling were held in oxygen-filled plastic bags at 5–8°C for 28–32 h (Taylor, 2005). Larvae regained activity within 15–20 min of being resuspended in filtered seawater at their destination and survival through metamorphosis to 45 days ranged from 0.1% to 19% (Taylor, 2005).

7.4.6.8. Early nursery culture

Early nursery culture of pearl oysters incorporates the period between settlement and transfer of spat from the hatchery to ocean-based culture systems. After settlement, spat are generally maintained on collectors in settlement tanks until transfer. Following metamorphosis, settlement tanks are generally provided with a flow-through water supply with the water itself being filtered to 1 µm (Southgate and Beer, 1997). There has been limited study of the optimal conditions for pearl oyster spat during early nursery culture.

Rose (1990) reported on the culture of *Pinctada maxima* spat in downwellers and noted an effect of density on growth rates. Spat held at a density of 4 per 100 cm^2 showed an average growth rate of 9.6 mm/month while those held at a density of 25 per 100 cm^2 recorded a growth rate of 6 mm/month. Taylor *et al.* (1998c) similarly reported that the density of *P. maxima* spat on plastic slat collectors influenced subsequent growth and survival. Highest growth rates and survival were associated with lower stocking density indicating competition for space and food at higher densities (Taylor *et al.*, 1998c). To maximize growth and survival of *P. maxima* spat, the authors recommended a density of around 70 spat per 100 cm^2.

Like larval rearing tanks, settlement tanks must be provided with micro-algae as food. Feeding rates used for spat during early nursery culture are significantly higher than rations used for larval culture (Tables 7.4 and 7.6). Mills (2000) reported that growth of *P. maxima* spat was optimal at a temperature of 26–29°C and at an algal concentration of 54 cells/µL. However, survival was at its highest at low algal concentrations and exceptional spat growth was recorded at algal cell densities as low as 12 cells/µL. The author concluded that to maximize growth, *P. maxima* spat should be cultured within a temperature range of 26–29°C and at algal cell densities of 12–54 cells/µL (Mills, 2000). The nutrient composition of the micro-algae used to feed spat is important in determining growth rates. Taylor *et al.* (1997f) assessed the nutritional value of five species of

TABLE 7.6
Feeding schedule for post-larvae and spat of *Pinctada maxima*.

Day	Shell length (mm)	Algal density (cells/mL)		Percentage of ration		
		AM	PM	T-Iso	*Chaetoceros*	*Tetraselmis*
43		26,000	34,000	30	60	10
48	1.60	30,000	35,000	30	60	10
53		30,000	40,000	30	60	10
58		30,000	50,000	35	60	5
63		35,000	50,000	35	60	5
68	5.10	40,000	50,000	35	60	5
73		45,000	50,000	40	55	5
78		45,000	55,000	40	55	5
83		45,000	50,000	40	55	5
88		45,000	55,000	45	50	5
93		50,000	50,000	45	50	5
98	14.80	52,000	53,000	45	50	5
103		53,000	55,000	50	45	5
108		55,000	55,000	50	45	5
113		55,000	55,000	50	45	5
118		55,000	60,000	55	40	5
123		55,000	60,000	60	35	5
128		60,000	60,000	65	30	5
135	41.70	60,000	60,000	70	25	5
138		60,000	60,000	75	20	5

Source: Rose (1990).

micro-algae for 75-day-old *P. maxima* spat. They reported relatively poor nutritional value of *Chaeterecos calcitrans* and *Palova lutheri*, but relatively high nutritional value of *Chaetoceros muelleri*, *Tetraselmis suecica* and *Isochrysis* (T-Iso).

Survival of 45-day-old *P. maxima* spat did not differ when held for 20 days at salinities ranging from 45‰ to 25‰ (Taylor *et al.*, 2004). However, spat growth was significantly depressed at salinities of 45‰, 40‰ and 25‰, with best growth at 30‰. The authors concluded that *P. maxima* spat are tolerant to a broad range of salinities and that reduced salinity (34‰ was ambient) was beneficial to spat growth. Rapid change in salinity has been shown to be an effective means of inducing detachment of *P. maxima* spat from settlement substrata. Spat with a mean shell height of 3.3 mm showed the greatest rate of detachment from spat collectors (92.3%) following 1 h exposure to hypersaline (45‰) seawater (Taylor *et al.*, 1997c). Exposure of spat to 45‰ seawater for 24 h did not result in any mortality (Taylor *et al.*, 1997c). Change in pH of culture water also brought about significant detachment with 86% of spat detaching after exposure to pH4 (Taylor *et al.*, 1997c). O'Connor and Lawler (2004) reported that survival of Akoya juveniles (17 mm) held within a water temperature range of 14–26°C was higher at salinities of 32‰ and 35‰ although a high rate of mortality occurred at salinities of 23‰ or less. Mortality of Akoya juveniles was found to be highest at the extremes of water temperatures tested, 14°C and 26°C (O'Connor and Lawler, 2004).

7.4.6.9. Transfer from the hatchery

Conditions in pearl oyster hatcheries are optimized for growth and survival of spat. The water may be heated, it is filtered to remove predators and there is an ample food supply. However, this changes when spat are transferred to the field and efforts are often made to reduce stress during this period. For example, the degree of water filtration in culture tanks was gradually reduced from 10 μm to 20 μm, and then to 30 μm, prior to transfer of *Pinctada maxima* spat to exposed spat to natural phytoplankton and silt (Rose, 1990). Furthermore, in an effort to minimize mortality, spat are usually held until they attain a certain size before they are transferred to the ocean. Akoya pearl oyster spat must generally reach a shell height of at least 3 mm before transfer (Alagarswami *et al.*, 1987; Chellam *et al.*, 1991; Dharmaraj *et al.*, 1991). Dybdahl *et al.* (1990) waited 2–3 months after settlement before transferring 6 mm *P. maxima* spat, while Rose and Baker (1994) transferred spat of the same species 3–4 weeks after settlement. Spat of the blacklip pearl oyster have been transferred to the ocean 3–7 weeks after settlement (Southgate and Beer, 1997; Alagarswami *et al.*, 1989).

Given the highly regulated environment within the culture tanks of land-based nursery systems (i.e. adequate and reliable food supply and relatively high water temperature) it is reasonable to assume that spat held in the hatchery for a longer period of time may be larger and more resistant to predation and nutritive stress, than those transferred from the hatchery at a younger age. However, research with *P. margaritifera* showed that spat transferred from the hatchery 3 weeks after settlement had a significantly greater shell height at grading (3.5 month of age) than spat from the same cohort transferred from the hatchery 5, 7 and 9 weeks after settlement (Pit and Southgate, 2000). Spat transferred earlier from the hatchery also had a greater proportion in larger size classes at grading. The authors suggested that these results reflected superior nutrition in the ocean and they recommended that spat be transferred from the hatchery as soon as possible after settlement in order to maximize growth rates.

At some time following transfer of spat collectors to the ocean, juveniles need to be removed from them to maximize subsequent growth and survival. Two approaches may be applied in commercial pearl oyster culture. The spat collectors may be sequentially harvested with larger juveniles being selectively removed from spat collectors on multiple occasions and smaller spat retained on collectors (Taylor *et al.*, 1997e). Alternatively, spat collectors can be removed from the culture system and all spat removed and graded (Southgate and Beer, 1997). Grading is important as it allows spat to be cultured in units with the largest possible mesh size. This facilitates water flow and subsequent growth rate, but it is also likely to reduce fouling and recruitment of predators to culture units. Southgate and Beer (1997) graded *P. margaritifera* spat at 3 month of age through three mesh sizes (15, 10 and 5 mm). They retained 0.2%, 8.9% and 67.3% of the spat, respectively, with a further 23.6% of spat being less than 5 mm in size. Subsequent research showed that the proportion of spat in the larger size categories at grading was significantly increased when spat were transferred to the ocean at an earlier age (Pit and Southgate, 2000).

The threat of predators needs to be considered following transfer of pearl oyster spat to the ocean and measures are normally taken to protect them by placing collectors in a protective culture unit such as mesh trays or within fine mesh sleeves (Southgate and Beer, 1997; Taylor *et al.*, 1998b). One of the major constraints in early nursery culture is that fine mesh containers, which prevent the loss of juveniles, encourage fouling and limit water flow and food supply to the oysters. Taylor *et al.* (1998b) assessed the impact of covering small (1.8 mm) *P. maxima* spat on PVC slats with mesh sleeves of varying pore size (0.75, 1.5 or 3.0 mm). After 2 weeks, spat that were covered with mesh had significantly greater survival than those without protective mesh. There was a significant increase in shell size of the spat with each increase in mesh size, resulting in mean dorso-ventral measurements of 3.7, 2.3 and 1.6 mm, respectively, for spat held under 3.0, 1.5 and 0.75 mm mesh (Taylor *et al.*, 1998b). The authors concluded that covering plastic slat spat collectors with mesh greatly improved survival of *P. maxima* spat. Furthermore, there was no advantage in using a mesh size small enough to retain dislodged spat (which perished anyway) because it rapidly fouled (Taylor *et al.*, 1998b).

7.5. CULTURE INFRASTRUCTURE

Pearl oysters are sub-tidal animals that can generally tolerate only minor emersion. Culture apparatus must account for this and pearl oysters are generally farmed using suspended-culture systems such as rafts or longlines, or bottom culture systems such as fencelines, racks or tressles. Gervis and Sims (1992) listed the following criteria as the basis for the choosing between these different culture systems:

- current speed
- water depth
- capital cost
- operational costs
- exposure to wind and waves
- ease of operation
- need for direct access from land
- security considerations
- tidal variation.

7.5.1. Rafts

Rafts are rigid floating structures composed of a framework from which culture apparatus containing pearl oysters can be hung (Fig. 7.9). They are commonly made from

FIGURE 7.9   Raft used for pearl culture in Li'an Bay, Hainan Island, China.

bamboo, timber or plastic pipe. Buoyancy is provided by attached floatation devices such as barrels, plastic container, drums and floats. Rafts commonly have 4–6 floats of 200–300 L capacity and, as a "rule of thumb", each 200 L barrel can support around 200 kg (Quayle and Newkirk, 1989). The material from which rafts are made generally reflects what is locally available. Raft size varies but the traditional Japanese raft measures 6.4 × 5.5 m and has four 0.6 × 1.05 m styrofoam floats (Gervis and Sims, 1992); it provides 100 points for attachment of oyster culture units.

Rafts may be moored to fixed structures such as jetties or anchored independently to maintain position. The former provides convenient access to culture stock and negates the need for a boat. Rafts are more commonly deployed in sheltered areas with relatively low wind and wave action where a number of rafts may be moored together. The relatively low-cost and simple construction of rafts are advantageous, and the work platform that they provide (Fig. 7.10) allows culture stock to be inspected and accessed without the use of divers. Perhaps the major disadvantage of rafts is the relatively close proximity of culture units holding pearls oysters which has implications relating to water flow and water quality, food availability and the spread of disease (Gervis and Sims, 1992).

7.5.2. Longlines

Longlines are composed of a large-gauge rope, or headline, supported by floats, that is anchored at either end to maintain position (Fig. 7.11). The headline provides an attachment point for "dropper" ropes to which culture apparatus containing pearl oysters are attached. Longlines are better suited to wind and wave exposure than rafts and their smoother movement under such conditions results in less wear and tear on the anchoring

FIGURE 7.10   Rafts in Li'an Bay, Hainan Island, China, provide a work platform for care and maintenance of pearl oysters (e.g. cleaning and grading) and cleaning, repair and storage of culture equipment.

FIGURE 7.11   Longline used for *Pinctada maxima* culture in northern Australia. Note the panel nets containing pearl oyster juveniles suspended from the longline. (Photo: Clare Flanagan.)

system and reduced impact on oyster growth. Longlines may be constructed to float on the ocean's surface ("surface longline") or may be deployed in mid-water ("sub-surface longline").

Surface longlines are extensively used for the culture of *Pinctada fucata* and *Pteria penguin* in Japan and for *Pinctada maxima* in Australia and Indonesia (Figs. 7.11 and 14.3). In Australia, surface longlines are generally 100 m long and deployed parallel to each other at 20–30 m intervals to avoid entanglement with adjacent lines (Wells and Jernakoff, 2006). Floats are placed at 1–2 m intervals and dropper ropes supporting culture apparatus (see Section 7.5.4) are attached at 1 m intervals (Fig. 7.11).

Both rafts and surface longlines provide a floating substrate from which culture units containing pearl oysters can be attached. However, potential problems associated with their use include surface movement that may affect the growth rate of pearl oysters, potential navigational hazard and potential security issues because of their visibility. Sub-surface longlines are less prone to these problems; however, they are more difficult to deploy and rely on scuba divers for regular maintenance and inspection. Sub-surface longlines are widely used in French Polynesia and the Cook Islands for culture of *P. margaritifera.*

7.5.3. Racks, tressles and fencelines

Racks or tressles are composed of a rigid framework fixed above the seabed using legs or other supports. They can be made from timber poles, or metal or plastic pipes to provide a platform onto which culture units containing pearl oysters can be placed or hung. Tressles are relatively low cost and simple to construct, require low maintenance and, being on the bottom, are less prone to adverse weather conditions and security risks. However, they are only assessable using SCUBA and cleaning is more difficult and labor intensive than with surface culture apparatus. Tressles have been used for pearl oyster culture in Australia, Sudan and Mexico and are the traditional culture system for culture of *Pinctada margaritifera* in French Polynesia and the Cook Islands (Coeroli *et al.*, 1984). In response to disease and oyster mortality resulting from overcrowding and poor water quality, tressles have been replaced by sub-surface longlines in Polynesia (Gervis and Sims, 1992) resulting in reduced oyster densities and improved water circulation (Sims, 1993a).

Fencelines utilize strong galvanized metal posts driven into the substrate, as supports for a line which is raised above the seabed. Culture units containing pearl oysters are hung from the line. Fencelines are generally used in relatively exposed sites in Western Australia (Wells and Jernakoff, 2006) and, while the capital cost of a fenceline is less than that of a surface longline, like tressles, they are more difficult to deploy and rely on divers for regular maintenance and inspection.

7.5.4. Culture units and methods

Various culture apparatus are used for ocean-based culture of pearl oysters. Most commonly used are pearl nets, panel nets, lantern nets and box nets. Nets are generally

FIGURE 7.12 Pearl nets used for culture of *Pinctada margaritifera* juveniles in the Cook Islands. (Photo: John Lucas.)

constructed from galvanized or, more commonly, plastic or powder-coated metal frames covered with polyethylene netting of varying mesh size appropriate for the size of cultured oysters. Pearl nets (Fig. 7.12) and lantern nets are commonly used to hold juvenile pearl oysters in large numbers, prior to their deployment to larger culture units (Gaytan-Mondragon *et al.*, 1993). Very often, they are used to hold oysters following their removal from spat collectors at grading. Panel nets, also known as pocket nets (Fig. 7.13), are composed of a strong frame covered with mesh that is sewn to form pockets to hold oysters. The size of the pocket and the size of the mesh can be altered, depending on species and the size of the cultured oysters. For example, panel nets used for small juveniles may have 48 pockets and be composed of 5 mm mesh, while those used for adult *Pinctada maxima* are likely to have six pockets made from 30–45 mm mesh. Clearly, the mesh size needs to be as large as possible to maximize water flow and food supply and minimize fouling.

Perforated plastic trays and baskets are also used for holding pearl oyster spat and juveniles. Plastic trays (55 × 30 × 10 cm) with lids have been used to hold spat collectors containing *P. margaritifera* following transfer from the hatchery until grading (Southgate and Beer, 1997) and have been used during nursery culture of *P. margaritifera* (Friedman and Southgate, 1999a; Southgate and Beer, 2000; Whitford, 2002).

FIGURE 7.13   Panel nets containing *Pinctada maxima* being inspected by a diver at a pearl farm in north Bali, Indonesia. (Photo: Joseph Taylor, Atlas Pearls.)

7.5.4.1. Ear-hanging

Pearl oysters are commonly cultured using the "ear-hanging" method in Polynesia (Fig. 7.14). It involves drilling a small hole (~2–3 mm) through the base of the shell in the dorsal-posterior region. Monofilament fishing line or wire can then be used to attach oysters to a rope that is itself attached to either a raft or longline. A number of oysters are usually attached to a single rope (Fig. 7.14) which is generally referred to as a chaplet (Friedman and Southgate, 1999b). Oysters can be attached singly or in pairs. The advantages of ear-hanging include greatly reduced outlay for culture equipment and, given that the cultured animals are not enclosed within a culture unit, there is greater flow of water and improved food availability. However, without any protective housing, they are prone to predation.

7.6. NURSERY CULTURE AND GROW-OUT

The longer established pearl culture industries in Japan, Australia and French Polynesia traditionally collect oysters from the wild, either as adults or relatively large juveniles.

FIGURE 7.14   *Pinctada margaritifera* attached to chaplets at a pearl farm in Takapoto Atoll, French Polynesia. (Photo: John Lucas.)

In Polynesia, for example, juveniles are usually held on spat collectors until they reach a size of around 60–90 mm (AQUACOP, 1982; Preston, 1990; Pouvreau and Prasil, 2001). Because of this, early nursery culture of pearl oysters, focusing on small spat and juveniles, had received little research attention until relatively recently. Increasing emphasis on hatchery production of pearl oysters (Gervis and Sims, 1992) and development of spat collection programs outside Polynesia where spat are removed from collectors at a much smaller size (e.g. Monteforte and García-Gasca, 1994; Friedman *et al.*, 1998), provided impetus for research to optimize nursery and juvenile culture methods for pearl oysters. The suitability of various culture units has been a particular emphasis of this research.

Growth rates of *Pinctada mazatlanica* juveniles (initial shell height 32.6 mm) over a 22-month period did not vary significantly when cultured in either lantern nets or pocket nets in suspended culture at a depth of 10 m, or in plastic cages attached to an iron platform at a depth of 10 m (Gaytan-Mondragon *et al.*, 1993). There were differences in survival, however, between oysters in cages (99%) and those in suspended culture equipment (65%). The same study reported significantly greater growth rates of *Pteria sterna* juveniles (initial shell height 37.8 mm) cultured for 18 months in pocket nets compared to those in cages (Gaytan-Mondragon *et al.*, 1993).

Friedman and Southgate (1999a) reported on a study in which lantern nets, panel nets and plastic mesh trays were assessed for early nursery culture of *P. margaritifera* in the Solomon Islands. Growth of small juveniles (16 mm) over a 3-month period was significantly greater in lantern nets than in panel nets; however, there was no differences in growth rates between culture units for two larger size classes of juveniles (24 and 33 mm) over the same period. Survival of small juveniles was greater in panel nets although, again, no significant differences in survival between culture units were evident for the two larger size classes (Friedman and Southgate, 1999a). Juvenile *P. margaritifera* (24 mm) held in panel nets showed significantly greater growth rates than those held in mesh trays for 3 months, but in a subsequent experiment, smaller *P. margaritifera* spat (11 mm) showed no significant differences in growth rate, wet weight or survival when held in the same culture units for 5 months (Friedman and Southgate, 1999a). Growth and wet weight of *P. margaritifera* juveniles were significantly greater in mesh trays when oysters were glued to the inner surface of the trays to prevent aggregation (Friedman and Southgate, 1999a).

A number of studies have reported on the tendency of pearl oyster juveniles to aggregate and form clumps (e.g. Crossland, 1957; Friedman and Southgate, 1999a; Southgate and Beer, 1997, 2000). Oysters grown under such conditions show reduced growth rates, shell deformity, and a higher proportion of smaller oysters compared to oysters grown in similar culture units where contact between individuals was prevented (Southgate and Beer, 1997; Taylor *et al.*, 1997b; Friedman and Southgate, 1999a; Southgate and Beer, 2000). The innermost individuals of these aggregations may become stunted or die (Crossland, 1957). Predictably then, culture units which hold individual oysters separately generally support greater growth rates. Maximizing growth rates of pearl oysters reduces the time required for them to reach operable size for pearl production. Furthermore, a narrower size range ensures that a greater proportion of oysters within a given cohort will reach operable size at a given time, thereby maximizing financial returns (Taylor *et al.*, 1997a; Southgate and Beer, 2000).

Friedman and Southgate (1999b) evaluated survival and growth rates of *P. margaritifera* juveniles (58–78 mm) cultured on chaplets in the Solomon Islands. They reported high survival of around 87% after 12 months in culture, and an annual growth rate of 60–64 mm, which compared favorably with those of *P. margaritifera* cultured on chaplets in Polynesia (Sims, 1993b; Coeroli *et al.*, 1984). Higher nutrient loads in the waters of the Solomon Islands compared to the atolls of Polynesia resulted in fouling by algae, and growth rates of oysters were significantly improved when algae were removed by brushing every 3–4 week (Friedman and Southgate, 1999b). Ear-hanging and culture in 24-pocket panel nets were superior to three other culture methods assessed for 8-month-old *P. margaritifera* juveniles in a 5-month study in Australia (Beer and Southgate, 2000). The small mesh size (5 mm) of some of the poorer culture units tested promoted rapid fouling which presumably limited food availability.

Friedman and Southgate (1999a) noted that pearl oyster growth depends on the method used to culture them; that is, oysters held under different conditions grow at different rates. While growth and survival are the major factors when assessing appropriate culture methods for pearl oysters, choice of culture method must also consider economic and practical factors, such as equipment cost, ease of construction, degree of fouling and, in developing countries, materials that are locally available (Southgate and

Beer, 2000). Furthermore, the criteria used to assess the suitability of a given culture unit may vary between sites depending on local conditions such as predator composition and abundance, and the degree and type of fouling (Southgate and Beer, 2000).

7.6.1. Stocking density

Taylor *et al.* (1997e) determined the effects of stocking density on growth rate and survival of 63-day-old ($\sim$6.2 mm) *Pinctada maxima* juveniles in suspended culture at a depth of 3 m. Oysters were held for 6 weeks on PVC slats within 3 mm mesh sleeves. Juveniles held at the lowest stocking density tested, 1.3/100 cm^2, had significantly greater shell length, shell height and wet weight than those at higher densities (6.7, 13.3 and 20 juveniles/100 cm^2) and juveniles were significantly smaller with each increase in stocking density from 6.7 to 20/100 cm^2 (Taylor *et al.*, 1997e). The authors also reported an increase in the incidence of growth deformity, expressed as a higher ratio of dorso-ventral height to antero-posterior length, with increasing stocking density, with more than 25% of juveniles held at a density of 20/100 cm^2 showing deformity. Juvenile survival was also significantly greater at a density of 1.3/100 cm^2 compared to other treatments (Taylor *et al.*, 1997e).

Panel nets with either 48 or 24 pockets were used to determine the effects of stocking density on growth and survival of larger (30 mm) *P. maxima* juveniles held for more than 5 weeks in either suspended culture at a depth of 2.5 m, or on the bottom at a depth of 20 m (Taylor *et al.*, 1997b). Survival of oyster held on the bottom was poor and their growth was significantly lower than that of oysters held in similar culture units in suspended culture. *P. maxima* held in 28 pocket panel nets (66 oysters/m^2) grew significantly larger than those held in 48 pocket panel nets (99 oysters/m^2) in both suspended and bottom culture (Taylor *et al.*, 1997b). The authors concluded that culture system had a greater influence on growth and survival than stocking density and this was influenced by reduced food availability at the bottom.

7.6.2. Control of fouling

The sub-tidal nature of pearl oyster culture facilitates fouling of oysters and culture equipment by a wide array of fauna and flora. Biofouling and its control are covered in detail in Chapter 15. However, given that biofouling control represents a major component of the activities undertaken by pearl farms, some of the basic methods used for this are outlined here.

Fouling causes a number of potential problems in pearl oyster culture including:

* interference with water flow-through culture units (and food availability);
* competition for space and food by filter-feeding fouling organisms;
* reduced growth rates of pearl oysters and possible deformities to oyster shells which influence their use for pearl production;
* increased weight and drag of culture equipment; and
* provision of substrate for further fouling and recruitment of predators.

Removal of fouling from culture stock and culture equipment is a routine, labor intensive and costly part of pearl farming which makes up around 25–30% of the operating costs (Crossland, 1957; Lewis 1994). Clearly, there is economic benefit in minimizing the frequency with which culture apparatus are cleaned and this frequency is likely to vary between latitudes, sites and seasons (see Section 15.4.2).

In West Papua, Indonesia, Taylor *et al.* (1997d) determined the effects of cleaning frequency on the degree of fouling and growth rates of *Pinctada maxima*. At the end of a 16-week period, oyster wet weight and shell length was significantly greater for oysters cleaned every 2 or 4 weeks compared to those cleaned after 8 or 16 weeks. Furthermore, a number of the latter showed shell deformities caused by invasion of the shell margin by other bivalves (particularly *Pteria* spp.). The authors concluded that oysters should be cleaned on a monthly basis to maximize growth rate and that more regular cleaning was unnecessary and would increase operational costs. The most common fouling organisms at this site in Indonesia were barnacles, other bivalves (*Pinctada* spp., *Pteria* spp., *Crassostrea* spp.) and polychaete worms.

Methods of fouling control at pearl farms are outlined in Section 15.4.2 and can vary greatly in their degree of mechanization. In China, for example, teams of around 6–8 (usually young female) workers pull pearl culture nets into small wooden boats and manually remove fouling from them using brushes and knives (Fig. 15.8). In Indonesia, floating pontoons that also function as work platforms, are pulled along longlines and the nets suspended from them are pulled onto the platform where they are cleaned using water jets (Fig. 14.3); hard-fouling may also be removed using knives. In Australia, a greater degree of mechanization is employed through the use of dedicated cleaning boats (Figs. 15.5 and 15.6). These boats have a mechanized pully system onto which a longline is fitted; as the boat moves forward, the longline is pulled clear of the water and the nets suspended from it are brought up onto a cleaning platform within the boat where they are cleaned with high-pressure water jets (Fig. 15.6).

7.6.3. Grow-out of seeded stock

Pearl oysters are seeded or grafted for pearl production using the methods outlined in Section 8.9.4. The days and weeks following pearl seeding are critical and special care of seeded oysters is required to maximize pearl yield (by minimizing nucleus rejection) and pearl quality. Great care is taken to minimize stress and oysters are held in calm conditions with minimum current and wave activity and appropriate and constants water quality parameters.

In Australia, seeded *Pinctada maxima* are returned to the ocean and generally placed in a holding area for a period of 7–8 days to facilitate recovery (Wells and Jernakoff, 2006). In Japan, newly operated Akoya pearl oysters are held in deep, calm waters for a period of 2–3 weeks to allow full recuperation, while in India, where coastal waters may be rough, they are generally held under controlled laboratory conditions for the period immediately following pearl grafting (Chellam *et al.*, 1991); they are then transferred to depths greater than 5 m in areas of high phytoplankton productivity. Following recuperation, seeded oysters in Japan may be examined using fluoroscopy to check for nucleus retention and to determine the position of the nucleus. Similarly, in Australia and French Polynesia seeded oysters are often checked for nucleus retention using X-ray.

In the Western Australian *P. maxima* industry, a "turning" program is initiated which is thought to assist in the development of a pearl-sac around the nucleus and in nucleus retention (see Sections 8.9.6). Oysters are laid horizontally for the first few days after seeding and are then turned so that they lie on the opposite valve. The frequency of turning varies between farms, but generally it is at 2–5 day intervals with frequency decreasing over time. The turning process lasts for 6–10 weeks and is assumed to increase the likelihood of obtaining perfectly spherical pearls. Following recuperation, seeded oysters can then be cultured along side the main farm stock. The assumed benefits of this process are, however, questionable (Gervis and Sims, 1992).

7.7. SUMMARY

The major pearl culture industries have traditionally relied on oysters collected from the wild for pearl nucleus insertion. Adult pearl oysters are collected or, alternatively, spat are obtained using various kinds of spat collectors in the field and then on-grown to a size suitable for nucleus insertion. The relatively recent development of hatchery culture methods for pearl oysters has reduced industry reliance on oysters taken from the wild. The methods used in hatchery culture for broodstock conditioning, spawning induction, fertilization, egg incubation, larval culture and inducing settlement are similar to those typically used in bivalve hatcheries. Larval culture is the most critical phase, with issues of water quality, water exchange, culling, stocking density, nutrition and food ration. Microalgae differ in their nutritional value for larvae and a mixture of several species is commonly used to provide a better balance of nutrients for larval development. Settlement is influenced by factors such as shape and surface properties of the substrate, its height in the water column, illumination and settlement inducing chemicals such as GABA. After settlement, the spat are retained for a further period of hatchery culture with temperature control and feeding with cultured microalgae to optimize growth and survival. The small juvenile oysters are then transferred to the field. Culture units enclosed in fine mesh are used to retain the juveniles and protect them from predators; however, the largest effective mesh size is used because of heavy fouling of fine meshes.

Various culture structures are used in pearl oyster farms: surface rafts and longlines; subsurface longlines; benthic racks, trestles and fencelines. The benthic structures are less commonly used. The oysters are cultured in units suspended from the structure. Most commonly the units consist of netting over frames, but in Polynesia the oysters are "ear-hung" by tying them to a vertical rope through a hole drilled in their hinge. Growth and survival are substantially influenced by the culture structure and culture units used. Small mesh size, fouling, high stocking density and aggregation of small oysters reduce water flow and hence growth rate.

Development of hatchery culture methods provides the industry with opportunities to develop selective breeding programs and develop techniques commonly used in other aquaculture industries, such as triploidy (see Section 12.4) and cryopreservation. There are unlikely to be major changes in the methods used for husbandry of juvenile and adult pearl oysters. However, hatchery selection programs and ploidy manipulation may result in significant improvements in the growth rates of pearl oysters and the quality of culture stock.

References

Achari, G.P.K., 1980. New designs of spat collectors breeding hapas, cages and improved technology of pearl farming. *Symposium on Coastal Aquaculture. Part 2: Molluscan culture.* January 12–18, 1980, Symp. Ser. Mar. Biol. Assoc, India *6*, p. 704 (Abstract).

Acosta-Salmón, H., Martínez-Fernández, E., Southgate, P.C., 2004. A new approach to pearl oyster broodstock selection: can saibo donors be used as future broodstock? *Aquaculture* 231, 205–214.

Acosta-Salmon, H., Southgate, P.C., 2005. Histological changes in the gonad of the blacklip pearl oyster (*Pinctada margaritifera* Linnaeus, 1758) during the reproductive season at two sites in north Queensland, Australia. *Molluscan Res.* 25, 71–74.

Acosta-Salmon, H., Jerry, D.J., Southgate, P.C., 2007. Effects of cryoprotectant agents and freezing protocol on motility of black-lip pearl oyster (*Pinctada margaritifera* L.) spermatozoa. *Cryobiology* 54, 13–18.

Alagarswami, K., Dharmaraj, S., Velayudhan, T.S., Chellam, A. Victor, A.C.C, 1983a. On the controlled spawning of the Indian pearl oyster *Pinctada fucata* (Gould). *Symposium on Coastal Aquaculture. Part 2: Molluscan Culture.* January 12–18, 1980, Symp. Ser. Mar. Biol. Assoc. India *6*, pp. 590–597.

Alagarswami, K., Dharmaraj, S., Velayudhan, T.S., Chellam, A., Victor, A.C.C., Gandhi, A.D., 1983b. Larval rearing and production of spat of pearl oyster *Pinctada fucata* (Gould). *Aquaculture* 34, 287–301.

Alagarswami, K., Dharmaraj, S., Velayudhan, T.S., Chellam, A., 1987. Hatchery technology for pearl oyster production. *Bull. Cent. Mar. Fish. Res. Inst.* 39, 62–71. CMFRI, Cochin, India.

Alagarswami, K., Dharmaraj, S., Chellam, A., Velayudhan, T.S., 1989. Larval and juvenile rearing of black-lip pearl oyster, *Pinctada margaritifera* (Linnaeus). *Aquaculture* 76, 43–56.

AQUACOP. 1982. French Polynesia – Country Report. In: Davy, F.B., Graham, M. (Eds.), *Bivalve culture in the Asia and the Pacific. Proceedings from a workshop*, February 16–19, 1982. Singapore. International Development Research Centre, Ottawa, pp. 31–33.

Beer, A.C., 1999. Larval culture, spat collection and juvenile growth of the winged pearl oyster, *Pteria penguin*. World Aquaculture '99. *The Annual International Conference and Exposition of the World Aquaculture Society*, April 26–May 2, 1999, Sydney, Australia. Book of Abstracts, p. 63.

Beer, A.C., Southgate, P.C., 2000. Collection of pearl oyster (Family Pteriidae) spat at Orpheus Island, Great Barrier Reef (Australia). *J. Shellfish Res.* 19, 821–826.

Bougrier, S., Rabenomanana, L.D., 1986. Cryopreservation of spermatozoa of the Japanese oyster *Crassostrea gigas*. *Aquaculture* 58, 277–280.

Brown, M.R., Jeffrey, S.W., 1992. The nutritional properties of microalgae used in mariculture: an overview. In: Allen, G.L., Dall (Eds.). *Proc. Aquaculture Nutrition Workshop*, April 15–17, 1991, Salamander Bay. NSW Fisheries, Brackish Water Fish Culture Research Station, Salamander Bay, Australia, pp. 174–179.

Chellam, A., Victor, A.C.C., Dharmaraj, S., Velayudhan, T.S., Satyanaryana Rao, T.S., 1991. *Pearl Oyster Farming and Pearl Culture. Training Manual 8.* Central Marine Fisheries Research Institute, Tuticorin. 104 pp.

Coeroli, M., De Gaillande, D., Landret, J.P.AQUACOP, 1984. Recent innovations in cultivation of molluscs in French Polynesia. *Aquaculture* 39, 45–67.

Coutteau, P., Hadley, N.H., Manzi, J.J., Sorgeloos, P., 1994. Effect of algal ration and substitution of algae by manipulated yeast diets on the growth of juvenile *Mercenaria mercenaria*. *Aquaculture* 120, 135–150.

Crossland, C., 1957. The cultivation of the mother of pearl oyster in the Red Sea. *Aust. J. Mar. Freshwat. Res.* 8, 111–130.

Dharmaraj, S., Velayudhan, T.S., Chellam, A., Victor, A.C.C., Gopinathan, G.S., 1991. *Hatchery production of pearl oyster spat, Pinctada fucata*. Central Marine Fisheries Research Institute, Cochin. Special Publication No. 49. CMFRI

Doroudi, M., Southgate, P.C., 2000. The influence of algal ration and larval density on growth and survival of blacklip pearl oyster (*Pinctada margaritifera* L.) larvae. *Aquaculture Res.* 31, 621–626.

Doroudi, M., Southgate, P.C., 2002. The effect of chemical cues on settlement behaviour of blacklip pearl oyster (*Pinctada margaritifera*) larvae. *Aquaculture* 209, 117–124.

Doroudi, M., Southgate, P.C., 2003. Embryonic and larval development of *Pinctada margaritifera* (Linnaeus, 1758). *Molluscan Res.* 23, 101–107.

Doroudi, M., Southgate, P.C., Mayer, R., 1999a. The combined effects of temperature and salinity on embryos and larvae of the black-lip pearl oyster, *Pinctada margaritifera* (L.). *Aquaculture Res.* 30, 1–7.

Doroudi, M., Southgate, P.C., Mayer, R., 1999b. Growth and survival of blacklip pearl oyster (*Pinctada margaritifera* L.) larvae fed different densities of micro-algae. *Aquaculture Int.* 7, 179–187.

Doroudi, M., Southgate, P.C., Mayer, R., 2002. Evaluation of partial substitution of live algae with dried *Tetraselmis* for larval rearing of black-lip pearl oyster, *Pinctada margaritifera*, (L.). *Aquaculture Int.* 10, 265–277.

Doroudi, M., Southgate, P.C., Lucas, J.S., 2003. Variation in clearance and ingestion rates by larvae of black-lip pearl oyster (*Pinctada margaritifera* L.) feeding on various micro-algae. *Aquaculture Nutr.* 9, 11–16.

Dybdahl, R., Harders, S., Nicholson, C., 1990. Developing on-growing techniques and disease prevention husbandry of pearl oysters in Western Australia (FIRTA Project 87/81) and On-growing mariculture techniques for the pearl oyster *Pinctada maxima* spat in Western Australia. FIRDTF Project 89/60. Final Report. Fisheries Department of Western Australia, 57 pp.

Friedman, K.J., Southgate, P.C., 1999a. Growout of blacklip pearl oysters, *Pinctada margaritifera* collected as wild spat in Solomon Islands. *J. Shellfish Res.* 18, 159–168.

Friedman, K.J., Southgate, P.C., 1999b. Grow-out of blacklip pearl oysters, *Pinctada margaritifera* (Linnaeus, 1758) on chaplets in suspended culture in Solomon Islands. *J. Shellfish Res.* 18, 451–458.

Friedman, K.J., Bell, J.D., Tiroba, G., 1998. Availability of wild spat of the blacklip pearl oyster, *Pinctada margaritifera*, from 'open' reef systems in Solomon Islands. *Aquaculture* 167, 283–299.

Gallager, S.M., Mann, R., Sasaki, G.C., 1986. Lipid as an index of growth and viability in three species of bivalve larvae. *Aquaculture* 56, 81–103.

Gaytan-Mondragon, I., Caceras-Martinez, C., Tobias-Sanchez, M., 1993. Growth of the pearl oysters *Pinctada mazatlanica* and *Pteria sterna* in different culture structures at La Paz Bay, Baja California Sur, Mexico. *J. World Aquaculture Soc.* 24, 541–546.

Gervis, M.H., Sims, N.A., 1992. *The Biology and Culture of Pearl Oysters (Bivalvia: Pteriidae)*. ICLARM. Metro Manila, Philippines/ODA, London, England.

Hart, A.M., Joll, L.M., 2006. Growth, mortality, recruitment and sex-ratio in wild stocks of silver-lipped pearl oyster *Pinctada maxima* (Jameson) (Mollusca: Pteriidae) in Western Australia. *J. Shellfish Res.* 25, 201–210.

Haws, M., Ellis, S., 2000. *Collecting Black-Lip Pearl Oyster Spat*. Centre for Tropical and Subtropical Aquaculture. Publication Number 144. CTSA, Honolulu.

Hayashi, M., Seko, K., 1986. Practical technique for artificial propagation of Japanese pearl oyster (*Pinctada fucata*). *Bull. Fish. Res. Ins.* 1, 39–62.

Ito, M., 1998. Major breakthrough in hatchery spat production of the silverlip pearl oyster. *Austasia Aquacult.* 12, 48–49.

Knuckey, I.A., 1995. Settlement of *Pinctada maxima* (Jameson) and other bivalves on artificial collectors in the Timor Sea, northern Australia. *J. Shellfish Res.* 14, 411–416.

Kuwatani, Y., 1965. Studies on the breeding of the Japanese pearl oyster *Pinctada martensii* (Dunker). I. Change in the maturation of the eggs obtained from the excised gonads during the spawning season. *Bull. Natl. Pearl Res. Lab.* 10, 1228–1243.

Lewis, T., 1994. Impact of biofouling on the aquaculture industry. In: Kjelleberg, S. Steinberg, P. (Eds.), *Biofouling: Problems and Solutions. Proceedings of an International Workshop*. University of New South Wales, Sydney, pp. 39–43.

Lucas, J.S., 2003. Bivalves. In: Lucas, J.S., Southgate, P.C. (Eds.), *Aquaculture: Farming Aquatic Animals and Plants*. Blackwell Publishing, Oxford, pp. 443–466.

Lyons, L., Jerry, D.R., Southgate, P.C., 2005. Cryopreservation of black-lip pearl oyster (*Pinctada margaritifera*, L.) spermatozoa: Effects of cryoprotectants on spermatozoa motility. *J. Shellfish Res.* 24, 1187–1190.

Mann, R., Gallager, S.M., 1985. Physiological and biochemical energetics of larvae of *Teredo navalis* L., and *Bankia Gouldi* (Bartsch). *J. Exp. Mar. Biol. Ecol.* 33, 211–228.

Martínez-Fernández, E., Southgate, P.C., 2007. Use of tropical microalgae as food for larvae of the black-lip pearl oyster *Pinctada margaritifera*. *Aquaculture* 263, 220–226.

Martínez-Fernández, E., Acosta-Salmón, H., Rangel-Dávalos, C., Olivera, A., Ruiz-Rubio, H., Romo-Pinera, A.K., 2003. Spawning and larval culture of the pearl oyster, *Pinctada mazatlanica* in the laboratory. *World Aquacult.* 34, 36–39.

Martínez-Fernández, E., Acosta-Salmón, H., Rangel-Dávalos, C., 2004. Ingestion and digestion of ten species of microalgae by winged pearl oyster *Pteria sterna* (Gould, 1851) larvae. *Aquaculture* 230, 419–425.

Martínez-Fernández, E., Acosta-Salmón, H., Southgate, P.C., 2006. The nutritional value of seven species of tropical micro-algae for black-lip pearl oyster (*Pinctada margaritifera*, L.) larvae. *Aquaculture* 257, 491–503.

Mills, D., 2000. Combined effects of temperature and algal concentration on survival, growth and feeding physiology of *Pinctada maxima* (Jameson) spat. *J. Shellfish Res.* 19, 159–166.

Minaur, J., 1969. Experiments on the artificial rearing of the larvae of *Pinctada maxima* (Jameson) Lamellibranchia. *Aust. J. Mar. Freshwat. Resour.* 20, 175–187.

Monteforte, M., García-Gasca, A., 1994. Spat collection studies of pearl oysters *Pinctada mazatlanica* and *Pteria sterna* (Bivalvia, Pteriidae) in Bay of La Paz, South Baja California, México. *Hydrobiologia* 291, 21–34.

Nayar, K.N., Mahadevan, S., Ranadoss, K., Sundaram, N., Rajan, C.T., 1978. Experimental study of the settlement and collection of pearl oyster spat from Tuticorin area. *Indian J. Fish.* 25, 246–252.

O'Connor, W.A., Lawler, N.F., 2004. Salinity and temperature tolerance of embryos and juveniles of the pearl oyster *Pinctada imbricata* Roding. *Aquaculture* 229, 493–506.

O'Foighil, D.O., Kingzett, B., O'Foighil, G., Bourne, N., 1990. Growth and survival of juvenile Japanese scallops, *Patinopecten yessoensis*, in nursery culture. *J. Shellfish Res.* 9, 135–144.

Passfield, R., 1989. *Basic requirements to set up a small pearl farm*. South Pacific Commission Library, Noumea, New Caledonia, 2 p.

Pit, J.H., Southgate, P.C., 2000. When should pearl oyster, *Pinctada margaritifera* (L.), spat be transferred from the hatchery to the ocean? *Aquaculture Res.* 31, 773–778.

Pouvreau, S., Prasil, V., 2001. Growth of black-lip pearl oyster, *Pinctada margaritifera*, at nine culture sites in French Polynesia; synthesis of several sampling designs conducted between 1994 and 1999. *Aqautic Living Resour.* 14, 153–202.

Preston, G., 1990. Spat collector construction costs. *SPC Pearl Oyster Information Bull.* 1, 7–9.

Quayle, D.B., Newkirk, G.F., 1989. *Farming bivalve molluscs: method for study and development*. World Aquaculture Society, Baton Rouge, 292 pp.

Rahma, I.H., Newkirk, G.F., 1987. Economics of tray culture of the mother-of-pearl shell *Pinctada margaritifera* in the Red Sea, Sudan. *J. World Aquacult. Soc.* 2, 156–161.

Ritchie, T.P., Menzel, R.W., 1969. Influence of light on larval settlement of American oysters. *Proc. Natl. Shellfish Assoc.* 59, 116–120.

Rose, R.A., 1990. *A Manual for the Artificial Propagation of the Silverlip or Goldlip Pearl Oyster, Pinctada maxima (Jameson) from Western Australia*. Fisheries Department, Perth.

Rose, R.A., Baker, S.B., 1989. Research and development of hatchery and nursery culture for the pearl oyster, *Pinctada maxima*. FIRTA Project 87/82, Final Report. Fisheries Department of Western Australia, 36 pp.

Rose, R.A., Baker, S.B., 1994. Larval and spat culture of the western Australian silver-or goldlip pearl oyster, *Pinctada maxima* Jameson (Mollusca): Pteriidae. *Aquaculture* 126, 35–50.

Saucedo, P.E., Rodriguez-Jaramillo, C., Aldana-Aviles, C., Monsalvo-Spencer, P., Reynoso-Granados, T., Villareal, H., Monteforte, M., 2001. Gonadic conditioning of the Calafia mother-of-pearl oyster, *Pinctada mazatlanica* (Hanley, 1856), under two temperature regimes. *Aquaculture* 195, 103–119.

Saucedo, P.E., Bervera-Leon, H., Monteforte, M., Southgate, P.C., Monsalvo-Spencer, P., 2005. Factors influencing recruitment of hatchery reared pearl oyster (*Pinctada matzalantica*; Hanley 1856) spat. *J. Shellfish Res.* 24, 215–219.

Shirai, S., 1970. *The Story of Pearls.* Japan Publications Inc., Japan. 132 pp.

Sims, N.A., 1993a. Pearl Oysters. In: Wright, A., Hill, L. (Eds.), *Near-Shore Marine Resources of the South Pacific.* FFA/ICOD, Suva, pp. 409–430.

Sims, N.A., 1993b. Size, age and growth of the black-lip pearl oyster, *Pinctada margaritifera* (L.) (Bivalvia; Pteriidae). *J. Shellfish Res.* 12, 223–228.

Southgate, P.C., 2003. Feeds and feed production. In: Lucas, J., Southgate, P.C. (Eds.), *Aquaculture: Farming Aquatic Animals and Plants.* Blackwell Publishing, Oxford, pp. 172–198.

Southgate, P.C., Beer, A.C., 1997. Hatchery and early nursery culture of the blacklip pearl oyster *Pinctada margaritifera* (L.). *J. Shellfish Res.* 16, 561–568.

Southgate, P.C., Ito, M., 1998. Evaluation of a partial flow-through technique for pearl oyster (*Pinctada margaritifera*) larvae. *Aquacult. Eng.* 18, 1–7.

Southgate, P.C., Beer, A.C., 2000. Growth of blacklip pearl oyster (*Pinctada margaritifera* L.) juveniles using different nursery culture techniques. *Aquaculture* 187, 97–104.

Southgate, P.C., Taylor, J.J., Ito, M., 1998a. The effect of egg density on hatch rate of pearl oyster (*Pinctada maxima* and *Pinctada margaritifera*) larvae. *Asian Fisheries Sci.* 10, 265–268.

Southgate, P.C., Beer, A.C., Duncan, P.F., Tamburri, R., 1998b. Assessment of the nutritional value of three species of tropical micro-algae, dried *Tetraselmis* and a yeast-based diet for larvae of the blacklip pearl oyster, *Pinctada margaritifera* (L.). *Aquaculture* 162, 247–257.

Strugnell, J.M., Southgate, P.C., 2003. Changes in tissue composition during larval development of the blacklip pearl oyster, *Pinctada margaritifera* (L.). *Molluscan Res.* 23, 179–183.

Tanaka, Y., Inoha, S., 1970. Studies on seed production of blacklip pearl oyster *Pinctada margaritifera* in Okinawa. IV. Resistibility of the larvae centrifugally-separated water from *Monochrysis* culture and to hypotonic sea water. *Bull. Tokai Reg. Fish. Res. Lab.* 63, 91–95.

Tanaka, Y., Kumeta, M., 1981. Successful artificial breeding of the silverlip pearl oyster, *Pinctada maxima* (Jameson). *Bull. Natl. Res. Inst., Aquacult.* 2, 21–28.

Tanaka, Y., Inoha, S., Kakazu, K., 1970. Studies on seed production I. Spawning induction by thermal stimulation. *Bull. Tokai Reg. Fish. Res. Lab.* 63, 75–78.

Taylor, J.J.U., 2005. Successful remote settlement of eyed silver- or gold-lip pearl oyster *Pinctada maxima* larvae. *World Aquaculture 2005*, May 9–13, Bali, Indonesia. Book of Abstracts. World Aquaculture Society, Baton Rouge, LA, p. 647.

Taylor, J.J., Rose, R.A., Southgate, P.C., 1997a. Byssus production in six age classes of the silver lip pearl oyster, *Pinctada maxima* (Jameson). *J. Shellfish Res.* 16, 97–101.

Taylor, J.J., Rose, R.A., Southgate, P.C., 1997b. Effects of stocking density on the growth and survival of juvenile silver-lip pearl oysters (*Pinctada maxima*, Jameson) in suspended or bottom culture. *J. Shellfish Res.* 16, 569–574.

Taylor, J.J., Rose, R.A., Southgate, P.C., 1997c. Inducing detachment of silver-lip pearl oyster (*Pinctada maxima*, Jameson) spat from collectors. *Aquaculture* 159, 11–17.

Taylor, J.J., Southgate, P.C., Rose, R.A., 1997d. Fouling animals and their effect on the growth of silver-lip pearl oysters, *Pinctada maxima* (Jameson) in suspended culture. *Aquaculture* 153, 31–40.

Taylor, J.J., Rose, R.A., Southgate, P.C., Taylor, C.E., 1997e. Effects of stocking density on growth and survival of early juvenile silverlip pearl oysters *Pinctada maxima* (Jameson), held in suspended nursery culture. *Aquaculture* 153, 41–49.

Taylor, J.J., Southgate, P.C., Wing, M.S., Rose, R.A., 1997f. The nutritional value of five species of microalgae for spat of the silver-lip pearl oyster, *Pinctada maxima* (Jameson) (Mollusca:Pteriidae). *Asian Fisheries Sci.* 10, 1–8.

Taylor, J.J., Southgate, P.C., Rose, R.A., 1998a. Assessment of artificial substrates for collection of hatchery-reared silver-lip pearl oyster (*Pinctada maxima*, Jameson) spat. *Aquaculture* 162, 219–230.

Taylor, J.J., Southgate, P.C., Rose, R.A., 1998b. Effects of mesh covers on the growth and survival of silver-lip pearl oyster (*Pinctada maxima*, Jameson) spat. *Aquaculture* 162, 241–246.

Taylor, J.J., Southgate, P.C., Rose, R.A., Keegan, A.J., 1998c. Effects of larval set density on subsequent growth and survival of the silver-lip pearl oyster, *Pinctada maxima* (Jameson). *J. Shellfish Res.* 17, 281–283.

Taylor, J.J., Southgate, P.C., Rose, R.A., 2004. Effects of salinity on growth and survival of silver-lip pearl oyster, *Pinctada maxima* (Jameson) spat. *J. Shellfish Res.* 23, 375–378.

Tranter, D.J., 1958. Reproduction in Australian pearl oysters (Lamellibranchia). IV. *Pinctada margaritifera* (L.). *Aust. J. Mar. Freshwat. Res.* 9, 509–523.

Vakily, J.M., 1989. The biology and culture of mussels of the genus *Perna*. *ICLARM Stud. Rev.* 17, 63.

Victor, A.C.C., Chellam, A., Dharmaraj, S., 1987. Pearl oyster spat collection. In: Alagarswami, K. (Ed.), *Pearl Culture. Bull. Cent. Mar. Fish. Res. Inst. No. 39*. Central Marine Fisheries Research Institute, Cochin, India, pp. 49–53.

Wada, S., 1942. Artificial insemination and development in *Pinctada maxima* (Jameson). *Kagaku-Nanyo* (Science of South Seas) 4(3), 26–32.

Webb, K.L., Chu, F.L., 1982. Phytoplankton as a food source for bivalve larvae. In: Pruder, G.D., Langdon, C.J., Conklin, D.E. (Eds.). *Proceedings of the Second International Conference on Aquaculture Nutrition: Biochemical and Physiological Approaches to Shellfish Nutrition*. Louisiana State University, Baton Rouge, LA, pp. 272–291.

Weiner, R.M., Walch, M., Labare, M.P., Bonar, D.B., Colwell, R.R., 1989. Effect of biofilms of the marine bacterium *Alteromonas colwelliana* (LST) on set of the oysters *Crassostrea gigas* (Thunberg, 1793) and *C. virginica* (Gmelin, 1791). *J. Shellfish Res.* 8, 117–123.

Wells, F.E., Jernakoff, P., 2006. An assessment of the environmental impact of wild harvested pearl aquaculture (*Pinctada maxima*) in Western Australia. *J. Shellfish Res.* 25, 141–150.

Whitford, J.R., 2002. Growth, survival and husbandry of black-lip pearl oyster juveniles (*Pinctada margaritifera*) in the Abaiang lagoon, Kiribati. Master's thesis, James Cook University, Australia, 148 pp.

Whyte, J.N.C., Bourne, N., Hodgson, C.A., 1987. Assessment of biochemical composition and energy reserves in larvae of the scallop *Patinopecten yessoensis*. *J. Exp. Mar. Biol. Ecol.* 113, 113–124.

Yankson, K., Moyse, J., 1991. Cryopreservation of the spermatozoa of *Crassostrea tulipa* and three other oysters. *Aquaculture* 97, 259–267.

CHAPTER 8

Pearl Production

Joseph Taylor and Elisabeth Strack

8.1. INTRODUCTION

Pearls are unique as the only gems produced by a living organism. The process of pearl formation, whether by natural means or human intervention, is a captivating process and has been the subject of myth, scientific speculation and research throughout human history. Even in this modern age, pearl culturing techniques differ little from those developed for round pearl culture over one hundred years ago by the pioneering Japanese pearlers Tokichi Nishikawa, Tatsuhei Mise and Kokichi Mikimoto (Gervis and Sims, 1992; Müller, 1997; Strack, 2006), and the precise manner in which a perfect pearl can be reliably formed is still unknown.

8.2. PEARL FORMATION

Pearls are the result of a mollusc's ability to produce shell material in response to an injury to the mantle tissue, and the process is identical to the laying down of the shell that protects the soft body tissues. This unique process has been termed biomineralization (Comps *et al.*, 2000; Che *et al.*, 2001; Chapter 3). The mantle of pearl oysters is responsible for both shell and pearl formation (Tsujii, 1960; Dix, 1972a, b; Jabbour-Zahab *et al.*, 1991; Garcia-Gasca *et al.*, 1994; Awaji and Suzuki, 1995). The mantle has several types of secretory cells responsible for the different types of shell material formed (Wang *et al.*, 2002; see Section 3.3, Table 3.1). In *Pinctada mazatlanica*, four different types of secretory cells were identified by Garcia-Gasca *et al.* (1994). In the mantle epithelium there were large secretory cells containing carbohydrates, acid proteins, sulphated acid mucopolysaccharides and calcium granules. Small cells, only in the middle mantle fold, were found to secrete acid mucopolysaccharides. The periostracal groove and shell epithelium contained acidophilic secretory cells that take part in protein synthesis. The central zone had large acidophilic secretory cells associated with glycogen synthesis. Garcia-Gasca *et al.* (1994) related the specialized secretory cells of the outer mantle epithelium and the presence of alkaline phosphatase and carbonic anhydrase with calcium deposition and suggested that this epithelium is the most suitable as graft tissue in pearl culture (see Table 3.1). Further details of the structure of pearl oyster mantle tissue and the processes of biomineralization are provided in Chapter 3.

It is commonly stated in the literature that natural pearls form when an irritant is enveloped in a cyst, i.e. the "grain of sand" or encapsulated parasite theory (Farn, 1986; Gervis and Sims, 1992; Strack, 2006). This common theory states that a foreign body that cannot be expelled by an oyster becomes encapsulated by the mantle, which forms a "sac" that subsequently coats the foreign body or "irritant" with conchiolin, thus forming a hard nucleus. The epithelial cells of the pearl-sac cover the nucleus with fine layers of nacre. If movement of the nucleus is unrestricted, the result is a spherical pearl (Farn, 1986). This defense mechanism has been called "nacrezation" and is common to all bivalves (Malek and Cheng, 1974).

However, a significant amount of research has been done to test this claim and no evidence has been found to support the encapsulated irritant theory (Hänni, 2006). Furthermore, pearl oysters have among the highest mantle cavity ventilation rates recorded in bivalve molluscs (see Section 4.2.2) and it is unlikely that they are unable to expel foreign particles that enter the mantle cavity.

The most recent body of evidence suggests that natural pearls are the result of an oyster's response to mantle tissue injury only, and that the presence of a foreign body is not required for pearl formation. Wounds to the mantle sometimes heal forming a cyst or pocket (the pearl-sac), which fills with calcium carbonate produced by the mantle's secretory cells. Hence the process does not require the encapsulation of a foreign body. In cultured pearls, this natural process is initiated by human intervention.

8.3. NATURAL PEARLS

The formation of a pearl is the result of a rather accidental occurrence within the normal life cycle of a mollusc. The various zones and folds of the pearl oyster mantle and their epithelial secretions are described in Section 3.3 (Fig. 3.14; Table 3.1). For natural pearls to form, it is essential that one or several epithelium cells of the upper epithelium layer of the mantle be transferred into the connective tissue, where they proceed to multiply by cell division and form a closed cyst, which is called a pearl-sac (Fig. 8.1).

The transfer of cells may result from a multitude of causes. The mantle tissue may suffer an injury when any type of attacker or intruder, such as small gastropod or crab, reaches the space between mantle and shell and causes damage to the mantle tissue. Small pieces of broken shell, lumps of conchiolin, small stones and the frequently mentioned "grain of sand" can be counted among the potential intruders. Egg cells or

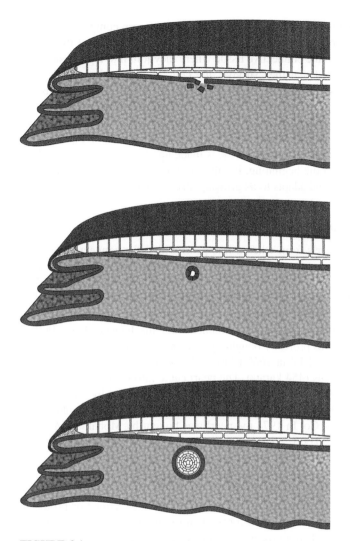

FIGURE 8.1 Schematic representation of the formation of a natural pearl. Mantle epithelial cells are displaced into connective tissue following damage to the mantle (top). They form a pearl-sac (middle) which secretes material forming a growing pearl (bottom). For greater detail of shell and mantle structures see Fig. 3.14. Redrawn from Strack (2006).

other remaining substances of the mollusc's own metabolism that were not ejected in the usual way may also act as intruders themselves.

The parasite theory, mainly favored in the 19th century, is probably true in so far as attacks by parasitic cestode larvae or parasitic copepods (see Section 11.7.6) on the mantle tissue may cause a transfer of epithelium cells, thus effecting the formation of a pearl-sac. The parasites are only several millimeters long and will, when a pearl-sac is formed, usually be encapsulated and dissolved. Abnormal growth of epithelium cells at high speed, caused by a biochemical process comparable to the growth of cancer, may also lead to the formation of a pearl, provided that the fast-growing layers sink down into the connective tissue where they join into a pearl-sac (Strack, 2006).

Provided that they survive, the cells of the pearl-sac will produce secretions, the type of which will depend on the age of the cells. Young cells will secrete conchiolin, similar to the periostracum of the shell. After some time they will begin to secrete prismatic calcite layers that correspond to the prismatic layer of the shell or ostracum (Fig. 3.14). Later they switch to producing nacre, corresponding to the hypostracum (see Fig. 3.14). If the pearl-sac is made up of older epithelium cells of the hypostracum right from the beginning, it will only secrete nacre. The pearl-sac surrounds the growing pearl and adapts to its growing size by producing new cells.

8.3.1. Natural blister pearls

Blister pearls form when a free pearl within the connective tissue, breaks through the mantle due to its size and weight and becomes pressed onto the shell. It will inter-grow with the shell while nacre layers are deposited over its surface. When such pearls are harvested, there is still a possibility that they can be used and sold as free pearls or that an experienced "pearl peeler" will be able to expose the original round pearl.

The famous Hope Pearl, named after the banker Henry Philip Hope, is a blister pearl. It was probably bought by Tavernier in India around 1650 and sold to the French King Louis XIV in 1669 (Strack, 2006). It measures $150 \times 83 \times 50$ mm and weights 1.816 grains (454 karats). The pearl came into the possession of Henry Philip Hope in London who died in 1839. In 1881, it was donated by A. J. Beresford-Hope as a permanent loan to the South Kensington Museum and was sold in 1886 by Christie's to Garrard & Co. In 1974 the pearl was bought by Mohamed Mahdi Altajiir from Dubai for US$200,000.

8.3.2. Blisters

Blisters, in contrast, are not pearls but are concretions or protuberances on the inner side of the shell. Blisters form when foreign objects or living marine organisms of the types described above get into the space between shell and mantle tissue, where they are trapped. The oyster may trap this intrusion by increasing the usual growth rate of the inner nacre layer of the shell (Fig. 8.2).

Blisters may also form when the pearl oyster responds to damage caused by polychaetes, sponges etc. boring a hole or holes through the shell from the external surface (see Sections 15.3.1; Fig. 15.1). The attack can occur either perpendicular or parallel to the length of the shell. Where holes penetrate through to the nacre layer, pearl oyster responds to damage by increasing the secretion of nacre in that region. The repair work may be so perfect as to leave no trace, but in most cases it will lead to a protuberance on the inside of the shell. Blisters are usually large, reaching up to a few centimeters in length. They have irregular shapes and are usually cut from the inside of the shell if they are to be used in jewelry.

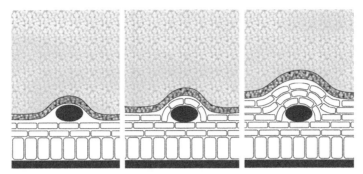

FIGURE 8.2   Schematic representation of the formation of a blister. A foreign object is trapped between shell and mantle (left) and is covered with successive layers of nacre (middle and right). Redrawn from Strack (2006).

8.4. THE STRUCTURE OF PEARLS

Sections of pearls reveal that their structure is similar to the structure of the shell, the difference lies only in the reversed sequence of layers, such that the inside of the shell corresponds in its layers to the outside of the pearl. The distinctive difference between shell and pearl lies in the varied structure of the individual layers. While the shell shows an evenly layered structure (Fig. 3.14), the structure of the pearl is concentric (Fig. 8.3).

The epithelium cells of the pearl-sac usually begin with the secretion of conchiolin in an ultra-thin lamellar structure, the so-called organic matrix, which represents a network of fine meshes that serve for the transport of calcium. Pearls often have an inner core of conchiolin, followed by a prismatic layer, made up of individual tapered prismatic crystals of calcite, arranged in a radial structure. The longitudinal axis of the prisms points towards the center of the pearl.

The outer layer of the pearl is made up of nacre, thus matching the hypostracum of the shell. The word "nacre" has the same meaning as "mother-of-pearl" (MOP) but is more appropriate when referring to pearls. The structure of the nacre layers can be compared with an onion. The individual sheets are studded with aragonite platelets of a polygonal, usually near hexagonal shape and are arranged on top of each other in such a way that they come to lie in the familiar brick structure known from the shell (Fig. 8.4a; see Section 3.3.4, Fig. 3.19). The sheets grow in stages and individual layers may twist and intertwine. The growth time is linked to the actual growth rate of the shell and depends on the time of the year. The average thickness of aragonite platelets is 0.5 microns in *Pinctada margaritifera*.

Layers on the surface of a pearl overlap in a terrace-like arrangement, reminiscent of roof tiles (Fig. 8.4a). Tiny ridges separate the platelets, combining with the terrace lines to form wavy or rounded relief patterns (Fig. 8.4b). The lines are visible under an optical microscope with magnification of 40x or more. The special structure of the

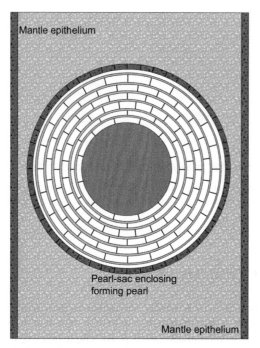

Mantle epithelium

Pearl-sac enclosing
forming pearl

Mantle epithelium

FIGURE 8.3   Schematic representation of a pearl during formation showing concentric layers of nacre surrounding an inner core of conchiolin. Redrawn from Strack (2006).

pearl's surface gives rise to interference and diffraction of light which lead to the phenomenon of iridescence that produces the orient and beauty of a pearl (Fig. 8.4c). The specific arrangement of aragonite platelets is thus considered a necessity for the definition of a pearl and it determines the appearance of a pearl. In other words: a pearl can only be defined as such when the outer layer consists of nacre. The inner layers, not visible to the naked eye but almost always present, do not affect beauty and value but they may have an influence on the durability of a pearl.

Concretions of conchiolin are quite often observed within both the prismatic and nacre layers of the pearl, where they show up as individual lumps, fine lines or accentuated rings, and are usually arranged parallel to the concentric structure. The color of a pearl will be influenced if the conchiolin core within it takes up more than three quarters of its size. An example of this rather rare case are the so-called "Blue Pearls" in the natural pearl trade which owe their bluish-gray color to a large inner core of conchiolin. The conchiolin substance is host to the coloring agents, usually organic pigments, carotinoids and porphyrins, which cause the various colors of pearls. Both blister and blister pearls differ from free pearls insofar as the MOP layers correspond to those of the shell. They have an even, parallel and not a concentric arrangement. In the case of blister pearls, the pearl within possesses a concentric structure.

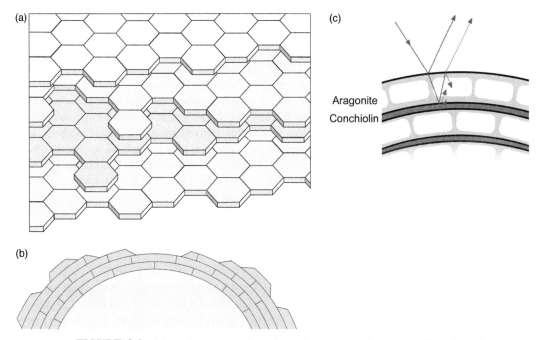

FIGURE 8.4 Schematic representations of (a) the aragonite platelets imbedded within the organic matrix within a pearl; (b) a cross-section of the incomplete outer nacre layers of a pearl which form contours on the pearl's surface; and (c) the interference on the surface of a pearl due to refraction and reflection of light from the alternating layers of aragonite and conchiolin Redrawn from Strack (2006).

8.5. PROPERTIES OF PEARLS

8.5.1. Chemical composition

Pearls are composed of calcium carbonate (91.50%) with traces of organic substances (3.83%), residual substances (0.01%) and water (3.97%). The residual substances include Na, Cl and K, and traces of other elements, such as Ba, Mg, P, Mn, Fe, Al, Cu, Zn, Ag, Hg, Li and Sr. The trace elements reflect the composition of the water at the location where a pearl was formed.

Acids attack calcium carbonate. A pearl will not dissolve completely, but will change into a soft, swollen mass that keeps its original outer shape, corresponding to the conchiolin frame. When a drop of diluted hydrochloric acid is applied to the surface of a pearl, there will be a slight effervescence that leaves a dull spot. When pearls are worn over a long period of time, they can be attacked by fatty acids from human perspiration.

The life span of pearls is mainly determined by the organic conchiolin compounds that may dry-out over time. At first, tiny fractures develop, and the pearl takes on a dull appearance, often followed by discoloration. This means that pearls should not be kept in a dry environment for long period, but should be worn regularly. Long exposure to high humidity may also have a damaging effect. It may lead to the decomposition of conchiolin. Cosmetic products of all types may contaminate the organic substance. The pearl trade of the past developed a number of remedies in order to save pearls from ageing. The pearl centers of the 19th century saw the so-called "pearl doctors", who had treatments involving alcohol or ether, bleaching agents like hydrogen peroxide and waxing or oiling agents such as lukewarm whale oil. The latter process, called "decraquelation" meant that fine fractures could no longer be seen by the naked eye. The methods of the pearl doctors were often derived from old Indian methods, and both India and China had methods for pearl enhancement going back to ancient times. One particularly inventive method, mentioned in the so-called Stockholm Papyrus, originating in Alexandria in 400 AD, involved feeding pearls to a rooster and trusting in the cleaning powers of the rooster's digestive acids.

8.5.2. Physical properties

8.5.2.1. Hardness

The hardness of pearls is in the range of 3.5–4.5 on Mohs' scale, it corresponds approximately with the hardness of aragonite (calcite has a hardness of 3, aragonite of 3.5–4). Although the hardness of pearls is, generally speaking, comparatively low, they are nevertheless tough due to their micro-crystalline structure. This is also partly due to the organic matrix holding the individual crystals together and resulting in stronger cohesion that makes it practically impossible to break pearls to pieces. The high elasticity of pearls is attributed to the same fact; it can be observed easily in the bouncing of pearls that are dropped.

8.5.2.2. Specific gravity

The specific gravity of calcite is 2.71, while that of aragonite is 2.93. The specific gravity of pearls is between 2.60 to 2.74, the actual figure depends on the individual composition of a pearl and above all on its conchiolin content. The average figure is quoted as 2.71, while around 80% of all pearls do not reach a figure higher than 2.72. The figure may fall to 1.91 or even lower if a large inner conchiolin core is present, the specific gravity of conchiolin is quoted as 1.34. All the above figures refer to pearls from *Pinctada*. Pearls fluoresce a light blue color under ultraviolet (UV) light, the fluorescence is more intense at longer wave lengths (366 nm) than under short wave UV light (254 nm).

8.6. CULTURED PEARLS

There are essentially four major types of cultured pearls:

1. Composite pearls (commonly called "mabé," half-pearls or blister pearls) that are grown directly on the internal nacreous surface of shells of both *Pinctada* and *Pteria* species.

2. Non-nucleated pearls that are produced in the mantle tissue following surgical implantation of sections of donor mantle tissue.
3. Nucleated pearls that are produced in the gonad of a variety of pearl oysters following surgical implantation of a spherical shell bead (the nucleus) and a section of donor mantle tissue.
4. Pearls that may develop when the bead or nucleus is rejected from the gonad, but the tissue graft continues to form a pearl-sac. This is also a non-nucleated cultured pearl and is generally referred to as a "keshi" pearl.

8.7. COMPOSITE PEARL PRODUCTION

Composite cultured pearls are commonly referred to as mabé and this term will be used to describe this type of pearl in the following text; they are analogous with natural blisters described in Section 8.3.2. Formation of mabé pearls takes advantage of an oyster's ability to cover concretions or protuberances on the inner side of the shell with nacre. Mabé pearls are formed by the adhesion of nuclei to the inner nacreous surface of pearl oyster shells; the nuclei are then covered with nacre to an appropriate thickness (ca. 1 mm) and harvested. The structure of the nacreous layer is equivalent to that of the MOP layer of the shell. This means that the aragonite platelets are arranged in parallel layers and not in concentric layers as within a true pearl (see Section 8.4, Fig. 8.3). At harvest mabé pearls are cut out of the shell and the nucleus is removed leaving a thin hollow half-sphere. The inside is cleaned and filled with resin and sealed with a flat base of MOP.

Mabé pearls were the earliest form of cultured pearls, with the technology having first been developed in China over 2,000 years ago (Dan, 2003). Lead effigies of Buddha were attached to the inside of river mussels in China (Farn, 1986). In modern times, plastic has become the material of choice for mabé pearl production (Gervis and Sims, 1992). Plastic shapes of various sizes are glued to the inner shell surface using a cyanoacrylate adhesive that is quick drying and waterproof (Acosta-Salmon et al., 2005) (Fig. 8.5). The traditional species used for mabé pearl production is the winged pearl oyster, Pteria penguin. However, the term mabé is now applied generically to this type of pearl produced from a range of pearl oyster species and even abalone (Matlins 1996; Müller, 1997). Very often mabé pearls are produced by pearl oysters which have already successfully produced round pearls or keshi pearls and are used for a final time before the end of their lives.

The key technical factor in the process of adhering mabé pearl nuclei to the inner surface of an oyster's shell is positioning. Nuclei need to be placed such that they do not interfere with the closing of the pearl oyster's shells and that there is sufficient space to allow for nacre deposition without interference from neighboring growing mabé pearls. In fast growing oysters, the adductor muscle may begin to cover the growing mabé pearls. This leads to differences in nacre thickness and reduced luster (Ruiz-Rubio et al., 2006) and, occasionally, a clear line between the covered area and clear areas of the mabé pearl. This effect can significantly reduce the value of the resulting pearls. Appropriate nucleus implantation is often facilitated by anaesthetizing the oysters

FIGURE 8.5   White hemi-sperical plastic nuclei glued to the inner shell surface of *Pinctada margaritifera* for mabé pearl production. (Photo: Paul Southgate.)

prior to implantation (Acosta-Salmon and Rangel-Davalos, 1997; Ruiz-Rubio *et al.*, 2006; Southgate *et al.*, 2006).

The number of mabé that can be cultured in an individual pearl oyster varies greatly depending on species. In general, *Pteria* species may accommodate a maximum of three mabé pearl nuclei: two on the dorsal valve and one on the ventral valve (Saucedo *et al.*, 1998; Ruiz-Rubio *et al.*, 2006). *Pinctada margaritifera* and *P. albina* can accommodate 4–5 nuclei (Southgate *et al.*, 2006; Taylor, unpublished data, 1990) while large *P. maxima* may be implanted with as many as seven nuclei.

The period between implantation and pearl harvest is usually much shorter for mabé pearls than for round pearls, and is generally between 6–12 months, depending on species, geographical location and culture system (Farn, 1986; Gervis and Sims, 1992; Ruiz-Rubio *et al.*, 2006). A culture period of 6–12 months, generally results in mabé pearls with a nacre thickness between 0.7–2.5 mm (Shirai, 1981; McLaurin *et al.*, 1997, 1999). Ruiz-Rubio *et al.* (2006) reported that a commercial nacre thickness in mabé pearls produced from *Pteria sterna* was reached within 5 months, (ca. 0.75 mm); however, pearl quality was highest after 9 months. Regardless of species, the process of harvesting results in the death of the pearl oyster. Mabé pearls are removed using a diamond tipped hole-saw. The back of the pearl (the shell of the pearl oyster) is ground off and the plastic nucleus is removed. The hollow space that remains is filled with a plastic resin and a MOP backing is glued in place. Finished pieces can be set into jewelry (Fig. 8.6).

The economic importance of mabé pearls declined significantly during the late 1980s as the production of Chinese freshwater pearls (non-nucleated mantle pearls) and of round marine pearls from culture began to expand rapidly (Müller, 1997). Very few mabé pearls are produced from *Pinctada* species and Australian production of mabé

FIGURE 8.6   Processed mabé pearls produced from *Pteria sterna*. (Photo: Douglas McLaurin, Perlas del Mar de Cortez.) (See Color Plate 10)

pearls using *P. maxima* has ceased altogether. However, there is still interest in producing mabé pearls in developing nations, as the process is far simpler and less costly than that for round pearls (Ruiz-Rubio *et al.*, 2006) and their production offers opportunity for income generation in coastal communities (Southgate *et al.*, 2006). MOP as jewelry has become increasingly fashionable, in recent years, particularly among young people. This is leading to a renewed interest in associated shell material, including mabé pearls. Mabé have comparatively low, but stable, prices. They represent a reasonable alternative to fully round pearls in sizes of 12–25 mm and they are also produced in fancy shapes. The pearls have to be handled with special care as the thin outer nacreous layer may crack easily or become damaged or separated from the filling underneath.

8.8.   NON-NUCLEATED PEARL PRODUCTION

This type of pearl cultivation most closely replicates the natural process and is the principal method of producing cultured freshwater pearls. The technique involves surgically implanting mantle sections from a donor oyster into the living mantle of a host oyster (Hänni, 2006). In the Japanese Akoya pearl industry, these pearls are termed "keshi" the Japanese word for "poppy seed," a term now used more generally in the marine cultured pearl industry to describe non-nucleated pearls that sometimes result in the gonad of pearl oysters following the expulsion of the nucleus (or bead) after nucleation (see Section 8.9). Marine production of non-nucleated pearls is insignificant and production figures are not available. However, freshwater non-nucleated pearls make up the vast majority of global pearl production with over 1,000 t produced annually (Strack, 2006). They account for >90% of all cultured pearl production.

8.9. NUCLEATED PEARL PRODUCTION

Most nucleated pearl production is from *Pinctada* species. The general technique involves surgical implantation of a shell-based nucleus (or nuclei) together with a section of mantle tissue removed from a selected donor oyster. There are several key steps in the nucleation process which is commonly called grafting or seeding. First and foremost, both the donor that provides mantle tissue and the hosts that will receive the graft must be in excellent health. The oysters should be mature enough to have reached a size to enable surgery, but still of an age where growth is vigorous.

8.9.1. Development of the cultured pearl technique

Today's cultured pearl industry owes its founding and present status to Kokicki Mikimoto who is considered a national hero in Japan. Mikimoto took up attempts at pearl culture after his contact with Dr. Yoshikichi Mizukuri, one of the founders of modern Japanese zoology. Mikimoto started his experiments in Ago Bay by using the local Akoya pearl oysters that had been known as a pearl producer around the coast of Ise peninsula for centuries. Mikimoto introduced nuclei made of different materials like coral, abalone shell, bone, powdered fish scales, etc. onto the inside of the shell of the oysters which he spread out on the ocean floor or kept in baskets. Success came when he began to use MOP beads and the first successful pearl harvest took place on July 11, 1893, in Nikasahi Bay. He applied for a patent and immediately started mass production; by 1902 he kept one million oysters on his farm that centered around a small island near Toba named Tatokushima. Female divers, the so-called "amas" that have a long tradition on the southeastern coast of Japan did the necessary diving at the farms, in order to produce enough oysters for pearl production.

Mikimoto recognized at an early time that he would have to find a way of organizing the necessary publicity work to successfully market his cultured pearls. Just before 1900 he opened his first shop in the Ginza district of Tokyo. Newspapers in London and Paris wrote for the first time about "Mikimoto pearls" after they had been presented as a gift to King Edward VII at his coronation ceremony. In the years that followed, the cultured pearls became known in Europe and the United States as Mikimoto participated in a number of foreign exhibitions. In 1904 a journalist from the New York Herald visited Mikimoto's pearl farm for the first time. In 1908 he opened his second farm in Gokasho and operated on about 10 million oysters, employing more than a 1,000 workers. By now Mikimoto was no longer the only pearl farmer in Japan, but he was the most successful.

The most important event in the history of pearl culture was the introduction of the so-called Mise-Nishikawa-method that allowed regular production of cultured pearls. It did not take place before 1916. The western world learned of the origin of the method only in 1949 when a report was written for the American occupying forces following World War II. More extensive information followed after the investigations of the Australian C. Denis George.

Tokishi Nishikawa was a marine biologist who in 1901 was sent as a Fisheries Officer with the Arafura Pearling Fleet to Australia. On Thursday Island where the Fleet was stationed, Nishikawa met the stepfather of Tatsuhei Mise who worked as a Fisheries Inspector in the same unit. The two Japanese probably became acquainted with William Saville-Kent (at the time Fisheries Commissioner for the British Colonial Government on Thursday Island in Torres Strait) or at least some of his collaborators. Saville-Kent had more than 20 years experience in the production of cultured mabé pearls and had even been successful in producing round pearls. At the time, Saville-Kent was famous above all for his work on the Great Barrier Reef (Saville-Kent, 1893).

Both Nishikawa and Mise returned to Japan in 1902 and in the years that followed Mise's stepson, Tatsuhei and Nishikawa himself developed two very similar methods for cultured round pearl production. They were eventually merged into the "Mise-Nishikawa method" which is universally used today for production of nucleated cultured pearls. Both were probably strongly influenced by Saville-Kent's method as they did not even try to follow the old Chinese method known until then. Both methods used a nucleus and a small piece of mantle tissue that did not totally envelop the nucleus.

Tatsuhei Mise (1880–1924) lived in Matoya Bay within Mie Prefecture; he received all necessary information from his stepfather and started his first pearl culture project in 1902 with 15,000 oysters. His first harvests were in 1904 and in 1907 he applied for a patent for the pearl culture method. Tokisi Nishikawa (1874–1909) started only in 1905 on a scientific basis and harvested his first pearls in 1907 applying for a patent in the same year. Both Mise and Nishikawa had used nuclei made of lead, silver and gold and they had used nearly the same implantation process. A patent controversy followed in 1907/1908 that was settled by mutual agreement. Nishikawa, who had been married to Mikimoto's daughter died at only 35 years of age. His work was continued by the talented Fujita brothers, his former assistants who received his patent in 1916 and owned it together with Nishikawa's son Shinkichi, Mikimoto's grandson. Mikimoto quickly recognized that the Mise-Nishikawa method was more suitable than the method that he himself had developed for production of round cultured pearls; it consisted of completely sewing a nucleus into a piece of mantle tissue. He had developed this method from his own experiments and it had been completed by his friend Otokichi Kuwabara before Mikimoto applied for a patent. In 1916 Mikimoto made peace with the Fujita brothers and his grandson Shinkichi as he bought the rights for Nishikawa's method. The Fujita brothers moved to work with Mikimoto in 1916. This year can be considered the birth year of cultured round pearls as it was the year that Mikimoto began mass production on the basis of the Mise-Nishikawa method. 1916 was also the first year in which Mikimoto, aged 58, made his first trip overseas to Korea and to Shanghai where he opened a shop.

The 1920s saw the rise of the cultured pearl. In 1921 the first fully round pearls appeared on the world market. H. Lyster Jameson, the English zoologist issued a statement to the press in London in which he said that Japanese cultured pearls were, from a scientific point of view, identical with pearls formed in nature. The London Chamber of Commerce had a different opinion and declared the pearls to be "fakes". In France, Mikimoto had to face several court trials and the first International Jeweller's Congress in April 1926 decided that cultured pearls were to be called "Perles cultivees" (cultured pearls). In the same year Mikimoto visited the United States for the first time where he had a store in New York and where he was received with enthusiasm. When he returned to Japan in 1927 after 7 months of traveling to a number of European and Asian countries he had become the "Pearl King."

When Mikimoto died at the age of 96 in September 1954, he left a well-established enterprise to his family and to the Japanese nation and a sound basis for the development of a global cultured pearl industry.

8.9.2. Pre-operative condition

To ensure the best possible result during the operation to produce cultured round pearls, a pearl oyster's metabolism must be in a reduced state and the gonad must be free of gametogenic activity (Alagarswami, 1970; Mizumoto, 1976; Hollyer, 1984; Alagarswami, 1987a, b; Chellam *et al.*, 1991; Gervis and Sims, 1992). In more temperate waters, e.g. Akoya pearl oysters in Japanese farms, this can be achieved by nucleating the pearl oysters during the non-reproductive seasons when water temperature is cooler (Mizumoto, 1976; Alagarswami, 1987a, b). This is also the case for *P. maxima* in Australia. However, in equatorial regions where water temperature shows little variation and during the warmer months in temperate regions, an appropriate pre-operative condition may need to be induced (Taylor, 2002). Techniques that are commonly employed include the use of fine mesh covers to encase pearl oysters, placing them deeper in the water column or placing them flat on the sea bed (Alagarswami, 1987a, b; Taylor, 2002) or deliberately crowding pearl oysters in baskets (Miyauti, 1965; Alagarswami, 1987a, b; Gervis and Sims, 1992) (Fig. 8.7). The result in all cases is reduced metabolic rate.

To reduce the risk of damage to the mantle and adductor muscle, which can occur when pearl oysters are forced open prior to operation, a variety of chemicals have been assessed as anesthetics for pearl oysters (Alagarswami, 1987a, b; Norton *et al.*, 1996; Mills *et al.*, 1997). Propylene phenoxetol has shown particular promise as it induces rapid anesthesia in *P. margaritifera*, *P. albina*, *P. fucata* (Norton *et al.*, 1996; Acosta-Salmon *et al.*, 2005) and *P. maxima* (Mills *et al.*, 1997) with short recovery times. In tests with *P. albina* and *P. margaritifera*, no mortalities were observed 7 days after recovery (Norton *et al.*, 1996) and large-scale testing with *P. maxima* resulted in only negligible mortality (Mills *et al.*, 1997). Whilst the above chemicals have proven to be effective anesthetic agents for pearl oysters, there has been no published account of commercial use of anesthesia during nucleation of any species. In commercial operations several thousand oysters will be nucleated on a daily basis and use of anesthesia on this scale is likely to be impractical.

8.9.3. Pearl nuclei

The nuclei (or beads) used for cultured pearl production are almost exclusively manufactured from the shells of American freshwater mussels. In particular, the pigtoe, washboard, butterfly, three ridge and dove shells of the Family Unionidae provide ideal raw material for nucleus manufacture (Alagarswami, 1970; Fassler, 1991). The shells

FIGURE 8.7 Akoya pearl oysters held at high density in a plastic conditioning box prior to nucleation for pearl production. (Photo: Josiah Pit.)

of these mussels have very thick nacreous layers with a hardness, specific gravity and thermal conductivity that make them particularly suitable for use as pearl nuclei (Gervis and Sims, 1992). The process of nucleus manufacture involves cutting the shells into cubes, followed by tumbling in a lapping machine to form spherical beads. A final treatment in hydrochloric acid produces a polished finish (Gervis and Sims, 1992). Nuclei of up to 13.5 mm in size can be produced from these freshwater mussels (Roberts and Rose, 1989).

Unfortunately, stocks of American freshwater mussels are under serious threat due to loss of habitat and the spread of the introduced zebra mussel, which has displaced the larger indigenous species in many areas (Fassler, 1995). As a result, there has been considerable interest in alternative materials for the manufacture of pearl nuclei. Roberts and Rose (1989) produced nuclei from giant clam shells; however, difficulties were encountered when drilling the resulting pearls due to differences in thermal conductivity. Scoones (1990) reported that nuclei made from *P. maxima* MOP were a suitable alternative to traditional nuclei. Pearl industry representatives raised concerns, however, due to the darker coloration of the MOP beads. Several patents exist in Japan for artificial nuclei produced from materials such as ceramics (Fassler, 1995). In 1996, Biron Corporation Ltd. in Australia produced artificial nuclei from a calcium carbonate-based material with the same attributes as the American mussel nuclei (M. Snow, Biron Corporation Ltd., personal communication). These "bironite" nuclei were implanted in *P. maxima* in a series of experiments to test nacre adhesion and other properties and the resulting pearls were indistinguishable from pearls nucleated with standard shell nuclei (Taylor, unpublished data, 1998). Alternative materials used for pearl nuclei should be readily available, have a relatively low cost, a white-based color and a composition similar to that of freshwater mussel shell (Ventouras, 1999). Undesirable characteristics for such materials include extreme hardness, problems relating to cracking and fracturing

TABLE 8.1

Some characteristics of clam shell, MOP and the manufactured material "Bironite", as potential alternatives to natural freshwater mussel shell used for pearl nuclei (Ventouras, 1999).

Characteristic	Mussel shell	Clam shell	MOP	Bironite
Density	2.8	2.72	2.71	2.84
Hardness	135–223	237–283	181–209	172–204
Expansion	17.2×10	15.4×10	15×10	21.6×10
Appearance	Glassy	Glassy	Glassy	Icy
Color	White/Brown	White	Brown	White
Drillability	Excellent	Good	Good	Good

when being drilled and unsuitable coloring. Important characteristics of some of the materials investigated as potential pearl nuclei are shown in Table 8.1.

The major problem with artificial nuclei to date has been market acceptance. Some of the major pearl traders have stated that they will not buy or sell cultured pearls that utilize artificial nuclei (A. Müller, personal communication). This combined with a surprising drop in the price of nuclei over the last 5 years has all but removed the incentive to commercialize the use of artificial nuclei.

8.9.4. Saibo donors and nucleus implantation

Nucleation for round pearl production is a surgical technique (Kawakami, 1952a). Methods vary slightly depending on the species, but the general technique is as follows. Saibo (pieces of mantle tissue for grafting) donors are carefully selected based on the visible quality of their nacre as the donor tissue will influence the color and quality of resulting pearls (Wada 1985; Wada and Komaru, 1996; Taylor, 2002). Suitable donor oysters are sacrificed and the anterior section of the mantle is removed from each valve as a single strip. The outer section of mantle tissue responsible for secreting the periostracal shell layers (see Section 3.3; Fig. 3.14; Table 3.1) is removed and the remaining inner epithelial layers of mantle (i.e. pallial mantle, see Fig. 3.14) are cut into square sections (Fig. 8.8). The number of useable saibo sections will vary depending on species and pearl oyster size. Typically, larger pearl oysters such as *P. maxima* and *P. margaritifera* will provide 20–30 pieces of saibo per donor oyster. The cut tissue is usually placed on a clean piece of moistened disposable tissue or cloth laid onto glass, wood or plastic for use.

Saibo can be removed from live oysters, allowing the mantle to regenerate (Acosta-Salmon and Southgate, 2005). This technique could allow the reuse of high-quality donor oysters; however, the technique may best be suited at this stage to evaluating nacre quality from donor oysters that could later be used as breeding stock (Acosta-Salmon *et al.*, 2004). The use of liquid nitrogen to preserve mantle tissue for later use as saibo

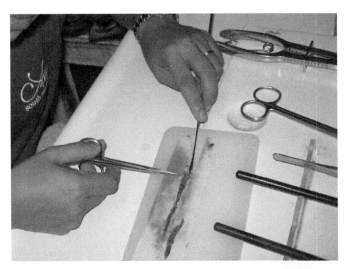

FIGURE 8.8 Strips of pallial mantle tissue from *Pinctada maxima* being prepared for use as saibo grafts for pearl nucleation. (Photo: Gustaf Mamangkey.)

FIGURE 8.9 Akoya pearl oysters are wedged open in preparation for nucleation. (Photo: Josiah Pit.)

grafts has been reported (Strack, 2006). These strips were used up to 14 months after freezing, but success was 40% lower than non-preserved mantle tissue.

Host pearl oysters are removed from the water and allowed to naturally open. Once open, a plastic or wooden wedge(s) is placed between the valves to keep the pearl oyster open (Fig. 8.9). The operating technician uses a speculum (ratcheted spreading tool) to open the pearl oyster's valves further and remove the wedge. The oyster is

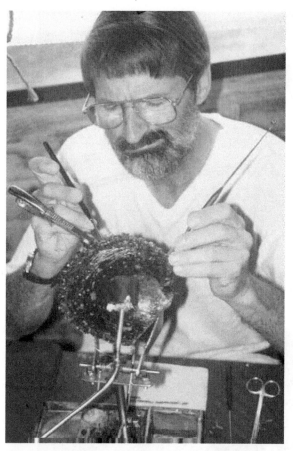

FIGURE 8.10   *Pinctada margaritifera* which is clamped open using a speculum and placed in the shell stand, undergoes nucleation for pearl production. The technician inserts a nucleus and a piece of mantle tissue into the gonad of the oyster. Note the row of mantle tissue sections on the moistened white paper in front of the technician. (Photo: John Lucas.)

placed in a clamp positioned to provide the grafting technician a view of the oyster's tissues between the shell valves (Fig. 8.10). The pearl oyster's gills are parted using a spatula and the foot is held firmly in place using a special tool with a hooked end. The technician first begins the incision near the join between the foot and gonad (Fig. 3.3). Using a specially shaped scalpel, the technician makes an incision within the gonad following the general line of the foot retractor muscles. Near the proximal end of the gonad, the technician twists the scalpel downwards in an arc that finishes at the desired location for placing the nucleus and saibo (Fig. 8.10).

The saibo and nucleus can be implanted in either order: the critical factor is ensuring that the saibo tissue is in secure contact with the nucleus. If the saibo implant is not in contact with the nucleus, the pearl-sac will not form around it and only a non-nucleated or "keshi" pearl may result (see Section 8.6). A range of specialized tools are used by the grafting technician (Fig. 8.11).

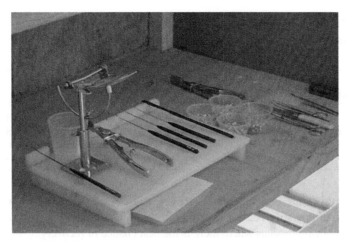

FIGURE 8.11 Specialized tools used by grafting technicians. On the left of the board is the shell stand used to hold the oyster to be operated, with the speculum (used to hold the oyster open) to the right of the stand. A range of different sized nuclei are held in the plastic containers to the right. Other tools are used for cutting and placing the nuclei and mantle tissue within the oysters, and the small board covered with white absorbent paper at the front is used to hold the mantle tissue sections. (Photo: Paul Southgate.)

8.9.5. Pearl formation

The process of pearl formation is identical to that of shell formation (Farn, 1986). The pearl-sac develops during the first weeks of recuperation. Initially, the inner epidermis and mesodermal layers of the mantle tissue graft degenerate leaving only the outer epidermal layer (Kawakami, 1952a, b). The pearl-sac is the result of the epithelial cells of the mantle epidermis growing around the nucleus to completely encase it. After 2 weeks, the pearl-sac begins to lay down periostracal material followed later by the prismatic layer. After approximately 40 days, the nacreous layers begin to form (Farn, 1986). This process of pearl formation through deposition of successive layers onto the nucleus is identical to that of shell formation (Farn, 1986).

8.9.6. Post-operative care

Pearl oysters are returned to the farm after the operation. Post-operative husbandry varies depending on species and historical development of techniques. In some cases, these techniques have little to do with scientific research. For example, it is common practice to "turn" *P. maxima* on a regular basis after the operation for periods up to 3 months (Scoones, 1990; Wells and Jernakoff, 2006). On most farms, *P. maxima* are placed in a horizontal plane with the left valve uppermost immediately after operation. On Australian farms this is achieved by placing oysters flat on the seabed. Once the wound has healed (usually after 5–7 days), "turning" commences. Every few days for a period lasting up to 3 months, the oyster is turned over. The theory is that this process will help centralize the developing pearl within the gonad, thus increasing the likelihood of a spherical pearl. However, this has not been confirmed experimentally (Scoones, 1990)

TABLE 8.2

Mortality and retention rates of pearl grafts post-operation for commercially important pearl oysters.

Species	Mortality	Retention of implant post-operation	Source
Akoya (Japan)	10–20%	>40%	Strack (2006)
P. margaritifera	Up to 30%	60%	Strack (2006)
P. maxima	2–10%	70–90%	Taylor, unpublished data, (1993–2006)

and many farms do not turn their pearl oysters and still achieve the same result. In either case, post-operative pearl oysters require careful handling and observation and must be placed in culture areas where there is the best chance of recovery.

Retention of growing pearls after the operation is variable. Reports of different commercial percentages for pearl retention are shown in Table 8.2. Retention is determined by a number of means. The most accurate technique is to X-ray the oysters; however, this equipment is expensive and can be difficult to maintain (a typical X-ray unit used for *P. maxima* or *P. margaritifera* costs ca. $US55,000). Many pearl farmers determine retention by placing a "catcher" mesh (usually a fine mesh bag) around the pearl oyster following the nucleation procedure. Should the nucleus be rejected, it will be caught in the "catcher" bag and help the farmer identify oysters which no longer retain a nucleus.

8.9.7. Culturing and harvest

The period from operation to harvest varies somewhat depending on the species. Akoya pearls grown mostly in Japan and China are harvested within 1 year of nucleation, whereas pearls from *P. maxima* and *P. margaritifera* are typically grown for between 18 months and 3 years depending on local environmental conditions (Gervis and Sims, 1992).

Most pearl farms favor suspended culture from rafts or long-lines (Gervis and Sims, 1992; Figs. 7.9 and 7.11). Some *P. maxima* farmers, however, favor bottom farming (see Section 7.5.3). Scoones (1990) reported better quality pearls produced using bottom farming practices in Roebuck Bay, Broome, Western Australia. Some small-scale farmers in China and Myanmar grow their pearl oysters (*P. maxima*) exclusively in baskets placed on the seabed.

A strict cleaning program must be maintained to promote the health of the pearl oysters and this is discussed in detail in Chapters 7 (Section 7.6.2) and 15 (Section 15.4.2). Cleaning methods vary depending on species and location. In countries such as Indonesia and China, where labor costs are comparatively low, smaller farms tend to hand clean their pearl oysters using scrapers and brushes. In Australia, Japan and Tahiti and most large-scale farms, pearl oysters are mechanically cleaned using high-pressure water pumps (e.g. Fig. 15.6). Experimental use of paraffin based wax coatings on live oysters and equipment are showing positive results both in Australia and Indonesia.

P. maxima treated in this way have been left without physical cleaning for more than 3 months with no detrimental effect on oyster health.

Pearl harvesting is usually done at the coolest time of the year. This is to promote a finer finish and better luster to the pearls as nacre deposition is slower in cooler water (Alagarswami, 1987b). In Indonesia, pearl growth is sometimes slowed during the last months prior to harvest by lowering the pearl oyster to greater depths or by crowding them on culture rafts. Careful monitoring of such activity is essential to maintain good health of the oysters in these last critical stages before harvest.

The number of pearls produced from a single oyster varies with species. Akoya pearl oysters can accommodate two nuclei at different positions within the gonad and subsequently can produce two pearls, but the pearl oyster is sacrificed at harvest (Gervis and Sims, 1992). *P. maxima* and *P. margaritifera* are nucleated with a single nucleus, but, following the first harvest can be reimplanted to produce another pearl. As a result, the harvest is another surgical procedure and the same degree of care is taken as with the initial operation. Unlike the initial operation, however, a saibo or mantle graft is not required, as the pearl-sac has already been formed. Some Australian pearl farmers produce up to three pearls sequentially from a single oyster.

8.10. PEARL QUALITY CRITERIA AND GRADING

Unlike diamonds, there is still no recognized world standard for the grading of pearls and various organizations apply different nomenclature (Matlins, 1996; Strack, 2006). In general, pearl value is based on what are termed the "five virtues" of pearl quality: size, shape, color, luster and surface.

The larger, rounder, smoother and brighter a pearl is, the higher its value. Color is a somewhat subjective indicator of value and will depend on the species. Tahitian "Black" South Sea pearls (from *P. margaritifera*) with overtones of purple, peacock green or pinks are more highly prized than pearls that appear gray or brown. "White" South Sea pearls (from *P. maxima*) that are a deep golden color carry the highest value of any marine pearl of comparative size, luster, shape and smoothness (Fig. 8.12).

Grading systems will vary depending on the pearl company involved. In general pearls are first graded into basic shapes. The most common shape grades are:

- Round: A pearl that is near to perfectly spherical.
- Near round: A pearl that appears very round without obvious deviations in symmetry.
- Drop: An elongated pearl that is generally tear shaped.
- Button: A pearl that is flat on one side but rounded on the other.
- Semi-baroque: A pearl lacking symmetry but approximating to the above shape categories.
- Baroque: A pearl without symmetry that cannot be placed in any obvious shape category.
- Circle: A symmetrical pearl with concentric lines running around the pearl.

FIGURE 8.12   High-quality gold and silver pearls produced from *Pinctada maxima* in Indonesia. (Photo: David Ramsay, Atlas Pearls) (See Color Plate 11)

Pearls of various shapes can be further divided by size. Size is determined by the maximum diameter of the pearl in millimeters or *bu* (a Japanese measure equivalent to approximately 3.1 mm). Marine cultured pearls vary in size depending on species. Those from Akoya oysters are 5–9 mm in size, whilst those from *P. margaritifera* are 8–16 mm and those from *P. maxima* are 9–20 mm (Müller, 1999; Strack, 2006). Further grading is completed on the basis of color before a final division by luster and surface perfection. An example of a typical description of a graded pearl parcel would be:

Shape:	Round
Size:	10–14 mm
Color:	White/Pink
Grade:	A

The grade here reflects the relative quality of the pearl's luster and surface. In this case, the pearls would be virtually free of any but the most minor blemishes (and these would be small enough to be lost through the drilling and setting process) with a very high reflective index in terms of luster. Pearls are often sold in two basic grades with further internal division. The main grade of pearls is generally termed "sellable" with 5–6 grades from gem quality (near perfect) to a lower grade with blemishes, imperfections and diminished luster. The other main grade is termed "commercial" goods, which are pearls with considerable blemishes, but which have some value for manufacture into relatively low cost jewelry. Significant proportions of harvested pearls have severe defects, i.e. the nucleus incompletely coated with nacre, and have no commercial value.

There are differing standards used in weighing pearls. The most widely used system by pearl producers, wholesale pearl traders and industry journals is based on the Japanese system of *momme* (Shor, 2007). One *momme* is equivalent to 3.75

g; 1000 *momme* is termed a *kan.* An alternative measure of weight, generally used in the jewelry trade, is carats (1 Carat = 0.200 g) while in the past they were weighed in grains (4 grains = 1 carat). During the 17th century, Jean Baptiste Tavernier established a rule for the evaluation of natural pearls that was still in use a few years ago. It is based on a size factor, named "One times the weight" (1 × the weight), which is obtained by multiplying the weight in grains by itself. The quality factor considers the four remaining grading factors by estimating a figure between 0.5 and 30. This multiplying figure for a pearl of very good quality may be, for example, between 15 and 20. A pearl with a diameter of 5.5 mm may weigh, for example, 4.68 grains. When the multiplying figure is 20×, it results in the following equation:

$$4.68 \times 4.68 \times 20 = 438$$

The price of the pearl would be US$438. The formula changes when pairs or necklaces are evaluated, as the grain weights are summed.

8.11. FACTORS INFLUENCING PEARL QUALITY

8.11.1. Saibo donors

One of the most critical factors in determining pearl quality is the saibo donor. The general quality of nacre deposition in the shells of donor pearl oysters will be reflected in the resulting pearls. In particular, it is clear that the color of the nacre from a donor will dominate in harvested pearls. Considerable research has been done in recent years demonstrating this effect (Alagarswami, 1987a, b; Wada and Komaru, 1996; Taylor, 2002). Other factors affecting the donor's suitability are growth rate and age. In general younger oysters are better as saibo tissue donors as they are rapidly producing high luster nacre. Older pearl oysters tend to produce nacre at a slower rate and the resulting nacre will tend to have lower luster.

8.11.2. Surgical technique and hygiene

The skill of the operating technician and the knowledge employed by them in selecting which pearl oysters should undergo nucleation is a contributing factor to pearl quality. Technicians nucleating similar oysters under almost identical conditions will still demonstrate significant differences in terms of nucleus retention and ultimate pearl quality. Hygiene is also recognized as an important factor in insuring success of the surgery (Cochennec-Laureau *et al.*, 2004). Various studies have looked at the use of antiseptics (Norton *et al.*, 2000; Cochennec-Laureau *et al.*, 2004) and anesthetics (Norton *et al.*, 1996, 2000; Mills *et al.*, 1997; Monteforte *et al.*, 2004; Acosta-Salmon *et al.*, 2005), but the effect on pearl quality is yet unproven.

A variety of antibiotic coatings are available for use on nuclei. The most common ingredient is tetracycline. These products are promoted as a means of improving postoperative recovery and consequently, pearl quality. At this time there are no published scientific data supporting this, but the products are widely used in the industry.

8.11.3. Pearl growth rate

The rate of pearl growth has a direct effect on quality. Very rapid growth often results in a pearl with a beaten surface appearance commonly referred to as "hammering." This can dramatically reduce a pearl's value despite it having excellent size, shape, color and luster. Very slow growing pearls rarely reach large sizes and often have a dull appearance. Ideally, pearls should grow at a rate resulting in consistent regular crystal formation for a very smooth and bright finish (Snow *et al.*, 2004). Several environmental and genetic factors interplay in terms of a pearl oyster's growth rate. Whilst donor tissue has a major influence on quality factors such as color, the host appears to determine the rate of pearl growth (Wada and Komaru, 1996).

8.11.4. Husbandry

Culture system and culture method will influence the health and growth of pearl oysters and ultimately pearl quality. Stocking density, water depth and cleaning systems and cycle will affect food availability, which in turn influence health and several studies have shown the need to reduce stocking density as pearl oysters grow (e.g. Taylor *et al.*, 1997a). Biofouling and its control are a significant cost for production and have a major influence on health and pearl quality (de Nys and Ison, 2004; Chapter 15).

A Chinese study (Wu *et al.*, 2003) reported that polyculture of the red sea weed *Kappaphycus alvarezii* with Akoya pearl oysters led to improved pearl quality with nacre deposition thicker and more homogenous in the polycultured pearl oysters compared to those in monoculture.

As in any farming venture, there is a trade off between maximizing health and pearl quality and control of costs. Putting this to one side, it is obvious that in any given farming operation the costs of producing a high-quality pearl will be the same as producing a pearl that cannot be sold and every farmer must strive to maximize quality within the scope of an individual operation.

8.11.5. Pearl enhancement

There are a variety of pearl enhancement methods available to improve the look of pearls (Table 8.3); however, the exact means employed are carefully kept industrial secrets (Strack, 2006). For certain types of pearls, notably freshwater and Akoya products, enhancement is the rule rather then the exception (Müller, 1997; Anon, 2005; Hänni, 2006; Strack, 2006). In general, South Sea pearls are not treated in their countries of origin and are sold initially in their original colors; however, South Sea pearls are treated by a number of Japanese pearl companies (Strack, 2006). Akoya pearls in their natural state are mainly cream, yellow and green, but are sold in a wide range of colors (Müller, 1997). Bleaching agents including chemical treatments and UV radiation are used to "whiten" pearls. Some low value, yellow South Sea pearls have been treated to appear gold in an effort to improve value.

A variety of dyes can be used to transform or enhance pearl color (Table 8.3). Traditionally, Japanese Akoya pearls were treated with the vegetable dye eosin to impart a pink hue. In modern times manufactured chemicals have replaced this traditional

TABLE 8.3
The effect of various enhancement methods for pearls. SSP (Black) = black South Sea pearls from *Pinctada margaritifera*. SSP (White) = white South Sea pearls from *P. maxima*.

Type of pearl	Effect	Enhancement method	Source
SSP (Black)	Black	Silver nitrate + UV light and/or hydrogen sulphate	Strack (2006)
	Chocolate color	Silver nitrate Hair dye extracts	Anon (2005) Hänni (2006)
SSP (White)	Gold color	Organic dyes Inorganic dyes	Strack (2006)
Akoya	Pink	Eosin (Traditional) Cobalt salts	Strack (2006)
	Whitening	Hydrogen peroxide + UV light	Strack (2006)
	Black	Silver nitrate + UV light and/or hydrogen sulphate	Strack (2006)

dye (Strack, 2006). The practice of artificially dyeing pearls black using silver salts did not come into fashion until the late 19th century and new color treatments have appeared in recent years. Large so-called "chocolate" pearls have been sold recently (Anon, 2006; Hänni, 2006). These pearls have been tested and found to be South Sea pearls (mostly Tahitian "Black") and have been treated with stains similar to that used in hair dye (Hänni, 2006) or, in some cases, silver nitrate (Anon, 2006).

Luster is enhanced through mechanical polishing and sometimes chemical treatment. Traditional luster enhancement for Akoya pearls is termed *maeshori* and involves the use of solvents such as methyl alcohol (Shor, 2007). Initial polishing usually involves a rotating drum filled with the pearls and a polishing material such as bamboo, walnut or cork beads. Paraffin or bees wax may be added. Final polishing can involve the use of diamond, gold or platinum dust (Strack, 2006). South Sea pearls can be polished manually using felt discs and bees wax polish.

In terms of marketing there are different perceptions of pearl enhancement depending on the species or type of pearl involved. South Sea pearls, both black and white, are strongly promoted as having minimal enhancement (washing and polishing only). Articles appear regularly in industrial journals condemning the use of color treatment on these pearls. In contrast, it is widely acknowledged, although not necessarily divulged to consumers, that all Akoya pearls are enhanced (Strack, 2006).

8.12. *IN VITRO* PEARL CULTURE

The idea of culturing pearls in the laboratory and thus abandoning the vagaries of farming has been an attractive (occasionally fearful) idea for many in the pearling industry. There is considerable literature demonstrating that it is possible to produce nacre using *in vitro* culturing of mantle cells (Suzuki and Mori, 1991; Awaji and Suzuki, 1998; Dharmaraj and Suja, 2001; Shi *et al.*, 2002; Barik *et al.*, 2004; Li *et al.*, 2005; Sugishita *et al.*, 2005). Several researchers have demonstrated an ability to produce nacre under laboratory conditions (Dharmaraj and Suja, 2001; Shi *et al.*, 2002; Barik *et al.*, 2004). However, there is as yet no reported production of a commercial size pearl. Wang *et al.* (2002) successfully cultured epithelial cells from Akoya pearl oysters for several weeks; however, after 31 days of culture cellular secretion began to weaken and by 41 days the cells began to die.

At this stage, the technology is best placed to look at specific biochemical interactions involved in the production of nacre in order to better improve our understanding of processes involved in nacre secretion. This line of research is being pursued (Pereira-Mouries, 2003; Tsukamoto *et al.*, 2004; Zhang *et al.*, 2004, 2006; Chen *et al.*, 2005).

8.13. SUMMARY

Contrary to the popular conception that natural pearls are formed by an irritant in the pearl oyster's tissues, recent evidence suggests that pearls result from mantle tissue damage. Pearls form when the mantle is damaged and one or more epithelial cells from the upper epithelial layer of the mantle are transferred into the connective tissue. The cells multiply and form a closed cyst, a pearl-sac. Young cells lining the pearl-sac secrete conchiolin, then a prismatic layer of calcite crystals and then nacre layers. The pearl has similar structure to the shell but the layers are in reverse order and concentric. The nacre layer has the same polygonal shapes in a brick-like structure as in shell nacre. This special structure of the pearl's surface gives rise to interference and diffraction of light, which lead to the iridescent properties of pearls.

As well as the typical natural pearl, there are natural blister pearls where pearls breaks free from the connective tissue and become pressed against the nacreous layer of the shell where nacre deposition fuses the pearl to the shell. Natural blisters form where foreign objects or organisms penetrate through the shell and the outer mantle surface secrete nacre to seal off the hole. This repair work may leave no trace, but, typically, it leaves protuberances (blisters) on the inside of the shell.

There are four major types of cultured pearls: mabé, non-nucleated, nucleated and keshi pearls. Mabé pearls are the oldest type of cultured pearls and involve fixing shapes on the inside of the shells. The shapes are subsequently coated with nacre by the outer surface of the mantle, like a natural blister pearl. Mabé are then cut out of the shell. Non-nucleated pearl production is by surgically implanting donor mantle sections into the mantle of the host. Production by this technique with pearl oysters is insignificant, but non-nucleated pearls from freshwater mussels account for >90% of all global pearl production. Nucleated pearl production involves surgical implantation of a shell-based bead and a section of mantle tissue (saibo) from a donor oyster into a host: a process known as grafting. A keshi pearl may form incidentally in a pearl-sac subsequent to the implanted nucleus being expelled.

There is no recognized world standard for grading pearls. The "five virtues" of pearl quality are size, shape, color, luster and surface. Significant proportions of pearl harvests have severe defects and are discarded. The other pearls are often sold in two basic grades: "sellable" (near perfect to some blemishes) and "commercial" (with considerable blemishes). The latter have some value for low cost jewelry. Harvested pearls may be enhanced by processes such as bleaching, dyeing and polishing.

Nucleated pearl production involves a series of stages: pre-operative oyster conditioning; saibo selection and insertion into the host's tissues with a nucleus; post-operative care; oyster culture during pearl formation; pearl harvest. The development of this method in the early 1900s provided pearl farmers with the means to produce cultured pearls on a reliable basis. The technique has changed little since it was developed. There is much potential for improving pearl yields and quality and this has stimulated research. A number of studies, for example, have looked at improving the nucleation procedure through the use of antiseptics, antibiotics and closure of the resulting wound. However, although such research is beneficial, significant advances in improved pearl yield and pearl quality will only be made through better understanding of the factors that directly influence nacre secretion, nacre quality and nacre color. The relative roles of genetics (of host and donor) and environment on pearl quality are unclear. Further research in these areas, including targeted pearl oyster breeding programs, is needed to improve pearl yield and pearl quality: the goals of pearl oyster farming.

References

Acosta-Salmon, H., Rangel-Davalos, C., 1997. Benzocaine (ethyl-*p*-aminobenzoate) as anesthetic for surgical implants of nucleus in the pearl oyster, *Pteria sterna* (Gould 1851). World Aquaculture Society, World Aquaculture '97. Book of Abstracts, Seattle, WA, USA. Feb 19–23, 1997.

Acosta-Salmon, H., Southgate, P.C., 2005. Mantle regeneration in the pearl oysters *Pinctada fucata* and *Pinctada margaritifera*. *Aquaculture* 246, 1–4.

Acosta-Salmon, H., Martinez-Fernandez, E., Southgate, P.C., 2004. A new approach to pearl oyster broodstock selection: Can saibo donors be used as future broodstock?. *Aquaculture* 231, 1–4.

Acosta-Salmon, H., Martinez-Fernandez, E., Southgate, P.C., 2005. Use of relaxants to obtain saibo tissue from the blacklip pearl oyster (*Pinctada margaritifera*) and the Akoya pearl oyster *Pinctada fucata*. *Aquaculture* 246, 167–172.

Alagarswami, K., 1970. Pearl culture in Japan and its lessons for India. In: Proceedings of the Symposium on Mollusca Part III. Mar. Biol. Assoc. India, pp. 975–998.

Alagarswami, K., 1987a. Technology of cultured pearl production. In: Alagarswami, K. (Ed.), *Pearl Culture. Bull. CMFRI No 39*. Central Marine Fisheries Research Institute, Cochin, India, pp. 98–106.

Alagarswami, K., 1987b. Cultured pearls – production and quality. In: Alagarswami, K. (Ed.), *Pearl Culture. Bull. CMFRI No. 39*. Central Marine Fisheries Research Institute, Cochin, India, pp. 107–111.

Anon., 2005. More dyed black Chinese pearls available. *Jewellery News Asia* 252, 58.

Anon., 2006. GIA identifies three types of brown pearls. *Jewellery News Asia* 265, 80.

Awaji, M., Suzuki, T., 1995. The pattern of shell proliferation during pearl-sac formation in the pearl oyster. *Fish. Sci.* 61, 747–751.

Awaji, M., Suzuki, T., 1998. Monolayer formation and DNA synthesis of the outer epithelial cells from pearl oyster mantle in coculture with amebocytes. *In Vitro Anm Cell Dev Biol* 34(6), 486–491.

Barik, S.K., Jena, J.K., Janaki, K.R., 2004. *In vitro* explant culture of mantle epithelium of freshwater pearl mussel. *Indian Exp. Biol.* 42, 1235–1238.

Che, L.M., Golubic, S., Le Campion-Alsumard, T., Payri, C., 2001. Development aspects of biomineralisation in the Polynesian pearl oyster *Pinctada margaritifera* var. *cumingii*. *Oceanol. Acta* 24, S37–S49.

Chellam, A., Victor, A.C.C., Dharmaraj, S., Velayudahan, T.S., Rao, K.S., 1991. Pearl oyster farming and pearl culture. FAO/UNDP Reg. Seafarming Development and Demonstration Project. Bangkok.

Chen, H.T., Xie, L.P., Yu, Z.Y., Xu, G.R., Zhang, R.Q., 2005. Chemical modification studies on alkaline phosphatase from pearl oyster (*Pinctada fucata*): A substrate reaction course analysis and involvement of essential arginine and lysine residues at the active site. *Int. J. Biochem Cell Biol.* 37, 1446–1457.

Cochennec-Laureau, N., Haffner, P., Levy, P., Saulnier, D., Langy, S., Fougerouse, A., 2004. Development of antiseptic procedure to improve cultured pearl formation in *Pinctada margaritifera*. *J. Shellfish Res.* 23, 285.

Comps, M., Herbaut, C., Fougerouse, A., 2000. Abnormal periostracum secretion during the mineralization process of the pearl in the blacklip pearl oyster *Pinctada margaritifera*. *Aquat. Living Resour.* 13, 49–55.

Dan, H., 2003. Fresh water pearl culture and production in China. *J. Shellfish Res.* 22(1), 325.

de Nys, R., Ison, O., 2004. Evaluation of antifouling products developed for the Australian pearl industry. FRDC Project No. 2000/254. Fisheries Research and Development Corp., Canberra, 113 pp.

Dharmaraj, S., Suja, C., 2001. Biotechnological approach in *in vitro* pearl production. In: Menon, N., Pillai, P. (Eds.), *Perspectives in Mariculture*. The Marine Biological Association of India, Cochin, India, pp. 405–412.

Dix, T.G., 1972a. Histology of the mantle and pearl-sac of the pearl oyster *Pinctada maxima* (Lamellibranchia). *J. Malac. Soc. Aust.* 2, 365–375.

Dix, T.G., 1972b. Histochemistry of mantle and pearl-sac secretory cells of the pearl oyster *Pinctada maxima* (Lamellibranchia). *Aust. J. Zool.* 20, 359–368.

Farn, A.E., 1986. *Pearls Natural, Cultured and Imitation.* Butterworth Gem Books, London, 150 pp.

Fassler, C.R., 1991. Farming jewels: The aquaculture of pearls. *Aquaculture Mag* 17(5), 34–52.

Fassler, C.R., 1995. New developments in pearl farming. *World Aquaculture* 26(3), 5–10.

Garcia-Gasca, A., Ochoa-Baez, R.I., Betancourt, M., 1994. Microscopic anatomy of the mantle of the pearl oyster *Pinctada mazatlanica* (Hanley, 1856). *J. Shellfish. Res.* 13(1), 85–91.

Gervis, M.H., Sims, N.A., 1992. The biology and culture of pearl oysters (Bivalvia: Pteriidae). ICLARM Stud. Rev. 21, ODA (Pub.), London, 49 pp.

Hänni, H., 2006. Study shows chocolate pearls are "stained". *Jewellery News Asia* 265, 78.

Hollyer, J., 1984. Pearls – jewels of the sea. *Infofish Marketing Digest* 5, 32–34.

Jabbour-Zahab, R., Chagot, D., Blanc, F., Grizel, H., 1991. Mantle histology and ultrastructure of the pearl oyster *Pinctada margaritifera* (L.). *Aquat. Living Resour.* 5, 287–298.

Kawakami, I.K., 1952a. Studies on pearl-sac formation. On the regeneration and transplantation of the mantle piece in the pearl oyster. *Mem. Fac. Kyushu Univ. (Ser. E.)* 1, 83–89.

Kawakami, I.K., 1952b. Marine regeneration in pearl oyster (*Pinctada martensii*). *J. Fuji Pearl Inst.* 2(2), 1–4.

Li, S., Xie, L., Ma, Z., Zhang, R., 2005. cDNA cloning and characterization of a novel calmodulin-like protein from pearl oyster *Pinctada fucata*. *FEBS J.* 272, 4899–4910.

Malek, F.A., Cheng, T.C., 1974. *Medical and Economic Malacology.* Academic Press Inc, New York pp. 155–263.

Matlins, A.L., 1996. *The Pearl Book: The Definitive Buying Guide.* Gemstone Press, Woodstock, Vermont, 198 pp.

McLaurin, D., Arizmendi, E., Farell, S., Nava, M., 1997. Pearls and pearl oysters from the Gulf of California, Mexico. *Australian Gemmologist* 19, 497–501.

McLaurin, D., Arizmendi, E., Farell, S., Nava, M., 1999. Pearls and pearl oysters from the Gulf of California, Mexico. *Australian Gemmologist* 20, 239–245. An update.

Mills, D., Tlili, A., Norton, J., 1997. Large-scale anaesthesia of the silver-lip pearl oyster, *Pinctada maxima* Jameson. *J. Shellfish Res.* 16(2), 573–574.

Miyauti, T., 1965. Studies on artificial spawning in pearl culture. I. On shell movement and oxygen consumption of the Japanese pearl oyster, *Pteria* (*Pinctada*) *martensii* (Dunker) inside the spawning-basket and dissolved oxygen of seawater inside and outside the spawning-basket. *Seiri Seitai* 13, 81–87. (in Japanese with English summary)

Mizumoto, S., 1976. Pearl farming – a review. In: Pilay, T.V.R. (editor) FAO *Technical Conference on Aquaculture*, Kyoto, Japan, 26 May, 1976. FAO-FIR: AQ/Conf. 76 R.13. pp. 381–385.

Monteforte, M., Bervera, H., Saucedo, P., 2004. Response profile of the Calafia pearl oyster, *Pinctada mazatlanica* (Hanley, 1856), to various sedative therapies related to surgery for round pearl induction. *J. Shellfish Res.* 23, 121–128.

Müller, A., 1997. *Cultured Pearls – The First Hundred Years.* Golay Buchel USA Ltd., 142 pp.

Norton, J.H., Dashorst, M., Lansky, T.M., Mayer, R.J., 1996. An evaluation of some relaxants for use with pearl oysters. *Aquaculture* 144, 39–52.

Norton, J., Lucas, J., Turner, I., Mayer, R., Newnham, R., 2000. Approached to improve cultured pearl formation in *Pinctada margaritifera* through the use of relaxation, antiseptic application and incision closure during bead insertion. *Aquaculture* 184, 1–17.

Pereira-Mouries, L., 2003. Etude des composants de la matrice organique hydrosoluble de la nacre de l'huître perlière *Pinctada maxima*. Bioactivité dans le processus de regeneration osseuse. Biologie te Biochimie Appliquées thesis, Museum Natl. d'Histoire Naturelle, Paris, 164 pp.

Roberts, R.B., Rose, R.A., 1989. Evaluation of some shells for use as nuclei for round pearl culture. *J. Shellfish Res.* 8(2), 387–389.

Ruiz-Rubio, H., Acosta-Salmon, H., Olivera, A., Southgate, P.C., Rangel-Davalos, C., 2006. The influence of culture method and culture period on quality of half-pearls ("mabé") from the winged pearl oyster *Pteria sterna*, Gould, 1851. *Aquaculture* 254, 269–274.

Saucedo, P., Monteforte, M., Blanc, F., 1998. Changes in shell dimensions of pearl oysters, *Pinctada matzalantica* (Hanley 1856) and *Pteria sterna* (Gould 1851) during growth as criteria for mabé pearl implants. *Aquaculture Res* 29, 801–814.

Saville-Kent, W., 1893. *The Great Barrier Reef: Its Products and Potentialities.* W.H. Allen, London.

Scoones, R.J.S., 1990. Research on practices in the Western Australian cultured pearl industry. Final Report to Fishing Industry Research and Development Council – Project BP 12, July 1987 – June 1990. Broome Pearls Pty Ltd, Perth 74 pp.

Shi, Z. et al. 2002. Comparison of the tissue culture and cell culture from the mantle epithelium of *Hyriopsis cumingii*. *J. Shanghai Fisheries University/Shanghai Shuichan Daxue Xuebao* 11(1), 27–30.

Shirai, S., 1981. *Pearls* 168. Shunposha Photoprinting Co. Ltd., Yamaguchi, Japan, 168 pp.

Shor, R., 2007. From single source to global free market: the transformation of the cultured pearl industry. *Gems & Gemology* 43, 200–226.

Snow, M.R., Prong, A., Self, P., Logic, D., Shaper, J., 2004. The origin of the color in pearls in iridescence from nano-composite structures of the nacre. *Am Mineral* 89, 1353–1358.

Southgate, P., Rubens, J., Kipanga, M., Smumi, G., 2006. Pearls from Africa. *SPC Pearl Oyster Information Bulletin* 17, 16–17.

Strack, E., 2006. *Pearls.* Rühle-Diebener-Verlag GmbH & Co., KG, Germany, 706 pp.

Sugishita, Y., Hirano, M., Tsutsumi, K., Mobin, S.M.A., Kanai, K., Yoshikoshi, K., 2005. Effects of exogenous lipid peroxides on mortality and tissue alterations in Japanese pearl oysters *Pinctada fucata martensii*. *J. Aquat. Anim. Health* 17, 233–243.

Suzuki, T., Mori, K., 1991. Immunolocalization and *in vitro* secretion of hemolymph lectin of the pearl oyster, *Pinctada fucata martensii*. *Zool. Sci.* 8, 23–29.

Taylor, J.J., 2002. *Producing Golden and Silver South Sea Pearls from Indonesian Hatchery Reared Pinctada maxima.* World Aquaculture 2002, Beijing, P. R. China. April 23–27, 2002. World Aquaculture Society, Book of Abstracts, Beijing, P.R. China, p. 754.

Taylor, J.J., Rose, R.A., Southgate, P.C., Taylor, C.E., 1997a. Effects of stocking density on growth and survival of early juvenile silver-lip pearl oysters, *Pinctada maxima* (Jameson), held in suspended nursery culture. *Aquaculture* 153, 41–49.

Taylor, J.J., Southgate, P.C., Rose, R.A., 1997b. Fouling animals and their effect on the growth of silver-lip pearl oysters, *Pinctada maxima* (Jameson) in suspended culture. *Aquaculture* 153, 31–40.

Tsukamoto, D., Sarashina, I., Endo, K., 2004. Structure and expression of an unusually acidic matrix protein of pearl oyster shells. *Biochem. Biophys. Res. Commun.* 320, 1175–1180.

Tsujii, T., 1960. Studies on the mechanism of shell and pearl formation in mollusca. *J. Fac. Fish. Pref. Univ. Mie-Tsu* 5, 1–70.

Ventouras., G. 1999. Nuclei alternatives – the future for pearl cultivation. World Aquaculture '99. Book of abstracts. p.793.

Wada, K.T., 1985. The pearls produced from the groups of pearl oysters selected for colour of nacre in shell for two generations. *Bull. Natl. Res. Inst. Aquacult. Yoshokukenho* 7, 1–7.

Wada, K., Komaru, A., 1996. Color and weight of pearls produced by grafting the mantle tissue from a selected population for white shell color of the Japanese pearl oyster *Pinctada fucata martensii* (Dunker). *Aquaculture* 142(1–2), 25–32.

Wang, A., Yan, B., Su, Q., Ye, L., 2002. Electron microscopic observations on the cells from cultured mantle of the pearl oyster, *Pinctada martensii. D. Mar. Sci./Haiyang Kexue* 26, 4–9.

Wells, F.E., Jernakoff, P., 2006. An assessment of the environmental impact of wild harvest pearl aquaculture (*Pinctada maxima*) in Western Australia. *J. Shellfish Res.* 25, 141–147.

Wu, M., Mak, S.K.K., Zhang, X., Qiian, P.Y., 2003. The effect of co-cultivation on the pearl yield of *Pinctada martensii* (Dunker). *Aquaculture* 221, 347–356.

Zhang, Y., Xiao, R., Ling, L., Xie, L., Ren, Y., Zhang, R., 2004. Extraction and purification of water-soluble matrix protein from nacre and its effect on CaCO$_3$ crystallization. *Mar. Sci./Haiyang Kexue* 28, 33–37.

Zhang, C., Xie, L.P., Huang, J., Liu, X.L., Zhang, R.Q., 2006. A novel matrix protein family participating in the prismatic layer framework formation of pearl oyster, *Pinctada fucata. Biochem. Biophys. Res. Commun.* 344, 735–740.

CHAPTER 9

Exploitation and Culture of Major Commercial Species

Paul C. Southgate, Elisabeth Strack, Anthony Hart, Katsuhiko T. Wada, Mario Monteforte, Micheline Cariño, Sandra Langy, Cedrik Lo, Hector Acosta-Salmón and Aimin Wang

9.1. INTRODUCTION

This chapter presents information on the history of exploitation and culture of the major commercial species of pearl oysters. Some of these species have very long and colorful histories of exploitation that, in some cases, are inexorably linked with economic and cultural development of a region. For example, the lure of the "Pearl Myth" and exploitation of *Pinctada mazatlanica* in the Gulf of California provided the major impetus for Spanish colonization and exploitation of the region. While exploitation of natural pearl oyster populations led in many cases to over-fishing and the collapse of these populations, development of techniques for cultured round pearl production in the early 1900s provided the basis for a global cultured pearl industry with a current value of about US$0.5 billion per annum. This chapter provides information on cultured pearl production from the major commercial species and outlines problems encountered by these industries and bottlenecks to production. This chapter also presents broad information on the culture methods used for pearl oysters and the reader is referred to Chapter 7 for more detail of these methods.

9.2. AKOYA PEARL OYSTER

As outlined in Chapter 2 (Section 2.3.1.13) the Akoya pearl oyster is, at this stage, best considered a species complex encompassing *Pinctada fucata, martensii, radiata* and *imbricata* and reference to Akoya pearl oysters in the following section of this chapter refers to oysters within this complex.

9.2.1. History of exploitation

Exploitation of Akoya pearl oysters for natural pearls has a long history with the earliest record of a pearl fishery being in India in 400 BC. Since then fishers have collected wild pearl oysters and harvested natural pearls throughout the entire distribution of the Akoya pearl oyster (Fig. 2.4) in the Middle East, India, Sri Lanka (Ceylon), Atlantic, South and North Pacific Oceans, including Japan and China (Wada, 1991; see Chapter1). However, natural oyster populations have declined as a result of overfishing, which results from the low yield of natural pearls from wild oysters. For example, Ogushi (1938) reported that, in Japan, natural pearls with a weight of only 9 *rin* (1 *rin* = ca. 3.75 mg which is international units of pearls for trading) could be harvested from the visceral parts of about 200 Akoya pearl oysters.

Pearl culture using the Akoya pearl oyster began in Japan ca. 1916 and has subsequently become established in China and India (Kobayashi and Watabe, 1959; Wada, 1973a, 1991; Chellam *et al.*, 1991; Ikenoue, 1992). The potential of pearl production from this 'species' has also been investigated on the Atlantic coast of South America (Urban, 2000a, b; Lodeiros *et al.*, 2002; MacKenzie *et al.*, 2003), in Australia (O'Connor *et al.*, 2003; Pit, 2004), Korea (Choi and Chang, 2003) and in the Arabian Gulf (Narayana and Michael, 1968; Almatar *et al.*, 1993; Behzadi *et al.*, 1997), but information on commercial production of cultured pearls from these regions is not yet available.

9.2.2. Culture and production trends – Japan

The Mise-Nishikawa method for cultured round pearl production was developed using the Akoya pearl oyster and is outlined in Section 8.9.1. It allowed regular production of cultured pearls and became established in Japan ca. 1916 when Mikimoto began mass production of pearls using this method (see Section 8.9.1). By 1926 there were 33 pearl farms in Japan, which had produced nearly 7,000 pearls; Mikimoto owned 9 of these farms. By 1938 the number of pearl farms in Japan had increased to 360 (Fig. 9.1) and these had collectively produced more than 10 million pearls. Production of cultured pearls was recorded in official statistics during the initial stage of the industry (1926–1945); they were, for example, 669 pearls in 1926, 7,749 pearls in 1935 and 10,883 pearls in 1938. Estimated maximum weight of annual production was 2,500–3,000 *kan* (1 *kan* = 3.75 kg) during this initial stage of the industry.

The number of pearl farms slumped drastically during World War II (Fig. 9.1). Pearl production was ca. 50 *kan* when culture efforts resumed after the war in 1946, and it increased to ca. 1,000 *kan* in 1950. The early 1950s saw quality become the primary consideration of the Japanese cultured pearl industry. Women working in the pearl sorting rooms selected no more than five 'graded' necklaces per day. The diameter of the pearls ranged from 3 to 5 mm with a center pearl of ca. 7 mm. The trade called these necklaces "3.5 *momme* graduates" as they usually had a weight of 3.5 *momme* (one *momme* = 3.75 g). In 1952, harvests of cultured pearls amounted to almost 10 t, sold

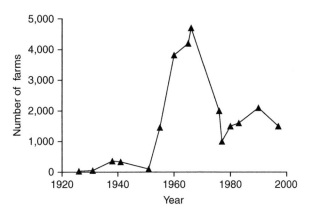

FIGURE 9.1 Changes in the number of Japanese pearl farms between 1920 and 2000. (*source*: Strack, 2006).

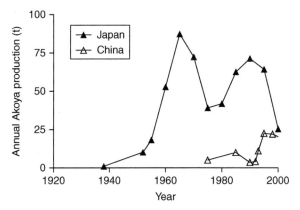

FIGURE 9.2 Annual production of Akoya pearls in Japan and China 1920–2000. (*source*: Strack, 2006).

almost exclusively to the United States, and by 1953 they were also exported to Europe. There was a remarkable increase in production from 1955 and by 1960 the number of cultured pearl farms in Japan had risen to 3,817 (Fig. 9.1). There were also 4,528 farms kept as breeding stations for production of juveniles. The cultured pearl industry at this time was estimated to have employed 200,000 people. Pearl production increased from 6,543 *kan* (24.5 t) in 1955 to 13,867 *kan* (52 t) in 1960 (Fig. 9.2).

Cultured pearls from Japan became a symbol for the western world. After a minor slump in 1961, production recovered quickly and the size of the pearls produced also increased to about 9 mm with choker necklaces, composed of similar sized pearls, gradually replacing the graded necklace. In 1966, pearl production reached its peak of 91.5 t produced from 4,700 farms (Figs. 9.1 and 9.2); however, a disease crisis for the industry in 1968 forced many companies to close (Fig. 9.1). The decrease in pearl quality that accompanied the crisis led to a slump in pearl exports from Japan by 50% and harvests continued to decline for the next 6 years. By this time, however, the domestic Japanese market had increased in importance and a third of all pearls produced by the industry were sold in Japan.

Prices rose again from 1972 onwards, although demand for pearls at first declined, as they did not go well with the new colorful fashions of the 1970s. However, when more conservative tastes returned after 1975, pearls became more popular again. Only 2,000 pearl farms remained in 1976 and pollution of pearl farming sites became an increasing problem. By 1977 only 1,000 pearl farms remained and these collectively produced only 35 t of pearls (Figs. 9.1 and 9.2). The number of pearl farms rose again to 1,500 in 1980, but production could not meet demand for high quality pearls and, as a result, large quantities of low quality pearls flooded the market. The Japanese cultured pearl industry was also influenced by strong competition from China's increasing production of pearls (Fig. 9.2). At the beginning of the 1990s, pearl production slumped and prices rose again as demand exceeded supply, particularly for 7–9 mm pearls. Japanese pearl farmers turned increasingly to producing large pearls to satisfy this demand and pearls in the 8–9 mm size range represented 25–30% of harvested pearls compared to about 5% in 1985. As a result of increasing Chinese freshwater pearl production, the supply of <6 mm Akoya pearls dropped.

Prices increased by 40% in 1994 and production focused on ever-more larger and better quality pearls to supply export markets in southeast Asia. A severe earthquake struck Kobe, the major Japanese pearl trading center, in January 1995. This affected the pearl industry and by the end of the year prices had dropped by 20–50%. In the summer of 1996 an epidemic claimed vast numbers of pearl oysters in Japan (see Section 9.2.6). This was the beginning of the worst catastrophe in the history of the Japanese cultured pearl industry. Oyster mortality soon reached 1 million animals per day and by the end of 1996, ca. 200 million oysters had been lost. This situation persisted and in 1999 pearl production was down to less than 20 t/year with a value of ca. US$130 million. This compares with the annual value of US$550–600 million in the early 1990s. The crisis of the late 1990s destroyed 75% of oysters in Japanese pearl farms. Production figures have since remained at about 20–25 t/year and the Japanese Akoya industry has had to focus on producing pearls of larger size and better quality to remain competitive. The most popular size range for Akoya pearls is within the 7–8 mm size range.

Ago Bay in Mie Prefecture is perhaps the most famous of the pearl farming areas in Japan (Fig. 1.13). Ago Bay and surrounding areas on the Ise peninsula (Ise Shima) produced nearly all of Japan's cultured pearls until 1938, but today produce about 33% of the total (Strack, 2006). Ehime and Kochi Prefecture have produced cultured pearls since before World War II and their main pearl farming sites today are in the Uwashima area of Ehime Prefecture which has around 35% of farms. Ehime was the most important pearl producing area until 1996 with greater production than both Mie Prefecture and Kyushu Island, which is the third major pearl producing area in Japan. However, Kyushu Island has produced slightly greater volumes of pearls than Mie and Ehime Prefectures since 1996, with about 40% of total production coming from Nagasaki Prefecture (Strack, 2006). Pearl farming sites at Kyushu Island experience relatively stable year-round water temperatures. The relatively high growth rates that result support production of large pearls which may exceed 9–10 mm. Most areas are typified by small family-run farms with around 200,000 oysters, although larger farms with several million oysters are also found in all pearl farming areas.

Recent years have seen China begin to dominate Akoya production (see Section 10.3.2). Nevertheless, Kobe remains the center of the pearl world and the Japanese Akoya cultured pearl still leads the market despite often being declared moribund. It still ranks first in meeting the needs of millions of women for round, white and lustrous pearls.

9.2.3. Culture and production trends – China

Marine pearl oyster cultivation began in China in 1961 in the Hepu district where pearl oysters are abundant, and has gradually spread (Yakushi, 1991; Wang *et al.*, 2007). Cultured Akoya pearl production in China was initially about 150 kg/year, but production increased rapidly during the 1980s when private farms became established (O'Connor and Wang, 2001). By the end of the 1980s and 1990s annual production of marine pearls was estimated to be greater than 2 t and 20 t, respectively (Wang *et al.*, 2007). The major culture areas are in the southern provinces of Guangxi, Gaungdong and Hainan (Fig. 1.11), and the major culture sites within these provinces are shown in Table 9.1.

TABLE 9.1
Major sites for marine pearl culture in southern China.

| Province | Culture sites | | Species | Culture scale[1] |
	City	Town		
Guangxi	Fangchenguang	Jiangshan	Akoya	Medium
	Beihai	Yingpan	Akoya	Large
Guangdong	Leizhou	Wushi	Akoya	Large
			Pteria penguin	Small
		Tandou	Akoya	Large
	Xuwen	Xilian	Akoya	Large
		Jiaowei	Akoya	Large
	Shenzhen	Nan'ao	Akoya	Small
Hainan	Wenchang	Puqian	Akoya	Small
	Lingshui	Li'an	Akoya	Large
			Pteria penguin	Small
		Xinchun	Akoya	Small
	Sanya	Yacheng	*Pinctada maxima*[2]	Small
		Anyou	Akoya	Small
			Pteria penguin	

[1]See text for production values.
[2]Very small scale or experimental culture.

Guangxi Province produces about 8–9 t of pearls annually with the majority originating from farms in the Beihai area. In Guangdong Province there are only two farms in Nan'ao which each produce 200–300 kg of pearls each year. However, there are over 1,000 pearl farms along the coast of Leizhou in Guangdong Province which, collectively with farms in Xuwen, annually harvest ca. 9–10 t of pearls. In Hainan Province there are single pearl farms in Puqian, Wenchang, in Xinchun, Lingshui and in Anyou, Sanya (Table 9.1). They collectively produce ca. 300 kg of pearls annually. There are a further six farms in Li'an, Lingshui which annually produce ca. 500 kg of pearls, so annual pearl production from Hainan Province is less than 1,000 kg.

Marine pearl production in China gained initial impetus in 1958 when Professor Xiong Daren began the first official pilot project on behalf of the government. During the early 1960s these experiments led to the first regular production of cultured pearls of good quality. Professor Xiong worked in Japan as a marine biologist before World War II where he probably became familiar with the Japanese culture methods. The first four official pearl farms were established in the Bay of Haikang in the South China Sea near the city of Zhanjiang. High growth rate resulted in a nacre thickness of over 2 mm within a culture period of about 2 years. The size of the resulting pearls generally ranged from between 6 and 9 mm and shapes were nearly perfectly round. In subsequent years pearl farms have spread from the China/Vietnam border along the southern coastline of China.

In the 1970s and 1980s, Chinese cultured marine pearls were primarily exported to Japan from where they reached the world market. However, the international market did not become aware of China as a producer of marine cultured pearls until 1992. The first 10,000 strings were offered in February 1992 by a Japanese company on behalf of the "China Pearl, Diamond, Gem and Jewelry Import and Export Corporation". But the large quantities and the relatively low quality of the pearls offered immediately earned them a mediocre reputation. The worldwide pearl industry had been unaware for more than three decades that part of the Japanese Akoya pearl production had actually come from China (see Section 10.3.2). After 1993, when China produced 5–6 t of marketable cultured marine pearls, Japanese companies began to invest in Chinese pearl farms and pearl factories. Pearls with larger size and of better quality are likely to come from these farms with foreign investment partners and they probably enter the world market via Japan. Processing is done either in Japan or in Japanese-supported pearl factories in China (Fig. 9.3). Higher quality Chinese marine pearls are generally sold overseas, although there are established local markets for both pearls and pearl products. Mother-of-Pearl (MOP) from pearl shells may be used in handicrafts and decorative work, and it is used as an ingredient in cosmetics; while oyster meat is sold fresh or dried at local markets.

9.2.4. Culture and production trends – other countries

India began Akoya pearl culture research at the Central Marine Fisheries Research Institute (CMFRI) at Tuticorin in 1972 and the first experimental round pearl production occurred in 1973. Between 1973 and 1978, CMFRI, in association with the

FIGURE 9.3 Processing and grading cultured Akoya pearls in a factory in Guangdong Province, China.

Department of Fisheries, Government of Tamil Nadu, conducted research into aspects of pearl oyster biology and culture as well as pearl production by multiple implantations. Importantly, CMFRI succeeded in developing methods for hatchery production of pearl oyster seed providing independence from wild stocks for culture operations. The Tamil Nadu Fisheries Development Corporation and the Southern Petro-chemical Industries Corporation Ltd. established a company for commercial pearl production in 1983 (Chellam *et al.*, 1991). The farm was at Krusadai and the nucleus implantation center at nearby Mandapam. Despite the subsequent establishment of a number pearl farms in India, particularly along the southeastern coast, commercial pearl farming has not become established on a large scale (Upare, 2001). Such development has been hampered by the relatively 'thin' shells of native Akoya pearl oysters, which restricts the size of the nuclei that can be implanted and the size of resulting pearls (Victor *et al.*, 2003; Mohamed *et al.*, 2006). As a result, commercially cultured Akoya pearls from India generally have a diameter of less than 5–6 mm (Mohamed *et al.*, 2006; Kripa *et al.*, 2007).

The Bay of Halong in the Gulf of Tonking in Vietnam (Fig. 1. 11) has been famous for its natural pearls for many centuries. Production of cultured Akoya pearls was first attempted in Vietnam in the 1960s and, since 1990, more than twenty companies have established farms along the coast from the Chinese border town to Nha Trang in the southeast. The farms belong to both Vietnamese and foreign companies and there are numerous joint ventures. Cultured Akoya pearls from Vietnam are generally within a size range of 2–6 mm and they have a better nacre thickness than Akoya cultured pearls from both Japan and China because of good water conditions and a longer growth period. Production exceeded 1,000 kg in 2001 and greater quantities are expected on the market in coming years.

Akoya pearl culture has also been investigated on the Atlantic coast of South America (Urban, 2000a, b; Lodeiros *et al.*, 2002; MacKenzie *et al.*, 2003), in Australia

(O'Connor *et al.*, 2003; Pit, 2004), Korea (Choi and Chang, 2003) and in the Arabian Gulf (Narayana and Michael, 1968; Almatar *et al.*, 1993; Behzadi *et al.*, 1997). However, there has been limited development towards commercial pearl production in many of these areas.

9.2.5. Akoya Pearl culture methods

9.2.5.1. Japan

Spat or juvenile oysters are generally obtained from spat collection or hatchery culture (see Chapter 7). The latter has now become popular due to recent declines and annual fluctuations in the abundance of oysters obtained using spat collection methods and because of demand for more stable production of superior seed. Cedar sprigs are the most popular material used as spat collectors in Japan and this is the traditional method for collection of wild spat. They are bunched together with rope and suspended in the sea from floating rafts during the spawning season (June) at the depth of 1.5–3.0 m. Shells of oysters or abalone, or plastic nets, are sometimes used as spat collectors. Collection of wild spat is not popular in China and India where hatchery seed production is preferred (Chellam *et al.*, 1991; Wang *et al.*, 2003, 2004). Hatchery production is also the basis for Akoya pearl production in Vietnam. In the Caribbean, Urban (2000a, b) reported on seasonal changes in the occurrence of larvae and spat in the wild population of Akoya pearl oysters in Colombia. He concluded that the natural availability of spat was sufficient to initiate and support aquaculture by using onion nets as spat collectors.

Techniques for hatchery production of Akoya pearl oyster seed are based on those used for other species of bivalves such as edible oysters, clams and scallops. There are few differences in the hatchery protocols among the species of bivalves including pearl oysters (Wada, 1973a,b; Qizeng *et al.*, 1981; Alagarswami *et al.*, 1983; Hayashi and Seko, 1986; Southgate and Beer, 1997) and the general methods used for pearl oysters are outlined in Section 7.4. After settlement, juveniles with a dorso-ventral measurement (DVM, see Section 5.7.4) of 10–20 mm are removed from collectors and transferred to rearing nets generally made of nylon. Oysters are commonly held in lantern nets (see Section 7.5.4) and suspended from rafts (Fig. 14.4) or longlines (Fig. 9.4). Frequent changes of nets are needed as the young oysters grow rapidly and fouling organisms and other substances disturb the growth of oysters.

9.2.5.2. China

The Chinese marine pearl industry is based almost exclusively on hatchery-produced spat that are available from the large number of commercial hatcheries. Hatcheries generally operate between March to May to enable spat to be available for the spring–summer growing season. Briefly, broodstock are sacrificed for gamete collection and larvae are cultured in large volume concrete tanks (Fig. 9.5). Larvae are settled onto plastic plates from which they are removed when they reach a size of ca. 1 mm DVM (O'Connor and Wang, 2001). Spat are then cultured in fine mesh bags to a size of

FIGURE 9.4 Longlines used for suspended culture of Akoya pearl oysters, Li'an Bay, Hainan Province, China.

FIGURE 9.5 Large volume concrete tanks used for hatchery culture of Akoya pearl oysters, Li'an Bay, Hainan Province, China.

5–8 mm DVM. Culture methods for pearl oysters vary in China, but generally oysters are held in cages suspended from supports that are driven into the substrate (O'Connor and Wang, 2001; Fig. 14.7), although some surface longlines are also used (Fig. 9.4). Oysters are generally cultured for ca. 1.5 years and on reaching 50–60 mm DVM are grafted for pearl production (see Section 8.9). Grafting most commonly takes place in early spring and resulting pearls are harvested the following winter.

9.2.6. Constraints to Akoya pearl production

A number of constraints have affected production of cultured Akoya pearls in Japan. Oyster parasites have been known since the early stages of the Japanese pearl industry. Damage to the pearl industry by parasites has been of particular significance since 1952 and has been facilitated by the translocation of oysters and lack of biological data on the parasites. Mud worms, sponges and a trematode are the main parasites affecting pearl production (see Chapters 11 and 15). Mud worms of the genus *Polydora* produce a mud tube and blister inside bivalve shells (Fig. 15.1) and *Polydra ciliata* is recognized as the most important of these (Mizumoto, 1975). Damage to the industry by mud worms has occurred since 1960 in all pearl culturing areas in Japan and infection results in heavy mortality. Research in 1970 confirmed that more than 50% of cultivated pearl oyster were infected in the main pearl farming areas of Mie, Wakayama, Ohita, Kumamoto, Miyazaki and Kagoshima, in the central and southern waters of Japan. To exterminate mud worm, the industry uses brine treatment in the mud worm larval setting season (June to October). Biological control, using natural enemies of mud worm has been explored (Funakoshi 1964; Mizumoto 1975). In Kuwait in the Arabian Gulf, *Polydora vulgaris* has been reported to be more prevalent in Akoya oysters cultured on the bottom compared to those in suspended culture (Mohammad, 1976). In India, immersing oysters in freshwater or 78% brine for 6 or 8 h has been shown to kill polychaetes, although the effects of these treatments on the oysters were not reported (Chellam *et al.*, 1991).

A major constraint to Akoya pearl oyster culture in Japan is the periodic abnormal blooming of toxic dinoflagellate algae or 'red tide' (see Section 11.6.2.1). As well as the aptly named, *Gymnodinium mikimotoi*, other species linked to mass mortality of oysters from red tide events include *Heterocapsa circularsquama*, *Gonyalax* sp. and *Cocurodium* sp. Most Japanese pearl farms are located in semi-closed coastal estuaries or small bays, which is a major factor in the development of such blooms. A large-scale blooming of *H. circularsquama* in the summer of 1992 in Ago Bay, which reached a maximum cell density of 87,420 cells/mL, caused mass mortality of pearl oysters that had not been seen before with *G. mikimotoi* or other species (Matsuyama *et al.*, 1995). Many studies in Japan have reported on the mechanism of algal blooming, physiological damage and mortality of pearl oysters, prediction of blooming on the basis of the field observation, and experimental exposure of animals to red tide dinoflagellates (e.g. Honjo, 1994; Nagai *et al.*, 1996; Iwata *et al.*, 1997). Nagai *et al.* (1996) reported that the LD_{50} for *H. circularsquama* was approximately 20,000 cells/mL after 24 h of exposure. They suspected that the cause of death was the direct action of cells because immediately after exposure to the algal cells, oysters rapidly contracted their mantles, closed their shell valves and contracted their gills before the heartbeat became irregular and eventually stopped. Further details on the occurrence, epidemiology and pathogenesis associated with toxic algal blooms are given in Section 11.6.2.1. Transportation of pearl nets holding pearl oysters from the blooming area to a non-blooming site is a common procedure to avoid the mortality associated with red tides in Japan.

Mass morality of pearl oysters in western Japan from summer to autumn may also be caused by changes in seawater temperature and reduced food availability. Tomaru *et al.* (2002) found that respiration rate increased consistently with food density and decreased at food densities $>10^7\,\mu m^3/mL$, suggesting inactivation of the digestion system at high food density. On this basis, eutrophicated or turbid environments may be unfavorable for the growth of Akoya oysters.

Since 1994, mass moralities of oysters have resulted in significant economic losses to the pearl culture industry in western Japan (see Section 9.2.2). The disease responsible for these mortalities occurred from summer to autumn and affected oysters showed yellowish-red coloration of the adductor muscle. Further details of "Akoya mortality syndrome" are provided in Section 11.7.9.1. Details of the occurrence, pathogenesis and histopathology of viral diseases of Akoya pearl oysters in Japan are detailed in Section 11.7.1.1. Management practices have been recommended to pearl farmers to reduce the risk of mass mortalities of pearl oysters and genetic programs to breed resistant strains of oysters have been initiated (Uchimura *et al.*, 2005).

Production and quality of Chinese marine pearls have been variable over recent years, generally as a result of environmental problems associated with poor husbandry techniques, over-crowding and lack of environmental awareness (O'Connor and Wang, 2001). A number of Chinese research institutions are now actively involved in research to remedy these problems. Research programs are focused on developing superior strains of oysters, as well as triploid and tetraploid oyster stocks, and development of culture methods for other species of pearl oyster including *P. maxima*, *P. margaritifera* and *Pteria penguin* (O'Connor and Wang, 2001; Wang *et al.*, 2007).

9.3. SILVERLIP/GOLDLIP PEARL OYSTER

The silverlip or goldlip pearl oyster, *Pinctada maxima,* is the largest species of pearl oyster (Shirai, 1994) with the greatest shell thickness. It can consequently be used to produce the largest cultured pearls. It is distributed within the central Indo-Pacific region, bounded by the Bay of Bengal to the west, Solomon Islands to the east, the Philippines to the north, and northern Australia to the south (Fig. 2.7). Originally distributed from the shallow sub-tidal, it occurs in depths in excess of 50 m. Some early reports from the Sulu Islands in the Philippines suggested that *P. maxima* live as deep as 120 m (Talavera, 1930). Strong tidal currents appear to be a common environmental feature of both past and currently important areas of wild stocks. This has resulted in many fatalities during the early years of the various fisheries (Bach, 1955), but also in unique technological solutions for safe diving to access these stocks (Lulofs and Sumner, 2002).

Historically, the most productive fisheries have been within the Arafura Sea between Australia and Indonesia (Fig. 1.12), although there are early reports of fisheries around the Mergui Islands in Myanmar (Kunz and Stevenson, 1908), and the Sulu Sea of the Philippines (Seale, 1910). The only currently viable wild fishery for *P. maxima* is in Western Australia and there are farming ventures to produce pearls within most of the areas where *P. maxima* is endemic.

The Japanese used the term "South Sea cultured pearl" for pearls produced in marine waters south of Japan (Strack, 2006). This name has become synonymous

TABLE 9.2

Production of cultured white South Sea pearls from *Pinctada maxima* in 2005 (*source*: Henricus-Prematilleke, 2005).

Country	Volume (kg)	% Total	Value (US$ in millions)	% Total
Indonesia	3,750	40.1	85	34.2
Australia	3,000	32.1	123	49.6
Philippines	1,875	20.1	25	10.1
Myanmar	563	6.0	13	5.2
Malaysia	75	<1.0	2	<1.0
Papua New Guinea	75	<1.0	Unknown	<1.0
Total	9,338		248	

with large cultured pearls produced from both *P. maxima* and the blacklip pearl oyster, *P. margaritifera*. Since the 1970s, however, the international market distinguishes between white and black South Sea cultured pearls. The white pearls are produced from *P. maxima*, almost exclusively in a geographical area situated between the Indian and Pacific oceans. The description "white" includes yellow, greenish, golden and light-gray hues.

The major producers of cultured pearls from *P. maxima* are Indonesia, Australia and the Philippines with approximately 40%, 32% and 20% of total production, respectively (Table 9.2). The total value of pearl production from *P. maxima* in 2005 was US$248 million of which Australia contributed $123 million or almost 50% of the total (Table 9.2). Pearl production from *P. maxima* has increased rapidly over recent years with an approximate 260% increase between 1999 and 2005 (Henricus-Prematilleke, 2005). It is interesting to note that while the level of pearl production from *P. maxima* in 1999 was only about 38% that of 2005, the value of the crop was about 90% that of the 2005 crop (Henricus-Prematilleke, 2005). South Sea cultured pearls from *P. maxima* are now the leading pearl category (see Section 10.3).

9.3.1. History of exploitation

It is likely that *Pinctada maxima*, like other pearl oysters, has been exploited since ancient times. However, its remoteness from the centers of ancient civilizations has afforded the wild stocks more protection than those of *P. mazatlanica* along the east coast of central America (Saucedo *et al.*, 1998; Section 9.5) and Akoya pearl oysters from the Gulf of Manaar in Sri Lanka, which have been harvested since antiquity (Herdman, 1903). Earliest records of exploitation of *P. maxima* are of fisheries in the Sulu Sea (Seale, 1910) and the Mergui Islands (Kunz and Stevenson, 1908) (Fig. 1.11). The use of pearl shell by Australian Aborigines was noted from the earliest times of contact with European settlers (Akerman and Stanton, 1994). For example, The Karadji of Eighty Mile Beach practiced totemic rituals designed to tame the weather and invoke

FIGURE 9.6   Large shells of *Pinctada maxima* being sorted, weighed and packed on board a pearling ship (*source*: Kunz and Stevenson, 1908).

the assistance of the bower bird, known for its attraction to shiny bright objects, to ensure a plentiful harvest (Elkin, 1933). The modern *P. maxima* pearling industry in Australia was initiated in 1861 with pearl shell traded by Aborigines to an explorer's team; however pearling *per se* began in the 1850s with the discovery of the *P. albina* stocks in Shark Bay (Hancock, 1989) (Fig. 1.12). Wada (1953) considered 1861 to be the first record of the discovery of the Arafura Sea fishing grounds, which he considered as stretching from northwestern Australia to the original fisheries based in Dobo in the Aru Islands of Indonesia (Fig. 1.11). In terms of production, however, it was the Australian stocks that supplied 80–90% of the harvest from the Arafura Sea fisheries (Wada, 1953).

The early fleets were dominated by Japanese vessels operating in international waters beyond the 3 nautical mile zone, together with Australian pearling enterprises using indentured labor primarily from Japan, but also from other countries bordering the Arafura Sea. The target of these fisheries was MOP shell, with pearls being a by-product (Fig. 9.6). Average yield of MOP from a large *P. maxima* (>180 mm DVM) is ca. 1 kg (Hart and Friedman, 2004), and this equation has been broadly applied to catch statistics.

There were three main fisheries in Australia: in Western Australia, Northern Territory and Queensland (Fig. 1.12). The Western Australian fishery began in 1867, initially as a wading fishery utilizing mainly Australian Aborigines, and later South Sea Islanders and Malays (Bach, 1955). The Queensland, or Torres Strait fishery was discovered in 1868. By 1875, it extended over 3,000 miles (Bach, 1955). The "Old Grounds", to the west of Cape York were discovered in 1881, and with it came the development of big, capitalized pearling fleets to work the remote locations, a trend which was exported to Western Australia (Fig. 9.7) and persists in the modern fishery. By the early 1890s, surface assisted or "hard-hat" (or helmet and suit) diving (Fig. 9.8),

FIGURE 9.7   Historical photo of pearling luggers and crew at low tide near Broome, Western Australia (ca. 1915).

FIGURE 9.8   *Pinctada maxima* (MOP) "hard hat" diver getting ready to descend. Broome, Western Australia (ca. 1915).

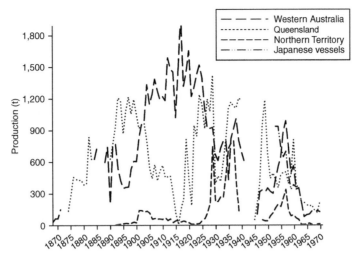

FIGURE 9.9 Production (tonnes) of *Pinctada maxima* (MOP) from wild fisheries within different Australian states between 1870 and 1970 including catches from Japanese vessels within the 1950s.

had been implemented across all fisheries, and eventually became the iconic image of the early *P. maxima* industry. Production of MOP from Queensland was nearly double that of the Western Australian grounds in the 6 years subsequent to 1881 (Fig. 9.9); however in 1886, there was a mass exodus of pearling fleets from Torres Strait, first to Western Australia, and then to the Aru Islands. Although never fully understood, the main reason for this was put down to stock depletion (Bach, 1955); at least enough to prevent the larger pearling fleets from being profitable.

Hard-hat diving allowed divers to work in deeper water and stay on the seabed much longer than free diving. It was possible to harvest from pearl oyster beds that were previously inaccessible. Diving in hard hats was a skilled and dangerous occupation, and the diver was the most highly paid person on board the pearl lugger (Edwards, 1994). The divers were often from Japan and Malaysia, being regarded as among the best at this occupation. Unfortunately, there was no understanding of decompression sickness or the 'bends' and, as the pearl beds were progressively depleted, it became necessary to work in progressively deeper water. A technological change also made it possible to work deeper. With the initial hand-driven air pumps on the luggers there was insufficient hose pressure and airflow for divers to operate much below 20m. The subsequent introduction of engine-driven air pumps made it possible to work at three times that depth (Edwards, 1994). There was a sudden escalation in diver deaths as there was no appreciation of the need for steady decompression to gradually release the large amounts of nitrogen gas that had dissolved in the diver's blood under pressure. Divers spent too long at depth and were brought up to the surface too rapidly. At the surface, the diver might experience agonizing pains, convulsions, paralysis, unconsciousness and death. From 1909 to 1917, 145 hard-hat divers died working out of Broome and 33 died in a single year (Edwards, 1994). Up to a thousand divers may have died during this early period of hard-hat diving in northern Australia and a similar number would have been crippled and paralyzed (Edwards, 1994). Large numbers of Japanese grave markers in the Japanese sections of the cemeteries of northern Australian pearling centers are stark reminders of this cost of pearling.

Fishing for *P. maxima* in the Northern Territory began substantially later than in the other states, and never achieved their scale (Fig. 9.9). Substantial exploration began about 1885 with an influx of pearling fleets from the other fisheries, presumably caused by depletions of stocks. Initial harvests were from Darwin Harbour and the nearby Melville Island. The "East" and "West" grounds (relative to Darwin) were subsequently discovered in the early 1900s; however, the pattern of exploitation was essentially that of serial depletion, a pattern that duly concerned the government authorities in the Northern Territory and other states. The pearlers generally rejected conservative measures on their own initiative, and this ultimately led to serious consideration of the possibility of culturing (or cultivating) *P. maxima*, and eventually pearls (Bach, 1955).

Original interest in culturing *Pinctada maxima* stemmed from concerns over stock depletion, but was generally focused on the notion of stock enhancement, rather than aquaculture. The idea was to transport immature oysters to pearl leases and grow them up for sale, subsequent breeding and natural spat settlement. William Saville-Kent, Australia's first professional fisheries scientist (Harrison, 1997) proposed the scheme in the 1890s, and Australia's first pearl farm was initiated in 1894 in the Torres Strait with 150,000 live pearl oysters transported to the Friday Island lease (Bach, 1955). The trial was largely unsuccessful due to ignorance of the basic ecology of the oyster, such as the inability to tell the difference between newly settled spat of *P. maxima* and *P. albina*, and inappropriate care of the oysters during transport, an issue which persisted into the modern era of *P. maxima* farming and was only subsequently solved in the 1980s with the discovery of bacterial infection build up in response to the stressing conditions during transport (see Section 11.7.2.4) (Pass *et al.*, 1987). The industry's use of the term 'shell' for pearl oysters has had unfortunate consequences. 'Shells' do not need oxygen, clean water and stable temperature; they can be transported in soupy conditions. However, the trial provided the initiative for Saville-Kent to go a step further and investigate the possibility of culturing pearls. Although the process is generally attributed to Japanese pioneers, a case has been made that Saville-Kent discovered how to culture pearls using *P. maxima*, first for blister pearls (half-pearls or mabé) in the late 1880s, and then spherical pearls in 1905, and unwittingly passed on the knowledge (Harrison, 1997; see Section 8.9.1). Ironically, the increasing resistance to the pearl culturing process by Australian pearlers of the time led to the banning of pearl culture in Western Australia in 1921 (Bach, 1955). The development of pearl culture moved to other countries bordering the Arafura Sea, facilitated by Japanese interest. The ban remained until 1949, after which pearl farming began again with the assistance of Japanese expertise and access to wild *P. maxima* oyster beds. Over time, the industry has shifted away from the "hard-hat" diving methods of the old pearling luggers and Japanese divers, to modern drifting vessels (Fig. 9.10), which deploy up to 8 divers simultaneously and fish using unique dive profiles developed specifically by the PPA (Pearl Producers Association) of Western Australia (Lulofs and Sumner, 2002).

FIGURE 9.10   Modern pearling vessel used for pearl diving, transport and pearl seeding within the *Pinctada maxima* industry of Western Australia.

9.3.2. Current wild fisheries management for *Pinctada maxima* in Australia

There are licenses to fish *Pinctada maxima* officially in four fisheries, the East Coast Pearl Fishery (Anon, 2006), the Torres Strait Pearl Fishery, the Northern Territory wild-pearl fishery (Souter, 2006) and the Western Australian fishery (Hart and Murphy, 2006a). However, the harvest from the East Coast, Torres Strait and Northern Territory fisheries is negligible (<1,000 oysters) and statistics are not reported. The Western Australian fishery is managed by a TAC (Total Allowable Catch) to ensure the long-term sustainability of the resource (Hart and Murphy, 2006a). The fishery is a "gauntlet fishery", with the target size-class being the young, fast-growing, protandrous males that hold the greatest potential for pearl production. There is a minimum legal size of 120 mm shell length, and maximum legal sizes and area-specific TACs have been set where appropriate. Oysters above 170 mm DVM are rarely taken, despite average maximum sizes (L_∞) being in excess of 200 mm DVM (Hart and Joll, 2006). The fishery is integrated into the pearling industry, which has three components: the collection of pearl oysters from the wild, production of hatchery-reared pearl oysters, and grow-out of pearls on pearl farm leases. Sale of MOP has been a by-product of the pearl farming process since about 1980 (Fig. 9.11).

In addition to detailed management controls of the resource, significant advances in biology and oceanography have improved management. A method for measuring *P. maxima* spat settlement in the wild (Hart and Murphy, 2006b) is providing the ability to forecast recruitment into the fishery up to 4 years ahead of time. A recently developed oceanographic model of larval transport of *P. maxima* on the main fishing grounds (80 Mile Beach, Broome, northwestern Australia; Fig. 1.12) has provided insight into how the pearling beds have sustained themselves over a century of fishing (Condie *et al.*, 2006). These factors contributed to the *P. maxima* fishery being one of the first fisheries in Australia accredited with export approval under the

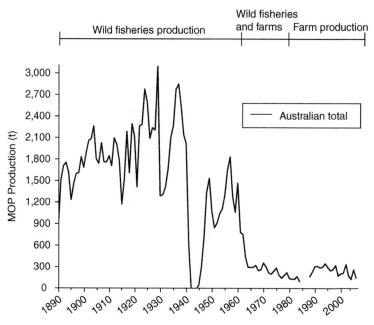

FIGURE 9.11   Production (tonnes) of *Pinctada maxima* (MOP) from Australian waters between 1890 and 2006. Time periods of the different phases (wild fisheries/farm production) are shown (*source*: Australian Bureau of Statistics).

biological sustainability conditions imposed by the Australian *Environmental Protection and Biodiversity Conservation* (EPBC) Act of 1999. The 5-year export accreditation license is due for renewal in 2008.

9.3.3. Management of white South Sea pearl production in Australia

Pearl production in Australia is managed through a quota management framework, underpinned by a policy of controlled expansion to maintain market position and avoid oversupply in what is primarily an export market (Anon, 2005). The policy is coordinated between the two major pearl producing states of Western Australia and Northern Territory, with the common principles being set out within a "Memorandum of Understanding" between the state governments. There are 1,342 pearl quota units overall, 922 in Western Australia (wild and hatchery), and 420 (hatchery) in the Northern Territory, allocated amongst 22 pearling licenses (Anon, 2005). The Western Australia quota is split between 572 wild-stock quota units and 350 hatchery quota units. The value of a hatchery quota unit is 1,000 pearl oysters, however the value of a wild-stock quota units varies up and down from 1,000 oysters, depending on status of wild stocks, which are reviewed each year to set TAC for each zone of the fishery (Hart and Murphy, 2006a). In 2007, the value of wild-stock quota unit in the major fishery (Zone 2) was 1,200 oysters as a result of predicted increased recruitment to the fishery

(Hart and Murphy, 2006a), arising from successful quantification of spat settlement rates. A recent policy review has proposed that the regulatory basis of management must change focus to "first operations seeding rights" of all pearl oysters, regardless of whether they are produced by wild stocks or within hatcheries (Anon, 2005). This action would de-couple the pearl production process from wild-stock sustainability assessment, which was the historical basis for management, but is no longer relevant with the advances in hatchery and pearl culture technology, and the globally compet- itive culture industry for South Sea pearls. However, the policy of controlled, rather than deregulated expansion, will still be the guiding principle of pearl production in Australia (Anon, 2005), despite its contentious nature.

9.3.4. Culture and production trends – Australia

9.3.4.1. Mother-of Pearl production

There are three broad historical phases to MOP production from *P. maxima* within Australia (Fig. 9.11). These are the wild-fishery phase (1880–1960), the wild-fishery/ farm production phase (1861–1980) and the farm production phase (1981–Present). Although occurring over an extended period, the wild-fishery phase was buffeted by a number of crises that ultimately caused the decline of the fisheries, particularly those in Northern Territory and Queensland. These were:

- localized and serial stock depletions;
- the restricted markets of the war years;
- the world economic depression of the late 1920s/early 1930s;
- the flooding of the world market after 1937 caused by the combined Japanese and Australian production (Bach, 1955);
- the development of the plastics industry for making buttons.

In 1956, pearl culture from *P. maxima* was introduced to Western Australian farms, and MOP production from farmed pearl oysters began in earnest in the 1960s (Malone *et al.*, 1988). By the early 1980s, MOP production from wild oysters was negligible, and the harvesting of MOP was officially prohibited in 1985 (Malone *et al.*, 1988), thus enabling the build-up of a substantial broodstock in the ensuing years. The prohibition on harvest of MOP still exists in Western Australia, however recent studies suggest that recruitment into the broodstock is likely to exceed natural mortality, and the potential exists for a sustainable harvest of MOP at low levels (Hart and Friedman, 2004). It is unlikely however, that such an activity would be economically viable under current markets and no future harvesting of wild stocks of MOP is expected at this time.

9.3.4.2. Pearl production

Pearl production officially began in Australia in 1956, with the establishment of the Kuri Bay pearl farm in the Kimberly Region of Western Australia. By 1965, exports had reached in excess of 300,000 pearls (Fig. 9.12), although the majority of earlier

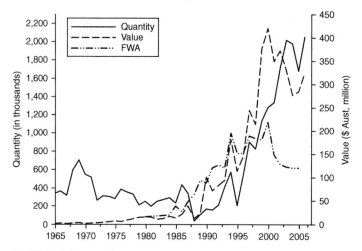

FIGURE 9.12   Total exports (quantity and value) of Australian cultured pearls between 1965 and 2006. Post 1995 export data have been reduced by 10% from the actual export figures to account for re-exports of pearls (*source*: Australian Bureau of Statistics). FWA-Production values from the Department of Fisheries, Western Australia.

exports were half-pearls, which are substantially lower in value than round pearls. Total pearl exports from 1965 to 1995 varied between 200,000 and 600,000 pearls per year. Major changes included the decline in the mid-late 1980s from 500,000 to 50,000 per year as a result of high oyster mortalities (Pass *et al.*, 1987). Once this problem was solved, export statistics show a 10-fold increase in pearl exports since 1995, from 200,000 to 2 million pearls/year (Fig. 9.12). Initially this was accompanied by a similar increase in value, although this declined substantially from a peak in 2000, whereas production stabilized. A cautionary note however, is that, while the general trend in value and production is considered true when compared with value of production statistics of the Western Australian Department of Fisheries (FWA), which show the same general trend (Fig. 9.12), the export statistics are likely to be inflated due to re-exports, i.e., pearls previously exported, imported for further processing, then re-exported.

Unit value of Australian cultured pearls rose slowly until the mid-1980s, and then jumped 10-fold from $50 to $500 per pearl by 1990 (Fig. 9.13). This appears to have resulted primarily from an increase in the proportion of round pearls sold, from 15% in 1986 to greater than 90% in 1988 and beyond (Fig. 9.13). The statistical break-up between round/baroque pearls and half-pearls has not been recorded since 1995; however, the situation is likely to have remained similar, with round pearls being the predominant export. Comparing the unit value with total quantity sold since 1988 shows prices decreasing with the increase in quantity exported (Fig. 9.14). The unit value of exported Australian pearls was at a 20 years low during 2004–2006. Increased production appears to be the likely predominant cause; however, factors such as increasing currency exchange rates, the Asian economic crisis of the late 1990s, and the SARS virus of the early 2000s, have all exacerbated the declining trend in unit value.

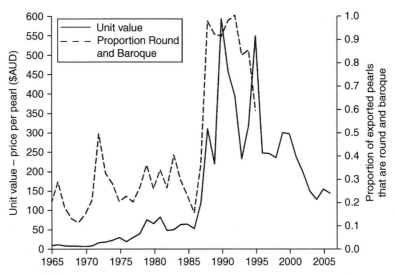

FIGURE 9.13 Unit value (price per pearl) of exported *Pinctada maxima* pearls from Australia, and proportion of exports that are round or baroque pearls (*source*: Australian Bureau of Statistics).

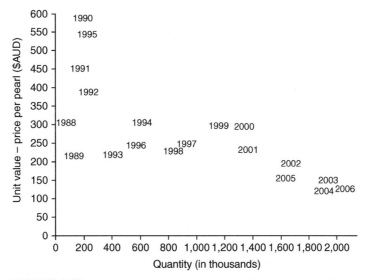

FIGURE 9.14 Unit value (price per pearl) of exported *Pinctada maxima* pearls as a function of quantity exported from Australia. Data are from the years 1988 to 2006 in which exports were predominately round and baroque pearls.

Future production of pearls from *Pinctada maxima* in Western Australia will require careful monitoring if prices and market position are to be maintained. Australian South Sea pearls (ASSP) have enjoyed a privileged position in the market on the basis of luxury, scarcity, and quality. Although the quality remains high, the perception of scarcity may have been eroded in recent times. Furthermore, production of pearls from

P. maxima is likely to increase in other countries of the Indo-Pacific region, particularly as the low volume/high value nature of the product lends itself to operations in remote, less-developed areas of the Indo-Pacific. The future will require careful branding of the ASSP in order to distinguish it from its rivals, and highlighting the unique aspects of the ASSP, e.g. production from a sustainably managed wild fishery.

9.3.5. Culture and production trends – Indonesia

The first attempts at pearl culture in Indonesia were made on the small island of Buton near Sulawesi Island under Dutch supervision as early as 1918. The required capital came from Baron Iaasaki of the Mitsubishi Company who had been forced to abandon his pilot farm on the Philippines 2 years earlier. The scientific director of the farm had been Sukeyo Fujita, the second of the talented Fujita brothers (see Section 8.9.1). In 1920, he founded the South Seas Pearl Company on Buton that used *Pinctada maxima* from the Arafura Sea. Between 1928 and 1932 Fujita was able to harvest between 8,000–10,000 pearls/year, while the total production amounted to 36,670 pearls between 1935 and 1938. The market responded with astonishment to the "Mitsubishi pearls" of 8–10mm in size, because the average size of Japanese cultured pearls at that time was still <5mm. In 1941 the Company was closed by the Japanese military forces at the onset of the Pacific War.

In the late 1930s a pearl culture project emerged at Dobo in the northern Aru Islands of Wokam (Fig. 1.11). The harvest of the "Dobo pearls" was interrupted by World War II, while the Boeton Pearl Company, having imported Australian *Pinctada maxima* from 3,000km away, successfully marketed 26,000 pearls in the years leading up to World War II.

As the first Indonesian President, Sukarno, was against Japanese involvement, in 1952 the Indonesian government confiscated the first farm owned by Tokuichi Kuribayashi, proprietor of the Arafura Pearling Fleet, whereupon Kuribayashi continued his work with great success in Australia. In 1964 the Fuji Pearl Company established a new farm on the site of Dr. Fujita's pre-war farm; however, the project was soon abandoned.

The real beginning of the Indonesian cultured pearl industry was in the 1970s after two laws from 1970 enabled foreign companies to invest in Indonesia. Four large Japanese companies that had previously invested in Australia set up farms on the Aru Islands. They were the Nippo Pearl Company, the Tayio Gyogyo Ltd., the Arafura Pearl Company and the Kakuda Pearl Company. The ships of the Arafura pearling fleet were still used and they brought huge quantities of *Pinctada maxima* from the Arafura Sea. The Indonesian government did not approve officially until the late 1970s. The 1990s brought the much-needed modernization of pearl farms when foreign companies began investing once more and several large companies from the international cultured pearl trade entered partnerships in Indonesia.

An earthquake in December 1992 caused the loss of more than half of all *Pinctada maxima* banks and from 1995 onwards the industry became reliant on hatchery-produced oysters. In 1995 the Indonesian Pearl Culturer's Association (ASBUMI) was founded, which aimed to develop its own marketing strategies. In 1999 Indonesia supplied more than a third of the world's production of South Sea cultured pearls and, in the years since 2000, production is in the range of 2–2.5 t/year. There are currently about 107 pearl farms in Indonesia, but only 32 of these are registered with the ABSUMI. There are at least 36 hatcheries in Indonesia and all commercial pearl production is hatchery based. Indonesia is currently the largest producer of cultured pearls from *P. maxima*, but is second to Australia in terms of the value of its pearls (Table 9.2).

9.3.6. Culture and production trends – Philippines

The roots of the Philippines cultured pearl industry go back to about 1900, when Dr. Alvin Seale of the Philippine Fisheries Department of the Bureau of Science, began his study on the biology of *Pinctada maxima*. The first Japanese attempts followed in 1916 when Sukeyo Fujita developed a farm near Zamboanga on Mindanao Island. Two years later the project, which had been supported by Baron Iaasaki of the Mitsubishi Company, was continued on Buton Island in Indonesia. Then towards the end of the 1950s, when the Philippine government showed an interest in the creation of a cultured pearl industry, a few farms were started by foreign and local companies. They produced mainly blister pearls (mabé pearls) and were supported by the Japanese. In March 1964, Kikiro Takashima from the South Seas Pearl Company came to the Philippines after his farm in Myanmar (Burma) had been nationalized. He built a new farm on Sacol Island near Zamboanga in a joint venture with the local Zamboanga Pearl Farms Company. A number of further farms were founded during the 1960s, on Bohol Island, Cebu Island and Mindanao Island, always according to the principles of the "Diamond Policy"[1]. Success was slow and blister pearls were mainly produced. The Philippines did not manage to establish itself as a serious producer of cultured pearls until the 1980s. The Jewelmer Company founded its first farm on the then uninhabited island of Bugsuk in the south of Palawan in 1979. In May 1996 the Philippine Association of Pearl Producers and Exporters (PAPPE) was founded as the national Philippine organization of the pearl industry that cooperates closely with government. Production has risen from ca. 0.5 to 2 t/year since 1999 from about 30 farms. Much of the production still goes to Japan (most farms still have Japanese partners), while a few of the larger companies market pearls themselves. The largest producer, Jewelmer International, organizes its own public relations and sells directly to large foreign companies. About 10% of its pearls are worked into jewelry and sold in exclusive hotel shops in Manila and other large cities in the country. The pearl farms are centered to the north of Palawan Island and the adjoining Calamian Group. Farms are also situated on Samar and Cebu Island around the southern tip of Palawan and on Mindanao Island. Only naturally occurring

---

[1]Written in 1953 by the Japanese government stating that: (1) pearl cultivating techniques shall remain secret to all but the Japanese; (2) the production objectives shall be controlled and regulated to safeguard home pearl production; and (3) all pearl production shall be exported to Japan (George, 1978). The principles of the "Diamond Policy" were eventually eroded by the Australian and French Polynesian pearl culture industries which challenged the Japanese monopoly (Gervis and Sims, 1992).

P. maxima were used for pearl production until about 1990; however, hatchery-produced oysters have played an increasingly important role since the end of the 1990s.

9.3.7. Culture and production trends – other countries

Other countries producing significant quantities of cultured South Sea pearls include Myanmar, Malaysia and Papua New Guinea (Table 9.2). Myanmar has a long history and a reputation for fine quality pearls (see Section 1.5.1). Cultured pearls were first produced in the 1960s and in 2005 production was ca. 563 kg, which made up 6% of total production of cultured 'white' South Sea pearls (Table 9.2). Malaysia and Papua New Guinea each produce ca. 75 kg/year of cultured South Sea pearls from *Pinctada maxima* each making up less than 1% of total production (Table 9.2). The attempts by C. Denis George in the 1960s to establish a cultured pearl industry on Daga Daga Island near Samarai for the people of Papua New Guinea was not successful. A few years ago, a new farm was established on Ungan Island in New Ireland province. Malaysia had a farm established in 1962 by the Kaya Pearl Company. The farmers built on the east coast of Sabah and round pearls were produced from the early 1970s onwards.

The whole western coast of Thailand has banks of *Pinctada maxima* and the first pearl farm was established in 1965 on Naganoi Island off the southwestern coast in the Andaman Sea. The Thai owner entered a joint venture with the South Seas Company of Kikiro Takashima and the Swiss Samourai S.A. Company. The farm was worked by Japanese technicians and used *P. maxima* from the area bordering Myanmar. The farm stopped working about 1995. Further farms were built on small islands off Phuket and Ranong Islands, although pearl production from these farms is low and are all are exported to Japan (Bussawarit, 1995).

Pinctada maxima occurs in the Gulf of Tongking along the coasts of Guangxi and Guangdong provinces and on Hainan Island in southern China (Fig. 1.11). Farms were established in the 1960s in the south of Hainan Island and more farms followed in the 1970s. There is little information available on the farms themselves, production figures and quality of the pearls produced. A number of pilot projects with *P. maxima* have taken place in northern Vietnam in connection with the production of Akoya pearls (see Section 9.2.4).

9.3.8. The market for 'white' South Sea cultured pearls

The large cultured pearls from Myanmar and Australia appeared on the European and American markets about 1960. By 1970 prices had reached astronomical heights. This was especially true for pearls from Myanmar, although most European and American jewelers did not yet know very much about the new cultured pearls. The pearls finally rose to prominence during the 1980s and prices increased accordingly. The 1990s brought the most important development, which was initiated by the Australians, namely improving production methods and quality, and developing independence from the Japanese. Nevertheless, more than half of all South Sea cultured pearls are still marketed around the world via Japan today. The beginning of the 21st century saw an increase in production of medium to low quality pearls from Indonesia and the Philippines, and prices began to decrease. Prices for good quality South Sea pearls have more or less remained stable, while prices for very good and fine qualities have remained stable or shown an upward trend.

South Sea cultured pearls are no longer only a luxury item. Commercial qualities have increasingly replaced Akoya cultured pearls and the number of companies dealing in South Sea cultured pearls has increased on the international market by 100–200%. The pearls are sold at several auctions worldwide, particularly in Hong Kong, but also in Kobe and Australia (see Section 10.3.2). Since the late 1990s the public relations activities of the South Seas Pearl Consortium (SSPC) have produced results especially in the United States.

South Sea cultured pearls are sized from 7–20 mm, while the average size is 12–14 mm. Size is the most striking characteristic of these pearls, but it varies between countries: sizes in Australia range from 10 to 20 mm, in Indonesia from 7 to 16 mm and in the Philippines from 9 to 16 mm. Small pearls from Indonesia are sometimes called indicator or baby pearls; they are often of commercial quality. The shapes of South Sea cultured pearls are divided into round, symmetrical and baroque (see Section 8.10). Only a few percent of the pearls produced are really round. While 'nearly round' shapes represent about 20% of Australian production, this is lower in other countries. About half the production is of symmetrical and baroque shapes, but this proportion increases to about 70% in Indonesia and the Philippines.

The body color of South Sea cultured pearls show mainly neutral and nearly neutral hues like white, light to dark gray, silver and cream. White pearls from Australia do not generally have a cream hue, but often show a silvery-white or silver hue. Indonesian and Philippine pearls almost always have a cream hue to which a yellowish or greenish hue may be added. Gray hues vary from light to dark gray and there is also silvery-gray and grayish-blue. A rather small proportion of the pearls have distinct body colorations like pink, golden, yellow, green and blue. Overtones are mainly pink, green, gray and blue. Colors generally speaking can be divided into three groups: the first group commands the highest value; it includes pearls with a pink body color and pink overtone golden pearls. They range from a deep honey golden to a dark orange or reddish body color that is called "burnt orange", "burnt gold" or "imperial gold" in the trade (Plate 10). The "silver pinks" with a silvery-white body color and a pink overtone also belong to the first group. The second group is made up of pearls with a white silvery-white or silver body color and overtones other than pink. They are immediately followed by pearls with a gray, silver gray or blueish gray body color. Golden yellow which does not match the golden hue of the first group is also included here. The third group includes cream, yellow and green hues. The combination of green and yellow is sometimes described as champagne and may fetch above average prices. The luster of South Seas cultured pearls is usually not as bright as the luster of Akoya cultured pearls. Nacre thickness usually ranges from 1.5 to 5 mm for good to very good qualities. However, pearls with a lower nacre thickness have been seen on the market for a few years. Surface features are quite common and spotless surfaces command a high premium price. The pearls are not bleached or dyed in their countries of origin, but treatments are applied in Japanese factories with artificial dyeing focusing on yellow and golden hues.

9.4. BLACKLIP PEARL OYSTER

The blacklip pearl oyster, *Pinctada margaritifera*, is famed for its production of 'black' South Sea cultured pearls. As outlined in Section 9.3., the term "South Sea cultured pearls" is generally used to describe pearls produced by both *P. maxima* and *P. margaritifera* with distinction between them made on the basis of color. *P. margaritifera* has a wide geographical distribution from the Red Sea and east Africa to eastern Polynesia (Fig. 2.6), but it is the atoll lagoons of French Polynesia that provide the primary region for cultured pearl production from this species.

9.4.1. History of exploitation

There is a long history of exploitation of *Pinctada margaritifera* for the MOP trade and this has occurred throughout the species range. *P. margaritifera* has also been used for cultured pearl production since before World War II when pearls farms were established in southern Japan and Palau. However, it is only since the 1970s that large-scale commercial production of black South Sea cultured pearls has developed in eastern Polynesia. Today, cultured black pearl production is dominated by French Polynesia with smaller-scale production in the Cook Islands and other Pacific nations.

9.4.2. Culture and production trends – Japan

Kokichi Mikimoto (see Section 8.9.1) established a pearl farm at Ishigaki, Okinawa, in 1914 and a second in Palau in 1923 from where he succeeded in producing round pearls from *Pinctada margaritifera* (Hisada and Fukuhara, 1999). After World War II, Okinawa became the only source of cultured black pearls and in 1951 there were nine companies in Okinawa investigating production of black pearls. Only one of these, the Ryukyu Pearls Co., survived and it reports annual production of ca. 2,000–3,000 pearls (Hisada and Fukuhara, 1999). Mikimoto and Ryukyu Pearls introduced cultured black pearls to the Japanese market emphasizing their scarcity and exclusivity. Cultured black pearls were also sold in New York. This exclusivity provided market acceptance of cultured black pearls in Japan and in the mid-1970s, when French Polynesia became the dominant producer, their product found a ready market in Japan.

9.4.3. Culture and production trends – French Polynesia

9.4.3.1. History

Shells of *Pinctada margaritifera* have been collected since before 1850 in French Polynesia to supply MOP to the button industry. Diving-based collection led to

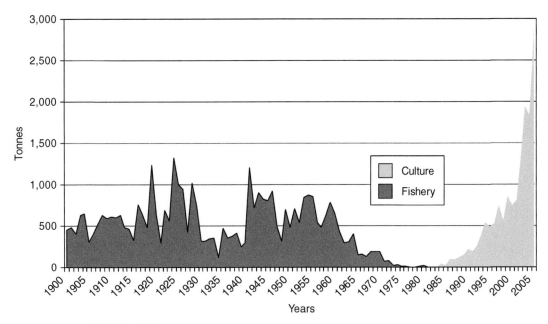

FIGURE 9.15 Exports of *Pinctada margaritifera* shells (MOP) from French Polynesia from 1900 to 2005 (*source*: French Polynesian Customs and ISPF).

over-exploitation and decimated natural shallow-water stocks of *P. margaritifera*. Then in 1885, Dr Bouchon-Brandeley suggested methods for sustainable resource management including introduction of a specified season for MOP collection and restrictions on collecting techniques (Bouchon-Brandeley, 1885). These measures allowed continued harvest of significant quantities of MOP well into the 1970s (Fig. 9.15).

The first pearl farms were built in French Polynesia in the late 1950s for production of mabé pearls (see Section 8.7). In 1961 the Fisheries Service secured the help of two Japanese companies, Nippo Pearl Company and Tayio Gyogyo Limited. They were both pioneers of pearl culture in Australia. The chief Japanese pearl seeding (grafting) technician came from Friday Island in the Torres Strait to Bora Bora. Other experiments followed under the guidance of Jean-Marie Domard, a French veterinarian who was Director of the Ministry of Agriculture and Fisheries at the time. During the 1960s, the Tahitian Fisheries Department began to conduct studies to develop pearl culture activities in remote islands (Domard, 1962). The first pearl seeding experiments were carried out in 1963 with a Japanese technician, Chiroku Muroi, on Hikueru, a Tuamotu atoll, and on the island of Bora Bora. In 1966, the brothers Jacques and Hubert Rosenthal built the first pearl farm on Manihi Atoll in the Tuamotu Archipelago (Fig. 1.14). The government aimed at developing a cultured pearl industry in the atolls and hoped to serve a number of ecological and sociological problems this way.

The first 71 black pearls were harvested on the Rosenthal farm in 1972. Other farms followed and the harvest of pearls in 1977 amounted to 28,000 in total. The pearl

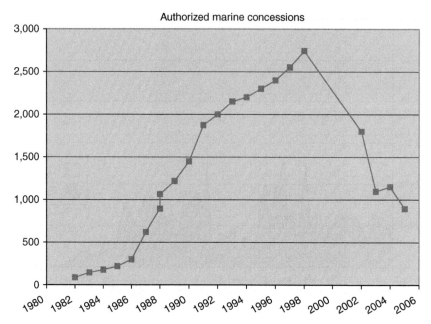

FIGURE 9.16 Authorized marine concessions (leases) for farming of *Pinctada margaritifera* in French Polynesia from 1982 to 2005.

market was hardly aware of the new dark-colored pearls in the early 1970s. However, this changed when Salvadore Assael, a pearl dealer from New York, began marketing the pearls after he became a business partner of Jean Claud Brouillet in a new pearl farm on Marutea Island, 2,000 km southeast of Tahiti. Assael created the "Tahitian black cultured pearl" and soon became the most successful promoter selling the pearls to well known jewelers in New York. The stage was set for the rise of the French Polynesian black pearl industry.

Due to the excellent profits generated by high value pearls, a large number of people, islanders and others, were attracted by this undemanding business. Using natural spat collection, it was easy for island residents to develop their own farms throughout the archipelagoes of French Polynesia. There was an exponential rise in the number of authorized marine concessions (leases) for pearl farms throughout the 1980s and into the late 1990s and, by the year 2001, the number had reached more than 2,500 (Fig. 9.16). Not surprisingly, there was a concomitant rise in production of MOP (Fig. 9.15) and exports of cultured pearls (Fig. 9.17).

Exports of raw pearls peaked in 2000 at more than 11 t with a value of approximately 170 million Euro (~US$200 million) (Fig. 9.17). This point, however, marked a turning point for the French Polynesian pearl industry. A flood of lower-grade pearls brought international prices for black pearls down and market demand declined. Evidence of low profits began in 1998 when for the first time the volume of loose

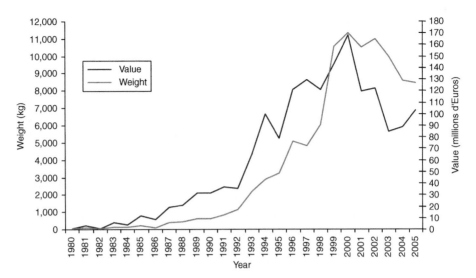

FIGURE 9.17   Export of raw pearls from French Polynesia (kg) and their value (millions of Euro) from 1980 to 2005 (*source*: French Polynesian Customs and ISPF).

black pearls exports increased at a higher rate than the value of exports (Fig. 9.17). As well as over-production, mean weights and quality of pearls were declining, and these factors lead to a dramatic crash in the market prices (Fig. 9.17). Pearl exports from French Polynesia between 2000 and 2005 declined by more than 20% and the value of pearl exports declined by about 40%. Total production in 2005 was in the range of 8–9 t and currently represents ca. 20% of total pearl market value. Despite government regulatory measures (see Section 9.4.6.2) there are still issues relating to the quantity and quality of pearls produced by French Polynesia (see Section 10.3.2).

Recent years have seen a decline in the number of pearl farms in French Polynesia to 516 in 2006. They vary from small (ca. <5 ha. in area) to large (>40 ha. in area) (see Section 13.3; Table 13.1). Most farms are situated on the islands of the Tuamotu Archipelago and the Gambier Archipelago (Fig. 1.14). The first farms were on the islands of Manihi and Takapoto, but farms spread to more than half of all atolls of Tuamotu, of which Narutea Sud atoll has become well known for having one of the largest pearl farms. The Gambier group situated far to the south is composed of only a few islands and it was on Mangareva Island where the Robert Wan Company established the largest and a very well organized pearl farm.

Pearl culture is French Polynesia's second largest economic resource after tourism and the first in terms of exports, with a value of more than US$120 million per year. This sector also generates employment for thousands of families spread over 30 islands in French Polynesia. For this reason, black pearl culture represents an essential part of the social and economic life of the country. In 2001, the Polynesian government became concerned about the black pearl industry in response to the falling profits of producers and a Ministry and government agency were created for the pearl culture industry. The President of French Polynesia was himself the Minister of Pearl Culture.

9.4.3.2. Organization of the French Polynesian pearling industry

In 1978, small cooperatives were grouped together to form the GIE Poe Rava Nui (the last three words meaning "large black pearl" in Polynesian) to provide reserve funds to pearl producers to establish their businesses (Hisada and Fukuhara, 1999). GIE Poe Rava Nui organized the first auction for its members in Papeete in Tahiti in 1979 and regular auctions have taken place in October ever since. Harvested pearls are delivered to GIE by producers where they are valued. Pearls are sold at the annual auction and the profits acquired from this are proportionally divided on the basis of estimated value (Hisada and Fukuhara, 1999). Saleable lots at auctions represent mixtures of pearls from various producers and this is the basis for pre-valuation. The GIE takes a commission of 8% of the total auction sales and a further portion of sales revenue is paid to the bank to service bank loans, where appropriate. The remaining amount is paid to the pearl producer (Hisada and Fukuhara, 1999). Despite the GIE, around 75% of pearls produced in French Polynesia were sold privately to large overseas buyers in 1999. As a result only 15–20% of production was available at auction (Hisada and Fukuhara, 1999). Since 2005, GIE Poe Rava Nui producers group has organized international auctions in China.

At the beginning of the new millennium, about 1,000 small farmers responsible for about 10% of pearl production were members of GIE Poe Rava Nui, while the GIE Tahiti pearl producers, established in 1996, had about 30 members with medium-sized farms that together accounted for about 20% of production. These organizations hold both separate and common auctions. The remaining 70% of pearl production from French Polynesia comes from approximately 15 members of the Syndicat Professionel des Producteurs des Perles (SPPP) of which the Tahiti Perles Company of Robert Wan holds the largest share. The company holds its own auctions in Hong Kong and, since 2003, also in Kobe, Japan. The major markets for cultured pearl exports from French Polynesia in 2005 were Hong Kong and China (42%), Japan (40%), Thailand (5.6%), USA (3.8%) and France (2.3%).

9.4.4. Culture and production trends – Cook Islands

Until the early 1970s, the inhabitants of the northern Cook Islands lived almost exclusively from fishing and from the sale of *Pinctada margaritifera* shells that were exported to English button manufacturers. There were trial pearl culture projects in the 1950s using the small local *P. maculata*, which were known for producing small yellow natural pearls called "Pipi pearls". However, pearl culture trials based on the Tahitian method and using *P. margaritifera*, began with the arrival of Peter Cummings from Australia in 1972.

Yves Tchen-Pan, a Tahitian of Chinese descent, constructed a pearl farm at Manihiki Atoll in 1986 and soon a number of farms owned by Cook Islanders followed, supported by the Ministry of Marine Resources. The pearl farms are located in the lagoon where they are built on top of natural platforms of coral rock known locally as "kaoa". People live and work on the kaoas where they keep chickens, pigs, plants and vegetables, and where they are surrounded by rafts or longlines holding their pearl oysters. Each kaoa represented something of a biosphere, and because of their low elevation, they are vulnerable to outside weather influences.

The grafting technicians used by these Manihiki farms came from Japan and in 1991 the Cook Islands Pearl Farmer's Association held its first auction on the main island, Rarotonga, offering 30,000 pearls for sale. In November 1997, intense typhoon Martin swept Manihiki atoll. Most of the underwater longlines with oysters remained intact, but the surface facilities were destroyed. The industry recovered and peaked in 2000 with export revenue of US$18 million, accounting for 20% of the country's gross domestic product. However, poor farming practices, particularly overstocking, meant that the oysters were susceptible to disease and the industry was virtually decimated by a disease outbreak towards the end of 2000. A rise in water temperature resulting from limited flushing of the Manihiki lagoon with oceanic water, combined with a mass spawning of oysters, triggered a rapid rise in the levels of pathogenic bacteria, *Vibrio harveyi* leading to the outbreak (Heffernan, 2006). To help ensure the long-term sustainability of the Cook Islands pearl industry and avoid further problems with disease, on-going monitoring of water quality and a greater understanding of the bathymetry and hydrodynamics in Manihiki lagoon have been critical in the development of a Pearl Farming Management Plan for Manihiki (Heffernan, 2006).

There were 205 pearl farms in the Cook Islands in 2003 with an estimated 1 million cultured adult oysters. However, as a result of increasing pearl production in French Polynesia, low international pearl prices and the continuing impacts of the 2000 disease outbreak, pearl export revenue from the Cook Islands declined to about US$2 million in 2005. Currently, 78% of the Cook Islands black pearl farms are within the lagoon of Manihiki Atoll where pearl production comes from 90 farms which seed ca. 900,000 pearl oysters annually, producing 300,000 saleable pearls. The remaining 20% of pearl culture occurs on Penrhyn Atoll where pearl culture began in 1994.

Pearl production in the Cook Islands amounts to about 5% of world production of black South Sea cultured pearls and benefits from the assistance of international development programs. It further benefits from the extensive public relation campaigns of the Tahitian cultured pearl industry. However, the high hopes for the industry of the early 1990s have not yet been fulfilled.

9.4.5. Culture and production trends – other countries

The Fiji Islands had Japanese pearl culture joint ventures as early as the 1950s and, from 1996 until the late 1970s, the Japanese Asia Pearl Company had a farm on Gau Island in Vukanicula Bay. In 2000, a pearl farm was established at Savusavu on the island of Vanua Levu. The farm is situated in a deep bay on a high island, and subject to nutrient-rich upwelling that supports abundant phytoplankton and rapid growth rates of pearl oysters. This situation is very different to that of pearl farms in the oligotrophic atoll lagoons of eastern Polynesia (French Polynesia and Cook Islands). The farm currently holds ca. 500,000 oysters and uses Japanese technicians to seed ca. 120,000 oysters annually. Approximately 80% of the farmed oysters are obtained from spat collectors; however, with a recently completed hatchery, the proportion of hatchery-produced spat will no doubt increase. The farm employs village-based spat collection, which provides much-needed income to communities close to the farm. The farm is operated by J. Hunter Pearls and the pearls, marketed as "Fiji pearls", have already gained a reputation for quality and their wide variety of interesting colors. The first auction of "Fiji pearls" in Japan in 2007 offered 30,000 pearls (Anon, 2007).

Between 1905 and 1922, Dungunab Bay in the Red Sea (Fig. 1.5) was the site of a most successful experiment with *Pinctada margaritifera* under the guidance of Cyril Crossland whose purpose was to support the thriving MOP industry. The so-called Red Sea cultured pearls have subsequently come from a farm in Dungunab Bay that was founded in 1997 by Rosario Autore and his brother Nino. The first harvest of 12,500 pearls from the farm in Dungunab Bay was sold in 1999/2000. Production has since been reduced and shows numerous keshis and circle pearls. Green and yellow hues dominate and there are pistachio and apricot colors as well as golden and champagne. Further down the east African coast, *P. margaritifera* has been used for trial blister (mabé) pearl production in Tanzania in a project to determine the potential of small-scale pearl production in generating income for coastal communities (Southgate *et al.*, 2006).

A repopulation program for *Pinctada margaritifera* in Micronesia was launched in the mid-1990s by Black Pearls Inc. on Majuro Atoll. The program was followed by a pearl farm that in 1999/2000 had exceeded an annual output of 50,000 pearls. Small privately owned pearl farms have also been developed in Micronesia as part of the College of Micronesia Land Grant Program. Two farms are planned for development in Pohnpei with a further two or three in outer islands (Dennis, 2006). The goal for this development is to harvest between 10,000 and 30,000 pearls by 2010.

Black South Sea cultured pearls from a number of research and pilot projects have reached the market during the last 15 years. Many are the result of international development programs to help create a new source of income for developing island nations of the Pacific. In 1999, a total of 891 black pearls produced on a farm near Gizo in the western province of the Solomon Islands were sold over the internet by an auctioneer from Sydney. Similar black pearl production trials have occurred in Kiribati and Micronesia (Fassler, 2002; Ito *et al.*, 2004; Southgate, 2004).

9.4.6. Marketing black (Tahitian) South Sea cultured pearls

9.4.6.1. Pearls and promotion

The atolls of French Polynesia are spread over an area the size of Europe, and they differ in environmental factors and in the genetic make-up of their resident pearl oyster populations. As a result, Tahitian cultured pearls show a variety of luster and color. The Tahitian government does not include color in its classification system as the variety of body colors, overtones and various combinations and saturations amount to a total of >80 possibilities and only individual grading is possible. This does not mean that there are not certain color combinations that command higher prices due to their rarity. The general rule is that pearls with overtones are valued higher than pearls without an overtone. A black body color with a green or "peacock" overtone is valued highest (peacock known as "poe rava" in Tahiti describes a combination of green, pink, red and yellow overtones). A purplish-pink overtone known as "aubergine" in French is valued next highest followed by a combination of pink and blue overtones. The Gemmological Institute of America developed a "color reference chart" which can be consulted for the description of body colors.

In general, cultured black South Sea pearls range in size from 9 to 20 mm; however, the so-called Robert Wan pearl, which is exhibited in the Perles de Tahiti museum in Papeete, has a diameter of 20.92 mm and weighs 62.5 carats. It represents a rare exception. The average and most popular size for Tahitian cultured pearls is 8–12 mm, although it has varied over recent years. Sizes between 8–11 mm and 13–17 mm sell best. The shapes of Tahitian pearls described by the official Tahitian classification include round to semi-round; semi-baroque; baroque to circles (pearls which show distinct circle lines on the surface); this sequence reflects diminishing rarity and is used as a criterion for evaluation (see Section 8.10).

The Ministry of Pearl Culture, the Pearl Culture Agency and affiliated producers constitute a non-profit organization named "Groupement d'Intérêt Économique GIE Perles de Tahiti". GIE Perles de Tahiti was founded in 1993 to pursue new marketing and promotional strategies for the French Polynesian pearl industry. The organization focuses on extensive global public relations, promotion in international markets and assisting producers in commercializing their products. Japan and Hong-Kong represent the major buyer countries and the Japan Black Pearl Promotion, which cooperates with GIE Perles de Tahiti, has an annual budget of about $US1.7 million (Hisada and Fukuhara, 1999). Promotion overseas is mostly focused on special events, such as jewelry fairs. GIE Perles de Tahiti communicates information about activities in French Polynesia to overseas partners and similarly reports on international events and market trends to local producers through a newsletter and a website. Recently, GIE Perles de Tahiti and the Agency of Pearl Culture collaborated to distribute an official brochure with guidelines for buyers.

9.4.6.2. Quality control

The Tahiti cultured pearl industry has benefited from successful government control and the promotion campaigns by GIE Perles de Tahiti, which had a promotional budget of US$5 million in 2004. Quality control remains a continuing problem for small pearl producers in Tahiti despite efforts by the French Polynesian government to improve technologies through teaching programs and by establishing control systems. In June 2000, the French Polynesian government sent out a strong message by throwing 35,000 pearls of low quality into the sea during a symbolic ceremony. In 1999, an official classification system was introduced for pearls that serve as the export guideline. Quality control measures are strictly applied to cultured pearls exported from French

Polynesia. The French Polynesian classification system does not include nacre thickness, but pearls with a nacre thickness <0.8 mm are not permitted to be exported. All pearls are checked using X-ray machines at the Pearl Culture Agency of the Ministry of Pearl Culture of French Polynesia (Fig. 9.18) and low-grade pearls without the minimum nacre thickness or which failed to meet the surface quality criteria, are rejected and destroyed (Fig. 10.1). The French Polynesian system groups pearls according to a classification from A to D which accounts for increasing irregularities on the surface and a decreasing amount of light reflected from the surface. Only grades A–C are permitted to be exported. In Japan, the major market for Tahitian pearls, the Japan Black Pearl Promotion, proposed to attach a guarantee mark to each piece of black pearl merchandise guaranteeing natural color and that the item had not been dyed (Hisada and Fukuhara, 1999).

Public organizations, including GIE Perles de Tahiti (which includes the Ministry of Pearl Culture and Pearl Culture Agency) as well as several producers' organizations (e.g. GIE Poe Rava Nui) that provide technical and financial assistance to pearl farmers, are dedicated to establishing and applying regulations to maintain quality production. The Agency's missions are broad and include:

- enforcing standards of exports, e.g. nacre thickness,
- assisting producers in meeting regulatory requirements,
- controlling marine leases,
- conducting scientific studies in collaboration with international and national research organizations such as Ifremer, CRIOBE (Ce Centre de Recherches Insulaires et Observatoire de l'Environnement) and IRD (Institut de Recherché pour le Développment).

In 1993, the Pearl Culture Agency established training courses for young people to teach them techniques for pearl oyster culture and pearl seeding. Two training schools

FIGURE 9.18   Black pearls being X-rayed to determine nacre thickness as part of French Polynesian government measures to maintain quality of exported pearls. The operator views X-ray images of the pearls (in the X-ray machine on the right) which clearly show pearl nuclei and nacre thickness.

for theory and cultural applications, and an experimental hatchery for scientific studies are situated in the Tuamotu islands.

9.4.7. Black South Sea pearl culture methods

By showing that spat collection was possible within the lagoons of French Polynesian atolls, Reed made it possible for islanders to set up pearl culture activities at their home islands (Reed, 1965). One of the major advantages that led French Polynesia to become the leader in black pearl culture was the availability of abundant natural spat for spat collection activities (see Section 7.3). Natural spat resources are not found in all the islands and atolls of French Polynesia. However, spat transfer between islands has enabled all archipelagos to become potential pearl producers, although this has had unfortunate genetic and fouling consequences (see Section 12.2.6). About fifteen atolls in French Polynesia undertake spat collecting activities, but natural spat production can vary from year to year due to biotic and abiotic factors.

Various techniques are used for natural spat collection in French Polynesia, with various substrates such as artificial plastic netted mesh and rope, and branches of a local plant, *Penphis acidula*, or "miki miki" being used. Once the spat are settled and are ca. 6 months old, they are removed from spat collectors to avoid spatial competition. They are then placed in a variety of culture units consisting mostly of pocket nets (see Section 7.5.4) and chaplets (Fig. 9.19) suspended from longlines. Frequent cleaning is necessary to avoid epibionts (see Section 15.4.2). An example of biosecurity management failure is the anemone pest, *Aiptasia pallida*, which has become a major fouling organism of pearl culture equipment throughout French Polynesia because of uncontrolled oyster transfers between islands.

Grafting is carried out by technicians from French Polynesia, China and Japan. Pearl growth is dependent on many factors and an average of 16 months to

FIGURE 9.19 Cultured *Pinctada margaritifera* held on chaplets hung from a longline at Takapoto Atoll, French Polynesia (Photo: John Lucas).

2 years of growth in the pearl farm is required to produce the minimum 0.8 mm nacre thickness required by export legislation. Both the government and producers are aware that management of farm density has to be a balance between profitability and environmental care.

9.4.7.1. Farm management

Since 2001, the French Polynesian government has established legislation to control pearl production in response to over-production, which contributed to the dramatic fall in pearl value that began in the late 1990s (Fig. 9.17). There are policies to strictly manage marine concessions (leases) and regulate producer numbers, and consequently pearl export volume. A software application based on GIS was developed and is used to manage marine concessions; a system similar to that used at Manihiki in the Cook Islands (Heffernan, 2006). Similarly, technical assistance is provided to French Polynesian pearl producers in order to help them to improve techniques and farm management. In return, producers must follow strict regulations. For example, it is required that rearing density does not exceed 10,000 oysters on a 200 m long line. A stock register must also be kept to ensure that producers act as good farm managers. To improve optimal growth, a maximum of 2,400 grafted oysters on a 200 m long line is allowed.

9.5. PANAMANIAN PEARL OYSTER

The Panamanian or Calafia pearl oyster, *Pinctada mazatlanica*, is distributed along the Pacific coasts of Central and South America (Fig. 2.6). Exploitation of this species for pearls and MOP has modeled the history of Baja California, from pre-hispanic times until the exhaustion of natural oyster beds by the middle of the 20th century. The presence of the resource was the principal incentive for colonial exploration of the region. The pearl oyster populations of this region established an economic base that allowed the colonization and construction of the first regional socio-economic structure in the 18th century. Exploitation of MOP and pearls, starting from the 1830s, provided the basis for founding the port of La Paz, Baja California, which was renowned as an important pearling area from then on. In fact it was in Bahía de La Paz where the first successful mass cultivation of a pearl oyster was carried out, using *P. mazatlanica*. The "Compañía Criadora de Concha y Perla de Baja California" (CCCP) was the first pearl emporium and it managed more than ten million pearl oysters. Yet despite this distinguished pearling history, this region and *P. mazatlanica* are no longer associated with MOP production or commercial pearl cultivation.

9.5.1. History of exploitation

The Californios Indian groups (Guaycuras, Pericúes, Cochimíes) were the first residents of Baja California (del Barco, 1780) and the marine fauna was more important in their diet than terrestrial foods. Pearl oysters provided both nutrition and ornament.

Analysis of the *concheros* (ancient shell deposits coinciding with Californios settlements) has shown that there was selection of specific oyster sizes, with only mature adults being killed, to assure conservation of the resource (Castellanos and Cruz, 1995). In 1533 the Spanish conquerors arrived at La Paz for the first time, in search of the "Pearl Myth". The sight of Californios with long hair braided with hundreds of pearls, and astonishingly bright large nacre shells adorning their bodies, caused the Spaniards to believe that they had found the mythical reign of Queen Calafia and her amazons, and their treasure of gold, gems and pearls.

The greed for pearls strongly motivated exploration of Baja California and the intent to conquer the Peninsula. Hernán Cortés commanded the first attempt to colonize the region. He reached the coast in 1535, naming Bahía de Santa Cruz, the bay that today is Bahía de La Paz, and confirmed the wealth of the pearl oyster banks. During the following 170 years, Spanish explorers tried in vain to establish a colony and to exploit the *placeres perleros* of the Gulf of California. The Viceroyalty of Nueva España established a strategy to promote the demarcation and exploration of the Californias coast without burdening the Royal Treasury. It consisted of granting licenses for pearl fishing, on condition that the beneficiaries must realize explorations of the peninsular coasts and establish a port-refuge for the Galleon of Manila. They must also pay the *quinto de perlas*, a tax of a fifth of the value of the pearls obtained. The combination of exploration and colonization with pearl fisheries consolidated the fame of the Gulf as one of the most important pearling regions in the world. By the end of the 17th century, King Carlos II ordered another expedition commanded by Admiral Isidoro Atondo y Antillón, who reported that the exploitation of pearl oysters in past decades had drained the *placeres perleros*.

During the missionary era lead by the Jesuits between 1697 and 1740, some pearl oysters beds had a period of recovery due to prohibition of fishing for oysters enforced by the Jesuits. In spite of such prohibition, Manuel de Ocio began exploiting the pearl oyster grounds in the central coasts of the Gulf of California. The fishing was so intensive that in 8 years those sites were depleted. The Marquis José de Gálvez was the Visitor (royal supervisor) that King Carlos III sent to drive out the Jesuits from Baja and to apply the modern reformation of the Borbón state. Gálvez based all development on the exploitation and commercialization of the pearl oyster resource. He planned the creation of an Asian-Mexican Company that would export large quantities of pearls to the Orient. However, the resource was extremely scarce due to the intensive exploitation that Ocio and other empresarios had exerted on the pearl beds.

The *Courts of Cádiz*, in April 1811, a few months after starting the independence movement, published an ordinance to promote the development of the Californias. It considered that the prosperity of the region would be dependent on the development and exploitation of marine resources. It declared "totally free the fishing of pearls in all the domains of the Indies, for all the subjects of the Monarchy" and "entirely free contracts made between *armadores* and divers". The young independent Mexican government summoned the *Junta de Fomento de Las Californias* for rebuilding and planning the economy and government of Baja California. This included establishing a trade with Asia to exchange pearls, fine fish and handcrafts in leather. The *Junta de Fomento* gave pearl oyster stocks as the major and most valuable resource. It is not surprising that, again in 1830, the general commander of Baja California, Manuel Victoria, proposed, among other measures to remedy the deplorable situation of living in the Territory, to dedicate ships to the coastal traffic trade, fishing of pearls, and hunting of whales and otters. The pearl oyster populations were in a phase of abundance, thus supporting the resurgence of the regional economy and establishing La Paz as the most important pearl center of México.

9.5.1.1. Mother-of-pearl and socio-economic development in La Paz (1830–1879)

New commercial opportunities developed in the early decades of the 19th century and the MOP of the Gulf of California became appreciated and sought by the world market. Until then, pearl oyster shells were considered as waste and were abandoned on the beaches. But in 1830, the value of MOP as a new and important natural resource brought an extraordinary turn to the development of the pearl oyster fishery and regional socio-economic structure (Cariño, 1998).

In 1830, the French merchant Cipryen Combier observed the great quantity of pearl oyster shells abandoned on the beaches and had the idea of ballasting his ship with them for selling later. Subsequently, MOP constituted the main target of the pearl oyster fisheries during the 19th century, as well as the main regional resource for export. The revitalization of this activity had an important impact on the regional socio-economic structure and increasing human population. Between 1838 and 1868 about a hundred *armadas* requested a license to fish pearl oysters (Valadéz, 1963). An *armada* was a fleet formed by a large ship – brig, frigate or sloop – and a certain number of canoes (Fig. 9.20 and Fig. 9.21). Generally, the *armador* was not the proprietor of the fleet but an employee of richer empresarios. The armadas traveled along the eastern coast from southern to northern Baja California, from Cabo Pulmo to Bahía Mulegé and the island of San Marcos. The skill of the divers – Mestizo Indians, Yaquis and Californian – was a matter of admiration. They were able to free dive for up to 2 minutes as deep 9–11 m; they made descents up to 40 times a day. The *armador* provided food for the divers and their family, considering it as payment in advance. Divers repaid the *armador* once the fishing season was finished. These debts were paid with a portion of the oysters that had been collected. They were allowed to freely sell any pearls that they found within the oysters, but the *patrón* had priority in purchasing them. In this way the divers paid their debt and the owners kept the entire product (Esteva, 1857).

Over-exploitation and the increase of fisheries activity during the 19th century resulted in the gradual decline of pearl oyster populations. Aware of the socio-economic disaster that this would bring to La Paz, José María Esteva proposed in 1857 regulations relating to management of the pearl oyster fisheries in the Gulf of California. Nevertheless, these regulations concerned mainly the work of the divers and not so much the conservation of the species. Toward 1870, the *placeres* began to show signs of exhaustion and the expeditions of the *armadas perleras* became less profitable. With pearl oyster populations on the brink of exhaustion, several laws were applied: in 1874 the coast was divided into four sections and fishing was restricted to

FIGURE 9.20   An *armada perlera* (pearl oyster fishing fleet). This spectacle was common during the late 1920s in Bahía de La Paz.

FIGURE 9.21   *Buzos de chapuz.* Native head divers in a canoe. Some were employed in smaller *armadas*, or worked independently.

every second year in each section. In 1878, there was another law which intended that fisheries should be carried out only every 4 years in each section. However, in general, the empresarios did not obey these restrictions and continued to exploit the oyster beds without control. Despite this, the application of these measures allowed fishing to prevail for some more decades.

The exploitation of pearl oysters was modernized in 1874 with the introduction of the 'hard-hat' or helmet and diving suit (Fig. 9.8). Territorial concessions for the exploitation of natural resources in Baja California were later granted to foreigners and Mexicans, and there were frequent commercial associations. In fact, the whole

Peninsula was a concession. The concession of marine areas for the exploitation of pearl oysters violated the first two articles of the 1874 legislation, which had declared these areas free of fishing. Officials received constant complaints from the *armadores* who were refused the right to fish in concession areas that the government had granted exclusively to the pearling companies.

In the politics of leasing the pearl areas, the federal government had fundamental goals: conservation of the pearl oyster resource, obtaining part of the revenue generated through taxes, and beneficial socio-economic effects for the region. From the 26 contracts of lease to the pearl fishery areas signed over a 22-year period, only 10 were actually put into operation.

Under the regime of Porfirio Díaz and policies of exclusive territorial concessions to pearl companies, the federal government favored the participation of rich capitalists who in turn imposed their conditions on the *armadas* for exploiting the region's pearling areas. The tendency was to concentrate capital, depriving the *armadores* and local divers of access to marine resources. As a consequence, several companies amalgamated to create the Compañía Perlífera de Baja California Sucesores in 1893. This new company had a short life because in the same year it gave up all its rights to The Mangara Exploration Limited Company, which had been founded in London and now had almost absolute control of the Mexican pearl resources.

La Mangara never fulfilled its obligations, under the conditions of its fishing concession, to establish places for the cultivation of pearl oysters and to establish new *placeres* every year to cultivate at least 10,000 pearl oysters. On the contrary, they used devastating fishing methods, including dynamite fishing. Despite this the government extended the concession for 16 more years beginning in 1916.

9.5.2. Culture and production trends

The consistently growing demand for pearls and MOP, and introduction of the diving suit, worsened the exploitation of the pearl oyster banks in the Gulf of California, driving them toward exhaustion. In all the countries of the world involved in the fishing or trading of pearl shell and pearls, merchants and empresarios pressed the scientists to develop techniques for their cultivation. The expectation of this scientific discovery was expressed by Lyster Jameson: "The man that solves the problem of pearl oyster cultivation, will have not only the privilege of making science and industry to progress, but rather his name will deserve the honor of being included among the founders of empires" (Jameson, 1901).

Gastón Vives was the first scientist in the world to achieve mass cultivation of a pearl oyster. In 1903, after years of field investigation, Vives founded the *Compañía Criadora de Concha y Perla de Baja California, S.A.* (CCCP), which became the first pearl emporium in the world (Cariño and Monteforte, 1999). The technology he developed had three stages: spat collection, nursery culture, and grow-out. It was necessary to build dedicated apparatus and facilities to complete the production cycle.

The operation center of CCCP was established at Espíritu Santo Island near La Paz and hundreds of workers were employed. To capture *Pinctada mazatlanica* spat, huge wooden boxes with metallic mesh were manufactured. They were filled with live pearl oysters and shells, and with branches of a local bush that is very resistant to rotting. These 'incubators' remained submerged during the summer months which are the pearl oyster's reproductive season. In autumn (fall) they were removed from the sea and the young oysters were individually removed by hand and placed in baskets divided into compartments where they would grow until reaching a mature size (Fig. 9.22). To carry out this second stage of the cultivation, Vives built an embankment 1 km long that divided the coastal lagoon from Bahía San Gabriel (Fig. 9.23). There was a system of 36 channels with ports which allowed the water to flow, providing food and oxygen to the juvenile pearl oysters, and protection against currents and predators. After 9 months, each adult pearl oyster was placed inside metallic armor (a metal container that fitted around the outside of the shell for protection) and moved to locations where the third and last stage of cultivation took place. This was developed on rocky substrates selected by Vives because of their favorable characteristics for the production of MOP and natural pearls.

Nine years after its foundation, the CCCP had multiplied its capital and became the largest exporter of high quality MOP and natural pearls in the world. This pearl emporium, which achieved the largest production of cultivated pearl oysters in history, was plundered and destroyed in 1914 during the Mexican Revolution. In spite of the obvious benefits to the region that would have resulted from re-establishing his pearl culture operations, Vives could not recover his company. Without the reproductive contribution of 10 million oysters in cultivation, intensive fishing in the Gulf of California over a period of two decades was more than enough to exhaust the resource,

FIGURE 9.22 Don Gaston Vives (right) supervising wire cages with thousands of *Pinctada mazatlanica* juveniles, ready to be placed in the nursery culture areas at Bahía San Gabriel.

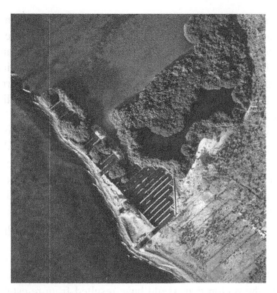

FIGURE 9.23   Aerial view of the remains of the Compañia Criadora de Concha y Perla de Baja California Sur, S.A. (CCCP) at Bahía San Gabriel, Espiritu Santo Island near La Paz, Baja California Sur. The nursery culture channels were constructed within an artificial lagoon and used to house wire cages containing *Pinctada mazatlanica*.

marking the end of Mexican pearl wealth. Both *P. mazatlanica* and *Pteria sterna* were officially declared 'in danger of extinction' in 1940 and fisheries were banned. This was the first measure for natural resource protection taken in Latin America (Cariño, 1998). Their status changed in 1994 to 'species under special protection', but the prohibition on fisheries is still in place.

9.5.2.1. Modern day culture of *Pinctada mazatlanica*

The decline of pearl oyster populations as a result of unsustainable fisheries, and the need to reliably obtain good quality pearls, compelled governments and scientists across the world to develop methods for pearl oyster husbandry and pearl cultivation. While such pioneering research resulted in development of large cultured pearl industries in many countries (see Sections 9.2, 9.3, 9.4), in México the pathway has been slower, discontinuous and complex (Cariño and Monteforte, 1999, 2006).

After the destruction of the Compañia Criadora in 1914, further efforts towards pearl culture did not start until 1939 (Cariño, 1998). To 1986, there were 22 failed attempts at pearl farming in the Gulf of California, 21 of them in the La Paz region. There has been significant research effort directed towards the culture of *P. mazatlanica* by Mexican scientists since 1988 and information is now available on reproductive seasonality and recruitment (Monteforte and López-López, 1990; Monteforte and Bervera, 1994), recruitment of *P. mazatlanica* spat to spat collectors and optimization of substrate for spat collectors (Bervera, 1994), growth and survival during nursery and juvenile culture (Bervera, 1994; Morales-Mulia, 1996; Monteforte and Morales, 2000), the effect of culture density and pearl production (Monteforte *et al.*, 1994). However, much of the research in Mexico has given priority to pearl production using the scarce wild-oyster population rather than development of sustainable culture techniques.

As shown in other parts of the world, pearl culture can provide significant socio-economic benefits for coastal communities (see Chapter 13) and is a form of low-impact aquaculture that can be conducted in accord with marine conservation efforts (e.g. Southgate *et al.*, 2006). Recently the SEMARNAT (Secretaría del Medio Ambiente y Recursos Naturales, Mexico) published the *Plan de Ordenamiento Ecológico para el Golfo de California* (SEMARNAT 2006), where low-impact mariculture was identified as the best option for coastal zones that have not yet been developed. The scenario has not changed much since 1985 when the advantages of pearl farming in Bahía de La Paz were highlighted (Monteforte, 1994, 1996). Coming years will hopefully see greater emphasis on pearl culture as a means of supporting sustainable socio-economic development in the Gulf of California, and re-establishment of this region as a significant producer of pearls and MOP.

9.6. WINGED PEARL OYSTERS

Numerous species of 'winged' pearl oysters belonging to the genus *Pteria* (from the Greek: *pteryga* or *ptero* = wing) occur in tropical and subtropical waters of the world (Fig. 2.8). The common name 'winged pearl oyster' relates to their elongated hinge that resembles a wing (see Section 2.2.2, Fig. 2.1).

In spite of the number of species and their wide distributions, only two species of *Pteria* are commercially cultured; the 'mabé' pearl oyster, *Pteria penguin*, is cultured throughout Southeast Asia, in Australia and in some Pacific island nations (Beer and Southgate, 2000) and the 'concha nácar' or rainbow lip pearl oyster, *Pt. sterna*, which is commercially cultured in Mexico in some areas in the Gulf of California (Kiefert *et al.*, 2004; Ruiz-Rubio *et al.*, 2006) (Fig. 2.8). A third species with potential for culture and pearl production is the Caribbean winged pearl oyster, *Pt. columbus*, which has recently been the subject of experimental culture trials (Lodeiros *et al.*, 1999). *Pteria* spp. are not yet widely utilized for pearl production. Many species are small and with narrow mantle cavity thickness, thus limiting the size of cultured pearls that can be obtained from them. Those *Pteria* species with the greatest potential for pearl production share geographical distributions with the larger species of *Pinctada*, which have greater commercial potential. Furthermore, *Pteria* spp. are common fouling organisms of culture equipment and oysters at pearl farms (Taylor *et al.*, 1997) and are often considered 'pest' species.

9.6.1. History of exploitation

Pteria spp. are generally cultured for production of mabé pearls (see Section 8.6). Their method of production is described in Section 8.7 and has been used since at least the 1950s. Although nucleated 'round' cultured pearls can be produced from *Pteria* spp., it is generally acknowledged that this is more difficult to achieve with *Pteria* spp. than *Pinctada* spp. as a result of morphological differences between genera. Only in recent years has steady successful production of round pearls from *Pteria* spp. been reported (Farell *et al.*, 1998; Yu and Wang, 2004). This is an important development and one that may provide an impetus for greater utilization of *Pteria* spp. for cultured pearl production. It would diversify the type of cultured round pearls available on the international market and may help many developing nations in the Pacific, Caribbean and Central and South America, to develop more lucrative aquaculture industries.

9.6.2. Culture and production trends – Japan

In the 1950s Japanese companies began using *Pteria penguin* on the Ryukyu Islands for production of mabé pearls. The shells of *Pt. penguin* can reach a length of up to 25 cm and they are called "mabé gai" in Japanese. The word probably comes from an old dialect that is spoken on the Ryukyu Islands. The nacre layer of *Pt. penguin* shows a distinct iridescence in which light-pink, blue-pink and green dominate. The rather broad rim is black and is underlined by a golden color. The center of production in the Ryukyu Islands is Amami-Oshima and at the start of the new millennium there were three or four companies in Amami-Oshima, producing commercial quantities of mabé pearls. Their annual yield is ca. 200,000 pearls (Hisada and Fukuhara, 1999) and they obtain their oysters from hatcheries because natural banks are almost exhausted.

9.6.3. Culture and production trends – Mexico

Successful production of round cultured pearls from *Pt. sterna* was achieved in 1995 at the Instituto Tecnologico y de Estudios Superiores de Monterrey (ITESM) in Guaymas. Regular production of cultured pearls started in 1996 and both mabé and round pearls have been produced in significant quantities. They are sold by the company Perlas del Mar de Cortez that developed from the ITESM research group led by Sergio Farell. Production is based on *Pt. sterna* and the mabé pearls produced are in the range of 12–15 mm (Fig. 8.6). Round pearls are sized 6.5–8.5 mm, although individual pearls may reach up to 14 mm. They are usually off-round in shape and show pleasing colors: gray, golden, bronze-colored, reddish-brown, black and purple. Overtones are pink, green and purple and sometimes there is a peacock overtone. The price of the pearls overlaps with those of white and black South Sea cultured pearls. Production of mabé and round pearls from *Pt. sterna* in one of the Mexican farms (Perlas del Mar de Cortez) began about 1995 at a pilot or experimental level and has increased to ca. 4,000 cultured pearls and 8,000 mabé annually. Approximately 50% of their production, mounted in jewelry, is sold at the farm (Kiefert *et al.*, 2004). Production is expected to gradually increase to 10,000 cultured pearls and 10,000 mabé by the year 2010 (D. McLaurin, personal communication).

9.6.4. Culture and production trends – China

While the mainstay of the Chinese marine cultured pearl industry is Akoya pearl production (see Section 9.2.3), recent years have seen production of pearls from other species including *P. maxima*, *P. margaritifera* and *Pt. penguin*. *Pteria penguin* is widely distributed along the southern Chinese coast and has been used for hatchery-based pearl culture in southern China (Yu and Wang, 2004). Three companies have been established at Hainan Island and Leizhou peninsula for profitable cultivation of mabé pearls from *Pt. penguin* (Yu and Wang, 2004). Furthermore, round pearls have also been successfully cultured from *Pt. penguin* at Hainan Island.

9.6.5. Culture and production trends – other countries

This species was used for production of mabé in Vava'u, Tonga. *Pteria penguin* was introduced to Tonga in 1975 by the Tasaki Pearl Company of Japan (Malimali, 1995; Finau, 2005). Subsequent research conducted by the Ministry of Fisheries in Tonga and supported by FAO South Pacific Aquaculture Development Project, attracted interest from local investors and there were 25 small pearl farmers in Tonga at the end of 2000. The current value of the industry is not known, as the majority of pearls are sold locally; however, the Ministry of Fisheries exported pearls to the value of ca. US$10,000 in 1996. Japanese specialists visiting Vava'u in the mid-1990s estimated that an area of approximately 850 ha could be farmed for half-pearl production. They estimated that such an area could support annual production of about 750,000 pearls with potential annual revenue of ca. US$7.5 million (Finau, 2005). Pearl production in Tonga relies on natural spat collection.

At Phuket Island in Thailand, the two major pearl farms were reported to hold 30,000 *Pt. penguin* for mabé production in addition to a number of smaller family farms that also produce mabé from *Pt. penguin* (Bussawarit, 1995).

9.6.6. *Pteria* spp. culture methods

9.6.6.1. *Pteria penguin*

Pteria penguin is the most widespread cultured winged oyster and is cultured in Japan, Australia, the Philippines, Indonesia, China, Thailand, Vietnam (Fig. 14.6) and the Kingdom of Tonga. Other nations with interest in the culture of *Pt. penguin* are the Seychelles, the Solomon Islands, the Maldives and Mayotte. Interestingly, despite its wide distribution, *Pt. penguin* has received relatively little research attention.

There are a few reports on the breeding cycle and hatchery production of this species. Beer and Southgate (2000) reported a very discrete pattern of recruitment, towards the end of summer, for *Pteria penguin* in northern Australia. A similar pattern of summer–autumn recruitment was reported in Tonga (Malimali, 1995; Tanaka, 1997). This pattern differs from the year-round reproduction reported for *Pt. penguin* in Thailand (Arjarasirikoon *et al.*, 2004), which is similar to patterns reported for other *Pteria* spp. *Pt. penguin* spat also show depth preference with greater recruitment onto spat collectors at a depth of 6 m than at a depth of 2 m (Beer and Southgate, 2000).

No detailed study has yet reported on larval culture of *Pteria penguin*; however, Beer (1999) reported that protocols used for the larval culture of other pearl oyster species

(*Pinctada* spp.) were suitable for *Pt. penguin*. Larval development required 23–25 days at a water temperature of 26.9–29.5°C and larvae showed relatively high survival (~16%) (Beer, 1999) compared to those of some *Pinctada* spp. (e.g. Alagarswami *et al.*, 1989; Southgate and Beer, 1997). Yu and Wang (2004) also reported development of methods for successful spawning induction and larval rearing of *Pt. penguin* in China.

There is a similar dearth of information relating to nursery and grow-out culture methods for *Pt. penguin*. Beer (1999) reported severe predation of *Pt. penguin* juveniles by fish when held on substrates without protective mesh; yet those held within 3 mm mesh showed slower growth (3.44 mm/month) than those held on unprotected ropes (6.67 mm/month), which reached a DVM of 100 mm within 15 months. Successful culture of hatchery produced *Pt. penguin* to adult size, and their subsequent use for pearl production, has also been undertaken in China (Yu and Wang, 2004).

9.6.6.2. *Pteria sterna*

This species has received more research attention than the more widely distributed *Pteria penguin* with all major aspects of its culture, including pearl production, being targeted. Hatchery production of *Pt. sterna* has been studied in some detail (McAnally-Salas and Valenzuela-Espinoza, 1990; Araya-Nuñez *et al.*, 1991, 1995). These earlier studies focused on descriptions of embryos and larvae and also on establishing some basic information on larval rearing protocols. Larval development in *Pt. sterna* ranges from 24 to 37 days, depending on water temperature (Serrano-Guzman and Salinas-Ordaz, 1993; Araya-Nuñez *et al.*, 1995) and, like those of other pearl oyster species, *Pt. sterna* larvae can only ingest small microalgae species during most of their development (Martínez-Fernández *et al.*, 2004). Research into spat collection and nursery culture studies of this species (Monteforte and García-Gasca, 1994) have allowed further development of this new aquaculture industry in Mexico. Studies on hatchery production, grow-out culture methods and mabé culture methods have also contributed to improve the efficiency of this industry (e.g. Martínez-Fernández *et al.*, 2004; Ruiz-Rubio *et al.*, 2006). A newly developed seeding technique for round pearls was a breakthrough for the cultured pearl industry in Mexico, and brought a new kind of cultured pearl to the international pearl market (Nava *et al.*, 2000).

Although hatchery production of *Pteria sterna* is feasible, natural spatfall is still sufficient to satisfy the demand of the two commercial pearl farms in Mexico. This species is preferred for pearl production to the other pearl oyster species occurring in Mexican waters (*Pinctada mazatlanica*; Section 9.5) because of its unique nacre coloration that seems to vary from year to year and also between culture areas (Kiefert *et al.*, 2004; Ruiz-Rubio *et al.*, 2006).

Pteria sterna is cultured to produce both, mabé and round pearls in Mexico (Nava *et al.*, 2000; Ruiz-Rubio *et al.*, 2006). This relatively small pearl oyster can hold a maximum of two to three hemispherical nuclei for mabé production before pearl quality is affected (Saucedo *et al.*, 1998; Ruiz-Rubio *et al.*, 2006). Other countries with potential to culture *Pt. sterna* include Guatemala, El Salvador, Honduras, Costa Rica, Nicaragua, Panama, Colombia, Ecuador, Peru and northern Chile. However, this species prefers colder waters and may be a challenge to culture in waters of Central American countries due to increased mortalities during the warmer summer months (Ruiz-Rubio et al., 2006).

9.6.6.3. *Pteria colymbus*

This species has only recently been subject to research in Venezuela and Colombia where it is relatively abundant. *Pteria colymbus* breeds continuously during the year but shows a major peak during winter and a secondary peak during summer months (Marquez et al., 2000). Although small, *Pt. colymbus* is a relatively fast-growing species that can reach a DVM of ca. 70mm in approximately one year (Lodeiros et al., 1999). This species may be able to produce cultured round pearls of a size similar to those produced by the Akoya pearl oyster. With appropriate broodstock management and breeding programs it may be possible to achieve a significant increase in maximum height in cultured animals. This would directly benefit the pearl production potential of this species and facilitate seeding with larger nuclei. Furthermore, with an increase in maximum DVM it may be possible to produce up to two high quality mabé per oyster.

9.7. SUMMARY

Commercial exploitation of pearl oysters goes back many centuries. Four *Pinctada* spp. have been the main targets: Akoya, *P. maxima*, *P. margaritifera* and *P. mazatlanica*. The initial industries were generally based on the collection of shells for MOP, with natural pearls being found incidentally. The history of pearl fisheries is of over-exploitation leading to depauperate stocks, which may not recover. In some cases the pearl beds were progressively overfished to increasing depths as a matter of necessity and as permitted by developing diving technology. Over-fishing of pearl oyster resources was a major factor in considering culture of pearl oysters for restocking and pearl production. Mass cultivation of pearl oysters was first achieved by Gastón Vives in 1903 with *P. mazatlanica* in the Gulf of California. This large and successful pearl oyster farm, producing shells and natural pearls, was subsequently destroyed and pearl oyster culture ceased until it emerged in Japan. Production of round cultured pearls from Akoya pearl oysters in Japan began in 1916 and the number of farms increased rapidly. The technology for producing cultured pearls subsequently spread to other countries and other species during the 20th century, with Japanese companies and Japanese technicians being strongly involved. Japan has also remained a major center for pearl marketing despite this diversification of the cultured pearl industry. Pearl fisheries and pearl culture industries have been characterized by large fluctuations in production and value. Fisheries have been affected by stock exhaustion and the development of plastic buttons. Cultured pearls are a luxury item and market fluctuations in demand and value reflect contemporary conditions of economic prosperity, supply versus demand, fashion and pearl quality. The Japanese cultured Akoya pearl industry has also been hugely affected by disease and adverse environmental conditions. Pearl production in

Japan rose from 24.5 t in 1955 to a peak of 91.5 t in 1966, produced from 4,700 farms, then pearl production declined to <20 t/year in 1999 after the onset of disease, etc. The Akoya pearl oyster, *P. maxima* and *P. margaritifera* have been cultured for decades using similar techniques. One difference is whether the pearl oyster spat are derived by hatchery production or from natural spatfall. The trend is towards the former. There are moves to expand the sources of cultured pearls as industries based on *P. mazatlanica* and some *Pteria* spp. are developing. In the immediate future, Australia, China, French Polynesia, Indonesia, Japan and the Philippines will be the major producers of cultured pearls, although a number of other countries around the world in tropical and subtropical latitudes where pearl oysters occur are also involved.

References

Akerman, K., Stanton, J., 1994. Riji and Jakuli: Kimberly Pearl Shell in Aboriginal Australia. Northern Territory Museum of Arts and Sciences, Monograph Series No. 4.

Alagarswami, K., Dharmaraj, T.S., Velayudhan, A., Chellam, A., Victor, A.C.C., Gandhi, A.D., 1983. Larval rearing and production of spat of pearl oyster *Pinctada fucata* (Gould). *Aquaculture* 34, 287–301.

Alagarswami, K., Dharmaraj, S., Chellam, A., Velayudhan, T.S., 1989. Larval and juvenile rearing of the black-lip pearl oyster, *Pinctada margaritifera* (L.). *Aquaculture* 76, 43–56.

Almatar, S.M., Carpenter, K.E., Jackson, R., Akhazeem, S.H., Al-Saffar, A.H., Abdul Ghaffar, A.R., Carpenter, C., 1993. Observations on the pearl oyster fishery of Kuwait. *J. Shellfish Res.* 12, 35–40.

Anon, 2005. Pearl Oyster Hatchery Policy: Phase II – Policy Direction. Fisheries Occasional Publication No. 27. Department of Fisheries, Western Australia, 40 pp.

Anon, 2006. Annual Status Report: East coast pearl fishery, March 2006. Department of Primary Industries and Fisheries, Queensland, 6 pp.

Anon, 2007. Justin Hunter and Pearls Fiji. *Pearl World: Inter. Pearling J.* 16(4), 10–15.

Araya-Nuñez, O., Ganning, B., Bueckle-Ramirez, F., 1991. Gonad maturity, induction of spawning, larval breeding, and growth in the American pearl-oyster (*Pteria sterna*, Gould). *Calif. Fish Game* 77, 181–193.

Araya-Nuñez, O., Ganning, B., Bueckle-Ramirez, F., 1995. Embryonic development, larval culture, and settling of American pearl-oyster (*Pteria sterna*, Gould) spat. *Calif. Fish Game* 81, 10–21.

Arjarasirikoon, U., Kruatrachue, M., Sretarugsa, P., Chitramvong, Y., Jantataeme, S., Upatham, E.S., 2004. Gametogenic processes in the pearl oyster *Pteria penguin* (Roding, 1798) (Bivalvia, Mollusca). *J. Shellfish Res.* 23(2), 403–409.

Bach, J.P.S., 1955. *The Pearling Industry of Australia: An Account of its Social and Economic Development.* Department of Commerce and Agriculture Commonwealth of Australia.

Beer, A., 1999. Larval culture, spat collection and juvenile growth of the winged pearl oyster, *Pteria penguin*. World Aquaculture '99. The Annual International Conference and Exposition of the World Aquaculture Society 26th April – 2nd May 1999, Sydney, Australia. Book of Abstracts, p. 63.

Beer, A.C., Southgate, P.C., 2000. Collection of pearl oyster (family Pteriidae) spat at Orpheus Island Great Barrier Reef (Australia). *J. Shellfish Res.* 19, 821–826.

Behzadi, S., Parivak, K., Roustaian, P., 1997. Gonadal cycle of pearl oyster, *Pinctada fucata* (Gould) in Northeast Percian Gulf, Iran. *J. Shellfish Res.* 16, 129–135.

Bervera, L.H., 1994. Evaluación de la captación de semilla de *Pinctada mazatlanica* (Hanley 1856) en diferentes células colectoras durante el período 1991–92, y tratamiento de juveniles en prengorda a partir de 1992 en Bahía de La Paz, B.C.S., México. B.Sc. Tesis de Licenciatura en Biología Marina. UABCS, Area de Ciencias del Mar. La Paz, B.C.S, México. 93 pp.

Bouchon-Brandeley, G., 1885. Les pêcheries des îles Tuamotu. *Journal des Etablissements Français d'Océanie*, 206–226.

Bussawarit, S., 1995. Marine pearl farms at Phuket Island. Phuket Marine Biology Centre, Special Publication 15, 41–42.

Cariño, M., 1998. Les mines marines du golfe de Californie, Histoire de la région de La Paz à la lumière des perles, Ph.D. Thesis in History and Civilizations, Ecole des Hautes études en Sciences Sociales, Paris.

Cariño, M., Monteforte, M., 1999. El primer emporio perlero sustentable del mundo: La Compañía Criadora de Concha y Perla de la Baja California, S.A. y perspectivas para México. Ed. UABCS-CONACULTA-FONCA (México D.F.), 325 pp.

Cariño, M., Monteforte, M., 2006. Histoire mondiale des perles et des nacres: peche, culture, commerce. France: Editions L'harmattan, Collection Maritime, 250 pp.

Castellanos, J.F., Cruz, A., 1995. Aprovechamiento de los moluscos en la dieta aborigen. In: Martha Micheline, Cariño-Olvera et al., *Ecohistoria de los Californios*. UABCS, La Paz.

Chellam, A., Victor, A.C.C., Dharmaraj, S., Velayudhan, T.S., Satyanaryana Rao, T.S., 1991. *Pearl Oyster Farming and Pearl Culture. Training Manual 8*. Central Marine Fisheries Research Institute, Tuticorin, India, pp. 104.

Choi, Y.H., Chang, Y.J., 2003. Gametogenic cycle of the transplanted–cultured pearl oyster, *Pinctada fucata martensii* (Bivalvia: Pteriidae) in Korea. *Aquaculture* 220, 781–790.

Condie, S.A., Mansbridge, J.V., Hart, A.M., Andrewartha, J.R., 2006. Transport and recruitment of silver-lip pearl oyster larvae on Australia's North West Shelf. *J. Shellfish Res.* 25, 179–185.

del Barco, M., 1780. Historia Natural y Crónica de la Antigua California. (Preliminary study, and notes by Miguel León-Portilla, Instituto de Investigaciones Históricas, México: UNAM.1988.

Dennis, K., 2006. Pohnpei pearl project enters commercial phase. *Center for Tropical and Sub-tropical Aquaculture Regional Notes* 1, 1–7.

Domard, J., 1962. Les bancs nacriers de la Polynesie Francaise. Leur exploitation, leur conservation, leur reconstitution. *Comm. Pac. Sud. Conf. Tech.* 1, 13–14.

Edwards, H., 1994. *Pearls of Broome and Northern Australia*. Hugh Edwards, Swanbourne, Western Australia, 84 pp.

Elkin, A.P., 1933. Totemism in north-western Australia. *Oceania* 3, 257–296.

Esteva, J.M., 1857. Memoria sobre la Pesca de la Perla en la Baja California (1857). In: *Las Perlas de Baja California*. Departamento de Pesca, México, pp. 30–45. 1977.

Farell, S., Arizmendi, E., McLaurin, D., Nava, M., 1998. "Perlas de Guaymas": An update on the first commercial marine pearl farm on the American continent, *"Aquaculture '98" Book of Abstracts*. World Aquaculture Society, p. 171.

Fassler, R., 2002. Recent developments in selected Pacific and Indian Ocean pearl projects. *"Aquaculture 2002" Book of Abstracts*. World Aquaculture Society, p. 218.

Finau, M.W., 2005. Tonga country report. SPC sub-regional technical meeting on pearl culture. Nadi, Fiji. 30 November- 2 December 2005. Secretariat of the Pacific Community (SPC), Noumea.

Funakoshi, S., 1964. Studies on the method using saturated salt solution to kill Poychaeta. I. On killing Polychaeta inhabiting mother oyster shells. *Bull. Natl. Pearl Res. Lab.* 9, 1156–1160. (in Japanese)

George, C.D., 1978. *The pearl. A report to the Government of Papua New Guinea, the Food and Agriculture Organisation of the United Nations and the Asia Development Bank*. Samarai, Milne Bay Province, Papua New Guinea, 169 pp.

Gervis, M.H., Sims, N.A., 1992. *The biology and culture of pearl oysters (Bivalvia: Pteriidae)*. ICLARM Metro Manila, Philippines/ODA, London, England.

Hancock, D.A., 1989. A review of the Shark Bay pearling industry. Fisheries Management Paper No. 27. Department of Fisheries, Western Australia, 93 pp.

Harrison, A.J., 1997. *Savant of the Australian Seas: William Saville-Kent (1845–1908) and Australian Fisheries*. Tasmanian Historical Research Association, Hobart, 173 pp.

Hart, A.M., Friedman, K., 2004. Mother of pearl shell (*Pinctada maxima*): Stock assessment for management and future harvesting. FRDC Project No: 1998/153, Fisheries Research Contract Report No. 10, Department of Fisheries, Western Australia, 84 pp.

Hart, A.M., Joll, L., 2006. Growth, mortality, recruitment, and sex ratio in wild stocks of the silver-lipped pearl oyster *Pinctada maxima* (Jameson)(Mollusca: Pteriidae) in Western Australia. *J. Shellfish Res.* 25, 201–210.

Hart, A.M., Murphy, D., 2006a. Pearl Oyster Managed Fishery status report. In: W.J. Fletcher, F. Head, State of the Fisheries Report 2005/06, Department of Fisheries, Western Australia, pp. 170–175.

Hart, A.M., Murphy, D., 2006b. Predicting and assessing recruitment variation – a critical factor for management of the *Pinctada maxima* fishery in Western Australia. Final Research Report. FRDC. Project No. 2000/127. Department of Fisheries Research Division, Western Australia, 63 pp.

Hayashi, M., Seko, K., 1986. Practical techniques or artificial propagation of Japanese pearl oyster *Pinctada fucata*. *Bull. Fish. Res. Inst. Mie* 1, 39–68. (in Japanese)

Heffernan, O., 2006. Pearls of Wisdom. *The Marine Scientist* 16, 20–23.

Henricus-Prematilleke, J., 2005. South Sea pearl production seen exceeding 2,000 kan. *Jewel. News Asia* December 2005, 75–80.

Herdman, W.A., 1903. *Report to the Government of Ceylon on the pearl oyster fisheries of the Gulf of Mannar. Part I.* The Royal Society, London, 146 pp.

Hisada, Y., Fukuhara, T., 1999. *Pearl Marketing Trends with Emphasis on Black Pearl Market. FAO Field Document No. 13.* FAO, Rome, 31 pp.

Honjo, T., 1994. The biology and prediction of representative red tides associated with fish kills in Japan. *Rev. Fish. Sci.* 2, 225–253.

Ikenoue, H., 1992. Pearl oyster (*Pinctada fucata*). In: Ikenoue, H., Kafuku, T. (Eds.), *Modern Methods of Aquaculture in Japan*, 2nd ed. Elsevier Science Publishers, Amsterdam, pp. 95–205.

Ito, M., Jackson, R., Singeo, S., 2004. Development of pearl aquaculture and expertise in Micronesia. *"Aquaculture 2004" Book of Abstracts.* World Aquaculture Society, p. 279.

Iwata, Y., Sugawara, I., Kimura, T., Silapajan, K., Sano, M., Mizuguchi, T., Nishimura, A., Inoue, M., Takeuchi, T., 1997. Growth potential of *Gymnodinium mikimotoi* in Gokasho Bay. *Nippon Suisan Gakkaishi (Bull. Jap. Soc. Sci. Fish)* 63, 578–584.

Jameson, H.L., 1901. On the identity and distribution of the Mother-of-pearl oysters; with a revision of the subgenus *Margaritifera*. *Proc. Zool. Soc. London* 1, 372–394.

Kiefert, L., McLaurin-Moreno, D., Arizmendi, E., Hanni, H.A., Elen, S., 2004. Cultured pearls from the Gulf of California, Mexico. *Gems Gemmol.* 40, 26–38.

Kobayashi, S., Watabe, N., 1959. *Studies on Pearl Culture.* Gihodo, Tokyo, 280 pp. (in Japanese)

Kripa, V., Mohamed, K.S., Appukuttan, K.K., Velayudhan, T.S., 2007. Production of Akoya pearls from the southwest coast of India. *Aquaculture* 262, 347–354.

Kunz, G.F., Stevenson, C.H., 1908. *The Book of the Pearl: Its History, Art, Science and Industry.* Dover Publications, New York, 548 pp.

Lodeiros, C.J., Rengel, J.J., Himmelman, J.H., 1999. Growth of *Pteria colymbus* (Roding, 1798) in suspended culture in Golfo de Cariaco, Venezuela. *J. Shellfish Res.* 18, 155–158.

Lodeiros, C., Pico, D., Prieto, A., Narvaez, N., Guerra, A., 2002. Growth and survival of the pearl oyster *Pinctada imbricata* (Roding 1758) in suspended and bottom culture in the Golfo de Cariaco, Venezuela. *Aquac. Res.* 10, 327–338.

Lulofs, H.M.A., Sumner, N.R. (2002). Historical diving profiles for pearl oyster divers in Western Australia. Fisheries Research Report No. 138, Department of Fisheries. Western Australia, 20 pp.

MacKenzie, C.L., Troccoli, L., Leon, S.L.B., 2003. History of the Atlantic pearl-oyster, *Pinctada imbricata*, industry in Venezuela and Colombia, with biological and ecological observations. *Mar. Fish. Rev.* 65, 1–20.

Malimali, S., 1995. Pearl oyster culture in Tonga. Present and future of aquaculture research and development in the Pacific island countries. Proceedings of an International Workshop held from November 20–24, 1995 at Ministry of Fisheries, Tonga. pp. 367–370.

Malone, F.J., Hancock, D.A., Jeffriess, B., 1988. Final report of the pearling industry review committee. Fisheries Management Paper No. 17. Department of Fisheries, Western Australia, 216 pp.

Marquez, B., Lodeiros, C., Jimenez, M., Himmelman, J.H., 2000. Disponibilidad de juveniles por captacion natural de la ostra *Pteria colymbus* (Bivalvia: Pteriidae) en el Golfo de Cariaco, Venezuela. *Rev. Biol. Trop.* 48(Suppl. 1), 151–158.

Martínez-Fernández, E., Acosta-Salmón, H., Rangel-Dávalos, C., 2004. Ingestion and digestion of ten species of microalgae by winged pearl oyster *Pteria sterna* (Gould, 1851) larvae. *Aquaculture* 230, 419–425.

Matsuyama, Y., Nagai, K., Mizuguchi, T., Fujiwara, M., Ishimura, M., Yamaguchi, M., Uchida, T., Honjo, T., 1995. Ecological features and mass mortality of pearl oysters during red tide of *Heterocapsa* sp. in Ago Bay in 1992 (in Japanese with English summary). *Nippon Suisan Gakkaishi (Bull. Jap. Soc. Sci. Fish.)* 61, 35–41.

McAnally-Salas, L., Valenzuela-Espinoza, E., 1990. Growth and survival of larvae of the pearl oyster *Pteria sterna* under laboratory conditions. *Cienc. Mar.* 16, 29–41.

Mizumoto S., 1975. Parasites affecting the pearl industry in Japan. Proceedings of 3rd US-Japan Meeting Aquaculture, Tokyo, October 15–16, 1974. Special Publication of Fishery Agency of Japan Government. 1997–1999. Fish Pathol. 34, 207–216 (in Japanese with English abstract).

Mohammad, M-B.M., 1976. Relationship between biofouling and growth of the pearl oyster *Pinctada fucata* (Gould) in Kuwait, Arabian Gulf. *Hydrobiologia*, 129–138.

Mohamed, K.S., Kripa, V., Velayudhan, T.S., Appukuttan, K.K., 2006. Growth and biometric relationships of the pearl oyster, *Pinctada fucata* (Gould) on transplanting from the Gulf of Mannar to the Arabian Sea. *Aquacult. Res.* 37, 725–741.

Monteforte, M., 1994. Perspectives for the installation of a pearl culture enterprise in Bahía de La Paz, South Baja California, México. *J. Shellfish Res.* 13, 339–340.

Monteforte, M., 1996. Cultivo de Ostras Perleras y Perlicultiura. In: Ponce, G., M. Casas (Eds.), Diagnóstico Pesquero y Acuícola del Estado de Baja California Sur. Convenio interinstitucional CIBNOR/CICIMAR/ UABCS/CET-MAR/SEMARNAP/FAO. Vol II, 571–613.

Monteforte, M., López-López, S., 1990. Captación masiva y prengorda de Madreperla *Pinctada mazatlanica* (Hanley 1856), en Bahía de La Paz, Sudcalifornia, México. IV Congreso de la Asociación Mexicana de Acuicultores. CICTUS-Universidad de Sonora. Hermosillo, Son. AMAC'90. Selected works, Vol. 1, 10 pp.

Monteforte, M., Bervera, H., 1994. Spat collection trials for pearl oysters *Pinctada mazatlanica* and *Pteria sterna* at Bahía de La Paz, South Baja California, México. *J. Shellfish Res.* 13, 341–342.

Monteforte, M., García-Gasca, A., 1994. Spat collection studies of pearl oysters *Pinctada mazatlanica* and *Pteria sterna* (Bivalvia, Pteriidae) in Bay of La Paz, South Baja California, México. *Hydrobiologia* 291, 21–34.

Monteforte, M., Bervera, H., Morales, S., Pérez, V., Saucedo, P., Wright, H., 1994. Results on the production of cultured pearls in *Pinctada mazatlanica* and *Pteria sterna* from Bahía de La Paz, South Baja California, México. *J. Shellfish Res.* 13, 342–343.

Monteforte, M., Morales, S., 2000. Growth and survival of the Calafia mother-of-pearl oyster *Pinctada mazatlanica* (Hanley 1856) under different sequences of nursery culture-late culture at Bahía de La Paz, Baja California Sur, México. *Aquacult. Res.* 31, 901–915.

Morales-Mulia, S.M., 1996. Crecimiento y supervivencia de la Madreperla *Pinctada mazatlanica* (Bivalvia : Pteriidae) en diferentes secuencias operativas del ciclo prengorda-cultivo durante el proceso de cultivo extensivo. BSc. In Biology. Facultad de Ciencias Biológicas, UNAM.

Nagai, K., Matsuyama, Y., Uchida, T., Yamaguchi, M., Ishimura, M., Nishimura, A., Akamatsu, S., Honjo, T., 1996. Toxicity and LD_{50} levels of the red tide dinoflagellate *Heterocapsa circularisquama* on juvenile pearl oysters. *Aquaculture* 144, 149–154.

Narayana, K.R., Michael, M.S., 1968. On the relation between age and linear measurements of the pearl oyster, *Pinctada vulgaris* (Schumacher) of the Gulf of Kutch. *J. Bombay Nat. Hist. Soc.* 65, 441–452.

Nava, M., Arizmendi, E., Farell, S., McLaurin, D., 2000. Evaluation of success in the seeding of round nuclei in *Pteria sterna* (Gould, 1851), a new species in pearl culture. *SPC Pearl Oyster Information Bulletin*, Noumea, New Caledonia, 14, 12–16.

O'Connor, W.A., Wang, A., 2001. Akoya pearl culture in China. *World Aquacult.* 32, 18–20.

O'Connor, W., Lawler, N.F., Heasman, M.P., 2003. *Trial farming the Akoya pearl oyster, Pinctada imbricata in* Port Stevens, NSW. New South Wales Fisheries, New South Wales Government, Sydney, 170 pp.

Ogushi, J., 1938. Study of Pearls, Tokyo. 217 pp. (in Japanese).

Pass, D.A., Dybdahl, R., Mannion, M.M., 1987. Investigations into the causes of mortality of the pearl oyster, *Pinctada maxima* (Jameson), in Western Australia. *Aquaculture* 65, 149–169.

Pit, J.H., 2004. Feasibility of Akoya pearl oyster culture in Queensland. Ph.D. Thesis, James Cook University, Queensland, Australia, 208 pp.

Qizeng, J., Yiyao, W., Weigo, J., 1981. On the artificial rearing of the larvae and juvenile pearl oyster *Pinctada f martensii* (Dunker). I. The artificial fertilization. *Nanhai Studia Marina Sinica* 2, 107–115. (in Chinese with English summary)

Reed, W., 1965. Marine fisheries development and mother of pearl oyster culture and recommendations for future development. Final technical report to Sudan Government, FAO, Rome, 27 pp.

Ruiz-Rubio, H., Acosta-Salmón, H., Olivera, A., Southgate, P.C., Rangel-Dávalos, C., 2006. The influence of culture method and culture period on quality of half-pearls ('mabé') from the winged pearl oyster *Pteria sterna*, Gould, 1851. *Aquaculture* 254, 269–274.

Saucedo, P., Monteforte, M., Blanc, F., 1998. Changes in shell dimensions of pearl oysters, *Pinctada mazatlanica* (Hanley 1856) and *Pteria sterna* (Gould 1851), during growth as criteria for Mabe pearl implants. *Aquacult. Res.* 29, 801–814.

Seale, A., 1910. The fishery resources of the Philippine Islands. Pt. 3. Pearls and pearl fisheries. *Philippines J. Sci.* 5, 87–100.

Serrano-Guzman, S.J., Salinas-Ordaz, D., 1993. Cultivo de larvas y produccion de semilla de *Pteria sterna* (Mollusca: Bivalvia) en un criadero comercial. *Rev. Inv. Cient.* 4, 81–90.

Shirai, S., 1994. *Pearls and Pearl Oysters of the World*. Marine Planning Co., Japan, 95 pp. (in Japanese and English)

Souter, A., 2006. Pearling Industry Status Report 2005. State of the Fisheries Report 2005, Fishery Report No. 85. Department of Primary Industry, Fisheries, and Mines, Northern Territory, Australia, pp. 99–175.

Southgate, P.C., 2004. Progress towards development of a cultured pearl industry in Kiribati, central Pacific. World Aquaculture Society. "Aquaculture 2004" Book of Abstracts p. 556

Southgate, P.C., Beer, A.C., 1997. Hatchery and early nursery culture of the black-lip pearl oyster, *Pinctada margaritifera* (L.). *J. Shellfish Res.* 16, 561–568.

Southgate, P., Rubens, J., Kipanga, M., Msumi, G., 2006. Pearls from Africa. *SPC Pearl Oyster Inform. Bull.* 17, 16–17.

Strack, E., 2006. *Pearls*. Ruhle-Diebener-Verlag GmbH & Co., Stuttgart, 707 pp.

Talavera, F., 1930. Pearl fisheries of Sulu. *Philippines J. Sci.* 43, 483–500.

Tanaka, Y., 1997. Potential of commercial development of mabe pearl farming in Vava'u islands, Kingdom of Tonga. FAO South Pacific Aquaculture Development project (Phase II), Filed Document No. 5. FAO, Rome, 26 pp.

Taylor, J.J., Southgate, P.C., Rose, R.A., 1997. Fouling animals and their effect on the growth of silver-lip pearl oysters, *Pinctada maxima* (Jameson) in suspended culture. *Aquaculture* 153, 31–40.

Tomaru, Y., Ebisuzaki, S., Kawabata, Z.I., Nakano, S.J., 2002. Respiration rates of the Japanese pearl oyster, *Pinctada fucata martensii*, feeding on *Pavlova lutheri* and *Chaetoceros gracilis*. *Aquac. Res.* 33, 33–36.

Uchimura, Y., Nishikawa, S., Hamada, K., Hyodou, K., Hirose, T., Ishikawa, K., Sugimoto, K., Nakajima, N., 2005. Production of strains of the Japanese pearl oysters *Pinctada fucata martensii* with low mortality through the less damages caused by infectious disease. *Fish Genet. Breed. Sci.* 34, 91–97. (in Japanese with English summary)

Upare, M.A., 2001. New horizon of fisheries development – mariculture through credit. In: Menon, N.G., Pillai, P.P. (Eds.), *Perspectives in Mariculture*. The Marine Biological Association of India, Cochin, India, pp. 421–428.

Urban, H.-J., 2000a. Culture potential of the pearl oyster (*Pinctada imbricata*) from the Caribbean. I. Gametogenic activity, growth, mortality and production of natural population. *Aquaculture* 189, 361–373.

Urban, H.-J., 2000b. Culture potential of the pearl oyster (*Pinctada imbricata*) from the Caribbean. II. Spat collection, and growth and mortality in culture systems. *Aquaculture* 189, 375–388.

Valadéz, A., 1963. Temas Históricos de la Baja California, Ed. Jus, México.

Victor, A.C.C., Jagadis, I., Ignatius, B., Chellam, A., 2003. Perspectives and problems of commercial scale pearl culture – An indicative study at Mandapam Camp, Gulf of Mannar. Abstracts of the First Indian Pearl Congress and Exposition. Central Marine Fisheries Research Institute, Cochin, India, pp. 73–75. 5–8 February 2003.

Wada, S.K., 1953. Biology and fisheries of the silver-lip pearl oyster. Unpublished Report contained within CSIRO library, Hobart.

Wada, K., 1973a. Modern and traditional methods of pearl culture. *Underwat. J.* 5, 28–33.

Wada, K.T., 1973b. Growth of Japanese pearl oyster larvae fed with three species of micro-algae. *Kokoritsu Shinju Kenkyusho Hokoku (Bull. Natl. Pearl Res. Lab.)* 17, 2075–2083. (in Japanese with English Summary)

Wada, K.T., 1991. Japanese pearl oyster, *Pinctada fucata* (Gould) (Family Pteridae). In: Menzel, W. (Ed.), *Estuary and Marine Bivalve Mollusks Culture*. CRC Press, Boca Raton, USA, pp. 245–260.

Wang, A., Yan, B., Ye, L., Lang, G., Zhang, D., Du, X., 2003. Comparison on main traits of F_1 from matings and crosses of different geographical populations in *Pinctada martensii*. *J. Fish. China/Shuichan Xuebao* 27, 200–206. (in Chinese with English abstract)

Wang, A., Shi, Y., Zhou, Z., 2004. Morphological trait parameters and their correlations of the first generation from matings and crosses of geographical populations of *Pinctada martensii* (Dunker). *Mar. Fish. Res./Haiyang Shuichan Yanjiu* 25, 39–45. (in Chinese with English abstract)

Wang, A., Shi, Y., Wang, Y., Gu, Z., 2007. Present status and prospect of Chinese pearl oyster culturing. World Aquaculture Society, San Antonio, Texas. Aquaculture 2007, February 26 to March 2, 2007.

Yakushi, K., 1991. *The New Pearl Topic – Chinese Akoya Pearl. Pearls of the World (II)*. Shinsoshoku Co. Ltd., Tokyo, 272 pp.

Yu, X., Wang, M, 2004. *The farming of and pearl cultivating from wing oyster Pteria penguin in southern China*. World Aquaculture Society. "Aquaculture 2004" Book of Abstracts. p. 665.

CHAPTER 10

The Pearl Market

Richard D. Torrey and Brigitte Sheung

10.1. INTRODUCTION

Pearls are fashion items and as such their markets are influenced by fickle demand trends. Pearl markets and pearl prices are also influenced by levels of production and supply, quality control and market perceptions of particular products. This chapter provides a brief overview of recent production trends and demands for different pearl products. It highlights some of the factors that are likely to influence pearl production, pearl markets and the viability of the pearling industry over coming years.

10.2. RECENT MARKET TRENDS

After the three lean years prior to 2004, the pearl industry has recently begun to realize recovery. Gains were seen in 2005 and 2006 with both producers and wholesalers reporting stronger sales. After too many years of cutting back on stock and hoping for an upturn in the market, many wholesalers are now in the process of rebuilding stocks thanks to firming prices and higher turnover. Major pearl auctions of South Sea and Tahitian pearls in the latter months of 2006 showed price increases from 10% to 40% compared to the year prior. It is notable that this increase included commercial qualities. Prices for Akoya pearls, notably for first-grade pearls, also registered a rise between 5% and 10% during the Japanese *hamaage* (raw pearls) auctions in 2006. Medium to good quality Akoya pearls of 9 mm and above continued to command higher prices than in previous years.

The current pearl market shows strong demand for the following products[1]:

- White South Sea pearls (8–12 mm) in commercial to fine quality
- South Sea baroque pearls of all sizes
- Golden South Sea pearls of various qualities, and "champagne" South Sea pearls with high luster
- Round and drop-shaped Tahitian pearls from 7 mm up
- Large size Tahitian pearls of 12 mm up with good quality (15 mm up are in very short supply)
- Tahitian keshi pearls and Tahitian baroques
- Tahitian pearls in treated chocolate color (see Section 8.11.5)
- Akoya pearls of good nacre and luster of 7–9 mm
- Multicolor necklaces with large (12 mm up) baroque pearls, mixing South Sea, Tahitian and freshwater pearls
- Long pearl sautoirs (80, 120, 150 cm or more) of small Tahitian pearls of mixed shades of silver-gray and or freshwater pearls (white or multicolor).

10.3. PRODUCTION OVERVIEW

10.3.1. Recent years

The value of world pearl production increased 7% in 2005 compared to 2004, and returned to 2001 levels, at US$640 million. South Sea pearl production registered record levels in 2005 and made up the lion's share of 37% of the total production value. Tahitian pearl production came under control with volume down and export value up, while Akoya pearl production in 2005 registered a slight decline in volume and in value.

While market demand recovered in 2005/2006, production of certain categories of pearls decreased compared with previous years; particularly exports of Tahitian cultured pearls and production of Akoya cultured pearls. The production of South Sea cultured pearls, however, has seen a record increase, due mainly to surging Indonesian harvests.

South Sea pearl production volume reached a record high in 2005 and marked a historic year for the development of the South Sea pearl industry. For the first time in history, the annual harvest of white South Sea pearls surpassed that of black South Sea pearls with total production of all South Sea pearls estimated at between 9 and 10 t, with an estimated value of US$248 million. Australia and Indonesia remain the top two white South Sea pearl producing countries, although the Philippines also contributed significantly to South Sea pearl harvests (Table 9.2), particularly in gold and yellow colors. Fully exploiting the use of hatchery produced oysters in Indonesia and the Philippines, and the solid financial base of established producers in these countries (plus those in Australia and Myanmar), South Sea pearls have become the leading cultured pearl category.

---

[1]See Chapter 8 (Section 8.10) for descriptions of terms relating to pearl characteristics and quality.

The supply of Tahitian pearls declined after their peak annual production above 10 t in 2002 and 2003. Annual exports dropped to 8.5 t in 2004 and then to 8.1 t in 2005. This reduced quantity is due to a series of government regulatory measures imposed to uphold the quality and market value of the Tahitian pearl (see Section 9.4.6.2, Fig. 10.1).

FIGURE 10.1   Low-grade black South Sea pearls being crushed as part of government regulatory measures to maintain export pearl quality in French Polynesia. (Photo: Sandra Langy.)

This reform has yielded positive results: export values rebounded to US$115 million in 2004, followed by a rise to US$126 million in 2005 (compared with US$95 million in 2003 when production peaked). In 2006, the value continued to climb for Tahitian pearl exports, and looks to be stabilized. The black pearl industry (Tahitian pearls as well as black pearls cultivated in other regions) had an overall estimated value of US$125 million, which represents around 20% of total pearl market value.

Akoya pearl production dropped in both Japan and China with efforts to produce better quality at the expense of quantity, and farming attempts in Vietnam and eastern Australia are yet to yield commercial quantities of pearls. In Japan, production dropped slightly to between 22 and 24 t in 2005, from over 25 t in 2004. Annual Chinese Akoya pearl production also dropped from 15 to 10 t. Nonetheless, the Akoya pearl industry remains a major pearl category, with a 2005 production value of US$128 million, making up around 20% of the total pearl market in value.

10.3.2. Production in 2007 and beyond

In 2007 some trends were seen which shall probably shape the future. One of these is the Tahitian pearl harvest. As with other pearl producing areas of the world, the farmers of French Polynesia are tempted toward producing quantity over quality, and letting quality fall by the wayside. This, one may note, was the downfall of the Japanese Akoya pearl industry which began when Japanese farmers began shortening the culturing time of their oysters in order to produce more and more of them in a shorter period of time. Naturally, larger sizes and higher quality goods suffered, and the end result was that China began to dominate Akoya pearl production.

The results of Tahitian producers' getting away from what brought them to prominence is reflected in an August 2007 verbatim report from a major purveyor of *Pinctada margaritifera* (i.e. "black pearl") goods:

- "Tahitian sizes are getting smaller and smaller while their problems are getting larger and larger."
- "It is now difficult to buy Tahitian goods which can produce a strand 12–14 mm in good color, that is really clean, and which has good luster."
- "All the larger sizes (11 mm and up) are from second and third operations[2], and, by nature, these goods usually lack the necessary luster."
- "Nice quality 14 mm or larger drops, rounds or nugget-shaped baroques have become so scarce that there is hardly any price anymore for them."

This rather negative view of the present and probable future of black pearl production in the South Pacific was ratified by the results of the 35th Robert Wan Auction in Hong Kong, in June 2007, in which Wan (the premier Tahitian black pearl producer)

---

[2]Oysters that produce a good quality pearl are often re-seeded for further pearl production (see Section 8.9.7)

sold only 65% of goods offered, and withdrew lots on the auction block because prices were soft, and because he wanted to maintain current price levels.

One must take note that, historically, it is difficult if not impossible to retreat from the cycle of producing smaller, lower quality pearls (oftentimes termed "commercial goods") to satisfy consumer demand for more affordable baubles. At another South Sea pearl auction, the 5th Poe Rava Nui Tahiti Pearl Auction, it was reported that "unsold goods were generally lower qualities (that is, below B-grade) with prices set higher than buyers' offers" (*Jewellery News Asia*, August 2007) which augurs a coming market glut of smaller, less attractive Tahitian goods.

If this is the probable future for black South Sea pearls, what is happening for other pearl categories?

The same issue of *Jewellery News Asia* reported on two recent auctions of white South Sea pearls. A June 2007 auction in Myanmar (formerly Burma) sold only 56% of the goods on offer "because we put up at the auction some of the low quality pearls that are normally sold at local auctions" it was reported by a participant.

This, too, is a growing trend which has historically been the sole province of the larger producers such as Wan and Paspaley (see below); selling off less attractive goods at auctions while saving the better material for privileged clients. Now even the smaller producers and groups have reached plateaus where they can be more selective in their offerings. Gone are the days when the early birds at auctions got the best worms.

The 35th Paspaley Pearl Auction in Hong Kong (which tends to sell only white South Sea pearls) sold over 83% of their offered goods, with average prices climbing by 6% overall, and with a 15–20% increase for similar pearl quality found at the previous March auction. An improved global economy and a strong Euro helped boost buying from non-Asian countries. This augurs well for white South Sea pearl goods, as Paspaley is regarded as "the bell cow" for that part of the industry, and even his detractors who produce and sell similar goods, appreciate his maintaining consistently high price levels for Australian cultured pearls.

As for Japan's production and sales of Akoya pearls, the *hamaage* (raw pearl) auctions in the early months of 2007 showed an 8% decline in volume and a 5% decline in value compared to the preceding year. Usually about 80% of Japan's Akoya pearl harvests are sold this way, with the remaining portion sold directly to processors and wholesalers. But a trend is emerging with a greater amount of the harvest being sold outside of the auction process. Smaller portions of the harvest becoming available to bidders will most likely encourage greater competition amongst participants, and therefore higher prices and profits, it is widely hoped. Thus, this deeper mining of the relatively static Japanese Akoya pearl volume may have a deleterious effect on this segment of the market by causing higher prices to be passed on to wholesalers, retailers, and, ultimately, the consumer.

And this will not in any way curtail the prevalent Japanese practice of taking more and more of Chinese Akoya pearl production and passing it off as their own (it is widely held that some 80% of Akoya pearl strands sold as "Product of Japan" are actually pearls of Chinese origin). This will undoubtedly continue to flummox the consumer who feels that the Japanese cachet is more valuable than that of the Chinese name, despite the fact that the Chinese Akoya pearl is becoming at least as good as, and often better than, the Japanese Akoya pearl, and considerably less expensive.

As for natural pearls, these days these precious orbs are relegated primarily to top auction houses. In April 2007, Christie's New York Auction sold the Baroda Pearls, a two-strand natural pearl ensemble (necklace, ear pendants, brooch and ring) that had been part of a famous seven-strand necklace of the Royal Treasury of the Maharaja of Baroda in India for a whopping US$7.09 million to an Asian private collector. Natural pearl sales continue to be a small but important niche of the pearl market in many parts of the world.

10.4. FRESHWATER PEARL PRODUCTION

Although this book focuses on the Pteriidae and therefore on marine pearl production, an overview of the current pearl market would not be complete without some consideration of freshwater pearl production.

Over recent years, freshwater pearl production, primarily from China, has continued at very high levels. Production of freshwater pearls from China peaked in 2004 and 2005, characterized by increased availability of larger sizes and innovative shapes. An estimated 1,500 t of Chinese freshwater pearls were put onto the market in 2005. The drastic price crash in 2001/2002 and the closing of a significant number of farms in 2002, 2003 and 2004 have led to an overall reduced scale of investment. Some predict a probable decline in the freshwater pearl supply for 2007 and 2008, due to the cultivation cycle of 4 years, which should prove positive for the Chinese industry. Meanwhile, the sizeable freshwater pearl industry represents over 23% of the total pearl market value, with an estimated production of US$150 million in 2005.

After years of drastic price drops and fluctuations, also showing a bottoming-out for the lower qualities and a firming-up for prices of higher qualities, some categories of freshwater pearls have experienced significant price hikes. These include top-quality rounds and near rounds of all sizes, nice baroques from 8 mm upward (a particularly popular item at present), and better quality pearls above 10 mm in all shapes. Major demand is for the following products:

- Round to near round freshwater pearls of all sizes.
- Freshwater pearls in baroque shapes and large sizes from 8 mm up.
- Freshwater pearls in fancy shapes: petals, coins, sticks, spikes, crosses, tubes.

Chinese freshwater pearls, produced primarily by implanting tissue pieces into the mantle of the mussel, *Hyriopsis cumingii*, are improving in size and quality. Not only are they a rival to both South Sea pearls and Akoya pearls in looks, their lower price has gained them an important advantage in the mass market and international gem shows are often awash in them. One of their strong selling points is that they are composed of virtually 100% nacre (as they are most often nucleated with just a sliver of mantle tissue). Oftentimes only cognoscenti can differentiate a top Chinese freshwater pearl strand from even a fine Australian South Sea pearls strand without seeing the vast difference of the two price tags, and the same goes for Akoya pearls. Recent developments in Chinese freshwater pearl production include bead nucleation and the use of another species of mussel, *H. schlegelii*, introduced from Japan (Fiske and Shepherd, 2007). With respect to pearl culture, the hybrid between *H. cumingii* and *H. schlegelii* is thought to be superior to either species (Xie *et al.*, 2006).

10.5. CHALLENGES

Compared with two decades ago, cultured pearls have become much more accessible due to increased supply and more affordable pricing. Pearls are now a major ingredient in jewelry, with very elastic consumer price ranges covering all jewelry segments, including fashion accessories, basic designs for the mass market, designer jewelry, fine jewelry brands and *haute couture* masterpieces.

The awareness of pearls has significantly increased among consumers in various countries and cultures. The image of pearls has dramatically evolved, from the previously conventional image of classic pearl necklaces reserved for mature women to taking on a fresh, contemporary image with a cutting edge chic for a far wider age group. Recently the pearl sautoir has become a most popular and indispensable accessory. The internet and television buying continue to accelerate sales, particularly in the lower price categories.

Pearl producers have faced up to heavy challenges in the past 20 years relating to farm management, development of hatchery methods, genetic research, financial considerations, cost controls to achieve economy of scale, and varying environmental issues. Striving to achieve a feasible and profitable operation amid drastic economic changes, as well as political and social instabilities in certain pearl producing countries, are common problems for all producers.

However, while the industry advances toward sophistication in terms of quality and supply, the consumer market has also evolved into a highly complex and competitive world in which there is a profusion of products for an increasingly discerning and demanding clientele. Indeed, advantageous quality/price ratios and attractive designs remain crucial factors conducive to sales. But shoppers are showing a switch in preference to favor an emotional shopping experience, requiring sophistication in shop designs and an internationally recognized brand often backed-up by celebrity endorsement.

The pearl world, like other industries in consumer goods, finds itself in a frantic race for communication and visibility. How many consumers first think of pearls when they look for a gift? How many consumers can tell the difference between good quality pearls and low quality pearls? How many understand the various categories of pearls and their price differentials? The latest study from the Jewelry Consumer Opinion Council in the United States revealed that 49% of respondents to a major pearl survey were not cognizant of the kind of pearls they themselves own, and 54% had no idea of which kinds of pearl are most desirable. How many consumers can name any pearl jewelry brand?

For years, pearl organizations and companies have injected significant funds into education, product development and institutional communications on pearls. Still, the overall pearl jewelry world is lagging behind when compared to leading automobile, watch, fashion, leather accessories and cosmetics brands. The pearl industry has a great deal to learn from other segments of consumer goods to bring pearl sales to a matching level of sophistication and to rival that of diamonds as a leader of jewelry marketing.

It is regrettable that the pearl industry/trade has so far failed to set up a unique and unified international organization to promote the global awareness of pearls, dating back to the failed formation of the International Pearl Association which was still-born at "Pearls '94" in Hawaii, to the collapse of efforts to form the International Pearl Organization in 2006 through the work of Martin Coeroli of CIBJO[3]. Though many industry members realize the importance of cooperation, few (especially major producers) have put their money where their mouth is. The challenges ahead for the pearl world are multiple and complex, but by combining competition and cooperation the pearl industry should continue to evolve if it is to continue to meet fast changing, global consumer demand.

10.6. FUTURE DEVELOPMENTS

Where does the pearl market go from here? This is a question that many are asking (and have been asking for quite some time). One might well consult a fortune teller, palm reader or diviner of tea leaves to accurately prognosticate the future of this vast global enterprise which is estimated to generate some US$5 billion annually. However, some reasonable assumptions may be made:

1. Larger producers will continue to buy up smaller producers, as Paspaley has done over the years in Australia. This will also happen in other emerging producing areas, such as Indonesia and Vietnam. Those in "the catbird seat" will work feverishly to maintain their dominance in the production of their countries and regions.
2. China will continue to expand its pearl production. The only thing that might confound them is pollution of their waterways, and other environmental catastrophes. There is little chance of any centralized organization taking control of overall marketing.
3. Tahiti will continue at about the same pace as at present. Governmental oversight is unconcerned about falling sizes and quality, and will continue to insist upon a minimum 0.8 mm nacre thickness.
4. Golden pearls will continue to be a focus for the Philippines' Jewelmer, as these South Sea pearls continue to generate premium prices due to their relative scarcity

---

[3]Confédération Internationale de la Bijouterie, Joaillerie et Orfèvrerie, also known as the World Jewellery Confederation, an international confederation of jewellery, gemstone, horology, and silverware trade organisations. Based in London, its primary purpose is as a decision-making body, to create and maintain standards and promote cooperation within these interconnected organisations; it seeks to cement nomenclature and set best practice guidelines to better engender consumer confidence. Including representatives of 36 countries, the CIBJO was renamed the World Jewellery Confederation in 2000, but still appears to retain its former name.

(many producers elsewhere prefer to aim their harvests toward white South Sea pearls, as these are more commonly considered more salable in major markets).

5. There is virtually no hope for a unified, DeBeers type of oversight organization for the cultured pearl industry. Too many conflicting interests preclude any type of a coordinating committee to evolve a consistent pearl grading standard, much less to police policies on exports, treatments and industry ethics. Many have gallantly tried, all have abjectly failed.

6. Finally, there seems to be no end to the capability of the consumer to admire, buy, collect and wear pearls. This is what drives the inherent desire of producers worldwide to cultivate more pearls faster, and what induces wholesalers and retailers to focus on price points to satisfy this demand: "How little can I pay for pearls and how much can I sell them for?" This two-edged sword of commerce is, unfortunately, the driving rationale of the marketplace; precious few in the industry have time to reflect on the history, beauty and mystique of the pearl.

References

Fiske, D., Shepherd, J., 2007. Continuity and change in Chinese freshwater pearl culture. *Gems and Gemology* 43, 138–145.

Xie, N., Li, Y., Zheng, H., Wang, G., Li, J., Oi, N., Yuan, W., 2006. Comparison of culture and pearl performances among *Hyriopsis schlegelii*, *Hyriopsis cumingii* and their reciprocal hybdrids. *Journal of Shanghai Fisheries University* 15, 264–269.

CHAPTER 11

Disease and Predation

John D. Humphrey

11.1. INTRODUCTION

Pearl oysters are not spared from disease which is frequently a limiting factor in the successful farming of pearl oysters as a consequence of:

- deaths;
- reduced growth, productivity and product value;
- poor pearl quality;
- costs of preventative or control measures;
- rehabilitation costs following mass mortality events;
- negative perceptions affecting investment in the industry.

Notable examples of diseases and their catastrophic social and economic consequences include Akoya oyster[1] disease syndrome (see Section 9.2.6), mortalities associated with dinoflagellate blooms, invasion of shell by polychaetes and sponges (see Section 15.3.1) and mass mortality events. Although numerous disease conditions are known in pearl oysters, in many cases the cause or causes are not well understood. Occasional reviews covering limited aspects of pearl oyster disease have been published (Humphrey *et al.,* 1998; Humphrey and Norton, 2005; Jones and Creeper, 2006),

---

[1]Throughout this text the term 'Akoya' is used for pearl oysters in the currently unresolved complex of *Pinctada fucata-martensii-radiata-imbricata* (see Section 2.3.1.13)

but there is no comprehensive review of those pathogens, parasites, diseases and associates impacting on pearl oysters. There are major deficiencies in knowledge of the cause of disease, the development of disease within the host, the spread of disease in the environment and the interactions between stress, the environment and the host in the ultimate expression of disease.

A taxonomically diverse spectrum of infectious microbial and parasitic organisms, symbionts and associates occur in pearl oysters, which may or may not be associated with disease. Environmental parameters outside the physiological range of the oyster may initiate or exacerbate disease, and organic and inorganic toxins are known to compromise pearl oyster health (see Section 6.9). Furthermore, a number of important disease syndromes are known for which the cause is obscure or multifactorial. The genesis of disease in pearl oysters often involves a complex of infectious, non-infectious factors and husbandry factors and the diagnosis of disease is frequently problematic.

A sound understanding of the occurrence, distribution and prevalence of infectious agents, and the nature of non-infectious and environmental factors that may impact adversely on the production and management of pearl oysters is a pre-requisite on which to develop sound industry practice. Such knowledge is also central to the development and implementation of quarantine policies and strategies to protect against disease incursions. Importantly, new infectious organisms and non-infectious conditions continue to emerge and impact on pearl oyster aquaculture and pearl production. The emergence of new diseases in association with decreased pearl oyster production and reduced pearl quality should be anticipated by farm managers, fisheries biologists and molluscan pathologists. For the purposes of this chapter, disease is taken to include abnormalities of structure or function and includes sub-clinical disease in which there may be no tangible signs of disease, but which is manifest by physico-chemical abnormalities including retarded growth and production. Sub-clinical disease may only be apparent where accurate production records are maintained. This chapter describes those diseases and disease syndromes in the genus *Pinctada* with reference to their etiology, clinical signs, pathogenesis, epidemiology and pathology. Also described are those symbionts and associates identified in *Pinctada* spp., which may not be associated with disease or whose pathogenic role remains uncertain. The nature and role of biofouling organisms and predators in disease are also discussed.

11.2. CLINICAL RESPONSE TO DISEASE

The pearl oyster has limited capacity to respond to disease or to produce signs referable to specific disease states or conditions, and the examination of living, farmed oysters presents difficulties. Restricted visual access inside the shell cavity limits direct observation of the living animal. In the living animal viewed undisturbed in water, the observer is generally dependent on relatively subjective and variable criteria to evaluate the health of individual animals. Adductor muscular tone assessed on strength and speed of closure of the shell, and mantle retraction are two common criteria evaluated in response to external stimuli. Retraction of the mantle and failure to lay down new shell may suggest disease.

Clinical signs of disease in an individual animal or in a population as a whole include deaths, empty shells, gaping shells, reduced growth, shell valve anomalies, mantle retraction, reduced adductor muscle strength and reduced rate of closure of the shell valves. Sub-clinical disease, in which no visible abnormalities are evident, may be manifest by sub-optimal shell growth, decreased body-weight gain, poor condition indices, poor pearl quality and quantity of pearl production. The detection of sub-clinical disease necessitates the maintenance of detailed farm production records, and the importance of maintaining such records is emphasized if optimal productivity is to be attained and the spatial and temporal source of losses identified.

An assessment of the general nutritional condition of the body and mantle of the oyster may assist in the evaluation of disease states. Poorly nourished animals appear wasted, with transparent mantle tissues and watery, atrophic visceral masses. Prolonged retraction of the mantle is indicative of disease states. Mantle retraction may result in fouling of the nacreous surface of the shell normally covered by the expanded mantle, together with anomalous deposition of conchiolin and nacre by the retracted mantle, leading to "shell disease." This process may result in a growth check, with the deposition of new shell on previously normal shell, giving rise to the name "double backs" for such oysters (Fig. 11.1). Shells with such lesions are good indicators of a pre-existing insult to the animal or population. Anomalous deposition of conchiolin and failure of mineralization at the margins of the retracted mantle, so-called shell disease (Fig. 11.2) is also a good indicator of pre-existing damage of disease to the oyster and may arise under different disease states.

FIGURE 11.1   "Double-back" shell of *Pinctada maxima* showing distinct ridge of new shell growth (arrow) following period of mantle retraction and cessation of growth associated with disease event.

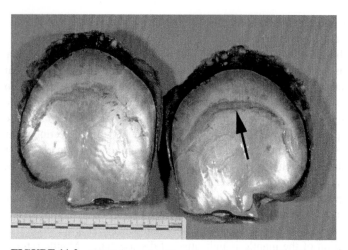

FIGURE 11.2   "Shell disease" in *Pinctada maxima* characterized by anomalous deposition of conchiolin on the inner shell valve (arrow) and failure of mineralization at the margins of the retracted mantle, typical of earlier or pre-existing disease or injury to the oyster.

11.3. PATHOPHYSIOLOGICAL RESPONSES TO DISEASE

There is no doubt that pearl oysters possess humoral and cell-mediated inflammatory and immune capabilities and can mount prompt and effective responses to injury or infection. These responses have, however, been poorly investigated when compared to other molluscan species, especially the edible oysters. Nevertheless, diverse aspects of the pathophysiological response to disease or injury have been investigated that demonstrate the capacity of pearl oysters to survive and recover from disease or injury.

11.3.1. Wound healing

Physical wounding and injury to pearl oysters may result from storm events, rough handling and predation. Additionally, pearl oysters are routinely subject to wounding as a result of seeding and pearl harvesting operations. Potential exists for the introduction of micro-organisms and foreign matter into the pearl sac or surrounding tissues at this time, and for subsequent scar formation at the site of operation, with anomalous pearl sac formation and possibly poor quality pearls.

Pearl oysters appear to have considerable capacity to recover from wounds, at least from simple injuries. Regeneration of shell following damage has long been recognized (Herdman, 1903), but only recently has the underlying cellular mechanisms of wound healing been investigated. A rapid response to experimental mantle excision in Akoya pearl oysters was demonstrated by Acosta-Salmon and Southgate (2006), characterized by hemocytic and connective tissue responses by 3–6h after excision, hemocytic plugging of the wound by within 6 h and an epithelial regenerative response within 25 h after excision. In a similar study, mantle tissues of Akoya pearl oysters and *Pinctada margaritifera* were shown by Acosta-Salmon and Southgate (2005) to regenerate following experimental excision, with an initial hemocytic infiltration at the excision site, followed by progressive proliferation of the connective tissue, vascular and epithelial elements. By 15 days post-excision, the internal, middle and external lobes of the marginal zone had regenerated, with functional secretion of conchiolin and progressive epithelial pigmentation leading to complete recovery within 90 days.

Suzuki *et al.* (1991) investigated early responses to wound healing in Akoya pearl oysters and demonstrated production by amoebocytes of an extracellular matrix containing collagen and possibly proteoglycan as a pre-requisite for epithelial migration and colonization. Further, Suzuki and Funakoshi (1992) described a fibronectin-like molecule in the hemolymph of Akoya pearl oysters and its incorporation into extracellular matrix produced by amoebocytes, suggesting that the fibronectin-like molecule supports epithelial migration in wound healing.

11.3.2. Cellular immunity and inflammation

Although little is published on the cellular inflammatory response of *Pinctada* species, the ability of pearl oysters to mount intense cell-mediated inflammatory responses to injury or infection is well recognized (Humphrey and Norton, 2005) and is described as part of the histopathology in a number of diseases.

The cellular immune response of Akoya pearl oysters was shown by Zhou *et al.* (2003) to be enhanced following administration of proteoglycan, with increased circulating hemocyte numbers and increased phagocytic activity. In contrast, phagocytic ability was decreased in Akoya pearl oysters following pearl seeding operations, suggesting compromised immune function after operations which might lead to increased post-operative mortalities. Lower water temperatures were considered to be a factor in reducing immune function and rendering oysters more susceptible to marine vibrionic infection (Pass *et al.*, 1987).

11.3.3. Humoral immunity

Zhou *et al.* (2003) described enhanced humoral immunity in Akoya pearl oysters fed proteoglycans, with increased agglutination ability and increased antibacterial activity of hemolymph. Following pearl seeding operations in Akoya pearl oysters, Zhou *et al.* (2001) described increased superoxide dismutase activity in the hemolymph. Yamaguchi and Mori (1988) described hemagglutinating ability of hemolymph of Akoya pearl oysters against horse and sheep erythrocytes, but not human erythrocytes, suggesting that pearl oyster hemolymph may agglutinate certain biochemical moieties.

Humoral immune function may be reduced under certain circumstances. Following pearl seeding operations, Zhou *et al.* (2001) reported a decrease in bacteriolytic activity of hemolymph of Akoya pearl oysters.

11.4. NON-SPECIFIC DISEASES AND CONDITIONS IN LARVAE

11.4.1. Hydropic degeneration

Hydropic degeneration or the accumulation of water in cells of the developing digestive gland may occur in larval oysters in association with stress or intercurrent disease (Humphrey and Norton, 2005). The change appears to be a reversible response to environmental stress or unsuitable water quality. Hydropic degeneration is seen as clear, fluid-filled cytoplasmic vacuoles in the digestive gland epithelium.

11.4.2. Larval bacillary necrosis

Deaths in hatchery-reared larvae associated with colonization and invasion by saprophytic micro-organisms occur commonly and are under-reported. The causes are not well understood or described. Larval *Pinctada maxima* are particularly susceptible to bacterial infections and adverse environmental conditions that may predispose the larvae to bacterial infections. Larval bacillary necrosis and mortalities have been described at least in Australia (Humphrey and Norton, 2005) and in Thailand (Nugranad *et al.*, 1999). The condition is associated with poor hatchery hygiene and poor water quality, resulting in debilitation of larvae and secondary colonization and invasion by marine Vibrionic bacteria, protozoa and fungi. Thus, the mortalities are commonly associated with poor water quality and hatchery hygiene.

Larvae are usually found dead. High mortalities are the norm. Bacterial invasion of larval oysters, especially associated with marine vibrionic organisms, may occur rapidly in populations and is accompanied by necrosis of tissues. Advanced bacterial necrosis results in dissolution and fragmentation of the larvae, and may be associated with colonization by saprophytic ciliated protozoa. Typically, numerous bacteria may be seen proliferating within the body cavity or intestine of infected larvae.

11.4.3. Anomalous development

Tanaka *et al.* (1970) described anomalous development of *P. margaritifera* larvae characterized by anomalous apposition of shell valves. Toxicity of the algae *Monochrysis lutheri* used as feed and bacterial proliferation at higher temperatures were associated with the anomalies.

11.5. NON-SPECIFIC DISEASES AND CONDITIONS IN SPAT TO ADULT STAGES

11.5.1. Mantle (Figs. 3.1 and 3.3)

11.5.1.1. Edema

Generalized edema may be associated with prolonged withdrawal from water and associated stress and is characterized by marked dilation of hemolymph sinusoids and

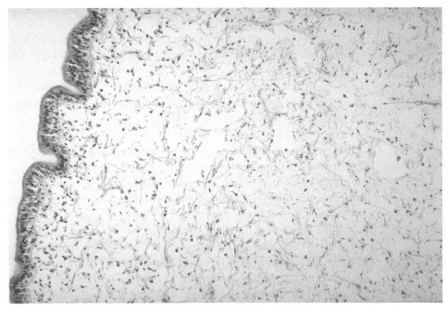

FIGURE 11.3 Generalized edema of mantle tissue in *Pinctada maxima* characterized by dilation of hemolymph sinusoids and irregular spongiform change with separation of fibers of stromal tissues by clear fluid-filled spaces.

irregular spongiform change in stromal tissues (Fig. 11.3). General edema is a feature of Oyster Edema disease in *P. maxima* (see Section 11.7.9.4). Focal edema of the mantle is seen commonly in populations of *P. maxima* associated with the crater-like depressions containing pea crabs *Pinnotheres* sp. (Humphrey and Norton, 2005).

11.5.1.2. Non-specific inflammation

Non-specific inflammatory changes characterized by focal or focally extensive infiltrations of hemocytic cells in the stroma of the mantle, in some cases extending through the epithelium, occur occasionally in spat and mature *P. maxima*. This inflammation is generally considered a response to bacterial infection with marine vibrionic organisms. Often, however, no organisms are evident on histological examination. Focal or multifocal, sub-epithelial accumulations of hemocytic cells in the stroma of the marginal and central mantle without an obvious etiological agent, may be related to mantle retraction and wedge-shaped deposition of conchiolin on the nacreous surface. This change is generally considered an indication of earlier or pre-existing localized bacterial infection.

Discrete, focal, granulomatous cellular accumulations are seen occasionally in the mantle stroma, which resemble the focal granulomas seen with metacestode infections but may show no evidence of a parasite. They may represent earlier parasitic invasion.

11.5.2. Shell disease

History, occurrence and distribution: The term "shell disease" is a non-specific term applied to discoloration of nacre caused by irregular, focal or extensive deposits of conchiolin and anomalous mineralization on the nacreous surface of the shell (Fig. 11.2). These anomalous deposits may be compounded by colonization by marine algae and other fouling organisms. The term shell disease is also loosely applied to invasion of the shell by boring or burrowing organisms resulting in lysis of shell structure and discoloration, which may also lead to anomalous mineralization (Fig. 15.1). Shell disease occurs in a diverse range of marine bivalve molluscs (Perkins, 1996). The presence of shell disease indicates current or earlier pathological insult to the oyster and should be considered a non-specific response to mantle pathology. It is not pathognomonic for any particular disease. Discoloration of the nacre may reduce the value of the shell as a by-product of pearl oyster farming.

Shell disease in pearl oysters has been recognized at least since the mid-1980s in French Polynesia and was a major clinical feature in mass mortalities in *P. margaritifera* (Marin and Dauphin, 1991; Comps *et al.*, 2001). Shell disease has also accompanied mortality events in *P. maxima* in Australia (Pass *et al.*, 1987) and is common is shell affected by disease processes. Shell disease is generally considered to be due to irritation to the mantle, mantle pathology and mantle dysfunction.

Although the cause has been attributed to mechanical trauma and chemical stress (Comps *et al.*, 2001), there is compelling clinical evidence suggesting that shell disease results from disease or disability causing mantle retraction, inflammation and anomalous deposition of conchiolin on the previously normal nacreous surface of the shell valves. Commonly, shell disease is associated with systemic vibriosis or other systemic disease. Biochemically, the malformations in the nacreous layer of the shell are associated with failure of the mantle epithelium to mineralize the organic matrix in a normal sequence of calcite crystals followed by aragonite crystals (Cuif and Dauphin, 1996; Marin and Dauphin, 1991) (see Section 3.3). Marin and Dauphin (1991) demonstrated irregular, anomalous deposition of nacre and differences in amino acid composition and content of the organic matrix in oysters with shell disease, compared to normal. These authors, together with Dauphin and Cuif (1990), showed decreased levels of aspartic acid, and glycine and alanine in the organic matrix of diseased shell, suggesting that reduced aspartic acid may be involved in failure of mineralization, and that reduced glycine and alanine may cause anomalous organic matrix composition accounting for the irregular patterns of mineralization.

Shell disease is characterized by the deposition of a yellow-brown organic matrix on the nacreous surface of the shell valves. Deposits may be focal, generalized, nodular or flattened and are commonly associated with retraction of the mantle. With time, the deposits may become covered with nacre. Comps *et al.* (2001) described yellow coloration and swelling of the mantle adjacent to the discolored nacre. Dauphin and Cuif (1990) described bilateral involvement of up to one-third of the shell area. In the earliest stages, anomalous conchiolin deposits occur as wedge-shaped or broad zones of thin conchiolin, which may subsequently be covered by a layer of nacre (Perkins,

1996). In more severe cases, conchiolin is deposited in broad, thicker brittle layers and is often multilayered (Perkins, 1996). The anomalies may be associated with prolonged local or regional mantle retraction and dysfunction (Fig. 11.2).

Histopathology: Perkins (1996) in examining mantle tissue of *P. maxima* affected with shell disease and normal oysters could not correlate epithelial sloughing or hemocytic infiltrations with the presence or absence of shell disease. Histological examination of thin sections of shell disease by Comps *et al.* (2001) have shown a homogenous layer of periostracum containing small lacunae and cellular debris is deposited on the inner surface of the shell. This layer is overlaid by lamina of periostracum alternating with layers of granulomatous tissue.

Ultrastructure: Comps *et al.* (2001) reported necrotic hemocytes in the deposits that contained paraspherical vacuoles 200–500 nm in size, some of which contained electron-dense membranous elements and spherical electron-dense particles approximately 45 nm in diameter. Anomalous biomineralization characterized by failure of sequential calcite and aragonite crystal deposition on organic matrix and the exuberant deposition of organic matrix resulting in protruding non-mineralized surfaces have been described by Cuif and Dauphin (1996).

Dauphin and Cuif (1990) using SEM described architectural disruption of the organic matrix with disorientation and loss of crystalline structure, together with anomalous organic matrix deposition (see Section 3.3). Dauphin and Denis (1987) described ultrastructural changes in the nacreous layer similar to proteolysis and decalcification, with architectural disorientation of crystallites and fracture of the vertical crystals with lacunae formation. These authors noted other changes associated with disruption of the nacreous layer including anomalous growth associated with loss of outer growth lamellae, local thickening of the shell and detachment of the adductor muscle.

11.5.3. Gills (Figs. 3.2, 3.5 and 3.6)

11.5.3.1. Cysts and mineralization

Only rarely, cysts or cystic structures of unknown origin appear in gill tissues and focal areas of mineralization may occasionally be observed in the gill of *P. maxima* (Humphrey and Norton, 2005).

11.5.3.2. Non-specific inflammation

Focal or regional areas of non-specific inflammation characterized by hemocytic infiltrations or accumulations are seen occasionally in the gills of clinically normal *P. maxima* (Humphrey and Norton, 2005).

11.5.3.3. Uncharacterized protozoa

Ovoid, eosinophilic organisms approximately 7–10 μm in diameter, resembling the protozoan *Haplosporidium*, have infrequently been visualized in the gill tissues of

adult *P. maxima* in Australia. The organisms were associated with epithelial degenera-
tion and an intense focal inflammatory cell response (Humphrey and Norton, 2005).
Ancistrocoma-like ciliates have occasionally been reported in the gills of spat of
P. maxima from Australia (Humphrey and Norton, 2005).

11.5.3.4. Metazoa

A variety of metazoan agents including pea-crabs and shrimp have been associated
with the external surfaces or epithelium of the gills of *P. maxima* in Australia, generally in the absence of inflammatory or degenerative responses. Histological sections of
gill may inadvertently include a section through one of these symbionts. Occasionally,
uncharacterized metazoa may invade and localize within the stromal tissues of the gill,
often without inciting an inflammatory or other host reaction (Fig. 11.4).

11.5.4. Labial palps (Figs. 3.3 and 3.9)

11.5.4.1. Edema

Edematous changes characterized by marked dilation of hemocytic sinusoids and
irregular spongiform change in stromal tissues have occasionally been observed in
the palps of *P. maxima* in Australia (Humphrey and Norton, 2005). The significance is
uncertain.

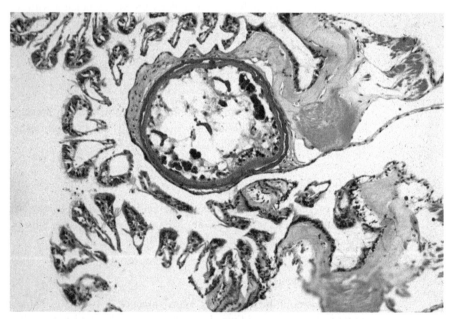

FIGURE 11.4   Uncharacterized metazoan localized in the stromal tissues of the gill of *Pinctada maxima*.
Note absence of inflammatory response. (Photo: John Norton.)

11.5.4.2. Non-specific inflammation

Non-specific inflammatory cellular responses in the stroma of the palps occur infrequently in mature *P. maxima* in Australia in the absence of observable etiological agents. Focal or focally extensive hemocytic infiltrations involving both the stroma and epithelium are likely to represent a response to bacterial infection. Discrete, focal or multifocal granulomas also occur in the sub-epithelial stroma and are similar in character to those initiated by metacestode invasion. These granulomas may be in the stroma of the fold of the palp and may significantly expand the fold (Fig. 11.5). In many cases, no agent is visible and serial sections may not reveal any such agents. These granulomas may represent earlier invasion by metazoan larvae.

11.5.4.3. Inclusions

Viral-like inclusions have occasionally been recorded from the epithelium of the palp of *P. maxima* in Western Australia (Humphrey and Norton, 2005). The inclusions differ to the papovavirus-like inclusions (see Section 11.7.1.3). Their nature and significance remains uncertain.

11.5.4.4. Metazoa

Turbellarian-like ciliated metazoan agents approximately 300–400 μm in length colonize the surface of the palp epithelium of wild-harvested oysters and have been found

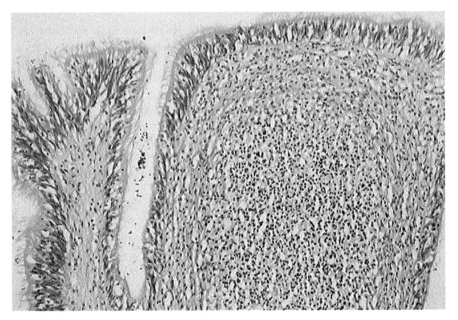

FIGURE 11.5 Focal granulomatous inflammatory cell accumulation in the sub-epithelial stroma of the palp of *Pinctada maxima* causing expansion of the fold. Granulomas commonly occur as a result of invasion by metacestodes and in their absence, may represent earlier invasion.

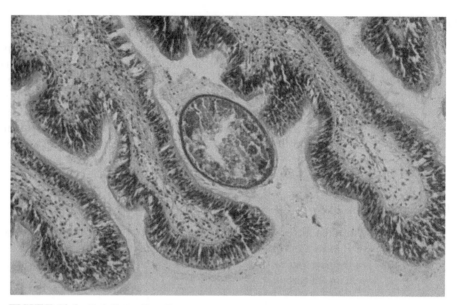

FIGURE 11.6   Turbellarian-like ciliate occupying the surface of the palp epithelium of wild-harvested *Pinctada maxima*. These organisms appear saprophytic and cause no appreciable tissue damage. (Photo: John Norton.)

at a prevalence of up to 12% of *P. maxima* in Australia (Humphrey *et al.,* 1998). No tissue damage is apparent with these agents (Fig. 11.6).

A variety of metazoan agents is occasionally found external to the palp epithelium of *P. maxima* in Australia. Generally, there are no associated inflammatory or degenerative responses associated with these agents and they are considered to be incidental occurrences of free-living or commensal organisms.

Single or multiple focal concentric hemocytic accumulations or granulomas peripheral to metacestodes in the stroma underlying the palp epithelium may be prominent in populations of mature oysters (see Section 11.7.6.1). Older lesions show well-developed fibrous encapsulation. Unidentified metazoa may occasionally localize in the stroma of the palp, forming small granulomas.

11.5.5. Mouth and esophagus (Figs. 3.3 and 3.7)

11.5.5.1. Metazoa

The copepod *Anthessius pinctadae* may on occasions be found inhabiting the superficial epithelium of the mouth region and the lumen of the esophagus (see Section 11.7.7.1). Metazoa typical of metacestodes are commonly associated with discrete, focal or multifocal granulomas in stromal tissues below the epidermis and occur

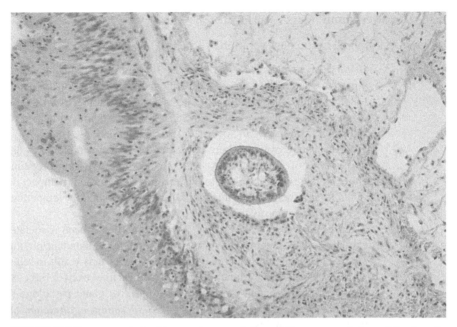

FIGURE 11.7   Low-grade focal hemocytic accumulation and fibrous encapsulation peripheral to a lecanicephalid metacestode in the stroma of the stomach of *Pinctada maxima.*

regularly in populations of *P. maxima* in Australia. (see Metacestodiasis, Section 11.7.6.1).

11.5.6. Stomach (Figs. 3.3 and 3.13)

11.5.6.1. Non-specific inflammation

The stromal tissues surrounding the gastric epithelium in mature *P. maxima* normally contains low numbers of hemocytes. Focal or regional hemocytic inflammatory cell infiltrations occur infrequently in the stroma of the stomach, associated with bacterial infections. Ulceration of stomach epithelium of unknown cause, with an underlying stromal cellular inflammatory response is rarely observed in *P. maxima* in Australia.

11.5.6.2. Metazoa

Focal hemocytic accumulations peripheral to metazoan agents consistent with larval lecanicephalid cestodes, often with fibrous encapsulation, are occasionally located in the stroma of the stomach of *P. maxima* in Australia (Fig. 11.7) (see Section 11.7.6.1). Occasionally, unidentified crustacea are found which may elicit an intense focal hemocytic inflammatory response. Molluscan larvae are commonly present in the lumen of the stomach of *P. maxima* in Australia.

11.5.7. Digestive gland (Figs. 3.3 and 3.7)

11.5.7.1. Dilation and cystic degeneration

History, occurrence and distribution: A syndrome characterized by dilation of the digestive gland diverticula and flattening or attenuation of the epithelium, sometimes accompanied by irregular cystic degeneration resulting from breakdown of the walls of the gland and coalescence of adjoining diverticula with loss of glandular integrity occurs commonly in spat of *P. maxima* in Australia (Fig. 11.8). This change may occur in mature oysters where it is generally focal or regional in distribution. In spat, the severity of the syndrome suggests that digestive gland dysfunction must occur, although there have been no studies that quantify the actual impact on the oyster.

Etiology: The cause is uncertain: toxic phytoplankton, temperature stress and starvation have been postulated as the cause.

Epidemiology: The occurrence of the lesions has been associated with starvation, extremes of water temperature and the presence of abundant phytoplankton, suggesting that toxic elements in certain plankton species may result in reduced feeding activity and a decrease in epithelial height, or they may exert a direct toxic effect. Negri *et al.* (2004) associated mortalities and digestive gland pathology in *P. maxima* with exposure to the bloom-forming alga *Trichodesmium erythraeum,* although its role in the syndrome remains uncertain.

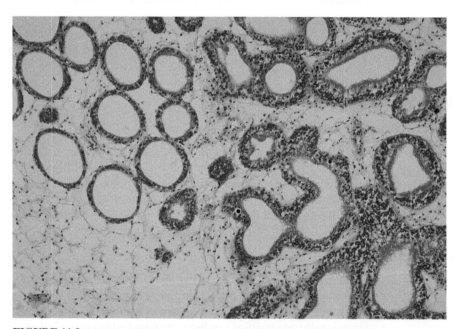

FIGURE 11.8   Dilation of the digestive gland diverticula and attenuation of the lining epithelium in spat of *Pinctada maxima*. This change is sometimes accompanied by irregular cystic degeneration resulting from breakdown of the walls of the gland and coalescence of adjoining diverticula with loss of glandular integrity. The change has been associated with starvation, water temperatures and toxic phytoplankton.

Clinical findings and gross pathology: Affected spat appear to grow and develop normally and the changes are generally seen in the absence of clinical signs. On occasions, the changes have been seen in spat undergoing mortality events.

Histopathology: In early cases, the epithelium of individual or multiple diverticula is attenuated and the gland dilated. Breakdown of the integrity of the gland is seen, with regression of the stroma and epithelium resulting in irregular multilocular cystic cavities. Degeneration of the epithelium is frequently seen, with loss of apical cytoplasm and the formation of ovoid, eosinophilic cytoplasmic fragments which stream out of the affected gland. High numbers of residual bodies may be present in the affected epithelium or free in the lumen, associated with degeneration and fragmentation of the epithelium.

11.5.7.2. Intracellular inclusions

Large intracellular amphophilic inclusions showing internal granular structure occasionally occur in the digestive diverticular epithelium of *P. maxima* in Australia. Smaller eosinophilic inclusions with granular internal structure are also seen occasionally. The nature and significance of these inclusions are uncertain.

11.5.7.3. Non-specific inflammation

Regional or focal infiltrations of hemocytes are occasionally found in the stroma of the digestive gland of apparently normal oysters in the absence of an obvious etiological agent (Fig. 11.9). Intense, focally extensive hemocytic infiltrations associated with

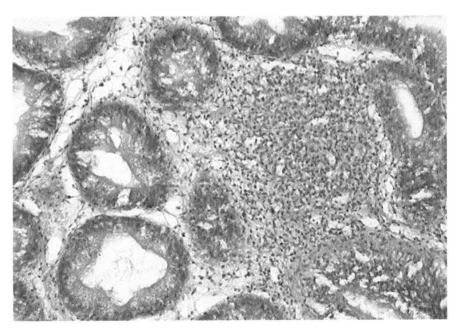

FIGURE 11.9 Focal hemocytic infiltration of the stromal tissues of the digestive gland of *Pinctada maxima* with distortion of glandular structure. No obvious etiologic agent is seen.

degeneration of digestive gland epithelium have also been seen in apparently clinically healthy oysters. No causal agent, however, has been found. It is likely that, in the absence of etiological agents or tissue damage, that the inflammatory response is associated with bacterial invasion.

11.5.7.4. Enigmatic bodies

Eosinophilic ovoid granular bodies, approximately 10–15 μm in diameter, closely associated with or in the digestive diverticular epithelium have been observed in populations of mature *P. maxima* from Australia at a prevalence of up to 3.3%. The bodies were associated with a moderately intense localized granulomatous inflammatory response and epithelial degeneration. Some of the bodies showed fine basophilic granular internal structure. No definite association with clinical disease could be established and the possibility that they represent degenerate cellular components cannot be excluded.

11.5.7.5. Internal metazoa

Morphologically and taxonomically diverse metazoan agents may occasionally be observed in the lumen of the digestive gland tubules or diverticula or in the interstitial tissues (Fig. 11.10). In general, these agents appear to incite little if any tissue damage

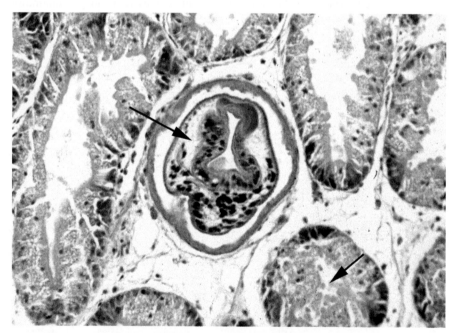

FIGURE 11.10   Unidentified metazoan (arrow) localized in the interstitial tissues of the digestive gland of a *Pinctada maxima* in the absence of host inflammatory response. (Photo: John Norton.)

or inflammatory response. Occasionally, metazoa in the digestive gland are associated with atrophy of the epithelium, hemocytic response and mineralization. A low proportion of *P. maxima* spat examined from Western Australia have been recorded with the copepod *Anthessius pinctadae* encysted in the digestive gland (B. Jones, personal communication).

On occasions, metazoa, including lecanicephalid metacestodes may invade the interstitial tissues peripheral to the digestive gland, inciting an intense hemocytic inflammatory response.

11.5.7.6. Residual bodies

Residual intracellular bodies, which are small spherical to ovoid, slightly refractile and up to approximately 6 μm in diameter, and containing golden-brown pigment, are commonly encountered in the epithelium of the digestive diverticula of *P. maxima* (Fig. 11.11). These bodies have been described as "protistan parasites" by Wolf and Sprague (1978) who considered them to be parasitic and associated their presence with degenerative changes and mortalities. Similar bodies were associated with a mass mortality of *P. margaritifera* in the Red Sea by Nasr (1982). Pass and Perkins (1985) considered these bodies to be "residual bodies," normal constituents of digestive cells comprising storage or secretory products bound within lysosomes, a view which is now widely accepted. Although the residual bodies are not considered to be of pathological significance, their presence may indicate feeding activity and, on at least one occasion, high

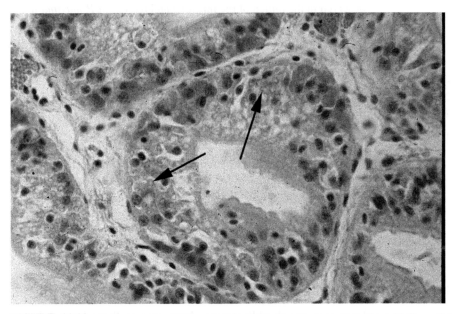

FIGURE 11.11 Golden-brown, slightly refractile residual bodies (arrows) in the epithelium of the digestive diverticula of *Pinctada maxima* are considered to be normal lysosomal degradation products.

numbers of residual bodies have been associated with mortality events in Australia, possibly reflecting a toxic insult to the digestive gland.

11.5.8. Intestine (Figs. 3.3 and 3.8)

11.5.8.1. Siliceous obstipation

Ingestion of fine silica spicules resulting in intestinal impaction with erosion of cilia is recorded in *P. maxima* in Australia resulting in secondary bacterial infection and death. This condition has occurred when oysters have been placed on seabeds containing degenerating siliceous sponges (Humphrey and Norton, 2005).

11.5.8.2. Non-specific inflammation

Diffuse infiltrations of hemocytes of light to moderate intensity commonly occur adjacent to the intestinal epithelium, especially in response to bacterial infections. Erosion and ulceration of the intestinal epithelium may occur, with intense inflammatory cell infiltrations extending into and partially occluding the lumen of the intestine or surrounding the intestine especially in spat. Intense focal inflammatory cell accumulations may occupy the stroma of the infolded epithelium of the distal intestine.

11.5.8.3. Metazoa

A range of morphologically diverse metazoan agents, including mollusc larvae may be observed in the lumen of the intestine and rectum of *P. maxima* in Australia. In general, these appeared to incite no tissue damage and no inflammatory response. While the normal food particles of pearl oysters obtained through filter feeding are *ca.* $<10\,\mu m$ in diameter, large particles are sometimes trapped and pass through the digestive system undigested and even alive (see Section 4.2.5).

11.5.9. Heart and vascular system (Figs. 3.3 and 3.12)

11.5.9.1. Non-specific inflammation

Non-specific inflammatory changes in the heart, characterized by mild to intense, focal or regional infiltrations of hemocytes in the stroma, sometimes accompanied by degeneration and necrosis may be observed occasionally in apparently normal oysters in the absence of obvious causative agents. Hemocytic accumulations on the valves may occur occasionally and may be granulomatous in nature with peripheral fibrous encapsulation. Vasculitis or inflammation of the hemolymph vessels may accompany inflammatory responses in any tissue or organ.

11.5.9.2. Bacterial colonization and microabscessation

Focal or regional infiltrations of hemocytes in the heart may be observed in diseased animals with bacterial septicemia, in which Gram negative bacilliform bacteria typical of marine vibrionic organisms may be seen in the lesion. The heart appears to be a primary site for bacterial localization in oysters with bacterial septicemia, where abscess or granuloma formation with central necrosis is a frequent finding. The presence of focal hemocytic inflammation in the myocardium, even in the absence of obvious microbial organisms, is typical in bacterial septicemia associated with *Vibrio* spp. Valvular inflammatory lesions are not uncommon, especially in oyster spat with vibriosis.

11.5.9.3. Protozoa and metazoa

An unidentified apicomplexan has been reported on one occasion in the heart within the myocardial fibers of spat of *P. maxima* from Western Australia.

Uncharacterized metazoan agents occasionally occur in the sinuses of the heart and in hemolymph sinusoids in *P. maxima* in Australia. There have been no reports of clinical disease associated with these agents.

11.5.10. Muscles (Figs. 3.9 and 3.10)

The muscles of pearl oysters are well defined and include the adductor muscle, retractors and levators of the foot, branchial muscles, pallial muscles and the heart (see Chapter 3). Details of the pathology and diseases of the heart are discussed in Section 11.5.9.

11.5.10.1. Edema

Edema of the adductor muscle is occasionally recorded in *P. maxima* in Australia. This may occur in oysters removed from the water and subject to high air temperatures.

11.5.10.2. Mineralization

Focal areas of concentric lamelliform mineralization have been seen occasionally in adductor muscle in *P. maxima* from Australia.

11.5.10.3. Non-specific inflammation

Non-specific foci of inflammation are common, probably associated with bacterial infections, especially in younger oysters. Focally extensive or diffuse hemocytic infiltrations of adductor muscle fibers occur in *P. maxima* in Australia associated with tissue damage caused by excessive force when opening the shell (see Section 8.9.4). More intense, focal hemocytic accumulations in the adductor muscle may occur in bacterial septicemia.

11.5.10.4. Adductor muscle necrosis, myositis and abscessation

Syndromes of necrosis, inflammation and abscessation of the adductor muscle have been described in *P. margaritifera* from French Polynesia in association with mass

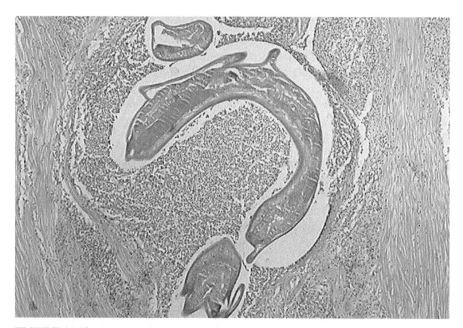

FIGURE 11.12   Uncharacterized metazoan parasite invading the adductor muscle of *Pinctada maxima* and inciting intense hemocytic inflammatory response.

mortalities and in Akoya pearl oysters from Japan as a characteristic lesion in Akoya oyster disease syndrome. Myofibrillar degeneration, myofiber necrosis and an associated granulomatous inflammatory response resulting in central abscesses were described in *P. margaritifera* by Comps *et al.* (1999, 2001) who reported viral-like particles in degenerate cells.

11.5.10.5. Metazoa

On rare occasions, metazoa may invade the muscle tissues, probably as incidental infections. An unidentified metazoan is recorded in the adductor muscle of *P. maxima* in Australia eliciting an intense inflammatory response (Fig. 11.12). Invasion of the levator of the foot by an unidentified metazoan is also recorded in *P. maxima* in Australia. The larval nematode *Cheiracanthus uncinatus* may be encysted in the adductor muscle (see Section 11.7.4.1).

11.5.11. Kidney

11.5.11.1. Non-specific inflammation

Non-specific inflammatory changes, characterized by focal or regional hemocyte infiltrations occur infrequently in the nephridia of *P. maxima* in Australia.

11.5.11.2. Metazoa

Occasionally, uncharacterized metazoa have been observed in the lumen of the kidney tubules in the absence of tissue damage in *P. maxima* in Australia.

11.5.11.3. Hyperplasia

Hyperplasia of the nephridial tubular epithelium, characterized by thickening and proliferation of the nephridial epithelium in the absence of gland formation, is occasionally observed in *P. maxima* in Australia. The cause and pathogenic significance is unknown.

11.5.12. Byssal organ (Figs. 3.3, 3.7 and 3.11)

11.5.12.1. Non-specific Inflammation

Diffuse hemocytic cell infiltrations of the byssal organ may occur as part of a systemic reaction or in response to local bacterial colonization. Focal or focally extensive hemocytic infiltrations of the byssal gland have been recorded in *P. maxima* in Australia. In some cases, focal abscessation may occur, probably a result of trauma and bacterial infection associated with tearing or pulling on the byssal threads during harvesting or removal from culture panels. The byssal threads are removed entirely leaving a cavernous space within the gland, which is susceptible to infection by marine vibrionic bacteria and subsequent hemocytic accumulation.

11.5.13. Foot (Figs. 3.2 and 3.9)

11.5.13.1. Edema

Commonly, edematous dilation and swelling of the foot is seen in *P. maxima* in Australia taken straight from the water. This is thought to represent a physiological escape response mechanism. Gross edema of the foot is seen in individual populations of mature *P. maxima*, associated with prolonged time out of water between collection and examination or fixation.

11.5.13.2. Mineralization

Foci of concentric, lamelliform, mineralization occur occasionally in the foot of mature *P. maxima* in Australia.

11.5.13.3. Non-specific inflammation

Focal or regional inflammation of the foot characterized by hemocytic infiltrations occurs uncommonly in clinically normal mature *P. maxima* in Australia.

11.5.13.4. Metazoa

Occasional metazoan agents may be encountered internally in the hemolymph sinuses of the foot and externally in the pedal groove. Commonly, no tissue damage or inflammation is apparent with these agents.

11.5.14. Gonad (Figs. 3.3 and 3.7)

11.5.14.1. Mineralization

Concentric, lamelliform, mineralized foci are occasionally observed in the gonad of *P. maxima* in Australia.

11.5.14.2. Non-specific inflammation

Regional hemocytic infiltrations are associated with regressing gonads as part of the normal cycle of regression. Multinucleate cells may occur in the regressing gonad. Occasional non-specific hemocytic accumulations have been recorded in the gonadal stroma of mature *P. maxima* in Australia.

11.5.15. Interstitial tissue

11.5.15.1. Non-specific inflammation

Focal or regional inflammation of the interstitium may involve adjacent organs or tissues and is characterized by focal, regional or generalized hemocytic infiltrations. These occur relatively commonly in clinically normal mature *P. maxima*, and may be associated with stress and bacterial invasion. On occasion, "brown cells" are seen, phagocytes containing brown pigment consistent with lipofuscins suggesting earlier cell damage. Granulomatous inflammatory foci are associated with *Perkinsus* infection in *P. maxima* in Australia (see Section 11.7.3.4).

11.5.16. The pearl sac

The process of creating a pearl sac, whereby a pearl nucleus and mantle graft are inserted into a lobe of the visceral mass (the "pearl pocket." see Section 3.2.1, Figs. 3.2 and 3.8), has potential for infection through cutting into the lobe and inserting the foreign bodies of a pearl nucleus and mantle graft. See Section 8.9.4 for details of this procedure.

11.5.16.1. Pearl sacculitis

History, occurrence and distribution: Inflammatory cell infiltrations of the pearl sac are not uncommon. Non-specific inflammation of the pearl sac is recorded in *P. maxima* from Australia and in *P. margaritifera* from French Polynesia. The inflammation has been associated with anomalous conchiolin production resulting in inferior pearl quality.

Etiology: The cause is generally unapparent although in some occasions, bacteria have been reported in association with the inflammatory response. Cochennec-Laureau *et al.* (2004) described *Vibrio harveyi* and *Vibrio alginolyticus* from the pearl sac of *P. margaritifera* following nucleus insertion.

Pathogenesis: It is probable that micro-organisms or detritus introduced during seeding operations is the primary cause.

Histopathology: An intense hemocytic infiltration of the epithelium and sub-epithelium, with extensive folding of the sac is reported in *P. maxima* (Dix, 1973a). Abscess formation with a granulomatous inflammatory cell infiltration, tissue debris and epithelial degeneration adjacent to the pearl nucleus has been recorded in *P. margaritifera* from French Polynesia (Comps *et al.*, 2001).

11.5.16.2. Accretions and anomalous mineralization

History, occurrence and distribution: A variety of accretions and mineralization anomalies are known to occur in the pearl sac. Comps *et al.* (2000) described lamelliform periostracal accretions, small (0.5–2 mm), macroscopically visible, para-spherical whitish bodies in the pearl sac of *P. margaritifera* from French Polynesia that were associated with surface flaws in pearls. Similar anomalies are recognized in these Australian *P. maxima*.

Etiology: The lamelliform accretions in *P. margaritifera* may be a response to earlier trauma or foreign matter introduced into the pearl sac during grafting.

Pathogenesis: The lamelliform accretions in *P. margaritifera* are believed to result from abnormal secretion of ostracum (conchiolin layer) and a local disturbance of mineralization of the secreted protein matrix.

Histopathology: The accretions described by Comps *et al.* (2000) in *P. margaritifera* comprise, on thin section: an amorphous basophilic layer in contact with the pearl; an inner fractal material; and a homogenous purple-stained material in contact with the pearl sac epithelium. These occurred within a cavity formed by infolding of the epithelium of the pearl sac. Histological examination of the pearl sac in *P. maxima* shows irregular folding or protrusions of the pearl sac epithelium associated with mineralized material (Fig. 11.13).

Ultrastructure: Ultra-structurally, the lamelliform periostracal bodies consist of lamelliform aggregates of the organic conchiolin matrix.

11.5.16.3. Post-operative mortalities and nucleus rejection

History, occurrence and distribution: Mortalities following grafting and seeding operations and rejection of implanted nuclei are well recognized in farmed pearl oysters but there is little information on the actual cause or causes.

Etiology: Poor hygiene and sepsis have been associated with post-operative mortalities and nucleus rejection. Sound antiseptic technique is a pre-requisite for successful operations. Norton *et al.* (2000), however, did not find any improvements in oyster survival, bead rejections and pearl quality after treatment of the wound site with an antiseptic solution. Cochennec-Laureau *et al.* (2004) reported that the use of antiseptic

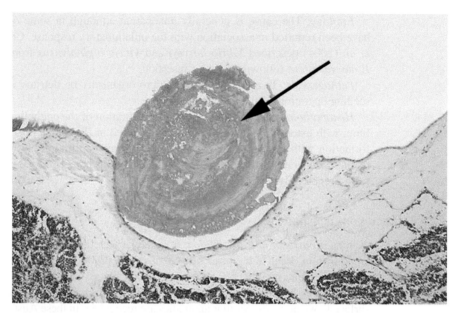

FIGURE 11.13   Anomalous mineralized ovoid accretion (arrow) in the pearl sac of *Pinctada maxima*.

during the operation procedure reduced the number of bacteria present in the pearl sac; however, no effect on mortality or nucleus rejection was observed. It is likely that post-operative mortalities and nucleus rejection result form a complex of factors including sepsis, failure of wound healing and insertion of the nucleus in and inappropriate location. Norton *et al.* (2000).

11.6. NON-INFECTIOUS AGENTS CAUSING DISEASE

A diverse range of physical, nutritional and environmental factors impact on pearl oysters in favorable and unfavorable ways. They are reviewed in Chapter 6. Unfavorable levels of these factors stress the oysters so the metabolism, feeding, growth, reproduction, etc., are adversely affected. Levels of some factors may stress the pearl oysters to the point of mortality, but it is more common that the stress renders them more susceptible to infectious agents, which might otherwise remain innocuous.

Some environmental factors will be treated here where there is evidence of infections being linked to stressful environmental conditions.

11.6.1. Low temperature

Decreased activity of host inflammatory processes at lower environmental temperature is considered the primary mechanism for growth of marine vibrionic bacteria and mortalities in transported *P. maxima* (Dybdahl and Pass, 1985; Pass *et al.*, 1987). Mannion

(1983) examined the influence of temperature on experimental infection with the bacteria *V. harveyi* and *Pseudomonas putreficiens* in *P. maxima* at 19°C and 29°C. Oysters held at the lower temperature showed a markedly greater incidence of disease and mortality.

11.6.2. Phytoplankton blooms

11.6.2.1. Dinoflagellate blooms ("Red tides") in Japan

History, occurrence and distribution: Phytoplankton blooms resulting from exuberant proliferation of dinoflagellate algae, cyanobacteria (blue-green algae) and other phytoplankton taxa may cause direct or indirect disease and mortalities in pearl oysters, although the mechanisms by which disease and deaths occur are generally unclear. Although "Red tides" have been recognized since the 16th century (Cahn, 1949), not all such events are toxigenic. The Japanese pearl oyster industry in particular has suffered severe losses as a result of phytoplankton blooms on numerous occasions (Sparks, 1985).

Seasonally recurrent large-scale red tides of the toxic dinoflagellate, *Heterocapsa circularisquama,* have been recognized in Japan since 1988, occurring particularly during summer and autumn and resulting in mass mortality of Akoya pearl oysters, together with numerous other commercially important bivalve species (Matsuyama *et al.*, 1995; Matsuyama, 2003a, b). Cahn (1949) described algal blooms of *Gymnodinium* in Japan in 1917 and in 1934, which caused extensive damage to fish and shellfish, including pearl oysters. Extensive blooms of the cyanobacterium, *Trichodesmium erythraeum*, have been associated with mortality events in *P. maxima* in Australia (Negri *et al.*, 2004) (see Section 11.6.2.2). Chellam and Alagarswami (1978) associated heavy mortalities in pearl oysters in the Gulf of Mannar exposed to water from a bloom of *Trichodesmium thiebautii*. Not all phytoplankton blooms result in adverse affects. Dharmaraj *et al.* (1987) reported widespread seasonal blooms of the blue-green alga *T. thiebautii* in India, without any apparent effect on pearl oysters and Miyamoto *et al.* (2002) described a major red tide in western Japan caused by the dinoflagellate *Gyrodinium* sp. without significant mortalities in Akoya pearl oysters.

Etiology: Numerous species of dinoflagellate have been identified in blooms impacting adversely on pearl oysters. In Japan, species associated with disease and mass mortalities in pearl oysters include the dinoflagellate genera *Gymnodinium*, *Gonyaulax, Peridinium Ceratium Cocurodium, Prorocentrum, H. circularisquama* and *Trichodesmium,* the distomacian genus *Chaetoceros* and the cyanobacteria or blue-green algae (Cahn, 1949; Sparks, 1985; Wada, 1991; Matsuyama *et al.*, 1995; Negri *et al.*, 2004). *H. circularisquama* is recognized as a major toxigenic dinoflagellate associated with mortalities in pearl oysters and it has been extensively studied (Kim *et al.*, 2000; Matsuyama, 2003a, b).

Epidemiology: The genesis of phytoplankton blooms is complex. They may be natural events or may be events potentiated or exacerbated through anthropogenic activities. Blooms may be associated with increasing water temperatures, stratification and calm seas and may persist for several weeks (Chellam and Alagarswami, 1978). A clear relationship exists in some cases between heavy rainfall, subsequent nutrient input and phytoplankton blooms (Miyamoto *et al.*, 2002).

The toxic dinoflagellate *H. circularisquama* is well studied as the cause of seasonally recurrent red tides and mass mortalities of Akoya pearl oysters in Japan since 1988 (Matsuyama *et al.*, 1995; Matsuyama *et al.*, 2000; Matsuyama, 2003a, b) and mortalities can readily be reproduced experimentally (Yamatogi *et al.*, 2005) These blooms, which are prevalent during the warmer months, July–November, may be preceded by heavy rainfall and/or vertical mixing of anoxic seawater associated with storm events (Matsuyama *et al.*, 1995) and occur under conditions of elevated water temperature (>23°C) and salinity (>30‰). It is postulated that temporary or sustained mixing of water may provide nutrients which promote the proliferations of *H. circularisquama*. Dense assemblages of *H. circularisquama* occur inshore, but not offshore or in channels suggesting their distribution is strongly affected by water exchange. *H. circularisquama* is considered a tropical species that does not grow below 11°C. Its emergence in Japan is believed to be due to:

- an ability to over-winter as a consequence of increased sea-water temperatures caused by global warming;
- adaptation to oligotrophic conditions of low phosphorous and nitrogen;
- dispersal associated with movements of shellfish for aquaculture.

Although *H. circularisquama* is clearly toxic for bivalves, no harmful effects on wild or cultured fish or other marine invertebrates have been recorded (Matsuyama, 2003a). High cell densities may occur during phytoplankton blooms. Cell densities may exceed 80,000 cells/mL (Miyamoto *et al.*, 2003b).

Pathogenesis: The mechanisms whereby disease and death occurs during or following exposure to phytoplankton blooms is not clear in the majority of cases. It is presumed that losses may be due to one or more of decreased oxygen, suffocation associated with clogging of gills or direct toxic effects through waterborne or ingested toxins (Cahn, 1949; Sparks, 1985). In certain situations, oysters close their shells tightly and die despite a high concentration of dissolved oxygen. It appears likely that more than one mechanism may be involved.

The pathogenesis of the toxigenic dinoflagellate *H. circularisquama* has been studied in detail. The lethal effect of *H. circularisquama* is due to a toxin and toxicity is related to density of cells in the water (Nagai *et al.*, 1996, 2000; Kim *et al.*, 2000; Matsuyama, 2003b). Toxicity varies amongst strains of *H. circularisquama* and is in part dependent on temperature, salinity and nitrogen, with higher toxicity at elevated temperatures and salinities, and lower toxicity in nitrogen limiting conditions (Matsuyama, 2003b). Toxicity is mediated by an innate chemical agent in intact cells and is not due to extracellular metabolites or disrupted cells (Kim *et al.*, 2000; Matsuyama, 2003b). A hemolysin specific to *H. circularisquama* was isolated by Oda *et al.* (2001a), which caused degenerative changes in unfertilized eggs of Pacific oysters, *Crassostrea gigas*, and was lethal for rotifers *Brachionus plicatilis*. Oda *et al.* (2001a, b) demonstrated species-specific hemolytic activity for rabbit, mouse and guinea pig erythrocytes, and associated its activity with intracellular K^+ ion. Although the role of hemolysin in the pearl oyster remains to be elucidated, a causative role in *H. circularisquama* mediated deaths in pearl oysters appears likely. The biochemical mechanisms of toxigenesis in *H. circularisquama* are discussed in detail by Matsuyama (2003b).

Under natural conditions, cells densities reached 6×10^6–8×10^7 cells/L resulting in death of exposed oysters within several days (Nagai *et al.*, 1996; Matsuyama, 2003b). Experimentally Matsuyama *et al.* (2000) showed that clearance rate of bivalves was reduced when exposed to 2×10^4 cells/L and 5×10^6 cells/L of *H. circularisquama* was lethal. Nagai *et al.* (1996) determined an LD_{50} of 2×10^7 cells/L at 24 h and 1×10^7 cells/L at 48 h using laboratory reared cells. In contrast, Nagai *et al.* (2000) using naturally occurring cells found an LD_{50} of 5×10^6 cells, suggesting loss of toxigenicity during artificial rearing. When exposed to naturally produced cells of *H. circularisquama*, deaths of pearl oysters were initiated by 24 h at cell densities of 2.2×10^7 cells. Both naturally and experimentally exposed oysters showed a similar disease pattern, with shell closure despite a high concentration of dissolved oxygen, rapid mantle contraction, contraction of gills, irregular heartbeat patterns and ultimate cessation of heart contractions leading to death (Matsuyama *et al.*, 1995; Matsuyama *et al.*, 2000; Nagai *et al.*, 1996). This process might well lead to high mortalities in the absence of significant histopathological changes. It is noteworthy that *H. circularisquama* toxin appears specific for bivalves, gastropods, ascidians, hydrozoans and certain other invertebrates whereas vertebrates including fish, crustaceans, starfish and sea urchins appear refractory (Matsuyama, 2003b).

Clinical findings and gross pathology: There is little information on clinical signs and gross pathology. Red tides of *H. circularisquama* cause mass mortality with shell closure, mantle and gill retraction, discoloration of the gut, cardiac arrhythmias and paralysis in Akoya pearl oysters (Matsuyama *et al.*, 1995; Nagai *et al.*, 1996; Matsuyama, 2003b).

Histopathology: Despite the biological and economic significance of phytoplankton blooms, including *H. circularisquama*, there is scant data on the histopathology of affected pearl oysters. Heavy blooms of the algae *Trichodesmium* have been associated with the dilation of the digestive gland diverticula and flattening or attenuation of the epithelium, together with a cellular inflammatory response in *P. maxima* (Negri *et al.*, 2004).

Diagnosis: Establishing a causal role in pearl oyster mortalities in the face of phytoplankton blooms is problematic and depends on:

1. the presence high numbers of dinoflagellates, cyanobacteria or other phytoplankton in the water column;
2. the taxonomic identification of such organisms which is a specialized discipline in itself;
3. the demonstration of toxigenic properties of the isolated organisms.

Given the highly toxic nature of some phytoplankton and the ephemeral nature of phytoplankton populations in marine environments, it is possible that significant numbers of organisms may no longer be present in the water column by the time mortalities in pearl oysters are observed.

A novel approach to the early detection of *H. circularisquama* is described by Nagai *et al.* (2006) who utilized a sensor in the shell to monitor contractions of the adductor muscle induced in the presence of the dinoflagellate to detect low levels of cells of *H. circularisquama* in real time.

11.6.2.2. Cyanobacterium blooms in Australia

Farmed *P. maxima* underwent high mortalities in the Dampier Archipelago, Western Australia in 1996. Juvenile and mature oysters were affected. The mortality event was coincided with blooms of the cyanobacterium, *Trichodesmium erythraeum*; however, subsequent studies could not attribute the mortalities to *T. erythraeum* (Negri *et al.*, 2004). Saxitoxin was detected in some of the affected oysters, but saxitoxin was not detected in the *T. erythraeum* and feeding experiments failed to elicit mortalities. It was proposed that the *T. erythraeum* was not a suitable food for *P. maxima* and that starvation may have contributed to the mortalities. There was no evidence of anoxic conditions. Histopathological examination of adult oysters from affected farms and of juvenile oysters exposed to *T. erythraeum* showed dilation of digestive diverticula, sloughing of epithelial cells and a granulocytic inflammatory cell infiltration.

11.7. INFECTIOUS DISEASES AND ORGANISMS

Infectious diseases and organisms described in pearl oysters include viral, bacterial, fungal, protozoan and metazoan organisms, which are directly or indirectly transmissible and which may result in disease, together with a spectrum of symbionts and associates.

11.7.1. Viruses and viral diseases

There are few reports of viruses or viral-like agents in *Pinctada* species, although viruses representing members of the families Birnaviridae, Herpesviridae, Iridoviridae, Papovaviridae, Reoviridae and Retroviridae, as well as several viral-like agents have been reported in molluscs other than the genus *Pinctada* (Sindermann, 1990).

11.7.1.1. Akoya pearl oyster viral disease in Japan

Mass mortalities have occurred in farmed Akoya oysters in Japan resulting in losses of hundreds of million oysters and huge economic loss. Various causative agents have been suggested, none conclusively, although there is good evidence that at least a icosahedral virus is involved. A number of agents estimated the loss to be over 30 billion

Japanese Yen, including loss of markets due to poor quality pearls. The disease raised fears that the Akoya pearl oyster may be driven to extinction in Japan (Miyazaki *et al.*, 1999). Although the disease has only been recorded in Akoya pearl oysters, Miyazaki *et al.* (1999) speculated that its origin was infected Chinese pearl oysters.

Etiology: Compelling evidence that a virus was a causative agent of Akoya oyster disease was presented by Miyazaki *et al.* (1999) and Kobayashi *et al.* (1999), who visualized a small (30 nm) inclusion in tissues of affected oysters, isolated the virus in cell culture and experimentally reproduced typical disease by inoculation of virus.

Epidemiology: Mortalities appeared to be restricted to the western oyster growing areas of Japan with an annual mortality rate of over 400 million oysters in 1996 and 1997 representing losses of over 50% of oyster stocks. Mortalities were associated with water temperature with high weekly mortalities at temperatures of 25°C and greater, and marked decreases in mortality below 25°C. Paradoxically, juvenile oysters examined from colder water showed histological evidence of muscle damage (Miyazaki *et al.*, 1999). Experimentally, mortalities of 50–100% occurred 10 days after inoculation.

Miyazaki *et al.* (1999) proposed that infected juvenile oysters carried the virus in winter and subsequently developed clinical disease as mature oysters when water temperatures increased above 25°C, initiating horizontal spread of the disease and the resulting mass mortalities.

Miyazaki *et al.* (1999) reported the isolation of a virus resembling the Akoya oyster virus from Chinese pearl oysters transported without quarantine and speculated that this was the source of the infection in Akoya pearl oysters.

Kurokawa *et al.* (1999) demonstrated the transmission of the disease after cohabitation of healthy and diseased oysters supporting the presence of an infectious agent.

Pathogenesis: The studies of Miyazaki *et al.* (1999) suggest that viral infection of the muscles of the adductor, mantle, foot, gills and heart resulted in the clinical syndrome, with functional disturbances of feeding, respiration, excretion and glycogen metabolism, and death due to cardiac muscle necrosis and heart failure. Miyazaki *et al.* (2000) examined the healing process in experimentally infected oysters and showed that phagocytosis of degenerate muscle tissue was performed by agranulocytes with new collagen produced by myofibroblasts in repairing the tissue. The administration of feline interferon-ω to Akoya pearl oysters prior to experimental challenge with the virus resulted in decreased mortalities and fewer muscular lesions (Miyazaki *et al.*, 2001). In contrast to natural infections, feline interferon-ω was found to promote both phagocytosis and collagen production by agranulocytes, and it was found to reduce mortality rates in experimentally infected oysters when administered prophylactically. Miyazaki *et al.* (2000) postulated that recombinant feline interferon-ω, but not human interferon-α, because of its tertiary structure, may bind to specific hemocyte receptors initiating or enhancing hemocytic antiviral activity. Miyazaki *et al.* (2002) further demonstrated that receptors for recombinant feline interferon-ω were present on agranulocytes of Akoya pearl oysters and that these receptors specifically bound recombinant feline interferon-ω, presumably enhancing phagocytosis, collagen production and the healing process.

Clinical findings and gross pathology: The disease affected mature and immature oysters. Clinically there were functional disturbances of shell valve movements and sluggish closure of shell valves on stimulation of the mantle margin. Reduced growth at the shell valve margin was seen. Affected oysters had atrophy of the adductor muscle, atrophy of the mantle margin and body, and transparency of the mantle. The adductor appeared yellowish brown, especially during periods of higher water temperature (Miyazaki *et al.*, 1999; Kobayashi *et al.*, 1999). Affected oysters produced poor quality pearls (Kobayashi *et al.*, 1999). Experimentally, adductor dysfunction similar to that observed in natural cases of disease was seen approximately 7 days after inoculation, with widely opened valves shortly before death.

Histopathology: Histopathological changes were predominantly associated with muscle fibers. Affected oysters showed atrophy, necrosis, swelling and vacuolization of adductor muscle fibers accompanied by a hemocytic infiltration of varying intensity. Necrotic fibers showed nuclear pyknosis, karyorrhexis and marginal hyperchromatosis. Staining by the periodic-Schiff method showed depleted glycogen in affected muscle fibers. Feulgen positive inclusions were demonstrable in affected fibers. Muscle fibers transecting the mantle lobes were also atrophic with necrosis and a hemocytic infiltration accompanied by fibrosis. Cardiac muscle fibers in affected oysters showed atrophy, necrosis and vacuolization accompanied by a hemocytic infiltration. On occasions, fibrosis peripheral to myonecrosis in mature oysters indicated a healing process. Hemocytic infiltrations of the mantle were also a feature. Decreased numbers of cellular phagosomes and less absorbed matter resulting in attenuation of the epithelium and dilation of the lumen in the digestive gland of affected oysters was recorded, together with depletion of interstitial glycogen storage cells, interstitial edema and markedly fewer hemocytes compared to normal (Miyazaki *et al.*, 1999). A similar range of histopathological changes to those seen in natural cases were induced by experimental inoculation.

Ultrastructure: Membranous inclusions bodies containing numerous round, non-enveloped viral particles 25–33 nm were present in the sarcoplasm of necrotic adductor and pallial muscle fibers. Affected fibers showed myofibrillar thinning or fragmentation, sarcoplasmic vacuolization and mitochondrial destruction.

Virology: Miyazaki *et al.* (1999) demonstrated cytopathic effects (CPE) following inoculation on EK-1 and EPC cell lines incubated at 24°C and characterized by karyopyknosis, vacuolation, rounding up and lifting within 7–14 days. Replication of the CPE was attained on second and third serial passage. Viruses were isolated at a prevalence of 76–88% of diseased mature oysters but were not recovered from normal oysters. Electron microscopic examination of infected cells showed viral particles morphologically identical to those in the muscle tissues of affected oysters.

11.7.1.2. Intranuclear viral inclusions in *P. maxima* in Australia

History, occurrence and distribution: Intranuclear viral inclusions have been commonly reported in the digestive gland epithelium in immature and adult *P. maxima* from diverse geographic regions across Queensland, Northern Territory and Western

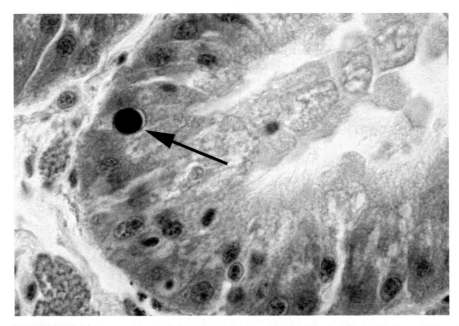

FIGURE 11.14 Intranuclear viral inclusion body (arrow) in the digestive gland epithelium of *Pinctada maxima* from Australia showing dense amorphous nature of the inclusion with margination of the nuclear membrane. Electron microscopic examination demonstrates arrays of icosahedral viral particles consistent with herpesviridae or adenoviridae.

Australia (Pass *et al.*, 1988; Humphrey *et al.*, 1998; Hine and Thorne, 2000; Humphrey and Norton, 2005).

Etiology: Electron microscopic examination of the inclusions shows arrays of regular icosahedral viral particles morphologically consistent with herpesvirus or adenovirus.

Epidemiology: The inclusions in the digestive gland are common in natural populations of mature *P. maxima* but rare in farmed oysters. A prevalence of infection in individual populations of up to 53% has been recorded (Humphrey *et al.*, 1998).

Clinical findings and pathology: No disease or gross pathology was attributed to the presence of the agent. Inclusions are round to ovoid, basophilic or amphophilic, occupying most of the nucleus (Fig. 11.14). Individual oysters may have high numbers of inclusions: in excess of 20 inclusion bodies per high power field in some areas. Inclusions generally appear randomly distributed in the nuclei of cells of the digestive diverticula and are not usually associated with any inflammatory, degenerative or proliferative changes. Localized hyperplasia and degeneration of digestive gland epithelium are occasionally associated with the presence of these bodies, however, especially where individual glands are heavily infected.

11.7.1.3. Papova-like virus in *P. maxima* in Australia

History, occurrence and distribution: Papovavirus-like infections have been described in the epithelium of the palps and gills from populations of juvenile and adult

P. maxima. The distribution appears restricted to oysters of Torres Strait, Queensland. A survey of *P. maxima* across northern Australian waters failed to demonstrate infection in oysters from the Northern Territory or Western Australia (Norton *et al.*, 1993a; Humphrey *et al.*, 1998; Humphrey and Norton, 2005).

Etiology: Non-enveloped, icosahedral viral-like particles approximately 60 nm in diameter consistent with the family Papovaviridae were demonstrable in tissues by transmission electron microscopic examination (Norton *et al.*, 1993a).

Epidemiology: Infections occurred in both wild and captive populations at a prevalence of 4.0–50.0%. No seasonal variation was apparent. Both spat and mature oysters are susceptible. The inclusion bodies and the epithelial lesions have been recorded only from Queensland (Humphrey *et al.*, 1998).

Clinical findings and pathology: Infection was detected in apparently healthy oysters. There were no gross lesions. Infection is characterized by hypertrophy of infected cells and the presence of large intranuclear inclusion bodies, especially in the ciliated columnar epithelium of the palp, but also in the epithelium of the gill. Large numbers of inclusions may be seen in some cases (Fig. 11.15). There is nuclear enlargement up to seven times normal size, with margination of chromatin and the presence of an amorphous central eosinophilic inclusions. The hypertrophied nucleus substantially replaces the cell. Loss of cilia and cytoplasm accompanies the nuclear hypertrophy. Expansive regions of the palp may be involved and the extent of involvement, together

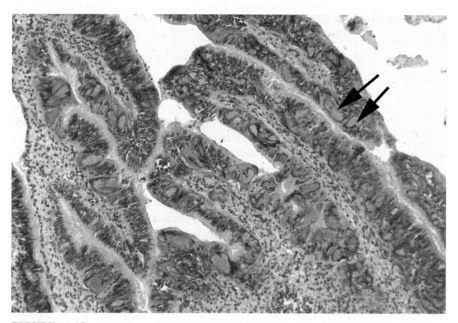

FIGURE 11.15   Severe generalized papovavirus-like infection of the palp of juvenile *Pinctada maxima* showing numerous inclusions (arrows) throughout the epithelium. There is marked nuclear hypertrophy of infected cells with loss of cilia and cytoplasm. A moderately intense hemocytic inflammatory response is associated with the hypertrophied cells in the stroma. (Photo: John Norton.)

with the marked epithelial hypertrophy and loss of cilia in affected cells, suggests the agent may be pathogenic, especially in younger animals. In spat, the large intranuclear inclusions may also occur in cells within the underlying stromal tissues. An intense hemocytic infiltration of the palp tissues and loss of epithelial ciliation may occur.

11.7.1.4. Viral-like particles in *P. margaritifera* in French Polynesia

History, occurrence and distribution: In 1985, high mortalities involving up to a million pearl oysters *P. margaritifera* were initially attributed to a mysterious "viral" disease. The disease was initially reported in the Gambier group of islands and later in Takapoto Atoll in French Polynesia (SPC, 1985). Viral-like particles were subsequently described in diseased oysters collected from several atolls in French Polynesia in the mid-late 1990s (Comps *et al.*, 1999). They were from populations that had earlier undergone mass mortality events. There is little evidence to support a viral etiology in these mortalities and the term virus appears to have been used in a generic sense to imply a spreading disease with high mortality.

Comps *et al.* (2001) noted a similarity of the lesions to those of Akoya oyster viral disease in Akoya pearl oysters, however, the viral-like particles were larger than those described in the Japanese condition.

Clinical findings and gross pathology: Oysters showed excess mucus production and grossly visible abscesses were present in the adductor muscles.

Histopathology: Progressive myofibrillar degeneration and adductor muscle necrosis, a diffuse hemocytic inflammatory cell infiltration and centralized areas of myonecrosis and granulomatous cell accumulation were reported.

Electron microscopy: Electron microscopic examination identified the inflammatory cells in the granuloma as granulocytes and macrophages, the latter containing lysosomes, membranous vesicles and cellular debris. Glio-interstitial cell proliferation was noted. Para-spherical or polygonal viral-like particles 40 nm in diameter with an outer membranous envelope and an inner electron-dense core 35 nm in diameter were common in degenerative tissues.

11.7.1.5. Birnavirus in Akoya pearl oysters in Japan

History, occurrence and distribution: A marine birnavirus was detected by PCR assay and was isolated in cell culture from Akoya pearl oysters from the Uwajima and Uchiumi fishing grounds in Japan in 1996. The virus was present in healthy oysters and in oysters showing signs of Akoya oyster disease syndrome (Suzuki *et al.*, 1998a). The virus appears widespread in clinically normal fish and molluscs (Suzuki, 1997). The same virus was subsequently identified from farmed Akoya pearl oysters and from the surrounding seawater by Kitamura *et al.* (2002). Although a causal role for the virus in Akoya oyster disease syndrome has not been established, Suzuki *et al.* (1998b) experimentally induced low-grade mortalities in oysters and recorded lesions similar to those seen in natural cases of Akoya oyster disease syndrome.

Epidemiology: The virus was present at a prevalence of up to 28.6% in Akoya pearl oysters and was present in surrounding sea-water, suggesting horizontal transmission. The prevalence of the virus in oysters and in the surrounding water increased during autumn and winter, when temperatures were low (Kitamura *et al.*, 2002). In addition to Akoya pearl oysters, the birnavirus was also detected by Suzuki (1997) in four species of finfish and one other species of mollusc.

Clinical findings and gross pathology: Suzuki *et al.* (1998a) isolated the birnavirus from oysters showing signs of Akoya oyster disease syndrome with red coloring of the mantle cavity and adductor muscle. The virus was also recovered from clinically normal animals.

Diagnosis: Suzuki (1997) described a sensitive PCR assay for birnavirus in Akoya pearl oysters which was more sensitive than virus isolation in cell culture.

11.7.2. Bacteria and bacterial diseases

11.7.2.1. Bacillary larval necrosis

History, occurrence and distribution: Necrosis of larvae associated with invasion and colonization by marine bacteria, primarily *Vibrio* spp., appears to be a common and under-reported condition in hatcheries.

Etiology: There are few data on the spectrum of organisms present and diagnosis is commonly made on histopathological grounds. It is generally assumed that marine Vibrionic bacteria are the principal organisms involved.

Clinical findings and pathology: Infection may result in high mortalities. Bacterial invasion of larval oysters, especially associated with marine vibrionic organisms, may occur rapidly in populations and is usually accompanied by necrosis of tissues. Advanced bacterial necrosis results on dissolution and fragmentation of the larvae, and may be associated with colonization by saprophytic ciliated protozoa. Typically, numerous bacteria may be seen proliferating within the body cavity or intestine of infected larvae.

11.7.2.2. Pseudomonas disease of larval Akoya pearl oysters from China

Mortalities in hatchery-reared larval Akoya pearl oysters were associated with infection by *Pseudomonas* sp. (Liu *et al.*, 1998). Two strains of *Pseudomonas* were isolated from affected larvae and disease was reproduced experimentally by bath transmission. Affected larvae showed dark coloration of the digestive diverticula and lack of ability to digest food with altered phototaxis, weakness and death.

11.7.2.3. Bacterial septicemia

History, occurrence and distribution: Mortalities in mature and immature pearl oysters associated with invasion by marine organisms, particularly marine vibrionic bacteria occur commonly. Such infections are probably under-reported. Infection may be localized or generalized. Infection may resolve or may result in mortalities. Lesions consistent with bacterial septicemia in populations of pearl oyster contra-indicate further stressful procedures, for example, transport or seeding or harvesting operations.

Etiology: Marine vibrionic bacteria are common inhabitants of marine environments and can be readily recovered from healthy oysters. A spectrum of bacteria has been isolated from diseased *P. maxima* during cultural examinations in Western Australia. These include *Vibrio* spp., *V. pelagicus*, *V. mediterranei*, *V. alginolyticus*, *V. anguillarum*, *V. splendidus*, *V. parahaemolyticus*, *Photobacterium* sp. Other organisms recovered from pearl oysters include *Cornebacterium* sp. and *Erwinia hebicola* (Mannion, 1983). Stress, including low salinity, is considered to play a major role in such septicemic infections.

Clinical findings and gross pathology: Acute infections, commonly seen in spat, may be manifest as dead oysters. Sub-acute or chronic disease, typically in mature shell, may be seen as gaping, weakness, loss of muscle tone and retraction of mantle. Segmental mantle retraction may be seen, with localized, often wedge-shaped areas of conchiolin deposition and anomalous mineralization.

Histopathology: Bacterial infection of spat may cause intense hemocytic infiltrations in or adjacent to organs and tissues including palps, intestinal tract and myocardium. Focal or regional infiltrations of hemocytes in the heart may be observed in diseased animals with bacterial septicemia, in which bacilliform bacteria typical of marine vibrios may be visible. The heart appears to be a primary site for bacterial localization in oysters with bacterial septicemia, where abscess or granuloma formation with central necrosis is a frequent finding. The presence of focal hemocytic inflammation in the myocardium in the absence of obvious etiological agents, is typical in bacterial septicemia, especially associated with *Vibrio* spp. Valvular inflammatory lesions are not uncommon, especially in oyster spat with vibriosis.

11.7.2.4. Post-transportation vibriosis in *P. maxima* in Australia

History, occurrence and distribution: Transportation *per se* imposes undefined physiological stresses on mature and immature pearl oysters. Such transportation of live pearl oysters is frequently an integral part of management and husbandry. Infections with *Vibrio* spp. have subsequently been the cause of post-transport mortalities in pearl oyster *P. maxima* in Australia in the 1980s. There were losses of up to 80% of harvested pearl oysters following removal from collecting grounds in Western Australia.

Similarly, in the early 1990s, there were mortalities of up to 90% in *P. maxima* in Torres Strait within 6 weeks following harvest of wild pearl oysters and transportation to farms. Joll (1994) described changes to handling practices which mitigated against stress, low oxygen and low water temperature, and the consequent invasion by vibrionic bacteria.

Etiology: Marine vibrionic bacteria were consistently isolated from the hemolymph of diseased oysters, with the predominant organism being *V. harveyi*, and a lower prevalence of *V. alginolyticus*. The bacterial colonization is considered secondary and related to environmental stressors.

Epidemiology: Infections were associated with poor water circulation and decreased water temperature during post-collection transport (Dybdahl and Pass, 1985; Pass *et al.*, 1987; Lester, 1990). Overcrowding of uncleaned oysters in transportation tanks, with low water circulation, was considered to be the cause of mortality from bacterial infections in the Torres Strait incidents (J. Norton, personal communication).

Pathogenesis: Marine vibrios tend to localize in the heart, inciting hemocytic inflammatory lesions (Pass *et al.*, 1987), an observation explained by Suzuki (1995), who demonstrated that fixed phagocytes of the pearl oyster in the auricle are a primary site of antigen localization, with entrapment of antigen within 6 h. Experimentally, Mannion (1983) investigated the relationship of water temperature to infection and mortalities of *P. maxima* by *V. harveyi* and concluded that lower temperature enhances bacterial disease by suppressing cellular defense mechanisms and promoting proliferation of bacteria, with subsequent invasion of oyster tissues *by V. harveyi* at 19°C, a conclusion reached by Pass *et al.* (1987) in studies of mortalities in transported oysters at 21°C. Pathogenic isolates of *V. harveyi* are toxigenic and invasive (Mannion, 1983).

Clinical findings and gross pathology: Diseased oysters showed gaping, mantle retraction and staining of nacre. Closure response was slow and closure was not complete. Experimentally, the clinical signs and lesions induced by inoculation of *V. harveyi* are the same as observed in natural cases of the disease.

Histopathology: Histologically, multiple focal accumulations of hemocytes were seen in the connective tissue of the digestive gland and mantle. Focal or diffuse infiltration of the heart by hemocytes was also a feature. Erosion of the outer fold and marginal zone of the epithelium of the mantle occurred, with infiltration of the subepithelial tissues by phagocytic hemocytes. Gram negative bacteria were occasionally observed within phagocytes. Atrophy of the digestive gland was a common feature.

11.7.2.5. Post-operative bacterial disease in Akoya pearl oysters

Septicemic bacterial disease associated with Gram negative bacilliform and diplococcoid organisms were recorded following seeding operations in Akoya pearl oysters in Japan and there were unusually high mortalities (Kotake and Miyawaki, 1954). The organisms were recoverable from the sea bed and water at the farm sites. Infection was suspected to be due to disturbance of sediments on moving cages as part of seeding operations, with infection through the incision wounds (Kotake and Miyawaki, 1954).

The disease was reproducible by inoculation of the Gram negative bacillus, leading to rapid death and putrefaction.

11.7.2.6. Pseudomoniasis in *P. maxima*

Pseudomoniasis has been induced experimentally in *Pinctada maxima* in Australia. Mannion (1983) demonstrated that an isolate of *Pseudomonas putrefaciens*, isolated from a diseased *Pinctada maxima* in the laboratory and inoculated into the same species induced a variably fatal septicemia from which the organism could be recovered from hemolymph.

11.7.2.7. Rickettsiosis in Chinese pearl oysters

History, occurrence and distribution: Severe systemic disease associated with Rickettsia-like organisms (RLOs) has been described in farmed Akoya pearl oysters and *P. maxima* from Hainan Island, China in 1993 and 1994 (Wu and Pan, 1999a).

Etiology: There is some debate regarding the validity of the observations and descriptions of RLOs. Wu and Pan (1999b) described the histological and ultrastructural morphology of their RLOs in diseased or moribund *P. maxima* and recorded their presence in epithelial cells, connective tissue and endothelial cells. Hine (2002), however, considered that the apparent high prevalence of infection, at least in *P. maxima*, was due to the misidentification of normal eosinophilic granular epithelial cells as rickettsial inclusions. Nevertheless, it is clear that a mortality event associated with RLOs occurred in Akoya pearl oysters and *P. maxima* and considerable detail exists on the disease syndrome associated with these bodies. The putative role of RLOs in the mortalities, however, appears unresolved. The fine structure of the RLOs is described in detail by Wu and Pan (1999b).

Epidemiology: Studies by Wu *et al.* (2001) failed to demonstrate RLOs in oocytes, fertilized ova or early embryonic phase larvae of *P. maxima*, with RLOs initially occurring in 24 h D-shaped larvae and mortalities occurring approximately 7 days after fertilization. These studies suggested that transmission occurred horizontally, associated with contact infection with the emergent gills and digestive tract, and that trans-ovarian transmission was unlikely. Wu *et al.* (2003) described outbreaks of mortality in juvenile populations of farmed *P. maxima* in which peak mortalities occurred in 4- to 6-months-old oysters and which correlated with severity of infection by RLOs. Infection rates in all populations examined were 100% with mortality rates up to 79.6%. Peak mortalities occurred in oysters 10–30 mm mean body length and mortalities declined in oysters over 50 mm mean body length. Mortalities could not be correlated with water temperature of salinity.

Pathogenesis: Wu and Pan (1999a) believed that the RLOs parasitize the host cytoplasm forming intracytoplasmic inclusions and described two propagative modes leading to cell pathology: transverse binary fission and budding.

Clinical findings and pathology: Wu and Pan (1999a) reported heavy mortalities in infected oysters. Wu and Pan (1999a, 1999b) described RLOs as highly pleomorphic, eosinophilic inclusions in epithelial tissues, connective tissue and endothelium, and described pre-granular and granular stages of the RLOs. These authors reported that the RLOs stained variably negative by the Gram method and were acid fast by the Ziehl-Neelson method. Acute and chronic phases of infection with intracellular infection by RLOs were described by Wu and Pan (1999c). An acute phase characterized by acute inflammatory lesions, with cellular necrosis and lysis involving mantle, gills, alimentary tract, digestive gland, gonad and vasculature. There was a reparative chronic phase with epithelial end endothelial cellular hyperplasia and fibrosis.

Ultrastructure: The RLOs possessed trilamellar membranes characteristic of bacteria, were located either free in the cytoplasm or within lysosomes, and existed in two forms: a ribosome rich small cell variant and a large cell variant (Wu and Pan, 1999a).

A process of nuclear pyknosis, karyohexis and karyolysis, mitochondrial and endo-plasmic reticular swelling and vacuolation responsible for cell degeneration and lysis was associated with the presence of RLOs in affected oysters (Wu and Pan, 1999d).

11.7.2.8. Rickettsiosis in Australian pearl oysters

History, occurrence and distribution: Infections with Rickettsiales-like organisms have been recorded from comprehensive surveys of Australian *Pinctada maxima* in tropical northern Australian waters (Humphrey *et al.*, 1998; Hine and Thorne, 2000). RLOs have also been recorded in *P. albina* and in *Pteria penguin* from Western Australia (Hine and Thorne, 2000). The pathogenic significance of these organisms remains uncertain. Although clinical disease has not been associated with the presence of these agents, an inflammatory cell infiltration may occur, associated with the presence of RLOs in the sub-epithelial tissues and it appears likely that the organisms may have potential to induce sub-clinical disease at least.

Etiology: Two morphologically different forms of RLO have been recorded in *P. maxima* in Western Australia (B. Jones, personal communication): a large form and a small form, varying only in the size of the intracellular inclusion. Whether or not the organisms are genetically related remains to be determined.

Epidemiology: In *P. maxima*, the overall prevalence of infection was found to vary with geographic location from 2.9% in Queensland to 0.3% in Western Australia (Humphrey *et al.*, 1998). Regional differences between populations of oysters were found, with up to 17% of individuals within a population infected. Infection was present in the majority of wild populations, whereas less than 50% of farmed popula-tions were positive (Humphrey *et al.*, 1998). Hine and Thorne (2000) also recorded RLOS in 0.4% of 469 *P. albina* and in 1.1% of 88 *Pteria penguin* from Western Australia.

Pathogenesis: There are few data on the means whereby the organism infects and colonized the tissues. There appears to be a predilection for different tissues. In the survey of a total of 4,502 *P. maxima* by Humphrey *et al.* (1998), the overall prevalence of infection was 0.40%, 0.40% and 1.5% in the gill, palp and digestive gland, respec-tively. In the case of palp infections, no RLOs were detected in the palp epithelium of oysters from the Northern Territory or Western Australia.

Clinical findings and pathology: No disease or gross pathology has been associated with the RLOs.

Histopathology: In *P. maxima*, RLOs have been seen in or immediately below the epithelium of the palps, gills, digestive diverticula and rarely the gonad. They are characterized by finely granular, basophilic intracellular inclusions approximately 20–40µm in diameter occupying or intimately associated with epithelial cells. The inclusions contain numerous small, fine, Gram negative bodies. There is minimal appar-ent tissue damage. On occasions, a mild inflammatory cell infiltration is associated with the presence of the bodies in the sub-epithelial tissues of the palp. Similarly, RLOs in the digestive gland colonize and occupy the epithelial cells of the digestive diverticula or they are intimately associated with these cells. Generally there is no tissue damage or inflammatory response associated with these organisms. Rarely an intense inflammatory cell infiltration and necrosis of epithelium is seen. Smaller RLO inclusions have been recorded in the digestive gland in spat of *P. maxima*. Sporadically, these organisms appear to initiate an intense inflammatory response. RLOs are rarely present in the gonad.

11.7.2.9. Rickettsiosis in *P. margaritifera* from French Polynesian

History, occurrence and distribution: Infections with RLOs were recorded by Comps *et al.* (1998) during routine examinations of *P. margaritifera* from French Polynesia.

Etiology: Ultrastructural examination by Comps *et al.* (1998) showed pleomorphic prokaryotic organisms 500–800 nm in length and 200–300 nm in diameter containing paracrystalline bodies, consistent with Rickettsiales.

Clinical findings and gross pathology: No disease or gross pathology has been associated with the RLOs.

Histopathology: Intracellular microcolonies of RLOs were occasionally observed in the epithelium of the digestive tubules as basophilic inclusions 5–25 μm in diameter. No inflammatory response was recorded; however, fibrosis peripheral to the inclusions, presumed to be a host response to the infection, was noted in some cases.

11.7.3. Protozoa and protozoal diseases

11.7.3.1. Gregarine infections of *P. maxima* in Australia

History, occurrence and distribution: Gregarine protozoa are known to colonize the epithelium of the stomach, intestine and digestive gland in mature *P. maxima* in Queensland. They have rarely been recorded from other pearl farming regions in Australia (Humphrey *et al.*, 1998). No disease is attributed to the organisms and they appear innocuous, causing no apparent damage or host response.

Etiology: Cui (1997) reported gregarine organisms within and between host digestive gland cells and identified two forms of the organism: Type I, a larger form with conspicuous longitudinal folds of the membrane and Type II, a smaller form which lacked folds in its limiting membrane and a showed a typical trilaminal membrane.

Epidemiology: Where present, gregarine protozoa have been reported from both wild and farmed populations at a prevalence of up to 41% (Humphrey *et al.*, 1998).

Clinical findings and histopathology: Clinically, the infection appears innocuous: no clinical signs have been reported. By light microscopy, there is no apparent tissue damage or inflammatory response. The gregarines present as non-ciliated, indented, ovoid organisms, approximately 8–20 μm in length lying on or within the host epithelium. Gregarines are particularly prevalent in the digestive gland with up to 41% of individuals in a population infected (Fig. 11.16). A lower prevalence of infection occurs in the stomach, up to 7%, and proximal intestine, up to 0.3%.

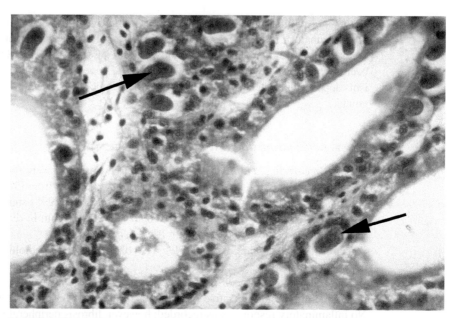

FIGURE 11.16   Gregarine protozoa (arrows) colonizing the epithelium of the digestive gland of *Pinctada maxima*. These organisms appear innocuous, causing no apparent tissue damage or host response. (Photo: John Norton.)

11.7.3.2. Gregarine infections of *P. margaritifera* in French Polynesia

History, occurrence and distribution: In 1990, Chagot *et al.* (1993) undertook pathological examinations on *P. margaritifera* from atolls in French Polynesia that had some 5 years earlier experienced high mortality (see Section 11.7.9.2). They described an elongate parasitic gregarine identical to a gregarine in *P. margaritifera* from the Red Sea, observed by the same authors. Although the parasite was associated with local tissue destruction, no link between the presence of the parasite and the mass mortality event in 1985 could be established and there was no effect on normal growth of the oyster (Chagot *et al.*, 1993).

Etiology: The gregarine was confirmed on ultrastructural examination to be characteristic of the Apicomplexan Eugregarinoida (Chagot *et al.*, 1993).

Epidemiology: Although 100% infection was reported in the examined oysters, a causative role in the earlier high mortality events was not established (Chagot *et al.*, 1993). The organism was reported to be common in a subsequent study of *P. margaritifera* from the same region by Comps *et al.* (2001).

Clinical findings and gross pathology: There appear to be no abnormal signs or gross lesions referable to infection with the organism.

Histopathology: The parasite colonized the distal intestine or rectum and extended into the proximal intestine and digestive gland in severe cases. Although primarily intracellular in the intestinal epithelium, its occurrence in sub-epithelial connective tissue in heavily parasitized individuals was noted. The parasite was associated with local destruction and sloughing of parasitized cells with severe erosion of the rectal mucosa.

11.7.3.3. Perkinsosis

History, occurrence and distribution: Infections of diverse molluscan species by *Perkinsus* spp. have been recorded in Asia, North America and Australia, and *Perkinsus* spp. are common parasites of bivalve molluscs including pearl oysters. Although *Perkinsus* sp. was nominated as a possible cause of Akoya oyster disease in Japan (Muroga, 2002), a causal role was subsequently discounted. Park *et al.* (2001) failed to identify *Perkinsus* in a survey of Akoya pearl oysters in Korea, even though a pathogenic *Perkinsus* sp. was known in Manila clams *Ruditapes phillippinarum* in Korean waters. The only confirmed report of *Perkinsus* as a pathogen of pearl oysters is in *Pinctada maxima* in Australia.

11.7.3.4. Perkinsosis in Australian pearl oysters

History, occurrence and distribution: A *Perkinsus* sp. was isolated from at least two species of pearl oyster, *Pinctada margaritifera* and *Pinctada albina (sugillata)*, and from a diverse range of apparently healthy bivalve mollusc on the Great Barrier Reef, Australia (Goggin and Lester, 1987). *Perkinsus*-like protozoa were described by Norton *et al.* (1993b) in tissues of adult *Pinctada maxima* from a population undergoing a high mortality in Torres Strait, Australia. In Western Australia, Hine and Thorne (2000) recorded a single schizont, considered to be *Perkinsus* sp., at the base of the gills in a single *Pinctada maxima* and they also recorded *Perkinsus* schizonts in *Pinctada albina*. No *Perkinsus* spp. have been recorded in pearl oysters from the Northern Territory.

Pathogenicity: Although affected oysters showed multifocal granulomatous systemic lesions which contained the protozoan, a causative role in the mortalities could not be established Norton *et al.* (1993b).

Epidemiology: Although *Perkinsus* spp. appear to be widespread in bivalve molluscs, heavy infections were recorded only in the Tridacnidae, Spondylidae, Arcidae and Chamidae (Goggin and Lester, 1987), suggesting that pearl oysters may be somewhat refractory to infection. Transmission studies on Queensland *Perkinsus* sp. suggest that the parasite is not host specific. Goggin *et al.* (1989) induced an experimental infection with *Perkinsus* sp. in the pearl oyster *Pinctada sugillata* by exposure to motile zoospores obtained from diverse molluscan species including *Tridacna gigas*, *Anadara trapezia*, *Haliotis scalaris*, *Haliotis laevigata*, *Haliotis cyclobates* and *Chama pacificus*. Conversely, zoospores from *Pinctada sugillata* were infective for other molluscs. Further, the origin of the parasite with respect to locality did not affect its transmissibility and isolates from gastropods were transferable to bivalves and vice versa.

Histopathology: *Perkinsus*-like protozoa were described by Norton *et al.* (1993b) in focal granulomatous lesions in the tissues of adult *Pinctada maxima* from a population undergoing a high mortality in Torres Strait, Australia. *Perkinsus* sch-izonts have been described at the base of the gill in *Pinctada maxima* and in tissues of *Pinctada albina* (Hine and Thorne, 2000).

11.7.3.5. Haplosporidiosis in *Pinctada maxima* in Australia

History, occurrence and distribution: *Haplosporidium* sp. has been described in *Pinctada maxima* from Western Australia in small (5–10 mm), hatchery-reared spat on several occasions (Hine, 1996; Hine and Thorne, 1998). The agent is distinct from *Haplosporidium nelsoni* (Hine, 1996). Infected oysters were considered to have acquired the infection from the local environment as testing prior to release from the hatchery failed to demonstrate the organism. Testing of other spat from the same batch reared elsewhere showed no evidence for infection.

Infected populations were destroyed on detection and thus the natural course of infection is unknown. The high numbers of organisms and the extent of infection sug-gest that clinical disease and death are probable sequelae (B. Jones, personal commu-nication) and the organism is considered to represent a serious potential threat to the pearling industry in Australia.

Etiology: Electron microscope examination of the organism has confirmed that the parasite is an undescribed species of the genus *Haplosporidium* (Hine and Thorne, 1998).

Epidemiology: The prevalence of haplosporidiasis in hatchery-reared populations of *P. maxima* is low, 0.1–5.7% (Hine, 1996; Hine and Thorne, 1998; B. Jones, personal communication). The low and sporadic occurrence of the parasite in *P. maxima* in what is clearly an endemic area suggests other molluscs may be a more suitable host for the parasite. A haplosporidian is associated with mortalities in the rock oyster *Saccostrea cuccullata* in north Western Australia (Hine and Thorne, 2002) and the possibility that the *Haplosporidium* in *P. maxima* is a coincidental infection with same organism can-not be excluded.

Pathogenesis: Hine and Thorne (1998) recorded three stages of the parasite:

1. a bi-nucleate stage containing haplosporosomes,
2. spores containing a sporoplasm,
3. dense, possibly senescent, spores.

Clinical findings and gross pathology: Although no mortalities have been associated with the infection, the heavy infections and tissue pathology strongly suggest that spat would not have survived.

Histopathology: Infected spat show sporulation and pre-sporulation stages replac-ing connective tissue (Leydig cells) peripheral to the digestive diverticula (Fig. 11.17). Light or moderate infections were also observed in the stromal tissues of the heart, foot, gills, adductor muscle and mantle: parasites were never observed in epithelial cells. Although granulocytic hemocytes were associated with the parasites, there was no evi-dence of phagocytosis. Mature spores have a yellow, refractile operculate wall enclos-ing an eosinophilic sporoplasm (Hine and Thorne, 1998). In some cases, the almost

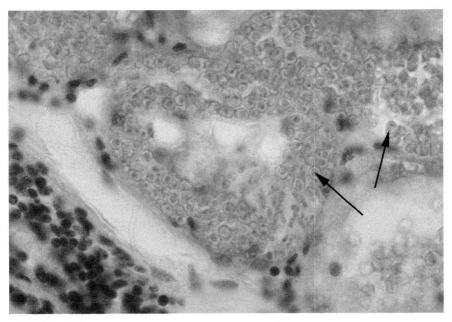

FIGURE 11.17   Intense infection of stromal tissues of spat of *P. maxima* by *Haplosporidium* sp. (arrows). (Photo: Brian Jones.)

total replacement of digestive gland stroma with spores in all stages from immature to mature spores strongly suggested the spat would not have survived the infection.

11.7.3.6. Rhynchodid-like ciliates in *Pinctada maxima* in Australia

History, occurrence and distribution: *Rhynchodid*-like ciliated have been reported by Jones and Creeper (2006) inhabiting the epithelium of the digestive gland of juvenile *P. maxima* from Western Australia, in oysters situated near the southern limits of their distribution.

Clinical findings and gross pathology: On occasions, infection is associated with an intense hemocytic response; however, they have not been associated with clinical disease (B. Jones, personal communication). The prevalence of infection is higher in smaller oysters, 20–50 mm.

Histopathology: The ciliate typically occupies a luminal or intraepithelial location in the digestive gland but may penetrate the basal lamina to enter the hemolymph sinuses or interstitial tissue. Commonly, infection is associated with an intense hemocytic response.

11.7.3.7. Ancistrocoma-like ciliates in Australian pearl oysters

History, occurrence and distribution: *Ancistrocoma*-like ciliates have been recorded from the intestine of mature and immature *P. maxima* from tropical Australian waters

(Humphrey *et al.*, 1998, Hine and Thorne, 2000). Hine and Thorne (2000) also recorded Ancistrocomid ciliates in 1.5% of 469 *P. albina* in Western Australia.

Etiology: The ciliates morphologically conform to the family Ancistrocomidae; however, there is no information regarding their classification to genus or species level. It is unknown if the ancistrocomids described comprise one or more species.

Epidemiology: *Ancistrocoma*-like ciliates were recorded at a high prevalence in *P. maxima* from Queensland (18.6%) and at lower prevalence in the Northern Territory (0.1%) and Western Australia (2.2%) (Humphrey *et al.*, 1998; Hine and Thorne, 2000).

Clinical findings and histopathology: No disease has been associated with the presence of the organism. Elongate, *Ancistrocoma*-like ciliated protozoa, approximately 25–30 µm in length, may be common in the intestine of mature *P. maxima*. These organisms are non-invasive and are readily seen free in the lumen or associated with the superficial epithelial border. There is no associated inflammatory response or tissue damage. *Ancistrocoma*-like ciliates should be distinguished from those saprophytic ciliated protozoa that commonly invade moribund or dead oysters.

11.7.3.8. Uncharacterized ciliates in Australian *Pinctada maxima*

Gut ciliates: Jones and Creeper (2006) record the presence of ciliates in the gut of *P. maxima* from Western Australia. No association with disease was made.

Apicomplexans: An uncharacterized Apicomplexan parasite has been reported on two occasions in the heart of *P. maxima* in Australia (B. Jones, personal communication).

Coccidians: A coccidian oocyst was recorded in the heart of a single *P. maxima* in Western Australia by Hine and Thorne (2000).

Cryptosporidium – like organisms: Multiple, small basophilic bodies *ca.* 2–3 µm in size and resembling *Cryptosporidium*, have been recorded on one occasion in or closely associated with the epithelium of the digestive gland diverticula of a single mature oyster from the Northern Territory in the absence of disease (Humphrey *et al.*, 1998). Cytoplasmic enlargement and foamy degeneration of the epithelium were associated with the bodies. Comps *et al.* (2001) reported an earlier occurrence of a protistan parasite in *P. margaritifera*, in French Polynesia, characterized by ovoid sporozoan spores encysted in the mantle. The pathogenic significance is unknown.

11.7.3.9. Uncharacterized Protistan parasites

History, occurrence and distribution: Comps *et al.* (2001) reported an occurrence of a protistan parasite in *P. margaritifera*, characterized by ovoid sporozoan spores encysted in the mantle. The pathogenic significance of this organism is unknown.

Jones and Creeper (2006) reported a single *P. maxima* from Western Australia containing an enigmatic, unidentified protistan parasite, the sporoblasts of which developed within the epithelium of the digestive tubules, filling the lumen of the tubules. An associated hemocytic response was present. Jones and Creeper (2006) also described and elongate proctistan in *P. maxima*, approximately 30 × 20 µm, adhering to the surface of digestive gland epithelial cells by several attachments. This proctistan was thought to have induced multinucleation in underlying epithelial cells.

A protistan tentatively identified as a member of the Thraustochytridea was reported by Jones and Creeper (2006) in a moribund *P. maxima* previously exposed to a cyanobacterial bloom with *Trichodesmium*. Affected oysters showed necrosis of the epithelium of the palps and mantle and invasion of the tissues by segmented, unicellular organisms 10–15 μm in diameter and smaller (5 μm) basophilic cells.

11.7.4. Nematodes and nematode diseases

11.7.4.1. Nematodiasis

History, occurrence and distribution: Nematodes parasitizing pearl oysters are rarely recorded and do not appear to be a significant cause of disease. Shipley and Hornell (1904, 1905) described the larval nematode *Ascaris meleagrina* encysted in the gonad or mantle of Akoya pearl oysters from the Gulf of Mannar. Thirty of the 534 oysters examined (5.6%) were infected. They noted that file fishes, *Balistes* spp. harbor the adult stage in the intestine. Similarly, Shipley and Hornell (1904, 1905) described the larval nematode *Cheiracanthus uncinatus* encysted in the adductor muscle in 41 of 542 oysters examined (7.6%). They also noted that file fishes, *Balistes* spp., harbor the adult stage in the intestine. These authors also described an *Oxyurus* sp. in the intestine of Akoya pearl oysters in only two occasions.

11.7.5. Trematodes and trematode diseases

As early as the 17th century, platyhelminth parasites were described in pearl oysters and were attributed with pearl formation (Shipley and Hornell, 1904). The association between parasitic invasion of pearl oyster tissues by platyhelminths and the formation of pearls was further recognized in 1857 when Dr. Kelaart and Mr. Humbert studied pearl oysters from Sri Lanka and described "parasitical worms infesting the viscera and other parts of the pearl oyster" and "that these worms play an important part in the formation of pearls" (Shipley and Hornell, 1904). The role of platyhelminths was subsequently confirmed by Thurston who in 1894 found "larvae of some platyhelminthian (flat worm)" in the tissues and alimentary tract of the pearl oyster. Southwell (1912) in reviewing the role of platyhelminths in pearl formation concluded that the metacestodes described earlier by Shipley and Hornell (1904) initiated pearl formation upon encystment and death within the tissues of the oyster. Southwell (1912) also described studies to elucidate the life cycle of these metacestodes.

11.7.5.1. Trematodiasis in Akoya pearl oysters

Shipley and Hornell (1904) reported three immature trematodes inhabiting the tissues of Akoya pearl oysters from the Gulf of Mannar: *Mutta margaritifera*, *Musalia herdmani* and *Aspidogaster margaritifera*. They described their morphology in detail.

Shipley and Hornell (1906) described the collection of *A. margaritifera* from Akoya pearl oysters in Sri Lanka. No disease was recorded in association with the parasites.

11.7.5.2. Bucephaliasis in Akoya pearl oysters

History, occurrence and distribution: Infection of pearl oysters by the bucephalid trematode *Bucephalus margaritae* is recorded in Akoya pearl oysters as a parasite causing seriously adverse affects on pearl development. It invades the gonad and digestive gland, and further invades the mantle, gills, palps and adductor muscle in advanced cases (Ozaki and Ishibashi, 1934). Wada (1991) described the bucephalid *Bucephalus varicus* in Japan, also in Akoya pearl oysters, and observed that oysters infected by *Bucephalus* are unsuitable for pearl production. Khamdan (1998) reported a *Bucephalus* sp. in the gonads of Akoya pearl oysters in the Persian Gulf, but ascribed no pathogenic significance to it.

Epidemiology: Wada (1991) considered Akoya pearl oysters to be the first intermediate host of the larval cercaria and found up to 40% of oysters infected with *B. varicus*. Khamdan (1998) reported a maximum incidence of 4.5% in 69–74 mm *P. radiata* with no apparent seasonal occurrence.

11.7.5.3. *Proctoeces* sp. in Akoya pearl oysters

Sakaguchi *et al.* (1970a) described trematode metacercaria of the genus *Proctoeces* parasitic in the heart, gonad, intestine and digestive gland of Akoya pearl oysters in Japan. Both farmed and natural oysters were infected, with a higher prevalence and higher numbers of parasites in natural oysters. Experimentally, Sakaguchi *et al.* (1970b) determined that the black porgy *Mylio macrocephalus*, a recognized predator of pearl oysters, was a final host for the parasite. Sakaguchi (1972) further identified the parasite as *P. ostreae*, noting that *M. macrocephalus* was found on pearl oyster farms.

11.7.5.4. Uncharacterized trematodes

Mizumoto (1979) identified invasion by trematode worms occasionally causing serious damage to pearl oysters in Japan, however, little detail as to the species and specific pathology is given. In India, CMFRI (1991) noted that three species of trematodes have been isolated from the tissues of pearl oysters.

11.7.6. Cestodes and cestode diseases

11.7.6.1. Metacestodiasis

History, occurrence and distribution: Larval cestodes have been long recognized as parasites of pearl oysters. Herdman (1903) associated a larval cestode, *Tetrarhynchus* with pearl formation. Shipley and Hornell (1906) described *Tetrarhynchus unionifactor* invading the tissues of Akoya pearl oysters. Seurat (1906) identified metacestode cysts, designated *Tylocephalum margaritifer,* in *Pinctada margaritifera* from French Polynesia and Shipley and Hornell (1906) described metacestodes in Akoya pearl oysters from Sri Lanka. Postulated as the invasion site of natural pearl formation (Herdman, 1903; Seurat, 1906; Shipley and Hornell, 1906; Sparks, 1985), it was subsequently claimed that invasion by larval trematodes was more important in natural pearl formation than larval cestodes (Sindermann, 1990), although data to substantiate these claims appear to be lacking.

Shipley and Hornell (1904) described in detail the morphology and development of cestode larvae in Akoya pearl oysters from the Gulf of Mannar, to which, at least in more developed stages, the identity *T. unionifactor* was assigned. Infections of pearl oysters by larval cestodes of the genus *Tylocephalum* have been recorded from diverse locations in diverse species. Metacestodes recorded in pearl oysters are shown in Table 11.1. In high numbers, the periesophageal localization and induction of inflammatory granulomas by the metacestodes, especially in smaller spat, may impede passage of food. In spat, parasitic invasion may be accompanied by bacterial infection. In mature oysters, the parasites and associated granuloma are generally considered an incidental finding of little consequence to the health of the oyster. The intense inflammatory reaction to the parasites suggests that *P. maxima* may not be a particularly suitable intermediate host for the cestode parasite.

Etiology: Larval cestodes are well recognized in bivalve and gastropod molluscs. The parasites are generally considered to be lecanicephalid metacestodes of the genera *Tylocephalum* or *Polypocephalus*, although other species may be involved. Although commonly referred to as metacestodes of *Tylocephalum*, at least in histological section, it is perhaps more correct to refer them as lecanicephalid metacestodes consistent with the genera *Tylocephalum* and *Polypocephalus*.

TABLE 11.1

Occurrence and distribution of metacestodes in pearl oysters.

Host	Parasite	Location	Author(s)
Pinctada margaritifera	*Tylocephalum*	Red Sea	Nasr (1982)
		French Polynesia	Seurat (1906); Comps *et al.* (2001)
		Sri Lanka	Southwell (1912)
Akoya pearl oysters	*Tylocephalum*	Red Sea	Herdman (1903); Crossland (1957)
	Tetrarhynchus unionifactor	Sri Lanka	Shipley and Hornell (1905, 1906)
	Cestode larvae	Sri Lanka	Shipley and Hornell (1905)
	Metacestodes	Sri Lanka	Shipley and Hornell (1906)
P. maxima	*Tylocephalum*	Australia	Humphrey *et al.* (1998); Hine and Thorne (2000)
P. albina	*Tylocephalum*	Australia	Hine and Thorne (2000)
Pteria penguin	*Tylocephalum*	Australia	Hine and Thorne (2000)

Epidemiology: Infection with *Tylocephalum* metacestodes is common in marine molluscs including pearl oysters. Larval *Tylocephalum*, which are not considered to be host specific, may occur at high prevalence and intensities in pearl oysters and edible oysters (Sindermann, 1990). The adult stages of these lecanicephalid tapeworms reside in the intestine of sharks or rays. Mature eggs pass from the shark or ray host in its feces and hatch to form a motile coracidium. The coracidium invades the tissues of the intermediate molluscan host. In suitable intermediate hosts, the coracidium develops into a metacestode which, if eaten by the shark or ray definitive host, develops into a mature tapeworm, completing the life cycle. Seurat (1906) appears to be the first to ascribe the life cycle of *Tylocephalum* in pearl oysters and the presence of the definitive host, the ray *Rhinoptera quadrilobata*. A high prevalence of infection may occur in the population, especially if spat have been reared on the sea floor and exposed to fecal contamination from sharks or rays.

Systematic surveys of pearl oysters in Australia have shown the parasite to be relatively common and geographically widespread. In Western Australia, Hine and Thorne (2000) recorded *Tylocephalum* metacestodes in 0.9% of 4,670 hatchery-reared *Pinctada maxima*; in 14.8% of 88 *Pteria penguin;* and in 12.2% of 469 *Pinctada albina*, with up to 51.7% of individual populations infected. Humphrey *et al.* (1998) recorded metazoa consistent with lecanicephalid metacestodes in eight populations of *Pinctada maxima*. *Tetrarhynchus* sp. has been reported from Akoya pearl oysters in Sri Lanka and the same species was recorded by Crossland (1957) in *P. margaritifera* in the Red Sea.

Pathogenesis: Although infections of mantle and gill occur, the palps, esophagus and stomach are the more common sites of invasion and localization of parasitic metacestodes, suggesting that the coracidium is drawn into alimentary tract via ciliary movements. Clearly viable metacestodes have been commonly observed within the sub-epithelial tissues of the host. They have on occasions been reported free in the lumen of the alimentary tract (Shipley and Hornell, 1906). It is uncertain how long infection persists. In spat, the intense inflammatory reaction to the parasites suggests that *P. maxima* may not be a particularly suitable intermediate host.

Clinical findings and gross pathology: Although infections have been commonly seen in the absence of clinical signs of gross lesions, Sindermann (1990) reported that heavy infections may reduce the condition of their molluscan host.

Histopathology: Lecanicephalid metacestodes have been recorded in the stromal tissues supporting the gills, palps, mantle, esophagus and stomach (Nasr, 1982; Humphrey *et al.*, 1998). Generally, the parasite evokes an intense inflammatory response seen as focal or multifocal granulomas with a central parasite. Older lesions may show marked fibrosis. Single or multiple focal granulomas may occur, each containing a single parasite. Not uncommonly, multiple granulomas subtend the esophageal epithelium (Fig. 11.18) and may localize in the stromal tissue near the junction of the esophagus and the stomach especially in smaller spat. Infection may be accompanied by bacterial invasion of tissues and an associated cellular inflammatory response. Commonly, focal granulomas have been seen identical to those induced by metacestodes, but which on serial section are shown not to contain parasites. The genesis of

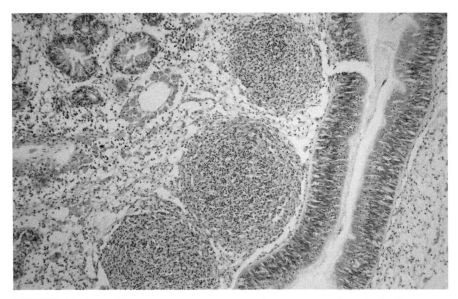

FIGURE 11.18   Multiple metacestode granulomas subtending the esophageal epithelium in spat of *Pinctada maxima.*

these granulomas is uncertain. Possibly they represent a host response to a parasite that has died or has migrated elsewhere.

11.7.6.2. Tetrabothriate cestodes

Hine and Thorne (2000) recorded immature tetrabothriate cestodes in the intestine of 6.8% of 88 *Pteria penguin* from Western Australia. The pathogenic significance of these cestodes is unknown.

11.7.7. Crustacea and crustacean diseases

11.7.7.1. *Anthessius pinctadae*

The copepod *Anthessius pinctadae* has been reported at high prevalence and intensity in the mouth and esophageal lumen in certain populations of mature *P. maxima* from the Northern Territory (Humphrey *et al.*, 1998). Occlusion of the lumen by large numbers of parasites and epithelial erosion and ulceration associated with feeding of the parasite have been observed in some animals.

11.7.7.2. Pinnotherid crabs

History, occurrence and distribution: Pea crabs, *Pinnotheres* spp, are common inhabitants of the mantle cavity of bivalve molluscs. They "steal" food from the host's filter

FIGURE 11.19   Pea crab *Pinnotheres* sp. (arrow) inhabiting a crateriform cavity in the mantle of *Pinctada maxima.*

feeding mechanism on the gills. Pea crabs identified as *Pinnotheres villosulus* are commonly found in the mantle of *Pinctada maxima* in Australia (Dix, 1973b; Humphrey *et al.*, 1998). Pinnotherids do not appear to have been recorded in pearl oysters elsewhere, which is surprising.

Clinical findings and gross pathology: The crab induces and resides in crateriform cavities with peripheral thickening in the mantle tissue without any apparent harm to the host (Fig. 11.19).

Histopathology: The areas of mantle inhabited by the crab show an edematous stroma with fibrosis and displacement and re-arrangement of pallial muscle fibers. There is no evidence of cellular inflammation.

11.7.7.3. Pontoniine shrimp

Pontoniine shrimps of the genus *Conchodytes* appear to be common commensal associates of bivalves. In Australia, *Conchodytes meleagrinae* was described from *P. margaritifera* by Bruce (1977). Subsequently Bruce (1989) described *Conchodytes maculatus* as a bivalve associate of *P. maxima* from the Northwest Shelf. Shrimps *C. maculatus* were found to be common inhabitants of the mantle cavity of *P. maxima* from the Northern Territory and *Conchodytes* spp were found in a population of *P. maxima* from Queensland. Shrimps commonly occurred in pairs at a prevalence of up to 72% (Humphrey *et al.*, 1998). *C. meleagrinae* was typically found as one small male and one large female in the mantle of *P. margaritifera* in the Red Sea at a prevalence of approximately 50% (Nasr, 1982). Doroudi (1996) considered *C. meleagrinae* as a fouling organism of *Pinctada* sp. in the Persian Gulf, Iran. A shrimp identified as *C. tridacnae* was recorded from *P. maxima* from the Palau Islands (Miyake and Fijino, 1968). Pontoniine shrimps also steal food from the host's filter feeding mechanism on the gills, but, unlike pinnotherids, do not cause pathological conditions.

11.7.8. Neoplasia

11.7.8.1. Polypoid mesenchymal tumors in *Pinctada margaritifera*

History, occurrence and distribution: Dix (1972) described two polypous mesenchymal tumors as an incidental finding in a single mature, wild-harvested *P. margaritifera* from Queensland, Australia in a survey of 131 oysters from six reefs.

Clinical findings and gross pathology: Both tumors were firm and attached to the visceral mass by a polypoid flexible stalk. One of the tumors was irregular wrinkled and lightly pigmented, approximately 1.3 cm. The other was a bi-lobed, smooth and poorly pigmented, approximately 1.0 cm.

Histopathology: Both tumors showed similar histological features, characterized by a heavily vascularized proliferative mass of stromal collagenous matrix containing isolated muscle fibers covered by a normal epithelium.

11.7.8.2. Neurofibroma in *Pinctada maxima*

Two lesions consistent with neurofibroma were recorded in a comprehensive histopathological survey of mature *P. maxima* in Australia (Humphrey *et al.*, 1998).

11.7.9. Diseases of uncertain etiology

Diseases and associated mass mortalities of pearl oysters of uncertain have been reported from numerous geographic locations. In some cases, these events have been investigated in detail, however, others remain poorly defined and understood.

11.7.9.1. Akoya oyster mortality syndrome in Japan

History, occurrence and distribution: A syndrome characterized by seasonally recurrent mass mortalities in farmed Akoya pearl oysters in western Japan was first recognized in 1994 and continued unabated over several subsequent years (Muroga, 1999, 2002). Miyazaki *et al.* (1999) isolated a virus from affected oysters and experimentally reproduced the disease, providing convincing evidence for a viral etiology (see Section 11.7.1.4). Other researchers, however, failed to demonstrate a viral etiology and suggested alternative aetiologies. Disparate views on the cause remain. The possibility of a disease syndrome with multiple aetiologies should not be excluded, especially given differences in histopathology reported by different authors.

Etiology: Some authors consider non-infectious factors to be the cause. Hirano *et al.* (2002) could not transmit the disease by cohabitation. Hirano *et al.* (2002, 2005) suggested the syndrome was associated with increased water temperatures, increased oxygen demand, the proximity of fish farms and pollution resulting in decreased feeding activity and debilitation. The disease showed a marked seasonality, occurring during the warmer months of summer and autumn (Morizane *et al.*, 2001; Hirano *et al.*, 2005). Hirano *et al.* (2005) reported a seasonally recurrent occurrence of the disease over three consecutive years, with normal growth between

December–May/June preceding a marked growth reduction and the onset of mortalities in July extending to December, with a peak in mortality in September. Cumulative mortalities varied from 56% in 2000 to 79% in 1998 (Hirano *et al.*, 2005). Mortalities decreased with decreases in water temperature (Morizane *et al.*, 2001). Sugishita *et al.* (2005) associated ingestion of lipid peroxides in oxidized feedstuffs, arising from pollution from fish farms, with the disease and experimentally induced lesions similar to those described in Section 11.7.1.1. Other suggested aetiologies included blooms of the toxic dinoflagellate *H. circularisquama,* perkinsosis and birnavirus infection, but these were subsequently discounted (Muroga *et al.*, 1999; Muroga, 2002). Tomaru *et al.* (2001) associated increased numbers of the inedible *Nitzschia* spp. with the mortality events and suggested that oysters were predisposed to disease through starvation. It appears likely that an infectious agent is involved, which may or may not be a virus, and which may be influenced by high water temperatures and/or other environmental stressors.

11.7.9.2. Mass mortalities of *Pinctada margaritifera* in French Polynesia

History, occurrence and distribution: There were mass mortalities in farmed and wild populations of *P. margaritifera* in 1985 and 1986 from a number of atoll lagoons in the Tuamotu Archipelago, French Polynesia. Mortalities were recorded in mature and juvenile oysters, and in spat (Chagot *et al.*, 1993; Remoissenet, 1995). The disease resulted in severe socio-economic losses in the region and was named Syndrome 85 by Comps *et al.* (2001), who conducted a detailed retrospective investigation into the disease. Changes to management practices, including replacement of galvanized platforms by rope lines, decreased stocking densities and cultivation in deeper water appeared to be successful in overcoming the disease syndrome.

Etiology: Despite considerable efforts to identify a cause, the etiology remains uncertain (Le Moullac *et al.*, 2003). Various investigations have attributed the mortalities to a wide variety of factors. Bernadac *et al.* (1980) considered the decreased numbers of oysters in this group to be due to over-exploitation and not to algal blooms as was initially suggested by some workers. Cabral (1989a) considered the etiology to be a complex of uncharacterized factors. Chagot *et al.* (1993) reported a sporozoan gregarine parasite from affected animals, although a causal relationship appears unlikely. It was not considered a primary cause of the disease. Remoissenet (1995) in reviewing the mass mortalities presented evidence suggesting that zinc eluted from galvanized platforms may have had a causative role, a hypothesis supported by DeNardi (1989) (cited by Remoissenet, 1995) who identified high levels of zinc in shellfish tissues. Vacelet *et al.* (1996) studied the nutrient concentrations of the aquatic environment of Takapoto Lagoon and found both nitrogen and phosphorus to be limiting. These authors, together with Cabral (1989b) considered that the biomass of the pearl oysters appeared to exceed the nutritional potential of the lagoon, inducing oligotrophic conditions under which phytoplankton and even bacteria were unable to sustain their nutrition, suggesting that nutritional inadequacy was associated with the disease in *P. margaritifera*. Virus-like particles were described by Comps *et al.* (1999) in *P. margaritifera* collected from several atolls in French Polynesia in the mid-late 1990s.

These atoll lagoon populations had earlier undergone mass mortality events; however, a primary microbial or parasitic etiology was never established. A virus was alleged to be the cause, but no such agent was isolated or visualized (Comps *et al.*, 2001). They hypothesized that environmental parameters or stress and wounds occasioned by handling and grafting were the possible causes of the mortalities.

Epidemiology: The mortalities were initially reported in 1985 in the Gambier group of islands and subsequently on Takapoto Atoll, Tuamotu Group. Losses of 50–80% in adult, juvenile and spat were recorded (Chagot *et al.*, 1993; Remoissenet, 1995; Comps *et al.*, 2001). The disease appeared to spread in cultured *P. margaritifera* in the Tuamotu Group following transfer of stocks among atolls (SPC, 1985, 1988; Cabral, 1989a, b; Braley, 1991; Eldredge, 1994). Braley (1991) and Cabral (1989b) also reported the disease in other species of bivalve molluscs in the Tuamotu Group during the same period. Signs referable to the 1985 outbreak were still present, albeit at a reduced prevalence, in 1996–1997 (Comps *et al.*, 2001).

Pathogenesis: Paraspherical or polygonal, membrane bound virus-like particles 40 nm in diameter were reported by Comps *et al.* (1999; 2001) in what were described as highly damaged cells from the granulomatous tissue. Although the exact nature and significance of these particles remains uncertain, Comps *et al.* (2001) speculated on similarities to the virus identified by Miyazaki *et al.* (1999) as the cause of Akoya oyster viral disease.

Clinical findings and gross pathology: Affected oysters in the initial outbreaks of disease showed excessive mucus secretion, mantle retraction, localized growth checks and abnormal biomineralization. Marin and Dauphin (1991) reported microstructural and biochemical alterations of the shell associated with changes in the amino acid content of soluble and insoluble organic matrices of the nacreous layer. Affected oysters typically exhibited "shell disease," irregular brownish deposits of unmineralized conchiolin on the nacreous surface of the shell valves (Marin and Dauphin, 1991; Remoissenet, 1995; Cuif and Dauphin, 1996). In addition to the shell disease, Comps *et al.* (2001) subsequently reported weakness, excess mucus production, lesions of the mantle and necrosis of the adductor muscle in oysters considered to be exhibiting similar signs to those in the initial outbreak.

Histopathology: Abscessation of the adductor muscle showed focal myonecrosis, central granulomatous inflammatory cell accumulation and necrosis, and a peripheral hemocytic response. Dense fibrous tissue was seen in some cases, suggesting healing. Comps *et al.* (2001) demonstrated phagocytosis of damaged muscle fibers, muscle tissue regeneration and fibrosis.

11.7.9.3. Mass mortalities of *Pinctada margaritifera* in the Red Sea

History, occurrence and distribution: Recurrent mass mortality events affecting *P. margaritifera* in Dongonab Bay, Red Sea in 1969, 1973 and 1993 have been recorded (Nasr, 1982; Suliman, 1995). Mortalities were confined to *P. margaritifera*. At least in the 1973 event, mortalities were restricted to areas at Dongonab Village and did not occur in the nursery area (Nasr, 1982).

Etiology: The cause of the 1969 mortality event was never determined. Although Nasr (1982) suggested an etiological role for the ovoid bodies described in the digestive gland, there is no clear evidence that they were not in fact residual bodies and a component of the normal cell. Starvation, overcrowding and heavy metal toxicities were excluded by Nasr (1982) as the cause of the mortality event, at least in the 1973 occasion. Thus the etiology has never been determined.

Clinical findings and gross pathology: Oysters affected in the 1973 event showed secretion of copious quantities of mucus, which completely covered the gills. Sudden closure of valves was reported with expulsion of mucus containing fine sediment and subsequently, a slow response to sensory stimulation. Engorgement of ventricles with hemolymph, swelling of the rectum, absence of the crystalline style and absence of food in the stomach were noted (Nasr, 1982).

Histopathology: Nasr (1982) reported the presence eosinophilic granule cells surrounding hemolymph vessels and the presence of putative "brown cells" in the digestive gland. Nasr (1982) also recorded the presence of spherical bodies similar to those described by Wolf and Sprague (1978) in the digestive gland of affected oysters, later considered by Pass and Perkins (1985) to be residual bodies, lysosomes-bound storage or secretory products of cellular metabolism.

11.7.9.4. Oyster edema disease in *Pinctada maxima* from Australia

History, occurrence and distribution: A disease syndrome characterized by mass mortalities in farmed pearl oysters *P. maxima* occurred in Exmouth Gulf, Western Australia, in October 2006. It is believed that at least 2.8 million oysters died, representing considerable economic loss and industry disruption. Approximately 60% of the recently seeded oysters died along with mortality rates up to 90% or higher in smaller oysters, with all class sizes affected. The disease spread to all lease sites in the Exmouth Gulf. The disease syndrome has been named oyster edema disease (OOD).

Etiology: Based on transmission studies, epidemiological data and electron microscopic examination of diseased oysters in which viral-like particles were demonstrated, a hypothesis has been developed that OOD is due to a transmissible, filterable agent, most likely to be a virus. No infectious agent has, however, been unequivocally demonstrated as the cause of the syndrome.

Epidemiology: Limited transmission and cohabitation studies are reported to have shown spread of disease to healthy oysters, suggesting an infectious cause, and a preliminary epidemiological study determined that an infectious agent is the probable cause of the syndrome, with apparent temporal spread from affected to unaffected farms due to transport of live oysters, fomites (e.g. culture equipment) or water movements. It appears that only *P. maxima* are affected and no disease has been recorded in other shellfish species, including other species of *Pinctada*.

In contrast, oysters with clinical OOD when translocated to areas outside Exmouth Gulf were found to recover, and healthy populations of *P. maxima* exposed to affected oysters under farm conditions have failed to develop the disease.

Clinical findings and gross pathology: OOD is characterized by mantle retraction, generalized tissue edema, reduced adductor muscle function with weak or delayed closure and high mortality. Highest mortalities were reported in smaller spat, 40–50 mm, although mortalities occurred in all age classes of oyster.

Histopathology: Histological changes described include mild inflammatory change, tissue edema and edematous separation of epithelial tissues from underlying stroma. These changes are not pathognomonic and can readily be confused with artifactual changes.

Diagnosis: Currently, there are no definitive diagnostic tests for OOD. Identification is based on a case definition of high mortalities with affected oysters showing edematous changes in the tissues, which are acknowledged as subjective.

11.7.9.5. Mass mortalities of *Pinctada maxima* in Australia

History, occurrence and distribution: Mass mortalities of larval, juvenile and mature *P. maxima* from diverse Australian pearl growing regions are well recognized, but the cause or causes of such losses have not been defined in the majority of cases. These mortalities are of considerable concern. Wolf and Sprague (1978) reported mass mortalities in some commercial pearl farms in the years 1968–1978, yet the cause remained unknown. Large numbers of loose shells of *P. maxima* have been reported from several oyster banks off northern Australia, indicating mass mortalities in wild oysters (BRS, 1987).

Recently, high mortalities at a farm in Northern Territory were reported to have a seasonal basis, possibly associated with decreased salinity and decreased water temperature. Unpublished accounts of massive mortalities in juvenile stock and occasional mass mortalities in larval stock have also been reported. Again, the cause or causes are largely undefined, but they are likely to involve both infectious and environmental/ husbandry factors. While these mass mortality events of undetermined or uncertain cause occur regularly in wild and farmed *P. maxima* in Australia, scant information is available on which to pursue an etiology and it is likely that the cause or causes may be long gone by the time the mortality event is investigated.

11.7.9.6. Mass mortality events of *Pinctada maxima* in Torres Strait, Australia

There were mass mortalities attributed to overcrowding in farms in Torres Strait in 1966 and 1968 (Pyne, 1972). In 1969, high mortalities were attributed to domestic sewage and chemical pollution that occurred adjacent to Thursday Island (Pyne, 1972).

Widespread mass mortalities occurred following the grounding of the oil tanker, *Ocean Grandeur*, on March 3, 1970 in Torres Strait. Mortalities of up to 80% were recorded in pearl oysters following transport to farms in the 1970 season. The oil spill and use of a non-biodegradable detergent were seen as significant sources of the mortalities by some observers (Yamashita, 1986), but other investigations suggested an uncharacterized infectious agent as the cause (Pyne, 1972).

Norton *et al.* (1993b) investigated a mortality event involving 85% of adult oysters in a farm at Torres Strait and identified a *Perkinsus*-like organism in affected oysters. Subsequent data indicated that mortalities continued on some farms. There was no clear evidence that *Perkinsus* was the cause and it was considered that the losses were caused by marine Vibrionic bacteria and deficiencies in hygiene, handling and transport procedures (J. Norton, personal communication).

11.7.9.7. Mass mortality in *P. maxima*, Cobourg Peninsula, Australia

History, occurrence and distribution: In 2002, a mortality event occurred in farmed *P. maxima* at Raffles Bay, Northern Territory, Australia. High mortalities were incurred in large spat and mature oysters. Oysters in adjacent embayments were not affected.

Etiology: The mortality event was strongly associated with a concurrent bloom of the diatom *Rhizosolenia*, which was identified at high abundance in the water column.

Clinical findings and gross pathology: A high proportion of oysters showed moderate to severe mantle retraction, discoloration of nacre and severe wasting.

Histopathology: Affected oysters showed degenerative changes of the digestive gland, characterized by localized or regional dilation of digestive gland diverticula and flattening of the epithelium; moderate to severe, generalized vacuolation; degeneration and lysis of the epithelium with a cellular effusion into the lumen; and breakdown of glandular architecture with the formation of dilated, irregular cystic cavities within the gland. Characteristically, moderate numbers of ovoid golden bodies were present within the epithelial cells or free in the lumen of the digestive diverticula, extending into the collecting ducts, stomach and intestine. These bodies stained strongly with PAS reagent. There was no cellular inflammatory change in the gland and there was no evidence of an infectious or parasitic causative agent or agents. The ovoid golden-brown bodies were 4–6 µm in diameter, appeared encapsulated and stained strongly with PAS. Although the identity of these bodies remained uncertain, they resembled hypertrophied "residual bodies," bodies resulting from lysosomal aggregation in degenerate cells.

Electron microscopy: Electron microscopic examination of samples of diseased shell failed to identify viral particles in the cells of the digestive diverticula. The ovoid golden-brown bodies present appeared to have no internal structure.

11.7.9.8. Mass mortalities of *Pinctada margaritifera* in the Cook Islands

An isolated report (SPC, 2002a) described a disease outbreak causing the deaths of young oysters in 2000. A "natural phenomenon" associated with mass spawning of shellfish, an algal bloom and murky conditions were associated with deaths (SPC, 2002b).

11.7.9.9. Miscellaneous mortality events of *Pinctada maxima*

Heavy losses of juvenile grow-out spat from pearl oyster hatcheries have been experienced on pearl farms in northern Australia. Causes of losses include the escape of spat from mesh baskets, the clumping of juvenile spat within the baskets causing secondary starvation, and reduced water flow into the baskets from the use of very small mesh that become fouled. Survivors from one batch of spat in Queensland had a papova virus-like infection of the gills and non-specific enteritis (J. Norton, unpublished data). Tlili (1995) described a mortality event of sudden onset, initially occurring in 3-month old farmed spat progressively occurring in 12-month old oysters in Bynoe Harbour, Northern Territory. Mortalities of 90% occurred within 1 week of spat entering the farm, however, mortalities in 12-month-shell were 3–4%. Mortalities appeared to be restricted to the farm and did not spread to adjacent farms. Although a *Vibrio* sp. was isolated from the hemolymph of affected oysters and ovoid parasite-like bodies observed in some spat samples, no clear identification of the cause was determined (Tlili, 1995). Affected 3-month-old oysters showed retracted mantles with demineralization of the nacreous surface and deposition of brown-yellow conchiolin (shell disease). The shell margins were colonized by zoo- and phytoplankton (Tlili, 1995).

11.8. PREDATORS AND PREDATION

A range of fishes, rays, octopus, gastropods, crustaceans, turbellarians and echinoderms commonly prey on farmed pearl oysters and on natural oyster beds, and may result in heavy losses (Dharmaraj *et al.*, 1987; Gervis and Sims, 1992; Pit and Southgate, 2003). Juvenile pearl oysters, by virtue of small size and corresponding thin shells, are particularly vulnerable to predation. For instance, Sarver *et al.* (1998) found that gastropod predation levels on *P. margaritifera* in culture in the Marshall Islands resulted in an attrition rate of up to 10% of juveniles per month. Predation usually declines with increasing oyster size, as is typically found in other bivalve molluscs and many other marine invertebrates.

Measures to reduce predation usually involve:

- meshes of appropriate size enclosing groups of pearl oysters;
- suspending oysters mid-water where they are inaccessible to benthic predators;
- regular removal of predators from within enclosures;
- combinations of mesh, suspension and removal.

Meshes, while reducing predation, bring problems of fouling. They reduce water circulation and hence food, according to mesh size and degree of fouling, so that the enclosed pearl oysters' growth may be impaired. Regular defouling programs are necessary, considerably adding to the cost of commercial farming (see Sections 7.6.2 and 15.4). Small meshes are more protective of small juveniles, but they are fouled more heavily and more rapidly. In choosing mesh size, there is need to balance level of protection versus reduction in water circulation and proneness to fouling.

There are, however, predators that cannot be excluded by the smallest practical mesh sizes or by suspending the meshed enclosures mid-water. These predators settle as larvae from the plankton either on the mesh or within the meshed enclosure. They metamorphose upon settling, grow rapidly and begin preying on the pearl oysters. They are among the most severe predators, while being particularly problematic since controlling them requires very regular inspection of the mesh enclosures and individual removal. There are three main groups of these predators of pearl oysters (Friedman and Bell, 2000):

- carnivorous gastropods,
- crustaceans,
- turbellarian flatworms.

It had been suggested that settlement is induced by the presence of the pearl oyster hosts (Gervis and Sims, 1992). This may well be the case as this phenomenon occurs in other marine planktonic larvae, which locate and then settle in environments that are appropriate for post-larval life.

There are numerous reports of various predators and predation on pearl oysters, but these are not systematic. The following text is organized into animal groups and countries where this is appropriate.

11.8.1. Fish, sharks and rays

Species from fish genera that are characterized by strong teeth and jaws, including *Lethrinus, Chrysophrys, Pagrus, Tetradon, Serranus,* and *Balistes* spp., are predators of pearl oysters, together with a variety of sharks and rays that feed by crushing. Young oysters <1-year of age are especially susceptible (Crossland, 1957; Dharmaraj *et al.*, 1987; Gervis and Sims, 1992). Rays are also predators of older pearl oysters (Dharmaraj *et al.*, 1987). In general, fish predators are not a problem in pearl farming if the oysters are enclosed (Gervis and Sims, 1992) or suspended mid-water. Furthermore, there is often an "escape size" for the larger species whereby the oysters have grown to a large enough size, and shell thickness, to be relatively free from fish predators.

Not all fish predators are lethal. The monocanthid fish, *Paramonocanthus japonicus*, preys on the soft ventral shell margins of *P. margaritifera* juveniles, reducing their growth rate but not causing mortality (Pit and Southgate, 2003).

11.8.1.1. Japan

The eel *Anguilla japonica*, the black porgy, *Acanthopagrus schlegeli (Sparus nilerocephalus),* and the globe fish *Spheroides* sp. are recognized predators of pearl oysters in Japan (Cahn, 1949; Wada, 1991), with the latter species feeding on young oysters.

11.8.1.2. India

Dharmaraj *et al.* (1987) identified the following species of fish as predators: *Balistes nitis, B. stellaris, B. maculatus* and *Lethrinus* sp. These authors noted that the fishes *Serranus* spp. and *Tetrodon* spp. preyed on young oysters <1-year old and also recorded the rays *Rhinoptera javanica* and *Ginglymostoma* spp. as predators.

11.8.1.3. Middle East

Doroudi (1996) reported *Tetradon stellatus* and *Balistes* spp. as predators of *P. marga-ritifera* and Akoya pearl oysters in the Persian Gulf, and noted that up to 35% mortality in Akoya pearl oysters may occur due to fish predation. In the Red Sea, Crossland (1957) described several species of fish, including *Lethrinus karwa*, *Chrysophrys bifasciatus* and *Pagrus* sp., as predators of *P. margaritifera*. Fish were reported to break the shells and eat oysters up to 18 months of age. Crossland (1957) also reported that larger fish including, *Balistes flavimarginatus*, could break the shells of older oysters, up to 6mm thick, and that rays could break open oyster shells.

11.8.1.4. Cook Islands

Sims (1992) described molluscivorous fishes as being abundant in certain atolls of the Cook Islands, south Pacific, with dead pearl oysters and shell fragments common.

11.8.2. Carnivorous gastropods

Predation of spat and juveniles by gastropods is a major source of mortality in pearl oyster farms and in pearl oyster beds. Fong *et al.* (2005) considered them the principal source of mortality in farmed *P. margaritifera* in the central Pacific. The predatory gastropods are from four families:

- Ranellidae (=Cymatidae), e.g. *Cymatium* spp., *Lintella caudate*,
- Muricidae, e.g. *Murex* spp., *Chicoreus ramosus*,
- Bullidae, e.g. *Gyrineus natator*,
- Bursidae, e.g. *Bursa rubeta*.

They prey on pearl oysters in natural oyster beds. The major problem for pearl oyster farming, however, is that many species settle as small larvae within the protective meshes (see above). They have a considerable capacity to rapidly consume pearl oysters by inserting their long proboscis between the oyster's valves and digesting the soft tissue *in situ* (Table 11.2).

11.8.2.1. India

Cymatium cingulatum is a voracious predator of pearl oysters in natural oyster beds in India (CMFRI, 1991) and of young Akoya pearl oysters in the Gulf of Mannar (Chellam *et al.*, 1983). Individuals may consume up to 20 pearl oysters in 19 days (CMFRI, 1991). Similarly, the gastropod *Murex virgineus* was found to be a major predator of natural oyster beds in India (CMFRI, 1991) and in beds of young Akoya pearl oysters in the Gulf of Mannar (Chellam *et al.*, 1983). Individual *M. virgineus* may consume up to 20 oysters in 49 days (CMFRI, 1991). Chellam *et al.* (1983) described severe mortalities in beds of Akoya pearl oysters from predation by the gastropods *C. cingulatum* and *M. virgineus*. Dharmaraj *et al.* (1987) also reported *C. cingulatum* and *M. virgineus*

TABLE 11.2

Predation and consumption of *Pinctada margaritifera* by various predators in aquarium studies.

Predator	Predator size (mm)	Killing method	No. of oysters consumed from three size classes (DVM, mm)			No. of oysters consumed per week
			<5	5–10	>10	
Fish:						
Paramonocanthus japonicus		(Shell trimming)				0
Xanthid crabs:						
Pilumnis curser	9[a]	Crushing	1	2	1	1
Pilumnis curser	22	Crushing	10	7	7	6
Stomatopod:						
Gonodactylus falcatus	40[b]	Smashing	10	10	10	7
Gonodactylus falcatus	60	Smashing	28	28	25	20
Gastropods:						
Cymatium muricinum	43[c]	Proboscis penetration	10	10	10	7
C. aquatile	80	Proboscis penetration	6	9	7	5
Turbellarians:						
Stylocus sp.	40[b]	Toxins	0	4	7	2
Stylocus sp.	60	Toxins	0	5	7	3

Source: Adapted from Pit and Southgate (2003). DVM = dorsoventral shell length.
[a]Carapace width.
[b]Total length.
[c]Shell length.

as the most serious predatory gastropods. They also included *Cymatium pileare*, *Murex ramosus*, *Bursa rubeta*, *Thais margariticola* and *Gyrineus natator*.

11.8.2.2. Vietnam

Chinh (2001) found that of 1,500 Akoya pearl oysters cultured in ten protective cages, only 67 remained after nearly a year. The massive mortality was from uncontrolled *C. pileare* and *Lintella caudate* infestations.

11.8.2.3. Red Sea

Murex virgineus preys on *P. margaritifera* in the Red Sea (Crossland, 1957). Nasr (1982) observed *Chicoreus ramosus* also preying on *P. margaritifera* in the Red Sea.

11.8.2.4. Okinawa

Gastropods of the family Ranellidae (Cymatidae) are serious pests in the culture of *P. margaritifera* and *P. maxima* in Okinawa (Gervis and Sims, 1992).

11.8.2.5. Solomon Islands

In laboratory studies in Solomon Islands, *Cymatium muricinum, C. aquatile, C. nicobaricum* and *C. pileare* were observed to prey on smaller *Pinctada* spp. (Gervis and Sims, 1992).

11.8.3. Cephalopods

Cahn (1949) identified the octopus *Octopus vulgaris* as a heavy consumer of Akoya pearl oysters in Japan and noted that cage culture is an effective barrier against these predators. Wada (1991) also reported *Octopus* sp. as a predator of Japanese pearl oysters. Dharmaraj *et al.* (1987) identified octopus as predators of pearl oysters in the Gulf of Mannar. Sims (1992) described octopuses (*Octopus* spp.) as abundant in certain atolls of the Cook Islands, south Pacific, with dead pearl oysters and shell fragments common.

11.8.4. Crustaceans

Portunid and xanthid crabs, including *Charybdis lucifer, Atergatis integerrimus, Leptodius exaratus, Portunus* spp., and *Thalamita* spp. were considered to be the worst predators of pearl oysters in India, entering the spat rearing cages during their larval stage and as they grow, crushing and feeding on the oyster spat (CMFRI, 1991). The crab *Charybdis* sp. was reported by Gervis and Sims (1992) as destroying entire cages of Akoya pearl oysters. In the Red Sea, Crossland (1957) reported several species of crabs and a hermit crab as attacking oysters of up to 100 mm in diameter, but noted that attacks were usually on very young stages up to approximately 12 mm in diameter. Dharmaraj *et al.* (1987) also described crabs as serious predators, with the larval stages entering cages, growing and feeding on the spat and causing high losses. *C. lucifer, Atergastis integerrisimus, L. exaratus, Portunus* spp. and *Thalamita* spp. were the most common species. In a study of predation on juvenile *Pinctada margaritifera.* Pit and Southgate (2003) tested a potential group of predators, including crabs, snapping shrimps, flatworms, gastropods, polychaetes, a fish, and a stomatopod. The most destructive predator was the stomatopod, *Gonodactylus falcatus*, with individuals consuming in excess of 20 juvenile pearl oysters per week (Table 11.2; Fig. 11.20). Stomatopods live in concealed benthic environments and it is likely that they would only be a problem in culture on the benthos.

11.8.5. Turbellarians

Newman *et al.* (1993) described a turbellarian predator, *Stylochus (Imogene) matatasi* as being consistently associated with mortalities of the cultured giant clam, *T. gigas* and the fouling pearl oyster *P. maculata* in the Solomon Islands. The stylochid flatworm, *Imogine mcgrathi*, preyed on Akoya pearl oysters in culture at Port Stevens, Australia (O'Connor and Newman, 2001). Individual flatworms consumed a mean number of 0.25–0.4 oysters/week. Stylochid turbellarians fed on juvenile *P. margaritifera*

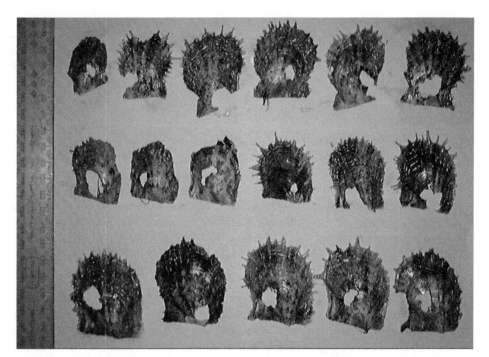

FIGURE 11.20 Damage caused to Akoya pearl oyster shells by the stomatopod *Gonodactylus falcatus*. Note that shells are not broken at the margins but at a point approximating the position of the adductor muscle. (Photo: Antoine Teitelbaum.)

at a higher rate (Table 11.2), probably reflecting the small size of the pearl oysters. The feeding process of these turbellarians is poorly known, but it seems that they secrete toxic mucus onto the soft tissues of the oyster.

11.8.6. Echinoderms

Dharmaraj *et al.* (1987) observed starfish feeding on adult pearl oysters in oyster banks.

11.8.7. Potential predators

Pit and Southgate (2003) tested the carnivorus polychaetes, *Palola* sp. and *Eunice* sp., and the snapping shrimp, *Alpheus* sp., in an aquarium study of predation of juvenile *P. margaritifera* (part of the study is presented in Table 11.2). These seemed potential predators of juveniles in benthic environments, as members of these groups are known to prey on bivalves, but there was no mortality.

11.9. SUMMARY

A broad range of disease causing and potential disease causing agents have been recorded in pearl oysters. There have been some spectacular examples of the impacts of major disease outbreaks on pearl culture industries. Perhaps most notable are those that affected Akoya pearl culture in Japan and the black pearl industry in French Polynesia. Such outbreaks can have devastating impacts, both functional and economic, which may influence the industry for many years. In many instances, the onset of disease in a population of cultured pearl oysters relates to inappropriate husbandry practices, such as overcrowding, or inadequate environmental conditions (e.g. water parameters or food). The occurrence of disease in pearl farming situations can be minimized through appropriate husbandry conditions and as a result of disease outbreaks, pearl industries have often developed more appropriate culture practices designed to minimize the chance of further disease outbreaks (e.g. Heffernan, 2006).

A large number of organisms have been recorded as predators of pearl oysters and many of these settle onto pearl oysters and pearl oyster culture equipment from the plankton. Many predators grow relatively rapidly in a pearl farm environment where there is plentiful food and, if not removed from culture equipment, can cause significant oyster mortality, with younger, smaller individual being particularly vulnerable. Again, the impact of predators in pearl oyster culture can be minimized using appropriate husbandry methods which may involve regular checking for and removal of predators from culture units.

References

Acosta-Salmon, H., Southgate, P.C., 2005. Mantle regeneration in the pearl oysters *Pinctada fucata* and *Pinctada margaritifera*. *Aquaculture* 246, 447–453.

Acosta-Salmon, H., Southgate, P.C., 2006. Wound healing after excision of mantle tissue from the Akoya pearl oyster, *Pinctada fucata*. *Comp. Biochem. Physiol. A* 143, 264–268.

Bernadac, M., Galenon, P., Secchi, F. Gallet de St. Maurin, D., 1980. Enquête ecologique sur le lagon d'un atoll des Tuamotu. Ministere Defense France. Dir. Cent. Serv. Sante des Armees. Travaux Scientifiques de Chercheurs du Service de Sante des Armees Durant L'annee 1980. No. 2, 404–407.

Braley, R.D., 1991. Pearl oyster introductions to Tokelau atolls: real potential or a long shot. *SPC Pearl Oyster Inform. Bull.* 3, 9–10.

BRS, 1987. An historical summary of pearling in the Northern Territory. *Bureau of Rural Science*, February 1987, 13 pp.

Bruce, A.J., 1977. Pontoniine shrimps in the collections of the Australian Museum. *Records of the Australian Museum* 312, 39–81.

Bruce, A.J., 1989. Notes on some Indo-Pacific Pontoniinae. XLV. *Conchodytes maculatus* sp. nov., a new bivalve associate from the Australian Northwest Shelf. *Crustaceana* 56, 182–192.

Cabral, P., 1989a. Problems and perspectives of the pearl oyster aquaculture in French Polynesia. Advances in Tropical Aquaculture, Workshop at Tahiti, French Polynesia, February 20–March 4, 1989. *Actes de Colloque* 9, 57–66.

Cabral, P., 1989b. Some aspects of the abnormal mortalities of the pearl oysters, *Pinctada margaritifera* L. in the Tuamotu Archipelago, French Polynesia. Advances in Tropical Aquaculture, Workshop at Tahiti, French Polynesia, February–March 4 1989. *Actes de Colloque* 9, 217–226.

Cahn, A.R., 1949. *Pearl Culture in Japan*. United States Department of the Interior. Fishery Leaflet 357, Fish and Wildlife Service, pp. 51–58.

Chagot, D., Fougerouse, A., Weppe, M., Marques, A., Bouix, G., 1993. Presence d'une gregarine (Protozoa Sporozoa) parasite de l'huitre perliere a levres noires *Pinctada margaritifera* (L., 1758) (*Mollusca Bivalvia*) en Polynesie Francaise. *Les Comptes Rendus de l'Academie des Sciencesi, Paris t 316 Serie lll*, pp. 239–244.

Chellam, A., Alagarswami, K., 1978. Blooms of *Trichodesmium thiebautii* and their effect on experimental pearl culture at Veppalodai. *Indian J. Fish.* 25, 237–239.

Chellam, A., Velayudhan, T.S., Dharmaraj, S., Victor, A.C.C., Gandhi, A.D., 1983. A note on the predation on pearl oyster *Pinctada fucata* (Gould) by some gastropods. *Indian J.Fish.* 39, 337–339.

Chinh, N., 2001. Harmful effects of the two pilate molluscan species *Cymatium pileare* and *Linatella caudata* on the culture of pearl oysters in Vung Ro sea water, Phu Yen, Viet Nam. *J. Shellfish Res.* 20, 1309–1310.

CMFRI, 1991. *Pearl Oyster Farming and Pearl Culture.* Training Manual 8. Pearl Oyster Farming and Pearl Culture Training Course, Central Marine Fisheries Research Institute, Tuticorin, India, Regional Seafarming Development and Demonstration Project RAS/90/002, 104 pp.

Cochennec-Laureau, N., Haggner, P., Levy, P., Saulnier, D., Langy, S., Fougerouse, A., 2004. Development of antiseptic techniques to improve cultured pearl formation in *Pinctada margaritifera. J. Shellfish Res.* 23, 285.

Comps, M., Fougerouse, A., Buestel, D., 1998. A procaryote infecting the black-lipped pearl oyster *Pinctada margaritifera. J. Invertebr. Pathol.* 72, 87–89.

Comps, M., Herbaut, C., Fougerouse, A., 1999. Virus-like particles in pearl oyster *Pinctada margaritifera. Bull. Eur. Assoc. Fish Pathol.* 19, 85–88.

Comps, M., Herbaut, C., Fougerouse, A., 2000. Abnormal periostracum secretion during the mineralisation process of the pearl in the blacklip pearl oyster *Pinctada margaritifera. Aquat. Living Resour.* 13, 49–55.

Comps, M., Herbaut, C., Fougerouse, A., Laporte, F., 2001. Progress in characterisation of Syndrome 85 in the black-lip pearl oyster *Pinctada margaritifera. Aquat. Living Resour.* 14, 195–202.

Crossland, C., 1957. The cultivation of the mother-of-pearl oyster in the Red Sea. *Aust. J. Mar. Freshw. Res.* 8, 111–136.

Cui, X.Z., 1997. *Ultrastructural Observations in Gregarines in Pearl Oysters Pinctada maxima and Polychaetes.* A Report Submitted in Partial Fulfilment of the Requirements for a Postgraduate Diploma in Science within the Department of Parasitology, University of Queensland, February 1977.

Cuif, J-P., Dauphin, Y., 1996. Occurrence of mineralisation disturbances in nacreous layers of cultivated pearls produced by *Pinctada margaritifera* var *cumingi* from French Polynesia. Comparison with reported shell alterations. *Aquat. Living Resour.* 9, 187–193.

Dauphin, Y., Cuif, J.P., 1990. Epizootic disease in the black lip pearl oyster *Pinctada margaritifera* in French Polynesia: ultrastructural alterations of the nacreous layer. In: Suga, S., Nakahara, H. (Eds.), *Mechanism and Phylogeny of Mineralization in Biological Systems.* Springer-Verlag, Tokyo. pp. 167–171.

Dauphin, Y., Denis, A., 1987. Alteration microstructurale de la nacre des huitres perlieres (*Pinctada margaritifera*) atteintes par l'epizootie en Polynesie française. *C. R. Acad. Sci. Paris, Ser. III* 305, 649–654.

Dharmaraj, S., Chellam, A., Velayudhan, T.S., 1987. Biofouling, boring and predation of pearl oysters. In: Alagarswami, K. (Ed.), *Pearl Culture.* CMFRI Bulletin, 39, pp. 92–97.

Dix, T.G., 1972. Two mesenchymal tumours in a pearl oyster, *Pinctada margaritifera. J. Invertebr. Pathol.* 20, 317–320.

Dix, T.G., 1973a. Histology of the mantle and pearl sac of the pearl oyster *Pinctada maxima* (Lamellibranchia). *J. Malacol. Soc. Aust.* 2, 365–375.

Dix, T.G., 1973b. Mantle changes in the pearl oyster *Pinctada maxima* induced by the pea crab *Pinnotheres villosulus. Veliger* 15, 330–331.

Doroudi, M.S., 1996. Infestation of pearl oysters by boring and fouling organisms in the northern Persian Gulf. *Indian J. Mar. Sci.* 25, 168–169, Springer-Verlag, Tokyo.

Dybdahl, R., Pass, D.A., 1985. *An Investigation of Mortality of the Pearl Oyster, Pinctada maxima, in Western Australia.* Report 71, Fisheries Department, Perth, Western Australia, 78 pp.

Eldredge, L.G., 1994. *Introductions of Commercially Significant Aquatic Organisms to the Pacific Islands.* Pacific Science Association, Honolulu, HI. South Pacific Commission, Noumea New Caledonia, 127 pp.

Fong, Q.S.W., Ellis, S., Haws, M.E., 2005. Economic feasibility of small scale black-lipped pearl oyster (*Pinctada margaritifera*) farming in the Central Pacific. *Aquacult. Econom. Manag.* 9, 347–368.

Friedman, K.J., Bell, J.D., 2000. Shorter immersion times increase yields of the blacklip pearl oyster, *Pinctada margaritifera* (Linne), from spat collectors in Solomon Islands. *Aquaculture* 187, 299–313.

Gervis, M.H., Sims, N.A., 1992. The biology and culture of pearl oysters (Bivalvia: Pteriidae). *ICLARM Study Rev.* 21, 1–49.

Goggin, C.L., Lester, R.J.G., 1987. Occurrence of *Perkinsus* species Protozoa, Apicomplexa, in bivalves from the Great Barrier Reef. *Dis. Aquat. Org.* 3, 113–117.

Goggin, C.L., Sewell, K.B., Lester, R.J.G., 1989. Cross-infection experiments with Australian *Perkinsus* species. *Dis. Aquat. Org.* 7, 55–59.

Heffernan, O., 2006. Pearls of wisdom. *The Marine Scientist* 16, 20–23.

Herdman, W.A., 1903. Observations and experiments on the life-history and habits of the pearl oyster. In Report Pearl Oyster Fisheries, Gulf of Mannar. In: Herdman, W.A. (Ed.), *Report to the Government of Ceylon on the Pearl Oyster Fisheries of the Gulf of Mannar*. Royal Society, London, pp. 125–146.

Hine, P.M., 1996. Southern Hemisphere mollusc diseases and an overview of associated risk assessment problems. *Rev. Sci. Tech. Off. Int. Epiz.* 15, 563–577.

Hine, P.M., 2002. Significant diseases of molluscs in the Asia-Pacific region. In: Lavilla-Pitogo, C.R., Cruz-Lacierda, E.R. (Eds.), *Diseases in Asian Aquaculture IV*. Fish Health Section, Asian Fisheries Society, Manila, pp. 187–196.

Hine, P.M., Thorne, T., 1998. *Haplosporidium* sp. (Haplosporidia) in hatchery-reared pearl oysters *Pinctada maxima* (Jameson, 1901), in north Western Australia. *J. Invertebr. Pathol.* 71, 48–52.

Hine, P.M., Thorne, T., 2000. A survey of some parasites and diseases of several species of bivalve molluscs in northern Western Australia. *Dis. Aquat. Org.* 40, 67–78.

Hine, P.M., Thorne, T., 2002. *Haplosporidium* sp. (Alveolata: Haplosporidia) associated with mortalities among rock oysters *Saccostrea cuccullata* in north Western Australia. *Dis. Aquat. Org.* 51, 123–133.

Hirano, M., Kanai, K., Yoshikoshi, K., 2002. Contact infection trials failed to reproduce the disease condition of mass mortality in cultured pearl oyster *Pinctada funcata martensii*. *Fish. Sci.* 68, 700–702.

Hirano, M., Sugishita, Y., Mobin, S.M.A., Kanai, K., Yoshikoshi, K., 2005. An investigation of the pathology associated with mass mortality events in the cultured Japanese pearl oyster *Pinctada fucata martensii* at four farms in western Japan. *J. Aquat. Anim. Health* 17, 323–337.

Humphrey, J.D., Norton, J.N., 2005. *The Pearl Oyster Pinctada maxima (Jameson, 1901) An Atlas of Functional Anatomy, Pathology and Histopathology*. Fisheries Research and Development Corporation, Australia, 111 pp.

Humphrey, J.D., Norton, J.H., Jones, J.B., Barton, M.A., Connell, M.T., Shelley, C.C., Creeper, J.H., 1998. *Pearl Oyster (Pinctada maxima) Aquaculture: Health Survey of Northern Territory, Western Australia and Queensland pearl Oyster Beds and Farms*. Project Report 94/079. Fisheries Research and Development Corporation, 108 pp.

Joll, L., 1994. Shell handling and mortality. *The Western Australian Experience in Developing the Torres Strait and Queensland East Coast Pearl Industry* 1994 Industry Workshop June 22–23, Torres Strait, pp 35–37.

Jones, J.B., Creeper, J., 2006. Diseases of pearl oysters and other molluscs: a Western Australian perspective. *J. Shellfish Res.* 25(1), 233–238.

Khamdan, S.A.A., 1998. Occurrence of *Bucephalus sp.* trematode in the gonad of the pearl oyster, *Pinctada radiata*. *Environ. Int.* 24, 117–120.

Kim, D., Sato, Y., Oda, T., Muramatsu, T., Matsuyama, Y., Honjo, T., 2000. Specific toxic effect of dinoflagellate *Heterocapsa circularisquama* on the rotifer *Branchionus plicatilis*. *Biosci., Biotechnol. Biochem.* 64, 719–2722.

Kitamura, S-I., Tomaru, Y., Kawabata, Z., Suzuki, S., 2002. Detection of marine birnavirus in the Japanese pearl oyster *Pinctada fucata* and seawater from different depths. *Dis. Aquat. Org.* 50, 211–217.

Kobayashi, T., Nozawa, N., Miyazaki, T., 1999. Studies on akoya-virus disease associated with mass mortalities in Japanese pearl oyster *Pinctada fucata martensii*. *Fourth Symposium on Diseases in Asian Aquaculture, Cebu City, Philippines*, November 22–26. Book of Abstracts.

Kotake, N., Miyawaki, Y., 1954. Experimental studies on the propagation of the pearl oyster, *Pinctada martensi* – ll. On the unusual high mortality of pearl oyster caused by bacteria – 1. *Bull. Jpn. Soc. Sci. Fish.* 19, 952–956.

Kurokawa, T., Suzuki, T., Okauchi, M., Miwa, S., Nagai, K., Nakamura, K., Honjo, T., Nakajima, K., Ashida, K., Funakoshi, S., 1999. Experimental infection of a disease causing mass mortalities of Japanese pearl oyster *Pinctada fucata martensii* by tissue transplantation and cohabitation. *Bull. Jpn. Soc. Sci. Fish.* 65, 241–251.

Le Moullac, G., Goyard, E., Salulier, D., Haffner, P., Thourard, E., Nedelec, G., Goguenheim, J., Rouxel, C., Cuzon, G., 2003. Recent improvements in broodstock management and larviculture in marine species in Polynesia and New Caledonia: genetic and health approaches. *Aquaculture* 227, 89–106.

Lester, R.J.G., 1990. Diseases of cultured molluscs in Australia. Advances in Tropical Aquaculture: Workshop held in Tahiti, French Polynesia, February 20–March 4, 1989. *Actes de Colloque* 9, pp. 207–216.

Liu, Z., Huang, H., Zhou, Y., 1998. Preliminary study on the pseudomonas disease of the larvae of *Pinctada martensi* (Dunker). *J. Zhanjiang Ocean Univ.* 18, 23–28.

Mannion, M.M., 1983. *Pathogenesis of a Marine Vibrio Species and Pseudomonas putrefaciens Infections in Adult Pearl Oysters, Pinctada maxima (Mollusca: Pelecypoda).* Thesis, Honours Degree in Veterinary Biology, Murdoch University, 130 pp.

Marin, F., Dauphin, Y., 1991. Diverse alteration in the amino–acid content of the nacreous organic matrices in the black lip pearl oyster (*Pinctada margaritifera*) in French Polynesia affected by an epizootic disease. C.R. Acad. Sci. *Serie III* 312, 483–488.

Matsuyama, Y., 2003a. Physiological and ecological studies on harmful dinoflagellate *Heterocapsa circularisquama* – 1 Elucisation of environmental factors underlying the occurrence and development of *H. circularisquama* red tide. *Bull. Fish. Res. Agency (Japan)* 7, 24–105.

Matsuyama, Y., 2003b. Physiological and ecological studies on harmful dinoflagellate *Heterocapsa circularisquama* – 2 Clarification on toxicity of *H. circularisquama* and its mechanisms causing shellfish kills. *Bull. Fish. Res. Agency (Japan)* 9, 13–117.

Matsuyama, Y., Nagai, K., Mizuguchi, T., Fujiwara, M., Ishimura, M., Yamaguchi, M., Uuchida, T., Honjo, T., 1995. Ecological features and mass mortality of pearl oysters during red tides of *Heterocapsa sp.* in Ago Bay in 1992. *Nippon Suisan Gakkaishi (Bull. Jap. Soc. Fish. Sci.)* 611, 35–41.

Matsuyama, Y., Uchida, T., Honjo, T., 2000. Impact of harmful dinoflagellate *Heterocapsa circularisquama* on shellfish aquaculture in Japan. *J. Shellfish Res.* 19, 636.

Miyake, S., Fujino, T., 1968. Pontoniinid shrimps from the Palau Islands (Crustacea, Decapoda, Palaemonidae). *J. Fac. Agr. Kyushu Univ.* 14, 401–431.

Miyamoto, M., Yoshida, Y., Koube, H., Matsuyama, Y., Takayama, H., 2002. The red tide of a dinoflagellate *Gyrodinium sp.* in Youkaka Bay in 1995: environmental features during the red tide and its effect on cultured finfish. *Nippon Suisan Gakkaishi (Bull. Jap. Soc. Fish. Sci.)* 68, 157–163.

Miyazaki, T., Goto, K., Kobayashi, T., Kageyama, T., Miyata, M., 1999. Mass mortalities associated with a virus disease in Japanese pearl oysters *Pinctada fucata martensii. Dis. Aquat. Org.* 32, 1–12.

Miyazaki, T., Nozawa, N., Kobayashi, T., 2000. Clinical trial results on the use of a recombinant feline interferon-ω to protect Japanese pearl oysters *Pinctada fucata martensii* from akoya-virus infection. *Dis. Aquat. Org.* 32, 1–12.

Miyazaki, T., Nozawa. N., Kobayashi, T., 2001. Efficacy of a recombinant feline interferon-omega in protecting Japanese pearl oysters *Pinctada fucata martensii* from akoya-virus infection. *Aquaculture 2001: Book of Abstracts*, p. 445.

Miyazaki, T., Taniguchi, T., Hirayama, J., Nozawa, N., 2002. Receptors for recombinant feline interferon-ω in hemocytes of the Japanese pearl oyster *Pinctada fucata martensii. Dis. Aquat. Org.* 51, 135–138.

Mizumoto, S., 1979. Pearl farming in Japan. In: Pillai, T.V.R., Dill, W.A. (Eds.), *Advances in Aquaculture.* Fishing News Book Ltd., Farnham, Surrey, pp. 431–435.

Morizane, T., Takimoto, S., Nishikawa, S., Matsuyama, N., Tyohno, K., Uemura, S., Fujita, Y., Yamashita, H., Kawakami, H., Koizumi, Y., Uchimura, Y., Ichikawa, M., 2001. Mass mortality of Japanese pearl oyster in Uwa Sea, Ehime, in 1997–1999. *Fish Pathol.* 36, 207–216.

Muroga, K., 1999. Diseases of maricultured gastropods and bivalves in Japan. *Fourth Symposium on Diseases in Asian Aquaculture: Aquatic Animal Health for Sustainability*, November 22–26, 1999, Cebu International Convention Centre, Cebu City, Philippines. Book of Abstracts (Unpaginated).

Muroga, K., 2002. Diseases of maricultured gastropods and bivalves in Japan. In: C.R. Lavilla-Pitogo, C.R., Cruz-Lacierda, E.R. (Eds.), *Diseases in Asian Aquaculture IV. Proceedings of the Fourth Symposium on Diseases in Asian Aquaculture*, November 22–26, 1999, Cebu City, Philippines. Fish Health Section, Asian Fisheries Society, Manilla, pp. 197–202.

Muroga, K., Inui, Y., Matsusato, T., 1999. Workshop "Emerging diseases of cultured marine molluscs in Japan". *Fish Pathol.* 34, 219–231.

Nagai, K., Matsuyama, Y., Uchida, T., Yamaguchi, M., Ishimura, M., Nishimura, A., Akamatsu, S., Honjo, T., 1996. Toxicity and LD sub50 levels of the red tide dinoflagellate *Heterocapsa circularisquama* on juvenile pearl oysters. *Aquaculture* 144, 149–154.

Nagai, K., Matsuyama, Y., Uchida, T., Akamatsu, S., Honjo, T., 2000. Effect of a natural population of the harmful dinoflagellate *Heterocapsa circularisquama* on the survival of the pearl oyster *Pinctada fucata*. *Fish. Sci.* 66, 995–997.

Nagai, K., Honjo, T., Go, J., Yamashita, H., Seok, J.O., 2006. Detecting the shellfish killer *Heterocapsa circularisquama* (Dinophyceae) by measuring bivalve valve activity with a Hall element sensor. *Aquaculture* 255, 395–401.

Nasr, D.H., 1982. Observations on the mortality of the pearl oyster, *Pinctada margaritifera*, in Dongonab Bay, Red Sea. *Aquaculture* 28, 271–281.

Negri, A.P., Bunter, O., Jones, B., Llewellyn, L., 2004. Effects of the bloom-forming alga *Trichodesmium erythraeum* on the pearl oyster *Pinctada maxima*. *Aquaculture* 232, 91–102.

Newman, L.J., Cannon, L.R.G., Govan, H., 1993. *Stylochus (Imogene) matatasi* n. sp. Platyhelminthes, Polycladida: Pest of cultured giant clams and pearl oysters from Solomon Islands. *Hydrobiologia* 2573, 185–189.

Norton, J.H., Shepherd, M.A., Prior, H.C., 1993a. Papovavirus-like infection of the golden-lipped pearl oyster, *Pinctada maxima*, from the Torres Strait, Australia. *J. Invertebr. Pathol.* 62, 198–200.

Norton, J.H., Shepherd, M.A., Perkins, F.P., Prior, H.C., 1993b. Perkinsus-like infection in farmed golden-lipped pearl oyster *Pinctada maxima* from the Torres Strait, Australia. *J. Invertebr. Pathol.* 62, 105–106.

Norton, J.H., Lucas, J.S., Turner, I., Mayer, R.J., Newnham, R., 2000. Approaches to improve cultured pearl formation in *Pinctada margaritifera* through use of relaxation, anticeptic application and incision closure during bead insertion. *Aquaculture* 184, 1–17.

Nugranad, J., Traithong, T., Promjinda, K., 1999. Hatchery seed production of gold-lipped pearl oyster *Pinctada maxima* (Jameson). *Proceedings of the Ninth Workshop of the Tropical Marine Mollusc Programme (TMMP), Indonesia*, August 19–29, 1988, Part 1, Vol. 19, No. 1, Special Publication, Phuket Marine Biological Centre, Phuket, pp. 247–248.

O'Connor, W.A., Newman, L.J., 2001. Halotolerance of the oyster predator, *Imogine mcgrathi*, a stylochid flatworm from Port Stephens, New South Wales, Australia. *Hydrobiologia* 459, 157–163.

Oda, T., Sato, Y., Kim, D., Muramatsu, T., 2001a. Hemolytic activity of *Heterocapsa circularisquama* (Dinophyceae) and its possible involvement in shellfish toxicity. *J. Phycol.* 37, 509–516.

Oda, T., Sato, Y., Kim, D., Matsuyama, Y., Honjo, T., 2001b. Species-specific haemolytic activity of *Heterocapsa circularisquama*, a newlty identified harmful red tide dinoflagellate. *Proceedings of the Seventh Canadian Workshop on Harmful Marine Algae*. British Columbia, Canada, pp. 69–78.

Ozaki, Y., Ishibashi, C., 1934. Notes on the cercaria of the pearl oyster. *Proc. Imp. Acad. Japan* 10, 439–441.

Park, K.I., Choi, K.S., Jeong, W.G., 2001. An examination for the protozoan parasite, *Perkinsus sp.* in the pearl oyster *Pinctada fucata martensii* (Dunker) from the southern coast of Korea. *Bull. Eur. Assoc. Fish Patholo.* 21, 30–32.

Pass, D.A., Perkins, F.O., 1985. "Protistan parasites" or residual bodies in *Pinctada maxima*. *J. Invertebr. Pathol.* 46, 200–201.

Pass, D.A., Dybdahl, R., Mannion, M.M., 1987. Investigations into the causes of mortality of the pearl oyster, *Pinctada maxima* (Jamson), in Western Australia. *Aquaculture* 65, 149–169.

Pass, D.A., Perkins, F.P., Dybdahl, R., 1988. Viruslike particles in the digestive gland of the pearl oyster *Pinctada maxima*. *J. Invertebr. Pathol.* 51, 166–167.

Perkins, F.O., 1996. Shell disease in the gold lip pearl oyster, *Pinctada maxima* and the Eastern oyster, *Crassostrea virginica*. *Aquat. Living Resour.* 9, 159–168.

Perkins, F.O., 1996. Shell disease in the gold lip pearl oyster *Pinctada maxima* and the Eastern oyster. *Aquat. Living Resour.* 9, 159–168.

Pit, J.H., Southgate, P.C., 2003. Fouling and predation; how do they affect growth and survival of the black-lip pearl oyster, *Pinctada margaritifera*, during nursery culture? *Aquacult. Int.* 11, 545–555.

Pyne, R.R., 1972. Interim synopsis – Pearl oyster *Pinctada maxima* mortality, Thursday Island, Unpublished report.

Remoissenet, G., 1995. From the emergence of mortalities and diseases on Pinctada margaritifera to their effects on the pearling industry. *Aquaculture Workshop on Present and Future of Aquaculture Research and Development in the Pacific Island Countries*. Technical Paper S7–6, JICA/ Ministry of Fisheries, November 20–24, 1995, Nuku'alofa, Tonga.

Sakaguchi, S., 1972. Studies on the trematode genus *Proctoeces* parasitic in pearl oyster, *Pinctada fucata*. lll. On the adult trematode, *Proctoeces ostreae* Fujita in Dolifus, 1925, found in the black porgy, *Mylio macrocephalus*, from Tanabe Bay. *Bull. Natl. Pearl Res. Inst.* 16, 2028–2037.

Sakaguchi, S., Hosina, T., Minami, T., 1970a. Studies on the trematode genus *Proctoeces* parasitic in pearl oyster, *Pinctada fucata*. 1. On the morphology, parasitic status and the distribution of pearl oyster infected with the parasite. *Kokoritsu Shinju Kenkyusho Hokoku (Bull. Natl. Pearl Res. Lab.)* 15, 1931–1938.

Sakaguchi, S., Minami, T., Yamamura, Y., 1970b. Studies on the trematode genus *Proctoeces* parasitic in pearl oyster, *Pinctada fucata*. 2. Experimental infection with the metacercaria to the final host. *Bull. Natl. Pearl Res. Inst.* 15, 1939–1947.

Sarver, D.J., Sims, N.A., Clarke, R.P., 1998. *Pinctada* nursery culture: overcoming the bane of *Cymatium*. *SPC Pearl Oyster Inform. Bull.* No. 11, 31.

Seurat, L.G., 1906. Sur un cestode parasite des huitres perlieres determinant la production des perles fines aux iles Gambier. C.R. Hebd. Seances Acad. Sci. *Serie D, Sciences Naturelles* 142, 801–803.

Shipley, A.E., Hornell, J., 1904. The parasites of the pearl oyster. In: Herdman, W.A. (Ed.), *Report to the Government of Ceylon on the Pearl Oyster Fisheries of the Gulf of Manaar, Part II*. Royal Society, London, pp. 77–106.

Shipley, A.E., Hornell, J., 1905. Further report on parasites found in connection with the pearl oyster fishery of Ceylon. In: Herdman, W.A. (Ed.). *Report to the Government of Ceylon on the Pearl Oyster Fisheries in the Gulf of Manaar*, Part II, Royal Society, London, pp. 49–56.

Shipley, A.R., Hornell, J., 1906. Report on the cestode and nematode parasites from the marine fishes of Ceylon. In: Herdman, W.A. (Ed.), *Report to the Government of Ceylon on the Pearl Oyster Fisheries of the Gulf of Mannar, Part V*. Royal Society, London, pp. 43–96.

Sims, N.A., 1992. Abundance and distribution of the black-lip pearl oyster, *Pinctada margaritifera* (L.), in the Cook Islands, South Pacific. *Aust. J. Mar. Freshw. at. Res.* 43, 1409–1421.

Sindermann, C.J., 1990. *Principal Diseases of Marine Fish and Shellfish. Volume 2. Diseases of Marine Shellfish*. Academic Press, New York, 516 pp.

Southwell, T., 1912. The Ceylon pearl inducing worm. A brief review of the work done to date. *Parasitology* 5, 27–36.

Sparks, A.K., 1985. *Synopsis of Invertebrate Pathology Exclusive of Insects* Elsevier Science Publishers B.V., Amsterdam, pp. 73.

SPC, 1985. Virus kills mother of pearl in French Polynesia. South Pacific Commission Fisheries Newsletter No. 34, p. 17.

SPC, 1988. *Workshop Report, Workshop on Pacific Inshore Fishery Resources, Noumea, New Caledonia*, March 14–15, 1978. Inshore Fisheries Research Project, South Pacific Commission, Noumea, New Caledonia.

SPC, 2002a. Disease outbreak costs Manihiki pearl farms millions. *Secretariat of the Pacific Community Pearl Oyster Information Bulletin*, No. 15, August 2002, 5.

SPC, 2002b. Penrhyn oyster killer not a disease, say scientists. *Secretariat of the Pacific Community Pearl Oyster Information Bulletin*, No. 15, August 2002, 5.

Sugishita, Y., Hirano, M., Tsutsumi, K., Mobin, S.M.A., Kanai, K., Yoshikoshi, K., 2005. Effects of exogenous lipid peroxides on mortality and tissue alterations in Japanese pearl oysters *Pinctada fucata martensii. J. Aquat. Anim. Health* 17, 233–243.

Suliman, A., 1995. Pathological studies on the pearl oysters *Pinctada margaritifera* cultivated in Dongonab Bay (Red Sea) with special reference to the recent mass mortality. Sustainable Development of Fisheries in Africa, Fisheries Society of Africa. 263 pp.

Suzuki, T., 1995. Accumulation of injected proteins in the auricles of the pearl oyster *Pinctada fucata. Bull. Natl. Res. Inst. Aquacult. Japan Yoshokukenho* 24, 23–32.

Suzuki, S., 1997. Molecular detection of aquatic birnaviruses from marine fish and shellfish. *Fourth International Marine Biotechnology Conference,* September 22–29, 1997, Rome. Abstract, p. 124.

Suzuki, T., Funakoshi, S., 1992. Isolation of a fibronectin-like molecule from a marine bivalve, *Pinctada fucata*, and its secretion by amoebocytes. *Zool. Sci.* 9, 541–550.

Suzuki, T., Yoshinaka, R., Mizuta, S., Funakoshi, S., Wada, K., 1991. Extracellular matrix formation by amoebocytes during epithelial regeneration in the pearl oyster *Pinctada fucata*. *Cell and Tissue Res.* 266, 75–82.

Suzuki, S., Kamakura, M., Kusuda, R., 1998a. Isolation of birnavirus from Japanese pearl oyster: *Pinctada fucata*. *Fish. Sci.* 64, 342–343.

Suzuki, S., Utsunomiya, I., Kasuda, R., 1998b. Experimental infection of marine birnavirus strain JPO-96 to Japanese pearl oyster *Pinctada fucata*. *Bull. Marine Sci. Fish., Kochi Univ.* 18, 39–41.

Tanaka, Y., Inoha, S., Kakazu, K., 1970. Studoes on seed production of black-lip pearl oyster *Pinctada margaritifera*, in Okinawa – V. Rearing of the larvae. *Bull. Tokai Reg. Fish. Res. Lab.* 63, 97–106.

Tlili, A., 1995. *Pearl Oyster Production Pinctada maxima (Jameson, 1901). Management and Economical Studies at Bynoe Harbour Pearl Farm, Northern Territory, Australia.* Department of Primary Industry and Fisheries, Northern Territory, Australia. 53 pp.

Tomaru, Y., Kawabata, Z., Nakano, S., 2001. Mass mortality of the Japanese pearl oyster *Pinctada fucata martensii* in relation to water temperature, chlorophyll a and phytoplankton composition. *Dis. Aquat. Org.* 44, 61–68.

Vacelet, E., Arnoux, A., Thomassin, B., 1996. Particulate material as an indicator of pearl-oyster excess in the Takapoto Lagoon Tuamotu, French Polynesia. *Aquaculture* 144, 133–148.

Wada, K.T., 1991. The pearl oyster *Pinctada fucata*. In: Menzel., W. (Ed.), *Estuarine and Marine Bivalve Mollusk Culture*, pp. 246–249.

Wolf, P.H., Sprague, V., 1978. An unidentified protistan parasite of the pearl oyster, *Pinctada maxima*, in tropical Australia. *J. Invertebr. Pathol.* 31, 262–263.

Wu, X., Pan, J., 1999a. Studies on Rickettsia-like organism disease of the tropical marine pearl oyster l: The fine structure and morphogenesis of *Pinctada maxima* pathogen Rickettsia-like organism. *J. Invertebr. Pathol.* 73, 162–172.

Wu, X., Pan, J., 1999b. Studies on the rickettsia-like organism diseases of tropical marine pearl oyster-the morphology, morphogenesis and ultrastructure of RLO inclusions, an agent for *Pinctada maxima*. *Oceanol. Limnol. Sin./Haiyang Yu Huzhao* 30, 73–80.

Wu, X., Pan, J., 1999c. Studies on rickettsia-like organism disease of tropical marine pearl oyster, *Pinctada maxima* and *P. fucata*. IV. On histo-cytopathology of RLO disease. *Acta Oceanologica Sinica* 21, 93–98.

Wu, X., Pan, J., 1999d. Studies on the Rickettsia-like organism disease of tropical marine pearl oyster: V. Ultrastructural pathology and pathogenesis of rickettsia-like organism disease. *Acta Oceanologica Sinica/Haiyang Xuebao* 21, 113–118, CRC Press, Boca Raton, Florida.

Wu, X., Li, D., Pan, J., 2001. Studies on rickettsia-like organism (RLO) disease of tropical marine pearl oyster – Epidemiological investigation of RLO disease in larvae populations of maricultured *Pinctada maxima*. *Acta Oceanologica Sinica* 20, 563–574.

Wu, X., Li, D., Pan, J., Jiang, J., 2003. Studies on rickettsia-like organism (RLO) disease of tropical marine pearl oyster – Epidemiological investigation of RLO disease in juvenile populations of maricultured *Pinctada maxima*. *Acta Oceanologica Sinica* 22, 421–435.

Yamaguchi, K., Mori, K., 1988. Hemagglutinin activities in pearl oyster, *Pinctada fucata martensii*, hemolymph. *Bull. Natl. Inst. Aquacul., Japan Yoshokukenho* 13, 25–32.

Yamashita, S., 1986. The Torres Strait Pearling Industry. In: Haines, A.K., Williams, G.C., Coates, C. (Eds.). *Torres Strait Fisheries Seminar*, Port Moresby, February 1985, Australian Government Publishing Service, Canberra, pp. 118–121.

Yamatogi, T., Sakaguchi, M., Matsuda, M., Iwanaga, S., Iwataki, M., Matsuoka, K., 2005. Effect on bivalve molluscs of a harmful dinoflagellate *Heterocapsa circularisquama* isolated from Omura Bay, Japan, and its growth characteristics. *Nippon Suisan Gakkaishi* 71, 746–754.

Zhou, C., Xie, L-P., Zhang, R-Q., 2001. Effect of pearl – nucleus-inserting operation on immune levels in *Pinctada fucata*. *J. Fish. China* 25, 419–423.

Zhou, Y., Chen, Y., Zeng, S., Long, L., Zhnag, S., 2003. Effects of protogrycan from pearl oyster on non-specific immunology of *Pinctada martensii*. *Mar. Sci.* 27, 47–52.

CHAPTER 12

Population Genetics and Stock Improvement

Katsuhiko T. Wada and Dean R. Jerry

12.1. INTRODUCTION

For the effective management of wild populations of pearl oysters, and their subsequent commercial exploitation through aquaculture, it is important to have both a sound appreciation of how genes are exchanged between populations and how genetic diversity is maintained or lost. Also required is knowledge of the underlying genetic basis behind the phenotypic expression of important traits. There has been a relative paucity of information on the genetics of pearl oysters compared to other important aquaculture groups. There are few examples where genetic techniques have been applied to increase aquaculture productivity. However, over the last decade there has been increased effort to acquire essential genetic information necessary to manage and exploit pearl oyster populations. This chapter summarizes current knowledge of the population genetics of pearl oysters, updates our understanding of the quantitative genetic basis of commercially important traits and highlights the application of emerging technologies to this group of bivalves.

12.2. POPULATION GENETICS

An understanding of how populations within a species interact and exchange genes is important for their sustainable management, as well as aiding in the identification of genetically diverse populations that may possess biological traits of interest to aquaculture. The following summarises current knowledge of the population genetics of the most important pearl producing species.

12.2.1. Silver/gold-lip pearl oyster

Despite the economic importance of this species, there have been limited population genetic studies on *Pinctada maxima* and there are currently no published studies that have examined the broad population structure across its entire distributional range. Population genetic surveys have primarily focused on determining connectivity and stock structure between populations in northern Australia and Indonesia. They show evidence of restrictions to gene flow in this species, particularly between eastern and Western Australian populations, and between Australian and Indonesian populations (Johnson and Joll, 1993; Benzie and Smith, 2003; Benzie and Smith-Keune, 2006). For instance, Johnson and Joll (1993) surveyed five northern Australian *P. maxima* populations from Exmouth Gulf, Western Australia to Thursday Island, Queensland (ca. 3,400 km apart) using six polymorphic allozyme loci. Significant heterogeneity in allele frequencies was found at three loci (*GOT, GPI, LTP*), with the greatest genetic differences observed between populations situated on opposite sides of the Gulf of Carpentaria (i.e. Oxley Island and Thursday Island – $F_{st} = 0.09$). Heterogeneity was also observed between close Northern Territory populations, indicating that genetic subdivision can occur over as little as a few hundred kilometers ($F_{st} = 0.08$) in this species. However, the pattern of genetic structuring found by Johnson and Joll (1993) was not solely a manifestation of geographical distance. Two populations from northwestern Australia, separated by 800 km, were essentially panmictic. This implied that *P. maxima* populations in this latter region could be managed as a single stock.

Given that the bulk of Australia's South Sea pearl production originates from this region (see Section 9.3) and the insensitivity of allozyme markers to detect low levels of population structuring, Benzie and Smith (2003) and Benzie and Smith-Keune (2006) applied both mitochondrial (mtDNA) and microsatellites to further examine the genetic structure of populations in Western Australia, and to establish the larger scale connectivity between Western Australian and two Indonesian populations (Java, Sumbawa). Both studies confirmed the single stock structure of northwestern Australian populations, although the level of genetic exchange among populations appeared to reduce over larger spatial scales as in an "isolation by distance" pattern of gene flow (Fig. 12.1). Significant genetic differences were found, however, between Australian and two Indonesian populations and *P. maxima* from these two regions can be considered as separate stocks.

A broader-scale population genetic survey by Lind (2008) using mitochondrial COI and microsatellite DNA data has shown that there are appreciable restrictions to gene flow among *P. maxima* populations from Vietnam, Indonesia, Australia and the Solomon Islands, with significant levels of genetic structuring evident (mtDNA COI $\theta_{ST} = 0.388$, microsatellite $R_{ST} = 0.036$) (Table 12.1). All populations were shown

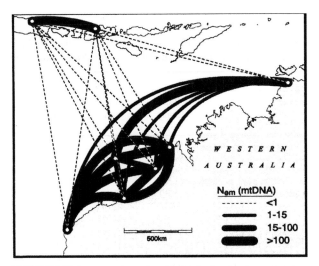

FIGURE 12.1 Map illustrating the relative levels of gene flow between *Pinctada maxima* populations in Australia and Indonesia estimated from mtDNA cytochrome oxidase 1 data. (Benzie and Smith, 2003).

to be genetically isolated, except for Bali and Raja Empat (West Irian) populations, where the microsatellite data suggest low levels of gene flow between these two populations. Numbers of detected alleles and genetic diversity estimates tend to be higher in Indonesian populations than further east suggesting that this region may have originally been the ancestral epicenter from which *P. maxima* evolved and subsequently expanded its geographical distribution.

Both allozyme and microsatellite studies have shown significant heterozygote deficits at many gene loci (Johnson and Joll, 1993; Benzie and Smith-Keune, 2006). This is a common occurrence in pearl oysters and it has been postulated that these deficits are a consequence of differential selection pressures acting against specific genotypes throughout the lifecycle (Durand and Blanc, 1989), temporal heterogeneity in recruitment (Johnson and Black, 1984), assortative mating (Smith, 1987) or null alleles (Zouros and Foltz, 1984). However, no evidence of temporal post-settlement selection was found by Benzie and Smith-Keune (2006), who examined microsatellite allele frequencies in two consecutive year classes of juveniles. They postulated that deficiencies at microsatellite markers may at least be partly due to null alleles. Stochastic levels of inbreeding due to variance in reproductive success and the genetic identity of individuals contributing to cohorts may also explain heterozygote deficiencies in this species (Benzie and Smith-Keune, 2006).

12.2.2. Black-lip pearl oyster

The population genetic structure of *Pinctada margaritifera* has been the most comprehensively examined of all the important pearl oyster species, perhaps due to its high

TABLE 12.1

Frequencies of cytochrome oxidase I (mtDNA) gene haplotypes observed in wild silver-lip pearl oyster (*Pinctada maxima*) populations from Vietnam, Indonesia, Australia and the Solomon Islands (Lind, 2008).

Haplotype No. n	Vietnam Phu Quoc 28	Indonesia Aru 55	Bali 52	Raja Empat 61	Australia Torres Strait 55	Western Australia 56	Hervey Bay 19	Solomon Islands 5
1	0.500	0.182	0.231	0.066	0.200	0.643	0.368	0
2	0.071	0	0.346	0.377	0.018	0	0	0
3	0	0.036	0	0.016	0.455	0.143	0	0
4	0	0.473	0	0	0	0.071	0	0
5	0	0.073	0.096	0.230	0	0	0	0
6	0	0	0.058	0.033	0.164	0	0.211	0
7	0	0.073	0.019	0	0.055	0.036	0	0
8	0	0.036	0	0	0.091	0.018	0	0
9	0	0.073	0	0	0	0.036	0	0
10	0	0.055	0	0.049	0	0	0	0
11	0	0	0	**0.082**	0	0	0	0
12	**0.179**	0	0	0	0	0	0	0
13	0	0	**0.077**	0	0	0	0	0
14	0	0	0	0	0	**0.054**	0	0
15	0	0	**0.038**	0	0	0	0	0
16	0	0	**0.038**	0	0	0	0	0
17	0	0	0.019	0.016	0	0	0	0
18	0	0	0	**0.033**	0	0	0	0
19	0	0	0	**0.033**	0	0	0	0
20	0	0	0	**0.033**	0	0	0	0
21	0	0	0	0	0	0	0	**0.400**
22	0	0	0	0	0	0	**0.105**	0
23	0	0	0	0	0	0	**0.105**	0
24	0	0	0	0	0	0	**0.105**	0
25	0.036	0	0	0	0	0	0.053	0
26	**0.071**	0	0	0	0	0	0	0
27	0	0	*0.019*	0	0	0	0	0
28	0	0	*0.019*	0	0	0	0	0
29	0	0	*0.019*	0	0	0	0	0
30	0	0	*0.019*	0	0	0	0	0
31	0	0	0	*0.016*	0	0	0	0
32	0	0	0	*0.016*	0	0	0	0
33	0	0	0	0	*0.018*	0	0	0
34	0	0	0	0	0	0	0	*0.200*
35	0	0	0	0	0	0	0	*0.200*
36	0	0	0	0	0	0	0	*0.200*
37	0	0	0	0	0	0	*0.053*	0
38	*0.036*	0	0	0	0	0	0	0
39	*0.036*	0	0	0	0	0	0	0
40	*0.036*	0	0	0	0	0	0	0
41	*0.036*	0	0	0	0	0	0	0
Frequency of unique	0.393	0.000	0.231	0.213	0.018	0.054	0.368	1.000

Bold text indicates a unique haplotype not observed in any other population; italic text indicates singleton haplotypes.

economic importance to scattered Pacific island communities and its spatially complex distribution. *Pinctada margaritifera* ranges across the Indian- and Pacific-Oceans (see Section 2.3.2.3). Its greatest abundance and region of greatest exploitation, however, are in the atoll lagoons of the Central Pacific (Gervis and Sims, 1992) and the majority of population genetic surveys have been undertaken in this region. Surveys of allozyme variation in populations of *P. margaritifera* throughout French Polynesia, Mauritius, Kiribati, Cook Islands, Japan and eastern Australia have all shown high levels of within-population variation, with reported mean observed heterozygosities ranging between 0.237 and 0.575 (Blanc, 1983; Blanc *et al.*, 1985; Durand and Blanc, 1986, 1989; Blanc and Durand, 1989; Durand *et al.*, 1993; Benzie and Ballment, 1994; Arnaud-Haond *et al.*, 2003a). Allozyme surveys, and later surveys based on anonymous nuclear DNA (anDNA) markers indicate complex population structuring throughout the Western and Central Pacific, with strong evidence for restrictions of gene flow between archipelagos, and, in some instances, islands within archipelagos (Blanc *et al.*, 1985; Durand and Blanc, 1989; Benzie and Ballment, 1994; Arnaud-Haond *et al.*, 2003a). For example, Benzie and Ballment (1994) showed significant genetic heterogeneity between populations within Western Pacific island groups, and between Great Barrier Reef (GBR) populations, and between the GBR populations and those from Kiribati and the Cook Islands. Rogers (1972) showed genetic distances ranging from an average of 0.136 between Cook Island populations to 0.303 between those from GBR and Kiribati. Similarly, genetic heterogeneity has been shown between atolls within the French Polynesian, Cook and Society Island archipelagos (Benzie and Ballment, 1994; Arnaud-Haond *et al.*, 2003a). Durand and Blanc (1986) found Nei's (1972) genetic distance values ranged from 0.0078 to 0.0423 between French Polynesian populations, and greater still between French Polynesia as a whole and Mauritius (Nei's D = 0.060).

Contrasting with the pattern of population structuring shown by nuclear markers, at least within the Central Pacific, mtDNA 12S rRNA and cytochrome oxidase 1 RFLP analyses did not detect significant heterogeneity between archipelagos (Arnaud-Haond *et al.*, 2003a). What is more, mtDNA, haplotype diversity (0.09–0.23) and sequence divergences were low (0.36–1.83%), suggesting a short coalescence time between DNA molecules. Short coalescence times are often indicative of recent genetic bottlenecks and Arnaud-Haond *et al.* (2003a) postulated that mtDNA data supports the hypothesis that many Polynesian *P. margaritifera* populations were extirpated following Pleistocene glacial sea level fluctuations which led to draining of atolls (Paulay, 1990). Consequently, many atoll populations can only be 8,000 years old and the low haplotype divergences reflect relatively recent colonization. In support of this hypothesis, *P. margaritifera* from the Marquesas Islands show appreciably higher levels of both haplotypic and anDNA gene diversity than those of central Polynesian populations. Due to its broad embayment topography, it may not have been affected to the same degree by Pleistocene sea level emersions and may have acted as a refuge area for inner reef species (Paulay, 1990). Based on the unique mtDNA signature of *P. margaritifera* from this group of islands, Arnaud-Haond *et al.* (2003a–c) suggested that this archipelago may have been a source population for recruits to other archipelagos once sea levels rose again around 8,000 years ago. The study of Arnaud-Haond *et al.* (2003a) highlight how genetic markers that have different effective sizes (i.e. mtDNA versus nuclear DNA) can often demonstrate differing patterns of differentiation in the face of varying demographic forces.

Of particular concern to the maintenance of historical genetic diversity among archipelago populations of the Central Pacific was the observation by Arnaud-Haond *et al.* (2004) of homogenization of previously genetically distinct stocks over as little as 20 years as a consequence of spat collection and translocation. Populations subject to translocations within and between the Tuamotu-Gambier and Society archipelagos became genetically homogenized over this period, while genetic structure was maintained in those populations in the same archipelagos that had never been stocked. The complex population structure in this region may be under threat from anthropomorphic processes.

There is currently no information on this species for genetic structuring in regions outside the Central and Western Pacific. However, Herbinger *et al.* (2006) published 10 microsatellite markers for this species and these, coupled with current genetic markers, provide unprecedented power to resolve fine-scale population genetic processes in *P. margaritifera*.

12.2.3. Akoya pearl oyster

The taxonomy of the group of pearl oysters that are collectively termed Akoya pearl oysters, i.e. *Pinctada fucata-martensii-radiata-imbricata* species complex is unresolved (see Section 2.3.1.13). Many species names have been given for these small pearl oysters, which are distributed widely over almost all tropical to warm-temperate regions of the world, except the western coasts of Africa and America. Because overall levels of genetic differentiation between populations appear to be similar regardless of which species the author considered they were examining, in the discussion below populations will be simply referred to as Akoya pearl oysters.

Population genetic variability of Akoya pearl oysters has been studied in populations from Japan, China, Bahrain and Australia initially using allozymes (Table 12.2) and later molecular DNA markers (Wang *et al.*, 2003b; Masaoka and Kobayashi, 2005; Yu and Chu, 2006a, b). These allozyme and molecular studies suggested that populations distributed from Japan to Australia regularly exchange high numbers of migrants between or among metapopulations resulting in significant levels of gene flow, often over large geographical distances. For example, based on allozymes, low levels of genetic differentiation were evident between Australian populations and three Japanese samples (Colgan and Ponder, 2002). Yu and Chu (2006a,b) also found low genetic differentiation among three Chinese, Japanese and Australian Akoya populations using amplified fragment length polymorphism (AFLP) markers. Across this broad geographical range 96% of the genetic variance present was contained within populations, compared to only 4% between populations. The pattern of genetic distance across this large geographical scale was consistent with isolation by distance (Mantel test $r = 0.96$, $P = 0.03$). However this correlation broke down once the Australian population was removed from the analyses, with no evidence of significant genetic differentiation evident between Chinese

TABLE 12.2

Studies that have estimated genetic variation in the *Pinctada fucata-martensii-radiata-imbricata* complex based on allozyme polymorphisms.

Nominal species/ subspecies	Scored loci	Locations	Number of samples	Source
Pinctada fucata martensii	1	Japan (2 locations)	97	Wada (1975a)
	4	Japan (4 locations)	112	Wada (1982)
	4	Japan (30 locations)	2,408	Wada (1984)
Pinctada fucata	8	China (1 location) (wild and hatchery)	270–274	Li *et al.* (1985)
Pinctada fucata martensii	4	Japan (4–6 selection lines)	232–398	Wada (1986a)
	17	Japan (4 locations)	709	Minami *et al.* (2000)
	5	Japan (2 locations)	99–140	Atsumi *et al.* (2004)
		China, Vietnam, Thailand (1 location each)	151–161	
Pinctada imbricata	10	Northern Australia (4 locations)	19–27	Colgan and Ponder (2002)
	10	Japan (3 locations)	19–30	Colgan and Ponder (2002)
	4	Northern Australia (3 locations)	19–23	Colgan and Ponder (2002)
	4	Southern Australia (12 locations)	236–267	Colgan and Ponder (2002)
	4	Japan (3 locations)	27–36	Colgan and Ponder (2002)
Pinctada radiata	2	Bahrain (3 location)	131	Beaumont and Khamdan (1991)

and Japanese samples (i.e. only 0.8% of the total genetic variance was explained by differences between these two northern Pacific populations). Similarly three populations from the Arabian Gulf were examined at two polymorphic allozyme loci suggesting that these are a single genetic stock (Beaumont and Khamdan, 1991).

Mean observed heterozygosities for Akoya pearl oysters range from 0.117 to 0.480 (Li *et al.*, 1985; Beaumont and Khamdan, 1991; Minami *et al.*, 2000; Colgan and Ponder, 2002). Given the small number of loci commonly surveyed in studies on Akoya pearl oysters, estimates of heterozygosity are in reasonable agreement within the genus and fall within the ranges observed for other bivalves (Fujio *et al.*, 1983; Beaumont and Zouros, 1991).

Another purpose of population genetics is aiding in the identification of genetically diverse populations that may possess biological traits of interest to aquaculture. Although biochemical or molecular methods have not detected much genetic differentiation among Asian Akoya populations, there is some suggestion of morphological or physiological genetic differentiations between northern and southern (subtropical) Japanese population. One of these was exemplified by the Japanese Akoya population from the small Namako lagoon isolated from the outer sea by sand banks through which seawater flows. The lagoon is inhabited by a large number of small pearl oysters, and their shell morphology has been shown to be genetically differentiated from other Japanese populations by mating experiments despite presumed selectivity neutral allozyme or molecular differentiation being low (Wada, 1975a, b, 1984; Masaoka and

Kobayashi, 2005). Another possible example of genetic differentiation in Akoya populations was reported in the physiological trait of reproduction. The subtropical Akoya population was reported to be mature and to spawn mainly in winter in contrast to the summer spawning of the Japanese northern population (Wada *et al.*, 1995). Tsutsumi (2002) reported a similar phenomenon of winter maturation and delayed onset of the reddening adductor muscle disease in subtropical Chinese Akoya strains raised in northern Japan when compared to those of the northern strain raised under the same conditions. It would be very interesting to determine whether there is an inherited relationship between this reproductive trait and disease resistance.

12.2.4. Panamanian pearl oyster

The Panamanian pearl oyster, *Pinctada mazatlanica*, occurs over a large latitudinal range from northwestern Mexico to northern Peru (Fig. 2.6). In recent times it has undergone substantial reductions in population size as a result of overfishing (Monteforte and Carino, 1992; Section 9.5). The distribution of this species is bisected by several surface equatorial currents and counter currents that probably limit gene flow between northern and southern populations (Fig. 12.2(a)). Genetic structure in this species has been assessed using mtDNA and nuclear markers, which give conflicting pictures of gene flow. Strong genetic structuring was shown at mtDNA 12S rRNA and COI ($\theta_{ST} = 0.18$, $P < 0.001$) (Arnaud-Haond *et al.*, 2001) with clear isolation by distance effects on genetic partitioning. However, no structuring was suggested at allozyme loci (Li and Hedgecock, 1991) and two anonymous nuclear markers ($\theta = 0.003$, $P > 0.05$) (Arnaud-Haond *et al.*, 2003b). *P. mazatlanica* is a protandrous hermaphrodite and, given that mtDNA is maternally inherited, discrepancies between the genetic marker sets may be explained by male-biased dispersal coupled with high mortality before the female phase of the life cycle. This would lead to structuring at mtDNA loci, but not nuclear loci. A lower effective population size for mtDNA also makes it a lot more sensitive to detecting population bottlenecks and this bottleneck effect would be emphasized if high variance in female reproductive success were also evident. High variance in female reproductive success would also accentuate the sex-ratio bias and increase the proportion of genetic contribution to a population by males (Li and Hedgecock, 1998; Arnaud-Haond *et al.*, 2003b). The other possibility for differences in genetic data may have been a lack of resolution by the nuclear markers.

Based on mtDNA, there is a clear pattern of population differentiation in *P. mazatlanica* in its northern distribution, with three distinct genetic stocks identified: Gulf of California-northern Mexico, southern Mexico and Panama (Arnaud-Haond *et al.*, 2001). Interestingly, haplotypic and nucleotide diversities are correlated with latitude, with northern populations more polymorphic than those of further south (Fig. 12.2(b)). This pattern is also present for gene diversity at nuclear markers (Table 12.3). Haplotypic diversity for this species at 12S rRNA and COI ranged between 0.000 in pearl oysters from Panama to 0.856 in Gulf of California samples. Gene diversity corresponded in a similar fashion (0.48–0.62). This suggests either a long-term lowering of effective population sizes in southern populations or, more likely, evidence of recent (re) colonization of southern Mexico and Panamanian populations in a step-by-step

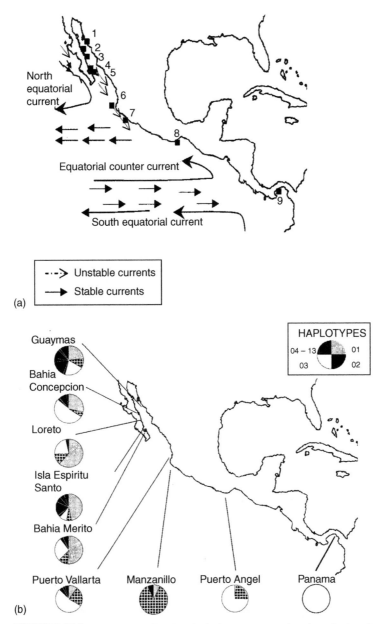

FIGURE 12.2 (a) Map showing the principal currents occurring along the American Pacific coast from Mexico to Panama. Numbers indicate locations sampled by Arnaud-Haond *et al.* (2001) in their study of the mtDNA diversity among *Pinctada mazatlanica* populations. 1: Guaymas; 2: Bahía Concepcíon; 3: Loreto; 4: Isla Espiritu Sancto; 5: Bahía Meritonillo; 8: Puerto Angel; 9: Panama. (b) mtDNA cytochrome oxidase subunit 1 haplotypic variation across this species natural distribution.

TABLE 12.3

Sample size, haplotype and nucleotide diversities estimated for mtDNA, gene diversity and heterozygous deficiencies estimated for nuclear DNA for each population of *Pinctada mazatlanica* sampled by Arnaud-Haond *et al.* (2001). F_{is} = inbreeding coefficient (Weir and Cockerham, 1984) and H = unbiased gene diversity (Nei, 1987).

Collection sites (north–south)	Sample size	Haplotype diversity (±SE)	Nucleotide diversity (%)	Gene diversity (H, nb ± SE)	F_{is} Pinuc1	F_{is} Pinuc2
Gulf of California						
Guaymas	53	0.86 (±0.02)	0.0063	0.60 (±0.02)	−0.00	0.12
Bahía Concepcíon	39	0.66 (±0.06)	0.0056	0.59 (±0.03)	−0.12	0.09
Loreto	37	0.57 (±0.08)	0.0044	0.55 (±0.03)	0.33**	0.14
Isla Espiritu Santo	36	0.75 (±0.07)	0.0044	0.61 (±0.03)	0.23*	0.09
Bahía Merito	38	0.66 (±0.07)	0.0052	0.59 (±0.03)	0.29*	0.24
Pacific coast						
Puerto Vallarta	39	0.66 (±0.07)	0.0039	0.62 (±0.02)	0.23	0.12
Manzanillo	14	0.28 (±0.15)	0.0017	0.48 (±0.07)	0.14	0.06
Oaxaca	11	0.41 (±0.13)	0.0014	0.55 (±0.07)	0.01	−0.01
Panama	21	0.00 (±0.00)	0.0000	0.49 (±0.05)	0.09	−0.23*

*$P < 0.05$ after 1,000 permutations.
**Value is significantly different from zero after sequential Bonferroni correction.

manner from northern migrants (Arnaud-Haond *et al.*, 2003b). Despite undergoing obvious reductions in population densities due to overfishing, nuclear and mtDNA data indicate that for most populations there are still abundant levels of genetic variation, similar to those in other pearl oyster species (i.e. average heterozygosity at 28 allozyme loci ranged between 0.124 and 0.153) (Li and Hedgecock, 1991).

12.2.5. Impact of culture and hatchery methods on genetic diversity

Maintaining high levels of genetic diversity is critical to the long-term success of re-stocking and selective breeding programs and ensures that populations have the capacity to adapt to changing environments and possess adequate levels of phenotypic variation (underpinned by allelic diversity) for targeted selection. It has been shown for many aquaculture species that routine production practices have the potential to significantly impede the maintenance of genetic diversity in captive populations (Cross and King, 1983; Taniguchi *et al.*, 1983; Beaumont, 1986; Sbordoni *et al.*, 1987; Busack and Currens, 1995; Evans *et al.*, 2004); often as a result of population founder effects, small effective broodstock numbers, differential contribution and survival among families (Frost *et al.*, 2006), grading and selection (Wada, 1986a). Loss of genetic diversity in aquaculture populations is usually greatest in those that are based on small broodstock founder numbers and have undergone several generations of closed breeding. If left unchecked, loss of genetic diversity may lead to inbreeding and the associated fitness consequences of inbreeding depression.

Understanding the impact of closed breeding on the long-term maintenance of genetic diversity in pearl oysters is only rudimentary. However, given the number of studies in other molluscs that have demonstrated adverse effects, it can be assumed that hatchery and culture practices will impact in a similar fashion (Hedgecock and Sly, 1990; Boudry et al., 2002; Li et al., 2004). Pearl oysters are extremely fecund and it is common practice for hatcheries to produce large numbers of progeny through the mass spawning of a limited number of broodstock. Although mass spawning generally ensures adequate numbers of offspring for culture purposes, the genetic contribution by individual broodstock to each cohort is not controlled and subsequently the genetic variability present in the cohort will be directly related to the number of parents that spawn and the percentage of their gametes that were viable and develop into juveniles. As for other mollusc species, the practice of mass spawning of pearl oysters (see Section 7.4) is likely to lead to small N_e/N ratios and temporal variance of allelic frequencies, which over time can lead to dramatic reductions in levels of genetic variability within cohorts (Hedgecock and Sly, 1990; Boudry et al., 2002).

Hatcheries are capricious environments, which, coupled with the fragility of pearl oyster larval stages, have the potential to induce frequent genetic bottlenecks. Additionally, normal practice in many hatcheries is to mass-spawn broodstock and this practice can often result in low effective breeding numbers and differential parental contributions to the ensuing cohort of progeny. The genetic impact has been examined in experimental generations of Japanese (Wada, 1986a) and hatchery produced Chinese (Su et al., 2002; Yu and Chu, 2006a) Akoya, and in the Japanese P. margaritifera produced in commercial hatcheries (Durand et al., 1993). Like other bivalve species most of these populations showed reduced levels of genetic diversity while in others genetic diversity was not as depressed, probably depending on the extent of inbreeding. For example, Wada (1986a) examined genetic variation at four protein loci in Japanese Akoya populations that had undergone up to six generations of selection for shell color and shape. Compared to a base cultured population, gradual reductions in the number of alleles were observed, with selected lines losing up to five out of seven alleles at some protein loci. While there were some indications of heterozygosity loss at many of the loci, the effect was not as dramatic as that observed for the loss of alleles. Similarly in P. margaritifera, Durand et al. (1993) observed a 17–18% reduction in allele number in second and third generation hatchery oysters compared to wild samples. Observed heterozygosity levels were, however, unaffected (H_o: wild = 0.237 and hatchery = 0.236–0.269). Reductions in total allele number without corresponding decreases in levels of heterozygosity are often observed in hatchery populations and it is generally accepted that heterozygosity levels are less sensitive to changes in genetic diversity of cultivated stocks (Hedgecock and Sly, 1990). Culture practices within an Indonesia P. maxima hatchery were shown to lead to the reduction in 50% or more in the number of mtDNA COI haplotypes and microsatellite alleles present in local populations (Lind, unpublished data).

Hatchery managers need to ensure that effective broodstock population sizes are as large as possible to combat losses in genetic diversity. Many studies of mass spawnings of oyster species have shown that the N_e/N ratio is often very low (Hedgecock and Sly, 1990; Boudry et al., 2002; Li et al., 2004). To increase N_e/N ratios, farm managers will need to increase the genetic contribution of each broodstock to the ensuing cohort, either through significantly increasing broodstock numbers or by having greater control over matings. Recently, Knauer et al. (2005) reported on a new protocol for

P. maxima that allows them to create numerous full-sib families through natural spawning of single mating pairs and therefore give a direct estimate of the number of families initially contributing to a cohort. However, even if a large number of families can be produced, hatchery managers also need to be aware that significant proportions of the initial genetic diversity sampled can be lost through differential family survival during development. This is the process where entire families may "crash" due to genetic factors or environmental stochasticity. For example, Taylor *et al.* (unpublished data) spawned 50 full-sib *P. maxima* families and reared them in individual culture vessels for 8 days. By day 2 of culture, 9 (18%) of the 50 families had crashed. Such family drop-outs, coupled with low N_e's, will significantly reduce overall genetic diversity in pearl oyster cohorts.

12.2.6. Spat collection and translocation

Culture techniques for many pearl oyster species (i.e. *P. margaritifera*) are often based on the rearing of wild caught spat (Gervis and Sims, 1992; see Section 7.3). It is only in more recent years that hatchery propagation techniques have been applied. Arnaud-Haond *et al.* (2003c) evaluated the impact natural spat collection has on genetic variability of both farmed and wild stocks of *P. margaritifera* in French Polynesia by comparing variation at four anDNA markers between four farms and adjacent wild populations. No genetic differences were found between farmed and wild populations (differentiation ranged between 0.0 and 0.2%), although the mean number of alleles/locus and heterozygosity were slightly lower in the farmed populations. Equally as important, no genetic differences were found between adjacent wild populations where there were no farms and those where farms were present. This shows that the collection of spat from populations close to where they are subsequently on-grown may only lead to very small losses in genetic variability in farmed stocks. However, caution needs to be exercised when interpreting the findings of Arnaud-Haond *et al.* (2003c) because even very small losses of genetic variation can result in significant changes to population structure over long periods of time. In fact the practice of spat collection, coupled with translocation between populations and archipelagos, has subsequently been shown to significantly impact the genetic structure of local populations over a 20 year period (Arnaud-Haond *et al.*, 2004). In the Tuamotu and Gambier Islands of French Polynesia, spatial and temporal variability in spat collection has led to importations from the neighboring Society Islands. Anonymous nuclear DNA data show that in the 1980s *P. margaritifera* from the two archipelagos were genetically differentiated ($\Phi_{ST} = 0.032$, $P < 0.05$). However, by the year 2000, translocations and subsequent spawning of non-indigenous individuals all but homogenized genetic variability between the two populations ($\Phi_{ST} = 0.006$). Such rapid breakdown of genetic structure due to the introduction of spat to natural populations will be exaggerated under high-density culture systems like those in French Polynesia. High density culture will lead to better synchronisation of gamete release and a subsequent effect on the lowering of variance in reproductive success. Given the generally low reproductive success observed in wild pearl oysters (Hedgecock, 1994; Li and Hedgecock, 1998), these effects would increase the opportunity for cultured pearl oysters (either native or introduced) to genetically contribute to the surrounding population.

12.3. QUANTITATIVE GENETICS

12.3.1. Background

Pearl oyster culture has not historically involved large-scale intensive domestica-tion like that of terrestrial livestock and many other aquaculture species. There were few in-depth studies into the genetic basis of commercially important traits until rela-tively recently. Studies aimed at quantifying the role genes play in the realization of pearl quality and oyster growth traits have progressed only slowly, primarily due to difficulties in controlled rearing and in the production of large numbers of synchro-nously spawned families (Wada, 1987, 1995, 1997, 1999, 2000; Beaumont and Zouros, 1991; Wada et al., 1995; Beaumont, 2000). Commercial hatchery production of spat, however, is common now in most pearl oyster industries and this has led to increased interest in applying genetic improvement techniques to improve productivity. Genetic techniques have led to enormous improvements in the productivity of farmed edible oyster species (reviewed by Sheridan, 1997) and the use of genetics to improve pro-ductivity of pearl farming in the future is only likely to increase as knowledge on the genetic basis of important traits is elucidated.

12.3.2. The conundrum – is pearl quality due to genetics or environment?

The production of a cultured pearl is a complex biological process potentially involv-ing genetic contributions by two oysters, i.e. those of the host and donor oysters (see Section 8.9), and modification of genetic expression by environmental influences (see Section 6.11). What is currently not understood for pearl oysters though, is the relative importance of the following factors in the realization of pearl traits: the genotype of the host; the genotype of the donor; the interaction between the genotypes of the host and donor; the environment; or the interaction between the various genotypes and the environment. It is also possible that the underlying genetic and environmental processes differ for the various important pearl traits. It is not known if the same processes are correlated with more than one trait; for example, are the same genetic or environmen-tal factors responsible for the realization of pearl color and roundness, for instance? It is imperative that the genetic basis of important commercial pearl quality and oys-ter growth traits are fully elucidated if high genetic gains are ever to be realized from selective breeding programs for this group of molluscs.

12.3.3. Heritability of commercial traits

The heritability of a trait broadly expresses that proportion of phenotypic variance within a population that is attributable to heritable genetic factors. Estimating the herit-ability of a trait before undertaking a selection program is important. It shows to what extent a trait has a genetic basis and allows the best improvement approach to maximize the selection response (Falconer and Mackay, 1996). There have been several trials to estimate the heritability of commercially important traits in Akoya pearl oysters, with perhaps the first being that of Wada (1975b) based on only three full-sib families at

TABLE 12.4

Realized heritability of shell traits of 3-year-old Akoya pearl oysters estimated from two selection experiments (Wada, 1984, 1986c).

			Heritability $(h^2)^a$		
Trait selected	Experiment	Generation	Large direction	Small direction	Both
Shell width	I	1st	–	0.245	–
		2nd	0.345	0.127	0.213
	II^b	1st	0.460	0.009	0.231
Shell convexity^c	I	1st–2nd	0.128	0.333	0.215
		2nd–3rd	0.095	–	–
	II	1st–2nd	0.368	0.274	0.318
		2nd–3rd	0.092	0.0735	–

^{a}Calculation of heritability (h^2) Large direction: $h^2 = (Mlo - M)/(Mlp - M)$. Small direction: $h^2 = (Mso - M)/(Msp - M)$. Both directions: $h^2 = (Mlo - Mso)/(Mlp - Msp)$, where M: Mean value of the base population, Mlp: Mean value of the parents selected for large direction, Msp: Mean value of the parents selected for small direction, Mlo: Mean value of the offspring from the parents selected for large direction, Mso: Mean value of the offspring from the parents selected for small direction.
^{b}Estimated at the age of 2 years.
^{c}Shell convexity = (Shell width)/(shell height + hinge line length + shell width).

1 and 2 years of age. Despite incompleteness in parameter estimation, this study encouraged the deliberate selection of cultured populations of pearl oysters. It was important in view of the few studies on heritability estimation of bivalves at that time (Lannan, 1972). After selective breeding had begun in this species, realized heritabilities of two shell traits, shell width and convexity, were estimated from selection responses (Wada, 1986c; Table 12.4). The realized heritabilities of shell width and convexity were 0.467 and 0.350, respectively. Velayudhan *et al.* (1996) estimated rather higher realized heritability of the same traits in Akoya pearl oysters from India over four generations.

12.3.4. Selective breeding and hybridization

Results from the selective breeding of pearl oysters have been reported from Japan, China and India, with the examples available restricted to Akoya pearl oysters (Wada, 1994; Wada and Komaru, 1990, 1991, 1994, 1996; Velayudhan *et al.*, 1995, 1996).

12.3.4.1. Shell size and shell shape

Shell size and shape are very important pearl oyster traits, as they determine the size of the nucleus that can be implanted during the seeding operation (see Section 8.9). As a result, these traits were included in the breeding objective of a selection trial for the Japanese pearl oyster (Wada, 1984, 1986c). In this trial, oysters were mass selected for three generations separately for shell width and convexity. Figure 12.3 shows the mean changes of shell width in the lines selected for shell width and shell convexity (Wada, 1986c). The mean shell width of the second generation in the line selected for large shell width (LII) was significantly greater than that of the first generation (LI)

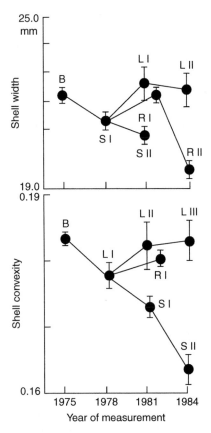

FIGURE 12.3 Changes in mean shell width and mean shell convexity in selected lines of Akoya pearl oysters over three generation. B, base population. L, selected for large. S, selected for small. R, random control. Roman numerals, I, II and III, are generations of lines of selected parents. Selection began in 1975, but only data for the small (S) line for shell width and the large (L) line for shell convexity were obtained in 1978. Vertical bars are 95% confidence intervals (from Wada, 1986a).

at 2 years (1983), but not at 3 years (1984). This decreased response of the large line at 3 years was attributed principally to adverse environmental conditions between the second and third years of this period, because the response of the random line (RI-RII) also markedly decreased over these 3 years. Changes in other traits such as shell height, hinge line length and shell weight showed the same tendency as shell width over this period (Wada, 1986c). Thus, high genetic correlations might be expected between these traits and shell width. Shell convexity of the large line from the second generation (LII) to the third generation (LIII) increased significantly at 2 years of age (1983), but was insignificant at 3 years (1984) (Fig. 12.3). Shell convexity of the small line (SI) decreased significantly at both the SI and SII generations. SI values for shell width and shell convexity were always smaller than the LI and RI values for the equivalent trait. Changes in other traits in this experiment were unpredictable and showed no relation to the responses measured for shell convexity.

12.3.4.2. Shell color

Nacre coloration is one of the traits that most influence pearl price. Selection experiments have shown that, of the various nacre colors, at least yellow coloration is inherited (Wada, 1984, 1986a). The amount of yellow pigments in the nacre of the shell varies in Akoya pearl oysters and it has been suggested that oysters without yellow nacre are useful as donors of mantle tissue for producing the valuable non-yellow pearls (Wada, 1969). It is, however, not practical for farmers or breeders to select oysters on the basis of nacre color, as yellow pigments are present in the nacre of most oysters (about 80%, see Fig. 12.4). The coloration of the inner shell is also very difficult to recognize, particularly without opening the shell valves. Even after opening the valves, it is not easy to eliminate shells with yellow pigments in the nacre. Reflection or interference of light makes observation of the color difficult. Selection trials for Japanese Akoya shells without yellow nacre pigment, i.e. white, resulted in an increase of this type of oyster from a starting percentage of 20% non-yellow nacre in the base population to 80% by the third generation (Fig. 12.4). Experimental transplantation of the mantle tissue to produce pearls was then carried out by using populations of second-generation animals as donors and hosts. White pearls were produced at a higher rate in the group of oysters selected for non-yellow nacre than in those of non-selected populations. No significant difference in mortality or other pearl characteristics were observed between the two groups and these results demonstrated the feasibility of selective breeding of oysters with non-yellow nacre to control pearl coloration (Wada, 1985, 1986b). Genetic mechanisms influencing pearl coloration for other colors have not been determined.

External color of the bivalve shell is mainly associated with the pigments contained in the prismatic layer of the shell. White Akoya oysters are rare in the wild in Japan, most are brown-purple with dark brown spots or stripes. In pearl oysters from hatchery production, white specimens became frequent and breeding experiments suggested that

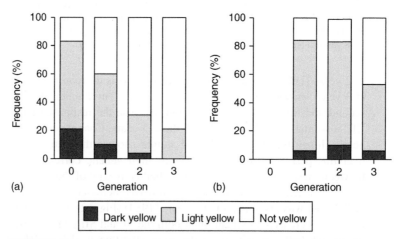

FIGURE 12.4   Changes of frequency of Akoya pearl oysters with different colors of shell nacre in two studies (a and b) with populations selected for shells without yellow pigments. Redrawn with data from Wada (1984, 1986b).

TABLE 12.5

Shell color of the parents and offspring of crossing experiments on the inheritance of white shell coloration in Japanese Akoya pearl oysters (Wada and Komaru, 1990).

Cross number	Parents		Offspring		Total number	Expected ratio[a]
	Female	Male	White	Brown		
1	White A	White B	282	0	282	1:0
2	White C	White D	839	23[b]	962	1:0
3	White E	Brown F	0	5,274	5,274	0:1
4[c]	Brown G	Brown H	117	638	755	1:3
5[c]	Brown I	Brown H	26	54	80	1:3

[a]Observed counts and expected ratios assuming Mendelian inheritance of two alleles at the shell color locus.
[b]Most of these animals were determined to be derived from other accidental matings on the basis of analysis of genotype at the two loci on zymogram patterns of LAP.
[c]All parents in crosses Nos. 4 and 5 were derived from offspring of the No. 3.

the white coloration is inherited under the control of a recessive gene or genes (Wada, 1983). All the progeny from crosses involving white specimens as parents were white, while those resulting from parental crosses between white females × brown males produced brown-shelled offspring. A 3:1 ratio of brown to white specimens was obtained in an F_2 generation cross of the F_1 brown-shelled individuals (Table 12.5). There was also less yellow pigment in the nacre of oysters with the white prismatic layer than in those with brown. Using white oysters as mantle tissue donors could be useful in producing more valuable pearls with less yellow pigment. Wada and Komaru (1996) conducted transplanting experiments using mantle tissue from specimens selected for white color in the prismatic layer to examine this possibility. The frequency of yellow pearls was significantly lower in the group produced by grafting mantle tissue from the inbred white line than from brown lines.

12.3.4.3. Weight of pearls

Pearl weight is an important market trait. In the experiment described in the previous section, pearl weight was significantly lower in the group of pearls produced from the transplantation of the mantle tissue of animals selected for non-yellow nacre (75.6 mg) than those from animals selected for yellow nacre (82.6 mg). Moreover, there was a similar trend of differences in pearl weight observed between yellow and non-yellow Akoya oysters within each line. Weights of dark yellow (103 mg), light yellow (83.4 mg) and non-yellow (74.0 mg), pearls were all significantly different from each other in the groups selected for non-yellow shell nacre. The same results were observed in the population selected for yellow shell nacre (dark yellow 110.6 mg; light yellow 88.9 mg; non-yellow 76.8 mg). It may thus be necessary to determine whether the phenotypic correlation between the weight of pearls and coloration is of genetic or environmental origin (Wada, 1985).

Which individual is more important in the production of thicker nacre and thus heavier pearls – the donor or the host oyster? Mantle tissue from the donor oyster produces the pearl-sac that secretes nacre from its epithelium (see Section 8.9). The host oyster provides the sources of energy and raw metabolites for pearl growth, and the physiological processes to recover from surgery and to maintain a healthy condition until pearl harvest. Genetic improvement of pearl weight, therefore, is an interesting objective of the scientific selective breeding of pearl oysters. Preliminary studies addressed this objective by using selected families for either the white or brown shell prismatic layer (Wada and Komaru, 1996). In this trial, a small positive correlation ($r = 0.451$, $P < 0.01$) was observed between pearl weight and the shell weight of host oysters, i.e. host oysters with heavier shells produced heavier pearls (Fig. 12.5). Interestingly, pearl weight did not differ due to the origin of mantle tissue; from white or brown donors. These results have implications for the selective breeding of host and donor pearl oysters. The shell weight of the host oyster may be a more important criterion in selecting a strain to produce heavier pearls than the shell weight of the donor oyster. This is because pearl weight is dependent on calcium metabolism in the pearl-sac and the physiological condition or activity of the host oyster determines whether the maximum potential for calcium metabolism in the pearl-sac epithelium is reached (Wada, 1972).

12.3.4.4. Inbreeding depression and hybridization

Although the genetic mechanisms are not well understood, extensive data on the effects of inbreeding depression after long-term selection have been presented not only in

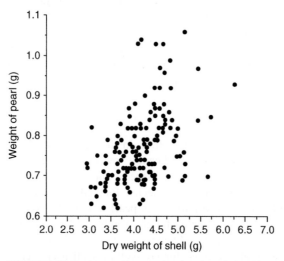

FIGURE 12.5 Scatter diagram showing the relationship between the weight of pearls and dry shell weight of the host Akoya pearl oyster from which pearls were harvested in the experiment of Wada and Komaru (1996). Each dot is the value of a single pearl harvested and the dry shell weight of the host ($r = 0.451$, $P < 0.01$).

terrestrial organisms but also in some aquaculture species. Selective breeding has the purpose of making desirable genes homozygous. However, selection of one trait can result in undesirable genes also becoming homozygous and may lead to the expression of recessive alleles, which normally remain concealed by the presence of their dominant allele variants in the heterozygous genotypic state. The number of parents that contribute to produce each generation in the bivalve hatchery is difficult to estimate because of differences in fertilization rates and mortality among gametes of each individual parent. Because this makes estimation of the extent of inbreeding impractical in pearl oyster hatcheries, it is suggested that hatcheries use as large a number of parents as possible, from different strains, and that the sperm and eggs should be obtained separately for fertilization and mixed afterwards to prevent inbreeding depression (Wada and Komaru, 1994).

Many breeding strategies, including hybridization, have been proposed and conducted in the breeding programs of aquaculture animals to avoid inbreeding depression. In pearl oysters, the results of some hybridization trials to detect and counter inbreeding depression have been reported in order to support a breeding program (Wada, 1984). Figure 12.6 shows survival of some inbred and crossbred strains of the Japanese Akoya

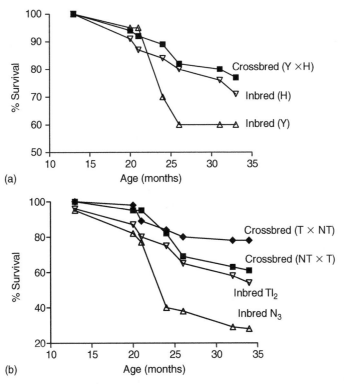

FIGURE 12.6 Survival rates of crossbred and inbred lines of the Japanese Akoya pearl oyster in two crossing experiments. H, N, NT, T, Tl and Y are the names of strains. The egg source is given first in crossbreds. N_3 refers to the 3rd generation of N strain. Tl_2 refers to the 2nd generation of a T strain selected for larger shell width (Wada, 1984).

pearl oyster and shows higher survival in crossbred than in inbred strains. Selected traits, like shell width, were also reported to be improved in crossbred strains than those of inbred strains. Increased variation in morphology of shell traits has been preliminarily reported in the first generation of crosses of three different local populations of the Chinese Akoya, compared with those of three inbred lines (Wang *et al.*, 2003a).

Recent prevalence of a new disease in the Japanese Akoya pearl oyster led researchers and farmers to start practical scientific breeding of culture stock. Since empirical evidence showed high resistance of the southern (subtropical) population derived from southern China to this new emergent syndrome, trials of cross breeding between the Japanese northern strains and southern strains were undertaken to develop hybrid strains with resistance to the new syndrome. In some cases the hybrid has high resistance (Uchimura *et al.*, 2005), but this is not so in all cases, with some problems in the quality of pearls produced also evident as a result of crossing.

12.4. CHROMOSOME MANIPULATIONS

Progress in the development of ploidy manipulation techniques in molluscs was reviewed by Beaumont and Fairbrother (1991) and Beaumont (2006). Interest in chromosome manipulations in pearl oysters has focused on gonad maturation and pearl quality in triploids.

12.4.1. Production of triploids and tetraploids

In bivalve molluscs, triploids can be produced by suppressing either meiosis I or II (Gosling, 2003) with chemical or physical treatment of fertilized eggs, and their production has been reported in many species. In the initial phase of chromosome manipulation studies, interest was focused on methods of induction, ploidy verification and practical evaluation of methods for mass production of triploids in the hatchery. In pearl oysters, attempts were made to induce triploidy using a variety of treatments including cytochalasin B (CB), combined heat and caffeine (CA) shock, hydrostatic pressure, and 6-DMAP (Jiang *et al.*, 1987; Wada *et al.*, 1989; Durand *et al.*, 1990; Komaru *et al.*, 1990a,b; Shen *et al.*, 1993; Fig. 12.7). Methods of ploidy verification have been much improved from the original method of chromosome counting air-dried cells of chopped larvae (Wada, 1976, 1978; Komaru and Wada, 1985; Wada and Komaru, 1985) and now generally consist of measuring relative amount of nucleic acids in eggs, embryo or larval cell nuclei with fluorescent dyes (Uchimura *et al.*, 1989; Komaru *et al.*, 1990a,b). Triploid production rates have been reported to reach 100%, but like triploid induction in most mollusc species, it is highly dependant on combinations and concentrations of chemicals, the timing of polar body interference after fertilization, duration of shock treatment and egg quality (Fig. 12.7; Wada *et al.*, 1989; Uchimura *et al.*, 1989). It seems, however, that the technology for mass production of triploids in pearl oysters is not being used commercially. Since high percentages of triploids are routinely produced

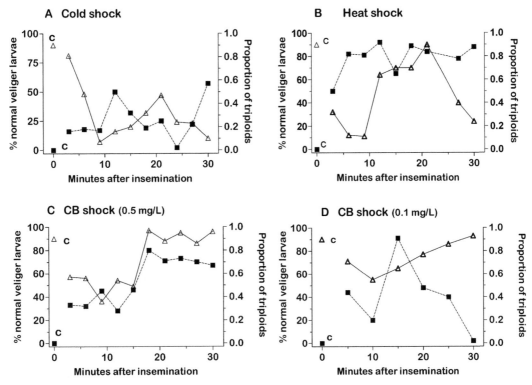

FIGURE 12.7 The effect of treatment for inducing triploidy represented by the percentage of eggs developing to normal veliger larvae (solid lines and triangles) and the proportion of triploid larval cells (dotted lines and squares) in the Japanese Akoya pearl oyster. A: Cold shock (5.0°C). B: Heat shock (35.2°C). C: Cytochalasin B (CB) treatment (0.5 mg/L). D: CB treatment (0.1 mg/L). Minutes after insemination shows the time of initiation of treatments (duration of each treatment = 15 min.). c = untreated controls. From 50 to 100 larvae or embryos were examined in each group. Temperature of incubation of fertilized eggs before treatment was 24°C in A–C, 25°C in D. Ploidy was estimated from the microfluorometry of relative DNA content of suspended cells from more than 20 larvae. (Wada *et al.*, 1989).

in hatcheries for a variety of other commercial bivalves, there seems to be no obstacle to their commercial production in pearl oyster hatcheries.

Tetraploid embryos can theoretically be induced in the pearl oyster when first polar body extrusion is inhibited (Komaru *et al.*, 1990b) or when triploid eggs are fertilized with diploid sperm following the inhibition of the first polar body (He *et al.*, 2000c). However, the stable production of tetraploids in pearl oysters through inhibition of the first polar body has proven elusive and it often results in the development of aneuploid embryos, i.e. with irregular chromosome numbers, more than in the case of second polar body inhibition (Komaru *et al.*, 1990a, b; He *et al.*, 2000a, b, 2002, 2004). Further research should focus on how to increase the survival rate of tetraploids.

12.4.2. Maturation and growth of triploids

The degree of gonad maturation in triploid Akoya pearl oysters varies with geographic population, age, farm environment and year (Uchimura *et al.*, 1989; Jiang *et al.*, 1990, 1993; Komaru and Wada, 1990; Matsuda and Yamakawa, 1993; He *et al.*, 1996). In most cases, triploid Akoya oysters were reported to produce more or less matured gametes, although gametogenesis obviously was much retarded in triploids derived from the same parents as those of control diploids. Experimental crosses between triploid females and diploid males produced viable larvae and juveniles in Japanese Akoya oysters, although many individuals developed abnormally (Komaru and Wada, 1994a). Oocytes from two female triploid oysters were artificially fertilized with spermatozoa from a diploid male ($3n \times 2n$ cross) to observe meiotic maturation in the triploid eggs and subsequent larvae and juveniles. Meiosis and development up to the four-cell stage appeared to be virtually identical to diploid controls, but karyological analyses revealed high levels of aneuploidy, with a broad range of chromosome numbers compared to the controls. Veliger larvae were highly abnormal and survival up to 4 months was very low in the $3n \times 2n$ group. Aneuploids in the $3n \times 2n$ cross died during the larval stage and only euploid ($2n$ and $3n$) juveniles survived (Komaru and Wada, 1994a). He *et al.* (1996) reconfirmed this finding with similar results for Chinese Akoya oysters and concluded that triploids were sterile because embryos from triploids did not develop normally. It is necessary to determine whether triploids produce abnormal gametes when they spawn on farms and, if so, what genetic and ecological risks do spawning triploids pose for wild stocks (Wada, 1992).

12.4.2.1. Growth

Several studies have reported improvements in growth rate at maturation in triploid pearl oysters, while others have reported no differences in growth. Jiang *et al.* (1991, 1993), Komaru and Wada (1994b) and Uchimura (1999) reported in both Chinese and Japanese Akoya populations that triploids became much larger than diploids after maturation (ca. 1 year-old), presumably because the available energy normally used for both growth and gametogenesis was directed purely to growth. Matsuda and Yamakawa (1993), however, observed no significant differences between the growth of triploid and diploid Japanese Akoya oysters.

Jiang *et al.* (1993) compared the height, length and width of shell, weight of body, and soft tissue of triploid pearl oysters produced by inhibition of the first polar body (Group I) and second polar body (Group II) with diploid siblings of Chinese Akoya oysters at 8 or 9 months and 23 or 24 months of age. Except for shell height, width and body weight at 8 months in Group II animals, all the traits were significantly superior in triploids compared with diploids. Differences in daily growth rate between the triploids and diploids were significantly higher during the reproductive season, but not significantly different in most months of the non-reproductive season. This supports the concept that retarded gonad development is a major cause of rapid growth in triploids (Jiang *et al.*, 1993). Komaru and Wada (1994b) obtained similar results in the Japanese

TABLE 12.6
Comparison of shell width and shell weight of diploid and triploid Japanese Akoya pearl oysters raised in Gokasho Bay, Mie prefecture Japan at 12, 21 and 26 month old of age.

Trait	Ploidy	12 month (June)	21 month (March)	26 month (August)
			Age (Season)	
Shell width (cm)	Diploid	1.45 (16)	81 (81)	2.19 (16)
	Triploid	1.30 (44)	1.97 (59)	2.57 (34)
Shell weight (g)	Diploid	5.07 (16)	10.21 (81)	15.81 (16)
	Triploid	4.16 (44)	12.44 (59)	23.30 (34)

Figures in the parenthesis refers to number of animals measured. All comparisons between diploid and triploid at the same age were significantly different ($P < 0.01$). Data from Komaru and Wada (1994b).

Akoya oyster (Table 12.6). In a large study comparing growth between triploid and diploid pearl oysters from Uwakai, Ehime prefecture, Uchimura (1999) found three kinds of growth curves for body weight between the ages of 7 and 17 months in 10 growth experiments. In six experiments, growth of triploids exceeded that of diploids during the adult stage, while in two others, growth of triploids was better than that of diploids during the whole period of evaluation. Finally, in the other two experiments, diploid growth exceeded that of triploids during both young and adult stages. They concluded that triploids generally had higher growth rates than diploids (Uchimura, 1999).

12.5. EMERGING TECHNOLOGIES AND FUTURE APPLICATIONS

Compared to most animal production industries, aquaculture has been relatively slow to embrace new genetic technologies as a means to improve profitability. In many cases, there has been a lag of 10–20 years between experimental evaluations and application by the industry (Hulata, 2001). The uptake of new genetic technologies by the pearl culture industry has been even slower and it is only relatively recently that even fundamental information has been obtained for the genetic basis of any important production traits. Consequently, there are few examples where newer molecular genetic technologies have been applied to pearl oysters. The integration of molecular genetic techniques into future selective breeding programs will, however, assist managers to achieve faster genetic responses than are achievable solely through traditional selection processes. Genes that have a major effect on trait expression can be directly targeted. It will also be possible to reduce the rate of inbreeding that can diminish genetic gains through the use of molecular pedigree based methods. Therefore, given the requirement for pearling companies to remain internationally competitive in the future, it is anticipated that genetic techniques like DNA parentage analyses, gene marker assisted selection and transgenics will become more widely utilized within the industry.

12.5.1. DNA parentage

Historically, the retention of pedigree information on aquaculture species throughout the entire production process has been one of the biggest challenges to researchers and farm managers alike. Due to the small initial size of early developmental stages and the large numbers of progeny that can be reared, individual tagging is often impossible and, as a consequence, the preferred genetic unit of pedigree management in most aquaculture breeding programs is that of the family (Boudry *et al.*, 2002). In those species that can be single-pair spawned, retention of family-level pedigrees is easily achieved by rearing families separately in tanks until individuals are large enough to be physically tagged and communally stocked. Although this separate rearing of families has the potential to introduce environmental heterogeneity between culture vessels and therefore bias in genetic analyses, it allows pedigree identification to be easily maintained. This approach has proven very successful for those species where individual full-sib families can easily be produced, but cannot be applied where cohorts of progeny result from mass spawning, as is common practice when propagating many pearl oyster species. The acquisition of initial family pedigree data in these species is more problematic because individual offspring belonging to a particular full-sib family cannot be identified.

Whether the goal of a pearl oyster breeding program is to produce spat for on-growing, or to produce improved genetic lines, preventing rapid inbreeding is essential to have some control over mating choices. Genealogical information must be retained; otherwise it is difficult to prevent consanguineous matings. The development of highly polymorphic DNA-based genetic markers such as microsatellites has revolutionized the way pedigree information can be obtained (Wilson and Ferguson, 2002; Liu and Cordes, 2004), and DNA parentage analyses are now being applied to retain pedigree information in many aquaculture breeding programs (Estoup *et al.*, 1998; Norris *et al.*, 2000; Boudry *et al.*, 2002; Walker *et al.*, 2002; Sekino *et al.*, 2003, 2004; Jerry *et al.*, 2004; McDonald *et al.*, 2004; Frost *et al.*, 2006). This approach has enormous applications to pedigree identification in pearl oysters, although there is limited published information on their use in these bivalves. However, with the recent interest in applying selective breeding practices to pearl oyster species it will be crucial to have a method of verifying pedigree so that maximum genetic gains can be achieved while at the same time controlling the rate of inbreeding. It is here where DNA parentage analyses will be applied. Currently, there is only one published example of a DNA parentage marker suite for a pearl oyster (Evans *et al.*, 2006), although microsatellite loci with potential application to DNA parentage analyses have been published for *P. maxima* (Smith *et al.*, 2003; Evans *et al.*, 2006) and *P. margaritifera* (Herbinger *et al.*, 2006). In the marker suite of Evans *et al.* (2005), simulations based on allele frequency information at six microsatellite loci were evaluated for their ability to assign progeny to a range of 2–50 families of *P. maxima* (Fig. 12.8). Assignment of progeny to the correct parents ranged from 100% if the cohort comprised only 4 families, to 75% where there were 50 families. Individual markers also had different levels of power to assign parentage (PIC ranged from 0.19–0.87). The study of Evans *et al.* (2006) illustrates

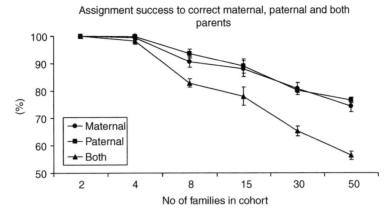

FIGURE 12.8 Simulation of power of six microsatellite markers developed by Evans *et al.* (2006) for *Pinctada maxima* to assign parentage of pearl oyster progeny to their correct maternal, paternal and both parents.

two important points in the development of future pearl oyster DNA parentage marker suites.

1. The power of the marker suite will depend largely on the polymorphism of each locus. The more alleles there are at an individual locus within the population, the higher its assignment power.
2. The more families that have to be assigned, the greater number of microsatellite markers required in the marker suite.

Additionally, microsatellites are notorious for null alleles and the presence of these in parentage data sets can severely impact on assignment success. Therefore, careful consideration has to be given to the choice of loci for inclusion in the final marker suite because not all loci prove to be useful for parentage determination (Jerry *et al.*, 2006).

12.5.2. Cryopreservation

While gamete cryopreservation techniques have been developed for some mollusc species (Kurokura *et al.*, 1990, Yankson and Moyse, 1991; McFadzen, 1995; Lin and Chao, 2000; Chao and Liao, 2001; Adams *et al.*, 2004), long-term storage of pearl oyster spermatozoa by freezing has received only cursory attention. The future development of specific breeding lines to improve traits like pearl quality and the subsequent requirement to store gametes from genetically superior animals will require development of cryopreservation techniques in this group of bivalves. For example, pearl quality traits like color are closely related to the quality of the mantle from donor oysters sacrificed during pearl grafting (Wada, 1969; Wada and Komaru, 1996; Acosta-Salmón *et al.*, 2004). Cryopreservation of spermatozoa from high quality mantle donors would then allow more efficient dissemination of genes from superior donor oysters, leading to an increase in the overall quality of donors in the commercial population.

Research on cryopreservation techniques for pearl oysters has so far been limited to *P. margaritifera* (Lyons *et al.*, 2005; Rouxel *et al.*, 2005; Acosta-Salmón *et al.*, 2007). Several cryoprotectant agents (CPA) and freezing protocols have been tried with varying success in this species. Of the CPAs evaluated to date, spermatozoa motility appears to be best preserved in 0.45–1.0M trehalose and 0, 5, 8 or 12% dimethyl sulfoxide (DMSO) and has been reported to be as high as 86% that of initial motility (Lyons *et al.*, 2005; Acosta-Salmón *et al.*, 2007). As has been observed for other oysters, slow-cooling rates retain spermatozoa motility better than direct plunging into liquid nitrogen. Lyons (unpublished data) also showed that pearl oyster spermatozoa retain significant levels of motility up to 10 days at 4°C raising the possibility that spermatozoa may be stored for short periods simply through strip spawning (with non-activation) and storing in a refrigerator. However, no studies have yet verified fertilization rate and subsequent possible effects of cryopreserved or cooled spermatozoa on embryos.

12.5.3. Genome mapping

Traditional animal breeding practices are based on assumptions of the infinitesimal model of quantitative genetic variation where small additive effects at numerous genetic loci sum over ensuing generations to lead to an improvement in a particular trait under selection (Falconer and Mackay, 1996). This model of quantitative genetic variation implies that traditional selection practices are a blind process where genotypes are modified without knowledge of which genes are actually involved. Their relative position in the genome to other genes is unknown or the effect of individual alleles on trait expression is not quantified. Recent advances in molecular genetics and the ability to fine-scale map the locations of individual gene regions, or even specific genes themselves, and examine their effects on the expression of traits has demonstrated that the infinitesimal model does not adequately explain all the variation observed. Commonly there are genes that contribute unequally to the expression of a phenotype. These genes have been termed as genes of "major effect", candidate genes, or quantitative trait loci (QTL) (Falconer and Mackay, 1996).

In most production systems for terrestrial animals, there is currently a huge investment into genome mapping and the identification of QTLs and candidate genes. Once QTLs or candidate genes that substantially contribute to the variation observed in a production trait are identified, individuals can be genotyped and the gene variant they possess are identified. They can then be directly selected if they possess favor variants of the gene (marker-assisted-selection – MAS). Once again, application of gene mapping techniques to aquaculture species has been slower, although there are a few QTLs that have been mapped and characterized for rainbow trout: upper thermal tolerance (Jackson *et al.*, 1998; Danzmann *et al.*, 1999; Perry *et al.*, 2001), spawning time (Sakamoto *et al.*, 1999) and embryonic development (Robison *et al.*, 2001). Then there is sex determination in tilapia (Lee and Kocher, unpublished data) and growth heterosis in Pacific oysters (Hedgecock *et al.*, unpublished data), to name a few. However, there has been no published gene mapping research in pearl oysters.

Despite the lack of research in this area with pearl oysters, several genes believed to be involved in nacre and shell prismatic layer biomineralization, and therefore of possible commercial interest, have been isolated. These include the shell matrix proteins, nacrein, N16, N66, N14 and pMSI1, pMSI2, MSI7, MSI31 and MSI60 (Miyamoto *et al.*, 1996; Sudo *et al.*, 1997; Samata et *al.*, 1999; Kono *et al.*, 2000; Zhang *et al.*, 2003; Samata, 2004), pearlin (Miyashita *et al.*, 2000), aspein (Tsukamoto *et al.*, 2004) and PFMG1 (Liu *et al.*, 2007). Although the exact contribution of these genes in determining pearl quality has not been publicly divulged, it demonstrates a recent stimulated interest in understanding the process of biomineralization in pearl oysters and the genetic factors underpinning this process. Future interest in this area, coupled with the application of genetic technologies like DNA microarrays, will facilitate our understanding of the genes determining the underlying phenotypic variation seen in pearl quality traits which in time, may lead to the development of perfect DNA markers for candidate genes that can be applied in MAS programs.

12.6. SUMMARY

Pearl oysters have a substantial period of planktonic larval life and therefore a seemingly high capacity for dispersion of their larvae. However, inferences from population genetic surveys in this group suggest that most species are relatively genetically structured throughout their distribution. This finding has important implications for long-term management of pearl oyster species, as genetic structure may also equate into local adaptation and variation in the expression of commercially important traits like pearl color. There is a history of movement of species within their natural distributions and, as selective breeding practices become more widespread in the future, these translocations may lead to similar levels of homogenization of genetic structure as those reported in French Polynesia for *P. margaritifera*. There is insufficient information on degrees of genetic structure across the entire distributions of the commercially important pearl oysters. This information will be essential for future management and exploitation of pearl oyster resources in a sustainable way.

The recent development of hatchery culture techniques for a number of pearl oyster species will lead to the use of quantitative genetic breeding techniques as a way to improve productivity. Despite limited information on inheritance of the characters on pearl formation, some shell traits in the Akoya pearl oyster are inherited with moderate estimated values of heritability. Traits studied were shell coloration (internal or external), shell width, shell shape (focusing on the ratios of shell width or convexity) and shell weight. These characters showed some evidence of inheritance and therefore could be improved through selective breeding and/or cross breeding. Chromosome manipulation has been trialed in Akoya pearl oysters. However, triploidy induction methodologies are not yet reliable or cost effective enough to permit wide-scale uptake by industry. The effects of triploidy on the productivity of commercial traits have also not been reliably reported. In addition, data are needed on whether triploid animals spawn naturally in the farm when raised together with diploid oysters and what impact this may have on local wild populations.

The pearl culture industry has been slow in adopting emerging genetic technologies, molecular biology and cryopreservation; and it is only relatively recently that even fundamental information on the genetic basis for any important production traits has been obtained. Consequently, there are few examples where new molecular genetic technologies have been applied to pearl oysters. However, several genes believed to be involved in nacre and shell prismatic layer biomineralization, and therefore of possible commercial interest, have been isolated. Future interest in this area, coupled with the application of genetic technologies like DNA microarrays, will greatly facilitate understanding of which genes determine the underlying phenotypic variation seen in pearl quality traits and in time may lead to the development of perfect DNA markers for candidate genes that can be applied in MAS programs. The integration of molecular genetic techniques into future selective breeding programs will assist managers to achieve faster genetic responses than are achievable solely through traditional selection. This will allow genes that have a major effect on trait expression to be directly targeted or reduce the rate of inbreeding that can diminish genetic gains. It is expected that genetic techniques like DNA parentage analyses, gene marker assisted selection, and transgenics will become more widely utilized within the cultured pearl industry in the future.

References

Acosta-Salmón, H., Martinez-Fernandez, E., Southgate, P.C., 2004. A new approach to pearl oyster broodstock selection: can saibo donors be used as future broodstock? *Aquaculture* 231, 205–214.

Acosta-Salmón, H., Jerry, D.R., Southgate, P.C., 2007. Effects of cryoprotectant agents and freezing protocol on motility of black-lip pearl oyster (*Pinctada margaritifera* L.) spermatozoa. *Cryobiology* 54, 13–18.

Adams, S.L., Smith, J.F., Rodney, D.R., Janke, A.R., Kaspear, H.F., Tervit, H.R., Pugh, P.A., Steven, C.W., King, N.G., 2004. Cryopreservation of sperm of the Pacific oyster (*Crassostrea gigas*): development of a practical method for commercial spat production. *Aquaculture* 242, 271–282.

Arnaud-Haond, S., Monteforte, M., Bonhomme, F., Blanc, F., 2001. Population structure and genetic variability of pearl oyster *Pinctada mazatlanica* along Pacific coasts from Mexico to Panama. *Conserv. Genet.* 1, 299–307.

Arnaud-Haond, S., Bonhomme, F., Blanc, F., 2003a. Large discrepancies in differentiation of allozyme, nuclear and mitochondrial DNA loci in recently founded Pacific populations of the pearl oyster *Pinctada margaritifera*. *J. Evol. Biol.* 16, 388–398.

Arnaud-Haond, S., Monteforte, M., Blanc, F., Bonhomme, F., 2003b. Evidence for male-biased effective sex ratio and recent step-by-step colonization in the bivalve *Pinctada mazatlanica*. *J. Evol. Biol.* 16, 790–796.

Arnaud-Haond, S., Vonau, V., Bonhomme, N., Boudry, P., Prou, J., Seaman, T., Veyret, M., Goyard, E., 2003c. Spat collection of the pearl oyster (*Pinctada margaritifera cumingii*) in French Polynesia: an evaluation of the potential impact on genetic variability of wild and farmed populations after 20 years of commercial exploitation. *Aquaculture* 219, 181–192.

Arnaud-Haond, S., Vonau, V., Bonhomme, F., Boudry, P., Blanc, F., Prou, J., Seaman, T., Goyard, E., 2004. Spatio-temporal variation in the genetic composition of wild populations of pearl oyster (*Pinctada margaritifera cumingii*) in French Polynesia following 10 years of juvenile translocation. *Mol. Ecol.* 13, 2001–2007.

Atsumi, T., Komaru, A., Okamoto, C., 2004. Genetic relationship among the Japanese pearl oyster *Pinctada fucata martensii* and foreign pearl oysters. *Fish Genet. Breed. Sci.* 33, 135–142. (in Japanese with English Summary)

Beaumont, A., 2006. Genetics. *Dev Aquacult Fish Sci.* 35, 543–594.

Beaumont, A.R., Khamdan, S.A.A., 1991. Electrophoretic and morphometric characters in population differentiation of the pearl oyster, *Pinctada radiata* (Leach), from around Bahrain. *J. Mollus. Stud.* 57, 433–441.

Beaumont, A.R., Fairbrother, J.E., 1991. Ploidy manipulation in molluscan shellfish: review. *J. Shellfish Res.* 10, 1–18.

Beaumont, A.R., Zouros, E., 1991. Genetics of scallops. In: Shumway, S.E. (Ed.), *Scallops: Biology, Ecology and Aquaculture*. Elsevier Science Publishers B.V., Amsterdam, pp. 585–623.

Beaumont, A.R., 1986. Genetic aspects of hatchery rearing of the scallop, *Pecten maximus* (L.). *Aquaculture* 57, 99–110.

Beaumont, A.R., 2000. Genetic considerations in hatchery culture of bivalve shellfish. In: Fingerman, M., Nagabhushanam, R. (Eds.), *Recent Advances in Marine Biotechnology, Vol. 4, Aquaculture Part A, Seaweeds and Invertebrates*. Science Publishers Inc., New Hampshire, USA, pp. 75–85.

Benzie, J.A.H., Ballment, E., 1994. Genetic differences among black-lipped pearl oyster (*Pinctada margaritifera*) populations in the western Pacific. *Aquaculture* 127, 145–156.

Benzie, J.A.H., Smith, C., 2003. Mitochondrial DNA reveals genetic differentiation between Australian and Indonesian pearl oyster *Pinctada maxima* (Jameson 1901) populations. *J. Shellfish Res.* 22, 781–787.

Benzie, J.A.H., Smith-Keune, C., 2006. Microsatellite variation in Australian and Indonesian pearl oyster (*Pinctada maxima*) populations. *Mar. Ecol- Prog. Ser.* 314, 197–211.

Blanc, F., 1983. Estimation du polymorphisme enzymatique dans trios p populations naturelle de nacre (*Pinctada margaritifera*) en Polynesie francaise. *C.R. Acad. Sci. Paris* 297, 199–202.

Blanc, F., Durand, P., Shinh-Milhaud, M., 1985. Genetic variability in populations of black pearl oyster *Pinctada margaritifera* (mollusc: bivalve) from Polynesia. *Proceedings of the 5th International Coral Reef Congress*, Vol. 4, Tahiti, pp. 113–118.

Blanc, F., Durand, P., 1989. Genetic variability in natural bivalve populations: the case study of the black-lipped pearl oyster *Pinctada margaritifera*. *La Mer* 27, 125–126.

Boudry, P., Collet, B., Cornette, F., Hervouet, V., Bonhomme, F., 2002. High variance in reproductive success of the Pacific oyster (*Crassostrea gigas*, Thunberg) revealed by microsatellite-based parentage analysis of multifactorial crosses. *Aquaculture* 204, 283–296.

Busack, C.A., Currens, K.P., 1995. Genetic risks and hazards in hatchery operations: fundamental concepts and issues. *Am. Fish. Soc. Symp.* 15, 71–80.

Chao, N.-H., Liao, I.C., 2001. Cryopreservation of finfish and shellfish gametes and embryos. *Aquaculture* 197, 161–189.

Colgan, D.J., Ponder, W.F., 2002. Genetic discrimination of morphologically similar, sympatric species of pearl oysters (Mollusca: Bivalve: *Pinctada*) in eastern Australia. *Mar. Freshwater Res.* 53, 697–709.

Cross, T.F., King, J., 1983. Genetic effects of hatchery rearing in Atlantic salmon. *Aquaculture* 33, 33–40.

Danzmann, R.G., Jackson, T.R., Ferguson, M., 1999. Epistasis in allelic expression at upper temperature tolerance QTL in rainbow trout. *Aquaculture* 173, 45–58.

Durand, P., Blanc, F., 1986. Divergence génétique chez un bivalve marin tropical: *Pinctada margaritifera*. In: *Biologie des Populations*. Coll. Natl. CNRS, Lyon, France, pp. 323–330.

Durand, P., Blanc, F., 1989. Diversite genetique chez un bivalve marin tropical: *Pinctada margaritifera* (Linne, 1758). *Bull. Soc. Zool. Fr.* 113, 293–304.

Durand, P., Wada, K., Blanc, F., 1993. Genetic variation in wild and hatchery stocks of the black pearl oyster, *Pinctada margaritifera*, from Japan. *Aquaculture* 110, 27–40.

Durand, P., Wada, K.T., Komaru, A., 1990. Triploidy induction by caffeine-heat shock treatments in the Japanese pearl oyster *Pinctada fucata martensii*. *Nippon Suisan Gakkaishi (Bull. Jap. Soc. Sci. Fish.)* 56, 1423–1425.

Estoup, A., Gharbi, K., San Cristobal, M., Chevalet, C., Haffray, P., Guyomard, R., 1998. Parentage assignment using microsatellites in turbot (*Scophthalmus maximus*) and rainbow trout (*Oncorhynchus mykiss*) hatchery populations. *Can. J. Fish. Aquat. Sci.* 55, 715–725.

Evans, B., Bartlett, J., Sweijd, N., Cook, P., Elliott, N.G., 2004. Loss of genetic variation at microsatellite loci in hatchery produced abalone in Australia (*Haliotis rubra*) and South Africa (*Haliotis midae*). *Aquaculture* 233, 109–127.

Evans, B.S., Knauer, J., Taylor, J.J., Jerry, D.R., 2005. A reliable DNA pedigree system for *Pinctada maxima* breeding studies. World Aquaculture Society Annual Conference, Bali, Indonesia.

Evans, B.S., Knauer, J., Taylor, J.J.U., Jerry, D.R., 2006. Development and characterisation of six new microsatellite markers for the silver or gold lipped pearl oyster, *Pinctada maxima* (Pteriidae). *Mol. Ecol. Notes* 6, 835–837.

Falconer, D.S., Mackay, T.F.C., 1996. *Introduction to Quantitative Genetics*, 4th Edition. Longman, Burnt Mill, England. 464 pp.

Frost, L.A., Evans, B.S., Jerry, D.R., 2006. Loss of genetic diversity due to hatchery culture practices in barramundi (*Lates calcarifer*). *Aquaculture* 261, 1056–1064.

Fujio, Y., Yamanaka, R., Smith, P.J., 1983. Genetic variation in marine molluscs. *Nippon Suisan Gakkaishi (Bull. Jap. Soc. Sci. Fish.)' Sci. Fish.)* 49, 1809–1817.

Gervis, M.H., Sims, N.A., 1992. *The Biology and Culture of Pearl Oysters* (*Bivalvia: Pteriidae*). ICLARM Stud. Rev. 21, 49 pp.

Gosling, E.M., 2003. *Bivalve Molluscs: Biology, Ecology and Culture*. Blackwell Publishing, Oxford, 456 pp.

He, M., Lin, Y., Jiang, W., 1996. Studies on the sterility of triploid pearl oyster, *Pinctada martensii* (D.). *Redai Haiyang (Trop. Oceanol.)* 15, 17–21. (in Chinese with English Abstract)

He, M., Lin, Y., Shen, Q., Hu, J., Jiang, W., 2000a. Production of tetraploid pearl oyster (*Pinctada martensii* Dunker) by inhibiting the first polar body in eggs from triploids. *J. Shellfish Res.* 19, 147–151.

He, M., Lin, Y., Shen, Q., Hu, J., Jiang, W., 2000b. Production of aneuploid *Pinctada martensii* (Dunker) in tetraploid induction. *Redai Haiyang (Trop. Oceanol.)* 24, 22–27. (in Chinese with English Abstract)

He, M., Shen, Q., Lin, Y., Hu, J., Jiang, W., 2000c. Inducement of tetraploid *Pinctada martensii* in eggs from triploids. *Shuichan Xuebao (J. Fish. China)* 24, 22–27. (in Chinese with English Abstract)

He, M., Jiang, W., Pan, J., 2002. Chromosome segregation in fertilized eggs from pearl oyster *Pinctada martensii* Dunker following the first polar body inhibition with cytochalasin B. *Shuichan Xuebao (J. Fish. China)* 26, 15–20. (in Chinese with English Abstract)

He, M., Jiang, W., Huang, L., 2004. Studies on aneuploid pearl oyster (*Pinctada martensii* Dunker) produced by using triploid females and a diploid male following inhibition of PB1. *Aquaculture* 230, 117–124.

Hedgecock, D., Sly, F., 1990. Genetic drift and effective population sizes of hatchery-propagated stocks of the Pacific oyster, *Crassostrea gigas*. *Aquaculture* 88, 21–28.

Hedgecock, D., 1994. Does variance in reproductive success limit effective population sizes of marine organisms?. In: Beaumont, A.R. (Ed.), *Genetics and Evolution of Aquatic Organisms*. Chapman and Hall, London, pp. 122–134.

Herbinger, C.M., Smith, C.-A., Langy, S., 2006. Development and characterization of novel tetra- and dinucleotide microsatellite markers for the French Polynesia black-lipped pearl oyster, *Pinctada margaritifera*. *Mol. Ecol. Notes* 6, 107–109.

Hulata, G., 2001. Genetic manipulations in aquaculture: a review of stock improvement by classical and modern technologies. *Genetica* 111, 155–173.

Jackson, T.R., Ferguson, M.M., Danzmann, R.G., Fishback, A.G., Ihssen, P.E., O'Connell, M., Crease, T.J., 1998. Identification of two QTL influencing upper temperature tolerance in three rainbow trout (*Oncorhynnchus mykiss*) half-sib families. *Heredity* 80, 143–151.

Jerry, D.R., Preston, N.P., Crocos, P.J., Keys, S., Meadows, J.R.S., Li, Y., 2004. Parentage determination of Kuruma shrimp *Penaeus* (*Marsupenaeus*) *japonicus* using microsatellite markers (Bate). *Aquaculture* 235, 237–247.

Jerry, D.R., Evans, B.S., Kenway, M., Wilson, K., 2006. Development of a microsatellite DNA parentage marker suite for black tiger shrimp *Penaeus monodon*. *Aquaculture* 55, 542–547.

Jiang, W., Li, G., Lin, Y., Xu, G., Qing, N., 1990. Observation on the gonad of triploidy in *Pinctada martensii* (D.). *Redai Haiyang (Trop. Oceanol.)* 9, 24–30. (in Chinese with English Abstract)

Jiang, W., Li, G., Xu, G., Lin, Y., Qing, N., 1993. Growth of the triploid pearl oyster, *Pinctada martensii* (D.). *Aquaculture* 111, 245–253.

Jiang, W.G., Li, G., Lin, Y.G., Qing, N., 1987. Induced polyploidization in pearl oyster. *Redai Haiyang (Trop. Oceanol.)* 6, 37–45. (in Chinese with English Summary)

Jiang, W.G., Xu, G., Lin, Y., Li, G., 1991. Comparison of growth between triploid and diploid of *Pinctada martensii* (D.). *Redai Haiyang (Trop. Oceanol.)* 10, 1–7. (in Chinese with English Summary)

Johnson, M.S., Black, R., 1984. Pattern beneath the chaos: the effect of recruitment on genetic patchiness in an intertidal limpet. *Evolution* 38, 1371–1383.

Johnson, M.S., Joll, L.M., 1993. Genetic subdivision of the pearl oyster *Pinctada maxima* (Jameson, 1901) (Mollusca: Pteriidae) in Northern Australia. *Aust. J. Mar. Freshwater Res.* 44, 519–526.

Knauer, J., Jerry, D.R., Taylor, J.J.U., Evans, B.S., 2005. Producing silver- or gold-lip pearl oysters (*Pinctada maxima*) in a commercial hatchery: mass spawning versus family lines. World Aquaculture Society Annual Conference, Bali, Indonesia.

Komaru, A., Wada, K.T., 1985. Karyotype of the Japanese pearl oysters, *Pinctada fucata martensii*, observed in the trochophore larvae. *Yoshoku Kenkyusho Kenkyu Hokoku (Bull. Natl. Res. Inst. Aquacult.)* 7, 105–107.

Komaru, A., Wada, K.T., 1990. Gametogenesis of triploid Japanese pearl oyster, *Pinctada fucata martensii*. In: Hoshi, H., Yamashita, K. (Eds.), *Advances in Invertebrate Reproduction*. Elsevier, Amsterdam, pp. 469–474.

Komaru, A., Wada, K.T., 1994a. Gametogenesis and growth of induced triploid Japanese pearl oyster, *Pinctada fucata martensii*. *Suisanzoshoku (J. Jpn. Soc. Aquacult.)* 42, 541–546. (in Japanese with English Summary)

Komaru, A., Wada, K.T., 1994b. Meiotic maturation and progeny of oocytes from triploid Japanese pearl oysters (*Pinctada fucata martensii*) fertilized with spermatozoa from diploids. *Aquaculture* 120, 61–70.

Komaru, A., Matsuda, H., Yamakawa, T., Wada, K.T., 1990a. Meiosis and fertilization of the Japanese pearl oysters eggs at different temperature observed with a fluorescence microscope. *Nippon Suisan Gakk.* 56, 425–430.

Komaru, A., Matsuda, H., Yamakawa, T., Wada, K.T., 1990b. Chromosome behavior of meiosis inhibited eggs with cytochalasin B in Japanese pearl oyster. *Nippon Suisan Gakkaishi (Bull. Jap. Soc. Sci. Fish.)* 56, 1419–1422.

Kono, M., Hayashi, N., Samata, T., 2000. Molecular mechanism of the nacreous layer formation in *Pinctada maxima*. *Biochem. Biophys. Res. Comm.* 269, 213–218.

Kurokura, H., Namba, K., Ishikawa, T., 1990. Lesions of spermatozoa by cryopreservation in oyster *Crassostrea gigas*. *Nippon Suisan Gakkaishi (Bull. Jap. Soc. Sci. Fish.)* 56, 1803–1806.

Lannan, J.E., 1972. Estimating heritability and predicting response to selection for the Pacific oyster, *Crassostrea gigas*. *Proc. Natl. Shellfish Assoc.* 62, 62–66.

Li, G., Hedgecock, D., 1991. Comparison of biochemical genetic variation between La Paz and Mazatlan populations of pearl oyster, *Pinctada mazatlanica* Hanley, in the Gulf of California, Mexico. *Redai Haiyang (Trop. Oceanol.)* 10, 56–61. (in Chinese with English Abstract)

Li, G., Hedgecock, D., 1998. Genetic heterogeneity, detected by PCR-SSCP, among samples of larval Pacific oysters (*Crassostrea gigas*) supports the hypothesis of large variance in reproductive success. *Can. J. Fish. Aquat. Sci.* 55, 1025–1033.

Li, G., Jin, Q., Jiang, W., 1985. Biochemical genetic variation in the pearl oysters, *Pinctada fucata* and *P. chemnitz*. *Acta Genet. Sin.* 12, 202–214. (in Chinese with English Abstract)

Li, Q., Park, C., Endo, T., Kijima, A., 2004. Loss of genetic variation at microsatellite loci in hatchery strains of the Pacific abalone (*Haliotis discus hannai*). *Aquaculture* 235, 207–222.

Lin, T.-T., Chao, N.-H., 2000. Cryopreservation of eggs and embryos of shellfish. In: Tiersch, T.R., Mazik, P.M. (Eds.), *Cryopreservation in Aquatic Species*. The World Aquaculture Society, Baton Rouge, Louisiana, pp. 240–250.

Lind, C., Evans, B.S., Taylor, J.J.U., Knauer, J., Jerry, D.R., 2007. Genetic diversity and spatial structure of a high-dispersal marine bivalve, *Pinctada maxima*, throughout the Indo-Australian Archipelago and Great Barrier Reef. *Mol. Ecol,* 16, 5193–5203.

Lind, C., 2008. Effects of aquaculture on genetic diversity in the silver-lipped pearl oyster *Pinctada maxima*. PhD thesis, James Cook University, Townsville, Queensland, Australia.

Liu, H., Liu, S.F., Ge, Y.J., Lie, J., Wang, X.Y., Xie, L.P., Zhang, R.Q., Wang, Z., 2007. Identification and characterization of a biomineralization related gene PFMG1 highly expressed in the mantle of *Pinctada fucata*. *Biochemistry* 46, 844–851.

Liu, Z.J., Cordes, J.F., 2004. DNA marker technologies and their applications in aquaculture genetics. *Aquaculture* 238, 1–37.

Lyons, L., Jerry, D.R., Southgate, P.C., 2005. Cryopreservation of black-lip pearl oyster (*Pinctada margaritifera,* L.) spermatozoa: effects of cryoprotectants on spermatozoa motility. *J. Shellfish Res.* 24, 1187–1190.

McDonald, G.J., Danzmann, R.G., Ferguson, M.M., 2004. Relatedness determination in the absence of pedigree information in three cultured strains of rainbow trout (*Oncorhynchus mykiss*). *Aquaculture* 233, 65–78.

McFadzen, I.R.B., 1995. Cryopreservation of sperm of the Pacific oyster *Crassostrea gigas. Meth. Mol. Biol.* 38, 145–149.

Masaoka, T., Kobayashi, T., 2005. Species identification of *Pinctada imbricata* using intergenicspacer of nuclear ribosomal RNA genes and mitochondrial 16S ribosomal RNA gene regions. *Fish. Sci.* 71, 746–837.

Matsuda, H., Yamakawa, T., 1993. Growth, survival, and sexual maturation of triploid Japanese pearl oyster *Pinctada fucata martensii. Bull. Fish. Res. Inst. Mie* 5, 1–8. (in Japanese with English Abstract)

Minami, H., Akera, S., Kijima, A., 2000. Genetic variability and genetic differences in and among localities in the Japanese pearl oyster, *Pinctada fucata martensii. Fish Genet. Breed. Sci.* 33, 135–142.

Miyamoto, H., Miyashita, T., Okushima, M., Nakano, S., Morita, T., Matsushiro, A., 1996. A carbonic anhydrase from the nacreous layer in oyster pearls. *P. Natl. Acad. Sci. USA* 93, 9657–9660.

Miyashita, T., Takagi, R., Okushima, M., Nakano, S., Miyamoto, H., Nishikawa, E., Matsushiro, A., 2000. Complementary DNA cloning and characterization of pearlin, a new class of matrix protein in the nacreous layer of oyster pearls. *Mar. Biotechnol.* 2, 409–418.

Monteforte, M., Carino, M., 1992. Exploration and evaluation of natural stocks of pearl oysters *Pinctada mazatlanica* and *Pteria sterna* (Bivalvia: Pteriidae): La Paz Bay, South Baja California, Mexico. *AMBIO* 21, 314–320.

Nei, M., 1972. Genetic distance between populations. *Am. Nat.* 106, 283–293.

Nei, M., 1987. *Molecular Evolutionary Genetics.* Columbia University Press, New York, 512 pp.

Norris, A.T., Bradley, D.G., Cunningham, E.P., 2000. Parentage and relatedness determination in farmed Atlantic salmon (*Salmo salar*) using microsatellite markers. *Aquaculture* 182, 73–83.

Paulay, G., 1990. Effects of late Cenozoic sea-level fluctuations on the bivalve faunas of tropical oceanic islands. *Paleobiology* 16, 415–434.

Perry, G.M., Danzmann, R.G., Ferguson, M.M., Gibson, J.P., 2001. Quantitative trait loci for upper thermal tolerance in outbred strains of rainbow trout (*Oncorhynchus mykiss*). *Heredity* 86, 333–341.

Robison, B.D., Wheeler, P.A., Sundin, K., Sikka, P., Thorgaard, G.H., 2001. Composite interval mapping reveals a major locus influencing embryonic development rate in rainbow trout (*Oncorhynchus mykiss*). *J. Hered.* 92, 16–22.

Rogers, J.S., 1972. Measures of genetic similarity and genetic distance. *Stud. Genet.VII. Univ. Tex. Publ.* 7213, 145–153.

Rouxel, C., Bambridge, R., Bellais, M., Pellan, A., Vonau, V., Lo, C.M., Cochard, J-C., 2005. Investigations on the cryopreservation of spermatozoa of pearl oyster *Pinctada margaritifera* in French Polynesia. World Aquaculture 2005, Bali, Indonesia, p.127 (Abstract).

Sakamoto, T., Danzmann, R.G., Okamoto, N., Ferguson, M.M., Ihssen, P.E., 1999. Linkage analysis of quantitative trait loci associated with spawning time in rainbow trout (*Oncorhynchus mykiss*). *Aquaculture* 173, 33–43.

Samata, T., Hayashi, N., Kono, M., Hasegawa, K., Horita, C., Akera, S., 1999. A new matrix protein family related to the nacreous layer formation of *Pinctada fucata. FEBS Lett.* 462, 225–229.

Samata, T., 2004. Recent advances in studies on nacreous layer biomineralization. Molecular and cellular aspect. *Thalassas* 20, 25–44.

Sbordoni, V., Rosa, G.L., Mattoccia, M., Cobolli-Sbordoni, M., De-Matthaeis, E., 1987. Genetic changes in seven generations of hatchery stocks of the Kuruma Prawn, *Penaeus japonicus* (Crustacea, Decapoda). In: Tiews, K. (Ed.), *Proceedings of the World Symposium on Selection, Hybridization and Genetic Engineering in Aquaculture, Bordeaux 27–30 May 1986*, Vol. 1. Heeneman Verlagsgesellschaft, Berlin, pp. 143–155.

Sekino, M., Saitoh, K., Yamada, T., Kumagai, A., Hara, M., Yamashita, Y., 2003. Microsatellite-based pedigree tracing in a Japanese flounder *Paralichthys olivaceus* hatchery strain: implications for hatchery management related to stock enhancement program. *Aquaculture* 221, 255–263.

Sekino, M., Sugaya, T., Hara, M., Taniguchi, N., 2004. Relatedness inferred from microsatellite genotypes as a tool for broodstock management of Japanese flounder *Paralichthys olivaceus*. *Aquaculture* 233, 163–172.

Shen, Y., Zhang, H., Ma, L., 1993. Triploid induction by hydrostatic pressure in the pearl oyster, *Pinctada martensii*. Dunker. *Aquaculture* 110, 221–227.

Sheridan, A.K., 1997. Genetic improvement of oyster production – a critique. *Aquaculture* 153, 165–179.

Smith, C., Benzie, J.A.H., Wilson, K.J., 2003. Isolation and characterization of eight microsatellite loci from silver-lipped pearl oyster *Pinctada maxima*. *Mol. Ecol. Notes* 3, 125–127.

Smith, P.J., 1987. Homozygote excess in the flounder, *Rhombosolea plebeia*, produced by assortative mating. Mar. Biol. 95, 489–492.

Su, T., Cai, Y., Zhang, D., Jiang, S., 2002. RAPD analysis of three cultured populations of *Pinctada martensii*. *Shuichan Xuebao (J. Fish. China)* 9, 106–109. (in Chinese with English Abstract)

Sudo, S., Fujikawa, T., Nagakura, T., Ohkubo, T., Sakaguchi, K., Tanaka, M., Nakashima, K., 1997. Structures of mollusc shell proteins. *Nature* 387, 563–564.

Taniguchi, N., Sumantadinata, K., Iyama, S., 1983. Genetic change in the first and second generations of hatchery stock of Black Seabream. *Aquaculture* 35, 309–320.

Tsukamoto, D., Sarashina, I., Endo, K., 2004. Structure and expression of an unusually acidic matrix protein of pearl oyster shells. *Biochem. Biophys. Res. Commun.* 320, 1175–1180.

Tsutsumi, M., 2002. Seasonal observation of color of adductor muscle and gonad in the hybrid strains of Akoa pearl oysters, China × China and China × Japan. *Zensinren- Gijutu- Kenkyukaiho (Tech. Rep. Pearl Farm. Assoc.)* 15, 19–30. (in Japanese)

Uchimura, Y., 1999. The pearl formation on physiological aspect in triploid Japanese pearl oyster, *Pinctada fucata martensii*. *Bull. Ehime Prefect. Fish. Exp. Stn.* 7, 1–61. (in Japanese with English figures and tables)

Uchimura, Y., Komaru, A., Wada, K.T., Ieyama, H., Yamaki, M., Furuta, H., 1989. Detection of induced triploidy at different ages for larvae of the Japanese pearl oyster, *Pinctada fucata martensii*, by micro-fluorometry with DAPI staining. *Aquaculture* 76, 1–9.

Uchimura, Y., Nishikawa, S., Hamada, K., Hyoudou, K., Hirose, T., Ishikawa, K., Sugimoto, M., Nakajima, N., 2005. Production of strains of the Japanese pearl oyster *Pinctada fucata martensii* with low mortality through less damage caused by infectious disease. *Fish Genet. Breed. Sci.* 34, 91–97.

Velayudhan, T.S., Dharmaraj, S., Victor, A.C.C., Chellam, A., 1995. Colour and thickness of nacre in four generations of Indian pearl oyster, *Pinctada fucata* (Gould) produced in the hatchery. *Indian Council Agricult. Res. Mar. Fish.* 137, 3–6.

Velayudhan, T.S., Chellam, A., Dharmaraj, S., Victor, A.C.C., Kasim, H.M., 1996. Comparison of growth and shell attributes of four generations of the pearl oyster *Pinctada fucata* (Gould) produced in the hatchery. *Indian J. Fish.* 43, 69–77.

Wada, K., 1969. Experimental biological studies on the occurrence or yellow color in pearl. *Kokoritsu Shinju Kenkyusho Hokoku (Bull. Natl. Pearl Res. Lab.)* 16, 1765–1820. (in Japanese with English Summary)

Wada, K., 1972. Relationship between metabolism of pearl sac and pearl quality. *Kokoritsu Shinju Kenkyusho Hokoku (Bull. Natl. Pearl Res. Lab.)* 16, 1949–2027. (in Japanese with English Summary)

Wada, K.T., 1975a. Genetic differentiation of the pearl oyster, *Pinctada fucata* (Gould) collected form Namako lagoon, Kamikoshikijima Island, Kagoshima Prefecture. *Yoshokukenho (Bull. Natl. Res. Inst. Aquacult.)* 19, 2169–2185. (in Japanese with English Summary)

Wada, K.T., 1975b. Experimental estimating heritability for shell attributes of the Japanese pearl oyster *Pincatada fucata* (Gould). *Kokoritsu Shinju Kenkyusho Hokoku (Bull. Natl. Pearl Res. Lab.)* 19, 2157–2168. (in Japanese with English Summary)

Wada, K.T., 1976. Number and gross morphology of chromosomes in the pearl oyster *Pinctada fucata* (Gould) collected from two regions of Japan. *Jpn. J. Malac. (Venus)* 35, 9–14.

Wada, K.T., 1978. Chromosome karyotypes of three bivalves: the oyster *Iognomon alatus* and *Pinctada imbricata*, and the bay scallop, *Argopecten irradians*. *Biol. Bull.* 155, 235–245.

Wada, K.T., 1982. Inter and intra-specific electrophoretic variation in three species of the pearl oysters from the Nansei Island of Japan. *Yoshokukenho (Bull. Natl. Res. Inst. Aquacult.)* 3, 1–10.

Wada, K.T., 1983. White coloration of the prismatic layer in inbred Japanese pearl oyster, *Pinctada fucata martensii. Yoshokukenho (Bull. Natl. Res. Inst. Aquacult.)* 4, 131–133. (in Japanese with English Abstract).

Wada, K.T., 1984. Breeding study of the Japanese pearl oyster, *Pinctada fucata martensii. Yoshokukenho (Bull. Natl. Res. Inst. Aquacult.)* 6, 79–157. (in Japanese with English Summary).

Wada, K.T., 1985. The pearls produced from the groups of pearl oysters selected for color of nacre in shells for two generations. *Yoshokukenho (Bull. Natl. Res. Inst. Aquacult.)* 7, 1–7. (in Japanese with English Summary)

Wada, K.T., 1986a. Genetic variability at four polymorphic loci in Japanese pearl oysters, *Pinctada fucata fucata,* selected for six generations. *Aquaculture* 59, 139–146.

Wada, K.T., 1986b. Color and weight of shells in the selected populations of the Japanese pearl oyster, *Pinctada fucata martensii. Yoshokukenho (Bull. Natl. Res. Inst. Aquacult.)* 9, 1–6. (in Japanese with English Summary)

Wada, K.T., 1986c. Genetic selection for shell traits in the Japanese pearl oyster, *Pinctada fucata martensii. Aquaculture* 57, 171–176.

Wada, K.T., 1987. Selective breeding and intraspecific hybridization – mollusks. In: Tiews, K. (Ed.), *Proceedings of the World Symposium on Selection, Hybridization and Genetic Engineering in Aquaculture, Bordeaux, 27–30 May 1986,* Vol. 2. Heeneman Verlagsgesellschaft, Berlin, pp. 313–332.

Wada, K.T., 1992. Gametogenesis of triploid bivalves with respect to aquaculture. *NOAA Tech. Rep. NMFS* 106, 19–20.

Wada, K.T., 1994. Genetics of pearl oyster in relation to aquaculture. *JARQ-Jpn. Agr. Res. Q.* 28, 276–282.

Wada, K.T., 1995. Quantitative genetics in molluscs. *Fish Genet. Breed. Sci.* 22, 15–26. (in Japanese with English Summary)

Wada, K.T., 1997. Recent research on genetic improvement of aquacultured species with reference to quantitative or population genetics in Japan. *Yoshoku Kenkyusho Kenkyu Hokoku (Bull. Natl. Res. Inst. Aquacult.)* 3, 25–28.

Wada, K.T., 1999. Aquaculture genetics of bivalve molluscs: a review. In: Xu, H.-S., Colwell, R.R. (Eds.), *Proceedings of the International Symposium on Progress and Prospect of Marine Biotechnology (ISPPMB '98). China Ocean Press, Beijing,* pp. 52–67.

Wada, K.T., 2000. Genetic improvement of stocks of the pearl oyster. In: Fingerman, M., Nagabhushanam, R. (Eds.), *Recent Advances in Marine Biotechnology. Vol. 4, Aquaculture Part A, Seaweeds and Invertebrates.* Science Publishers Inc, New Hampshire, USA, pp. 75–85.

Wada, K.T., Komaru, A., 1985. Karyotypes in five species of the Pteriidae (Bivalvia: Pteriomorphia). *Jpn. J. Malac. (Venus)* 44, 183–192.

Wada, K.T., Komaru, A., Uchimura, Y., 1989. Triploid production in the Japanese pearl oyster *Pinctada fucata martensii. Aquaculture* 76, 11–19.

Wada, K.T., Komaru, A., 1990. Inheritance of white coloration of the prismatic layer of shells in the Japanese pearl oyster *Pinctada fucata martensii* and its importance in the pearl oyster industry. *Nippon Suisan Gakk.* 56, 1787–1790.

Wada, K.T., Komaru, A., 1991. Estimation of genetic variation in shell traits of the Japanese pearl oyster. *Yoshokukenho (Bull. Natl. Res. Inst. Aquacult.)* 20, 19–24. (in Japanese with English Abstract)

Wada, K.T., Komaru, A., 1994. Effect of selection for shell coloration on growth rate and mortality in the Japanese pearl oyster, *Pinctada fucata martensii. Aquaculture* 125, 59–65.

Wada, K.T., Komaru, A., 1996. Color and weight of pearls produced by grafting the mantle tissue from a selected population for white shell color of the Japanese pearl oyster *Pinctada fucata martensii* (Dunker). *Aquaculture* 142, 25–32.

Wada, K.T., Komaru, A., Ichimura, Y.., Kurosaki, H., 1995. Spawning peak occurs during winter in the Japanese subtropical population of the pearl oyster, *Pinctada fucata fucata* (Gould, 1850). *Aquaculture* 133, 207–214.

Walker, D., Porter, B.A., Avise, J.C., 2002. Genetic parentage assessment in the crayfish *Orconectes placidus,* a high-fecundity invertebrate with extended maternal brood care. *Mol. Ecol.* 11, 2115–2122.

Wang, A., Yan, B., Ye, L., Zhang, D., Du, X., 2003a. Comparison on main traits of F1 from matings and crosses of different geographical populations in *Pinctada martensii. Shuichan Xuebao (J. Fish. China)* 27, 200–206. (in Chinese with English Summary)

Wang, A., Ding, X., Deng, F.-J., Ye, L., Yan, B., 2003b. The genetic diversity of the first filial generation from matings and crosses of two wild populations (Daya Bay, Guangdong and Sanya, Hainan) in *Pinctada martensii* (Dunker). *Haiyang Shuichan Yanjiu (Mar. Fish. Res.)* 24, 19–25. (in Chinese with English Abstract)

Weir, B.S., Cockerham, C.C., 1984. Estimating F-statistics for the analysis of population structure. *Evolution* 338, 1358–1370.

Wilson, A.J., Ferguson, M.M., 2002. Molecular pedigree analysis in natural populations of fishes: approaches, applications, and practical considerations. *Can. J. Fish. Aquat. Sci.* 59, 1696–1707.

Yankson, K., Moyse, J., 1991. Cryopreservation of the spermatozoa of *Crassostrea gigas* and three other oysters. *Aquaculture* 97, 259–267.

Yu, D.H., Chu, K.H., 2006a. Genetic variation in wild and cultured populations of the pearl oyster *Pinctada fucata* from southern China. *Aquaculture* 258(1–4), 220–227.

Yu, D.H., Chu, K.H., 2006b. Low genetic differentiation among widely separated population of the pearl oyster *Pinctada fucata* as revealed by AFLP. *J. Exp. Mar. Biol. Ecol.* 333, 140–146.

Zhang, Y., Xie, L., Meng, Q., Jiang, T., Pu, R., Chen, L., Zhang, R., 2003. A novel matrix protein participating in the nacre framework formation of pearl oyster, *Pinctada fucata. Comp. Biochem. Phys. B* 135, 565–573.

Zouros, E., Foltz, D.W., 1984. Possible explanations of heterozygote deficiency in bivalve mollusks. *Malacologia* 25, 583–591.

CHAPTER 13

Economics of Pearl Farming

Clem Tisdell and Bernard Poirine

13.1. INTRODUCTION

The pearl oyster industry has experienced substantial economic change particularly in the last 50 years or so. It has been transformed from an industry dependent solely on wild catch to one that depends mainly on the culture of oysters, either taken from the wild, then seeded and cultured (a form of ranching), or on oysters raised in hatcheries and then grown out (see Chapter 7). Moreover, the industry's structure has altered due partly to market developments and new technologies and the spread of knowledge about techniques for culturing pearls.

In this chapter, the market structure of the industry is discussed and related to new technologies, differences in the industry's socio-economic impacts are explored, sources of market supply are considered and features involved in the marketing of pearls are given particular attention. Most, but not exclusive attention, is given to the experiences of the Australian and French Polynesian pearl industries. Australia is the major global producer of the South Sea pearls and French Polynesia is the main global supplier of Tahitian black pearls (see Chapter 9).

According to Love and Langenkamp (2003), South Sea pearls obtained from *Pinctada maxima* and Tahitian black pearls, derived from *Pinctada margaritifera* together account for about a half of the world market by value. Japanese Akoya pearls and Chinese freshwater pearls, produced from mussels, each supply about a quarter of the world market by value. Twenty-five years ago, Japanese Akoya pearls supplied 90% of the value of the world market. However, Japan no longer dominates the global pearl market and Australia, French Polynesia, Indonesia and China have secured substantial shares of the world sales of pearls.

According to IEOM (2005), Japan is still the major world importer of raw pearls, i.e., US $241 million, with a 58.5% market share. It is also the top exporter of worked pearls (necklaces and other pearl made jewelry) with a 25.7% market share, followed by Australia (23.3%), China, Hong Kong and French Polynesia. In 2004, French Polynesia was the top producer of raw seawater pearls, with 26.9% of the total pearl market, followed by Indonesia (23.6%) and Australia (19.5%).

13.2. CHANGING TECHNOLOGIES AND THE INDUSTRY'S MARKET STRUCTURE

New technologies and more widespread access to technologies for cultivating pearls have played a major role in altering the economic structure of the cultured pearl industry. Japanese domination of the global pearl industry during most of the 20th century arose to a large extent as a result of its early development of methods for culturing pearls. Koichi Mikimoto played a key role in the development (see Section 8.9.1), was a very successful innovator and marketer and his firm is reputed at its height to have supplied about three-quarters of the world supply of pearls. Mikimoto Pearls still retains an important general market position but now there are several other important suppliers. It has been claimed that Japan dominated the cultured pearl industry for many years by keeping its implantation techniques secret.

The structure and nature of the pearl industry varies between countries but on a global scale, the pearl oyster industry is dominated by a few large vertically integrated companies. While some small producers have embarked on pearl production from oysters in developing countries, they are not vertically integrated and only contribute a small fraction of global output.

In order to appreciate the interdependence between new technologies in pearl production and market structure, consider the Australian and French Polynesian industries for illustrative purposes.

13.3. THE AUSTRALIAN AND TAHITIAN PEARL INDUSTRIES: CHANGING TECHNOLOGIES AND MARKET STRUCTURE

Australia is the world's major supplier of South Sea pearls. Most of its pearls originate from Western Australia's northern coastal areas (85%), with some contribution from the Northern Territory (14.6%). Queensland makes only a minor contribution to total production (0.2%).

The Australian pearl oyster industry developed as a captive industry in the second half of the 19th century. Mother-of-pearl shells used for buttons and inlay work became its mainstay and Australia supplied up to three-quarters of world output. Australian production peaked by the beginning of the 1920s and then began to decline. Initially this was because open access existed and over-harvesting occurred. Subsequently, by the 1930s the advent of plastic buttons and the Great Depression reduced demand for mother-of-pearl shells. Production virtually ceased during World War II and while there was some recovery in production thereafter, the market eventually faded away. In 1987, permits were no longer issued in Western Australia for the collection of mother-of-pearl shells (Fletcher *et al.*, 2006).

The main reason for this cessation was not only the decline in demand for mother-of-pearl shells but the switch of the Australian pearl industry to the culture of *Pinctada maxima* for pearls. Until 1949, the culture of pearls was prohibited in Western Australia (Fletcher *et al.*, 2006). In 1956, Nicholas Paspaley Snr. formed a joint venture company, Pearls Proprietary Ltd., in conjunction with a Japanese businessman and began culturing South Sea pearls at Kuri Bay, 420 km north of Broome in Western Australia. This, the first commercial venture to culture pearls in Australia, was a success and in 1963, the Paspaley Pearling Company in conjunction with another Japanese company began culturing pearls at Port Essington, east of Darwin in the Northern Territory. By the early 1980s, the Kuri Bay farm was culturing 200,000 oysters per year and the Port Essington farm 70,000 oysters per year.

Paspaley Snr. was a "first mover" in the culture of pearls in Australia. He had many years of experience in the pearl oyster industry prior to this. He was also able to draw on Japanese experience in pearl culture. Eventually, the Paspaley Pearling Company would come to dominate the supply of Australian pearls and account for more than two-thirds of Australian supply. It would become a major force in the global South Sea pearl market. The second major Australian producer of South Sea pearls is M.G. Kailis. Together, these firms dominate the Australian industry.

Whereas the Paspaley Pearling Company has basically specialized in the pearl industry, M.G. Kailis Group entered the industry as a means of diversifying its business interests in its existing marine industries, mainly seafood production and marine services. It entered the industry in 1974 and is the world's second largest global producer of South Sea pearls.

In Western Australia and the Northern Territory, the supply of South Sea pearls is indirectly controlled by government quotas on total allowable catch (TAC) of wild oysters for seeding. Around the mid-1980s these quotas became binding in Western Australia (Fletcher *et al.*, 2006, see Section 9.3.3). These quotas are important for ensuring sustainable catches and for helping to maintain South Sea pearl prices. At the same time, the introduction of quotas undoubtedly advantaged those already established in the industry. While quotas are transferable, there appears to be little trade in these. Greater detail of the quota system used to manage the Western Australian pearl industry and the zones employed by the industry are given in Section 9.3.3.

The second major development on the aquaculture side of relevance to the Australian pearl industry has been the development of hatchery-based production of *P. maxima*. Since 1992 in Western Australia, pearl producers have had an option of meeting some of their TAC by substituting hatchery-based oysters for wild oysters.

However, this substitution has basically been confined to the southernmost pearling zone (Zone 1) of Western Australia where wild recruitment and catches are erratic and where wild catch to effort ratios are much lower than in Zones 2/3 centered on Broome (Fletcher *et al.*, 2006). In 2001, total wild oyster catch per hour in Zone 2/3 was 41.7 whereas in Zone 1 it was only 7.3, and in 2000 it was, respectively, 54.2 compared to 11.3 (Fletcher *et al.*, 2006).

The decision not to use hatchery-bred oysters in Zones 2/3 but to use them in Zone 1 seems to hinge on the comparative cost of the wild catch. In Zone 2/3, it is cheaper to rely on wild catch. While about half of cultured oysters are obtained from hatcheries in Zone 1, overall more than 90% of Western Australia's pearl supply comes from wild catch. Without the use of hatchery-bred oysters in Zone 1, it is doubtful if pearl oyster farming there would be sustainable.

Apart from technological advances in the culture of Australian pearl oysters, advances have also occurred in the harvesting of wild seed oysters and their husbandry. Technological change has resulted in the industry becoming more capital intensive in Australia. When such economies in marketing are also taken into account, significant economies of scale seem to be experienced by this industry. Apart from barriers to entry created by the quota system, scale economies may be a significant barrier to entry into the Australian South Sea pearl industry.

In French Polynesia, pearl culture was initiated in 1963 when the head of the Fisheries Department, Jean Domard, decided to try to graft the local oyster *Pinctada margaritifera*, with the help of an Australian company and two Japanese grafters. He obtained what are now called black pearls or Tahitian pearls. But it took many years before private pearl farms began to produce significant quantities, thanks to a few pioneers such as Jean Claude Brouillet and David Rosenthal (see Section 9.4.3.1). At first the product (the black pearl) was unknown to jewelers, and it took a lot of effort to make it known and appreciated by jewelers and consumers worldwide.

Production really started to grow in the 1980s, and then accelerated in the 1990s, but came to a sudden halt in 2000–2001, as prices plummeted and world imports were affected by a series of adverse events including the Kobe earthquake, the September 11, 2001 attack, SARS in Asia and the war in Iraq. From 29 kg in 1980, pearl production reached 575 kg in 1990 (a 20-fold increase), 11,541 kg in 2000 (another 20-fold increase), and 11,161 kg in 2001, but has shown a regular decline since then, i.e., 7,304 kg of pearl produced in 2005 (see Section 9.4.3). The general pattern is evident from Figure 13.1.

The steep decline in the price per gram of black pearls is the result of a supply curve sliding much faster to the right than the demand curve in the 1990s. As a consequence, a 20-fold increase in the quantity supplied was met by an equivalent increase in the quantity demanded thanks only to the average price falling to one-fifth of its earlier level. While the world market share of the Tahitian pearl expanded rapidly from a negligible amount in the 1980s to around 25% in 2000, it has stagnated since. In the mid-1980s, the Tahitian pearl was seen as a rare, niche market upscale product, and benefited from the decline in the Japanese Akoya pearl supply. In the 1990s, its falling price led to it becoming a more widely marketed product. Moreover, while for a long time Tahiti had

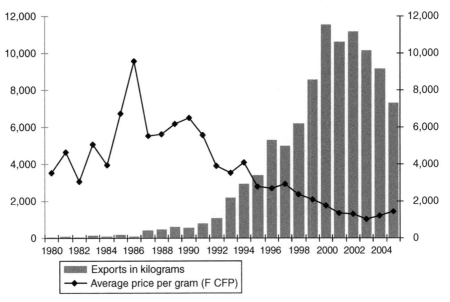

FIGURE 13.1   Pearl exports and price per gram of pearl exports from French Polynesia, 1980–2005. *Source*: ISPF.

a monopoly on the black pearl (obtained from *Pinctada margaritifera*), competitors began to emerge at the end of the 1990s, in the Cook Islands, Indonesia and elsewhere.

There are no oyster quotas in French Polynesia, and no hatchery, because spat can be collected in many lagoons, to provide the juvenile oysters producers need for grafting, without danger of depleting the wild stock (see Section 9.4). As a result, no limit has been set to the number of producers. However, since 2002, the local government requires producers to hold a professional card and proof of a maritime concession of at least 1 ha. The professional card is given to persons proving their aptitude and ownership of equipment for pearl production. As a result of falling prices, the number of producers has been declining since 2001, as well as production. As of November 2006, there were 409 oyster producers and 421 pearl producers (many engage in both activities). There were a total of 516 pearl farms on 31 islands, most of the islands (27) being in the Tuamotu and Gambier archipelagos.

The distribution of ownership of pearl farms by size in French Polynesia is shown in Table 13.1. Forty-one large farms (each with more than 40 ha of maritime concession) make up more than two-thirds of the total surface area of maritime concessions (and therefore, probably also account for about two-thirds of total production). A total of 505 small farms (less than 5 ha in size) make up only 5% of the total exploited surface area and 183 medium-sized farms (between 5 and 40 ha in size), account for 34.5% of the total area farmed with pearl oysters. Considerable inequality in the size of farms is evident.

TABLE 13.1
Structure of French Polynesia's pearl industry.

	Large	Medium	Small	Totals
Size of farms (ha)	>40	5–40	<5	
Total area of farms (ha)	6,601	2,609	505	9,715
Percentage of farmed area	67.9	26.9	5.2	100
Number of farms	41	183	307	531
Percentage of farms	7.7	34.5	57.8	100

Source: Service de la Perliculture, French Polynesia.

13.4. SOCIO-ECONOMIC IMPACT

Pearl culture has the potential to provide increased economic opportunities to remote marine communities. For such communities, pearls have the advantage that their value is high relative to their weight and they are easily stored. These two factors reduce transport difficulties. Several development organizations such as Worldfish Centre and the Australian Centre for Agricultural Research (ACIAR) have supported pearl projects in developing countries as a means to improve the livelihood of disadvantaged communities (e.g., Lane *et al.*, 2003; Southgate, 2004).

In Australia, the pearl industry is located in some of its more remote and sparsely settled areas. However, no farms are owned by Australian Aborigines who account for the major proportion of the population in these remote areas. Nevertheless, according to the Northern Territory Government: "various (Aboriginal) communities have significant involvement with the pearling industry. All the land for pearl farms around the coast is leased from traditional owners or land councils and the farms provide employment opportunities for indigenous workers".

Such communities usually have few economic opportunities. Nevertheless, in the Northern Territory the industry only directly employed about 300 persons in 2004. In Western Australia, it has been estimated that the industry employs directly about 1,500 people, most of whom are from Broome (Fletcher *et al.*, 2006). Taking into account indirect local employment and assuming a regional multiplier of around 1.5, the pearling industry may result in the employment of 2,500–3,000 persons in northern Australia.

It is interesting to note that even today Mikimoto Pearls claims as one of the achievements of Koichi Mikimoto, his contribution to economic development of remote island areas. Mikimoto America states:

> "While trying to meet the challenge of producing black lipped and silver lipped cultured pearls, he [Koichi Mikimoto] encouraged the development of local pearl industries on previously underdeveloped islands. He contributed so significantly to the development of these islands that the name "Mikimoto" is spoken of with reverence even to this day."

However, we cannot assume that the development of pearl culture is always beneficial to local communities.

In recent years, several Pacific Island communities have been given aid to develop the culture of pearl oysters, particularly *P. margaritifera*, as a means of assisting their economic development. Southgate *et al.* (2006) states:

> "As demonstrated in the Pacific, cultured pearl production can provide considerable opportunity for income generation for coastal communities. Pearl production may occur on small family-based pearl farms and individuals may enter the industry at a number of levels to produce oyster shells (mother-of-pearl), half-pearls or round pearls, or they may simply collect spat from the wild for sale to pearl farms. Furthermore, the pearl industry provides opportunity for the involvement of women and provides the raw materials for local handicraft manufacture which may include lower grade pearls or pearl shell."

In French Polynesia, pearl culture was successfully developed by private initiative, once Jean Domard, Head of the Fisheries Department, had proven that it was possible to obtain cultured pearls from *P. margaritifera* oysters in the 1960s. But it took a lot of courage for a few entrepreneurs to pioneer this all new product in the mid-1970s and make it known worldwide. Pearl culture took place in the remote Tuamotu and Gambier archipelagos that had been greatly depopulated during the 1960s because of French atomic testing in Mururoa: the center for atomic experiments needed labor to set up in military bases on Hao, Fangataufa and Mururoa, so most young people went from fishing and copra culture to well paid employment by the military. Once the military installations were completed, many were engaged as servicemen or went to work on the main island of Tahiti in the military bases there. Therefore, when the pearl industry picked up in the 1980s and 1990s, it led to a repopulation of both archipelagos: many people came back from Tahiti to their island of origin, those who had saved and still had land tried to set up pearl farms, those who could not went to work for large-scale farms. As a consequence, the population of atolls such as Takaroa, Fakarava and Manihi, shot up again, and the standard of living of their inhabitants was greatly increased, because prices in the 1980s were still high and pearl culture was very profitable, even for small-scale family farms, much more so than previous activities, such as copra culture.

In recent years, falling prices have made this activity less profitable than before, and production has declined, as well as the number of pearl farms, but small-scale family farms have not disappeared, since they do not have to pay wages, and their grafters (pearl seeding technicians) are paid piece rates (most of the grafters are from China now). Many families spread risks by engaging simultaneously in other activities, such as: fishing, making and selling pearl jewelry for tourists, gathering copra, making handicrafts, operating small hotels, setting up lagoon and pearl farm tours for cruise boat tourists (especially on the islands of Tahaa and Raiatea).

Medium-scale farms with paid employees were the most affected by the pearl price downturn: small family farms could survive because they had no payroll, large-scale farms also survived because they had lower unit cost than medium-scale farms. An unpublished study by Poirine and Kugelmann for the *Service de la perliculture* found evidence of significant economies of scale in French Polynesia's pearl farming.

French Polynesia showed that small-scale pearl farming was possible outside Japan. Manihiki in the Cook Islands is another example of successful small-scale pearl farming using *P. margaritifera*.

Yet, as is to be expected, not all such projects in the Pacific islands have been an economic success. In this regard, a study of the reasons for economic failures as well as success could improve future decisions about proposed projects for pearl oyster culture in developing countries.

Demonstrating the technical feasibility of culturing pearl oysters in a developing country is only an initial step towards establishing whether this culture will be an economic success. Amongst other things, techniques for culture need to be successfully transferred to locals, (e.g., the government of French Polynesia is actively promoting the training of Polynesian grafters by creating a public school of grafting, in order reduce its dependence on Japanese and Chinese grafters), the resources needed for pearl culture must be available or affordable to locals and they must have adequate access to markets for their produce (early on, the government of French Polynesia encouraged small producers to market their products through cooperatives called "groupements d'intérêt économique" or GIEs. Several GIEs hold auctions in Tahiti once a year to sell the pearls of their members).

In Manihiki, Cook Islands, initial expertise came from a part Chinese, part Tahitian, part Cook Islander entrepreneur, Yves Tchen Pan, who set up a large-scale farm in Manihiki. To maximize the benefits for the local families, the Island Council of Manihiki entered into an agreement with Cook Islands Pearls, Yves Tchen Pan's company: "Under the terms of this agreement, Manihikians dived for oysters, which were taken to Cook Islands Pearl's farm. The farmers, who tended their own oysters with advice from Cook Islands Pearls, gained knowledge of farm husbandry practices, access to management advice, access to technicians, and a market for their pearls. In effect oysters were managed, seeded, harvested and marketed on their behalf by Tchen Pan's company, Cook Islands Pearl Ltd., in return for 40% of the proceeds. With proceeds from sales of oysters, proceeds of their own crops' sales, and the above arrangement, Manihikians could become established on their own farms relatively easily" (Macpherson, 2000).

In addition, Tchen Pan's company helped family farms by renting out his grafting technicians, charging a small fee per oyster grafted. Experience in French Polynesia and the Cook Islands suggests that it is often in this way that locals acquire expertise: first by collecting and growing spat to juvenile size (which does not require much capital investment and is a relatively simple task). The proceeds from selling juvenile oysters can be reinvested over time by setting up oyster lines and getting the grafting task done by outside technicians hired by big farms or cooperatives (Lane *et al.*, 2003).

Furthermore, acceptable and enforceable property rights in the cultured pearl oysters must exist; otherwise, there will be a lack of economic incentive for their culture. In many developing areas, lack of "adequate" property rights in cultured species is a barrier to their commercial culture. The enforcement of property rights is a problem in many countries (Indonesia for example), and even in French Polynesia the frequent stealing of pearl oysters from the oyster lines is a problem for pearl farmers.

Another way to maximize the socio-economic benefits of pearl culture is to try to develop downstream activities adding value to the pearl, such as the sale of pearl necklaces (instead of raw pearls) and of pearl jewelry. This industry is worth more than the raw pearl trade worldwide. French Polynesia recently encouraged pearl jewelry for export by creating "free firms" (entreprises franches in French): jewelers working with Tahitian pearls may import inputs such as gold or silver, or gold or silver made jewels, free of tax, and also export free of tax (there is an export tax on pearls in French Polynesia). It is worth noting that Australia exported US $77 millions worth of pearls in 2003 (18% of the world market), but was the first exporter of worked pearls (necklaces made of pearls), with a value of US $153 millions, i.e., twice as much as the value of raw pearls exported the same year. It was for a long time Japan's privilege to import raw pearls from all over the world and transform them into necklaces through the lengthy process of pairing pearls of the same shape, quality, color and size that were re-exported to the rest of the world. This value added industry is now growing strongly outside Japan, in Australia, Hong Kong, and French Polynesia. The big producers, such as Paspaley in Australia or Robert Wan in French Polynesia, make necklaces and jewelry to reap the value added from marketing the finished product instead of the raw material.

Southgate *et al.* (2006) gave consideration to the possibility that the introduction of pearl culture to East Africa could assist the sustainable economic development of coastal communities. On the basis of early evidence from trials of the culture of *P. margaritifera* in the Mafia Island Marine Park in Tanzania, they find physical production from such culture to be very satisfactory, and claim that the transfer of techniques for half-pearl production to locals is not difficult. They see the main market for the produce as being for jewelry to be sold at resorts on Mafia Island and through retail outlets in Dar-es-Salaam and Zanzibar.

Despite some economic uncertainties, the project is to be expanded. Southgate *et al.* (2006) state the following:

> "The long term sustainability of this project will depend on reliable sources of culture stock. Expansion of current spat collection activities and development of local hatchery production are immediate goals for the project. Ongoing research will also investigate the potential for round pearl production within the [Mafia Island Marine Park] and development of local jewellery making skills."

In the article, Southgate *et al.* do not mention whether property-rights issues are likely to be a problem, nor what the impact on the local community would be if the project is in the end not economically viable.

In relation to small-scale pearl oyster farming in the Central Pacific, Fong *et al.* (2005) came to the view, on the basis of evidence from the Republic of Marshall Islands and the Federation States of Micronesia, that it is likely to be profitable. Over a 2 year period of operations, they suggest that an internal rate of return of 9.6% is realistically achievable. However, this rate of return is quite sensitive to market price and mortality rates (Fong *et al.*, 2005). Furthermore, the representative pearl oyster farm that they model has quite large negative cash flows during the first 5 years of its operation (see Fong *et al.*, 2005) and financing these could be a problem for many local communities. There may also be differences of opinion about whether the size of the farm they envisage is really very small. They assume that a stock of 25,000 oysters for seeding will be maintained once the farm reaches a steady state.

Economic modeling software was developed by the Secretariat of the Pacific Community (SPC) in association with the Queensland Department of Primary Industries and Fisheries (QDPIF), Australia, to help assess various scenarios for development of cultured pearl industries in Pacific island countries (Johnston and Ponia, 2003). Following inputs relating to farming costs (equipment, labor, maintenance, etc.), estimated production levels (yield) and product value, the software generates values for output parameters including annual gross revenue, annual production costs, production cost pear pearl, revenue per pearl, and for economic parameters including net present value, annual return, internal rate of return and benefit–cost ratio. The models generated allow assessment of the economic viability of pearl farms of various sizes producing various products (e.g., pearl shell, half-pearls, round pearls). Using this software, economic models for various black pearl farming scenarios were generated for Kiribati in the central Pacific: (1) a large sized, private sector owned farm with 100,000 oysters producing round pearls; (2) a medium sized, private sector owned farm with 30,000 oysters; (3) A small sized, private sector owned farm with 5,000 oysters; and (4) a medium sized, private sector owned farm with 30,000 oysters growing mabé (half round) pearls (Table 13.2). The models generated also allowed estimation of the major contributors to pearl productions costs for each farming scenario as well as discounted cumulative cash flow (Fig. 13.2).

Under the assumptions made in the model, a large sized, privately owned farm with 100,000 oysters producing round pearls was considered the best option (Table 13.2). The costs of establishing a large farm (scenario 1) in Kiribati was estimated to be around $300,000 and annual operating costs would average $300,000 (Southgate, unpublished data). Such a farm was estimated to require 6 years to break even but provided a reasonable risk profile, with a 67% chance of making a zero profit or less (Fig. 13.3). It would also be possible for four types of small business ventures to "spin-off" from such

TABLE 13.2

Output summary and economic indicators generated for various black pearl farming scenarios in Kiribati, central Pacific, using modeling software (see Johnston and Ponia, 2003). scenario 1: a large sized, private sector owned farm with 100,000 oysters producing round pearls; scenario 2: a medium sized, private sector owned farm with 30,000 oysters; scenario 3: a small sized, private sector owned farm with 5,000 oysters; and scenario 4: a medium sized, private sector owned farm with 30,000 oysters growing mabé pearls.

	Pearl farm options			
	scenario 1	scenario 2	scenario 3	scenario 4
Output summary				
Annual production (pearl)	9,673	2,902	484	5,892
Annual gross revenue	$496,879	$138,182	$23,030	$62,984
Annual production costs	$306,483	$113,852	$33,804	$50,747
Production costs per pearl	$31.69	$39.24	$69.90	$8.61
Revenue per pearl	$51.37	$47.62	$47.62	$10.69
Economic indicators				
Net present value	$1,869,337	$238,877	−$105,779	$120,145
Annual return	$190,396	$24,330	−$10,774	$12,237
Internal rate of return	28.33%	15.45%	NA	14.91%
Benefit–cost ratio	1.62	1.21	0.68	1.24

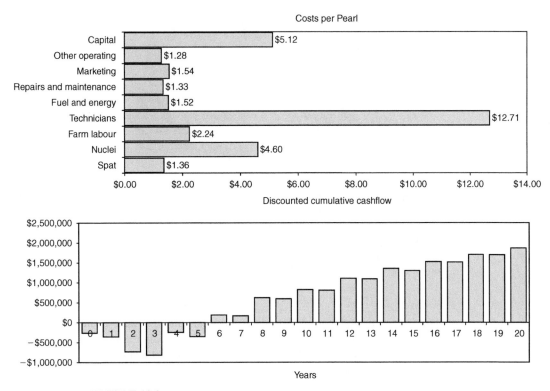

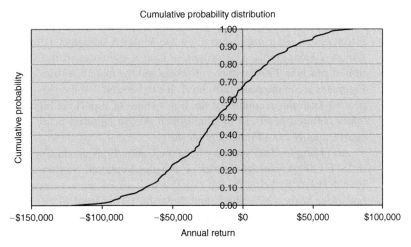

FIGURE 13.2 Relative contributions of the various components of pearl cost and discounted cumulative cashflow calculated using economic modeling software (Johnston and Ponia, 2003) for a large sized, private sector owned farm with 100,000 oysters producing round black pearls in Kiribati, central Pacific.

FIGURE 13.3 Predicted returns using economic modeling software (Johnston and Ponia, 2003) for a large sized, private sector owned farm with 100,000 oysters producing round black pearls in Kiribati, central Pacific. Summary of Risk Assessment indicates a 67% chance of zero profit or less with a lowest potential return of −$121,519 and a highest potential return of $78,677.

FIGURE 13.4   Handicrafts made from lower grade pearls and pearl shell offer opportunities for income generation in coastal and island communities in the Pacific islands. The photograph shows participants in a pearl handicraft training workshop held in Kiribati, central Pacific. (Photo: Antoine Teitelbaum.)

a farm: cleaning and gear making/repair services; culture of spat/juveniles to pearl producing size (to approximately 2 years old) and selling them to the central farm for pearl production; production of half-pearls (mabé) from oysters that have rejected nuclei; and use of pearl shell and lower grade round and half-pearls in handicrafts (Fig. 13.4).

13.5. PRODUCTION AND TRADE STATISTICS

It is very difficult to obtain accurate global statistics for pearl production and sales. However, some interesting statistics have been provided to *Pearl World: The International Pearling Journal* by the Golay Company, a leading trader in pearls (Anon, 2006a). Table 13.3 provides Golay's estimates of the value of world production of pearls in 2004 by types of pearls and the main countries producing these pearls. Estimates are at the pearl farm level. It can be seen that the value of South Sea cultured pearls (Australia accounts for the lion's share of supply) is highest followed by freshwater cultured pearls from China, then Akoya cultured pearls and Tahitian cultured pearls, mainly from French Polynesia.

According to 2004 statistics from the Centre Français du Commerce Extérieur, cited in ISPF (2006), worldwide sales of unworked cultured pearls were at US $412 million, French Polynesia being the top exporter with a 27% market share, Indonesia taking second place with a 24% share, and Australia taking third place. Japan and Hong Kong imported 79% of the total value of pearl imports in 2004. Total sales of worked pearls (pearls necklaces and earrings) amount to US $787 million in 2004 (a 33% increase compared to 2003). Japan and Australia are the main exporters of worked pearls with a market share of 26% and 23%, respectively. French Polynesia's market share is only 7%.

TABLE 13.3

Value at the pearl farm level of world production of cultured pearls in US$ in 2004 according to Golay's estimates.

Type of pearl and main producers	Value (US$ millions)	Percentage of supply
White South Sea pearls (Australia, Indonesia, The Philippines, Myanmar)	220	35
Freshwater pearls (China)	150	24
Akoya pearls (Japan, China)	135	22
Tahitian pearls (French Polynesia)	120	19
Total	625	100

Source: Based on Anon (2006a).

TABLE 13.4

Global share of each country or region of the world pearl jewelery market in 2004 according to Golay's estimates.

Country/region	Value of sales (US$ millions)	Percentage of total sales
USA	1,500	30
Japan	1,200	24
Europe	900	18
China	600	12
Southeast Asia	500	10
Other Countries	300	6
Total	5,000	100

Source: Based on Anon (2006a).

China's entry into the global pearl market has been important in recent years. It has become a major supplier of freshwater pearls. While South Sea pearls have increased their market shares and so have Tahitian cultured black pearls, there has been a significant reduction in the market share of Akoya pearls (see Section 10.2).

The major markets for pearl jewelery, as estimated by Golay, are presented in Table 13.4. The USA is the major market followed by Japan, Europe, China and Southeast Asia. Pearls are particularly popular in Asia. Table 13.4 implies that Asia purchased more than half (over 52%) of the global supplies of pearl jewelery in 2004.

The United Nations Commodity Trade Statistics for 2005 provides data on the exports of pearls, natural or cultured. The values of exports of the main exporting countries are shown in Table 13.5. These data indicate that Japan is the largest exporter, followed by Australia, China and French Polynesia. The value of exports by the Philippines and Indonesia are considerably smaller than the exports of the countries just mentioned and

TABLE 13.5

The value of exports of unworked pearls by the main exporting countries in 2005. The data are for category H52002-7101 in the COMTRADE statistics. United Nations, Statistics Division (2006) Commodity Trade Statistics Database (COMTRADE).

Country	Value of exports (US$ millions)	Percentage
Japan	263.6	33.2
Australia	219.5	27.6
China	146.5	18.4
French Polynesia	128.3	16.1
The Philippines	15.45	1.9
Indonesia	10.7	1.3
Total	784.05	100

the data for Myanmar were not available. In some cases, the statistics would include re-exports and some exports from some countries, e.g., the Philippines and Indonesia, may bypass official channels and not be recorded. However, the data may give an indication of the relative importance of nations in international trade in unworked pearls.

The value of Australian pearl exports since 2002 have shown an upward trend whereas those of Japan have been relatively stagnant. However, Australian production of pearls by volume remains fairly stationary due to the quota system limiting the number of pearl oysters seeded for pearl production. This system has the twin objectives of ensuring that the take of wild pearl oysters is ecologically sustainable and of restricting supply so as to maintain prices for South Sea pearls. Because Australia is the globally dominant supplier of these pearls, it is in a position to influence their price internationally. Its strategy has been to limit supply and concentrate on improving the quality of the product and its promotion. A review early in this decade by the Australian National Productivity Commission, as a part of ongoing reviews of business competition in Australia, supported the continuation of this policy. This policy seems to be in Australia's interest since most of Australia's pearl production is exported and pearls, as a product, have features that require their marketing to be evaluated in a different way to most economic goods.

13.6. MARKETING AND THE NATURE OF THE MARKET FOR PEARLS

13.6.1. The economic nature of pearls differs from the majority of commodities

Even though pearls do not satisfy any basic needs, they are highly valued. Their value derives from their inherent beauty and the social "messages" they convey when worn or given. Their relative scarcity, especially of sought after specimens, adds to their economic value.

Because social factors have such an important influence on the economic value of pearls, they can be classified as Veblen-type goods (Leibenstein, 1950; Tisdell, 1972). Veblen (1934) stressed the importance of social factors in determining the economic value of some types of goods. Demand for pearls may be enhanced by emphasizing their exclusiveness, associating pearl jewelery with desired life-styles and images, and promoting their quality. Demand for pearls may even rise to a point where there is an increase in their price.

13.6.2. Consequences of the characteristics of pearls for their marketing and market structure

Given the above attributes of pearls, considerable scope exists for major pearl suppliers to increase the demand for pearls by advertising and promotion and by securing the recognition of their brands. The earliest producer of cultured pearls, Mikimoto Pearls, recognized the importance of these aspects in its early development. It continues to promote its products heavily. A similar pattern has been followed by market leaders entering the industry later such as Paspaley Pearls, the leader in South Sea pearls, and Wan's Tahiti Perles, the major supplier of Tahitian pearls. In French Polynesia, in 1993, the government decided to set up a promotion board, the GIE Perles de Tahiti, financed by a tax on pearl exports. The tax is now set at 200 F CFP (French Pacific franc) per gram of exported pearl, but 35% of the receipts only go to the GIE Perles de Tahiti (50% before 2002). Promotion efforts at first went toward organizing trade events linked to jewelery and haute couture, and placing pearl necklaces in movies or television series in order to promote a luxury image for the Tahitian pearls. The United States, Japan and Europe were the main markets aimed at. From 1998 on, promotion was more geared toward the general public, and trying to create new markets (i.e., jewelery for young men and sales in the Middle East and Russia).

Major suppliers have an incentive to promote recognition of their brands. This is partly because many end-buyers are ignorant of the quality and market value of pearls and therefore, when purchasing more expensive pearls or pearl jewelery are likely to put their trust in well known brands. Branding and rigorous quality control help to build the reputation of major establishments in the industry. Brands also provide a focused means for promoting and advertising the social value of pearls. Furthermore, such a strategy may help the market to operate more efficiently by ensuring that the sale of poor pearls by less reputable sellers does not drive out producers of better quality pearls. This type of phenomenon is well recognized in the economics literature (Akerloff, 1970; Varian, 1987).

The downside of this phenomenon is that unbranded or little known suppliers of pearls are likely to be treated with suspicion by customers. Even when their product is of high quality, this is likely to be discounted by uncertain buyers. Consequently, small and relatively unknown suppliers of pearls and pearl products may obtain lower prices for the same quality product as that of market leaders.

To reassure buyers, the government of French Polynesian is considering the creation of a quality label for exported pearls. Quality controls already exist in French Polynesia for all exported pearls (see Section 9.4.6.2): before they can be exported, all pearl lots must go through an X-ray machine at the Service de la perliculture (Fig. 9.18). All pearls must have a minimum thickness of 0.8 mm of nacre around the nucleus. Pearls that do not qualify are rejected and crushed by the Service de la perliculture (Fig. 10.1). The producers are paid a small indemnity for the rejects, i.e., 50 F CFP per gram, with a limit of 500 g per year and per hectare of oyster grafted (arrêté N 1027 CM du 17 novembre 2005). The pearl lots qualifying for export are then put in a sealed bag and go directly to customs in their sealed bag.

It should also be noted that the quality and attributes of pearls are quite diverse. This may lead to market segmentation. Some types of pearls may be marketed to the high end of the market (e.g., high quality Akoya, Tahitian or South Sea pearls) whereas others may be sold to a lower market, e.g., Chinese freshwater pearls. Different marketing strategies may be used in different market niches or segments.

This raises the question of the extent to which pearls from different species or of different shapes and so on are substitutes. Where large suppliers of pearls specialize in supply from a particular species, they may in their promotion stress their special qualities compared to pearls from other species. This is intended to reduce substitution between pearls from different species. Furthermore, there is probably a low degree of substitution between lower quality pearls and those of higher quality. This would be consistent with significant segmentation of the market. Furthermore, this segmentation is likely to be promoted by the market leaders in the industry.

Thus, marketing considerations tend to favor large vertically integrated suppliers in the pearl industry. Therefore, for this and other reasons, the global industry appears to be characterized by a few large easily recognizable suppliers and by many small relatively unknown suppliers. The industry structure is dualistic or bipolar.

13.6.3. Further observations on the bipolar nature of the industry, constraints on new entrants, external costs in production due to the abuse of a common resource, and the existence of scale economies

The dominant suppliers in the industry appear to be those who made an early start in developing the culture of particular species of pearl oyster and which have, on the whole, specialized in these species. This applies to Mikimoto in Akoya pearls, Paspaley in South Sea pearls and Tahiti Perles in Tahitian black pearls. The dominance of these companies is reinforced by economies of scale in marketing, their considerable attention to quality control, their vertical integration and existing exclusive rights to utilize natural areas very suitable for pearl oyster collection and culture.

Efficient suppliers seem to benefit from an early entry advantage. It is extremely difficult for other firms entering later to emulate their success. Early entrants usually control the most suitable sites for pearl oyster culture, have benefited from learning by doing, are technologically sophisticated in their operations and have established market recognition and networks. For example, in French Polynesia, Robert Wan owns private atolls, such as Nengo Nengo and Marutea, where no one else may produce pearls. He may then maximize quality by limiting the extent of pearl culture in the lagoon to prevent over-exploitation. This is not true of many pearl producers in a public lagoon: there is a risk of over-exploitation if too many producers extend their maritime concession and the government does not limit or does not enforce effectively the limits of each exploitation. In other words, only the producer owning a private lagoon can "internalize" the external cost of over-exploitation in a given lagoon. The recent (2001) episode of high oyster mortality in the lagoon of Manihiki in the Cook Islands (see Section 9.4.4) illustrates the external costs arising from over-exploiting the lagoon. Economic analysis of such external effects is given in Poirine (2003) and in a general context in Tisdell (2003).

It is interesting to note that one year before the high mortality peak in Manihiki, a survey of the lagoon had determined that 30% of the lagoon surface was occupied by pearl farms. The study concluded that the number of oysters could be increased from 1.5 million at the time of the census to 2 millions: "Assuming space is properly managed in the lagoon then the portion of farmable strata occupied (30%) suggests that farming of two million pearl shells could be attained without density-dependent consequences on the health of the oyster" (Ponia *et al.*, 2000). Since then, however, the Government of the Cook Islands has enforced strict rules to prevent over-exploitation of the lagoons.

Similar high mortality episodes occurred in French Polynesia, at Takapoto atoll in 1985, at Hikueru atoll in 1994 and at Manihi atoll in 1997. It is not yet sure if a virus was responsible, or if over-exploitation might weaken the natural defenses of oysters against viruses already present in the lagoon. To prevent the reoccurrence of such events in French Polynesia, the government monitors maritime concessions and tries to respect a policy of not allowing more than 10% of the lagoon area to be occupied by pearl farms; it does not grant any more maritime concessions or extensions of existing ones when this limit is reached.

Another example of external costs arising from intensive exploitation is the use of high pressure hoses to clean the oyster lines on boats on pontoons (e.g., Fig. 14.3): it seems that as a result of these methods, sea anemones which are shredded to pieces and ejected into the lagoon multiply much faster and colonize oysters at neighboring pearl farms as a major component of biofouling. In French Polynesia, this "cost effective" method of cleaning oysters in large-scale pearl farms is viewed as the main cause of the increasing proliferation of sea anemones in the lagoons of French Polynesia.

It is very difficult for late entrants to become market leaders in the supply of pearls. It is interesting to observe that market leaders emerging after Japan's domination of the industry, did so by favoring different species of pearl oysters and by having access to different geographical areas suitable for their culture. Consider the entry of Paspaley Pearls through the culturing of South Sea pearls in Australia and Tahiti Perles via the culture of Tahitian black pearls in French Polynesia. In so doing, they were lucky to profit from the severe downturn of Akoya pearl production in Japan, first from the 1966 to the mid-1970s, then from 1995–1996 to 2000, due to the "Akoya virus", which encouraged Japanese buyers to look elsewhere for substitutes.

Another often overlooked advantage of early players in the game is the existence of significant economies of scale, giving a decisive advantage to large-scale farms over small-scale farms. An unpublished study by Kugelmann and Poirine (2003) has shown evidence of such scale economies in French Polynesia. According to their survey, the average cost of harvesting a pearl is halved when the average size of the farms quadruples from 25,000 to 200,000 oysters in stock. In fact, at the time of study (2002), prices were so low that on average the small farms (less than 25,000 oysters in stock) were operating at a loss (note that the study included an estimated opportunity cost of family labor in the economic costs). However, the study found no significant economies of scale beyond 200,000 oysters in stock. The study further showed that the cost of rearing pearls is the most sensitive to size. The average cost of grafting increases and then decreases with size. The other average costs (management, taxes, boats, diving equipment) also tend to decrease with size. The general relationship between average costs and profitability as a function of farm size as illustrated in Figure 13.5.

In many cases, small producers of pearls, particularly in developing countries, are unable to access world pearl markets economically. Their sales are often made to passing tourists and to middlemen. Because they lack market power and knowledge, the prices they receive may be low compared to those achieved by leading suppliers in the industry.

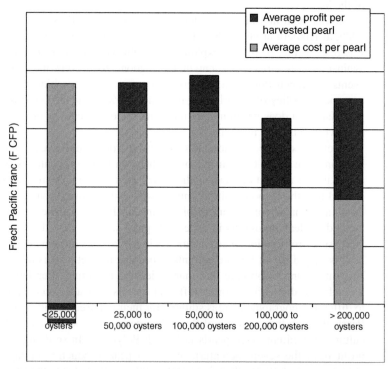

FIGURE 13.5 Average cost and profit for different sized pearl farms in French Polynesia (estimated with 2002 prices). The vertical axis is not graduated to preserve confidentiality of the profit and cost figures. Data from Kugelmann and Poirine (2003).

In French Polynesia, the middlemen buy with cash the crops of small producers, at a very high discount, even though small producers have the option to join a cooperative (groupement d'intérêt économique) to sell the pearls through an annual international auction. The cooperatives also routinely complain that small producers sell their best quality pearls to middlemen, leaving the rest of the crop to the cooperative to sell at the auction, with the result that the average quality of lots sold at the auction is not good enough to obtain a reasonable price (personal communication to Bernard Poirine).

13.6.4. Incomes and prices as influences on the demand for pearls

As noted above, social factors, advertising and promotion of pearls influence the demand for pearls. In addition, the price of pearls and the levels of income of consumers can be expected to affect the demand for them.

Few studies of the responsiveness of the demand for pearls to their price and income are available. However, Poirine (Tisdell and Poirine, 2000) estimated the price elasticity of demand for Tahitian pearls and found it to be inelastic (-0.36). Although no estimates are available, demand can also be expected to be price inelastic for Akoya pearls and for South Sea pearls, maybe even more so. The implication of this inelasticity is that an increase in the supply of pearls reduces the total revenue obtained by pearl producers.

This elasticity figure implies that a rise in the price of Tahitian pearls by 10% reduces demand for them by only 3.6%, everything else unchanged. On the other hand, a reduction in price of Tahitian pearls by the same amount would result in only a gain of 3.6% in the quantity sold. This matter can also be considered inversely. For example, the inverse of -0.36 is -2.7. This implies that a 10% rise in the supply of Tahitian black pearls would reduce their price on average by 27.7% and consequently cut the receipts of suppliers. In these circumstances, producers of Tahitian pearls would benefit by restricting their supply.

However, quite a different thing happened from 2002 to 2005 (see Fig. 13.1): the supply of Tahitian black pearls fell by -34.6%, the price per gram went up only by 11%, implying a price elasticity of -3.1. The Tahitian black pearl now has close substitutes from the Cook Islands and Indonesia, therefore the total supply of black pearls is no longer measured by the quantity exported from French Polynesia, and it is difficult now to draw conclusions on the value of price elasticity by looking only at figures from French Polynesia.

Because the price elasticity of demand for pearls is low, the incomes of pearl producers as a whole can be maintained by limiting their supply. This strategy also increases the long-term rarity value of pearls. Furthermore, because pearls are not a necessity and because of demand features associated with pearls, the public may not be very critical of this restrictive strategy from a social welfare point of view, especially if most buyers are foreign purchasers. In Australia's case, this may be why the Productivity Commission did not recommend a change in Australian policy which limits supply of South Sea pearls, even though hatchery techniques make it possible to increase the production of pearls significantly without depleting the wild stock of oysters (which was the main reason for limiting pearl grafting in Australia in the first place).

The demand for pearls, especially quality pearls, appears to be sensitive to income levels. Although no empirical estimates of income elasticities of demand for pearls seem to be available, some circumstantial evidence points to this sensitivity. As noted above, Asia is the major global market for pearls. The sharp fall in pearl prices in the period 1998 to 1999 was associated with the Asian financial crisis which reduced incomes for a time in Asia and created economic uncertainty. Again, it seems likely that the rising demand for pearls in China (see Table 13.4) is associated with rising incomes in China. The fact too that pearls are durable and their purchase can be deferred probably adds to the sensitivity of the demand for pearls to income variations.

No estimates of the cross price elasticities of demand for pearls are available. It was speculated above that their cross elasticity might not be high. This would imply that most consumers do not regard pearls from different species to be highly substitutable. This view is likely to be promoted by large suppliers primarily supplying pearls from a single species. Nevertheless, it is becoming more common for pearls from different species to be incorporated in jewelery (Anon, 2006b). This would foster complementarily in demand for pearls.

13.6.5. Other marketing aspects

Pearl auctions and jewelery fairs have become important wholesale outlets for pearls. The Paspaley Pearling Company conducts its own auctions as does Robert Wan's Tahiti Perles. Auctions are held in Hong Kong and Japan. These auctions help the major suppliers to gauge the market and, to some extent, control the price by setting reserve prices. If batches of pearls do not reach the reserve, a price may be negotiated with interested buyers following the auction or supply may be withheld (Anon, 2006a). The system opens the buying side of the pearl market to competition. However, small suppliers may find it difficult to access such opportunities for selling their pearls. Nevertheless, one Polynesian cooperative of small producers (GIE Poe Rava Nui), which held annual auctions in Papeete, is now considering holding its annual auction in Hong Kong.

As mentioned above, many small producers in developing countries rely on passing tourists and on middlemen to sell their pearls. They are at a disadvantage in accessing international markets and their pearls are likely to obtain a lower price than those supplied by large producers for several reasons outlined above. Small producers in developing countries are likely to be in a better position to market their pearls locally if there are skills in producing jewelery locally (Southgate et al., 2006). This increases value adding locally and can foster indirect employment.

This problem is recognized by Fong et al. (2005) who point out that a small-scale pearl farm may not compete well in the global pearl market with larger producers. They (Fong et al., 2005) recommend that:

> "farms in the Central Pacific may consider forming production and/or marketing cooperatives and/or partnerships to share resources to reduce monetary and non-monetary costs. Further, farms may consider different product differentiation strategies such as co-branding with wholesale/retail operations, mechanisms to ensure only high quality pearls enter the international market, and implement best management practices and use it as a marketing tool."

Even in more developed countries, tourists are targeted for pearl jewelery sales both generally and when they visit areas associated with pearling. In Australia, there are several pearl museums for example, one in Darwin and another in Broome. These provide background on the pearl oyster industry and promote it. They also sell some pearl jewelery. Some pearl farms have also supplemented their income by visits from tourists, such as the farm of Atlas-Pacific Ltd in Bali, Indonesia which intends to expand its pearl tourism operations (Fassler, 2006). While pearl-based tourism can be profitable for a pearl farm in a suitable location, it is not without costs. Each situation has to be individually assessed to decide whether pearl farm visits by tourists are likely to be profitable to a particular farm.

13.7. SUMMARY

The structure and nature of the cultured pearl industry shows variation between countries in which it has developed. This was illustrated by differences between the Australian industry, that of French Polynesia and the industry in the Cook Islands. Production is highly concentrated in Australia in very few hands. While industry concentration is also evident in French Polynesia, many small producers exist unlike in Australia. In the Cook Islands, a symbiotic or cooperative arrangement has evolved between small suppliers and a substantial producer and trader. Cultural and historical factors may help to explain the differences. However, developments in technology, and wider access to it, have and continue to play a role in shaping the structure of the industry which on a global scale appears to be bipolar.

On the production side, there is evidence of economies of scale which tend to favor larger production units. Furthermore, economies of size seem likely to be very marked in the global marketing of pearls. Large producers have integrated the production and marketing of their pearls and to a considerable extent are working these to add value to them. While smaller producers may be able to access local markets to sell their pearls and related products, they face difficulties in accessing global markets on their own. In French Polynesia, cooperative selling arrangements are being developed to reduce these barriers and standards are being set that must be met by exporters of pearls. An alternative strategy is for small producers to sell to a larger trader, as in the Cook Islands, who markets the product. As pointed out, pearls are not a standard product. The demand for them depends on their perceived social value and the image surrounding them. Keeping pearls scarce, particularly quality pearls, can add to their social worth.

In Australia, the supply of South Sea pearls is restricted by a quota system and presumably, the major Australian players try to limit supplies from other countries by obtaining leases etc. and buying unworked South Sea pearls from these sources. Sustaining the price of South Sea pearls requires supplies from new sources to be limited. On the other hand, supplies of black pearls have not been restricted in French Polynesia, and supplies have increased from other countries, such as the Cook Islands, in recent years. This has been a factor in the fall in the average price of black pearls in recent times. The demand for black pearls from French Polynesia is now more elastic than it used to be because of competition from other countries.

The market shares of pearls from different species of pearl oysters have changed considerably in recent decades, as have the shares of the main countries producing cultured pearls. In terms of the value of market shares, South Sea pearls have the highest share, followed by freshwater pearls, Akoya pearls and the black pearls. In previous times, Akoya pearls dominated the global pearl market, and gave Japan the dominant position in it. However, the Akoya virus cut Japanese pearl supplies and Japan had to search for alternative sources of supply. For example, joint ventures to culture South Sea pearls began in Australia. This helped to diffuse Japanese know-how and helped raise demand for pearls obtained from other species of oysters. Also, in earlier times, Japan had a dominant position in the freshwater pearl industry but as a result of water pollution in Japan, particularly in Lake Biwa, many freshwater mussels failed to survive. So, Japan moved some of its production offshore to other countries, such as China. This, together with the opening up of China to the outside world, resulted in China gaining a dominant position in the production of freshwater pearls. To some extent, environmental problems experienced in Japan assisted the global diffusion of Japanese technology for pearl culture and played a role in the changing geographical location of pearl culture and the altered composition of pearls marketed. This, however, is only part of the story because many different factors have played a role in the evolution of the pearl oyster industry.

On the socio-economic side, it was found that in some countries, substantial numbers of small-scale producers of pearl oysters have been able to survive economically. However, they are at an economic disadvantage compared to large producers. A positive feature of the industry is that it fosters decentralization and provides economic opportunities in remote areas. Sometimes, however, these opportunities are only realized by locals working for larger oyster producers, by the locals engaging in value adding by, for instance, producing and selling jewelery, or being involved in the tourist trade centered on the culture of pearls. Interesting examples of such activities were given for French Polynesia. In Australia, the pearl oyster industry appears to be much more capital intensive than in the Pacific islands. While pearl culture does provide employment and sources of extra income in remote areas of Australia, its production appears to be less integrated with local communities, except in very few centers such as in Broome, than in the Pacific islands. Pearl farming in Australia has more of an industrial element than in other countries and the grow-out of Australian pearl oysters often occurs well offshore and consequently rather distant from settlements onshore. Thus little interaction often occurs with the nearest onshore settlements.

References

Akerloff, G., 1970. The market for lemons: quality, uncertainty and the market mechanism. *Quar. J. Econ.* 84, 488–500.

Anon, 2006a. Golay's global view. *Pearl World: Int. Pearling J.* 15(1), 7–8.

Anon, 2006b. Pearl sales up at HK show. *Pearl World: Int. Pearling J.* 15(1), 1, 5, 8–9, 12, 14–16.

Fassler, C.R., 2006. Atlas-Pacific has a record year. *Pearl World: Int. Pearling J.* 15(1), 18–20.

Fletcher, W., Friedman, K., Weir, V., McCrea, J., Clark, R., 2006. *Pearl Oyster Fishery*, ESD Report Series, No. 5. Department of Fisheries, Western Australia.

Fong, Q.S.W., Ellis, S., Haws, M., 2005. Economic feasibility of small-scale black-lipped pearl oyster (*Pinctada margaritifera*) pearl fishing in the Central Pacific. *Aquacult. Econ. Manage.* 9, 347–368.

IEOM (Institut d'émission d'outre-mer) 2005). La Polynésie française en 2005, p. 64–65.

ISPF (Institut de la statistique de Polynésie française) 2006. Regards sur l'économie de l'année 2005, 30–33.

Johnston, B., Ponia, B., 2003. *Pacific Pearl Economic Model. Economic models for aquaculture and agriculture commodities.* Queensland Department of Primary Industries and Fisheries, Brisbane, and Secretariat of the Pacific Community, Noumea.

Kugelmann S., Poirine B., 2003. Etude économique des déterminants de la rentabilité des fermes perlières en Polynésie française, unpublished report, Service de la Perliculture.

Lane, C., Oengpepa, C., Bell, J., 2003. Production and Grow-out of the Black-Lip Pearl Oyster *Pinctada margaritifera. Aquacult. Asia* 8(1), 5–7.

Leibenstein, H., 1950. Bandwagon, snob and Veblen effects in the theory of consumer's demand. *Quar. J. Econ.* 64, 183–207.

Love, G., Langenkamp, D., 2003. Australian Aquaculture: Industry Profiles for Selected Species, ABARE eReport 03.8. Prepared for the Fisheries Research Fund, Canberra.

Macpherson, C., 2000. Oasis or mirage: the farming of black pearl in the northern Cook Islands. *Pacific Studies* 23(3/4), 33–55.

Ponia, B., Napara, T., Ellis, M., Tuteru, R., 2000. Manihiki Atoll Black Pearl Farm Census and Mapping Survey. *SPC Pearl Oyster Information Bulletin*, 14, 4–10.

Poirine, B., 2003. Managing the commons: an economic approach to pearl industry regulation. *Aquacult. Econ. Manage.* 7, 179–194.

Southgate, P.C., 2004. Progress towards development of a cultured pearl industry in Kiribati, central Pacific. *World Aquaculture Society Aquaculture 2004, Book of Abstracts*, World Aquaculture Society. p. 556.

Southgate, P., Rubens, J., Kiponga, M., Msumi, G., 2006. Pearls from Africa. *SPC Pearl Oyster Information Bulletin*, 17, 16–17.

Tisdell, C.A., 1972. *Microeconomics: The Theory of Economic Allocation.* John Wiley and Sons, Sydney, New York and London.

Tisdell, C.A., 2003. *Economics and Ecology in Agriculture and Marine Production.* Edward Elgar, Cheltenham, UK and Northampton MA, USA.

Tisdell, C.A., Poirine, B., 2000. Socio-economics of pearl culture: Industry changes and comparisons focusing on Australia and French Polynesia. *World Aquacult.* 11(2), 30–37. 58–61

Varian, H., 1987. *Intermediate Microeconomics: A Modern Approach.* W.W. Norton and Company, New York.

Veblen, T., 1934. *The Theory of the Leisure Class: An Economic Study in the Evolution of Institutions.* The Macmillan Company, New York.

CHAPTER 14

Environmental Impacts of Pearl Farming

Wayne A. O'Connor and Scott P. Gifford

14.1. INTRODUCTION

Several decades of burgeoning aquaculture production in response to increased demand for seafood or other related products has brought with it increasing scrutiny of the practices and potential impacts that may arise. Mollusc culture, and pearl culture in particular, are considered to have a low potential for environmental impact (Simpson, 1998; Gavine and McKinnon, 2002; Jernakoff, 2002; Yokoyama, 2002; Crawford *et al.*, 2003; Shumway *et al.*, 2003); however, there have been substantial adverse impacts in the pearl farming regions of Japan and China over some decades and the potential remains for adverse environment impacts from pearl farming industries. Many sectors of the pearl industry have recognized this and have responded accordingly:

> "It is clear that humans cause some effects on the environment as a direct, and often indirect, consequence of their presence. Whether those effects are evident at first glance, or after closer examination, is not necessarily relevant. It is necessary to accept that such effects must occur"
> *(AMWING Pearl Producers Association, 2003).*

Studies have been undertaken to measure the threat of particular impacts and risks in areas such as Japan, Australia and the Pacific. For example, the Australian pearl industries based on *Pinctada maxima* and other species have assessed environmental risks and have developed environmental management plans. At workshops held in

2001 and 2004, the *P. maxima* industry and community stakeholders identified the following "moderate grade" environmental risks to be the greatest facing the industry, and began to address them:

- water quality loss from the chemical treatment of vessel sewage;
- water quality loss from hydrocarbon spills;
- introduction of disease from seeding;
- introduction of exotic organisms;
- attraction of other fauna (Jernakoff, 2002; Pearl Producers Association, 2004).

While these risks are similar to those facing pearl producers in many countries and are closely mirrored by those raised as concerns for the other-species industries in Australia (AMWING Pearl Producers Association, 2003), they are not an exhaustive list for the international industry as a whole, nor do they necessarily reflect all community views. At the time of the *P. maxima* workshops in Western Australia, applications to establish Akoya[1] pearl culture on the opposite side of the country in more highly populated areas of New South Wales (NSW) highlighted an alternative set of concerns. Despite general similarities in culture methods between the east and west coast industries, the major public concerns with Akoya pearl culture were:

- degradation of the environment and the acknowledged natural beauty of the area;
- alienation of areas of public waterway by restricting the use of the waters;
- risk of entanglement of marine animals, particularly dolphins;
- damaging the important and growing tourism industry (Cleland, 2002).

These concerns were later echoed by the relevant NSW Government minister, who, when refusing the initial applications for pearl culture stated: "In such an ecologically important area, it is vital that any activity undertaken there is compatible with the environment." This area "is renowned for its natural beauty and attracts people from across NSW and Australia for holidays and as a place to live." The minister went on to say that even with stringent operation and management practices "I am not satisfied that the proposed development could be operated at all times with no risk to the environment." "There are concerns, even though the risk is low, that marine animals such as dolphins could become entangled" and "There is a risk, even though minimal, that the sediment plume associated with cleaning could have a detrimental impact on the immediate and broader environment." Although the ministerial decision was later reversed following hearings in the NSW Land and Environment Court, it highlighted the nature and breadth of community concerns.

A key for future development in the pearl industry is to develop an even greater understanding of the potential for impacts, to continue the process of monitoring for potential impacts in those areas considered to be of greatest risk and to ensure that the outcome of this research is effectively communicated to legislators and the broader

---

[1]Throughout this text the term, 'Akoya', is used for pearl oysters in the currently unresolved complex of *Pinctada fucata-martensii-radiata-imbricata* (see Section 2.3.1.13)

FIGURE 14.1 A reminder to staff and visitors at the entrance to a *Pinctada maxima* land base in Western Australia. (Photo courtesy Paspaley Pearl Co.)

community. The failure to do this can significantly and unnecessarily limited pearl culture development in some areas.

It is important, however, to acknowledge that the modern pearl industry is diverse. Today, there are a number of species cultivated across a wide range of environments on several continents using an array of culture techniques and apparatus. Potential impacts are dependent on these factors, on the socio-economic features of the culture area, and on other users and uses of the culture environment (Fig. 14.1). Finally, the perception of environmental impact is, in itself an impact, regardless of its reality or tangibility.

14.2. MECHANISMS OF IMPACT

The term "environmental impact" has been variously defined, but here it is considered to be any change in the environment, adverse or beneficial, that wholly or partially results from an industry's activities, products or services. While impacts are most often perceived to be negative, it is also important to recognize the positive environmental and socio-economic outcomes that can arise (see Chapter 13).

For the purposes of this chapter, potential impacts have been divided into classes based loosely on those used by GESAMP (1991), although considerable overlap can occur between classes, with processes such as bioaccumulation and sedimentation having biological, physical and esthetic impacts.

14.3. BIOLOGICAL IMPACTS

14.3.1. Genetic impacts: species translocations and artificial propagation

Of all the potential impacts of aquaculture, the introduction of a new self-sustaining species or the alteration of the gene pool of indigenous populations is potentially the most ecologically damaging. These are outcomes that can arise through the translocation of oysters and, to a lesser extent, the artificial propagation of oysters in hatcheries.

With respect to translocation, a distinction can be made between two types of species movements: introductions and transfers. "Introductions" are translocations intended to introduce a new species beyond its current geographical range. "Transfers" occur within the geographical range of a particular species and are intended to support stressed populations, enhance population genetics or re-establish a lost population (Welcomme, 1988). While the majority of impacts in pearling can be very effectively managed, any impacts arising from species translocations are usually irreversible (Arthington and Blüdhorn, 1996).

Historically, both species introductions and transfers have been a feature of the development of terrestrial agriculture and were inevitably extended to aquaculture. Oyster farming is now over 2,000 years old and many translocations have been made. Edible oysters were cultured in Italy as early as 75 BC and by 95 AD translocations of oysters were being made from Britain to Italy (Philpots, 1890). For much of the early twentieth century, species translocations of edible oysters occurred regularly and were advocated for experimental and commercial purposes (Stafford, 1913; Orton, 1937). Not surprisingly, marine pearl culture, with its roots in the late nineteenth century, included species translocations as a legitimate and valuable means to enhance fisheries and culture operations.

The species that dominate marine pearl production have relatively broad distributions (see Section 2.2.3), but the allure of a pearl industry or the attraction of "desirable" genetic characteristics has seen most of the economically important species translocated within and beyond their natural boundaries (Table 14.1). Translocations have been occurring since the beginning of last century (Saville-Kent, 1905) when the movement of stocks was seen as a potential means of dealing with constraints, such as variability in pearl fisheries. The approach to translocations of pearl oysters has changed more recently, but some countries have actively sought the importation of pearl oysters to establish a pearl industry (Tanaka, 1990). This practice risks the introduction of other non-native species, genetic pollution of native stocks and the introduction of pests, parasites and diseases.

Occasionally, inadvertent or accidental translocation of species may have occurred as sometimes species boundaries are blurred, either through the transient nature of populations in certain areas or the considerable taxonomic confusion among some pearl oyster species (see Section 2.3.1). However, there are cases where deliberate introductions of pearl oysters have markedly changed species boundaries. For instance, *Pinctada margaritifera* was deliberately introduced to the Mediterranean when stocks from the

TABLE 14.1

Translocations of pearl oysters.

Species	Source	Destination	Notes	Author
Akoya	Red sea	Mediterranean	Lessepsian migration through the Suez canal	Kinzelbach (1985)
Akoya	Japan	Marshall Islands and Palau	Introduced 1935–1936 by the Mikimoto company, abandoned 1942	Smith (1947); Humphrey (1995)
Akoya	Red Sea	Greece		Serbetis (1963)
Akoya		Hawaii	Introduced 1956. Now established and listed as a non-indigenous species (Ray 2005)	Kanayama (1967)
Akoya	Japan (Amami)	Tonga	1975–1977	Tanaka, 1990
Akoya	China (Hainan)	Japan	Putative source of Akoya virus in Japan	Hine (1999)
Akoya	China, Cambodia, Vietnam, Myanmar	Japan	Post 1994 in response to Akoya virus	Masaoka and Kobayashi (2005)
Akoya	Unknown	Kiribati	1902	Munro (1993)
Pinctada maxima	Australia (Torres strait)	Cook Islands	1904: 1,500 oysters to Suwarrow Atoll	Saville-Kent (1905)
Pinctada maxima	Australia and Indonesia	Palau	1935–1942: 77,460 oysters transported	Humphrey (1995)
Pinctada maxima	Japan (Amami)	Tonga	1975–1977	Tanaka (1990)
Pinctada maxima	Australia (Kuri Bay)	New Guinea	1977 Presumed *P. maxima*	Humphrey (1995)
Pinctada margaritifera	Red Sea	Mediterranean	1860	Bellet (1899)
Pinctada margaritifera	Japan (Amami)	Tonga	1975–1977	Tanaka (1990)
Pinctada margaritifera	Japan (Okinawa)	French Polynesia	1979	Millous (1980)
Pteria penguin	Japan (Amami)	Tonga	1975–1977	Tanaka (1990)

Source: Modified from Eldridge (1994).

Red Sea were imported about 1860 with the intention of establishing the species on the Calabrian coast (Bellet, 1899). Although this venture failed, there has been a subsequent report of the species on the Egyptian coast at Alexandria (Hasan, 1974). A similar case occurred with the Akoya pearl oyster, which, although not native, is now established and widespread in the Mediterranean. At least in this latter case there is some debate as to whether this arose from deliberate introduction (Serbetis, 1963) or it has been maintained through Lessepsian migration through the Suez Canal (Kinzelbach, 1985).

More recently, the attraction of "desirable" genetic traits has perhaps been a more significant driver for translocation than the desire to establish an industry. As might be expected from their broad distribution, genetic differences occur between populations of particular pearl oysters species (Wada, 1982; Blanc *et al.*, 1985; Johnson and Joll, 1993; Benzie and Ballment, 1994; Benzie *et al.*, 2003; Arnaud-Haond *et al.*, 2004; Section 12.2) and these differences are often of commercial interest. For example, species such as *P. margaritifera* and *P. maxima* exhibit differences in nacre color between populations (Durand, 1987; Taylor, 2002) and attempts have been made to exploit these differences commercially. For *P. maxima*, changes in the percentages of gold and silver-lip oysters between locations are well known and these differences have been shown to affect the ultimate color of the pearls produced and thus their commercial value (Knauer and Taylor, 2002). As a result of these phenotypic differences, translocations of pearl oysters over thousands of kilometers have been common within a number of species and have on some occasions been found to have negative impacts.

In French Polynesia, there have been translocations of spat between genetically distinct populations of *P. margaritifera*, leading to increased homogeneity among populations at two locations (Arnaud-Haond *et al.*, 2004). Interbreeding in this fashion can reduce the fitness of native populations through the loss of adaptations and the separation of beneficial gene combinations. Furthermore, these movements can also be commercially detrimental. Anecdotal reports have noted that, in the wake of large-scale *P. margaritifera* transfers between atolls of the Tuamotu Islands, the pre-existing differences in pearl color, luster and orient that were once characteristic of the various islands have now been lost (Lewis *et al.*, 2004).

In addition to translocation, the relatively recent proliferation of hatcheries to allow the artificial propagation of pearl oysters may also pose threats to the genetics of local pearl oyster populations. Arising from the initial desire by many pearl companies to ensure the timing and supply of pearl oyster seed, hatcheries are now commonplace. There are numerous hatcheries in Japan and China, where an area such as Beihai, Guangxi Province, alone has more than 50 small hatcheries (O'Connor and Wang, 2001). There are more than 30 hatcheries in Indonesia and the Pacific. Even in Australia, where the industry is still based largely on oysters collected from the wild, there are more than six pearl oyster hatcheries that are collectively capable of meeting total industry demand.

Hatcheries are potentially environmentally beneficial in that they reduce pressure on wild stocks as the source of oysters for culture as well as reducing the incentive for stock translocation. Indeed, some hatcheries have been set up specifically for this purpose. However, the increasing prevalence of hatcheries and their importance in culture means that some care is required to insure that they pose little threat to the genetics of local populations. Selective breeding, intentional or inadvertent, can occur in mollusc hatcheries (Fig. 14.2), and the major implications of this are outlined in Section 12.3.4. Studies with other bivalves have shown that hatchery produced stocks can have low effective populations sizes (Hedgecock *et al.*, 1992; Taris and Goulletquer, 2005). This results from factors such as variable fertility and genetic differences in larval survival, development rate, size and success at settlement (Boudry *et al.*, 2004). Simple hatchery

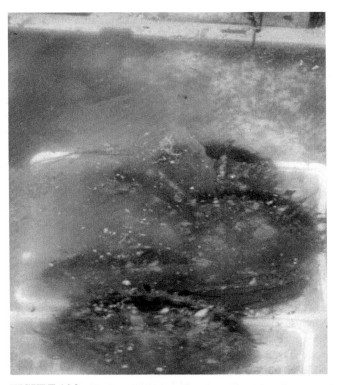

FIGURE 14.2 Hatchery cultivation of pearl oysters, such as these spawning *Pinctada margaritifera*, can result in loss of genetic diversity, inadvertent selection of non-desirable traits and large-scale production may influence gene frequencies in non-farmed stocks. (Photo: Paul Southgate.)

practices such as selectively sieving larger larvae and grading (see Section 7.4.6.3) can pose a substantial risk for diversity loss (Taris and Goulletquer, 2005). In addition, programs to select pearl oysters for various physiological and morphological characteristics are widespread (Wada, 1984; Taylor, 2002; Wang *et al.*, 2005; Chapter 12). As a result, some care is required by hatchery operators to ensure that inbreeding is reduced, that inadvertent selection of non-desirable traits is avoided and that large-scale production does not alter gene frequencies in non-farmed stocks. For example, McGinnity *et al.* (2003) showed that the fitness of wild Atlantic salmon (*Oncorhynchus tshawytscha*) was lowered by breeding with escaped aquaculture fish. Repeated escapes caused cumulative fitness depression in wild fish. In the case of pearl oysters, escape is not necessary for the animals to contribute to local gene pools.

An additional concern leveled at hatchery production in aquaculture has been the modification of the genome of an aquaculture species through the transfer of foreign genetic material. To date this technology has been limited largely to fish; however, the technology has been attempted with molluscs (Tsai *et al.*, 1997) and will presumably be available for use with pearl oysters in time. In response to public concern, some sectors of the pearl industry in Australia considered the question of genetically modified organisms (GMOs) and rejected their use (AMWING Pearl Producers Association, 2003).

14.3.2. Disease, parasites and pests

An unfortunate consequence of the translocation of molluscs has been the spread of a number of other aquatic organisms. Although transfer of a foreign organism does not imply that it will successfully establish within the new location, there are still numerous accounts of the introduction and establishment of diseases, parasites and pests (Eldridge, 1994; Humphrey, 1995). These introductions can have serious consequences for aquaculture, commercial fisheries and the aquatic ecosystem, and examples of pearl oyster mediated transfer have been recorded.

Those responsible for an inadvertent introduction can be the first to discover that it has occurred as the transfer of diseases or parasites has direct implications for aquaculture. The devastation of the Japanese Akoya industry is an example of the consequences of such transfers. Massive mortalities of pearl oysters occurred on Hainan Island off the coast of China in 1994 from an unidentified pathogen and, following the movement of some of these oysters to western Japan, more than 400 million Akoya oysters were lost in 1996 and 1997 (Hine, 1999). These mortalities continue and the Japanese industry has yet to recover (see Section 9.2.6).

In response to the Akoya disease, some Japanese farmers undertook programs to collect oysters from other areas including Thailand, Vietnam, Cambodia, Myanmar, China and the Red Sea (Masaoka and Kobayashi, 2005; W. O'Connor personal observation) in the hope that they would be resistant to the disease and that interbreeding with Japanese stocks would confer some degree of hybrid vigor. This in turn risked the introduction of other pathogens. A similar case was reported in which the movement of *P. margaritifera* spat between atolls in French Polynesia was implicated in the transfer of a "viral" disease (Cabral, 1989) and few areas were left unaffected (Cabral, 1992). This disease was also said to have been spread by the transfer of shells (Eldridge, 1994) and to have affected other unspecified bivalves (Braley, 1991).

Translocations of pest species of invertebrates and algae with commercial bivalve translocations are well documented. The movement of the American oyster, *Crassostrea virginica*, from the American Atlantic to the Pacific coast also introduced the Atlantic oyster drill, *Urosalpinx cinerea*, the slipper shells, *Crepidula fornicata* and *C. plana*, the polychaete, *Polydora cornuta*, and the salt marsh cord grass *Spartina alterniflora* (Ray, 2005). Likewise, a number of species have been introduced to European waters during translocations of oysters including *U. cinerea* and *C. fornicata* as well as the Japanese brown seaweed, *Sargassum muticum* (Wolff and Reise, 2002). These introductions were unintentional. However, unintentional introduction is probably the major source of entry for invasive species into new environments (Ray, 2005) and this warrants particular care for translocations.

There are few records of unintentional species introductions with pearl oysters. However, transfers of *P. margaritifera* have been blamed for the spread of anemones among some French Polynesian islands (Le Moullac *et al.*, 2003). These anemones occur as fouling organisms on the oysters and are increasing the cost of production. They are considered to be an important ecological problem.

The threat of translocations of some pest species can be overcome with careful cleaning, although some fouling organisms, particularly some algae, fungi, sponges and polychaetes that bore into the pearl oyster shell, may pose a greater threat (Mao Che *et al.*, 1996). In a study of farmed *P. maxima* in Western Australia and the Northern Territory, 62% contained the boring sponge, *Cliona dissimilis*, and two other species of sponge (Fromont *et al.*, 2005). A survey of Akoya pearl oysters in the Gulf of Mannar found evidence of polychaete infestation in more than 70% of shells (Alagarswami and Chellam, 1976). Surveys of adult oysters in NSW, Australia, found polychaete blisters (Fig. 15.1) in 23% of Akoya and 30% of *P. albina* shells (O'Connor, 2002a; O'Connor and Lawler, 2003). Given that these organisms cause considerable economic damage in pearl culture (Mao Che *et al.*, 1996) and are also variously capable of infesting other species such as corals and other bivalves (Fromont *et al.*, 2005), great care must be taken to ensure that there are no translocations of these species.

In accordance with the threat of introduced species, countries and specific industries have enacted legislation, developed environmental management systems and imposed industry-based codes of practice to prevent mishaps. Examples in Australia include: the adoption of the "National System for the prevention and Management of Marine Pest Incursions" in late 2003; the Pearl Oyster Translocation Protocols in states such as Western Australia, and industry codes of practice such as the Pearl Producers Association's "Recommendations for the sterilization of technicians equipment". The latter in recognition of the need for good husbandry practices as pearl technicians move between farms.

14.3.3. Impacts on native species

In certain circumstances, pearling operations can pose threats to local pearl oyster populations and other species within the general vicinity of the farm. Apart from the potential impacts of introduced species, threats can arise through direct pressures, such as the harvest of local pearl oyster stocks for seeding, or through indirect means, such as resource competition.

Overfishing and variability in pearl oyster resources were an impetus for pearl culture and have in part shaped the modern pearl culture industry (see Chapters 1 and 9). Many pearl industries now rely on hatchery propagation as a source of seed to reduce dependence on local pearl oyster stocks and local impacts. Furthermore, with the increase in interest in genetic selection for cultured stocks, the reliance on hatcheries is increasing. Perhaps the most notable exception is the Western Australian *P. maxima* industry, which retains the capacity to collect ca. 570,000 oysters from the wild annually (WA Pearling Act, 1990, see Section 9.3.3). The collection is strictly controlled. Allocations for collection are based on quota units with approval to take oysters of a restricted size from particular coastal zones. The numbers of oysters that may be collected and the fishing pressure in each zone are based on strictly controlled catch and fishing effort data. The data are regularly reviewed on the basis of extensive annual stock assessments. Any localized impact arising from fishing in particular areas seems unlikely, as estimates by Joll (1996) found that only 24–31% of the appropriate-sized oysters were removed from the environment during fishing.

Pearl culture may indirectly affect surrounding organisms through various environmental interactions. In measuring bivalve recruitment on *P. margaritifera* farms, Braley (1998) found that settlement of bivalves was greater in unfarmed areas and he suggested this might result from filtration of larvae from the water column. Pearl oysters ingest small zooplankters, including bivalve eggs and larvae, although it is not clear that they digest them (see Section 4.2.5). Alternatively, in oligotrophic waters such as those in the Cook Islands, where some concerns were already being voiced over stocking densities, resource competition may have also been a factor in reduced bivalve settlement. The literature cites a number of examples of the capacity of bivalves to remove suspended food material. Species such as mussels are capable of consuming between 35–60% of the food material available in the surrounding water (Figueras, 1989; Hickman, 1989). In the Cook Islands, farmed *P. margaritifera* were found to be capable of removing 40% of the phytoplankton from water passing through the farm (Anderson, 1998). In Takapoto lagoon, French Polynesia, pearl oyster culture was found to be capable of significant localized reductions in food availability; however, their overall consumption was found to be insignificant when compared to other bivalves in the lagoon (Pouvreau *et al.*, 2000).

Indirect pressures on other species can also include practices in the production of one species that impact on another. An often-cited example in aquaculture is the harvest of trash fish to produce diets for species of higher value. The demand for the shell nuclei commonly used to produce cultured pearls (see Section 8.9.3) has had significant impacts on some freshwater mussel species that are used to make the nuclei. In some states of the United States, such as Kentucky, Tennessee, Ohio, and Pennsylvania, "Mussel rustlers" have been responsible for the illegal harvest of mussels to provide nuclei for the pearl industry. Poachers can obtain from US$15 to $22 per kg for mussel shells and mussel harvesting is considered to be sufficiently threatening to have seen several southern states impose quotas or pass laws against mussel collection.

14.3.4. "Icon" species

The impacts of aquaculture are often assessed within the context of a particular ecosystem. Increasingly, however, the impacts on particular "icon" species are being considered, particularly if those species are themselves the subject of some economic activity. Some debate in Australia has focused on the impacts of longline mollusc culture on whales and dolphins, and to a lesser extent turtles and dugongs. The popular appeal of these species and their economic value to "ecotourism" ventures has led to a variety of claims of potential impact. Foremost among the impacts considered for pearl farming are: entanglement, increased risk of boat strike and injury, increased noise disturbance, and habitat loss and displacement (Paton *et al.*, 2002, cited in Cleland 2002).

The issue of whale entanglement has been one of the most emotive and its potential impacts can be readily demonstrated. In species such as the North Atlantic right whale, up to 57% of animals studied bore evidence of entanglement in fishing gear (Kraus, 1990) and 7% of known mortalities arose from entanglement in 1970–1999 (Knowlton and Kraus, 2001). Indeed, if as few as two human-caused deaths of North Atlantic right whale females could be prevented each year, population growth in this species could be restored to replacement level (Fujiwara and Caswell, 2001). However, despite a need for caution, there is little evidence to support any significant risk to whale populations from entanglement in pearl culture lines. A survey of Australian mollusc culture operations in 2002 found that there are more than 87,000 hectares of lease area licensed for longline and hanging line culture in Australian waters. The vast majority was for pearl oysters (O'Connor, 2002b). Many of these leases are located in areas frequented by whales and dolphins. For example, the waters off the NW Australian coast, north of Cape Leveque, are listed by Environment Australia as an area of "special interest" for whales. "This area is important for calving and breeding of the "Area IV" humpback whale population, with preliminary studies indicating that calving grounds extend north and east of Cape Leveque (16°24′ S) across the Buccaneer Archipelago and Camden Sound region" (Jenner and Jenner, 1996). Similar areas of "special interest" for cetaceans include the Albany area in the Great Australian Bight and Mercury Passage (Tasmania), both of which include significant areas of mollusc longline culture.

Over its 50 year history as a production method in Australia, there are only two substantiated records of whale entanglement in longlines and both were released. While any such entanglements are unfortunate, their impact is unlikely to affect whale populations and they are insignificant compared with other anthropogenic impacts. In 2004 in the state of Queensland alone, 54 whales, dolphins and dugongs were killed or injured as a result of human activities that included shark meshing, boat strikes, etc. (Greenland *et al.*, 2005; Greenland and Limpus, 2005).

14.3.5. Habitat exclusion and modification

Discussion of habitat exclusion and modification has featured dolphins among the "icon species." While there is no record of dolphin entanglement in pearl longlines and equipment, the limited observations of dolphin populations surrounding pearl leases have suggested that there may be changes in behavior. In Shark Bay, Western Australia, a long-term study of dolphins, including hand-fed wild individuals, provided the opportunity for incidental observations of dolphin behavior in conjunction with pearl leases. This study found that the ranging patterns of female dolphins appeared to alter, with those females with dependent calves rarely using pearl farm areas with longlines (Mann and Janik, 1999; Watson-Capps and Mann, 2005). The reasons for avoidance are unclear, particularly for mothers and calves. They "may" have been the direct consequence of pearling activities, although the authors acknowledged that tour operations (including shark baiting) and increased recreational activity might have contributed to the changes (Mann and Janik, 1999; Watson-Capps and Mann, 2005). In reviewing data on the Shark Bay dolphin population, Bedjar (2005) noted that harmful impacts are unlikely to result from single encounters, rather it is repeated, close, prolonged and persistent encounters, like those associated with tourist activities that can be detrimental. While pearl farming may not have the impact of tourism, there are some common features and procedures to reduce the potential for impact. Interactions with all marine mammals need to be avoided, especially feeding them.

In studies of the interaction between dolphins and other forms of bivalve culture, it has been suggested that culture infrastructure may impair foraging by visual or acoustic obstruction (Markowitz et al., 2004), by acting as physical obstacles to hunting (Wursig and Gailey, 2002) or by interfering with echolocation of prey (Au, 1993). On the other hand, there have been claims of positive interactions, with anecdotal reports of dolphins feeding and sheltering within pearl leases. Pearl culture infrastructure may act as an attractant for fish species that may in turn provide food for dolphins. While there is limited evidence for this, studies of mussel farms in Ireland have suggested that culture lines may act as a shelter for seals and sea birds by dampening out the effect of extreme wave action, and so creating more favorable foraging conditions (Roycroft et al., 2004).

14.4. PHYSICAL

14.4.1. Biodeposition and sedimentation

Bivalves are filter feeders that consume substantial quantities of suspended particulate matter from the water column (see Section 4.2.5). This food consumption can lead to significant biodeposition: the benthic accumulation of wastes from the bivalves themselves. The accumulation occurs predominantly in the form of feces and pseudofeces, but in some forms of bivalve culture it may be significantly augmented with organic material arising from stock losses, and through cleaning and harvest operations that dislodge fouling. In some instances, the process of biodeposition is exacerbated by hydrological changes induced by the bivalve culture apparatus. There will be greater settlement of heavier particles if the infrastructure severely retards currents flowing through the lease area.

Biodeposition and sedimentation may together lead to the accumulation of organically rich materials beneath bivalve leases (Dahlback and Gunnarsson, 1981; Mattson and Linden, 1983; Kaspar et al., 1985; Hatcher et al., 1994; Stenton-Dozey et al., 1999, 2001). This process can significantly alter benthic environments, both chemically and physically, which can in turn cause significant changes in the macrofaunal assemblages in the sediments (Mattson and Linden, 1983; Kaspar et al., 1985; Weston, 1990; Weston, 1991; Hatcher et al., 1994; Grant et al., 1995).

Chemically, biodeposition can cause increases in consumption of oxygen in the sediment beneath farms that can lead to anoxia and in extreme cases, the release of carbon dioxide, methane and hydrogen sulfide (Kaspar et al., 1985). Physically, the structure of the surficial sediment can also change. In some instances, the fine materials deposited smother benthic species. Alternatively, it has been suggested that the accumulation of lost stock and the associated fouling organisms beneath the farms does more to influence the environment than the deposition of feces and pseudofeces (Tenore *et al.*, 1982; Jaramillo *et al.*, 1992; Grant *et al.*, 1995; Stenton-Dozey *et al.*, 1999). Collectively, the chemical and physical changes can lead to changes in the species present, infaunal diversity and infaunal biomass. However, the literature on the effects of biodeposition on benthic communities is not consistent, with the measures such as infaunal biomass having been reported on occasions to both increase (Grant *et al.*, 1995) and decrease (Tenore *et al.*, 1982; Mattson and Linden, 1983), though biodiversity generally decreases.

The impacts of biodeposition and sedimentation have led to changes in management practices. In Japan, the relocation or temporary fallowing of oyster leases has been used to counteract impacts of the accumulation of wastes beneath leases (Imai, 1971). In areas such as the Cook Islands considerable effort has been made to determine appropriate densities for culture so that impacts on pearl productivity do not occur (Anderson, 1998; Braley, 1998; Ponia, 2000; Ponia and Okotai, 2000; Heffernan, 2006).

The effects of biodeposition can be a function of a number of factors including the type of infrastructure, stocking density, the type of material deposited, the assimilative ability of the particular environment and hydrology (water depth, current flows, etc.). Each culture operation needs to be considered in the light of these factors. Pearl culture operations therefore have varying abilities to cause impacts through biodeposition. Generally, most pearl culture operations, particularly in Australia and the Indo-Pacific work at markedly lower stocking densities than those used for edible bivalves such as mussels (Table 14.2). The low densities reduce the output of feces and pseudofeces as well reducing the amount of infrastructure present to slow water flows.

TABLE 14.2

A comparison of stocking densities used for pearl culture with those used for suspended culture of other bivalves.

Location	Species	Suspended culture method	Density (t/ha/year)	Author
Ago Bay, Japan	Akoya	Raft Culture	22.9	Uyeno *et al.* (1970)
Port Stephens, NSW, Australia	Akoya	Mid-water long-line	6.9*	O'Connor *et al.* (2003)
Penrhyn atoll, Cook Islands	*Pinctada margaritifera*	Surface longline	0.91**	Braley (1998)
Ria de Arousa, Spain	*Mytilus galloprovincialis*	Raft culture	48.0	Pérez Camacho *et al.* (1991)
Tjärnö, Sweden	*Mytilus edulis*	Raft culture	47.5	Dahlback and Gunnarsson (1981)

*Maximum allowed under lease permit conditions (average density approximately 2.2 t/ha/year)
**Assumes average individual oyster weight is 350 g.

Other factors implicated in bivalve farming impacts, such as those arising from lost stock (Stenton-Dozey *et al.*, 1999), are also lessened in pearl culture as a function of the comparatively high value of the pearl oysters. Unlike species such as mussels, pearl oysters are usually held in secure cages and in the rare event of stock losses, there are often attempts made to recover the animals and their cages. In the only known estimate of its type, the maximum expected loss for *P. maxima* farms was estimated at 160 panel nets (see Section 7.5.4.) in a year in which four cyclones occurred (Enzer Marine Environmental Consulting, 1998). At this rate, the likelihood of impact is negligible.

While the numbers of studies of the impacts of pearl culture on sediment quality and benthic flora or fauna are less than in edible bivalve culture, similarities between edible bivalve culture and pearl culture are nonetheless apparent. In Japan, Uyeno *et al.* (1970) investigated the contribution of biodeposition from Akoya pearl oysters, farmed at 22.9 t/ha/year, in the degradation of benthic sediments in Ago Bay. They concluded that hydrographic conditions, season and oyster stocking density all affected the accumulation of organic material in the sediment and recommended that the number of oysters under culture be reduced, at least during summer. More recently, Yokoyama (2002) contrasted the benthic environment below fish culture with that found below pearl culture areas and control locations in Gokasho Bay, Japan. Here, organic carbon and nitrogen loads from pearl farming were estimated from the work of Uemoto *et al.* (1978) to range from 0.6–$2.7\,gC/m^2/day$ and 0.06–$0.4\,gN/m^2/day$, respectively. At this level the benthic community structure below the pearl farm site remained similar to that of the control site.

Outside of Japan, there have also been investigations of the potential for benthic disturbance below pearl culture areas. Anderson (1998) measured sediment oxygen demand beneath *P. margaritifera* farms in Manihiki Atoll, Cook Islands, and concluded that at densities of $0.7\,oyster/m^2$ there were no discernable effects. More recently, four consecutive years of quarterly sediment monitoring beneath an Akoya farm on the Australian east coast found no significant changes in sediment nitrogen, phosphorous or total organic carbon when compared with other control sites (Gifford, 2006). A comparison of benthic faunal abundance and diversity beneath the same farm and at five other similar locations in the vicinity ($>500\,m$ but $<2\,km$ from the farm) found no significant impact on species numbers or diversity (Roberts, 2002). Finally, comparisons of the health and extent of seagrass beds adjacent to the farm with those made in the same area a decade earlier found no reductions, while the seagrass beds closest the farm ($150\,m$) were considered to be "healthy" and in a similar state to those more distant from the farm (Roberts, 2002).

14.4.2. Turbidity and shading

Turbidity, the properties of water that cause light to be scattered and absorbed, reflect the density of suspended particulate matter. Together with any shading of the seafloor that may arise from the culture apparatus, increases in turbidity can affect aquatic food webs by limiting primary productivity through the reduction or attenuation of light available to phytoplankton and benthic plant communities. Notwithstanding the

FIGURE 14.3   Cleaning of *Pinctada maxima* and their culture apparatus using high-pressure water jets on a wooden pontoon at Aljui, Iryan Jaya, Indonesia. (Photo: Joseph Taylor, Atlas Pearls.)

fact that bivalves contribute to a net reduction in turbidity as they filter particles from the water column (Newell, 2004), bivalve farming can cause short-term and localized decreases in light levels. When pearl oysters and their culture units are handled and cleaned, material is washed from them and it returns to the water and reduces light penetration (Fig. 14.3; see Section 15.4.2 and Figs. 15.4–15.6). As a function of comparatively low stocking density, turbidity and shading are not perceived to be significant problems (other than with respect to visual amenity). Indeed in many instances bivalve culture is being proffered as a means of lessening the impacts of other anthropogenic activities in coastal and estuarine areas.

14.4.3. Bioaccumulation and zooremediation

Filter feeding bivalves remove suspended material from the water column and in doing so reduce levels of organic material, silt, nutrients, bacteria and other dissolved components. This process can improve water clarity, remove pollutants and have significant environmental benefits (Dame, 1996; Newell, 2004), so much so that in some areas the restoration or introduction of bivalve communities is being encouraged to assist ecological balance (Mann and Evans, 2004). The largest of these programs is in the United States, where various stakeholders have committed US$100 million to increasing natural stocks of oysters 10-fold by 2010 in Chesapeake Bay (Chesapeake Bay Program Federal Agencies Committee, 2001). The use of bivalve aquaculture stock will most likely be

used to augment efforts to increase natural stocks of bivalve molluscs, with similar ecological benefits (Gifford *et al.*, 2004). Using bivalves in this fashion has been described
as "zooremediation" or animal-based bioremediation (Macfarlane *et al.*, 2004).

Like bivalve culture in general, pearl culture has the potential for remedial benefits
to the environment, but it is particularly suited to this role for a number of reasons
(Gifford *et al.*, 2004; Macfarlane *et al.*, 2005). The largest species of pearl oysters,
P. margaritifera and *P. maxima*, have amongst the highest pumping and filtering rates
per unit mass reported for bivalves. They process large volumes of water, removing
large quantities of suspended particulate matter and being exposed to large quantities
of potential pollutants. Pearl oysters also have a comparatively high protein content,
between 65–88% of dry tissue weight (Suzuki, 1957; Seki, 1972; Numaguchi, 1995),
which translates into a high requirement for nitrogen and phosphorous and hence a
great potential for remediation by taking up large quantities of these nutrients (Gifford
et al., 2004). Pearl oyster species have a wide global distribution in the tropics and
subtropics so that native organisms can be used in many areas. Finally, and possibly
most importantly, remediative programs can be expensive to establish and maintain.
Thus, the use of pearl oysters has advantages from their ability to produce a valuable
product, pearls, while not needing to be a product fit for human consumption.

Gifford *et al.* (2005a, b) provide an example of the potential for pearl oyster-based
zooremediation. In assessments of a small Akoya pearl oyster farm in NSW, Australia,
the harvest of each tonne of oysters resulted in 7.45 kg nitrogen, 0.55 kg phosphorus,
and up to 0.70 kg of metals (Zn, Al, Fe, Sr, Mn, As, Cd, V, Se, Cu, Pb, Ni, U and Cr)
being removed from the water (Gifford *et al.*, 2005a, b). At its existing harvest rate
(9.8 t/year), some 6.9 kg of heavy metals are currently being removed from the waters
of Port Stephens. They calculated that the small experimental farm would have to
increase current production 51 times, from 9.8 to 499 t/year, to balance the nitrogen
input of the existing sewage treatment plant located on the shores of the Port.

As a consequence of their ability to accumulate potential pollutants, pearl oysters
have been used as localized biomonitors of both heavy metals (Bou-Olayan *et al.*,
1995; Al-Madfa *et al.*, 1998) and hydrocarbons (Fowler *et al.*, 1993). Furthermore,
pearl oysters have also been shown to accumulate antibiotics such as sulfamethazine
and thus may serve as a useful biomonitor of antibiotic presence (MacFarlane *et al.*,
2004). The wide global distribution of pearl oyster species also lends them to being
a suitable candidate for regional or global environmental monitoring programs of
marine pollution (Sarver *et al.*, 2003; Gifford, 2006).

14.5. CHEMICALS

While the use of chemicals in terrestrial agriculture is widespread and often widely
accepted, their use in aquaculture attracts considerable scrutiny. This attention arises
in part because of the lack of effective barriers between the farms and the general
aqueous environment, but also because the fate of many chemicals and their impacts
are poorly understood. In this regard, the use of large quantities of chemicals is not

a feature of pearl farming; however, their use to some extent is unavoidable. Chemicals that pose a threat to aquatic life are present in some construction materials and can be used as antifoulants, while others such as fuel and oil are an integral requirement for farm operation. In some instances, antibiotics and antiseptics are used with certain pearl oyster species and the application of relaxants and hormones have been evaluated experimentally although not used commercially.

Plastics are commonly used on most pearl farms in ropes, floats and cages. They may contain stabilizers (fatty acid salts), pigments (chromates, cadmium sulfate), anti-oxidants (hindered phenols), UV absorbers (benzophenones), flame retardants (organo-phosphates), fungicides and disinfectants (GESAMP, 1991) that can leach into the environment and adversely affect aquatic life. The fuels and oils used in powerboats and generators contain toxic substances that can inadvertently enter the environment through spills or directly through motor discharges. Other chemicals, toxic to marine life, are present in antifoulants (see Section 15.6), which are deliberately designed for progressive release. However, none of these are unique to pearl oyster farming: some occur in each of the wide variety of commercial and recreational uses of the marine environment.

Intentional introduction of chemicals to culture water in pearl production is extremely limited. Antibiotics and antiseptics are used in small quantities, while disin-fectants are used for some applications. Among these, antibiotics in aquaculture prob-ably pose the greatest concern. Their use can lead to an increase in antibiotic resistant bacteria, their residues can accumulate in other organisms, they can impact on normal processes of biodegradation and they can impact on their users and seafood consumers (see Weston 1996 for review). Antibiotics have two applications in pearl culture: they are occasionally used in some pearl oyster hatcheries and also during pearl nuclei inser-tion operations.

Antibiotics have been used in Australian pearl hatcheries in the past in situations where water quality is poor or larval densities are high. They have also been evalu-ated for use with *P. fucata* in India (Subhash *et al.*, 2004), *P. martensii* in China (Liu *et al.*, 1998) and applied to *P. margaritifera* in Micronesia in situations where chronic failure of larval batches has occurred (Haws *et al.*, 2003). In these instances, antibiotic use is probably symptomatic of serious underlying problems with hatchery production that at least in one instance in Australia is likely to have arisen from poor site selection.

Antibiotics and antiseptics have been evaluated as a means of improving hygiene during nuclei insertion operations to reduce post-operative mortality and increase pearl quality (Norton *et al.*, 2000). In practice, antibiotics can be applied to the instruments and to oyster tissue during the operation, or the nuclei can be coated with an antibiotic. Neomycin (10% Oxytetracycline) is used in small quantities with Akoya pearl oys-ters in Japan and Australia. Neomycin solutions are used to soak instruments between operations on individual oysters. Total usage is low with estimates for a medium size farm (500,000 oysters operated per annum) using ca. 240 g/year. The bulk of this is actually retained following the operations and can be disposed of appropriately so that the quantities reaching waterways is negligible. Alternatively, antibiotic coated nuclei are also available and have been widely evaluated within the pearl industry, although they have not been accepted universally.

Disinfectants are used for the general maintenance of hygiene in hatcheries, oyster holding and transport facilities, and operating rooms and equipment. They vary from acid solutions and hypochlorite to commercially available disinfectants such as Pyroneg® and Virkon S®. The disinfectants are by necessity used carefully as their applications can impact directly upon the bivalves (Dove and O'Connor, 2007).

Hormones such as insulin promote growth in a variety of bivalves and have been shown to accelerate nacre deposition in *P. margaritifera* (Paynter and Haws, 2004), but they have not to the authors knowledge been applied commercially.

The invasive nature of the operative processes required to insert pearl nuclei has also encouraged the evaluation of relaxants to reduce stress and enhance post-operative recovery (Tranter, 1957; Hildermann *et al.*, 1974; Dev, 1994; Norton *et al.*, 1996, 2000). A number of chemicals have been tested including propylene phenoxetol, menthol, clove oil, benzocaine and MS222 (Norton *et al.*, 1996), evaluated with *P. margaritifera* (Hildermann *et al.*, 1974; Norton *et al.*, 1996), *P. albina* (Norton *et al.*, 1996; O'Connor and Lawler, 2003), *P. maxima* (Mills *et al.*, 1997), *P. mazatlanica* (Monteforte *et al.*, 2004) and Akoya oysters (O'Connor and Lawler, 2003; Acosta-Salmon and Southgate, 2004). Anesthetics are not in common use in pearl production. They are used in containment on land or on vessels and thus any broader ecological effects will be minimal.

14.6. ESTHETICS AND AMENITY

Increasingly, the impacts of aquaculture ventures are being considered in broader terms that include potential esthetic effects and amenity issues for other user and traditional owner groups. This is particularly so in areas considered to be "pristine" or having high population pressure or having value for cultural reasons, tourism or recreational pastimes. Aquaculture, including pearl culture, can be seen as an "invasion of the commons" (Stickney, 2003) and pearl culture may be considered to be incongruous with the community's perceptions of a particular area.

The extent of the incongruity of particular farming operations is a function of both individual and cultural preferences. In extreme examples, impacts considered by some to be excessive or obtrusive may well be acceptable to others (Fig. 14.4).

14.6.1. Indigenous communities

Indigenous and local communities have played an important role in the development of the various pearl industries, although the extent to which those communities have benefited from the pearling activities has varied. Commonly they have provided much of the labor necessary for the construction and operation of farms, but their role in the management and ownership of the farms has been limited. In many regions now there is legislation to ensure that local communities have greater input into industry development and thus the opportunity to protect areas of traditional cultural value.

FIGURE 14.4 Chairman's prizewinner in the 4th contest for beautiful scenes of Japanese villages. *"A beautiful scene is formed by the setting sun shining on the small islands of Kuroshima and Goitsumijima with the rafts for pearl and fish cultivation floating in orderly formations off shore."* (*Source*: MAFF, 2006.)

In many areas in Japan, prospective farmers have been required to negotiate with local communities, particularly fishing cooperatives, for the use of coastal waters when issues such as access to traditional fishing grounds will be considered. In the south Pacific, pearl industries are being encouraged for the benefit of many communities and in accordance there are legislative protective measures to establish local ownership.

In Australia, indigenous communities have long been involved in the pearl industry, providing the first pearl divers and then a labor force for the further development of the industry (Wright and Stella, 2003). But recently indigenous rights have been formalized to a much greater extent. In the first of its type, the "Croker Seas Native Title" claim saw indigenous rights to a 2,000 km^2 sea area used in a 1993 agreement that was brokered by the Northern Land Council between the company Barrier Pearls and traditional owners at Croker Island, Northern Territory. It is now common for the leading Australian pearl companies to negotiate access agreements with local indigenous communities.

14.6.2. Commercial and recreational users

Pearl culture equipment and the associated vessels can be restrictive for certain activities, such as recreational boating, indeed guidelines for locating pearl operations in Western Australia specifically state that areas of high recreational use should be avoided (Simpson, 1998). Pearl culture may also impinge on other commercial users such as commercial fishers or other aquaculture ventures.

14.6.3. Visual amenity

A common community concern with aquaculture is the loss of visual amenity. Complaints often focus on unsightly culture equipment, particularly surface buoys and the sediment plumes that can arise during cleaning operations. Buoys used to mark leases and to support various bivalve culture apparatus have been criticized and there have been studies undertaken to select sizes and colors to reduce visual impact (NSW Fisheries, 1997).

Sedimentation arising from cleaning can also have visual impacts and can be perceived to be pollution. The processes of sedimentation and biodeposition lead to the accumulation of fine material on the culture apparatus. This material, along with fouling removed during cleaning returns to the water during cleaning and maintenance operations causing discoloration. The ANZECC Guidelines (ANZECC, 2000) state that the natural visual clarity of recreational waters should not be reduced by greater than 20%, a level that can be exceeded during pearl cleaning operations.

Cleaning in *P. maxima* farms can cause a stream of discolored water that can be seen for up to 200 m on calm days, before it merges into the background (Enzer Marine Environmental Consulting, 1998). An investigation of the impacts of farming Akoya pearl oysters found that "During both the hauling and washing process the turbid plume generated extends to some depth – up to 7 m, approximating the depth from which the bags were being held when hauling commenced. The wash effluent plume is discharged at around 4 m and its behavior is less clear. The laboratory test of settleable solids (suspended solids: 1,140 mg/L, settleable solids: 1,020 mg/L) would suggest that much of this would settle toward the bottom in a period of several hours" (NSW EPA, 2003). Figure 14.5 provides a visual representation of the impact of cleaning on seawater

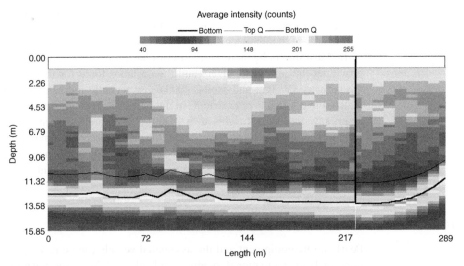

FIGURE 14.5 Acoustic Doppler Backscatter readings during cleaning operations on pearl lines in Port Stephens (NSW EPA, 2003). Increasing intensity indicates increased turbidity. The area in red shows the sediment plume around the cleaning boat. The localized zone of measurable impact is shown in yellow and can be seen to be limited to the upper 4–5 m of the water column and to extend for up to 70 m. The approximate position of the seafloor is indicated by the black line. (See Color Plate 12)

turbidity during pearl cleaning operations. While independent studies have failed to detect impacts of this process on sediment characteristics or benthic fauna beneath the farm, a plume was generated that was visible for up to 100 m (NSW EPA, 2003).

14.6.4. Noise

The remote nature of many pearling operations has meant that noise, as long as it poses no occupational hazard to pearl farm staff, has little or no esthetic consequence. However, in areas such as Japan, China and the Australian east coast, pearl farming can occur in close proximity to population centers. Regardless, there need not be excessive noise. Assessments of farm generated noise, predominantly from boats and cleaning machines, in Port Stephens, NSW, Australia, confirmed that activities could be undertaken within the relevant legislative noise guidelines (Cleland, 2002).

14.7. SOCIO-ECONOMIC IMPACTS

The socio-economic impacts of pearling are discussed in Chapter 13 and are only addressed briefly here. Pearling often occurs in relatively remote economically undeveloped areas and thus has the potential to have significant social impacts. Throughout Asia subsistence is involved in aspects of pearl culture (Fig. 14.6). In the Gambier and Tuamotu archipelagos of French Polynesia, one in four families earn a living from pearling and pearls form greater than 95% of French Polynesian exports (Tisdell and Poirine, 2000). Black pearls account for about 85% of the Cook Islands' export industry

FIGURE 14.6 Small-scale pearl oyster (*Pteria penguin*) culture on fish farming rafts in Vietnam to supplement income.

FIGURE 14.7   High-density Akoya pearl oyster culture at Beihai, Guangxi Province, China. (Photo: Aimin Wang.)

and the economic value of this industry is second only to tourism. In many of the outer islands of Polynesia, the growth of the pearl industry has reversed population declines and improved standards of living. In northern Australia, pearling has supplied work and various other indirect support mechanisms such as supplies and access to medical help (D. Mills personal communication.) for local indigenous communities. In Japan and Australia, local communities derive income from the use of resources. In Japan, pearl areas are leased from local fishing cooperatives providing direct financial benefit to local communities and in Australia access right agreements have been negotiated with traditional owners. On the downside, inequities can develop between those economically advantaged by pearling and, in some areas, pearling can encourage over exploitation of available resources. An example of the latter has occurred in China. Unregulated expansion of pearl farms in areas such as Beihai has seen reductions in productivity and pearl quality, which in turn has resulted in further expansion to replace the shortfall (O'Connor *et al.*, 2003; Fig. 14.7).

14.8. SUMMARY

Pearl culture is now more than a century old and its practices are constantly evolving. In many instances, the pearl industry is acutely aware of the importance of the environment in which they operate and have taken steps to protect that environment. Accordingly pearl culture is acknowledged to be among the most environmentally friendly forms of aquaculture (Jernakoff, 2002) with demonstrable positive environmental impacts

(Dame, 1996; Gifford *et al.*, 2004; Macfarlane *et al.*, 2004; Newell, 2004). This, however, is not universal across the global pearl industry and has not diverted criticism of particular practices. Unregulated pearl development in China has and will continue to affect the surrounding environment. It is important for all of the industry to acknowledge potential impacts and to continue to assess interactions with the environment. In areas of high public scrutiny, such as Australia, it is critical that the outcomes of these assessments continue to be open, transparent and effectively communicated to the public and legislators alike.

The acknowledgement by sectors of the pearl industry of the importance of the environment to the viability of their businesses and to the perception of their product in the market place has meant that many positive proactive steps for environmental protection have been taken. Accordingly, past practices that may have posed environmental risks, such as large-scale unregulated translocations of stock, have ceased in many areas. Although, this practice has not halted in Asia, despite the obvious example of direct catastrophic impact on the Japanese industry. Hatcheries are increasingly being developed to provide selected stock for industry and are serving to reduce pressures on wild populations and increase awareness of the potential genetic impacts.

Criticisms have often been leveled at aquaculture industries regarding the impacts of intensive culture, but while this is relevant to some sectors, it is now of declining relevance to the major pearl industries. The observed impacts associated with what was described as the "stupidity" of high-density cultivation on Japanese Akoya farms in the 1960s and 1970s were associated with production declines (Wada, 1991), and current production and farming densities in Japan are much reduced today (Yokoyama, 2002). In Australia and the Pacific, pearl farmers have historically used far lower stocking densities and where pearl farms are located in environments of good water flow with an average depth greater than 8 m, and stocking densities below 6 t/ha, the risk of benthic disturbance below pearl farm lease areas is greatly reduced (Wells and Jernakoff, 2006).

Well-managed pearl production requires few chemicals and the heightened awareness of the impacts of those that are used reduces the associated risks. This is also balanced by a greater understanding of the industry by legislators, which has increased the relevance of legislated controls. The esthetic impacts of all aquaculture ventures including pearl culture will require careful management to avoid community conflict. The geographical isolation of many farms will assist in this regard while other visual and noise impacts can be ameliorated through appropriate management protocols.

In the previously mentioned case in NSW, Australia, judge Cleland said the risks are real but can be adequately managed. With care, pearl culture can coexist happily, often synergistically, with other aquatic user groups without significant environmental costs.

References

Acosta-Salmon, H., Southgate, P.C., 2004. Use of a biopsy technique to obtain gonad tissue from the black-lip pearl oyster *Pinctada margaritifera* (L). *Aquacult. Res.* 35, 93–96.

Alagarswami, K., Chellam, A., 1976. On fouling and boring organisms and mortality of pearl oysters on the farm at Veppalodai, Gulf of Mannar. *Indian J. Fish.* 23, 10–22.

Al-Madfa, H., Abdel-Moati, M.A.R., Al-Gimlay, F.H., 1998. *Pinctada radiata* (pearl oyster): A biomonitor for metal pollution in Qatari waters (Arabian Gulf). *Bull. Environ. Contam. Toxicol.* 60, 245–251.

AMWING Pearl Producers Association, 2003. *Environmental Code of Practice*. Environment Australia and Aquaculture Council of Western Australia, Osborne Park, Western Australia, 139 pp.

Anderson, M., 1998. The ecological sustainability of pearl farming in Manihiki lagoon, Northern Cook Islands. *SPC Pearl Oyster Inform. Bull.* 11, 7–11.

ANZECC, 2000. *An Introduction to the Australian and New Zealand Guidelines for Fresh and Marine Water Quality*. Australian and New Zealand Environment and Conservation Council and the Agriculture and Resource Management Council of Australia and New Zealand, ISBN 0 9578245 2 1.

Arnaud-Haond, S., Vonau, V., Bonhomme, F., Boudry, P., Blanc, F., Prou, J., Seaman, T., Goyard, E., 2004. Spatio-temporal variation in the genetic composition of wild populations of the pearl oyster (*Pinctada margaritifera cumingi*) in French Polynesia following 10 years of juvenile translocation. *Mol. Ecol.* 13, 2001–2007.

Arthington, A.H., Blüdhorn, D.R., 1996. The effects of species interactions resulting from aquaculture. In: Baird, D.J., Beveridge, M.C.M., Kelly, L.A., Muir, J.F. (Eds.), *Aquaculture and Water Resource Management*. Blackwell Science, Oxford, 219 pp.

Au, W.W.L., 1993. *The Sonar of Dolphins*. Springer Verlag, New York, 277 pp.

Bedjar, L., 2005. Linking short and long term effects of nature based tourism on cetaceans. Ph.D. thesis, Dalhousie University. Halifax, NS, 158 pp.

Bellet, D., 1899. La culture des huîtres perlières en Italie. *La Nature* 27(1355), 375.

Benzie, J.A.H., Ballment, E., 1994. Genetic differences among black-lipped pearl oyster (*Pinctada margaritifera*) populations in the western Pacific. *Aquaculture* 127, 145–156.

Benzie, J.A.H., Smith, C., Sugama, K., 2003. Mitochondrial DNA reveals genetic differences between Australian and Indonesian pearl oyster *Pinctada maxima* (Jameson 1901) populations. *J. Shellfish Res.* 22, 781–788.

Blanc, F., Durand, P., Shin-Milhaud, M., 1985. Variabilitié génétique des populations noire perlière *Pinctada margaritifera* (Mollusque, Bivalve) de Polynésie. In: Delasalle, B., Galzin, R., Salvat, B. (Eds.), *Proceedings of the Fifth International Coral Reef Congress French Polynesian Coral Reefs, Vol. 1*. Antenne Museum-EPHE, Morea, French Polynesia, pp. 113–118.

Boudry, P., Degremont, L., Taris, N., Mcombie, H., Haffray, P., Ernande, B., 2004. Genetic variability and selective breeding for traits of aquacultural interest in the Pacific oyster (*Crassostrea gigas*). *Bull. Aquacult. Soc. Canada* 104(2), 12–18.

Bou-Olayan, A.-H., Al-Mattar, S., Al-Yakoob, S., Al-Hazeem, S., 1995. Accumulation of lead, cadmium, copper and nickel by pearl oyster, *Pinctada radiata*, from Kuwait marine environment. *Mar. Pollut. Bull.* 30(3), 211–214.

Braley, R.D., 1991. Pearl oyster introductions to Tokelau atolls: Real potential or a long shot. *SPC Pearl Oyster Inform. Bull.* 3, 9–10.

Braley, R.D., 1998. Sedimentation rates associated with the longline culture of the blacklip pearl oyster, *Pinctada margaritifera*, at Penrhyn Atoll, Cook Islands. *Aquaculture '98*, February 15–19, 1998, Las Vegas, NV (USA), p. 64.

Cabral, P. (1989). Some aspects of the abnormal mortalities of the pearl oysters, *Pinctada margaritifera* L. in the Tuamotu Archipelago (French Polynesia). *Advances in Tropical Aquaculture: Tahiti*, February 20–March 4, 1989. AQUACOP. IFREMER. Actes de Colloque 9, pp. 217–226.

Cabral, P., 1992. Inter-island transfers of pearl oysters to blame for contamination of many atolls in French Polynesia. *SPC Pearl Oyster Inform. Bull.* 5, 14–15.

Chesapeake Bay Federal Agencies Committee, 2001. Recommendations on Suminoe oyster (*Crassostrea ariakensis*) aquaculture in Chesapeake Bay. Chesapeake Bay Program, Annapolis, MD, 3 pp.

Cleland, K. (2002). Port Stephens Pearl Oyster Industry. Report to the Honourable Dr Andrew Refshauge, Deputy Premier, Minister for Planning, Minister for Aborigine Affairs, Minister for Housing. Office of the Commissioners of Inquiry for Environment and Planning, Sydney, NSW, 156 pp.

Crawford, C.M., Macleod, C.K.A., Mitchell, I.M., 2003. Effects of shellfish farming on the benthic environment. *Aquaculture* 224, 117–140.

Dahlback, B., Gunnarsson, L.A.II., 1981. Sedimentation and sulphate reduction under a mussel culture. *Mar. Biol.* 63, 269–275.

Dame, R., 1996. *Ecology of Marine Bivalves, an Ecosystem Approach.* CRC Press, Boca Raton, FL, 254 pp.

Dev, D.S., 1994. Pearl culture project in India. *Pearl Oyster Inform. Bull. South Pacific Commission, Noumea* 7, 5–6.

Durand, P., 1987. Biogéographie et différenciation génétique chez la nacre *Pinctada margaritifera* (L.). These de 3ème cycle USTL Montpellier, 94 pp.

Dove, M., O'Connor, W.A., 2007. Ecotoxicological evaluations of common hatchery substances and procedures used in the production of Sydney rock oysters, *Saccostrea glomerata* (Gould, 1850). *J. Shellfish Res.* 26, 501–508.

Eldridge, L.G., 1994. Introduction of commercially significant aquatic organisms to the Pacific. *IFRP Technical Document No. 7,* pp. 9–41.

Enzer Marine Environmental Consulting, 1998. The environmental impact of pearling (*Pinctada maxima*) in Western Australia. Report to the Pearl Producers Association Inc. Enzer Marine Environmental Consulting.

Figueras, A.J., 1989. Mussel culture in Spain and France. *World Aquacult.* 20(4), 8–17.

Fowler, S.W., Readman, J.W., Oregioni, B., Villeneuve, J.-P., McKay, K., 1993. Petroleum hydrocarbons and trace metals in nearshore Gulf sediments and biota before and after the 1991 war: An assessment of temporal and spatial trends. *Mar. Pollut. Bull.* 27, 171–182.

Fromont, J., Craig, R., Rawlinson, L., Aldaer, J., 2005. Excavating sponges that are destructive to farmed pearl oysters in Western and Northern Australia. *Aquacul. Res.* 36, 150–162.

Fujiwara, M., Caswell, H., 2001. Demography of the endangered North Atlantic right whale. *Nature* 414, 537–541.

Gavine, F.M., Mackinnon, L.J., 2002. Environmental monitoring of marine aquaculture in victorian coastal waters: A review of appropriate methods. Technical Report No. 46. Marine and Freshwater Resources Institute, Victoria.

GESAMP (Group of Experts on Scientific Aspects of Marine Environmental Protection), 1991. Reducing environmental impacts of coastal aquaculture. IMO/FAO/Unesco/WMO/WHO/IAEA/UN/UNEP, Reports and Studies No. 47, 35 pp.

Gifford, S., 2006. Environmental considerations associated with the development of an akoya pearl industry in NSW, Australia. Ph.D. Thesis, University of Newcastle, 210 pp.

Gifford, S., Dunstan, R.H., O'Connor, W.A., Roberts, T., Toia, R., 2004. Pearl aquaculture-profitable environmental remediation? *Sci. Total Environ.* 319, 27–37.

Gifford, S., Dunstan, H., O'Connor, W., Macfarlane, G.R., 2005a. Quantification of in situ nutrient and heavy metal remediation by a small pearl oyster farm at Port Stephens, Australia. *Mar. Poll. Bull.* 50, 417–422.

Gifford, S.P., Macfarlane, G.R., O'Connor, W.A., Dunstan, R.H., 2005b. Animals can do it too!: "Zooremediation" of aquatic environments. *3rd European Conference on Bioremediation,* Greece.

Grant, J., Hatcher, A., Scott, D.B., Pocklington, P., Schafer, C.T., Winter, G.V., 1995. A multidisciplinary approach to evaluating the impacts of shellfish aquaculture on benthic communities. *Estuaries* 18, 124–144.

Greenland, J.A., Limpus, C.J., Brieze, I., 2005. *Marine Wildlife Stranding and Mortality Database Annual Report. II. Cetacean and Pinniped.* State of Queensland Environmental Protection Agency, Brisbane, 30 pp.

Greenland, J.A., Limpus, C.J., 2005. *Marine Wildlife Stranding and Mortality Database Annual Report. I. Dugong.* State of Queensland Environmental Protection Agency, Brisbane, 27 pp.

Hasan A.K., 1974. Studies on bottom mollusca (Gastropods and Bivalves) in Abou-Kir Bay. M.Sc. Thesis, Faculty of Science, University of Alexandria, Egypt, 319 p.

Hatcher, A., Grant, J., Schofield, B., 1994. Effects of suspended mussel culture on sedimentation, benthic respiration and sediment nutrient dynamics in a coastal bay. *Mar. Ecol. Prog. Ser.* 115, 219–235.

Haws, M., Ellis, E., Ellis, S., 2003. Development of blacklip pearl oyster farming in Micronesia. Final Report to the Center for Tropical and Subtropical Aquaculture, Oceanic Institute, Oahu, Hawaii, 12 pp.

Hedgecock, D., Chow, V., Waples, R.S., 1992. Effective population numbers of shellfish broodstocks estimated from temporal variance in allelic frequencies. *Aquaculture* 108, 215–232.

Heffernan, O., 2006. Pearls of wisdom. *Mar. Sci.* 16, 20–23.

Hickman, R.W., 1989. Farming the green mussel in New Zealand. *World Aquacult.* 20(4), 20–28.

Hildermann, W.H., Dix, T.G., Collins, J.D., 1974. Tissue transplantation in diverse marine invertebrates. In: Cooper, E.L. (Ed.), *Contemporary Topics in Immunobiology, Vol. 4, Invertebrate Immunology*. Plenum Press, New York, pp. 141–150.

Hine, M., 1999. Significant diseases of molluscs in the Asia-Pacific region. *Fourth Symposium of Diseases in Asian Aquaculture*. Cebu City, Philippines, OP 34.

Humphrey, J.D. (1995). Introductions of aquatic animals to the Pacific Islands: Disease threats and guidelines for quarantine. SPC Inshore Fisheries Research Project Technical Document No. 8, 63 pp.

Imai, T., 1971. *Aquaculture in Shallow Seas: Progress in Shallow Sea Culture*. AA. Balkema, Rotterdam, 615 pp.

Jaramillo, E., Bertran, C., Bravo, A., 1992. Mussel biodeposition in and estuary in southern Chile. *Mar. Ecol. Prog. Ser.* 82, 85–94.

Jenner, K.C.S., Jenner, M-N., 1996. Group IV humpback whale calving ground and population monitoring program 1995. Unpublished Report for the Australian Nature Conservation Agency project # SCA01842.

Jernakoff, P., 2002. *Environmental Risk and Impact Assessment of the Pearling Industry*. Final Report to FRDC 2001/099. IRC Environment, Perth, Australia, 182 pp.

Johnson, M.S., Joll, L.M., 1993. Genetic subdivision of the pearl oyster *Pinctada maxima* (Jameson, 1901) (Mollusca: Pteriidae) in northern Australia. *Aust. J. Mar. Freshw. Res.* 44, 519–526.

Joll, L.M., 1996. Stock evaluation and recruitment measurement in the W.A. pearl oyster fishery. FDWA Final Report to Fisheries Research and Development Corporation for project 92/147, 62 pp.

Kanayama, R.K., 1967. Hawaii's aquatic animal introductions. *Forty-seventh Annual Conference of Western Association of State Game and Fish Commissioners*, Honolulu, 8 p.

Kaspar, H.F., Gillespie, P.A., Boyer, I.C., MacKenzie, A.L., 1985. Effects of mussel culture on the nitrogen cycle and benthic communities in Kenepuru Sound, Marlborough Sounds, New Zealand. *Mar. Biol.* 85, 127–136.

Kinzelbach, R., 1985. Lessps'sche wanderung: neue stationen von muscheln (Bivalvia: Anisomyaria). *Arch. Fur Molluskenkunde* 115, 273–278.

Knauer, J., Taylor, J.J.U. (2002). Production of silver nacred "saibo oysters" of the silver- or gold-lip pearl oysters *Pinctada maxima* in Indonesia. *World Aquaculture 2004*, April 23–27, 2002, Beijing, China, p. 355.

Knowlton, A.R., Kraus, S.D., 2001. Mortality and serious injury of northern right whales (*Eubalaena glacialis*) in the western North Atlantic Ocean. *J. Cetacean Res. Manag.* 2, 193–208.

Kraus, S.D., 1990. Rates and potential causes of mortality in North Atlantic right whales (*Eubalaena glacialis*). *Mar. Mammal Sci.* 6(4), 278–291.

Le Moullac, G., Goyard, E., Saulnier, D., Hafner, P., Thouard, E., Nedelec, G., Gougenhiem, J., Rouxel, C., Cuzon, G., 2003. Recent improvements in broodstock management and larviculture in marine species in Polynesia and New Caledonia: genetic and health approaches. *Aquaculture* 227, 89–106.

Lewis, T., Martin, C., Muir, C., Haws, M., Ellis, S., Ueanimatang, M., David, D., Nair, M. (2004). Population genetics of the black pearl oyster *Pinctada margaritifera*. *Aquaculture 2004*, March 1–5, Honolulu, Hawaii, p. 350.

Liu, Z., Huang, H., Zhou, Y., 1998. Preliminary study on the pseudomonas disease of the larvae of *Pinctada martensii* (Dunker). *J. Zhanjiang Ocean Univ.* 18, 23–28.

MacFarlane, G.R., Gifford, S.P., Dunstan, H.R., O'Connor, W.A., Linz, K., 2004. Zooremediation: animals make it crystal clear! Pearl oysters are the perfect animals for remediating coastal waters. *Aust. Sci.* November/December, 18–19.

MacFarlane, G.R., Markich, S., Linz, K., Gifford, S., Dunstan, R.H., O'Connor, W.A., Russel, R.A., 2005. The shell of the Akoya pearl oyster (*Pinctada imbricata*) as an archival indicator of lead exposure. *Environ. Pollut.* 143, 166–173.

McGinnity, P., Prodoehl, P., Ferguson, A., Hynes, R., Maoileidigh, N.O., Baker, N., Cotter, D., O'Hea, B., Cooke, D., Rogan, G., Taggart, J., Cross, T., 2003. Fitness reduction and potential extinction of wild populations of Atlantic salmon, *Salmo salar*, as a result of interactions with escaped farm salmon. *Proc. R. Soc. Lond., Ser. B: Biol. Sci.* 270, 2443–2450.

MAFF, 2006. www.maff.go.jp/soshiki/koukai/muratai/21j/english/no4/mura17.html.

Mann, J., Janik, V.M., 1999. Preliminary report on dolphin habitat use in relation to oyster farm activities Red Cliff Bay, Shark Bay. Unpublished Report to CALM and Fisheries WA, 6 pp.

Mann, R., Evans, D.A., 2004. Site selection for oyster habitat rehabilitation in the Virginia portion of the Chesapeake Bay: A commentary. *J. Shellfish Res.* 23, 41–49.

Mao Che, L., Le Campion-Alsumard, T., Boury-Esnault, N., Payri, C., Golubic, S., Be Zac, C., 1996. Biodegradation of shells of the black pearl oyster, *Pinctada margaritifera* var. *cumingi*, by microborers and sponges of French Polynesia. *Mar. Biol.* 126, 509–519.

Markowitz, T.M., Harlin, A.D., Wursig, B., McFadden, C.J., 2004. Dusky dolphin foraging habitat: overlap with aquaculture in New Zealand. *Aquatic. Conserv. Mar. Freshw. Ecosyst.* 14, 133–149.

Masaoka, T., Kobayashi, T., 2005. Species identification of *Pinctada imbricata* using intergenic spacer of ribosomal RNA genes and mitochondrial 16S RNA gene region. *Fish. Sci.* 71, 837–846.

Mattson, J., Linden, O., 1983. Benthic macrofauna succession under mussels, *Mytilus edulis* L., cultured on hanging long lines. *Sarsia* 68, 97–102.

Millous, O., 1980. Essais de production contrôlée de naissain d'huîtres perlières (*Pinctada margaritifera*) en laboratoire. *CNEXO/COP COP/AQ*, 80.017.

Mills, D., Tlili, A., Norton, J., 1997. Large scale anaesthesia of the silver lipped pearl oyster, *Pinctada maxima* Jameson. *J. Shellfish Res.* 2, 573–574.

Monteforte, M., Bervera, H., Saucedo, P., 2004. Response profile of the calafia pearl oyster, *Pinctada mazatlanica* (Hanley, 1856) to various sedative therapies related to surgery for round pearl induction. *J. Shellfish Res.* 23, 121–128.

Munro, J.L., 1993. *Aquaculture development and environmental issues in the Tropical Pacific*. In: Pullin, R.S.V., Rosenthal, H., Maclean, J.L. (Eds.), *Environment and Aquaculture in Developing Countries*, ICLARM *Conf Proc 31*, pp. 125–138.

Newell, R.I.E., 2004. Ecosystem influences of natural and cultivated populations of suspension feeding bivalve molluscs: A review. *J. Shellfish Res.* 23, 51–61.

Norton, J.H., Dashorst, M., Lansky, T.M., Mayer, R.J., 1996. An evaluation of some relaxants for use with pearl oysters. *Aquaculture* 144, 39–52.

Norton, J.H., Lucas, J.S., Turner, I., Mayer, R.J., Newnham, R., 2000. Approaches to improve cultured pearl formation in *Pinctada margaritifera* through the use of relaxation, antiseptic application and incision closure during bead insertion. *Aquaculture* 184, 1–18.

NSW EPA, 2003. Report for the Pearl Oyster Commission of Inquiry, 22 pp.

NSW Fisheries, 1997. *Mussel Aquaculture in Twofold Bay: Nominated Determining Authority's Report.* NSW Fisheries, Cronulla, 118 pp.

Numaguchi, K., 1995. Effects of water temperature on catabolic losses of meat and condition index of unfed pearl oyster *Pinctada fucata martensii*. *Fish. Sci.* 61, 735–738.

O'Connor, W.A., 2002a. Latitudinal variation in reproductive behaviour in the pearl oyster, *Pinctada albina sugillata*. *Aquaculture* 209, 333–346.

O'Connor, W.A., 2002b. Unpublished report to the Port Stephens Pearl Oyster Industry, Commission of Inquiry, June 2002.

O'Connor, W.A., Lawler, N., 2003. Reproductive condition of the pearl oyster, *Pinctada imbricata* (Roding) in Port Stephens, NSW, Australia. *Aquacult. Res.* 35, 1–12.

O'Connor, W.A., Wang, A., 2001. Akoya pearl culture in China. *World Aquacult.* September, 18–20.

O'Connor, W.A., Lawler, N.F. Heasman, M.P., 2003. Trial farming the akoya pearl oyster, *Pinctada imbricata*, in Port Stephens, NSW. Final Report to Australian Radiata Pty Ltd. NSW Fisheries Final Report Series N 42, 175 pp.

Orton, J.H., 1937. *Oyster Biology and Oyster Culture*. Edward Arnold & Co., London, 211 pp.

Paynter, K.T., Haws, M. (2004). Insulin accelerates shell growth and pearl nacre deposition in *Pinctada margaritifera*. *Aquaculture 2004*, March 1–5, Honolulu, Hawaii, p. 461.

Pérez Camacho, A., Gonzales, R., Fuentes, J., 1991. Mussel culture in Galicia (NW Spain). *Aquaculture* 94, 263–278.

Pearl Producers Association, 2004. *Workshop Report: 2004 Environmental Risk Assessment of the Pearling Industry*. Seafood Services, Ascot, Qld, Australia, 9 pp.

Philpots, J.R., 1890. *Oysters and All About Them*, Vols I and II. John Richardson and Co., London, 1370 pp.

Ponia, B., 2000. Proposed grid system to manage pearl farms at Manihiki Lagoon. Ministry of Marine Resources, Cook Islands, Misc. Report: 2000/2006.

Ponia, B., Okotai, T., 2000. Draft Manihiki Island Council pearl farming management guidelines. Ministry of Marine Resources, Cook Islands. Misc. Report: 2000/2005.

Pouvreau, S., Bacher, C., Heral, M., 2000. Ecophysiological model of growth and reproduction of the black pearl oyster, *Pinctada margaritifera*: potential applications for pearl farming in French Polynesia. *Aquaculture* 186, 117–144.

Ray, G.L., 2005. *Invasive estuarine and marine animals of Hawai'i and other Pacific Islands* ANSRP Technical Notes Collection (ERDC/TN ANSRP-05-3). US Army Engineer Research and Development Center, Vicksburg. MS, 19 pp.

Roberts, D.E., 2002. Seagrass meadows and benthic assemblages in the vicinity of the Wanda Head pearl oyster lease in Port Stephens, NSW. Report for the Pearl Oyster Commission of Inquiry, 22 pp.

Roycroft, D., Kelly, T.C., Lewis, L.J., 2004. Birds, seals and the suspension culture of mussels in Bantry Bay, a non-seaduck area in Southwest Ireland. *Estuar. Coastal Shelf Sci.* 61, 703–712.

Sarver, D., Sims, N.A., Harmon, V., 2003. Pearl oysters as a sensitive, sessile monitor for non-point source heavy metal pollution. *SPC Pearl Oyster Inform. Bull.* 16, 13–14.

Saville-Kent, W., 1905. Torres Straits pearl shell fisheries. Queensland Parliamentary Papers: Session 2 of 1905. Report to both Houses of Parliament. Government Printer, Brisbane, 2, pp. 1075–1078.

Seki, M., 1972. Studies on environmental factors for the growth of the pearl oyster, *Pinctada fucata*, and the quality of its pearl under the culture conditions. *Bull. Mie Pref. Fish. Exp. Stn.* 32, 143.

Serbetis, C.D., 1963. L'acclimatation de la *Meleagrina (Pinctada) margaritifera* (Lam.) en Grèce. *CIESM* 17, 271–272.

Shumway, S.E., Davis, C., Downey, R., Karney, R., Kraeuter, J., Parsons, J., Rheault, R., Wikfors, G., 2003. Shellfish Aquaculture – In praise of sustainable economies and environments. *World Aquacult.* 34(4), 15–18.

Simpson, C.J., 1998. Environmental guidelines and procedures in relation to CALM advice on the marine aspects of aquaculture and pearling proposals in Western Australia. Marine Policy and Coordination Report: MPC- 01/1998, 13 pp.

Smith, R.O., 1947. Survey of the fisheries of the former Japanese mandated Islands. *Fish. Leaflet* 273, 105.

Stafford, J., 1913. *The Canadian Oyster*. Mortimer Co., Ottawa, Canada, 159 pp.

Stenton-Dozey, J.M.E., Jackson, L.F., Busby, A.J., 1999. Impact of mussel culture on macrobenthic community structure in Saldanha Bay, South Africa. *Mar. Pollut. Bull.* 39, 357–366.

Stenton-Dozey, J., Probyn, T., Busby, A., 2001. Impact of mussel (*Mytilus galloprovincialis*) raft culture on benthic macrofauna, in situ oxygen uptake and nutrient fluxes in Saldanha Bay, South Africa. *Can. J. Fish. Aquat. Sci.* 58, 1021–1031.

Stickney, R., 2003. How did we get into this mess? Junk science vs real science. *World Aquacult.* 34(4), 4–7.

Subhash, S.K., Lipton, A.P., Raj, R.P., 2004. Antibiotic exposure to minimise microbial load in live feed *Isochrysis galbana* used for larval rearing of Indian pearl oyster *Pinctada fucata*. *Curr. Sci.* 87, 1339–1340.

Suzuki, K., 1957. Biochemical studies on the pearl oyster (*Pinctada martensii*) and its growing environments. I. The seasonal changes in chemical components of the pearl oyster, plankton and marine mud. *Bull. Natl. Pearl Res. Lab.* 2, 57–62.

Tanaka, H., 1990. Winged pearl; shell newly found in Tonga. *SPC Pearl Oyster Inform. Bull.* 2, 10.

Taris, N., Goulletquer, P., 2005. *Genetic Consequences of Intensive Production on Pacific Oyster Larval Stage: Drift and Selective Pressures Due to Rearing Practices*. IFREMER, Plouzane (France). 220 pp.

Taylor, J.J.U. (2002). Producing golden and silver South Sea pearls from Indonesian hatchery reared *Pinctada maxima. World Aquaculture 2004*, April 23–27, 2002, Beijing, China, p. 754.

Tenore, K.R., Boyer, L.F., Cal, R.M., Corral, J., Garcia-Fernandez, C., Gonzalez, N., Gonzalez-Gurmaran, E., Hanson, R.B., Iglesias, J., Krom, M., Lopez-Jamar, E., Maclain, J., Pamatmat, M.M., Perez, A., Rhoades, D.C., de Santiago, G., Tietjen, J., Westrich, J., Windom, H.L., 1982. Coastal upwelling in the Rias Bajos, NW Spain, contrasting benthic regimes of the Rias de Arosa and de Muros. *J. Mar. Res.* 40, 701–772.

Tisdell, C.A., Poirine, B. (2000). Socio-economics of pearl culture: Industry changes and comparisons focusing on Australia and French Polynesia. *World Aquaculture June 2000*, pp. 30–60.

Tranter, D.J., 1957. Pearl culture in Australia. *Aust. J. Sci.* 19, 230–232.

Tsai, H.J., Lai, C.H., Yang, H.S., 1997. Sperm as a carrier to introduce an exogenous DNA fragment into the oocyte of Japanese abalone (*Haliotis diversicolor*). *Transgenic Res.* 6, 85–95.

Uemoto, H., Inoue, H., Tanaka, Y., 1978. The load of organic substance in pearl culture ground. *Tech. Rep. Natl. Pearl Cult. Res. Lab.* 5, 50–54.

Uyeno, F., Funahashi, S., Tsuda, A., 1970. Preliminary studies on the relation between feces production of pearl oyster (*Pinctada martensii* (Dunker)) and bottom condition in an estuarine pearl oyster area. *J. Faculty Fish. Pref. Univ. Mie* 8, 113–137.

Wada, K.T., 1982. Inter- and intraspecific electrophoretic variation in three species of pearl oyster from the Nansei Islands of Japan. *Bull. Natl. Res. Inst. Aquacul.* 3, 1–10.

Wada, K.T., 1984. Breeding study of the pearl oyster, *Pinctada fucata*. *Bull. Nat. Res. Inst. Aquacult.* 6, 79–157.

Wada, K.T., 1991. *What is the Secret of the Pearls Beauty*. Les Joyaux, Special Edition, September, pp. 89–96.

Wang, A., Shi, Y., Gu, Z., Li, D., Wang, Y., Qu, Y., Ye, H., 2005. Early growth in family lines of the pearl oyster *Pinctada martensii*. *World Aquaculture 2004*, May 9–13, 2005, Nusa Dua, Bali, p. 689.

Watson-Capps, J.J., Mann, J., 2005. The effects of aquaculture on bottlenose dolphin (*Tursiops* sp.) ranging in Shark Bay, Western Australia. *Biol. Conserv.* 124, 519–526.

Welcomme, R., 1988. Proposed international regulations to reduce risks associated with transfers and introductions of aquatic organisms. In: Grimaldi, E., Rosenthal, H. (Eds.), *Efficiency in Aquaculture Production: Disease Control. Proceedings of the third International Conference on Aquafarming*. October 9–10, 1986, Verona, Italy, pp. 65–76.

Wells, F.E., Jernakoff, P., 2006. An assessment of the environmental impact of wild harvest pearl aquaculture (*Pinctada maxima*) in Western Australia. *J. Shellfish Res.* 25, 141–147.

Weston, D.P., 1990. Qualitative examination of macrobenthic community changes along an organic enrichment gradient. *Mar. Ecol. Prog. Ser.* 61, 232–244.

Weston, D.P., 1991. The effects of Aquaculture on indigenous biota. In: Bruno, D.E., Tomasso, J.R. (Eds.), *Aquaculture and Water Quality*, World Aquaculture Society, Baton Rouge, LA, pp. 534–567.

Weston, D.P., 1996. Quantitative examination of macrobenthic community changes along an organic enrichment gradient. *Mar. Ecol. Prog. Ser.* 61, 233–244.

Wolff, W.J., Reise, K., 2002. Oyster imports as a vector for the introduction of alien species into northern and western European coastal waters. In: Leppakoski, E., Gollash, S., Olenin, S. (Eds.), *Invasive Species of Europe – Distribution, Impact and Management*. Kluwer Academic Publishing, Dordrecht, pp. 193–205.

Wright, G., Stella, L., 2003. Pearling in the Pilbara 1860s–1890s. *Conference on Indigenous Fishing Rights: Moving Forward 2003*, Research Unit, National Native Title Tribunal, 43 pp.

Wursig, B., Gailey, G.A., 2002. Marine mammals and aquaculture: potential conflicts and resolutions. In: Stickney, R.R., McVey, J.P. (Eds.), *Responsible Marine Agriculture*. CAB International Press, New York, pp. 45–59.

Yokoyama, H., 2002. Impact of fish and pearl farming on the benthic environments in Gokasho Bay: Evaluation from seasonal fluctuations of the macrobenthos. *Fish. Sci.* 68(2), 258–268.

CHAPTER 15

Biofouling

Rocky de Nys and Odette Ison

15.1. INTRODUCTION

All pearl culture is impacted by biofouling which is a key operational issue and economic cost for the vast majority of pearl production. Pearl oysters are cultured by a variety of methods depending on the species and the location of the farm (see Chapter 9). The most common culture methods employ submerged panel nets, baskets or lantern nets suspended from longlines or rafts with alternative techniques including bottom culture on racks or "ear hanging" (see Section 7.5). While the method of culture is principally dependent on environmental factors specific to the location of the farm, including site, depth and season, all pearl culture is affected detrimentally by the biofouling process.

Biofouling is the settlement, metamorphosis and growth of plants and animals (Wahl, 1989) and it affects all marine industries ranging from marine infrastructure and shipping through to aquaculture (Yebra *et al.*, 2004; Braithwaite and McEvoy, 2005). The cost of biofouling on freshwater and marine infrastructure has been estimated in the billions of dollars (Abbott *et al.*, 2000; Champ, 2000). The true costs are unclear, but the impacts of biofouling in pearl aquaculture, are very significant.

Biofouling can be catastrophic for commercial pearl enterprises inducing mass mortality and significantly reducing pearl production (Korringa, 1951; Alagarswami and Chellam, 1976; Thomas, 1979). In addition to directly inducing oyster mortality, biofouling exerts sub-lethal effects on oyster growth and pearl yield (Nishii, 1961; Nishii *et al.*, 1961; Mohammad, 1972; Taylor *et al.*, 1997; Lodeiros *et al.*, 2002).

Consequently, control of fouling is imperative to ensure the growth and quality of both the oyster and the pearl. Biofouling control is one of the most difficult and costly production issues facing the pearl industry. However, the costs of controlling biofouling are not well documented due to the commercially sensitive nature of operations. The removal and prevention of fouling on both pearl oyster shells and culture equipment has been estimated at between 25% and 30% of the operating costs of a pearl farm (Crossland, 1957; Lewis, 1994), and these estimates appear to remain relevant to modern operational practices.

15.2. BIOFOULING COMMUNITIES

The ecological succession or development of biofouling in marine environments is well documented (reviewed in Little, 1984; Wahl, 1989; Maki and Mitchell, 2002). An organic conditioning film, of proteins, proteoglycans and polysaccharides, initially accumulates on all immersed surfaces. Primary colonization by prokaryotes and eukaryotes then commences on the organic film. Passive bacterial adsorption on the surface takes place and a complex biofilm develops. This can act as a cue for diatoms and other small eukaryotes (microfouling). More complex colonization occurs over time with the settlement of macroalgal spores and propagules, and the larvae of invertebrates including ascidians, sponges, bryozoans, barnacles, bivalves and polychaetes (macrofouling).

The development of biofouling depends on the availability of surfaces for colonization. The presence of large arrays of aquaculture structures provides an opportunity for a broad diversity of fouling organisms to settle and grow (Milne, 1975; Hodson and Burke, 1994; Hodson et al., 1997), and the diversity of structures, including pearl oyster shells, associated with pearl culture provides a ready habitat for fouling organisms. Surface fouling communities on pearl oysters and pearling equipment are generally dominated by barnacles, bivalves, bryozoans and tubiculous polychaetes (Alagarswami and Chellam, 1976; Dharmaraj et al., 1987; Wada, 1991, Monteforte and Garcia-Gasca, 1994; Doroudi, 1996; Taylor et al., 1997; Guenther et al., 2006). Ascidians, hydroids, sponges and algae are also common on the surface of pearl shells (Alagarswami and Chellam, 1976; Mohammad, 1976; Dharmaraj et al., 1987; Wada, 1991; Doroudi, 1996; Guenther and de Nys, 2006) (Table 15.1). However, the development and composition of fouling communities is often highly site specific due to regional variation in larval supply from benthic communities. Furthermore, the composition of fouling communities on pearl oysters in culture often differs from that found on the same species of oyster in their natural environments (Dharmaraj and Chellam, 1983; Dharmaraj et al., 1987; Doroudi, 1996).

15.3. THE IMPACT OF BIOFOULING

While fouling community structure is spatially and temporally variable, the impact of fouling is, in nearly all cases, highly detrimental to the productivity of pearl culture.

TABLE 15.1

An overview of the most common fouling organisms documented from pearl oysters, their impacts, and their distribution.

Fouling organism	Impact	Region	Pearl oyster species	Author(s)
Cyanobacteria	Erosion of shell		*P. margaritifera*	Mao Che *et al.* (1996)

Arthropoda
  Maxillopoda
    Balanus amphitrite variegates
    Balanus amphitrite communis
	Physical disruption to opening and closing of valves, recession of shell growth and mortality	Australia
		Indonesia
		India
		Persian Gulf
		Japan
		Arabian Gulf
			P. maxima
			P. fucata
			P. radiata
			P. martensii
				Taylor *et al.* (1997)
				Takemura and Okutani (1955)
				Alagarswami and Chellam (1976)
				Dharmaraj *et al.* (1987)
				Dharmaraj and Chellam (1983)
				Doroudi (1996)
				Miyauti (1968)
				Mohammad (1976)
				Wada (1991)

Mollusca
  Bivalvia
    Crassostrea sp.
    Pteria sp.
    Pinctada sp.
    Mytilus sp.
    Pinna sp.
    Lithophaga sp.
    Martesia sp.
    Saccostrea sp.
	Physical disruption to opening and closing of valves, damage to shell, recession of shell growth, shell deformity, mortality and competition for food and space	India
		Persian Gulf
		Australia
		Indonesia
		Red Sea
			P. fucata
			P. radiata
			P. maxima
			P. margaritifera
				Alagarswami and Chellam (1976)
				Dharmaraj *et al.* (1987)
				Doroudi (1996)
				Takemura and Okutani (1955)
				Taylor *et al.* (1997)
				Crossland (1957)
				Guenther *et al.* (2006)

Bryozoa *Membranipora* sp. *Bugula* sp. *Parasmittina* sp.	Not demonstrated	Japan Indian Ocean Australia	*P. fucata*	Alagarswami and Chellam (1976) Dharmaraj *et al.* (1987) Wada (1991) Guenther *et al.* (2006)

Chordata
  Ascidiacea
    Dicarpa sp.
    Didemnum sp.
    Diplosoma sp.
	Not demonstrated	Persian Gulf
		Japan
		Australia
			P. radiata
			P. maxima
			P. martensii
			P. fucata
				Doroudi (1996)
				Miyauti (1968)
				Takemura and Okutani (1955)
				Alagarswami and Chellam (1976)
				Mohammad (1976)
				Dharmaraj *et al.* (1987)
				Guenther *et al.* (2006)

Biofouling on oysters and culture equipment can inhibit the growth of oysters and the yield of pearls (Nishii, 1961; Nishii *et al.*, 1961; Miyauti, 1967; Mohammad, 1972; Wada, 1973; Mohammad, 1976; Taylor *et al.*, 1997; Lodeiros *et al.*, 2002) and heavy fouling can directly cause oyster mortality, with major impacts on pearl production (Korringa, 1952; Alagarswami and Chellam, 1976; Thomas, 1979). However, in some circumstances biofouling either does not affect oyster growth (Braley, 1984; Lodeiros *et al.*, 1999) or may have a beneficial effect by increasing the abundance of plankton (Mohammad, 1976; Lodeiros *et al.*, 2002), the primary food source for filter feeders including pearl oysters (see Section 4.1; Jørgensen, 1996). This is the exception and the control of fouling is a key operational parameter in the cost-effective production of pearls.

The major effects of the colonization of shell surfaces and culture equipment by fouling organisms fall into five major categories.

1. *Physical damage* by invasive organisms (endoliths) that bore into the shell causing shell deformities, impacting on nacre production and eventually resulting in the death of the host.
2. *Mechanical interference* of shell function by fouling organisms that settle on the hinge and lip of the oyster, disrupting the valve opening and closing, and leaving the oyster susceptible to predators.
3. *Biological competition* for resources such as phytoplankton (food) and space.
4. *Environmental modification* due to colonization of culture infrastructure, resulting in reduced water flow to the cultured animals and thereby creating waste build-up, and lower available oxygen and food for filter feeders. This has direct impacts on the growth and fitness of the pearl oysters.
5. *Increased friction and drag* on equipment such as panel nets, ropes and floats, through increased weight from biofouling. The maintenance and loss of equipment directly contributes to production costs.

15.3.1. Physical damage by invasive boring organisms (endoliths)

Boring organisms have a surprisingly destructive effect on pearl culture. Three ecologically distinct groups of boring (endolithic) organisms occur on pearl oysters. They are extremely effective at infecting pearl shells, and as a consequence inflict very significant damage. The first group are phototrophic micro-organisms including cyanobacteria (*Hyella* sp., *Plectonema* sp. and others) and green algae (*Phaeophila dendroides*, *Ostreobium quekettii*) (Mao Che *et al.*, 1996). The second are heterotrophic boring micro-organisms such as fungi (*Ostracoblabe implexa*) (Mao Che *et al.*, 1996), while the third, and most significant of the group of boring organisms, are the filter-feeding macro-borers including sponges, mussels and polychaetes (Mohammad, 1972; Alagarswami and Chellam, 1976; Thomas, 1979; Dharmaraj *et al.*, 1987; Mao Che *et al.*, 1996). Shell penetration and excavation of the oyster shell by these organisms results in cavities, burrows and tunnels deep within the nacreous layer of the shell (Cobb, 1969; Blake and Evans, 1973; Silina, 2006) (Table 15.2).

The process of pearl shell infestation commences with the micro-organisms that colonize and penetrate the external prismatic layers of the shell (Mao Che *et al.*, 1996). Photosynthetic endoliths can cause extensive damage to bivalve shells (Kaehler, 1999), which results in increased investment in shell regeneration and reduced reproductive output and mortality (Webb and Korrûbel, 1994; Kaehler and McQuaid, 1999). Heterotrophic boring micro-organisms, including the fungus *O. implexa*, tend to dominate the nacreous region of the shell where they form perforations (Mao Che *et al.*, 1996). In addition to directly causing damage, these microbial endoliths also represent a succession stage leading to the colonization of macro-borers (Mao Che *et al.*, 1996).

TABLE 15.2

An overview of the most common boring organisms documented from pearl oyster species, their impacts, and their distribution.

Boring organism	Impact	Region	Pearl oyster species	References
Porifera	Brittleness, hinge	Red Sea	*P. margaritifera*	Crossland (1957)
Callyspongia fibrosa	instability, blister	Persian Gulf	*P. margaritifera* var.	Alagarswami and
Cliona sp.	formation, damage to	French Polynesia	*cumingii*	Chellam (1976)
C. margaritiferae	MOP, shell damage, shell	Indian Ocean	*P. fucata*	Dharmaraj *et al.* (1987)
C. vastifica	deformity, and mortality	Australia	*P. radiata*	Dharmaraj and
C. celata				Chellam (1983)
C.dissimilis				Doroudi (1996)
C. orientalis				Mao Che *et al.* (1996)
Pione velans				Thomas (1979)
				Velayudhan (1983)
				Mohammad (1976)
				Fromont *et al.* (2005)
Annelida	Blisters in nacreous layer	Red Sea	*P. margaritifera*	Crossland (1957)
Polychaeta	Weakened shell	Arabian Gulf	*P. fucata*	Alagarswami and
Polydora ciliata		Indian Ocean		Chellam (1976)
Polydora vulgaris		Japan		Dharmaraj *et al.* (1987)
				Dharmaraj and
				Chellam (1983)
				Mohammad (1976)
				Mohammad (1972)
				Velayudhan (1983)
				Wada (1991)
				Doroudi (1996)
				Taylor *et al.* (1997)

Of all the macro-boring organisms, the clionid sponges (Alagarswami and Chellam, 1976; Thomas, 1979; Mao Che *et al.*, 1996, Fromont *et al.*, 2005) cause the most extensive damage to pearl oysters. Clionid sponges within the family Clionidae (Porifera: Demospongiae) are unique in that all the genera within the family, including the most destructive and widespread genus *Cliona*, bore into calcareous objects at some stage in their life cycle (Thomas, 1979; Fromont *et al.*, 2005). Clionid sponges will excavate cavities throughout the entire prismatic layer and etch deep into the nacreous layer (Mao Che *et al.*, 1996). Clionid sponge infestations also inhibit the pearl formation process and can cause mass mortality of pearl oysters (Korringa, 1952; Alagarswami and Chellam, 1976; Thomas 1979). At sub-lethal levels, infestations of pearl shells by clionid sponges result in decreases in pearl yield (Mohammad, 1972), and economic loss due to discolored and deformed pearls (Crossland, 1957; Moase *et al.*, 1999). As for all fouling species, untreated initial infection or fouling of oyster shells, can result in an increased spread of fouling within a population of oysters through an increased supply of larvae. However, clionid sponges can be treated to protect valuable oysters and to reduce larval supply for larger scale infestation through localized recruitment within intensive culture systems (see Section 15.6). While the taxonomy and biology of boring sponges is well documented (Rützler, 2002; Fromont *et al.*, 2005), there is little or no quantitative information on the rates of infestation and the epidemiology of boring sponge infestations in pearl culture. There is also a need to quantitatively assess the ecological and economic costs of increased predation, loses in fecundity, loss of pearl production, and loss of pearl quality, associated with infection by clionid sponges in pearl oysters.

In contrast to pearl oysters, quantitative data on fouling by boring sponges is more accurately reported for other bivalve industries and, paradoxically, the effects of clionid infestation are well addressed in the ecological literature. Clionid sponges have been well documented from a wide range of commercially important bivalves, including scallops (Evans, 1969; Barthel *et al.*, 1994) and edible oysters (*Ostrea edulis*) (Thomas, 1979; Pomponi and Meritt, 1990; Wesche *et al.*, 1997; Rosell *et al.*, 1999). Infestation rates of clionid sponges in bivalves can be very high. Case studies of infestation rates by *Pione* or *Cliona* spp. range from 51% for a commercial stock of the Australian Sydney rock oyster, *Saccostrea glomerata* (Wesche *et al.*, 1997), through 83% for scallops (*Chlamys islandica*) in Balsfjord, Norway (Barthel *et al.*, 1994), and up to 100% of flat oysters (*Ostrea edulis*) in the Mediterranean Sea (Rosell *et al.*, 1999). These studies highlight the broad geographic range of clionid sponges and their ability to settle on, and infest, cultured bivalve shells. Clionid sponges also affect gastropod molluscs, and the effects of infestation include increased vulnerability to predation due to structural changes in the shell, and decreased growth and fecundity due to diversion of energy to supplement interior layers of the shell (Stefaniak *et al.*, 2005).

Boring polychaete worms also constitute a major problem in many pearl growing regions. Boring polychaetes have been reported as a significant problem in Indian, French Polynesian, and Japanese waters (Mohammad, 1972; Alagarswami and Chellam, 1976; Dharmaraj and Chellam, 1983; Coeroli *et al.*, 1984; Dharmaraj *et al.*, 1987; Wada, 1991; Monteforte and Garcia-Gasca, 1994). In a survey of Akoya pearl oysters in the Gulf of Mannar, India, 70% of shells had one or more blisters from polychaete worm (*Polydora* sp.) infestations (Alagarswami and Chellam, 1976). However, significant infestation by polychaete worms has not been reported in cultured pearl oyster populations in Australia. As a consistent theme, there are few quantitative studies on polychaete infestations in pearl culture, although they are recognized as a costly problem. In contrast, there is a strong information base on polychaete infestations in other bivalve industries (Bishop and Hooper 2005; Silina, 2006) and in the ecological literature (Buschbaum *et al.*, 2007; Thieltges and Buschbaum, 2007).

Polychaete worms have been responsible for broad-scale losses in other bivalve aquaculture industries around the world. *Polydora* species are associated with mortalities of bivalves in Australian, English, Japanese and American shellfish industries. The polychaete worms *Polydora ciliata*, *P. hoplura*, *P. vulgaris* and *P. websteri* have been identified as the causative agents in most of these cases (Whitelegge, 1890; Roughley, 1922; Mohammad, 1972; Blake and Evans, 1973). A related polychaete, *Boccardia knoxi*, is a significant problem in abalone culture (Lleonart *et al.*, 2003), while the

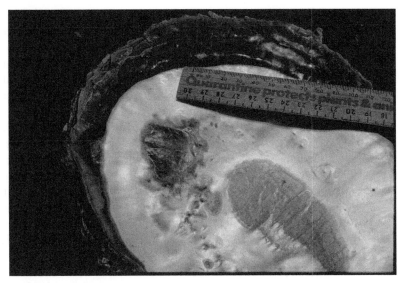

FIGURE 15.1   The inner surface of a pearl oyster shell showing the discoloured area resulting from polychaete (*Polydora* sp.) invasion (Photo: John Humphrey).

polyclad turbellarian flatworm, *Stylochus* sp., is a predator in pearl oyster (Newman *et al.*, 1993; Monteforte and Garcia-Gasca, 1994; Pit and Southgate, 2003; see Section 11.8.5) and edible oyster culture (Bailey-Brock and Ringwood, 1982; Littlewood and Marsbe, 1990).

Bivalves react to polychaete infestation by secreting a layer of conchiolin followed by a layer of nacre in an attempt to prevent the borer invading the body cavity (Blake and Evans, 1973). Shell pockets or blisters are formed under the nacreous layer during this process and the worm fills these with mud. These unsightly "mud blisters" are visible through the inner shell surface, resulting in deformation or disruption of the economically valuable mother of pearl shell (Fig. 15.1) (Mohammad, 1972; Blake and Evans, 1973; Thomas, 1979; Mao Che *et al.*, 1996).

In addition to blisters, the physical damage caused to pearl oysters by boring or eroding organisms includes hinge instability and disruption of the formation of the shell resulting in fragility, brittleness and a loss of thickness (Mohammad, 1972; Dharmaraj *et al.*, 1987; Doroudi, 1996; Mao Che *et al.*, 1996). The result of infestation, excavation and loss of calcareous material is a substantially weakened shell in pearl oysters (Hornell, 1904 in Thomas, 1979), other bivalves, and gastropods (Kent, 1981; Pomponi and Meritt, 1990; Wesche *et al.*, 1997; Kaehler and McQuaid, 1999; Stefaniak *et al.*, 2005). Bivalves and gastropods with weakened shells become more susceptible to crushing and predation (Bailey-Brock and Ringwood, 1982), particularly by crustaceans (Kent, 1981; Pomponi and Meritt, 1990; Stefaniak *et al.*, 2005; Buschbaum *et al.*, 2007), and are more susceptible to parasites such as intestinal copepods (Williams, 1968; Thieltges and Buschbaum, 2007).

Pearl yield and quality can also be affected by the activities of boring organisms, as the oyster diverts energy to repair damaged shell (Mohammad, 1972; Wada, 1973). Shell formation and regeneration in bivalves, including nacre deposition, is reported to be energy expensive, using up to one third of the total growth energy budget (Wilbur and Saleuddin, 1983). This cost of repair can result in reduced growth and fecundity (Stefaniak *et al.*, 2005). Ultimately, the damage can become so extreme that the infected animal may die and commercial operations may experience stock losses (Korringa, 1951). Damage to the shell of pearl oysters is specific to regions of the shell, and is also age dependent. Infestations by boring sponges generally begin around the hinge or apex region of the shell and progress to the mid-portion of the shell (Evans, 1969; Thomas, 1979; Barthel *et al.*, 1994; Mao Che *et al.*, 1996) (Fig. 15.2). Large parts of the oyster shell can be eroded and erosion of the calcareous substrate can occur within a very short period (Haigler, 1969). In *Pinctada margaritifera*, as much as 32% of the shell volume was removed from the hinge area of 3-year-old oysters collected as natural spat from Takapoto Island, French Polynesia (Mao Che *et al.*, 1996). The loss of hinge shell is particularly important as pearl oysters have a limited capacity to repair the hinge area (Watabe, 1983).

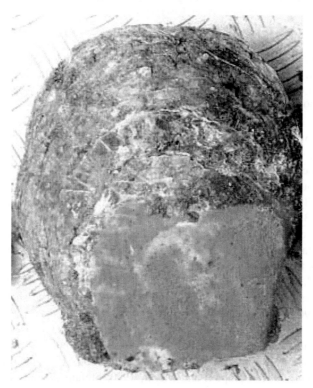

FIGURE 15.2   Pearl oyster, *Pinctada maxima*, infected with red sponge (*Cliona* sp.) also commonly known as "red arse". (See Color Plate 14)

The relationship between oyster age (or size) and fouling is well established in the more traditional literature. Pearl shells (165 mm in height) were not fouled in natural populations of *P. maxima* (Takemura and Okutani, 1955) and cultured shells were not infected by *Cliona*, which was considered to be a disease of adults (Hornell, 1904 in Thomas, 1979). Another early study also revealed a direct correlation between size class and polychaete worm blister infestation in pearl-carrying oysters (Mohammad, 1972). In a recent study this paradigm was reinforced, with the age of the shell, rather than the size of the shell *per se*, determining susceptibility to fouling in Akoya[1] pearl oysters on the Great Barrier Reef (Guenther *et al.*, 2006).

The relationship between age and susceptibility to fouling is a general feature of bivalves. The relationship between larger shell size (and age) and higher rates of infestations by boring organisms has also been reported in scallops (Evans, 1969; Barthel *et al.*, 1994) and mussels (Ambariyanto and Seed, 1991). It has been proposed that the age susceptibility to borers is due to erosion of the periostracum, a proteinaceous external layer of variable thickness, cover and morphology, present in the majority of shell bearing molluscs (see Section 3.3.1) (Bottjer and Carter, 1980; Harper, 1997). The periostracum of some molluscs, including some pearl oyster species, provides physical protection against fouling (Mao Che *et al.*, 1996; Guenther *et al.*, 2006) and erodes naturally over the life of the animal resulting in increased fouling with increasing age in some species (Scardino *et al.*, 2003; Guenther *et al.*, 2006).

The erosion of the periostracum in natural systems may be the result of mechanical wearing such as wave and sediment action (Kaehler, 1999; Ambariyanto and Seed, 1991). However, modern culturing techniques appear to have a significant impact on the rate of progression of biofouling. For example, Akoya pearl oysters in natural beds had very few blisters from boring polychaetes compared with farmed oysters (Dharmaraj and Chellam, 1983). Furthermore, rates of shell erosion were 36 times higher in 1-year-old cultured *P. margaritifera* than in natural populations (Mao Che *et al.*, 1996). The most plausible explanation is that the periostracum is being worn away more quickly through fouling removal practices (Velayudhan, 1983; Mao Che *et al.*, 1996) (see Section 15.4.2).

15.3.2. Mechanical interference with shell function

Mechanical interference with shell function has been attributed primarily to biofouling by the settlement and growth of bivalves such as mussels, and particularly barnacles (Yamamura *et al.*, 1969). Barnacles are the most problematic (non-endolithic) fouling organisms, and are the most common fouling organisms found on pearl oysters in Japan (Miyauti, 1968; Wada, 1991), northwestern Australia (de Nys and Ison, 2004), Indonesia (Takemura and Okutani, 1955; Taylor *et al.*, 1997), India (Alagarswami and Chellam, 1976; Dharmaraj and Chellam, 1983; Dharmaraj *et al.*, 1987), the Persian Gulf (Doroudi, 1996) and Venezuela (Lodeiros *et al.*, 2002). Barnacles are in most

---

[1]Throughout this text the term "Akoya" is used for pearl oysters in the currently unresolved complex of *Pinctada fucata–martensii–radiata–imbricata* (see Section 2.3.1.13).

FIGURE 15.3   A pearl oyster within a plastic culture panel heavily overgrown with barnacles.

cases the climax community after successional fouling and they can reach very high densities of up to 100% shell cover (Fig. 15.3). Barnacles adhere densely all over the shells. Fouling by the barnacle, *Balanus amphitrite variegates*, caused physical disruption to the opening and closing of the oyster valves and hinge on cultured Akoya pearl oysters in the Gulf of Mannar in India leading to high levels of mortality (27.5%) (Alagarswami and Chellam, 1976; Dharmaraj and Chellam, 1983; Dharmaraj *et al.*, 1987). Similarly, the presence of barnacles has been linked to the disruption of shell formation and hinge movement in Akoya pearl oysters (Miyauti, 1968). The effect of barnacles on hinge movement has a major effect on respiration rates in oysters. Pearl oysters fouled with barnacles had only half the oxygen consumption of oysters following shell cleaning (Miyauti, 1968). A similar effect was observed when pearl oysters are fouled by mussels, but surprisingly there was no effect on respiration when fouled by the soft-bodied ascidian, *Ciona intestinalis* (Miyauti, 1968).

Spat from bivalves, including unwanted pearl oyster species (sometimes known as "bastard shell") can also be a considerable component of the fouling assemblage found on pearl oysters. Spat from *Pteria* spp. and *Pinctada* spp. cause shell deformities, or a recession of shell growth, in the lip region of *P. maxima* in Indonesian waters (Taylor *et al.*, 1997). Even after removal of fouling the oyster lip growth is uneven ("double back"; Fig. 11.1) and subsequently more susceptible to infestation by boring sponges and polychaetes (Taylor *et al.*, 1997).

There is considerable scope to further investigate the mechanisms and impacts of settlement by barnacles and bivalves (hard fouling), as they are the most difficult and costly species to remove from shells. Understanding the larval and settlement biology of barnacles in particular, and determining key points for disruption of the settlement and metamorphosis process, will assist in developing new antifouling technologies to cost-effectively control their impacts.

15.3.3. Biological competition for resources

The preceding sections outline the negative impacts of boring organisms, and the mechanical disruption to oyster functions by bivalves and barnacles. However, other members of the biofouling community also impact on oysters, and this is often indirectly through biological competition for resources, in particular food. The strength of the competitive interaction between the pearl oyster and the fouling community depends on the species comprising the fouling community, and on the degree to which resources are limiting in the environment. Most fouling organisms are filter feeders and potentially compete for food with farmed bivalves. For example, ascidians can limit edible oyster growth by trophic competition for phytoplankton (Jackson, 1983; Stuart and Klumpp, 1984; Rissgård et al., 1995). Nevertheless, in the limited studies on competitive interactions between pearl oysters and other species it appears that fouling organisms do not necessarily compete directly for food resources (Mohammad, 1976; Lodeiros et al., 2002). This may in part be due to selective feeding by pearl oysters (Loret et al., 2000), and resource partitioning within the fouling community. Pearl oysters are extremely efficient feeders, even in limited food resource environments (Yukihira et al., 1998, 1999; Pouvreau et al., 2000). While biological competition for resources does not appear to be significant in many situations, these studies need to be interpreted in the context that fouling may limit food resources where environmental conditions are variable (Yukihira et al., 1999, 2000).

Surprisingly, in some circumstances biofouling may even have a net positive effect on pearl oyster (Mohammad, 1976) or bivalve (Ross et al., 2002) growth. The commensal relationship of bivalves and ascidians has been demonstrated by several studies. For example, overgrowth of the oyster, *Ostrea equestris*, by ascidians increased the growth of the oyster in two out of three field trials (Dalby and Young, 1993). Biofouling by encrusting organisms (including sponges, colonial tunicates, etc.) of the Akoya pearl oyster and the Pacific oyster, *Crassostrea gigas*, have been associated with an increased shell and weight growth rate, thought to be a response to competition (Tanita and Kikuchi, 1961; Arakawa, 1990). In these circumstances, it has been suggested that resource partitioning is occurring, with ascidians, which trap small particles efficiently, feeding on the picoplankton and thus allowing the oysters to feed more efficiently on larger particles (Lesser et al., 1992; Mazouni et al., 2001). Beneficial trophic interactions are also occurring, with ascidians contributing directly to the food resource available to the oysters through recycling large quantities of bio-deposits containing particles trapped within mucus sheets (Mook, 1981; Mazouni et al., 2001).

15.3.4. Environmental modification

Biofouling affects all aquaculture through the reduction of water flow thereby affecting the supply of food and oxygen, and the concentration of waste products in the environment (Cronin *et al.*, 1999; Braithwaite and McEvoy, 2005). This is also the case for pearl oyster culture, although the isolation and quantification of effects is less well documented than for many other aquaculture species. For example, *Pinctada maxima* and *P. margaritifera* spat cultured in small mesh, which was fouled rapidly, had reduced growth (Taylor *et al.*, 1997; Southgate and Beer, 2000). However, the effects were not partitioned between possible reductions in food, oxygen or removal of waste. Although pearl oysters are extremely efficient feeders (Yukihira *et al.*, 1998; Yukihira *et al.*, 1999; Pouvreau *et al.*, 2000; see Section 4.2.2), without fast water currents and large water exchanges, oysters have the potential to deplete water bodies of food in high-density culture situations (Yukihira *et al.*, 1998). A reduction in water flow can also lead to a build-up of toxic metabolites including ammonium, nitrate and nitrite. Nitrogenous waste has a significant impact in laboratory studies where high levels can be lethal over time. However, in more commercially relevant studies (*in-situ*) nitrogen also has a measurable impact. In *in-situ* culture a reduction in nitrogenous metabolites, through the co-culture of the red alga *Kappaphycus alvarezii* with the Akoya pearl oyster, resulted in a significant increase in growth of both pearl oysters and algae (Qian *et al.*, 1996). Furthermore, pearl yield and quality were consistently higher, with thicker and more homogeneous nacre on pearls being attributed to improved water quality through co-culture with *K. alvarezii* (Wu *et al.*, 2003).

While the decreased water flow through a culture system is relative to the level of biofouling, this reduction is not always detrimental. A reduction of water flow resulting in higher localized levels of phosphorous and nitrogen can encourage primary production of phytoplankton within scallop and mussel nets (Ross *et al.*, 2002; Le Blanc *et al.*, 2003). For example, there was no significant build-up of waste products despite reduced water flow through fouled nets of cultured scallop, *Pecten maximus*, and an increase in primary production may explain the growth of scallops despite fouling (Ross *et al.*, 2002). Given these findings there is scope to improve our understanding of the opportunities for polyculture of pearl oysters with other organisms, to either deter fouling or promote pearl yield and quality, based on their ability to improve the culture environment.

15.3.5. Increased friction and drag

Biofouling adds significant weight to aquaculture infrastructure and this can often be rapid and extreme. The most comprehensive data are for the fouling of cage systems for fish, where biomass increase ranges from 1 to 5 kg/m^2 of surface area depending on the site and time of immersion (Wee, 1979; Lee *et al.*, 1985; Chengxing, 1990; Cronin, 1995). This adds significantly to drag and structural fatigue for aquaculture materials (Milne, 1975; Swift *et al.*, 2006). Fouling also rapidly becomes a management issue on the equipment associated with bivalve culture, leading to increases in weight, and hence the need for additional floatation, and subsequent increases in operational costs. In one of the few studies quantitatively measuring fouling in pearl culture, barnacle settlement on nets used for culturing pearl oysters increased their weight 5-fold over a 6-month period (Dharmaraj and Chellam, 1983). As for most studies on biofouling in aquaculture there are strong data for biofouling in temperate regions, in particular for cage culture of fish, while there are fewer data for tropical regions, including pearl culture, and these are often more qualitative. Given the rapid and year-round growth of fouling in tropical regions, further details on the spatial and temporal development of communities on both shells and equipment will facilitate control through husbandry and intervention.

15.4. BIOFOULING CONTROL

The economic cost of biofouling in bivalve culture has been estimated at 30% of operational costs (Claereboudt *et al.*, 1994) and much of the workforce within the pearling industry is associated with the control of fouling on oysters and equipment using manual cleaning. There has been effort across aquaculture industries to prevent biofouling, however, the strategies and technologies used for major marine industries such as marine transport, oil and gas infrastructure are not directly transferable to pearl culture. There are, however, effective new strategies being developed for the large-scale prevention of fouling and these are discussed below.

15.4.1. Natural defences against biofouling

Biofouling has an ecological cost to marine organisms (de Nys and Steinberg, 1999; Stefaniak *et al.*, 2005) and in response marine organisms have evolved natural defense mechanisms against fouling. These can be physical, chemical or mechanical in nature. In bivalves the best example of a natural defense against fouling is the periostracum, which acts as a physical mechanism that prevents the settlement and growth of boring and fouling organisms (Bottjer and Carter, 1980; Wahl, 1989; Harper and Skelton, 1993; Mao Che *et al.*, 1996; Scardino *et al.*, 2003; Bers and Wahl, 2004; Scardino and de Nys, 2004). The periostracum of the shell strongly influences the nature and distribution of fouling on pearl oysters, and shells with an intact periostracum have a much stronger defense against fouling (Mao Che *et al.*, 1996; Scardino *et al.*, 2003; Guenther *et al.*, 2006). This is most obvious in the difference in fouling between younger shells with an intact periostracum, which are often less fouled, and older shells with a degraded or abraded periostracum where fouling is heavy (Mao Che *et al.*, 1996; Scardino *et al.*, 2003; Guenther *et al.*, 2006). While the periostracum does not act as a deterrent for all fouling species, it is broadly deterrent to boring organisms. Based on this there is an opportunity to manage husbandry and cleaning regimes to prevent the removal of the periostracum, and thereby optimize the fouling resistant properties of the shell, extending its efficacy in deterring boring organisms (Mao Che *et al.*, 1996; Guenther *et al.*, 2006).

15.4.2. Control methods

Given the prevalence of fouling in commercial-scale pearl culture, a diversity of biofouling control strategies have been applied to pearl and bivalve culture. Strategies for controlling biofouling rely on either the prevention of fouling or its removal. Technologies for the prevention of fouling in the marine environment mostly rely on the release of biocidal compounds, in particular heavy metals and organic biocides (Yebra *et al.*, 2004; Chambers *et al.*, 2006). However, the detrimental effect of biocides on the survival and growth of bivalves has meant that the control of biofouling in pearl culture has relied almost exclusively on the removal of fouling organisms, rather than the prevention of biofouling. A plethora of methods for biofouling removal have been trialed over time for commercial bivalve species, including exposure to air, freshwater dips, hypersaline dips and high temperature (Alagarswami and Chellam, 1976; Arakawa, 1990; Nell, 1993); biological control (Hidu *et al.*, 1981; Enright *et al.*, 1983; Minchin and Duggan, 1989; Cigarria *et al.*, 1998; Carver *et al.*, 2003; Lodeiros and García, 2004; Ross *et al.*, 2004); modifying culture methods (Adams *et al.*, 1991); and mechanical and high-pressure water cleaning (Arakawa, 1990; Wada, 1991; Taylor *et al.*, 1997).

Each of these fouling removal practices can have negative or positive impacts on the species being cleaned. Any negative impacts of cleaning need to be compared with those of simply allowing fouling to develop, and this is obviously not acceptable given the broad-scale implementation of cleaning regimes in pearl culture (see Section 7.6.2). The most important issue is therefore developing effective and efficient cleaning and husbandry practices taking into account the large spatial and temporal variation in fouling. Overall, the most commonly used method to prevent fouling in pearl culture is mechanical and high-pressure water cleaning.

In Australia and Indonesia, the pearl culture industry, predominantly culturing *P. maxima*, relies principally on regular cleaning using high-pressure water (Taylor *et al.*, 1997; Fig. 14.3). *P. maxima* generally benefits from regular cleaning. In experimental studies, *P. maxima* was larger when handled and cleaned every 2–4 weeks compared with those handled and cleaned every 8 or 16 weeks, demonstrating that the effect of removing fouling outweighs any negative effect of handling the animals (Taylor *et al.*, 1997). This benefit is clearly translated to the broader industry where cleaning is an integral part of the husbandry process. The cleaning process is usually done at the surface, where the pressurized water jets remove epiphytic algae and ascidians (soft fouling). High-pressure water cleaning may also remove other bivalve spat and oysters (hard fouling); however, these are often sequentially removed manually as they adhere strongly to the shell and cannot be removed by water cleaning (Fig. 15.4). In Australia, the majority of farms operate specifically designed cleaning boats which carry between 2 and 4 crew to clean oysters and culture nets on a rotating cycle (Figs. 15.5 and 15.6). The frequency and specifics of cleaning regimes vary between farms with the most intense cleaning regimes being implemented at lower latitudes where fouling is often consistent and heavy throughout the year (Fig. 15.7).

Biofouling is seasonal in most pearl oyster growing regions (Alagarswami and Chellam, 1976; Monteforte and Garcia-Gasca, 1994; Taylor *et al.*, 1997) and is correlated with the monsoonal wet season in low latitudes. In northern Australian and Indonesian waters, growers experience very high levels of fouling, particularly of barnacles, during the wet season between December and April (Taylor *et al.*, 1997). The variation of biofouling over the year means that farmers adapt management practices

FIGURE 15.4   Manual removal of fouling from pearl oyster culture nets using knives and chisels.

to control fouling as required with more frequent cleaning in summer than in winter. Cleaning cycles can vary from once every 8–10 days to once every 10–12 weeks depending on the season, the type of operation (bottom or longline culture; see Section 7.5), the geographic position of the farm, and the husbandry practices of individual companies.

The pearling industries in other countries including French Polynesia, Japan and China are generally more reliant on manual (non-mechanized) cleaning. In French Polynesia, culture is mostly based on spat collection (*P. margaritifera*) from the wild, with grow-out in baskets for the first year, and subsequent ear hanging of shells (Mao Che *et al.*, 1996). During grow-out fouling is mainly by boring sponges, polychaete worms and mussels, with algae also clogging rearing baskets (Coeroli *et al.*, 1984). Fouling problems are managed by manual cleaning of the baskets every 2 to 4 weeks (Coeroli *et al.*, 1984; Mao Che *et al.*, 1996). Older oysters (12–18 months) attached to ropes are also cleaned manually, using knives, brushes and water pressure, every 4 to 8 months (Mao Che *et al.*, 1996). In China, fouling is removed from oysters and equipment by small teams of local workers using brushes and knives (Fig. 15.8).

FIGURE 15.5   A cleaning boat cleans pearl oysters and culture equipment in Western Australia. As the boat moves along the longline (indicated by the row of floats), the line and the culture nets suspended from it are pulled onto the deck of the boat for cleaning (see Fig. 15.6).

FIGURE 15.6   A net containing pearl oysters is passed beneath a high-pressure water spray to remove fouling.

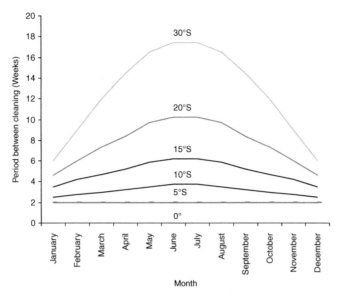

FIGURE 15.7 Theoretical cleaning frequency over a latitudinal gradient.

FIGURE 15.8 A team of young local women remove fouling from pearl oyster culture nets at Hainan Island, China.

15.5. ALTERNATIVE ANTIFOULING STRATEGIES

A number of unusual and somewhat risky methods have been used to reduce or remove fouling in bivalve culture and some of these have also been trialed, mostly with little success, in pearl culture. Clionid sponge infestations and polychaetes are not easily combated with standard fouling removal practices and are the target of many of these alternative treatments. These methods are reviewed for their novelty. In nearly all cases they are not applicable to large-scale commercial culture.

Both hypo- and hyper-tonic water have been trialed to control fouling on pearl oysters. Pearl oysters are tolerant of freshwater immersion to a limited degree, and this has been used to treat fouling on the principle that fouling organisms are more sensitive to treatment. Immersion of Akoya pearl oysters in freshwater for 6–10h was effective in controlling polychaetes and did not induce mortality in the oysters (Velayudhan, 1983). An alternative treatment to freshwater is the use of hypersaline solutions. Submerging Akoya pearl oysters in saturated solutions of salt for 15–40min has been used to kill the mud worm, *Polydora ciliata*, in Japan without significant oyster mortality (Waki and Yamaguchi, 1964; Gervis and Sims, 1992). However, other polychaete species, such as *Hydroides elegans*, are more resistant, requiring much longer treatment times and making the practice ineffective for broad spectrum fouling (Arakawa, 1973). Desiccation, a commonly used method to reduce biofouling in the edible oyster industry is not suitable for pearl oysters which are much more sensitive to air exposure (Gervis and Sims, 1992).

Among the intriguing treatments trialed for fouling removal is brushing with a 1% formalin solution. Akoya pearl oysters were then exposed to air for 15 minutes and subsequently washed in freshwater. The treatment killed all sponges and most polychaetes (87%) without affecting the oysters (Velayudhan, 1983). In another treatment that also has potential occupational health and safety aspects, fouling organisms have been burnt off oysters with an oil burner or torch in Japan. The process was noted to be dangerous and technically demanding, as the oysters all have to be evenly heated to the desired temperature (Arakawa, 1990).

In some cases of fouling, in particular clionid sponge infestations, the most appropriate strategy is to cull infected oysters to avoid the spread of the infestation through an increase in the localized supply of larvae (Fromont *et al.*, 2005). However, treatments that effectively suffocate fouling sponges have proven effective (Wang *et al.*, 2004) and this principle is the basis for coatings developed to treat clionid infestations in pearl oysters (de Nys and Ison, 2004).

One strategy to prevent or minimize fouling is to re-position stock according to depth. Fouling generally decreases with depth with decreased algal fouling below the photic zone, and the presence of a less diverse fouling community in deeper waters (Cronin *et al.*, 1999). This strategy relies on the trade-off between maximum food availability and growth in the subtidal region, and reduced biofouling deeper in the water column. Oysters can be temporarily moved away from the depth level favored by fouling organisms during the peak period of mussel or barnacle settlement, enabling fewer larvae to colonize and foul the oysters. This is a sensible and effective strategy for regions where fouling is predictable and seasonal, but is less effective in tropical regions where fouling pressure is consistent.

15.5.1. Biological control

Biological control utilizes natural species to control biofouling and a number of examples of its application have been reported. This approach is a current focus for research given impending restrictions on antifouling coatings and the potential to value-add to aquaculture industries when the species used for biological control also have commercial value. For example, successful biological control of fouling during culture of Akoya pearl oysters in nets from longlines has been demonstrated using the sea urchin, *Lytechinus variegates*. The urchins reduced fouling on the nets by 74% and fouling on the shells by 71% over a 3-month period (Lodeiros and García, 2004). *L. variegates* is an omnivore feeding on invertebrates, particularly bivalves, and algae (Watts *et al.*, 2007). It has commercial value and has a closed life cycle which can be completed on artificial feeds (George *et al.*, 2004) and is therefore a potential commercial by-product of pearl culture. An excellent example of the potential of biological control is in suspended scallop culture where urchins are highly effective at removing fouling while again providing a potential complementary product through polyculture (Ross *et al.*, 2004). Given the diversity of edible sea urchin species and their rapidly developing aquaculture (Lawrence, 2007) this opinion of biological control may provide long-term financial benefits.

15.5.2. Coatings for biofouling control

The technological development of coatings for shipping and infrastructure provides an opportunity to develop coatings technology to control biofouling in pearl culture on both equipment and shells. The industrial nature of pearl culture with the regular use of cleaning provides opportunities for the application of coatings using both biocides and foul-release coatings to shells and equipment (reviewed by Yebra *et al.*, 2004 and Chambers *et al.*, 2006). However, shells and equipment require markedly different technologies given the ability to coat equipment on land, while oysters need to be coated *ex-situ* and rapidly returned to their environment to avoid desiccation and death.

Coatings have been developed for the treatment of pearl oysters infected with clionid sponges, and to prevent fouling on pearl oysters. The coating PearlSafe®™, a biodegradable non-toxic coating, has been successfully used to treat *Pinctada maxima* grown in longline culture heavily infected with boring clionid sponges (*Pione* spp.). PearlSafe® is a wax-based emulsion that creates an impervious layer, effectively smothering and killing the boring sponge within 4 weeks of application, and preventing regrowth of the sponge (de Nys and Ison, 2004). Coating of the hinge area by dipping half of the shell (immersion for 1–2 s after air drying for 10 min) was the most effective treatment regime, resulting in 92% eradication of the sponge and mortality of only 2.5% in oysters with light to medium infestations (de Nys and Ison, 2004). In heavily infested shells, the success rate was 83%, with a mortality rate of 8.3%. This

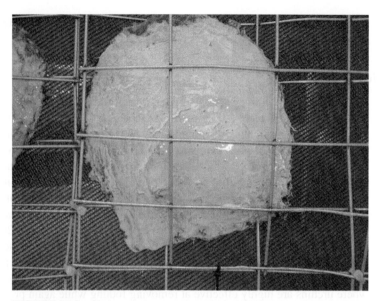

FIGURE 15.9   Modern coatings contain both biocidal ingredients and low surface energy additives such as silicone to prevent fouling.

increased mortality rate was attributed to the impact of high infestation on the oyster (de Nys and Ison, 2004). A second application eradicated the sponge completely. The wax-based coating only has surface integrity for 2–3 months after which it degrades and as such is unsuitable as a long-term antifouling control.

The development of a water-curing epoxy-based coating, provides a broader spectrum deterrence to fouling, including barnacle and oyster spat, for periods of up to 12 months depending on location (Fig. 15.9). This coating uses biocidal ingredients and low surface energy additives (silicones) to provide a two pronged defense against fouling. Firstly, biocides prevent fouling. Secondly, silicones facilitate the removal of any settled organisms, in particular barnacles and oysters, which are difficult to remove using current cleaning methods. The coating can be applied to oysters without heating or the need for air drying (de Nys and Ison, 2004). The coating remains under development and is not yet commercially available.

15.6. BIOFOULING CONTROL AND PROTECTION FOR EQUIPMENT

Treating pearl oyster shells for fouling is less effective when equipment (panels, nets) is also fouled. The feasibility of fouling prevention for culture equipment such as panel nets and trays has been investigated. Some methods have used traditional heavy metals including copper, nickel and tin (Huguenin and Huguenin, 1982; Changsheng *et al.*, 1990; Lee, 1992). For example, the use of 90–10 copper-nickel alloy in shellfish trays is effective in preventing biofouling of the tray mesh (Huguenin and Huguenin, 1982), and was reported to save 60–100% of the labor associated with antifouling control. The inorganic copper-organic tin antifouling agent JS-867 trialed as a coating for lantern nets used in scallop culture was also effective, with the weight of biofouling on the control nets being three times greater than the treated nets over a 6-month period (Changsheng *et al.*, 1990). Survival of scallops in the control and treated nets was equivalent ($>$97%), but the scallops in the treated net weighed almost 40% more than the untreated controls, and were also significantly larger. However, even though nets were air dried for 1 week after coating, and then soaked for 3 weeks to remove rapidly exuded toxic heavy metals prior to receiving scallops, tin and copper residues were detected in the scallop biomass, concentrated in the digestive gland (Changsheng *et al.*, 1990). Clearly, widespread application of antifouling agents based on tin and/or copper in close proximity to shellfish is unadvisable. Tributyltin oxide (TBT), a common ingredient in antifouling paints in the past is extremely toxic to molluscs with well-documented sub-lethal effects. TBT is banned due to its toxic effects on embryogenesis, adult shell growth and spat settlement (reviewed by Fent, 2006). Similarly, copper-based coatings are under review because copper can have negative impacts on invertebrates (Hodson *et al.*, 1997; Katranitsas *et al.*, 2003) and vertebrates, particularly during their early developmental stages (Oliva *et al.*, 2007).

An alternative is to use coatings with foul-release (low surface energy) coatings, or less toxic organic biocides. Foul-release coatings prevent the strong adhesion of fouling and given the incorporation of cleaning cycles in pearl culture, have a strong potential application if coating methods can be developed. The use of biocides is limited due to strong regulation and they are effective through toxicity, but their effects are well documented (reviewed by Konstantinou, 2006). Biocides have proven effective in limited trials. Oyster cages coated with a polymer film based on a hydrogel polymer film containing the antimicrobial agent alkyldimethyl-benzylammonium chloride (BCI) were only lightly fouled with diatoms and filamentous algae after 12 weeks (His *et al.*, 1996; Cowling *et al.*, 2000). There is scope to test efficacious and registered biocides for application to pearling infrastructure, and this may offer a readily available, cost-effective approach to prevent fouling. A combination of foul-release coatings and biocides would allow for both prevention and improved removal of fouling.

The development of antifouling technologies for aquaculture industries, including pearl culture, is underpinned by the development of new antifouling solutions for marine transport industries where costs for biofouling are very high. Future antifouling coatings will focus on non-toxic mechanisms (reviewed by Yebra *et al.*, 2004; Chambers *et al.*, 2006) and if successfully developed, they can be tailored for aquaculture industries alleviating the risk of using biocidal activity as the mechanism to prevent fouling.

15.7. SUMMARY

Biofouling has major implications for pearl farming. It is a major influence on the health and growth rates of pearl oysters, and the yield and quality of pearls they produce. It also affects the life and durability of equipment used to culture pearl oysters, and because of increased weight and drag may increase the vulnerability of pearl farming equipment to storm damage. Not surprisingly, regular removal of fouling organisms from pearl oysters and their culture equipment is a continuous activity at pearl farms and is a major component of farm operating costs. Biofouling control varies in its degree of mechanization ranging from removal by hand using brushes and knives, through to highly mechanized dedicated cleaning boats. Reducing the degree of fouling and the frequency of cleaning on pearl farms will result in considerable cost savings, and this has driven research into the potential of biological control of fouling, as well as development of chemical treatments and coatings to inhibit fouling. Many treatments have in the past relied on toxic chemicals and heavy metals which have significant environmental impacts. Modern coatings being developed use biocidal ingredients as well as low surface energy additives, and may deter all fouling for up to 12 months. The potential economic benefits that would result from a reduction of fouling in the pearling industry will continue to drive research in this field.

References

Abbott, A., Abel, P.D., Arnold, D.W., Milne, A., 2000. Cost-benefit analysis of the use of TBT: the case for a treatment approach. *Sci. Total Environ.* 258, 5–19.

Adams, M.P., Walker, R.L., Heffernan, P.B., Reinert, R.E., 1991. Eliminating spat settlement on oysters cultured in coastal Georgia: a feasibility study. *J. Shellfish Res.* 10, 207–213.

Alagarswami, K., Chellam, A., 1976. On fouling and boring organisms and mortality of pearl oysters in the farm at Veppalodai, Gulf of Mannar. *Indian J. Fish.* 23, 10–22.

Ambariyanto, A., Seed, R., 1991. The infestation of *Mytilus edulis* Linnaeus by *Polydora ciliata* (Johnston) in the Conwy Estuary, North Wales. *J. Mollus. Stud.* 57, 413–424.

Arakawa, K.Y., 1973. Biological studies on prevention and extermination of fouling organisms attached to cultured oyster. III. *Bull. Hiroshima Fish. Exp. Stn.* 4, 29–33.

Arakawa, K.Y., 1990. Competitors and fouling organisms in the hanging culture of the Pacific oyster, *Crassostrea gigas* (Thunberg). *Mar. Behav. Physiol.* 17, 67–94.

Bailey-Brock, J.H., Ringwood, A., 1982. Methods for control of the mud blister worm, *Polydora websteri*, in Hawaiian oyster culture. *Sea Grant Q.* 4, 1–6.

Barthel, D., Sundet, J., Barthel, K.G., 1994. The boring sponge *Cliona vastifica* in a subarctic population of *Chlamys islandica*. An example of balanced commensalism? In: van Soest, R.W.M., van Kempen, T.M.G., Braekman, J.C. (Eds.), *Sponges in Time and Space*. AA Balkema, Rotterdam, pp. 289–296.

Bers, V., Wahl, M., 2004. The influence of natural surface microtopographies on fouling. *Biofouling* 20, 43–51.

Bishop, M.J., Hooper, P.J., 2005. Flow, stocking density and treatment against *Polydora* spp.: influence on nursery growth and mortality of the oysters *Crassostrea virginica* and *C. ariakensis*. *Aquaculture* 246, 251–261.

Blake, J.A., Evans, J.W., 1973. *Polydora* and related genera as borers in mollusk shells and other calcareous substrates. *Veliger* 15, 235–249.

Bottjer, D.J., Carter, J.G., 1980. Functional and phylogenetic significance of projecting periostracal structures in the Bivalvia (Mollusca). *J. Paleont.* 54, 200–216.

Braley, R.D., 1984. Mariculture potential of introduced oysters *Saccostrea cucullata tuberculata* and *Crassostrea echinata*, and a histological study of reproduction of *C. echinata*. *Aust. J. Mar. Freshwat. Res.* 35, 129–141.

Braithwaite, R.A., McEvoy, L.A., 2005. Marine biofouling on fish farms and its remediation. *Adv. Mar. Biol.* 47, 215–252.

Buschbaum, C., Buschbaum, G., Schrey, T., Thieltges, D.W., 2007. Shell-boring polychaetes affect gastropod shell strength and crab predation. *Mar. Ecol. Prog. Ser.* 329, 123–130.

Carver, C.E., Chisholm, A., Mallet, A.L., 2003. Strategies to mitigate the impact of *Ciona intestinalis* (L.) biofouling on shellfish production. *J. Shellfish Res.* 22, 621–631.

Chambers, L.D., Stokes, K.R., Walsh, F.C., Wood, R.J.K., 2006. Modern approaches to marine antifouling coatings. *Surf. Coating. Tech.* 201, 3642–3652.

Champ, M.A., 2000. A review of organotin regulatory strategies, pending actions, related costs and benefits. *Sci. Total Environ.* 258, 21–71.

Changsheng, Z., Bingyin, S., Yunbi, H., Xianggui, X., 1990. Effects of JS-876, a copper-tin compounded anti-fouling agent, on growth of the scallop *Pecten maximus*. *J. Oceanogr. Huanguai Bohai Seas* 8, 40–46. (in Chinese with English abstract).

Chengxing, Z., 1990. Ecology of ascidians in Daya Bay. *Coll. Pap. Mar. Biol. Daya Bay* 2, 397–403.

Cigarria, J., Fernandez, J., Magadan, L.P., 1998. Feasibility of biological control of algal fouling in intertidal oyster culture using periwinkles. *J. Shellfish Res.* 17, 1167–1169.

Claereboudt, M.R., Bureau, D., Côté, J., Himmelman, J.H., 1994. Fouling development and its effect on the growth of juvenile giant scallops (*Placopecten magellanicus*) in suspended culture. *Aquaculture* 121, 327–342.

Cobb, W.R., 1969. Penetration of calcium carbonate substrates by the boring sponge, *Cliona. Am. Zoolog.* 9, 783–790.

Coeroli, M., De Gaillande, D., Landret, J.P., Coatanea, D., 1984. Recent innovations in cultivation of molluscs in French Polynesia. *Aquaculture* 39, 45–67.

Cowling, M.J., Hodgkiess, T., Parr, A., Smith, M.J., Marrs, S.J., 2000. An alternative approach to antifouling based on analogues of natural processes. *Sci. Total Environ.* 258, 129–137.

Cronin, E.R., 1995. An investigation into the effects of net fouling on the oxygen budget of southern bluefin tuna (*Thunnus maccoyii*) farms. Honours thesis, University of Adelaide, South Australia, pp. 72.

Cronin, E.R., Cheshire, A.C., Clarke, S.M., Melville, A.J., 1999. An investigation into the composition, biomass and oxygen budget of the fouling community on tuna aquaculture farm. *Biofouling* 13, 279–299.

Crossland, C., 1957. The cultivation of the mother-of-pearl oyster in the Red Sea. *Aust. J. Freshwat. Mar. Res.* 8, 111–130.

Dalby, J.E., Young, C.M., 1993. Variable effects of ascidian competitors on oysters in a Florida epifaunal community. *J. Exp. Mar. Biol. Ecol.* 167, 47–57.

de Nys, R., Steinberg, P., 1999. Role of secondary metabolites from algae and seagrasses in biofouling control. In: Fingerman, M., Nagabhushanam, R., Thompson, M-F. (Eds.), *Recent Advances in Marine Biotechnology*, Vol. 3. Science Publishers, Enfield, New Hampshire, USA, pp. 223–244.

de Nys, R., Ison, O., 2004. Evaluation of antifouling products developed for the Australian pearl industry. Project No 2000/254. Final Report to Fisheries Research and Development Corporation. James Cook University, Townsville, 114 pp.

Dharmaraj, S., Chellam, A., 1983. Settlement and growth of barnacle and associated fouling organisms in pearl culture farm in the *Gulf of Mannar. Proc. Symp. Coast. Aquacult.* 2, 608–613.

Dharmaraj, S., Chellam, A., Velayudhan, T.S., 1987. Biofouling, boring and predation of pearl oyster. In: Alagarswami, K. (Ed.), *Pearl Culture*. Central Marine Fisheries Research Institute special publication No 39. CMFRI, Cochin, India, pp. 92–97.

Doroudi, M., 1996. Infestation of pearl oysters by boring and fouling organisms in the northern Persian Gulf. *Indian J. Mar. Sci.* 25, 168–169.

Enright, C., Krailo, D., Staples, L., Smith, M., Vaughan, C., Ward, D., Gaul, P., Borgese, E., 1983. Biological control of fouling algae in oyster aquaculture. *J. Shellfish Res.* 3, 41–44.

Evans, J.W., 1969. Borers in the shell of the Sea Scallop, *Placopecten magellanicus. Am. Zoolog.* 9, 775–782.

Fent, K., 2006. World wide occurrence of organotins from antifouling paints and effects in the aquatic environment. In: Konstantinou, I. (Ed.), *Antifouling Paint Biocides. The Handbook of Environmental Chemistry 5.0.* Springer-Verlag, Berlin, pp. 71–100.

Fromont, J., Craig, R., Rawlinson, L., Alder, J., 2005. Excavating sponges that are destructive to farmed pearl oysters in Western and Northern Australia. *Aquacult. Res.* 36, 150–162.

Gervis, M.H., Sims, N.A., 1992. The Biology and Culture of Pearl Oysters (Bivalvia: Pteriidae). *ICLARM Stud. Rev.* 21, 49.

George, S.B., Lawrence, J.M., Lawrence, A.L., 2004. Complete larval development of the sea urchins *Lytechinus variegates* fed an artificial die. *Aquaculture* 242, 217–228.

Guenther, J., de Nys, R., 2006. Differential community development of fouling species on the pearl oyster *Pinctada fucata*, *Pteria penguin* and *Pteria chinensis* (Bivalvia, Pteriidae). *Biofouling* 22, 163–171.

Guenther, J., de Nys, R., Southgate, P.C., 2006. The effects of age and shell size of the Akoya pearl oyster *Pinctada fucata* (Bivalvia, Pteriidae) on the accumulation of fouling organisms. *Aquaculture* 253, 366–373.

Haigler, S.A., 1969. Boring mechanism of *Polydora websteri* inhabiting *Crassostrea virginica*. *Am. Zoolog.* 9, 821–828.

Harper, E.M., 1997. The molluscan periostracum: an important constraint in bivalve evolution. *Palaeontology* 40, 71–97.

Harper, E.M., Skelton, P.W., 1993. A defensive value of the thickened periostracum in the Mytiloidea. *Veliger* 36, 36–42.

Hidu, H., Canary, C., Chapman, S.R., 1981. Suspended culture of oysters: biological fouling control. *Aquaculture* 22, 189–192.

His, E., Beiras, R., Quiniou, F., Parr, A.C.S., Smith, M.J., Cowling, M.J., Hodgkiess, T., 1996. The non-toxic effects of a novel antifouling material on oyster culture. *Water Res.* 30, 2822–2825.

Hodson, S.L., Burke, C., 1994. Microfouling of salmon cages netting: a preliminary investigation. *Biofouling* 8, 93–105.

Hodson, S.L., Lewis, T.E., Burke, C., 1997. Biofouling of fish-cage netting: efficacy and problems of *in situ* cleaning. *Aquaculture* 152, 77–90.

Huguenin, J.E., Huguenin, S.S., 1982. Biofouling resistant shellfish trays. *J. Shellfish Res.* 2, 41–46.

Jackson, J.B.C., 1983. Biological determinants of present and past sessile animal distributions. In: Tevesz, M.J.S., McCail, P.L. (Eds.), *Biotic Interactions in Recent and Fossil Benthic Communities*. Plenum Press, New York, pp. 39–120.

Jørgensen, C.B., 1996. Bivalve filter feeding revisited. *Mar. Ecol. Prog. Ser.* 142, 287–302.

Katranitsas, A., Castritsi-Catharios, J., Persoone, G., 2003. The effects of a copper-based antifouling paint on mortality and enzymatic activity of a non-target marine organism. *Mar. Pollut. Bull.* 46, 1491–1494.

Kaehler, S., 1999. Incidence and distribution of phototrophic shell-degrading endoliths of the brown mussel. *Perna perna. Mar. Biol.* 135, 505–514.

Kaehler, S., McQuaid, C.D., 1999. Lethal and sub-lethal effects of phototrophic endoliths attacking the shell of the intertidal mussel *Perna perna. Mar. Biol.* 135, 497–503.

Kent, R.M.L., 1981. The effect of *Polydora ciliata* on the shell strength of *Mytilus edulis. J. Cons. Int. Explor. Mer.* 39, 252–255.

Konstantinou, I.K., 2006. *Antifouling paint biocides*. Springer-Verlag Publishing, Berlin, Germany.

Korringa, P., 1951. The shell of *Ostrea edulis* as a habitat. *Archiv. Neerland. Zool.* 10, 32–135.

Korringa, P., 1952. Recent advances in oyster biology. *Quart. Rev. Biol.* 27, 266–308.

Lawrence, J.M., 2007. *Edible sea urchins: Biology and ecology*, 2nd Edition. Developments in Aquaculture and Fisheries Science Volume 37, Elsevier Press.

Le Blanc, A.R., Landry, T., Miron, G., 2003. Fouling organisms of the blue mussel *Mytilus edulis*: their effect on nutrient uptake and release. *J. Shellfish Res.* 22, 633–638.

Lee, H.B., Lim, L.C., Cheong, L., 1985. Observations on the use of antifouling paint in netcage fish farming in Singapore. *Singapore J. Primary Ind.* 13, 1–12.

Lee, D.Y., 1992. The effect of antifouling paint against the fouling organisms in the pearl-oyster, *Pinctada fucata* farming. *Bull. Natl. Fish Res. Dev. Agency (Korea)* 46, 129–143. (in Korean with English summary)

Lesser, M.P., Shumway, S.E., Cucci, T., Smith, J., 1992. Impact of fouling organisms on mussel rope culture: Interspecific competition for food among suspension-feeing invertebrates. *J. Exp. Mar. Biol. Ecol.* 165, 91–102.

Lewis, T., 1994. Impact of biofouling on the aquaculture industry. In: Kjelleberg, S., Steinberg, P. (Eds.), *Biofouling: Problems and solutions. Proceedings of an international workshop*. University of New South Wales, Sydney, 939–943.

Little, B.J., 1984. Succession in microfouling. In: Costlow, J.D., Tipper, R.C. (Eds.), *Marine deterioration: An interdisciplinary study*. E and F N Spon Ltd, London, pp. 63–67.

Littlewood, D.T.J., Marsbe, L.A., 1990. Predation on cultivated oysters, *Crassostrea rhizophorae* (Guilding), by the polyclad turbellarian flatworm, *Stylochus* (Stylochus) *frontalis* Verrill. *Aquaculture* 88, 145–150.

Lleonart, M., Handlinger, J., Powell, M., 2003. Spionid mudworm infestation of farmed abalone (*Haliotis* spp.). *Aquaculture* 221, 85–96.

Lodeiros, C., García, N., 2004. The use of sea urchins to control fouling during suspended culture of bivalves. *Aquaculture* 231, 293–298.

Lodeiros, C.J., Rengel, J.J., Himmelman, J.H., 1999. Growth of *Pteria colymbus* (Roding, 1798) in suspended culture in Golfo de Cariaco, Venezuela. *J. Shellfish Res.* 18, 155–158.

Lodeiros, C., Pico, D., Prieto, A., Narvaez, N., Guerra, A., 2002. Growth and survival of the pearl oyster *Pinctada imbricata* (Roding 1758) in suspended and bottom culture in the Golfo de Cariaco, Venezuela. *Aquacult. Int.* 10, 327–338.

Loret, P., Pastoureaud, A., Bacher, C., Delesalle, B., 2000. Phytoplankton composition and selective feeding of the pearl oyster *Pinctada margaritifera* in the Takapoto lagoon (Tuamotu Archipelago, French Polynesia): in situ study using optical microscopy and HPLC pigment analysis. *Mar. Ecol. Prog. Ser.* 199, 55–67.

Mao Che, L., Le Campion-Alsumard, T., Boury-Esnault, N., Payri, C., Golubic, S., Bezac, C., 1996. Biodegradation of shells of the black pearl oyster, *Pinctada margaritifera* var. *cumingii*, by microborers and sponges of French Polynesia. *Mar. Biol.* 126, 509–519.

Maki, JC., Mitchell, R., 2002. Biofouling in the marine environment. In: Bitton, G. (Ed.), *Encyclopedia of environmental microbiology*. John Wiley & Sons, New York, pp. 610–619.

Mazouni, N., Gaertner, JC., Deslous-Paoli, J.M., 2001. Composition of biofouling communities on suspended oyster cultures: An *in situ* study of their interactions with the water column. *Mar. Ecol. Prog. Ser.* 214, 93–102.

Milne, P.H., 1975. Fouling of marine cages, Part 1. *Fish Farming Int.* 2, 15–19.

Minchin, D., Duggan, C.B., 1989. Biological control of the mussel in shellfish culture. *Aquaculture* 81, 97–100.

Miyauti, T., 1967. Studies on the effect of shell cleaning in pearl culture – II. The influence of fouling organisms upon the shell regeneration and byssus secretion in Japanese pearl oysters. *Nihon Seitai Gakkai Shi (Jap. J. Ecol.)* 17, 227. (in Japanese)

Miyauti, T., 1968. Studies on the effect of shell cleaning in pearl culture – III. The influence of fouling organisms upon the oxygen consumption in the Japanese pearl oysters. *Nihon Seitai Gakkai Shi (Jap. J. Ecol.)* 18, 40. (in Japanese with English abstract)

Moase, P.B., Wilmont, A., Parkinson, S.A., 1999. *Cliona*-an enemy of the pearl oyster, *Pinctada maxima* in the west Australian pearling industry. *SPC Pearl Oyster Inform. Bull.* 13, 27–28.

Mohammad, M.-B.M., 1972. Infestation of the pearl oyster *Pinctada margaritifera* (Linne) by a new species of *Polydora* in Kuwait, Arabian Gulf. *Hydrobiologia* 39, 463–477.

Mohammad, M.-B.M., 1976. Relationship between biofouling and growth of the pearl oyster *Pinctada fucata* (Gould) in Kuwait, Arabian Gulf. *Hydrobiologia* 51, 129–138.

Monteforte, M., Garcia-Gasca, A., 1994. Spat collection studies on pearl oysters *Pinctada mazatlanica* and *Pteria sterna* (Bivalvia, Pteriidae) in Bahia de La Paz, South Baja California, Mexico. *Hydrobiologia* 291, 21–34.

Mook, D.H., 1981. Removal of suspended particles by fouling communities. *Mar. Ecol. Prog. Ser.* 5, 279–281.

Nell, J.A., 1993. Farming the Sydney rock oyster (*Saccostrea commercialis*) in Australia. *Rev. Fish. Sci.* 1, 97–120.

Newman, L.J., Cannon, R.G., Govan, H., 1993. *Stylochus* (*Imogene*) *matatasi* n. sp. (Platyhelminthes, Polycladida): pest of cultured giant clams and pearl oysters from Solomon Islands. *Hydrobiologia* 257, 185–189.

Nishii, T., 1961. The influence of sessile organisms on the growth of pearl oyster and the quality of cultured pearls. *Kokoritsu Shinju Kenkyusho Hokoku (Bull. Natl. Pearl Res. Lab.)* 6, 684–687. (in Japanese with English Abstract)

Nishii, T., Shimizu, S., Taniguchi, M., 1961. Experiments on the cleaning of pearl shells and culture cages in relation to both the growth of pearl oyster and the quality of cultured pearls. *Kokoritsu Shinju Kenkyusho Hokoku (Bull. Natl. Pearl Res. Lab.)* 6, 670–675. (in Japanese with English Abstract)

Oliva, M., Carmen Garrido, M.D., Perez, E., Gonzalez De Canales, M.L., 2007. Evaluation of acute copper toxicity during early life stages of gilthead, *Sparus aurata. J. Environ. Sci. Health A* 42, 525–533.

Pit, J.H., Southgate, P.C., 2003. Fouling and predation: how do they affect growth and survival of the black-lip pearl oyster, *Pinctada margaritifera*, during nursery culture? *Aquacult. Int.* 11, 545–555.

Pomponi, S.A., Meritt, D.W., 1990. Distribution and life history of the boring sponge Cliona truitti in the upper Chesapeake Bay. In: Rutzler, K. (Ed.), *New Perspectives in Sponge Biology: Third International Conference on the Biology of Sponges, 1985*, Smithsonian Institution Press, Washington DC, pp. 384–413.

Pouvreau, S., Bodoy, A., Buestel, D., 2000. In situ suspension feeding behaviour of the pearl oyster, *Pinctada margaritifera*: combined effects of body size and weather-related seston composition. *Aquaculture* 181, 91–113.

Qian, P.Y., Wu, C.Y., Wu, M., Xie, Y.K., 1996. Integrated cultivation of the red sea alga *Kappaphycus alvarezii* and the pearl oyster *Pinctada martensi. Aquaculture* 147, 21–35.

Rissgård, H.U., Christensen, P.B., Olesen, N.J., Petersen, J.K., Moller, M.M., Anderson, P., 1995. Biological structure in a shallow cove (Kertinge Nor, Denmark). Control by benthic nutrient fluxes and suspension-feeding ascidians and jellyfish. *Ophelia* 41, 329–344.

Rosell, D., Uriz, M.J., Martin, D., 1999. Infestation by excavating sponges on the oyster (*Ostrea edulis*) populations of the Blanes littoral zone (north-western Mediterranean Sea). *J. Mar. Biol. Assoc. UK* 79, 409–413.

Ross, K.A., Thorpe, J.P., Norton, T.A., Brand, A.R., 2002. Fouling in scallop cultivation: Help or hindrance? *J. Shellfish. Res.* 21, 539–547.

Ross, K.A., Thorpe, J.P., Brand, A.R., 2004. Biological control of fouling in suspended scallop cultivation. *Aquaculture* 229, 99–116.

Roughley, T.C., 1922. Oyster culture on the George's River, New South Wales. *Tech. Educ. Ser. (Tech. Museum, Sydney)* 25, 1–69.

Rützler, K., 2002. Family Clionaidae D'Orbigny, 1851. In: Hooper, J.N.A., Van Soest, R.W.M. (Eds.), *Systema Porifera: A guide to the classification of sponges*. Kluwer Academic/Plenum Publishers, New York, pp. 173–185.

Scardino, A., de Nys, R., Ison, O., O'Connor, W., Steinberg, P., 2003. Microtopography and antifouling properties of the shell surface of the bivalve mollusks *Mytilus galloprovincialis* and *Pinctada imbricata. Biofouling* 19, 221–230.

Scardino, A., de Nys, R., 2004. Fouling deterrence of the bivalve shell *Mytilus galloprovincialis*: a physical phenomenon? *Biofouling* 20, 249–257.

Silina, A.V., 2006. Tumor-like formations on the shells of Japanese scallops *Patinopecten yessoensis. Mar. Biol.* 148, 833–840.

Southgate, P.C., Beer, A.C., 2000. Growth of black lip pearl oyster (*Pinctada margaritifera*) juveniles using different nursery culture techniques. *Aquaculture* 187, 97–104.

Stefaniak, L.M., McAtee, J., Shulman, M.J., 2005. The cost of being bored: Effects of a clionid sponge on the gastropod *Littorina littorea. J. Exp. Mar. Biol. Ecol.* 327, 103–114.

Stuart, V., Klumpp, D.W., 1984. Evidence for food-resource partitioning by kelp-bed filter feeders. *Mar. Ecol. Prog. Ser.* 16, 27–37.

Swift, M.R.F., Robinson Fredriksson, D.W., Unrein, A., Fullertonc, B., Patursson, O., Baldwin, K., 2006. Drag force acting on biofouled net panels. *Aquacult. Engin.* 35, 292–299.

Takemura, Y., Okutani, T., 1955. Notes on animals attached to the shells of the silver-lip oyster, *Pinctada maxima* (Jameson), collected from the "East" fishing ground of the Arafura Sea. *Nippon Suisan Gakkaishi (Bull. Jap. Soc. Fish. Sci.)* 21, 92–101. (in Japanese)

Tanita, S., Kikuchi, S., 1961. Studies on the organisms attaching to raft-cultured oysters. III. Effects on the growth and fattening of oysters. *Bull. Tohoku Reg. Fish. Res. Lab.* 19, 142–148. (in Japanese with English summary).

Taylor, J.J., Southgate, P.C., Rose, R.A., 1997. Fouling animals and their effect on the growth of silverlip pearl oysters, *Pinctada maxima* (Jameson) in suspended culture. *Aquaculture* 153, 31–40.

Thieltges, D.W., Buschbaum, C., 2007. Vicious circle in the intertidal: Facilitation between barnacles epibionts, a shell boring polychaete and trematode parasites in the periwinkle *Littorina littorea*. *J. Exp. Mar. Biol Ecol.* 340, 90–95.

Thomas, P.A., 1979. Boring sponges destructive to economically important molluscan beds and coral reefs in Indian seas. *Indian J. Fish.* 26, 163–200.

Velayudhan, T.S., 1983. On the occurrence of shell boring polychaetes and sponges on the pearl oyster *Pinctada fucata* and control of boring organisms. *Proc. Symp. Coastal Aquacult.* 2, 614–618.

Wada, K.T., 1973. Modern and traditional methods of pearl culture. *Underwater J. Feb.*, 28–33.

Wada, K.T., 1991. The pearl oyster, Pinctada fucata (Gould) (Family Pteriidae). In: Menzel, W. (Ed.), *Estuarine and Marine Bivalve Mollusk Culture*. CRC Press, Boca Raton, pp. 245–260.

Wahl, M., 1989. Marine epibiosis. I. Fouling and antifouling: some basic aspects. *Mar. Ecol. Prog. Ser.* 58, 175–189.

Waki, S., Yamaguchi, K., 1964. Extermination of mud worm, penetrating into the shell of the pearl oyster, by brine treatment. *Kaiho (Natl. Fed. Pearl Cult. Co-op. Assoc.)* 64, 15–25. (*in Japanese*).

Wang, A., Yaohua, S., Xing, W., 2004. The comparison among effects of four treatments to reduce polychaete infestation in pearl oyster, *Pinctada martensii. Mar. Fish. Res./Haiyang Shuichan Yanjiu* 25, 41–46.

Watabe, N., 1983. Shell Repair. In: Saleuddin, A.S.M., Wilbur, K.M. (Eds.), *The Mollusca. Vol. 4: Physiology*. Academic Press, New York, pp. 289–316.

Watts, S.A., McClintock, J.B., Lawrence, J.M., 2007. Ecology of Lytechinus. In: Lawrence, J.M. (Ed.), *Edible sea urchins: Biology and Ecology 2nd edition. Developments in Aquaculture and Fisheries Science*, Vol. 37. Elsevier, Amsterdam, pp. 473–491.

Wee, K.L., 1979. Ventilation of floating cages. M.Sc. Thesis, University of Stirling, pp. 42.

Webb, S.C., Korrûbel, J.L., 1994. Shell weakening in marine mytilids attributable to blue-green alga, *Mastigocoleus sp. (Nostrochopsidaceae) J. Shellfish Res.* 13, 11–17.

Wesche, S.J., Adlard, R.D., Hooper, J.N.A., 1997. The first incidence of clionid sponges (Porifera) from the Sydney rock oyster *Saccostrea commercialis* (Iredale and Roughley, 1933). *Aquaculture* 157, 173–180.

Whitelegge, T., 1890. Report on the worm disease affecting the oysters on the coast of New South Wales. *Rec. Aust. Mus.* 1, 41–54.

Wilbur, K.M., Saleuddin, A.S.M., 1983. Shell Formation. In: Saleuddin, A.S.M., Wilbur, K.M. (Eds.), *The Mollusca. Vol. 4: Physiology*. Academic Press, New York, pp. 235–287.

Williams, C.S., 1968. The influence of *Polydora ciliata* (Johnst.) on the degree of parasitism of *Mytilus edulis* L. by *Mytilicola intestinalis* Steuer. *J. Anim. Ecol.* 37, 709–712.

Wu, M., Mak, S.K.K., Zhang, X., Qian, P-Y., 2003. The effect of co-cultivation on the pearl yield of *Pinctada martensi* (Dunker). *Aquaculture* 221, 347–356.

Yamamura, Y., Kuwatani, Y., Nishii, T., 1969. Ecological studies of marine fouling communities in pearl culture grounds – I. Seasonal changes in the constitution of marine fouling communities at a pearl cultivating depth in Ago Bay. *Rep. Natl. Pearl Res. Lab.* 14, 1836–1861.

Yebra, D.M., Kiil, S., Dam-Johansen, K., 2004. Antifouling technology – past, present and future steps towards efficient and environmentally friendly antifouling coatings. *Prog. Org. Coatings* 50, 75–104.

Yukihira, H., Klumpp, D.W., Lucas, J.S., 1998. Effects of body size on suspension feeding and energy budgets of the pearl oysters *Pinctada margaritifera* and *P. maxima. Mar. Ecol. Prog. Ser.* 170, 119–130.

Yukihira, H., Klumpp, D.W., Lucas, J.S., 1999. Feeding adaptations of the pearl oysters *Pinctada margaritifera* and *P. maxima* to variations in natural particulates. *Mar. Ecol. Prog. Ser.* 182, 161–173.

Yukihira, H., Lucas, J.S., Klumpp, D.W., 2000. Comparative effects of temperature on suspension feeding and energy budgets of the pearl oysters *Pinctada margaritifera* and *P. maxima. Mar. Ecol. Prog. Ser.* 195, 179–188.

CHAPTER 16

Future Developments

Paul C. Southgate, John S. Lucas and Richard D. Torrey

16.1. INTRODUCTION

Preceding chapters of this book described the methods used for pearl oyster culture and pearl production, current levels of pearl production and the status of the marine pearl market. They also highlight opportunities for future industry development and potential constraints to such development. This final chapter will further explore some of these issues and discuss some of the contemporary research interests involving pearl oysters.

16.2. THE PEARL MARKET

Recent years have seen cultured pearls, both marine and freshwater, become increasingly accessible through greater supply and more affordable pricing. Pearls are now a major component in a wide range of contemporary jewelry with about 10% of all jewelry items sold having pearls as the principal gems (Anon, 2006). Cultured pearl jewelry has developed from the classic pearl necklaces reserved for mature women, and evolved a chic contemporary image that caters for a range of markets. The consumer market has also evolved into a highly complex and competitive world with a profusion of products for an increasingly discerning and demanding clientele. A recent review of the pearl market (Anon, 2006) attributed the growth in demand for pearl jewelry since the mid 1990s to four major factors:

1. Well developed pearl farming industries that benefit from accumulated experience and technological breakthroughs resulting in improved pearl quality and quantity.
2. Creative pearl jewelry designs in an extensive price range that target all market segments.
3. A fast-growing pearl jewelry retail network including traditional outlets and the internet.
4. High-profile local, national and international promotions conducted by pearl organizations and individual companies.

The review concluded with the statement that we now have "the largest and most beautiful pearls ever produced anytime in history and enjoy the widest choice of cultured pearls every offered" (Anon, 2006).

However, the pearl jewelry world generally lags behind in promotion of its products when compared to leading brand names of other sophisticated accessories. While acknowledging the advantages of a unique and unified international organization to promote the global awareness of pearls, the pearl industry has failed to establish such a body. The challenge for the pearl industry is to continue evolving to meet fast changing, global consumer demand.

16.3. PEARL-PRODUCING COUNTRIES

The majority of cultured marine pearls originate from a relatively small number of countries, particularly Australia, French Polynesia, Japan and Indonesia (Table 16.1).

Given the potential economic and socio-economic benefits of developing commercial pearl production (see Chapter 13), a number of countries have undertaken preliminary research and trial pearl production to determine the feasibility of such development. In the Pacific islands for example, countries such as the Solomon Islands, Kiribati, Marshall Islands and the Federated States of Micronesia have produced trial batches of cultured pearls from *P. margaritifera*. Fiji has also recently established commercial pearl production using *P. margaritifera* with every indication that this production will expand (Anon, 2007). Whilst cultured pearl production in such countries is likely to have a relatively minor impact on the world pearl market, collectively they may have an impact on the market for black South Sea pearls and potentially provide competition to the major producer, French Polynesia.

There is similar interest in development and expansion of small-scale commercial pearl production in Mexico, using *P. mazatlanica* and *Pteria sterna*, in countries within the Persian Gulf and western Indian Ocean, using Akoya and *P. margaritifera*, and in the Caribbean using the Akoya pearl oyster. Despite significant hurdles in developing a globally competitive cultured pearl industry, pearl production in many of these countries is likely to target local markets and the tourist market.

TABLE 16.1

Major cultured pearl producing countries and the value of world cultured pearl production at the pearl farm level.

Species	Pearl type	Major producers	Value US$ millions	Percentage of supply
Pinctada maxima	White South Sea Pearls	Indonesia Australia Philippines Myanmar	220	35
Pinctada fucata	Akoya pearls	Japan China	135	22
Pinctada margaritifera	Black South Sea Pearls or Tahitian Pearls	French Polynesia Cook Islands	120	19
Freshwater Mussels	Freshwater Pearls	China Japan	150	24
Total			625	100

Source: Anon (2006).

As well as new-comers to pearl production, future years are likely to see diversification of pearl production within traditional pearl-producing countries. In Australia, for example, there has been considerable research effort and small-scale commercial harvests of pearls from both Akoya pearl oysters and *P. margaritifera* (Love and Langenkamp, 2003; Strack, 2006). Given that Australia is a major pearl producer, these new industries are likely to benefit from an established reputation and marketing expertise already present. In 2002, the Aquaculture Council of Western Australia (ACWA) predicted that Australian pearl production may reach AUD$500 million by 2010 (ACWA, 2002). However, this figure may be over-optimistic. Pearl production from *P. maxima* in Australia declined to AUD$139.4 million in 2004/2005 and the value of products from other pearl oysters species was only AUD$0.55 million over the same period (O'Sullivan *et al.*, 2007).

16.4. FRESHWATER PEARL PRODUCTION

Perhaps the most significant challenge to the marine cultured pearl industry is growing competition from cultured freshwater pearls. Recent years have seen a substantial increase in volume (Table 16.2), and proportionally in value, of Chinese freshwater pearl production, which now makes up around 24% of cultured pearl production (Table 16.1). The increased availability of good quality Chinese freshwater pearls at affordable prices has improved access to pearls by the mass market (Anon, 2006).

TABLE 16.2
Estimated production of freshwater cultured pearls from China.

Year	Production (t/year)
1967	0.5
1972	11
1980	12
1987	120
1991	150
1993	200
1994	250–300
1995	500
1996	350–600
1997	200–400
1998	750–800
1999	800–900
2000–2004	1,000

Source: Strack (2006).

Increasing production of larger and higher-quality freshwater pearls from China has brought about increased competition with marine pearl producers. For example, approximately 1% of Chinese freshwater pearls are above 8 mm in size and another 1% are in the 7–8 mm size range; prices for these pearls have equalled those of Akoya pearls while prices for freshwater pearls of ≥10 mm are on a par with those of South Sea pearls (Strack, 2006). Annual production of 30–60 t of higher quality Chinese freshwater pearls has been predicted with the possibility that those in the 6–7 mm and 8–9 mm size ranges may replace Akoya cultured pearls in the short to medium term (Strack, 2006).

Chinese freshwater cultured pearls are traditionally produced by implanting tissue pieces from donor mussels into the mantle of host mussels (i.e. without a nucleus), and resulting pearls are harvested after 3–5 years. However, Chinese pearl farmers have developed techniques to produce large bead-nucleated pearls with a nacre thickness of at least 2 mm (Federman, 2006). This bead-nucleation technique was described in detail by Fiske and Shepherd (2007) and involves implanting mussels with a coin-shaped bead and donor tissue. After 2 years the "coin-pearls" can be harvested and a spherical bead or nucleus is then implanted into the existing pearl-sac. The oyster is grown for a further 1–2 years when the bead-nucleated pearls are harvested (Fiske and Shepherd, 2007). Freshwater mussels may deposit 0.75 mm of nacre per year onto implanted nuclei and this process results in baroque pearls that can measure up to 25 mm long, and round and near-round pearls that range from 10–15.5 mm (Fiske and Shepherd, 2007). These pearls often have excellent luster and variable colors and improvements in quality and greater consistency in shape are expected over the next few years. Production of Chinese bead-nucleated freshwater pearls is growing rapidly and pearls in the 15–20 mm size range are expected to be available to the market within the next few years (Federman, 2006; Fiske and Shepherd, 2007). Such pearls will provide direct competition with South Sea pearl products.

16.5. GENETICS AND BREEDING

Hatchery culture techniques are well established for all major pearl oyster species. This provides an opportunity for research aimed at generating the fundamental genetic information required to initiate selective breeding programs for desirable traits. Research has shown evidence of inheritance for traits such as shell dimension, growth rate and pearl production, which provides a basis for selective breeding programs (see Chapter 12). The importance with which this field of research is seen by the industry was demonstrated at the pearl oyster session of the World Aquaculture Society Conference in Bali, Indonesia in 2005 where more than 40% of the presentations related to the development of selective breeding and commercially important traits of pearl oysters (e.g. Evans *et al.*, 2005; Knauer *et al.*, 2005). A better understanding of the genetic basis of commercially important pearl quality traits is likely to result in improved yield (i.e. greater retention of nuclei), higher-quality pearls (i.e. a greater proportion of high-quality pearls, ability to use larger nuclei etc.) and a greater range of pearl products (i.e. greater color ranges and the development of specific natural colors and resulting branding opportunities). Although the outcomes of much of this research will not be seen for a number of years, there is no doubt of the increasing importance of breeding programs in commercial pearl production.

Other aspects of genetic-based research that may have significant impacts on pearl production are the development of technologies such as triploidy and cryopreservation of pearl oyster gametes (see Sections 12.4 and 12.5.2, respectively). These technologies are widely used in other forms of aquaculture, including bivalve culture. Triploidy, for example, is used routinely in the oyster industry, with significant benefits. Nell *et al.* (1994) reported that triploid *Saccostrea commercialis* were on average 41% heavier than diploids after 2.5 years of growth and reached market size 6–18 months faster than diploids. These technologies offer great potential to the pearling industry but have so far received limited research interest.

16.6. *IN VITRO* CULTURE OF PEARLS

An ultimate technical goal, but not necessarily an industry goal, for pearl production is to culture them *in vitro* under rigorously controlled conditions of culture medium and other parameters. It should overcome the sources of irregularities and failures in culturing pearls in pearl oyster farms, which occur even with the best technical procedures available. The natural pearl-sac is a foreign organ that is sustained by the environment of its host and the objective of *in vitro* pearl culture is to achieve an artificial equivalent. Cultured mantle or other epithelial cell lines could be used to spread over an appropriate nucleus *in vitro* creating the equivalent of a pearl-sac and secreting nacre onto the nucleus. The pearl-sac then needs to be sustained during this process by an appropriate culture medium. Genetically selected cell lines could be used to produce pearls of various colors and textures. Using nuclei of various shapes, it is likely that the resultant pearls would better reflect the shape of the nucleus, leading to designer shaped pearls. Finally, of crucial commercial importance, it should be possible to produce *in vitro* pearls at a fraction of the cost of those from pearl farms, but they may disrupt or saturate what has traditionally been a volatile market. Another major down-side would be that the "pearl mystique" is diminished. There is still a prevailing attitude at the retail level associating pearls with irritants in oysters, even though the modern market is dominated by cultured pearls from mantle-transplanted freshwater mussels and nucleus-inserted oysters. These methods of inducing cultured pearls have not detracted from pearls being regarded as highly desirable natural gems and not synthesized gems. It will be difficult to sustain the image of natural gems with *in vitro* production.

This ultimate technical goal for pearl production appears to be somewhat futuristic at this stage. The most advanced study has probably been with abalone, *Haliotis varia* (Suja and Dharmaraj, 2005). Pieces of abalone mantle tissue placed in a complex culture medium gave rise to individual granulocytes, hyalinocytes and fibroblast-like cells, which migrated from the tissue. Some granulocytes and hyalinocytes developed long pseudopodia in all directions and formed thick pseudopodial cell networks together with round granulocytes. The round granulocytes secreted intracellular granules, which induced nucleation and growth of crystals. During the process of crystal formation there was an amorphous stage of irregular shaped crystals and low Ca. Subsequently, the amorphous crystals grew together, forming rhombohedral crystals with high Ca content. They were apparently calcite crystals, but Suja and Dharmaraj (2005) suggested that deposition of calcite precedes aragonite. The study showed shell substances being crystallized after three processes: the formation of organic matrix as the basis of shell material; the fixation of Ca in this organic matrix; and the deposition of $CaCO_3$ crystal. An incidental finding of interest is that the mean survival of cells was 102 d in T 25 flasks.

Barik *et al.* (2004) also studied the nacre secreting capacities of dissociated mantle cells *in vitro*. They used pallial mantle tissue from the freshwater pearl-producing mussel, *Lamellidens marginalis*. Epithelial-like cells migrated out and formed a complete cell sheet surrounding the pallial tissue within 12–15 days. These epithelial cells were functionally viable and typical aragonitic "nacre" crystals of $CaCO_3$ could be observed throughout the culture plates after 38–40 d of culture.

Other *in vitro* and biochemical studies have focused on identifying and investigating the structures of matrix proteins. That is, proteins that are involved in or control the synthesis of nacre, i.e. "signal molecules". Some examples of these studies are given here. It seems that there is no comprehensive synthesis of these studies at this stage.

Samata *et al.* (1999) isolated a new matrix protein family (N16) that is specific to the nacreous layer of Akoya pearl oysters. They cloned and characterized the cDNA coding for its components. Analysis of the deduced amino acid sequence revealed that N16 had no definitive homology with other proteins. *In vitro* studies of the crystallization showed that N16 induced aragonite crystals when fixed on the substrate. Long intervals of incubation resulted in crystalline layers highly similar to the nacreous layer.

Kono *et al.* (2000) cloned the cDNAs that encode two matrix proteins, namely N66 and N14, in the nacreous layer of *P. maxima*. The molecular weights of these proteins were estimated as 59,814 and 13,734 Da, respectively. Kono *et al.* considered that the long repeated domain sequences of N66 and N14 may be responsible for the excellent calcification in *P. maxima*. *In vitro* crystallization experiments revealed that the mixture of N66 and N14 could induce platy aragonite layers highly similar to the nacreous layer, once adsorbed onto the membrane of the water-insoluble matrix.

Zhang *et al.* (2003) isolated a matrix protein, termed MSI7 because of its estimated molecular mass of 7.3 kDa, from Akoya. The novel protein shares similarity with MSI31, a prismatic framework protein of Akoya. MSI7 mRNA occurs specifically at the folds and outer epithelia of the mantle, indicating that MSI7 participates in the framework formation of both the nacreous layer and prismatic layer. *In vitro* experiment on the function of MSI7 suggested that it accelerates the nucleation and precipitation of $CaCO_3$.

Zhang *et al.* (2004) investigated the regulatory function of the water-soluble matrix and its components in nacre biomineralization, using Akoya nacre. Three protein fractions were isolated. The water-soluble matrix and these three fractions markedly affected $CaCO_3$ crystallization *in vitro*. However, two fractions stimulated $CaCO_3$ formation, the other suppressed $CaCO_3$. Zhang *et al.* (2004) suggested that $CaCO_3$ crystallization in nacre is under both positive and negative control.

Zhang *et al.* (2006) isolated a novel matrix protein from the nacreous layer of Akoya. The protein was designated p10 because of its apparent molecular mass of 10 kDa. *In vitro* experiments showed that p10 accelerates the nucleation of $CaCO_3$ crystals and induces aragonite formation. It plays a key role in both nacre biomineralization and in the differentiation of the cells involved in this process.

16.7. BIOREMEDIATION AND BIOSEQUESTRATION

The potential of pearl oysters to mediate anthropogenic nutrient enrichment and pollution in coastal ecosystems was suggested by Gifford *et al.* (2004) who estimated that a pearl farm with 100 t of pearl oysters could remove 300 kg of heavy metals, 24 kg of organic contaminants and 19 kg of nitrogen from surrounding waters into their tissues and shells. Because the primary reason for pearl oyster culture is pearl production, not human consumption, the accumulation of heavy metals and pollutants is not a threat to human health (Gifford *et al.*, 2004). In a subsequent study to quantify nutrient and heavy metal removal by Akoya pearl oysters in New South Wales, Australia, Gifford *et al.* (2005) reported that each tonne of harvested pearl oysters effectively removed 703 g of heavy metals, 7,452 g of nitrogen and 545 g of phosphorous from the environment. They noted that nutrient uptake is likely to show seasonal variation related to oyster condition and that biosequestration would be most effective if oysters were harvested during periods of high condition. The authors calculated that an increase in farm production from

9.8 t/year to 499 t/year would account for the nitrogen input of a local sewage treatment plant.

Sarver *et al.* (2004) proposed the use of pearl oysters as monitors for heavy metal contamination of tropical waters. They used aquarium studies to demonstrate the ability of *P. margaritifera* to bioaccumulate heavy metals in proportion to their concentration in surrounding water, and used *P. margaritifera* to establish field trials for heavy metal monitoring around the Hawaiian Islands. On the basis of this research, Sarver *et al.* (2004) suggested that *P. margaritifera* could be useful in monitoring levels of radioactive strontium and cobalt in Pacific island atolls previously used for nuclear bomb tests, a role of great importance in the remediation and repopulation of such areas.

These studies clearly demonstrate a potentially important role for pearl oyster culture in assisting with monitoring and biosequestration of marine pollutants and in bioremediation of coastal ecosystems. A major advantage of this approach is the potential commercial return from pearl production that could be used to off-set the cost of remediation (Gifford *et al.*, 2005). While the effects of relatively high levels of pollution on pearl quality are unknown (Gifford *et al.*, 2005) there is some doubt as to whether high pearl quality could be maintained in polluted waters (Wells and Jernakoff, 2006) and this may limit commercial returns.

16.8. BIOMEDICAL APPLICATIONS

Bone and nacre are highly organized biomineralized structures (Milet *et al.*, 2004). They are calcified structures that share an organic framework laid down by specialized cells, the bone cells in vertebrates and epithelial mantle cells in molluscs (see Section 3.3). The organic framework provides a scaffold for crystallization and directs mineralization. The similarity has led to pearl oyster nacre being proposed as a bone substitute. Furthermore, *in vivo* and *in vitro* studies have shown that shell nacre is a promising bioactive material for bone repair (Lamghari *et al.*, 1999; Moutahir-Belqasmi *et al.*, 2001; Shen *et al.*, 2006).

In an early *in vitro* study, Lopez *et al.* (1992) demonstrated that nacre can induce mineralization by human osteoblasts. Nacre chips were placed on a layer of human osteoblasts, which proliferated and attached to the chips. There was particular mineralization in bundles of osteoblasts surrounding the nacre chips and nodules were formed. The nodules contained foci with features of mineralized structures and bone-like structures. A complete sequence of bone formation is reproduced when human osteoblasts are cultured in the presence of nacre.

Cognet *et al.* (2003) found that human osteoprogenetic cells attached to nacre from *P. margaritifera* in an *in vitro* study. The cells proliferated, synthesized type I collagen and osteocalcin. The diffraction spectrum of the crystalline structure corresponded to crystallized $CaCO_3$ in the form of calcite for the outer part and in the form of aragonite for the inner part. Cognet *et al.* (2003) concluded that *P. margaritifera* nacre is cytocompatible *in vitro*, with mechanical properties similar to cortical bone.

An *in vitro* study on rat bone marrow explants cultured with a water-soluble extract of nacre organic matrix stimulated osteogenic bone marrow cells suggesting that nacre contains one or more "signal molecules" capable of activating osteogenic bone marrow cells (Lamghari *et al.*, 1999). Other *in vitro* studies have focused on the proposed "signal molecules". A water-soluble organic matrix was extracted from nacre of *P. maxima* (Lopez *et al.*, 2004). Three mammalian cell types, fibroblasts (human), bone marrow stromal cells (rat) and pre-osteoblasts (mouse), were used to characterize the effect of water-soluble nacre matrix on mammal cell recruitment and differentiation in the osteogenic pathway. The water-soluble organic matrix from nacre apparently contained diffusible signal molecules that were responsible for recruitment of mammal cells in the osteogenic pathway and bone cell activation, thus undergoing a complete sequence of mineralization.

Pereira-Mouries (2003) characterized the molecular composition of the water-soluble nacre matrix of *P. maxima*. It was highly hydrophobic and rich in glycine and alanine (more than 65% of the global amino acid composition). Similar data seemed to confirm the existence of molecular relationships between the nacre water-soluble matrix and the factors present in bone matrix. *In vitro* experiments on different mammalian cell types proved that the water-soluble matrix isolated from *P. maxima* nacre contains the signals responsible for the biological activity of the whole nacre. Indeed, this matrix acts particularly on bone cell differentiation, until the final stage of mineralization.

In the ultimate test, *in vivo* implants of *Pinctada maxima* nacre into a sheep did not invoked intolerance reactions. Experimental bone defects in the vertebrae of sheep were used to test the suitability of nacre as an injectable osteogenic biomaterial for treating vertebral bone loss (Lamghari *et al.*, 1999). Experimental bone cavities were filled with nacre powder, which slowly dissolved. Large active cell populations developed in the cavities and subsequently there were newly matured bone trabeculae in contact with or adjacent to the dissolving nacre. The functional new bone trabeculae were covered with osteoid lined with osteoblasts, indicating continuing bone formation. The recipient sheep bone underwent a sequence of bone regeneration within an osteoprogenitor rich cell layer (Lamghari *et al.*, 1999). In another study with sheep, newly formed bone and nacre welded into a dual biomineralized unit (Lopez *et al.*, 2004; Milet *et al.*, 2004).

Research in this field is limited and clearly needs pursuing with the potential for significant biomedical outcomes.

16.9. SUMMARY

Pearl oysters support a global cultured pearl industry with a value of almost $US 0.5 billion. This industry will face a number of challenges to maintain its position over coming years. In particular it must face increasing competition from cultured freshwater pearls which have long been dismissed by marine pearl producers on the basis of their small size and perceived lower quality. This is now changing and larger sized freshwater pearls with good quality provide direct competition to their marine counterparts. Techniques developed for the *in vitro* culture of pearls cannot be precluded, but these could result in cheap pearls lacking the 'pearl mystique'.

There are opportunities too for the industry. The number of countries producing pearls is likely to increase over coming years and diversification within existing pearl-producing countries will also increase production. As the industry matures, and pearl oyster become increasingly domesticated, more emphasis will be placed on the development of specific genetic lines which will allow increased yield, improved pearl quality and, potentially, the development of specific niche products.

Finally, aspects of pearl oyster biology have found application beyond the cultured pearl industry. Their high filtration rates and ability to accumulate heavy metals, pollutants and bacteria have been used to demonstrate their potential for bioremediation of polluted coastal environments. Furthermore, pearl oyster nacre has proven to be a bioactive material with potential for bone repair and future research may demonstrate considerable biomedical application in this field.

References

ACWA (Aquaculture Council of Western Australia), 2002. "State Report", in Australian Aquaculture Yearbook 2002. Perth, Western Australia.

Anon, 2006. Golay's global view. Pearl world: *Int. Pearling J.* 15, 7–8.

Anon, 2007. Justin hunter and pearls Fiji. Pearl world: *Int. Pearling J.* 16, 10–15.

Barik, S.K., Jena, J.K., Janaki, K.R., 2004. *In vitro* explant culture of mantle epithelium of freshwater pearl mussel. *Indian J. Exp. Biol.* 42, 1235–1238.

Cognet, J.M., Fricain, J.C., Reau, A.F., Lavignolle, B., Baquey, C., Lepeticorps, Y., 2003. *Pinctada margaritifera* nacre (mother-of-pearl): physico-chemical and biomechanical properties, and in vitro cytocompatibility. *Revue De Chirurgie Orthopedique Et Reparatrice De L Appareil Moteur* 89, 346–352.

Evans, B., Knauer, J., Taylor, J.J., Jerry, D.J., 2005. A reliable DNS Pedigree system for *Pinctada maxima* breeding studies. World Aquaculture 2005, May 9–13 Bali, Indonesia. Book of Abstracts, World Aquaculture Society, p. 181.

Federman, D., 2006. Fireball cultured pearls. *Modern Jeweler* 105, 51–52.

Fiske, D., Shepherd, J., 2007. Continuity and change in Chinese freshwater pearl culture. *Gems Gemol* 43, 138–145.

Gifford, S., Dunstan, R.H., O'Connor, W., Roberts, T., Toia, R., 2004. Pearl aquaculture – profitable environmental remediation? *Sci. Total Environ.* 319, 27–37.

Gifford, S., Dunstan, R.H., O'Connor, W., MacFarlane, G.R., 2005. Quantification of *in situ* nutrient and heavy metal remediation by a small pearl oyster (*Pinctada imbricata*) farm at Port Stephens, Australia. *Mar. Pollut. Bull.* 50, 417–422.

Knauer, J., Jerry, D., Taylor, J.J.U., Evans, B., 2005. Towards a selective breeding program for silver- or gold-lip pearl oysters *Pinctada maxima* in Indonesia. World Aquaculture 2005, May 9–13 Bali, Indonesia. Book of Abstracts, World Aquaculture Society, p. 320.

Kono, M., Hayashi, N., Samata, T., 2000. Molecular mechanism of the nacreous layer formation in *Pinctada maxima*. *Biochem. Biophys. Res. Commun.* 269, 213–218.

Lamghari, M., Almeida, M.J., Berland, S., Huet, H., Laurent, A., Milet, C., Lopez, E., 1999. Stimulation of bone marrow cells and bone formation by nacre: *In vivo* and *in vitro* studies. *Bone (N.Y.)* 25, 91S–94S.

Lopez, E., Vidal, B., Berland, S., Camprasse, S., Camprasse, G., Silve, C., 1992. Demonstration of the capacity of nacre to induce bone formation by human osteoblasts maintained *in vitro*. *Tissue Cell* 24, 667–679.

Lopez, E., Milet, C., Lamghari, M., Mouries, L.P., Borzeix, S., Berland, S., 2004. The dualism of nacre. *Bioceramics* 16, 733–736.

Love, G., Langenkamp, D., 2003. Australian Aquaculture: Industry Profiles for Related Species. ABARE eReport 03.8, Prepared for the Fisheries Resources Research Fund. Australian Bureau of Agricultural and Resource Economics. Canberra.

Milet, C., Berland, S., Lamghari, M., Mouries, L., Jolly, C., Borzeix, S., Doumenc, D., Lopez, E., 2004. Conservation of signal molecules involved in biomineralisation control in calcifying matrices of bone and shell. *Comptes Rendus Palevol.* 3, 493–501.

Moutahir-Belqasmi, F., Balmain, N., Lieberrher, M., Borzeix, S., Berland, S., Barthelemy, M., Peduzzi, J., Milet, C., Lopez, E., 2001. Effect of water soluble extract of nacre (*Pinctada maxima*) on alkaline phosphatase activity and Bcl-2 expression in primary cultured osteoblasts from neonatal rat calvaria. *J. Mate. Sci. Mater. Med.* 12, 1–6.

Nell, J.A., Cox, E., Smith, I.R., Maguire, G.B., 1994. Studies on triploid oysters in Australia. I. The farming potential of triploid Sydney rock oysters *Saccostrea commercialis* (Iredale & Roughley). *Aquaculture* 126, 243–255.

O'Sullivan, D., Savage, J., Fay, A., 2007. Status of Australian Aquaculture in 2004/2005. *Austasia Aquaculture Trade Directory 2007*. Turtle Press, Hobart, Tasmania.

Pereira-Mouries, L., 2003. Etude des composants de la matrice organique hydrosoluble de la nacre de l'huitre perlière *Pinctada maxima*. Bioactivite dans le processus de régénération osseuse. Biologie te Biochimie Appliquées thèses, Muséum Natl. D'Histoire Naturelle, Paris. 164 pp.

Samata, T., Hayashi, N., Kono, M., Hasegawa, K., Horita, C., Akera, S., 1999. A new matrix protein family related to the nacreous layer formation of *Pinctada fucata*. *FEBS Lett.* 462, 225–229.

Sarver, D.J., Ellis, A., Sims, N.A., Wise, D., 2004. Using pearl oysters as heavy metal monitors in tropical waters. Aquaculture 2004, "Aquaculture – An ecologically sustainable and profitable venture" March 1–5, Honolulu, Hawaii. Abstract book, World Aquaculture Society, p. 523.

Shen, Y.T., Zhu, J., Zhang, H.B., Zhao, F., 2006. *In vitro* osteogenetic activity of pearl. *Biomaterials* 27, 281–287.

Strack, E., 2006. *Pearls*. Ruhle-Diebener-Verlag GmbH & Co., Stuttgart. 707 pp.

Suja, C.P., Dharmaraj, S., 2005. *In vitro* culture of mantle tissue of the abalone *Haliotis varia*; Linnaeus. *Tissue Cell* 37, 1–10.

Wells, F.E., Jernakoff, P., 2006. An assessment of the environmental impact of wild harvest pearl aquaculture (*Pinctada maxima*) in Western Australia. *J. Shellfish Res.* 25, 141–150.

Zhang, G., Xie, X., Wang, D., Tong, Y., 2003. Scanning electron microscope study of nacre in pearl-culturing mollusc shells of China. *J. Trop. Oceanograhy/Redai Haiyang Xuebao* 22, 55–61.

Zhang, Y., Xiao, R., Ling, L., Xie, L., Ren, Y., Zhang, R., 2004. Extraction and purification of water soluble matrix protein from nacre and its effect on $CaCO_3$ crystallization. *Mar. Sci./Haiyang Kexue* 28, 33–37.

Zhang, C., Li, S., Ma, Z.J., Xie, L.P., Zhang, R.Q., 2006. A novel matrix protein p10 from the nacre of pearl oyster (*Pinctada fucata*) and its effects on both $CaCO_3$ crystal formation and mineralogenic cells. *Mar. Biotechnol.* 8, 624–633.

INDEX

PLATE 1 The so-called Imperial Crown of Konrad II. The front plate is 14.9 cm high. Weltliche und Geistliche Schatzkammer, Kunsthistorisches Museum, Vienna. (Photo: Kunsthistorisches Museum, Vienna.) (See Figure 1.1 on page 5)

PLATE 2 The "Kasel Cross" was embroidered around 1500 and is today part of on antependium at Ebstorf Monastery in North Germany. It measures 95 cm in height and 73.5 cm in width. The frames around the pictorial representations, the gowns of the persons depicted and the various symbols are embroidered with river pearls. The photograph shows a section only. (Photo: Elisabeth Strack. Reproduction with the permission of The Abbess, Kloster Ebstorf.) (See Figure 1.2 on page 6)

PLATE 3 Queen Elizabeth I. (*Source*: Kunz and Stevenson, 1908.) (See Figure 1.3 on page 7)

PLATE 4 Pendant: Sea dragon, late 16th century. Spanish or German, height 10 cm. (Photo: British Museum Publications Ltd. The Waddeston Bequest 3, Renaissance Jewellery, ML 82.) (See Figure 1.4 on page 7)

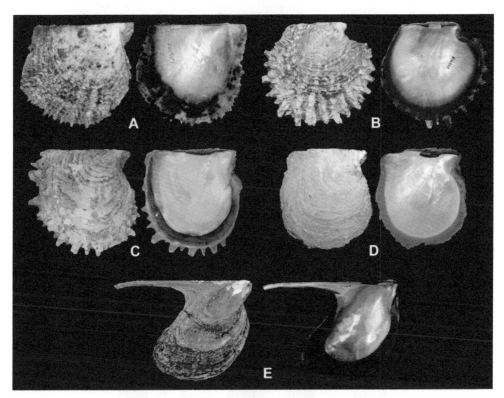

PLATE 5 Commercial species of pearl oysters. Size is expressed as shell length (measured as the longest distance between the extremities of the anterior and posterior shell margins parallel to the straight hinge line). A: specimen of *P. fucata/martensii/radiata/imbricata* species complex (Nada, Wakayama Prefecture, Japan; AMNH 307417, 54.76 mm); B: *Pinctada margaritifera* (Kontiki Lagoon, Vanua Levu Island, Fiji; AMNH 250418, 97.88 mm); C: *Pinctada mazatlanica* (Bahia de Bacochibampo, Guaymas, Mexico; AMNH 311788, 87.82 mm); D: *Pinctada maxima* (Vansittart Bay, Northern Territory, Australia; AMNH 232675, 146.16 mm); E: *Pteria penguin* (Amami Islands, Kagoshima Prefecture, Japan; AMNH 304802, 165.09 mm). (See Figure 2.1 on page 42)

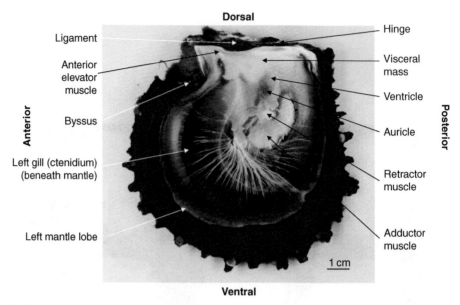

PLATE 6 Lateral view of *Pinctada margaritifera* after removal of the left valve (from Fougerouse-Tsing and Herbaut, 1994). (See Figure 3.1 on page 78)

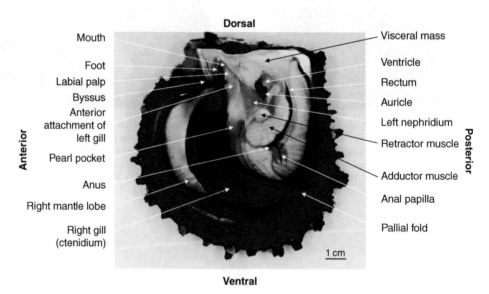

PLATE 7 Lateral view of *Pinctada margaritifera* after removal of the left valve, left mantle lobe, left ctenidium and pericardium (from Fougerouse-Tsing and Herbaut, 1994). (See Figure 3.2 on page 79)

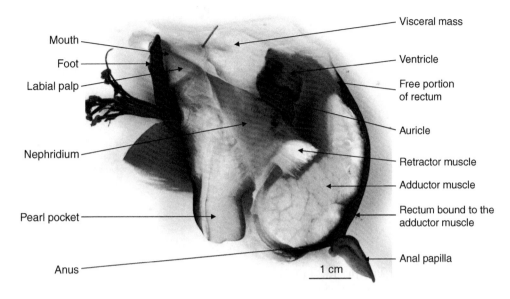

Mouth

Foot

Labial palp

Nephridium

Pearl pocket

Anus

Visceral mass

Ventricle

Free portion
of rectum

Auricle

Retractor muscle

Adductor muscle

Rectum bound to the
adductor muscle

Anal papilla

1 cm

PLATE 8 The visceral mass, adductor muscle and rectum of *Pinctada margaritifera* (from Fougerouse-Tsing and Herbaut, 1994). (See Figure 3.9 on page 85)

E

PLATE 9 Female *Pinctada maxima* spawning following removal from communal spawning tank (see Fig. 7.2). Note that this individual has been marked as a female following inspection to determine sex and reproductive condition. Spawned eggs (E) are negatively buoyant. (Photo: Clare Flanagan). (See Figure 7.3 on page 238)

PLATE 10   Processed mabé pearls produced from *Pteria sterna*. (Photo: Douglas McLaurin, Perlas del Mar de Cortez.) (See Figure 8.6 on page 283)

PLATE 11   High-quality gold and silver pearls produced from *Pinctada maxima* in Indonesia. (Photo: David Ramsay, Atlas Pearls) (See Figure 8.12 on page 294)

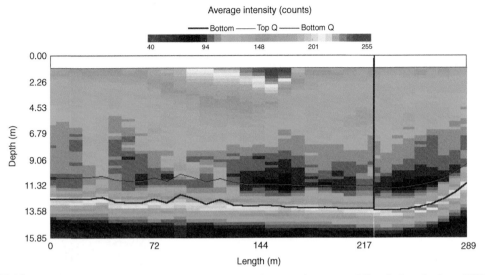

PLATE 12 Acoustic Doppler Backscatter readings during cleaning operations on pearl lines in Port Stephens (NSW EPA, 2003). Increasing intensity indicates increased turbidity. The area in red shows the sediment plume around the cleaning boat. The localized zone of measurable impact is shown in yellow and can be seen to be limited to the upper 4–5 m of the water column and to extend for up to 70 m. The approximate position of the seafloor is indicated by the black line. (See Figure 14.5 on page 516)

PLATE 13 Cultured *Pinctada margaritifera* held on chaplets hung from a longline at Takapoto Atoll, French Polynesia (Photo: John Lucas). (See Figure 9.19 on page 337)

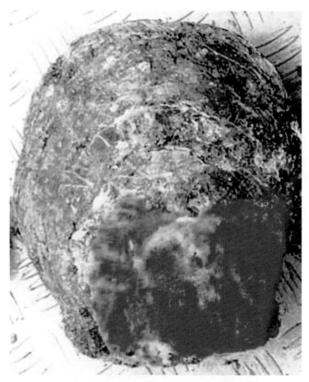

PLATE 14 Pearl oysters, *Pinctada maxima*, infected with red sponge (*Cliona* sp.), a condition commonly known as "red arse". (See Figure 15.2 on page 534)

Printed and bound by CPI Group (UK) Ltd, Croydon, CR0 4YY

03/10/2024

01040332-0004